Scheikunde

Custom Edition

Custom Edition for Leiden University

ISBN 13: 9781307051032

McGraw-Hill Custom Publishing

http://create.mheducation.com/uk/

Published by McGraw-Hill Education, 2 Penn Plaza, New York, NY 10121.

This McGraw-Hill Custom text may include materials submitted to McGraw-Hill for publication by the instructor of this course. The instructor is solely responsible for the editorial content of such materials.

Cover image: ©iStock

ISBN: 9781307051032

Scheikunde

CUSTOM EDITION

Contents

Credits

CHAPTER 1

Chemistry: The Central Science

The "Epidemic Memorial" masks, on display at the Washington State History Museum in Tacoma, Washington, were created by five Native American artists. They represent the effects of smallpox and other diseases on the Native American population.

Credit: *Washington State History Research Center.*

In This Chapter, You Will Learn

Some of what chemistry is and how it is studied using the scientific method. You will learn about the system of units used by scientists and about expressing and dealing with the numbers that result from scientific measurements.

Before You Begin, Review These Skills

- Basic algebra
- Scientific notation [⏭ Appendix 1]

How the Scientific Method Helped Defeat Smallpox

To advance understanding of science, researchers use a set of guidelines known as the scientific method. The guidelines involve careful observations, educated reasoning, and the development of hypotheses and theories, which must undergo extensive testing. One of the most compelling examples of the success of the scientific method is the story of smallpox.

Smallpox is one of the diseases classified by the Centers for Disease Control and Prevention (CDC) as a Category A bioterrorism agent. This disease has had an immeasurable impact on human history. During the sixteenth century, European explorers brought smallpox with them to the Americas, devastating native populations and leaving them vulnerable to attack—in effect, shaping the conquest of the New World. In the twentieth century alone, the disease killed an estimated half a *billion* people worldwide—leaving many more permanently disfigured, blind, or both.

Student Note: *Category A* agents are those believed to pose the greatest potential threat to the public and that have a moderate to high potential for large-scale dissemination.

Late in the eighteenth century, an English doctor named Edward Jenner observed that even during outbreaks of smallpox in Europe, milkmaids seldom contracted the disease. He reasoned that when people who had frequent contact with cows contracted *cowpox,* a similar but far less harmful disease, they developed a natural immunity to smallpox. He predicted that intentional exposure to the cowpox virus would produce the same immunity. In 1796, Jenner exposed an 8-year-old boy named James Phipps to the cowpox virus using pus from the cowpox lesions of a milkmaid named Sarah Nelmes. Six weeks later, when Jenner then exposed Phipps to the smallpox virus, the boy did *not* contract the disease. Subsequent experiments using the same technique (later dubbed *vaccination* from the Latin *vacca* meaning "cow") confirmed that immunity to smallpox could be induced.

The last naturally occurring case of smallpox occurred in 1977 in Somalia. In 1980, the World Health Organization declared smallpox officially eradicated. This historic triumph over a dreadful disease, one of the greatest medical advances of the twentieth century, began with Jenner's astute observations, inductive reasoning, and careful experimentation—the essential elements of the *scientific method.*

Student Note: Although naturally occurring smallpox was wiped out worldwide, samples have been kept in research laboratories in the United States and the former Soviet Union, and several countries are now thought to have unauthorized stockpiles of the virus.

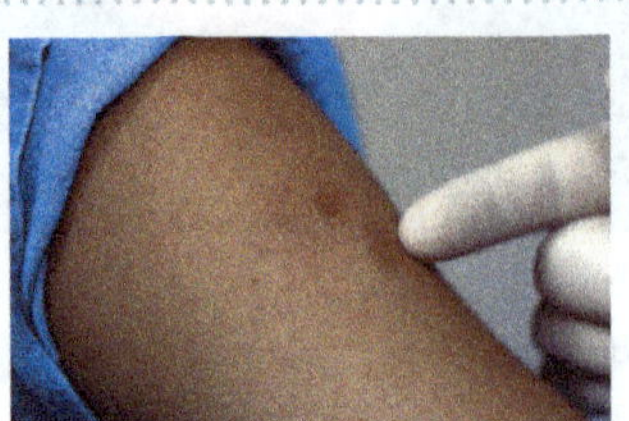

Until recently, almost everyone had a smallpox vaccine scar—usually on the upper arm.

Credit: © *Chris Livingston/Getty Images.*

At the end of this chapter, you will be able to answer several questions related to the smallpox vaccine [⏭ Applying What You've Learned, page 30].

1.1 The Study of Chemistry

Chemistry often is called the *central science* because knowledge of the principles of chemistry can facilitate understanding of other sciences, including physics, biology, geology, astronomy, oceanography, engineering, and medicine. ***Chemistry*** is the study of *matter* and the *changes* that matter undergoes. Matter is what makes up our bodies, our belongings, our physical environment, and in fact our universe. ***Matter*** is anything that has mass and occupies space.

Chemistry You May Already Know

You may already be familiar with some of the terms used in chemistry. Even if this is your first chemistry course, you may have heard of *molecules* and know them to be tiny pieces of a substance—much too tiny to see. Further, you may know that molecules are made up of *atoms,* even smaller pieces of matter. And even if you don't know what a chemical formula is, you probably know that H_2O is water. You may have used, or at least heard, the term *chemical reaction;* and you are undoubtedly familiar with a variety of chemical reactions, such as those shown in Figure 1.1.

The reactions in Figure 1.1 are all things that you can observe at the *macroscopic level.* In other words, these processes and their results are visible to the human eye. In studying chemistry, you will learn to visualize and understand these same processes at the *molecular level.*

Although it can take many different forms, all matter consists of various combinations of atoms of only a relatively small number of simple substances called *elements.* The properties of matter depend on which of these elements it contains and on how the atoms of those elements are arranged.

The Scientific Method

Experiments are the key to advancing our understanding of chemistry—or any science. Although not all scientists will necessarily take the same approach to experimentation, they all follow a set of guidelines known as the ***scientific method*** to add their results to the larger body of knowledge

(a)

(b)

(c)

(d)

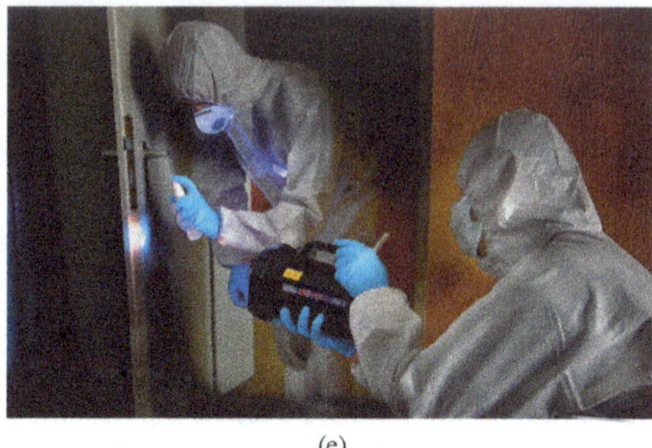

(e)

Figure 1.1 Many familiar processes are chemical reactions: (a) The flame of a gas stove is the combustion of natural gas, which is primarily methane. (b) The bubbles produced when Alka-Seltzer dissolves in water are carbon dioxide, produced by a chemical reaction between two ingredients in the tablets. (c) The formation of rust is a chemical reaction that occurs when iron, water, and oxygen are all present. (d) Many baked goods "rise" as the result of a chemical reaction that produces carbon dioxide. (e) The glow produced when luminol is used to detect traces of blood in crime-scene investigations is the result of a chemical reaction.

Credit: *a: © Steve Allen/Getty Images; b: © McGraw-Hill Education/Charles D. Winters, photographer; c: © Stockbyte/PunchStock; d: © Danilo Calilung/Corbis; e: © Jochen Tack/Alamy.*

What Do Molecules Look Like?

Molecules are far too small for us to observe them directly. An effective means of visualizing them is by the use of molecular models. Throughout this book, we will represent matter at the molecular level using *molecular art,* the two-dimensional equivalent of molecular models. In these pictures, atoms are represented as spheres, and atoms of particular elements are represented using specific colors. Table 1.1 lists some of the elements that you will encounter most often and the colors used to represent them in this book.

Molecular art can be of *ball-and-stick* models, in which the bonds connecting atoms appear as sticks [Figure 1.2(b)], or of *space-filling* models, in which the atoms appear to overlap one another [Figure 1.2(c)]. Ball-and-stick and space-filling models illustrate the specific, three-dimensional arrangement of the atoms. The ball-and-stick model does a good job of illustrating the arrangement of atoms, but exaggerates the distances between atoms, relative to their sizes. The space-filling model gives a more accurate picture of these *interatomic* distances but can obscure the details of the three-dimensional arrangement.

TABLE 1.1 Colors of Elements Commonly Used in Molecular Art

Hydrogen	Sodium
Boron	Phosphorus
Carbon	Sulfur
Nitrogen	Chlorine
Oxygen	Bromine
Fluorine	Iodine

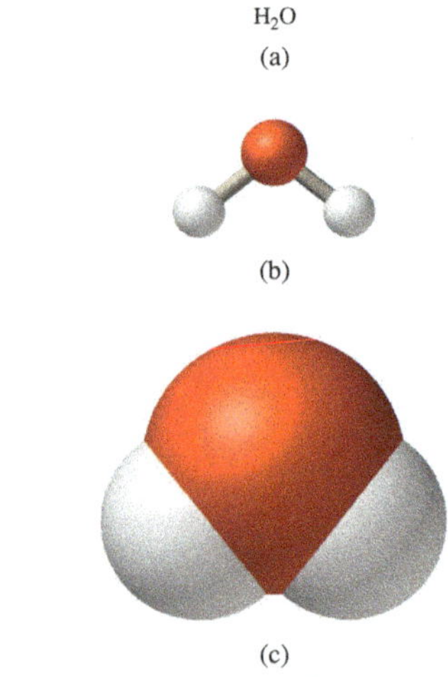

Figure 1.2 Water represented with a (a) molecular formula, (b) ball-and-stick model, and (c) space-filling model.

within a given field. The flowchart in Figure 1.3 illustrates this basic process. The method begins with the gathering of data via observations and experiments. Scientists study these data and try to identify *patterns* or *trends*. When they find a pattern or trend, they may summarize their findings with a ***law,*** a concise verbal or mathematical statement of a reliable relationship between phenomena. Scientists may then formulate a ***hypothesis,*** a tentative explanation for their observations. Further experiments are designed to test the hypothesis. If experiments indicate that the hypothesis is incorrect, the scientists go back to the drawing board, try to come up with a different

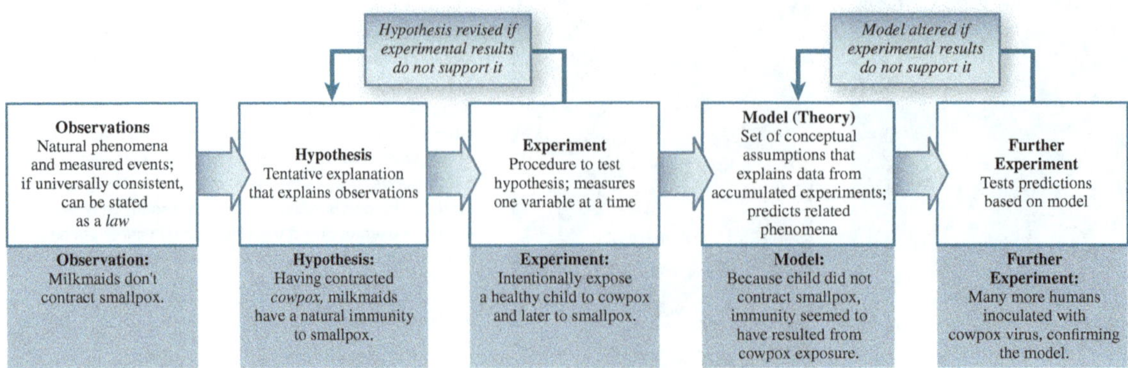

Figure 1.3 Flowchart of the scientific method.

interpretation of their data, and formulate a new hypothesis. The new hypothesis will then be tested by experiment. When a hypothesis stands the test of extensive experimentation, it may evolve into a theory. A ***theory*** is a unifying principle that explains a body of experimental observations and the laws that are based on them. Theories can also be used to predict related phenomena, so theories are constantly being tested. If a theory is disproved by experiment, then it must be discarded or modified so that it becomes consistent with experimental observations.

1.2 Classification of Matter

Student Note: Some books refer to *substances* as *pure substances*. These two terms generally mean the same thing although the adjective *pure* is unnecessary in this context because a substance is, by definition, pure.

Chemists classify matter as either a *substance* or a *mixture* of substances. A substance may be further categorized as either an *element* or a *compound*. A ***substance*** is a form of matter that has a definite (constant) composition and distinct properties. Examples are salt (sodium chloride), iron, water, mercury, carbon dioxide, and oxygen. Substances can be either elements (such as iron, mercury, and oxygen) or compounds (such as salt, water, and carbon dioxide). They differ from one another in composition and can be identified by appearance, smell, taste, and other properties.

States of Matter

All substances can, in principle, exist as a solid, a liquid, and a gas, the three physical states depicted in Figure 1.4. Solids and liquids sometimes are referred to collectively as the *condensed phases*. Liquids and gases sometimes are referred to collectively as *fluids*. In a solid, particles are held close together in an orderly fashion with little freedom of motion. As a result, a solid does not conform to the shape of its container. Particles in a liquid are close together but are not held rigidly in position; they are free to move past one another. Thus, a liquid conforms to the shape of the part of the container it fills. In a gas, the particles are separated by distances that are very large compared to the size of the particles. A sample of gas assumes both the shape and the volume of its container.

The three states of matter can be interconverted without changing the chemical composition of the substance. Upon heating, a solid (e.g., ice) will melt to form a liquid (water). Further heating will vaporize the liquid, converting it to a gas (water vapor). Conversely, cooling a gas will cause it to condense into a liquid. When the liquid is cooled further, it will freeze into the solid form. Figure 1.5 shows the three physical states of water.

Animation
Matter—three states of matter.

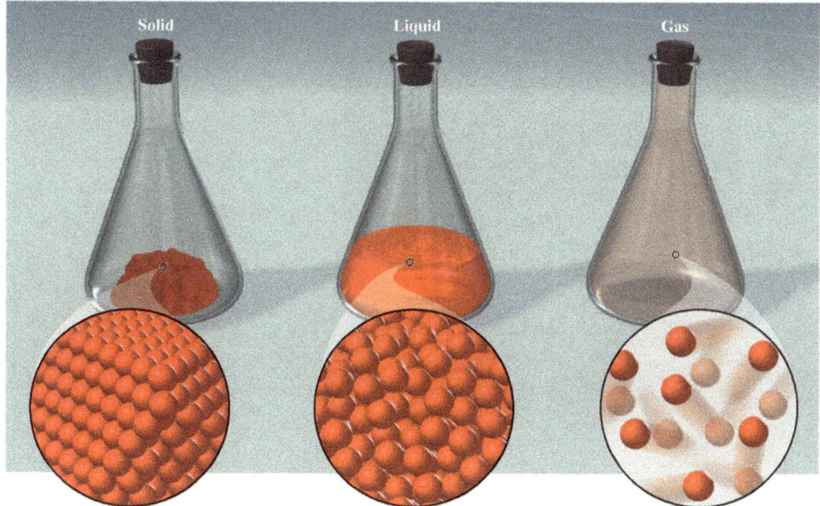

Figure 1.4 Molecular-level illustrations of a solid, liquid, and gas.

Figure 1.5 Water as a solid (ice), liquid, and gas. (We can't actually see water vapor, any more than we can see the nitrogen and oxygen that make up most of the air we breathe. When we see steam or clouds, what we are actually seeing is water vapor that has condensed upon encountering cold air.)

Credit: *© McGraw-Hill Education/Charles D. Winters, photographer.*

Elements

An ***element*** is a substance that cannot be separated into simpler substances by chemical means. Iron, mercury, oxygen, and hydrogen are just 4 of the 118 elements that have been identified. Most of the known elements occur naturally on Earth. The others have been produced by scientists via nuclear processes, which are discussed in Chapter 20. As shown in Figure 1.6(a) and (b), an element may consist of atoms or molecules.

For convenience, chemists use symbols of one or two letters to represent the elements. Only the first letter of an element's chemical symbol is capitalized. A list of the elements and their symbols appears on the inside front cover of this book. The symbols of some elements are derived from their Latin names—for example, Ag from *argentum* (silver), Pb from *plumbum* (lead), and Na from *natrium* (sodium)—while most of them come from their English names—for example, H for hydrogen, Co for cobalt, and Br for bromine.

(a)

(b)

(c)

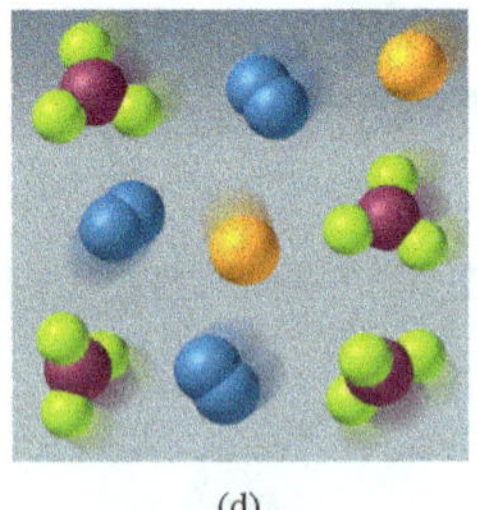

(d)

Figure 1.6 (a) Isolated atoms of an element. (b) Molecules of an element. (c) Molecules of a compound, consisting of more than one element. (d) A mixture of atoms of an element and molecules of an element and a compound.

Compounds

Most elements can combine with other elements to form compounds. Hydrogen gas, for example, burns in the presence of oxygen gas to form water, which has properties that are distinctly different from those of either hydrogen or oxygen. Thus, water is a ***compound,*** a substance composed of atoms of two or more elements chemically united in fixed proportions [Figure 1.6(c)]. The elements that make up a compound are called the compound's *constituent elements.* For example, the constituent elements of water are hydrogen and oxygen.

A compound cannot be separated into simpler substances by any physical process. (A physical process [⏭ Section 1.4] is one that does not change the identity of the matter. Examples of physical processes include boiling, freezing, and filtering.) Instead, the separation of a compound into its constituent elements requires a *chemical reaction.*

Student Note: A compound may consist of *molecules* or *ions*, which we will discuss in Chapter 2.

Mixtures

A ***mixture*** is a combination of two or more substances [Figure 1.6(d)] in which the substances retain their distinct identities. Like pure substances, mixtures can be solids, liquids, or gases. Some familiar examples are mixed nuts, 14-carat gold, apple juice, milk, and air. Mixtures do not have a universal constant composition. Therefore, samples of air collected in different locations will differ in composition because of differences in altitude, pollution, and other factors. Various brands of apple juice may differ in composition because of the use of different varieties of apples, or there may be differences in processing and packaging, and so on.

Mixtures are either *homogeneous* or *heterogeneous.* When we dissolve a teaspoon of sugar in a glass of water, we get a ***homogeneous mixture*** because the composition of the mixture is uniform throughout. If we mix sand with iron filings, however, the sand and the iron filings remain distinct and discernible from each other (Figure 1.7). This type of mixture is called a ***heterogeneous mixture*** because the composition is *not* uniform.

(a)

(b)

Figure 1.7 (a) A heterogeneous mixture contains iron filings and sand. (b) A magnet is used to separate the iron filings from the mixture.

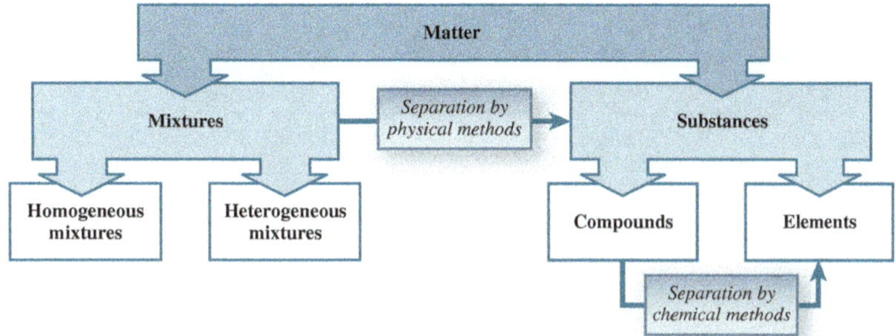

Figure 1.8 Flowchart for the classification of matter.

Mixtures, whether homogeneous or heterogeneous, can be separated by physical means into pure components without changing the identities of the components. Thus, sugar can be recovered from a water solution by evaporating the solution to dryness. Condensing the vapor will give us back the water component. To separate the sand–iron mixture, we can use a magnet to remove the iron filings from the sand, because sand is not attracted to the magnet [see Figure 1.7(b)]. After separation, the components of the mixture will have the same composition and properties as they did prior to being mixed. The relationships among substances, elements, compounds, and mixtures are summarized in Figure 1.8.

1.3 Scientific Measurement

Scientists use a variety of devices to measure the properties of matter. A meterstick is used to measure length; a burette, pipette, graduated cylinder, and volumetric flask are used to measure volume (Figure 1.9); a balance is used to measure mass; and a thermometer is used to measure temperature. Properties that can be measured are called *quantitative* properties because they are expressed using numbers. When we express a measured quantity with a number, though, we must always include the appropriate unit; otherwise, the measurement is meaningless. For example, to say that the depth of a swimming pool is 3 is insufficient to distinguish between one that is 3 *feet* (0.9 meter) and one that is 3 *meters* (9.8 feet) deep. Units are essential to reporting measurements correctly.

Student Note: According to the U.S. Metric Association (USMA), the United States is "the only significant holdout" with regard to adoption of the metric system. The other countries that continue to use traditional units are Myanmar (formerly Burma) and Liberia.

The two systems of units with which you are probably most familiar are the *English system* (foot, gallon, pound, etc.) and the *metric system* (meter, liter, kilogram, etc.). Although there has been an increase in the use of metric units in the United States in recent years, English units still are used commonly. For many years, scientists recorded measurements in metric units, but in 1960, the General Conference on Weights and Measures, the international authority on units, proposed a revised metric system for universal use by scientists. We will use both metric and revised metric (SI) units in this book.

SI Base Units

The revised metric system is called the ***International System of Units*** (abbreviated SI, from the French *Système Internationale d'Unités*). Table 1.2 lists the seven SI base units. All other units of measurement can be derived from these base units. The ***SI unit*** for *volume,* for instance, is derived by cubing the SI base unit for *length*. The prefixes listed in Table 1.3 are used to denote decimal fractions and multiples of SI units. This enables scientists to tailor the magnitude of a unit to a particular application. For example, the meter (m) is appropriate for describing the dimensions of a classroom, but the kilometer (km), 1000 m, is more appropriate for describing the distance between two cities. Units that you will encounter frequently in the study of chemistry include those for mass, temperature, volume, and density.

Mass

Although the terms *mass* and *weight* often are used interchangeably, they do not mean the same thing. Strictly speaking, weight is the force exerted by an object or sample due to gravity.

TABLE 1.2 Base SI Units

Base Quantity	Name of Unit	Symbol
Length	meter	m
Mass	kilogram	kg
Time	second	s
Electric current	ampere	A
Temperature	kelvin	K
Amount of substance	mole	mol
Luminous intensity	candela	cd

Student Note: Only one of the seven SI base units, the kilogram, itself contains a prefix.

TABLE 1.3 Prefixes Used with SI Units

Prefix	Symbol	Meaning	Example
Tera-	T	1×10^{12} (1,000,000,000,000)	1 teragram (Tg) = 1×10^{12} g
Giga-	G	1×10^{9} (1,000,000,000)	1 gigawatt (GW) = 1×10^{9} W
Mega-	M	1×10^{6} (1,000,000)	1 megahertz (MHz) = 1×10^{6} Hz
Kilo-	k	1×10^{3} (1,000)	1 kilometer (km) = 1×10^{3} m
Deci-	d	1×10^{-1} (0.1)	1 deciliter (dL) = 1×10^{-1} L
Centi-	c	1×10^{-2} (0.01)	1 centimeter (cm) = 1×10^{-2} m
Milli-	m	1×10^{-3} (0.001)	1 millimeter (mm) = 1×10^{-3} m
Micro-	µ	1×10^{-6} (0.000001)	1 microliter (µL) = 1×10^{-6} L
Nano-	n	1×10^{-9} (0.000000001)	1 nanosecond (ns) = 1×10^{-9} s
Pico-	p	1×10^{-12} (0.000000000001)	1 picogram (pg) = 1×10^{-12} g

Volumetric flask
(a)

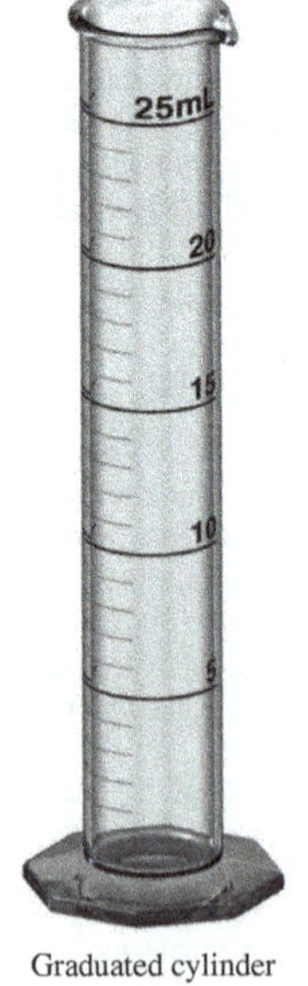

Graduated cylinder
(b)

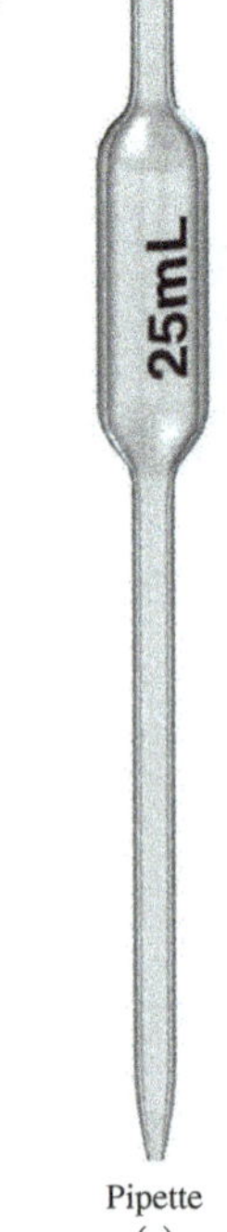

Pipette
(c)

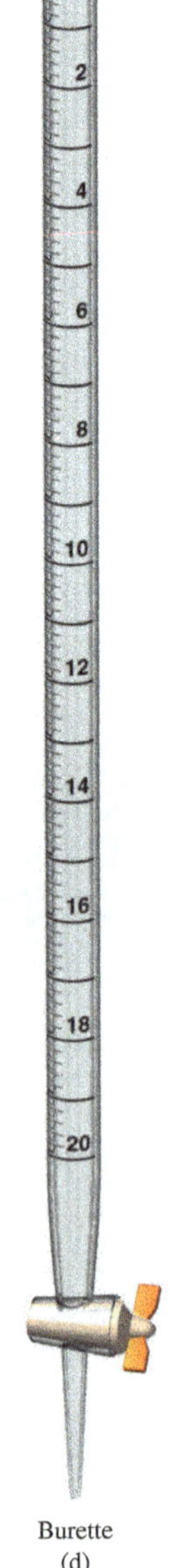

Burette
(d)

Figure 1.9 (a) A volumetric flask is used to prepare a precise volume of a solution for use in the laboratory. (b) A graduated cylinder is used to measure a volume of liquid. It is less precise than the volumetric flask. (c) A volumetric pipette is used to deliver a precise amount of liquid. (d) A burette is used to measure the volume of a liquid that has been added to a container. A reading is taken before and after the liquid is delivered, and the volume delivered is determined by subtracting the first reading from the second.

Mass is a measure of the amount of matter in an object or sample. Because gravity varies from location to location (gravity on the moon is only about one-sixth that on Earth), the weight of an object varies depending on where it is measured. The mass of an object remains the same regardless of where it is measured. The SI base unit of mass is the kilogram (kg), but in chemistry the smaller gram (g) often is more convenient and is more commonly used:

$$1 \text{ kg} = 1000 \text{ g} = 1 \times 10^3 \text{ g}$$

Temperature

There are two temperature scales used in chemistry. Their units are degrees Celsius (°C) and kelvin (K). The Celsius scale was originally defined using the freezing point (0°C) and the boiling point (100°C) of pure water at sea level. As Table 1.2 shows, the SI base unit of temperature is the ***kelvin.*** Kelvin is known as the *absolute* temperature scale, meaning that the lowest temperature possible is 0 K, a temperature referred to as "absolute zero." No *degree* sign (°) is used to represent a temperature on the Kelvin scale. The theoretical basis of the Kelvin scale has to do with the behavior of gases and is discussed in Chapter 10.

Units of the Celsius and Kelvin scales are equal in magnitude, so *a degree Celsius* is equivalent to *a kelvin*. Thus, if the temperature of an object increases by 5°C, it also increases by 5 K. Absolute zero on the Kelvin scale is equivalent to –273.15°C on the Celsius scale. We use the following equation to convert a temperature from units of degrees Celsius to kelvin:

Equation 1.1 $$\text{K} = {}^\circ\text{C} + 273.15$$

Depending on the precision required, the conversion from degrees Celsius to kelvin often is done simply by adding 273, rather than 273.15. Sample Problem 1.1 illustrates conversions between these two temperature scales.

SAMPLE PROBLEM 1.1

Normal human body temperature can range over the course of the day from about 36°C in the early morning to about 37°C in the afternoon. Express these two temperatures and the range that they span using the Kelvin scale.

Strategy Use Equation 1.1 to convert temperatures from the Celsius scale to the Kelvin scale. Then convert the range of temperatures from degrees Celsius to kelvin, keeping in mind that 1°C is equivalent to 1 K.

Setup Equation 1.1 is already set up to convert the two temperatures from degrees Celsius to kelvin. No further manipulation of the equation is needed. The range in kelvin will be the same as the range in degrees Celsius.

Solution 36°C + 273 = 309 K, 37°C + 273 = 310 K, and the range of 1°C is equal to a range of 1 K.

THINK ABOUT IT

Check your math and remember that converting a temperature from degrees Celsius to kelvin is different from converting a *difference* in temperature from degrees Celsius to kelvin.

Practice Problem ATTEMPT Express the freezing point of water (0°C), the boiling point of water (100°C), and the range spanned by the two temperatures using the Kelvin scale.

Practice Problem BUILD According to the website of the National Aeronautics and Space Administration (NASA), the average temperature of the universe is 2.7 K. Convert this temperature to degrees Celsius.

Practice Problem CONCEPTUALIZE If a single degree on the Celsius scale is represented by the rectangle on the left, which of the rectangles on the right best represents a single kelvin?

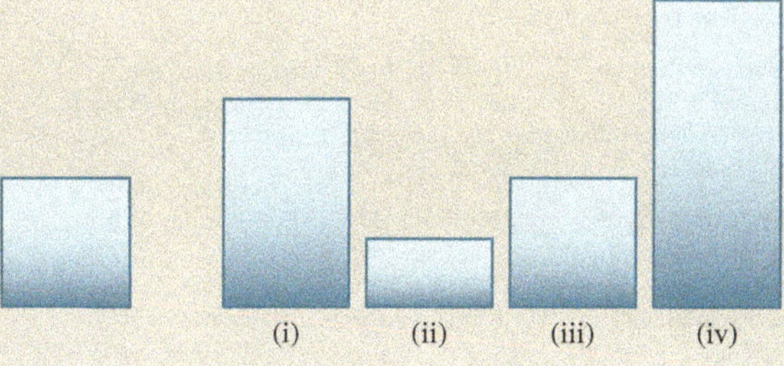

Bringing Chemistry to Life

Fahrenheit Temperature Scale

Outside of scientific circles, the Fahrenheit temperature scale is the one most used in the United States. Before the work of Daniel Gabriel Fahrenheit (German physicist, 1686–1736), there were numerous different, arbitrarily defined temperature scales, none of which gave consistent measurements. Accounts of exactly how Fahrenheit devised his temperature scale vary from source to source. In one account, in 1724, Fahrenheit labeled as 0° the lowest artificially attainable temperature at the time (the temperature of a mixture of ice, water, and ammonium chloride). Using a traditional scale consisting of 12 degrees, he labeled the temperature of a healthy human body as the twelfth degree. On this scale, the freezing point of water occurred at the fourth degree. For better resolution, each degree was further divided into eight smaller degrees. This convention makes the freezing point of water 32° and normal body temperature 96°. Today we consider normal body temperature to be somewhat higher than 96°F.

The boiling point of water on the Fahrenheit scale is 212°, meaning that there are 180 degrees (212° − 32°) between the freezing and boiling points. This separation is considerably more than the 100 degrees between the freezing point and boiling point of water on the Celsius scale [named after Swedish physicist Ander Celsius (1701–1744)]. Thus, the size of a degree on the Fahrenheit scale is only 100/180 or five-ninths of a degree on the Celsius scale. Equations 1.2 and 1.3 give the relationship between Fahrenheit and Celsius temperatures.

$$\text{temp in °C} = (\text{temp in °F} - 32\text{°F}) \times \frac{5\text{°C}}{9\text{°F}}$$ **Equation 1.2**

and

$$\text{temp in °F} = \frac{9\text{°F}}{5\text{°C}} \times (\text{temp in °C}) + 32\text{°F}$$ **Equation 1.3**

Sample Problem 1.2 illustrates the conversion between Celsius and Fahrenheit scales.

SAMPLE PROBLEM 1.2

A body temperature above 39°C constitutes a high fever. Convert this temperature to the Fahrenheit scale.

Strategy We are given a temperature in Celsius and are asked to convert it to Fahrenheit.

Setup We use Equation 1.3:

$$\text{temp in °F} = \frac{9\text{°F}}{5\text{°C}} \times (\text{temp in °C}) + 32\text{°F}$$ **Equation 1.3**

Solution

$$\text{temp in °F} = \frac{9\text{°F}}{5\text{°C}} \times 39\text{°C} + 32\text{°F} = 102.2\text{°F}$$

THINK ABOUT IT

Knowing that "normal" body temperature on the Fahrenheit scale is approximately 99°F (98.6°F is the number most often cited), 102.2°F seems like a reasonable answer.

Practice Problem ATTEMPT Convert the temperatures 45.0°C and 90.0°C, and the difference between them, to degrees Fahrenheit.

Practice Problem BUILD In Ray Bradbury's 1953 novel *Fahrenheit 451*, 451°F is said to be the temperature at which books, which have been banned in the story, ignite. Convert 451°F to the Celsius scale.

Practice Problem CONCEPTUALIZE If a single degree on the Fahrenheit scale is represented by the rectangle on the left, which of the rectangles on the right best represents a single degree on the Celsius scale? Which best represents a single kelvin?

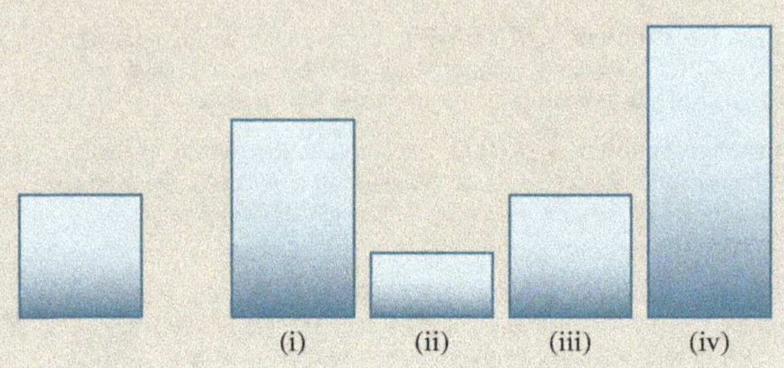

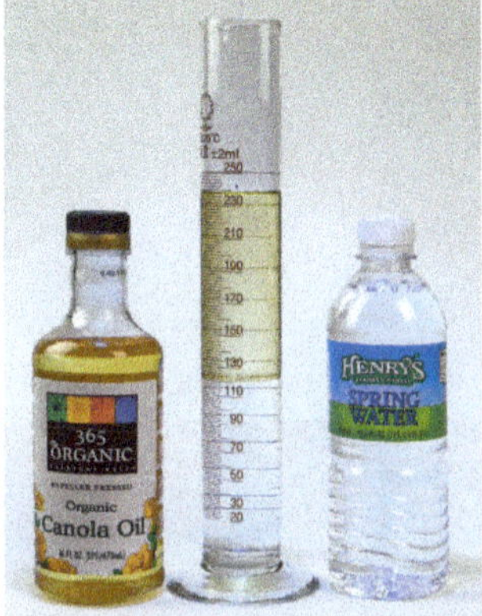

Oil floating on water is a familiar demonstration of density differences.
Credit: © *David A. Tietz/Editorial Image, LLC.*

Derived Units: Volume and Density

There are many quantities, such as volume and density, that require units not included in the base SI units. In these cases, we must combine base units to *derive* appropriate units for the quantity.

The derived SI unit for volume, the meter cubed (m^3), is a larger volume than is practical in most laboratory settings. The more commonly used metric unit, the *liter* (L), is derived by cubing the *decimeter* (one-tenth of a meter) and is therefore also referred to as the cubic decimeter (dm^3). Another commonly used metric unit of volume is the *milliliter* (mL), which is derived by cubing the centimeter (1/100 of a meter). The milliliter is also referred to as the cubic centimeter (cm^3). Figure 1.10 illustrates the relationship between the liter (or dm^3) and the milliliter (or cm^3).

Density is the ratio of mass to volume. Oil floats on water, for example, because, in addition to not mixing with water, oil has a lower density than water. That is, given *equal volumes* of the two liquids, the oil will have a *smaller mass* than the water. Density is calculated using the following equation:

Equation 1.4

$$d = \frac{m}{V}$$

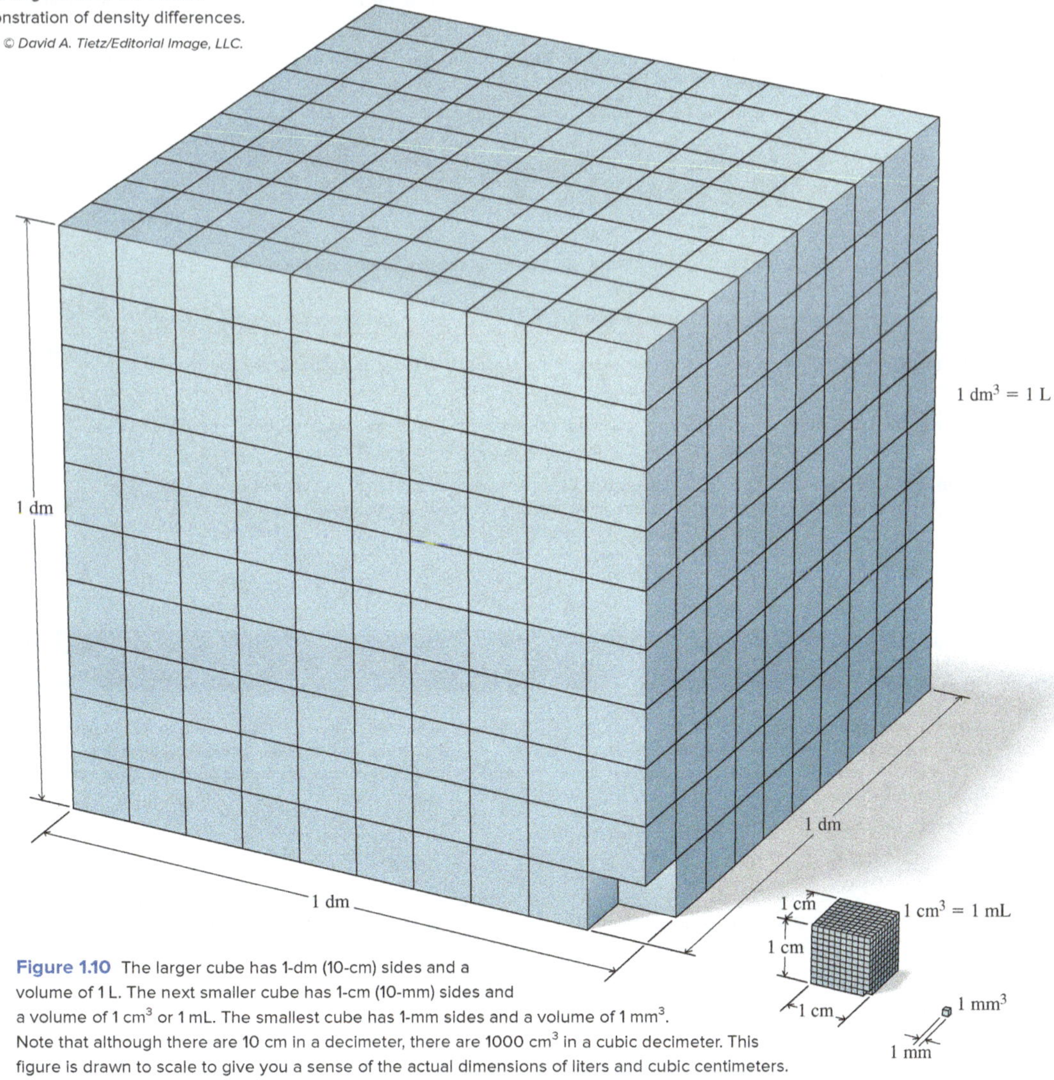

Figure 1.10 The larger cube has 1-dm (10-cm) sides and a volume of 1 L. The next smaller cube has 1-cm (10-mm) sides and a volume of 1 cm^3 or 1 mL. The smallest cube has 1-mm sides and a volume of 1 mm^3. Note that although there are 10 cm in a decimeter, there are 1000 cm^3 in a cubic decimeter. This figure is drawn to scale to give you a sense of the actual dimensions of liters and cubic centimeters.

where d, m, and V denote density, mass, and volume, respectively. The SI-derived unit for density is the kilogram per cubic meter (kg/m^3). This unit is too large for most common uses, however, so grams per cubic centimeter (g/cm^3) and its equivalent, grams per milliliter (g/mL), are used to express the densities of most solids and liquids. Water, for example, has a density of 1.00 g/cm^3 at 4°C. Because gas densities generally are very low, we typically express them in units of grams per liter (g/L):

$$1\ g/cm^3 = 1\ g/mL = 1000\ kg/m^3$$

$$1\ g/L = 0.001\ g/mL$$

Sample Problem 1.3 illustrates density calculations.

SAMPLE PROBLEM 1.3

Ice cubes float in a glass of water because solid water is less dense than liquid water. (a) Calculate the density of ice given that, at 0°C, a cube that is 2.0 cm on each side has a mass of 7.36 g, and (b) determine the volume occupied by 23 g of ice at 0°C.

Strategy (a) Determine density by dividing mass by volume (Equation 1.4), and (b) use the calculated density to determine the volume occupied by the given mass.

Setup (a) We are given the mass of the ice cube, but we must calculate its volume from the dimensions given. The volume of the ice cube is $(2.0\ cm)^3$, or 8.0 cm^3. (b) Rearranging Equation 1.4 to solve for volume gives $V = m/d$.

Solution

(a) $d = \dfrac{7.36\ g}{8.0\ cm^3} = 0.92\ g/cm^3$ or 0.92 g/mL (b) $V = \dfrac{23\ g}{0.92\ g/cm^3} = 25\ cm^3$ or 25 mL

THINK ABOUT IT

For a sample with a density *less* than 1 g/cm^3, the number of cubic centimeters should be *greater* than the number of grams. In this case, 25 (cm^3) > 23 (g).

Practice Problem ATTEMPT Given that 25.0 mL of mercury has a mass of 340.0 g, calculate (a) the density of mercury and (b) the mass of 120.0 mL of mercury.

Practice Problem BUILD Calculate (a) the density of a solid substance if a cube measuring 2.33 cm on one side has a mass of 117 g and (b) the mass of a cube of the same substance measuring 7.41 cm on one side.

Practice Problem CONCEPTUALIZE Using the picture of the graduated cylinder and its contents, arrange the following in order of increasing density: blue liquid, pink liquid, yellow liquid, grey solid, blue solid, green solid.

CHECKPOINT – SECTION 1.3 Scientific Measurement

1.3.1 The coldest temperature ever recorded on Earth was −128.6°F (recorded at Vostok Station, Antarctica, on July 21, 1983). Express this temperature in degrees Celsius and in kelvins.

a) −89.2°C, −89.2 K
b) −289.1°C, −15.9 K
c) −89.2°C, 183.9 K
d) −173.9°C, 99.3 K
e) −7.0°C, 266.2 K

1.3.2 What is the density of an object that has a volume of 34.2 cm^3 and a mass of 19.6 g?

a) 0.573 g/cm^3
b) 1.74 g/cm^3
c) 670 g/cm^3
d) 53.8 g/cm^3
e) 14.6 g/cm^3

1.3.3 A sample of water is heated from room temperature to just below the boiling point. The overall change in temperature is 72°C. Express this temperature change in kelvins.

a) 345 K
b) 72 K
c) 0 K
d) 201 K
e) 273 K

1.3.4 Given that the density of gold is 19.3 g/cm^3, calculate the volume (in cm^3) of a gold nugget with a mass of 5.98 g.

a) 3.23 cm^3
b) 5.98 cm^3
c) 115 cm^3
d) 0.310 cm^3
e) 13.3 cm^3

The following box illustrates the importance of using units carefully in scientific work.

Why Are Units So Important?

On December 11, 1998, NASA launched the 125-million-dollar Mars Climate Orbiter, which was intended to be the Red Planet's first weather satellite. After a 416-million-mile (mi) journey, the spacecraft was supposed to go into Mars's orbit on September 23, 1999. Instead, it entered Mars's atmosphere about 100 km (62 mi) lower than planned and was destroyed by heat. Mission controllers later determined that the spacecraft was lost because English measurement units were not converted to metric units in the navigation software.

Engineers at Lockheed Martin Corporation, who built the spacecraft, specified its thrust in pounds, which is an English unit of force. Scientists at NASA's Jet Propulsion Laboratory, on the other hand, who were responsible for deployment, had assumed that the thrust data they were given were expressed in *newtons,* a metric unit. To carry out the conversion between pound and newton, we would start with 1 lb = 0.4536 kg and, from Newton's second law of motion,

$$\text{force} = (\text{mass})(\text{acceleration}) = (0.4536\ \text{kg})(9.81\ \text{m/s}^2)$$
$$= 4.45\ \text{kg} \cdot \text{m/s}^2 = 4.45\ \text{N}$$

because 1 newton (N) = 1 kg · m/s^2. Therefore, instead of converting 1 lb of *force* to 4.45 N, the scientists treated it as a force of 1 N. The considerably smaller engine thrust employed because of the engineers' failure to convert from English to metric units resulted in a lower orbit and the ultimate destruction of the spacecraft.

Commenting on the failure of the Mars mission, one scientist said, "This is going to be the cautionary tale that will be embedded into introduction to the metric system in elementary school, high school, and college science courses till the end of time."

Mars Climate Orbiter during preflight tests.
Credit: © *Joe Skipper/Reuters/Corbis.*

1.4 The Properties of Matter

Substances are identified by their properties as well as by their composition. Properties of a substance may be ***quantitative*** (measured and expressed with a number) or ***qualitative*** (not requiring explicit measurement).

Physical Properties

Color, melting point, boiling point, and physical state are all physical properties. A ***physical property*** is one that can be observed and measured without changing the *identity* of a substance. For example, we can determine the melting point of ice by heating a block of ice and measuring the temperature at which the ice is converted to water. Liquid water differs from ice in appearance but not in composition; both liquid water and ice are H_2O. Melting is a ***physical change:*** one in which the state of matter changes, but the identity of the matter does not change. We can recover the original ice by cooling the water until it freezes. Therefore, the melting point of a substance is a *physical* property. Similarly, when we say that nitrogen dioxide gas is brown, we are referring to the physical property of color.

Chemical Properties

The statement "Hydrogen gas burns in oxygen gas to form water" describes a ***chemical property*** of hydrogen, because to observe this property we must carry out a ***chemical change***—burning in oxygen (combustion), in this case. After a chemical change, the original substance (hydrogen gas in this case) will no longer exist. What remains is a different substance (water, in this case). We *cannot* recover the hydrogen gas from the water by means of a physical process, such as boiling or freezing.

Every time we bake cookies, we bring about a chemical change. When heated, the sodium bicarbonate (baking soda) in cookie dough undergoes a chemical change that produces carbon dioxide gas. The gas forms numerous little bubbles in the dough during the baking process, causing the cookies to "rise." Once the cookies are baked, we cannot recover the sodium bicarbonate by cooling the cookies, or by *any* physical process. When we eat the cookies, we cause further chemical changes that occur during digestion and metabolism.

Extensive and Intensive Properties

All properties of matter are either *extensive* or *intensive*. The measured value of an ***extensive property*** depends on the amount of matter. *Mass* is an extensive property. More matter means more mass. Values of the same extensive property can be added together. For example, two gold nuggets will have a combined mass that is the sum of the masses of each nugget, and the length of two city buses is the sum of their individual lengths. The value of an extensive property depends on the amount of matter.

The value of an ***intensive property*** does *not* depend on the amount of matter. *Density* and *temperature* are intensive properties. Suppose that we have two beakers of water at the same temperature and we combine them to make a single quantity of water in a larger beaker. The density and the temperature of the water in the larger combined quantity will be the same as they were in the two separate beakers. Unlike mass and length, which are additive, temperature, density, and other intensive properties are not additive. Sample Problem 1.4 shows you how to differentiate chemical and physical processes.

SAMPLE PROBLEM 1.4

The diagram in (a) shows a compound made up of atoms of two elements (represented by the green and red spheres) in the liquid state. Which of the diagrams in (b) to (d) represent a physical change, and which diagrams represent a chemical change?

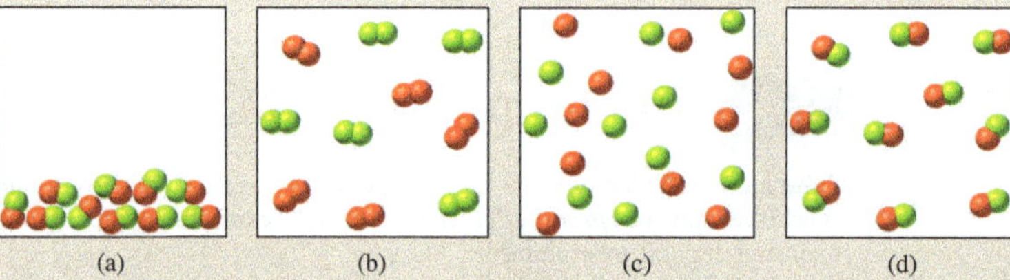

(a) (b) (c) (d)

Strategy We review the discussion of physical and chemical changes. A physical change does not change the *identity* of a substance, whereas a chemical change *does* change the identity of a substance.

Setup The diagram in (a) shows a substance that consists of molecules of a compound, each of which contains two different atoms, represented by green and red spheres. Diagram (b) contains the same number of red and green spheres, but they are not arranged the same way as in diagram (a). In (b), each molecule is made up of two identical atoms. These are molecules of *elements*, rather than molecules of a compound. Diagram (c) also contains the same numbers of red and green spheres as diagram (a). In (c), however, all the atoms are shown as isolated spheres. These are atoms of elements, rather than molecules of a compound. In diagram (d), the spheres are arranged in molecules, each containing one red and one green sphere. Although the molecules are farther apart in diagram (d), they are the same molecules as shown in diagram (a).

Solution Diagrams (b) and (c) represent chemical changes. Diagram (d) represents a physical change.

THINK ABOUT IT

A chemical change changes the *identity* of matter. A physical change does not.

Practice Problem ATTEMPT Which of the following processes is a physical change? (a) evaporation of water; (b) combination of hydrogen and oxygen gas to produce water; (c) dissolution of sugar in water; (d) separation of sodium chloride (table salt) into its constituent elements, sodium and chlorine; (e) combustion of sugar to produce carbon dioxide and water.

(Continued on next page)

Practice Problem BUILD The diagram on the left shows a system prior to a process taking place. Which of the other diagrams [(i) to (iv)] could represent the system after a *physical* process; which could represent the system after a *chemical* process; and which could not represent either?

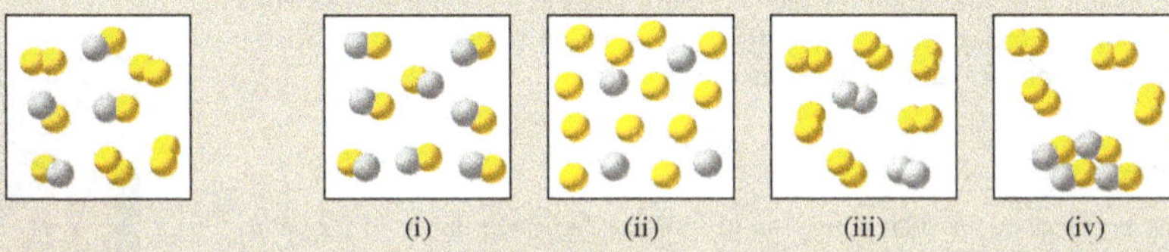

Practice Problem CONCEPTUALIZE The diagram on the left represents the result of a process. Which of the diagrams [(i) to (iii)] could represent the starting material if the process were physical, and which could represent the starting material if the change were chemical?

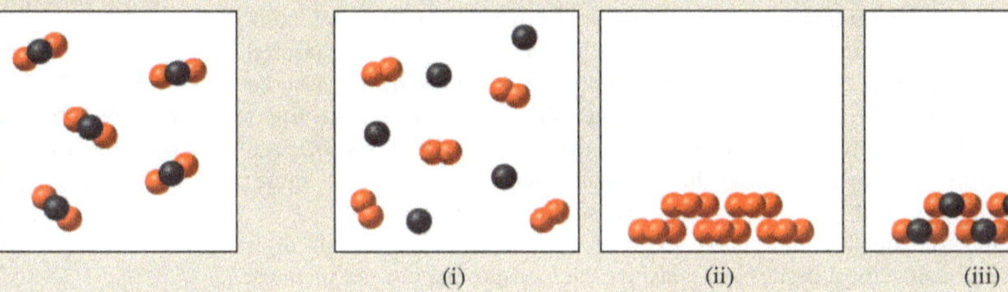

CHECKPOINT – SECTION 1.4 The Properties of Matter

1.4.1 Which of the following [(a)–(f)] represents a physical change? (Select all that apply.)

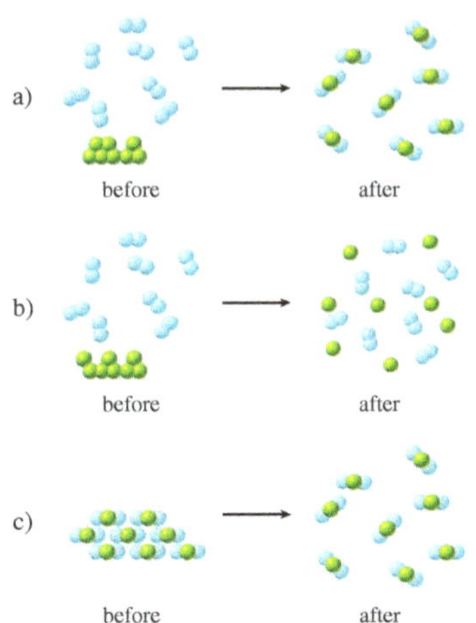

1.4.2 Which of the following [(a)–(f)] represents a chemical change? (Select all that apply.)

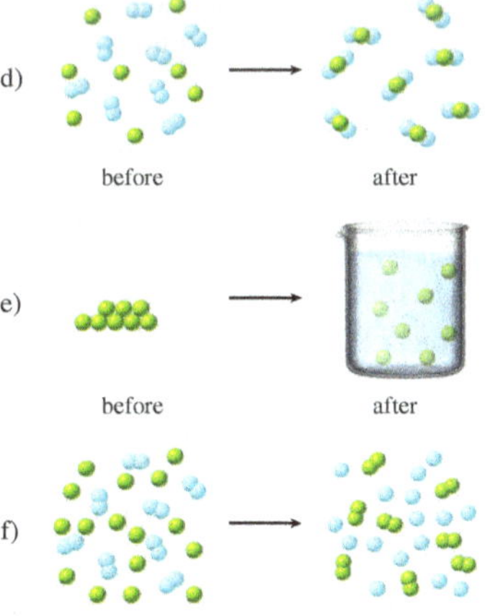

1.5 Uncertainty in Measurement

Chemistry makes use of two types of numbers: exact and inexact. *Exact* numbers include numbers with defined values, such as 2.54 in the definition 1 inch (in) = 2.54 cm, 1000 in the definition 1 kg = 1000 g, and 12 in the definition 1 dozen = 12 objects. (The number 1 in each of these definitions is also an exact number.) Exact numbers also include those that are obtained by counting. Numbers measured by any method other than counting are *inexact*.

Measured numbers are inexact because of the measuring devices that are used, the individuals who use them, or both. For example, a ruler that is poorly calibrated will result in measurements that are in error—no matter how carefully it is used. Another ruler may be calibrated properly but have insufficient resolution for the necessary measurement. Finally, whether or not an instrument is properly calibrated or has sufficient resolution, there are unavoidable differences in how different people see and interpret measurements.

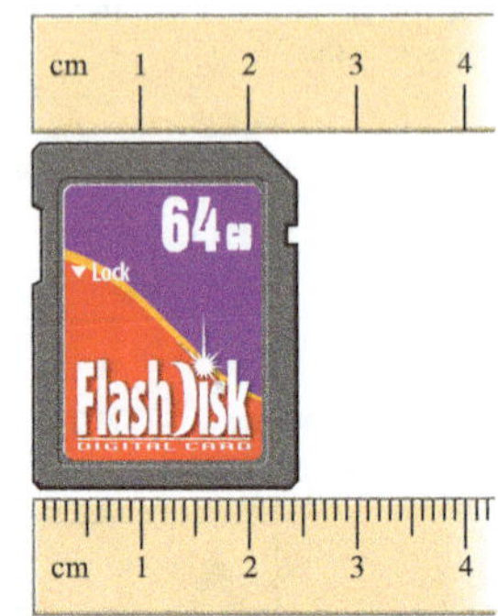

Figure 1.11 The width we report for the memory card depends on which ruler we use to measure it.

Significant Figures

An inexact number must be reported in such a way as to indicate the uncertainty in its value. This is done using significant figures. ***Significant figures*** are the *meaningful digits* in a reported number. Consider the measurement of the memory card in Figure 1.11 using the ruler above it. The card's width is between 2 and 3 cm. We may record the width as 2.5 cm, but because there are no gradations between 2 and 3 cm on this ruler, we are *estimating* the second digit. Although we are certain about the 2 in 2.5, we are *not* certain about the 5. The last digit in a measured number is referred to as the *uncertain digit;* and the uncertainty associated with a measured number is generally considered to be ± 1 in the place of the last digit. Thus, when we report the width of the memory card to be 2.5 cm, we are implying that its width is 2.5 ± 0.1 cm. Each of the digits in a measured number, including the uncertain digit, is a significant figure. The reported width of the memory card, 2.5 cm, contains two significant figures.

A ruler with millimeter gradations would enable us to be certain about the second digit in this measurement and to estimate a third digit. Now consider the measurement of the memory card using the ruler below it. We may record the width as 2.45 cm. Again, we estimate one digit beyond those we can read. The reported width of 2.45 cm contains three significant figures. Reporting the width as 2.45 cm implies that the width is 2.45 ± 0.01 cm.

Student Note: It is important not to imply greater certainty in a measured number than is realistic. For example, it would be inappropriate to report the width of the memory card in Figure 1.11 as 2.4500 cm, because this would imply an uncertainty of ± 0.0001.

The number of significant figures in any number can be determined using the following guidelines:

1. Any digit that is not zero is significant (112.1 has four significant figures).
2. Zeros located between nonzero digits are significant (305 has three significant figures, and 50.08 has four significant figures).
3. Zeros to the left of the first nonzero digit are not significant (0.0023 has two significant figures, and 0.000001 has one significant figure).
4. Zeros to the right of the last nonzero digit are significant if the number contains a decimal point (1.200 has four significant figures).
5. Zeros to the right of the last nonzero digit in a number that does not contain a decimal point may or may not be significant (100 may have one, two, or three significant figures—it is impossible to tell without additional information). To avoid ambiguity in such cases, it is best to express such numbers using scientific notation [⏭ Appendix 1]. If the intended number of significant figures is one, the number is written as 1×10^2; if the intended number of significant figures is two, the number is written as 1.0×10^2; and if the intended number of significant figures is three, the number is written as 1.00×10^2.

Sample Problem 1.5 lets you practice determining the number of significant figures in a number.

SAMPLE PROBLEM 1.5

Determine the number of significant figures in the following measurements: (a) 443 cm, (b) 15.03 g, (c) 0.0356 kg, (d) 3.000×10^{-7} L, (e) 50 mL, (f) 0.9550 m.

Strategy All nonzero digits are significant, so the goal will be to determine which of the zeros is significant.

Setup Zeros are significant if they appear between nonzero digits or if they appear after a nonzero digit in a number that contains a decimal point. Zeros may or may not be significant if they appear to the right of the last nonzero digit in a number that does not contain a decimal point.

Solution (a) 3; (b) 4; (c) 3; (d) 4; (e) 1 or 2, an ambiguous case; (f) 4.

(Continued on next page)

THINK ABOUT IT

Be sure that you have identified zeros correctly as either significant or not significant. They are significant in (b), (d), and (f); they are not significant in (c); and it is not possible to tell in (e).

Practice Problem ATTEMPT Determine the number of significant figures in the following measurements: (a) 1129 m, (b) 0.0003 kg, (c) 1.094 cm, (d) 3.5×10^{12} atoms, (e) 150 mL, (f) 9.550 km.

Practice Problem BUILD For each of the following numbers, determine the number of significant figures it contains, rewrite it without using scientific notation, and determine the number of significant figures in the result. (a) 3.050×10^{-4}, (b) 4.3200×10^{2}, (c) 8.001×10^{-7}, (d) 2.006080×10^{5}, (e) 1.503×10^{-5}, (f) 6.07510×10^{4}.

Practice Problem CONCEPTUALIZE Report the number of colored objects contained within each square and, in each case, indicate the number of significant figures in the number you report.

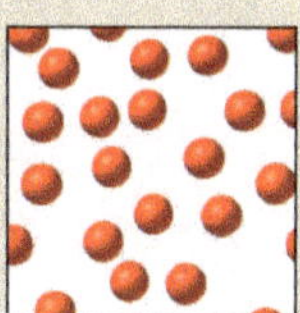

Calculations with Measured Numbers

Because we often use one or more measured numbers to calculate a desired result, a second set of guidelines specifies how to handle significant figures in calculations.

1. In addition and subtraction, the answer cannot have more digits to the right of the decimal point than the original number with the smallest number of digits to the right of the decimal point. For example:

   102.50 ← two digits after the decimal point
   + 0.231 ← three digits after the decimal point
   102.731 ← round to 102.73

   143.29 ← two digits after the decimal point
   −20.1 ← one digit after the decimal point
   123.19 ← round to 123.2

   The rounding procedure works as follows. Suppose we want to round 102.13 and 54.86 each to one digit to the right of the decimal point. To begin, we look at the digit(s) that will be dropped. If the leftmost digit to be dropped is less than 5, as in 102.13, we *round down* (to 102.1), meaning that we simply drop the digit(s). If the leftmost digit to be dropped is equal to or greater than 5, as in 54.86, we *round up* (to 54.9), meaning that we add 1 to the preceding digit.

2. In multiplication and division, the number of significant figures in the final product or quotient is determined by the original number that has the smallest number of significant figures. The following examples illustrate this rule:

   $1.4 \times 8.011 = 11.2154$ ← round to 11 (limited by 1.4 to *two* significant figures)

   $\frac{11.57}{305.88} = 0.037825290964$ ← round to 0.03783 (limited by 11.57 to *four* significant figures)

3. *Exact numbers* can be considered to have an infinite number of significant figures and do not limit the number of significant figures in a calculated result. For example, a penny minted after 1982 has a mass of 2.5 g. If we have three such pennies, the total mass is

   $$3 \times 2.5 \text{ g} = 7.5 \text{ g}$$

   Student Note: Note that it is the number of pennies (3), not the mass, that is an exact number.

   The answer should *not* be rounded to one significant figure because 3 is an exact number.

4. In calculations with multiple steps, rounding the result of each step can result in "rounding error." Consider the following two-step calculation:

   First step: $A \times B = C$
   Second step: $C \times D = E$

Suppose that $A = 3.66$, $B = 8.45$, and $D = 2.11$. The value of E depends on whether we round off C prior to using it in the second step of the calculation.

Method 1	Method 2
$C = 3.66 \times 8.45 = 30.9$	$C = 3.66 \times 8.45 = 30.93$
$E = 30.9 \times 2.11 = 65.2$	$E = 30.93 \times 2.11 = 65.3$

In general, it is best to retain at least one extra digit until the end of a multistep calculation, as shown by method 2, to minimize rounding error.

Sample Problems 1.6 and 1.7 show how significant figures are handled in arithmetic operations.

SAMPLE PROBLEM 1.6

Perform the following arithmetic operations and report the result to the proper number of significant figures: (a) 317.5 mL + 0.675 mL, (b) 47.80 L − 2.075 L, (c) 13.5 g ÷ 45.18 L, (d) 6.25 cm × 1.175 cm, (e) 5.46×10^2 g + 4.991×10^3 g.

Strategy Apply the rules for significant figures in calculations, and round each answer to the appropriate number of digits.

Setup (a) The answer will contain one digit to the right of the decimal point to match 317.5, which has the fewest digits to the right of the decimal point. (b) The answer will contain two digits to the right of the decimal point to match 47.80. (c) The answer will contain three significant figures to match 13.5, which has the fewest number of significant figures in the calculation. (d) The answer will contain three significant figures to match 6.25. (e) To add numbers expressed in scientific notation, first write both numbers to the same power of 10. That is, $4.991 \times 10^3 = 49.91 \times 10^2$, so the answer will contain two digits to the right of the decimal point (when multiplied by 10^2) to match both 5.46 and 49.91.

Solution

(a)
$$\begin{array}{r} 317.5\text{ mL} \\ +\ 0.675\text{ mL} \\ \hline 318.175\text{ mL} \end{array} \quad \leftarrow \text{round to } 318.2\text{ mL}$$

(b)
$$\begin{array}{r} 47.80\text{ L} \\ -2.075\text{ L} \\ \hline 45.725\text{ L} \end{array} \quad \leftarrow \text{round to } 45.73\text{ L}$$

(c) $$\frac{13.5\text{ g}}{45.18\text{ L}} = 0.298804781\text{ g/L} \quad \leftarrow \text{round to } 0.299\text{ g/L}$$

(d) $6.25\text{ cm} \times 1.175\text{ cm} = 7.34375\text{ cm}^2 \quad \leftarrow$ round to 7.34 cm^2

(e)
$$\begin{array}{r} 5.46 \times 10^2\text{ g} \\ +\ 49.91 \times 10^2\text{ g} \\ \hline 55.37 \times 10^2\text{ g} = 5.537 \times 10^3\text{ g} \end{array}$$

THINK ABOUT IT

It may look as though the rule of addition has been violated in part (e) because the final answer (5.537×10^3 g) has three places past the decimal point, not two. However, the rule was applied to get the answer 55.37×10^2 g, which has *four* significant figures. Changing the answer to correct scientific notation doesn't change the number of significant figures, but in this case it changes the number of places past the decimal point.

Practice Problem ATTEMPT Perform the following arithmetic operations, and report the result to the proper number of significant figures: (a) 105.5 L + 10.65 L, (b) 81.058 m − 0.35 m, (c) 3.801×10^{21} atoms + 1.228×10^{19} atoms, (d) 1.255 dm × 25 dm, (e) 139 g ÷ 275.55 mL.

Practice Problem BUILD Perform the following arithmetic operations, and report the result to the proper number of significant figures: (a) 1.0267 cm × 2.508 cm × 12.599 cm, (b) 15.0 kg ÷ 0.036 m^3, (c) 1.113×10^{10} kg − 1.050×10^9 kg, (d) 25.75 mL + 15.00 mL, (e) 46 cm^3 + 180.5 cm^3.

Practice Problem CONCEPTUALIZE A citrus dealer in Florida sells boxes of 100 oranges at a roadside stand. The boxes routinely are packed with one to three extra oranges to help ensure that customers are happy with their purchases. The average weight of an orange is 7.2 ounces, and the average weight of the boxes in which the oranges are packed is 3.2 pounds. Determine the total weight of five of these 100-orange boxes.

SAMPLE PROBLEM 1.7

An empty container with a volume of 9.850×10^2 cm^3 is weighed and found to have a mass of 124.6 g. The container is filled with a gas and reweighed. The mass of the container and the gas is 126.5 g. Determine the density of the gas to the appropriate number of significant figures.

Strategy This problem requires two steps: subtraction to determine the mass of the gas, and division to determine its density. Apply the corresponding rule regarding significant figures to each step.

Setup In the subtraction of the container mass from the combined mass of the container and the gas, the result can have only one place past the decimal point: 126.5 g − 124.6 g = 1.9 g. Thus, in the division of the mass of the gas by the volume of the container, the result can have only two significant figures.

Solution

$$\begin{array}{r} 126.5\text{ g} \\ -124.6\text{ g} \\ \hline \text{mass of gas} = \quad 1.9\text{ g} \end{array} \quad \leftarrow \text{one place past the decimal point (two significant figures)}$$

$$\text{density} = \frac{1.9\text{ g}}{9.850 \times 10^2\text{ cm}^3} = 0.00193\text{ g/cm}^3 \leftarrow \text{round to } 0.0019\text{ g/cm}^3$$

The density of the gas is 1.9×10^{-3} g/cm^3.

THINK ABOUT IT

In this case, although each of the three numbers we started with has *four* significant figures, the solution has only *two* significant figures.

Practice Problem ATTEMPT An empty container with a volume of 150.0 cm^3 is weighed and found to have a mass of 72.5 g. The container is filled with a liquid and reweighed. The mass of the container and the liquid is 194.3 g. Determine the density of the liquid to the appropriate number of significant figures.

Practice Problem BUILD Another empty container with an unknown volume is weighed and found to have a mass of 81.2 g. The container is then filled with a liquid with a density of 1.015 g/cm^3 and reweighed. The mass of the container and the liquid is 177.9 g. Determine the volume of the container to the appropriate number of significant figures.

Practice Problem CONCEPTUALIZE Several pieces of aluminum metal with a total mass of 11.63 g are dropped into a graduated cylinder of water to determine their combined volume. The graduated cylinder is shown before and after the metal has been added. Use the information shown here to determine the density of aluminum. Be sure to report your answer to the appropriate number of significant figures.

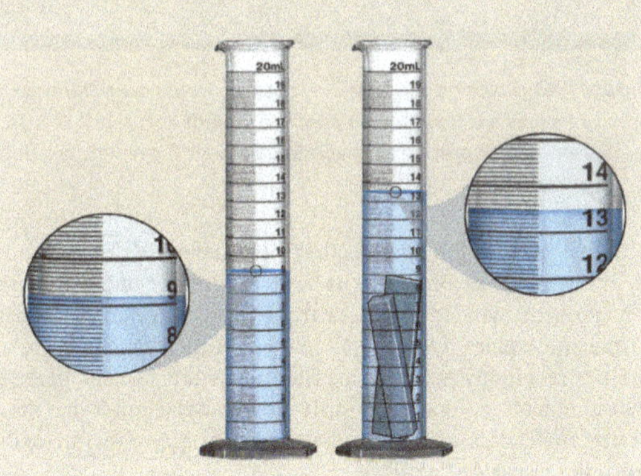

Accuracy and Precision

Accuracy and precision are two ways to gauge the quality of a set of measured numbers. Although the difference between the two terms may be subtle, it is important. ***Accuracy*** tells us how close a measurement is to the *true* value. ***Precision*** tells us how close multiple measurements of the same thing are to one another (Figure 1.12).

Suppose that three students are asked to determine the mass of an aspirin tablet. Each student weighs the aspirin tablet three times. The results (in grams) are

	Student A	Student B	Student C
	0.335	0.357	0.369
	0.331	0.375	0.373
	0.333	0.338	0.371
Average value	0.333	0.357	0.371

(a)

(b)

(c)

Figure 1.12 The distribution of papers shows the difference between accuracy and precision. (a) Good accuracy and good precision. (b) Poor accuracy but good precision. (c) Poor accuracy and poor precision.

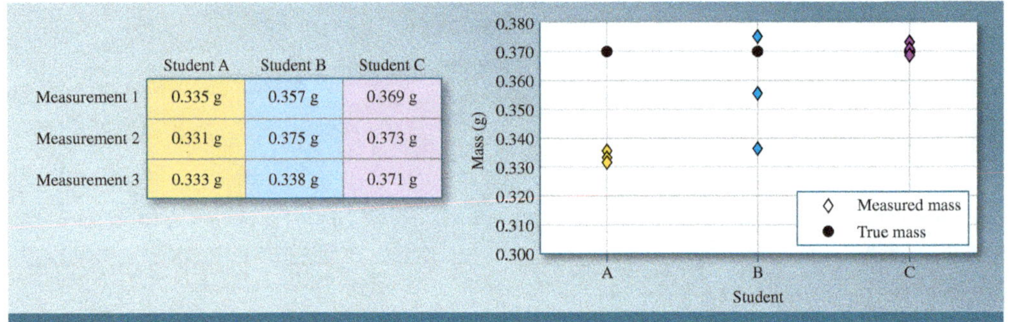

	Student A	Student B	Student C
Measurement 1	0.335 g	0.357 g	0.369 g
Measurement 2	0.331 g	0.375 g	0.373 g
Measurement 3	0.333 g	0.338 g	0.371 g

Figure 1.13 Graphing the students' data illustrates the difference between precision and accuracy. Student A's results are precise (values are close to one another) but not accurate because the average value is far from the true value. Student B's results are neither precise nor accurate. Student C's results are both precise and accurate.

The true mass of the tablet is 0.370 g. Student A's results are more precise than those of student B, but neither set of results is very accurate. Student C's results are both precise (very small deviation of individual masses from the average mass) and accurate (average value very close to the true value). Figure 1.13 shows all three students' results in relation to the true mass of the tablet. Highly accurate measurements are usually precise, as well, although highly precise measurements do not necessarily guarantee accurate results. For example, an improperly calibrated meterstick or a faulty balance may give precise readings that are significantly different from the correct value.

CHECKPOINT – SECTION 1.5 Uncertainty in Measurement

1.5.1 What volume of water does the graduated cylinder contain (to the proper number of significant figures)?

a) 32.2 mL
b) 30.25 mL
c) 32.5 mL
d) 32.50 mL
e) 32.500 mL

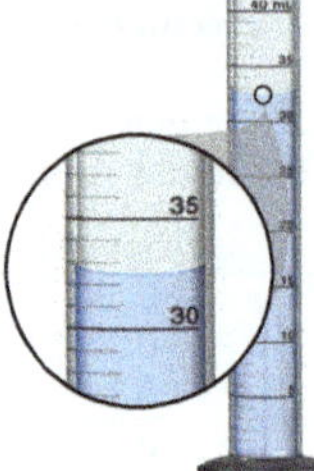

1.5.2 Which of the following is the sum of the following numbers to the correct number of significant figures?

$$3.115 + 0.2281 + 712.5 + 45 =$$

a) 760.8431
b) 760.843
c) 760.84
d) 760.8
e) 761

1.5.3 The true dependence of y on x is represented by the line. Three students measured y as a function of x and plotted their data on the graph. Which set of data has the best accuracy and which has the best precision, respectively?

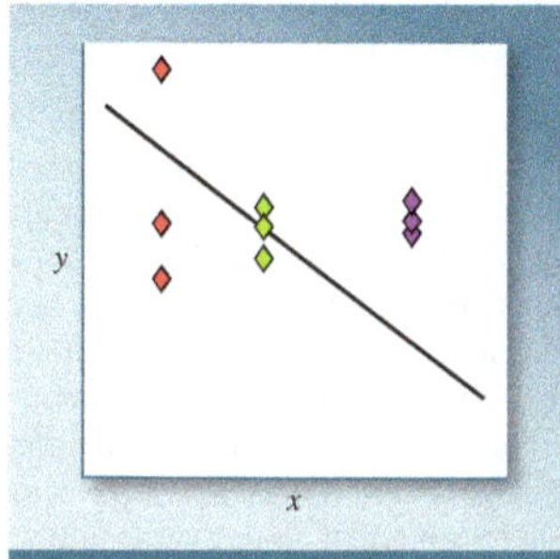

a) red, green
b) green, green
c) green, purple
d) purple, purple
e) purple, green

1.5.4 What is the result of the following calculation to the correct number of significant figures?

$$(6.266 - 6.261) \div 522.0 =$$

a) 9.5785×10^{-6}
b) 9.579×10^{-6}
c) 9.58×10^{-6}
d) 9.6×10^{-6}
e) 1×10^{-5}

1.6 Using Units and Solving Problems

Solving problems correctly in chemistry requires careful manipulation of both numbers and units. Paying attention to the units will benefit you greatly as you proceed through this, or any other, science course.

Conversion Factors

A ***conversion factor*** is a fraction in which the same quantity is expressed one way in the numerator and another way in the denominator. By definition, for example, 1 in = 2.54 cm. We can derive a conversion factor from this equality by writing it as the following fraction:

$$\frac{1\text{ in}}{2.54\text{ cm}}$$

Because the numerator and denominator express the same length, this fraction is equal to 1; as a result, we can equally well write the conversion factor as

$$\frac{2.54\text{ cm}}{1\text{ in}}$$

Because both forms of this conversion factor are equal to 1, we can multiply a quantity by either form without changing the value of that quantity. This is useful for changing the units in which a given quantity is expressed—something you will do often throughout this text. For instance, if we need to convert a length from inches to centimeters, we multiply the length in inches by the appropriate conversion factor.

$$12.00\text{ }\cancel{\text{in}} \times \frac{2.54\text{ cm}}{1\text{ }\cancel{\text{in}}} = 30.48\text{ cm}$$

We chose the form of the conversion factor that cancels the unit inches and produces the desired unit, centimeters. The result contains four significant figures because exact numbers, such as those obtained from definitions, do not limit the number of significant figures in the result of a calculation. Thus, the number of significant figures in the answer to this calculation is based on the number 12.00, not the number 2.54.

How Can I Enhance My Chances of Success in Chemistry Class?

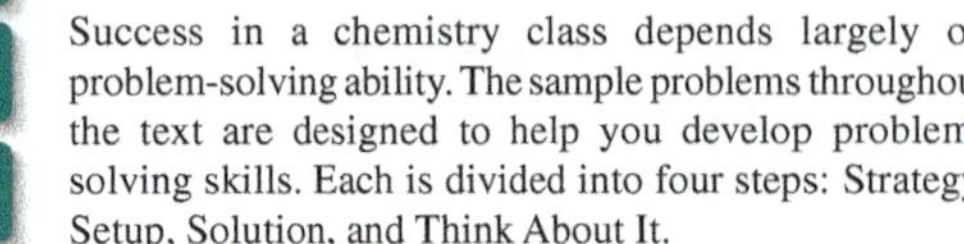

Success in a chemistry class depends largely on problem-solving ability. The sample problems throughout the text are designed to help you develop problem-solving skills. Each is divided into four steps: Strategy, Setup, Solution, and Think About It.

Strategy: Read the problem carefully and determine what is being asked and what information is provided. The Strategy step is where you should think about what skills are required and lay out a plan for solving the problem. Give some thought to what you expect the result to be. If you are asked to determine the number of atoms in a sample of matter, for example, you should expect the answer to be a whole number. Determine what, if any, units should be associated with the result. When possible, make a ballpark estimate of the magnitude of the correct result, and make a note of your estimate.

Setup: Next, gather the information necessary to solve the problem. Some of the information will have been given in the problem itself. Other information, such as equations, constants, and tabulated data (including atomic masses) should also be brought together in this step. Write down and label clearly all of the information you will use to solve the problem. Be sure to write appropriate units with each piece of information.

Solution: Using the necessary equations, constants, and other information, calculate the answer to the problem. Pay particular attention to the units associated with each number, tracking and canceling units carefully throughout the calculation. In the event that multiple calculations are required, label any intermediate results, but don't round to the necessary number of significant figures until the final calculation. Always carry at least one extra significant figure in intermediate calculations. Make sure that the final answer has the correct number of significant figures.

Think About It: Consider your calculated result and ask yourself whether or not it makes sense. Compare the units and the magnitude of your result with your ballpark estimate from the Strategy step. If your result does not have the appropriate units, or if its magnitude or sign is not reasonable, check your solution for possible errors. A very important part of problem solving is being able to judge whether the answer is reasonable. It is relatively easy to spot a wrong sign or incorrect units, but you should also develop a sense of magnitude and be able to tell when an answer is either way too big or way too small. For example, if a problem asks how many molecules are in a sample and you calculate a number that is less than 1, you should know that it cannot be correct.

Finally, each sample problem is followed by three practice problems. The first, "Attempt," typically is a very similar problem that can be solved using the same strategy. The second and third, "Build" and "Conceptualize," generally test the same skills, but require approaches slightly different from the one used to solve the preceding sample and practice problems.

Regular use of the sample problems and practice problems in this text can help you develop an effective set of problem-solving skills. They can also help you assess whether you are ready to move on to the next new concepts. If you struggle with the practice problems, then you probably need to review the corresponding sample problem and the concepts that led up to it.

Dimensional Analysis—Tracking Units

The use of conversion factors in problem solving is called ***dimensional analysis*** or the *factor-label method*. Many problems require the use of more than one conversion factor. The conversion of 12.00 inches into meters, for example, takes two steps: one to convert inches to centimeters, which we have already demonstrated, and one to convert centimeters to meters. The additional conversion factor required is derived from the equality

$$1 \text{ m} = 100 \text{ cm}$$

and is expressed as either

$$\frac{100 \text{ cm}}{1 \text{ m}} \quad \text{or} \quad \frac{1 \text{ m}}{100 \text{ cm}}$$

We must choose the conversion factor that will introduce the unit meter and cancel the unit centimeter (i.e., the one on the right). We can set up a problem of this type as the following series of unit conversions so that it is unnecessary to calculate an intermediate answer at each step:

$$12.00 \cancel{\text{in}} \times \frac{2.54 \cancel{\text{cm}}}{1 \cancel{\text{in}}} \times \frac{1 \text{ m}}{100 \cancel{\text{cm}}} = 0.3048 \text{ m}$$

Careful tracking of units and their cancellation can be a valuable tool in checking your work. If we had accidentally used the *reciprocal* of one of the conversion factors, the resulting units would have been something other than meters. Unexpected or nonsensical units can reveal an error in your problem-solving strategy.

Student Note: If we had accidentally used the reciprocal of the conversion from centimeters to meters, the result would have been 3048 cm^2/m, which would make no sense—both because the units are nonsensical and because the numerical result is not reasonable. You know that 12 inches is a foot and that a foot is not equal to *thousands* of meters!

Sample Problem 1.8 shows how to derive conversion factors and use them to do unit conversions.

SAMPLE PROBLEM 1.8

The Food and Drug Administration (FDA) recommends that dietary sodium intake be no more than 2400 mg per day. What is this mass in pounds (lb)? (1 lb = 453.6 g)

Strategy This problem requires a two-step dimensional analysis, because we must convert milligrams to grams and then grams to pounds. Assume the number 2400 has four significant figures.

Setup The necessary conversion factors are derived from the equalities 1 g = 1000 mg and 1 lb = 453.6 g.

$$\frac{1\text{ g}}{1000\text{ mg}} \text{ or } \frac{1000\text{ mg}}{1\text{ g}} \quad \text{and} \quad \frac{1\text{ lb}}{453.6\text{ g}} \text{ or } \frac{453.6\text{ g}}{1\text{ lb}}$$

From each pair of conversion factors, we select the one that will result in the proper unit cancellation.

Solution

$$2400\text{ mg} \times \frac{1\text{ g}}{1000\text{ mg}} \times \frac{1\text{ lb}}{453.6\text{ g}} = 0.005291\text{ lb}$$

THINK ABOUT IT

Make sure that the magnitude of the result is reasonable and that the units have canceled properly. Because pounds are much larger than milligrams, a given mass will be a much smaller number of pounds than of milligrams. If we had mistakenly multiplied by 1000 and 453.6 instead of dividing by them, the result (2400 mg × 1000 mg/g × 453.6 g/lb = 1.089×10^9 mg^2/lb) would be unreasonably large—and the units would not have canceled properly.

Practice Problem ATTEMPT The American Heart Association recommends that healthy adults limit dietary cholesterol to no more than 300 mg per day. Convert this mass of cholesterol to ounces (1 oz = 28.3459 g). Assume 300 mg has just one significant figure.

Practice Problem BUILD An object has a mass of 24.98 oz. What is its mass in grams?

Practice Problem CONCEPTUALIZE The diagram contains several objects that are constructed using colored blocks and grey connectors. Note that each of the objects is essentially identical, consisting of the same number and arrangement of blocks and connectors. Give the appropriate conversion factor for each of the specified operations.

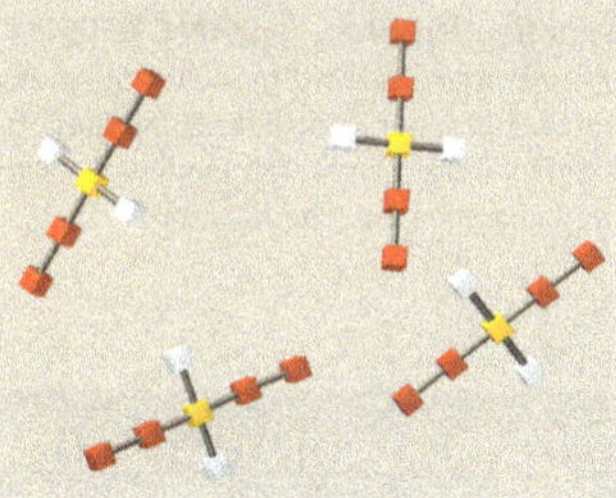

(a) We know the number of objects and wish to determine the number of red blocks.

(b) We know the number of yellow blocks and wish to determine the number of objects.

(c) We know the number of yellow blocks and wish to determine the number of white blocks.

(d) We know the number of grey connectors and wish to determine the number of yellow blocks.

Many familiar quantities require units raised to specific powers. For example, an area may be expressed in units of *length* squared (e.g., square meters, m^2, or square inches, in^2). Volumes sometimes are expressed in units of length *cubed* (e.g., cubic feet, ft^3, or cubic centimeters, cm^3). More often, though, volumes are expressed in liters (L) or milliliters (mL). It's important to remember that these are the common names given to specific units of length cubed. The liter is defined as a deciliter (dL) cubed: 1 L = 1 dL^3; and the milliliter is defined as the centimeter cubed: 1 mL = 1 cm^3. (See Figure 1.10.) When units are squared or cubed, special care must be taken when using them in dimensional analysis. For example, converting from cubic meters to cubic centimeters requires the following operation:

$$1\text{ m}^3 \times \frac{100\text{ cm}}{1\text{ m}} \times \frac{100\text{ cm}}{1\text{ m}} \times \frac{100\text{ cm}}{1\text{ m}} = 1.00 \times 10^6\text{ cm}^3$$

or

$$1\text{ m}^3 \times \left(\frac{100\text{ cm}}{1\text{ m}}\right)^3 = 1.00 \times 10^6\text{ cm}^3$$

Failing to raise the conversion factor to the same power as the unit itself is a common error—and one that can happen easily when the units L or mL appear—because they do not explicitly show the power 3.

Sample Problem 1.9 shows how to handle problems in which conversion factors are squared or cubed in dimensional analysis.

SAMPLE PROBLEM 1.9

An average adult has 5.2 L of blood. What is the volume of blood in cubic meters?

Strategy There are several ways to solve a problem such as this. One way is to convert liters to cubic centimeters and then cubic centimeters to cubic meters.

Setup $1\ L = 1000\ cm^3$ and $1\ cm = 1 \times 10^{-2}\ m$. When a unit is raised to a power, the corresponding conversion factor must also be raised to that power in order for the units to cancel appropriately.

Solution

$$5.2\ L \times \frac{1000\ cm^3}{1\ L} \times \left(\frac{1 \times 10^{-2}\ m}{1\ cm}\right)^3 = 5.2 \times 10^{-3}\ m^3$$

THINK ABOUT IT

Based on the preceding conversion factors, $1\ L = 1 \times 10^{-3}\ m^3$. Therefore, 5 L of blood would be equal to $5 \times 10^{-3}\ m^3$, which is very close to the calculated answer.

Practice Problem ATTEMPT The density of silver is 10.5 g/cm^3. What is its density in kg/m^3?

Practice Problem BUILD The density of mercury is 13.6 g/cm^3. What is its density in mg/mm^3?

Practice Problem CONCEPTUALIZE Each diagram [(i) or (ii)] shows the objects contained within a cubical space. In each case, determine to the appropriate number of significant figures the number of objects that would be contained within a cubical space in which the length of the cube's edge is exactly five times that of the cube shown in the diagram.

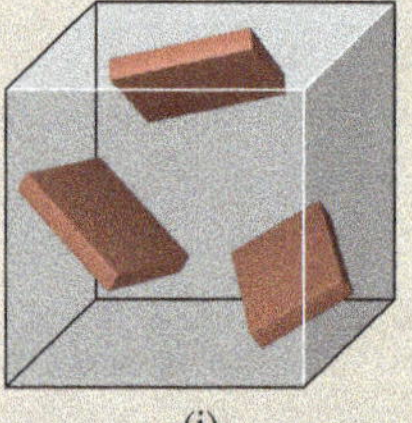

(i)

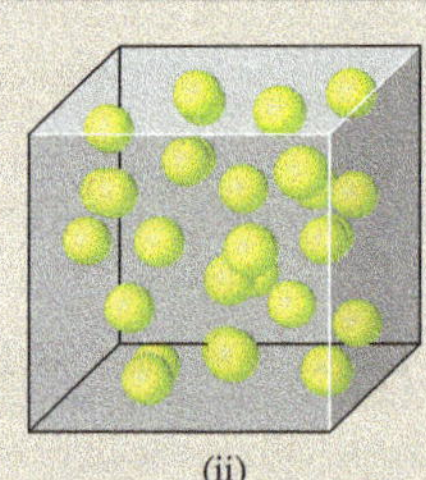

(ii)

CHECKPOINT – SECTION 1.6 Using Units and Solving Problems

1.6.1 The density of lithium metal is 535 kg/m^3. What is this density in g/cm^3?

a) 0.000535 g/cm^3
b) 0.535 g/cm^3
c) 0.0535 g/cm^3
d) 0.54 g/cm^3
e) 53.5 g/cm^3

1.6.2 Convert 43.1 cm^3 to liters.

a) 43.1 L
b) 43,100 L
c) 0.0431 L
d) 4310 L
e) 0.043 L

1.6.3 What is the volume of a 5.75-g object that has a density of 3.97 g/cm^3?

a) 1.45 cm^3
b) 0.690 cm^3
c) 22.8 cm^3
d) 0.0438 cm^3
e) 5.75 cm^3

1.6.4 How many cubic centimeters are there in a cubic meter?

a) 10
b) 100
c) 1000
d) 1×10^4
e) 1×10^6

Chapter Summary

Section 1.1

- ***Chemistry*** is the study of ***matter*** and the changes matter undergoes.
- Chemists go about research using a set of guidelines and practices known as the ***scientific method,*** in which observations give rise to ***laws,*** data give rise to ***hypotheses,*** hypotheses are tested with experiments, and successful hypotheses give rise to ***theories,*** which are further tested by experiment.

Section 1.2

- All matter exists either as a ***substance*** or as a mixture of substances. Substances may be ***elements*** (containing only one kind of atom) or ***compounds*** (containing two or more kinds of atoms). A ***mixture*** may be ***homogeneous*** (a solution) or ***heterogeneous.*** Mixtures may be separated using physical processes. Compounds can be separated into their constituent elements using chemical processes. Elements cannot be separated into simpler substances.

Section 1.3

- Scientists use a system of units referred to as the ***International System of Units*** or ***SI units.***
- There are seven *base* SI units including the kilogram (for ***mass***) and the ***kelvin*** (for temperature). SI units for such quantities as volume and ***density*** are derived from the base units.

Section 1.4

- Substances are identified by their ***quantitative*** (involving numbers) and ***qualitative*** (not involving numbers) properties.
- ***Physical properties*** are those that can be determined without the matter in question undergoing a chemical change. A ***physical change*** is one in which the identity of the matter involved does not change.
- ***Chemical properties*** are determined only as the result of a ***chemical change,*** in which the original substance is converted to a different substance. Physical and chemical properties may be ***extensive*** (dependent on the amount of matter) or ***intensive*** (independent of the amount of matter).

Section 1.5

- Measured numbers are *inexact*. Numbers obtained by counting or that are part of a definition are *exact* numbers.
- ***Significant figures*** are used to specify the uncertainty in a measured number or in a number calculated using measured numbers. Significant figures must be carried through calculations such that the implied uncertainty in the final answer is reasonable.
- ***Accuracy*** refers to how close measured numbers are to a *true* value. ***Precision*** refers to how close measured numbers are to *one another*.

Section 1.6

- A ***conversion factor*** is a fraction in which the numerator and denominator are the same quantity expressed in different units. Multiplying by a conversion factor is *unit conversion*.
- ***Dimensional analysis*** is a series of unit conversions used in the solution of a multistep problem.

Key Words

Key Equations

1.1 $K = °C + 273.15$	Temperature in kelvins is determined by adding 273.15 to the temperature in Celsius. Often we simply add 273, depending on the precision with which the Celsius temperature is known.
1.2 $\text{temp in °C} = (\text{temp °F} - 32°F) \times \frac{5°C}{9°F}$	Temperature in Fahrenheit is used to determine temperature in Celsius.
1.3 $\text{temp in °F} = \frac{9°F}{5°C} \times (\text{temp in °C}) + 32°F$	Temperature in Celsius is used to determine temperature in Fahrenheit.
1.4 $d = \frac{m}{V}$	Density is the ratio of mass to volume. For liquids and solids, densities are typically expressed in g/cm^3.

KEY SKILLS Dimensional Analysis

Solving problems in chemistry often involves mathematical combinations of measured values and constants. A conversion factor is a fraction (equal to one) derived from an equality. For example, 1 inch is, by definition, equal to 2.54 centimeters:

$$1 \text{ in} = 2.54 \text{ cm}$$

We can derive two different conversion factors from this equality:

$$\frac{1 \text{ in}}{2.54 \text{ cm}} \quad \text{or} \quad \frac{2.54 \text{ cm}}{1 \text{ in}}$$

Which fraction we use depends on what units we start with, and what units we expect our result to have. If we are converting a distance given in centimeters to inches, we multiply by the first fraction.

$$37.6 \text{ cm} \times \frac{1 \text{ in}}{2.54 \text{ cm}} \Rightarrow 14.8 \text{ in}$$

If we are converting a distance given in inches to centimeters, we multiply by the second fraction.

$$5.23 \text{ in} \times \frac{2.54 \text{ cm}}{1 \text{ in}} \Rightarrow 13.3 \text{ cm}$$

In each case, the units cancel to give the desired units in the result.

When a unit is raised to a power to express, for example, an area (cm^2) or a volume (cm^3), the conversion factor must be raised to the same power. For example, converting an area expressed in square centimeters to square inches requires that we square the conversion factor; converting a volume expressed in cubic centimeters to cubic meters requires that we cube the conversion factor. The following individual flowcharts converting an area in cm^2 to m^2 show why this is so:

$$48.5 \text{ cm}^2 \Rightarrow 48.5 \text{ cm} \times \text{cm} \qquad \left(\frac{1 \text{ in}}{2.54 \text{ cm}}\right)^2 \Rightarrow \frac{1 \text{ in}}{2.54 \text{ cm}} \times \frac{1 \text{ in}}{2.54 \text{ cm}}$$

$$48.5 \text{ cm} \times \text{cm} \times \frac{1 \text{ in}}{2.54 \text{ cm}} \times \frac{1 \text{ in}}{2.54 \text{ cm}} \Rightarrow 7.52 \text{ in}^2$$

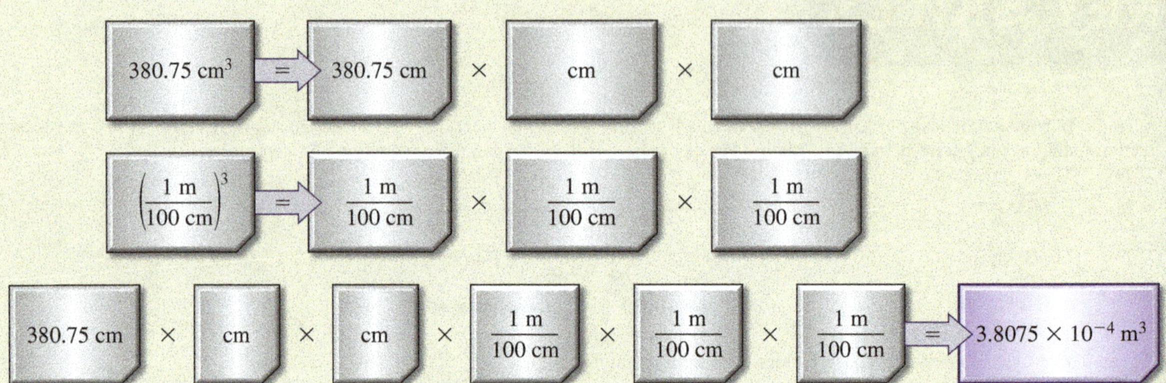

Failure to raise the conversion factor to the appropriate power would result in units not canceling properly.

Often the solution to a problem requires several different conversions, which can be combined on a single line. For example: If we know that a 157-lb athlete running at 7.09 miles per hour consumes 55.8 cm^3 of oxygen per kilogram of body weight for every minute spent running, we can calculate how many liters of oxygen this athlete consumes by running 10.5 miles (1 kg = 2.2046 lb, 1 L = 1 dm^3).

$$157\ \text{lb} \times \frac{1\ \text{kg}}{2.2046\ \text{lb}} \times \frac{1\ \text{h}}{7.09\ \text{mi}} \times \frac{55.8\ \text{cm}^3}{\text{kg}\cdot\text{min}} \times \frac{60\ \text{min}}{1\ \text{h}} \times \left(\frac{1\ \text{dm}}{10\ \text{cm}}\right)^3 \times 10.5\ \text{mi} = 353\ \text{dm}^3$$

$$353\ \text{dm}^3 = 353\ \text{L}$$

Key Skills Problems

1.1
Given that the density of gold is 19.3 g/cm^3, calculate the volume (in cm^3) of a gold nugget with a mass of 5.98 g.

(a) 3.23 cm^3 (b) 5.98 cm^3 (c) 115 cm^3 (d) 0.310 cm^3 (e) 13.3 cm^3

1.2
The SI unit for energy is the joule (J), which is equal to the kinetic energy possessed by a 2.00-kg mass moving at 1.00 m/s. Convert this velocity to mph (1 mi = 1.609 km).

(a) 4.47×10^{-7} mph (b) 5.79×10^{6} mph (c) 5.79 mph (d) 0.0373 mph (e) 2.24 mph

1.3
Determine the density of the following object in g/cm^3. A cube with edge length = 0.750 m and mass = 14.56 kg.

(a) 0.0345 g/cm^3 (b) 1.74 g/cm^3 (c) 670 g/cm^3 (d) 53.8 g/cm^3 (e) 14.6 g/cm^3

1.4
A 28-kg child can consume a maximum of 23 children's acetaminophen tablets in an 8-h period without exceeding the safety-limit maximum allowable dose. Given that each children's tablet contains 80 mg of acetaminophen, determine the maximum allowable dose in mg per pound of body weight for one day.

(a) 80 mg/lb (b) 90 mg/lb (c) 430 mg/lb (d) 720 mg/lb (e) 3.7 mg/lb

Questions and Problems

Applying What You've Learned

Although naturally occurring, smallpox was eradicated by a superbly coordinated effort including the World Health Organization and health care providers worldwide, the classification of smallpox as a Category A bioterrorism agent has renewed interest in its treatment and prevention. Moreover, although vaccination against the disease is considered relatively safe for most individuals, it is not entirely without risk.

The CDC estimates that 14 to 52 out of every million people who are vaccinated for smallpox will suffer serious, potentially life-threatening reactions to the vaccine. In these cases, immediate medical attention is required. The first course of treatment is with *vaccinia immune globulin* (VIG). If a patient does not respond to treatment with VIG, a second option is cidofovir, a drug that currently is approved by the Food and Drug Administration (FDA) to treat specific viral infections of the eye in individuals with compromised immune systems. Both drugs are available for use in the treatment of a serious reaction to the smallpox vaccine only through the FDA's Investigational New Drug (IND) protocol.

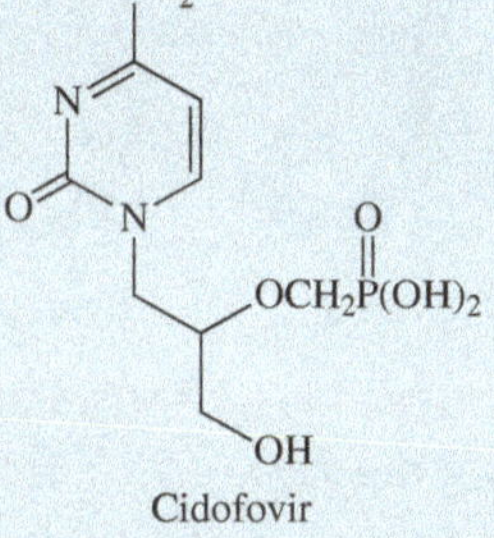

Cidofovir

Cidofovir, marketed under the name Vistide, is distributed in vials containing 375 mg of the drug dissolved in 5 mL of water. The manufacturer specifies that the drug should be kept at room temperature (68°F–77°F). The vial contents are first diluted with saline and then administered intravenously with a recommended dosage of 5 mg cidofovir per kilogram of body weight. (a) Convert cidofovir's recommended storage-temperature range to the Celsius scale [◀◀ Sample Problem 1.2]. (b) If the fluid in a single vial of cidofovir has a volume of 5.00 mL and a mass of 5.89 g, what is the density of the fluid [◀◀ Sample Problem 1.3]? (Report the density to the appropriate number of significant figures [◀◀ Sample Problem 1.6]). (c) What mass of cidofovir should be administered to a 177-lb man (1 lb = 0.4536 kg) [◀◀ Sample Problem 1.8]? (d) Convert the density in part (b) to g/L and to kg/m^3 [◀◀ Sample Problem 1.9].

SECTION 1.1: THE STUDY OF CHEMISTRY

Review Questions

1.1 Define the terms *chemistry* and *matter*.

1.2 Explain what is meant by the scientific method.

1.3 What is the difference between a hypothesis and a theory?

Computational Problems

1.4 Classify each of the following statements as a hypothesis, law, or theory. (a) Beethoven's contribution to music would have been much greater if he had married. (b) An autumn leaf gravitates toward the ground because there is an attractive force between the leaf and Earth. (c) All matter is composed of very small particles called atoms.

1.5 Classify each of the following statements as a hypothesis, law, or theory. (a) The force acting on an object is equal to its mass times its acceleration. (b) The universe as we know it started with a big bang. (c) There are many civilizations more advanced than ours on other planets.

Conceptual Problems

1.6 Identify the elements present in the following molecules (see Table 1.1).

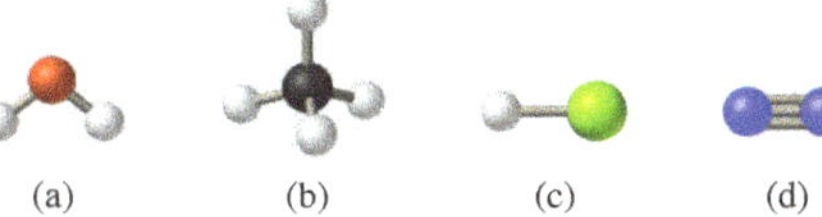

(a) (b) (c) (d)

1.7 Identify the elements present in the following molecules (see Table 1.1).

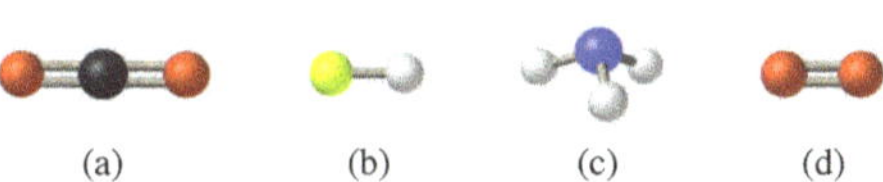

(a) (b) (c) (d)

SECTION 1.2: CLASSIFICATION OF MATTER

Review Questions

1.8 Give an example for each of the following terms: (a) matter, (b) substance, (c) mixture.

1.9 Give an example of a homogeneous mixture and an example of a heterogeneous mixture.

1.10 Give an example of an element and a compound. How do elements and compounds differ?

1.11 What is the number of known elements?

Computational Problems

1.12 Give the names of the elements represented by the chemical symbols Li, F, P, Cu, As, Zn, Cl, Pt, Mg, U, Al, Si, Ne (see the table inside the front cover).

1.13 Give the chemical symbols for the following elements: (a) potassium, (b) tin, (c) chromium, (d) boron, (e) barium, (f) plutonium, (g) sulfur, (h) argon, (i) mercury (see the table inside the front cover).

1.14 Classify each of the following substances as an element or a compound: (a) hydrogen, (b) water, (c) gold, (d) sugar.

1.15 Classify each of the following as an element, a compound, a homogeneous mixture, or a heterogeneous mixture: (a) seawater, (b) helium gas, (c) sodium chloride (salt), (d) a bottle of soft drink, (e) a milkshake, (f) air in a bottle, (g) concrete.

Conceptual Problems

1.16 Identify each of the diagrams shown here as a solid, liquid, gas, or mixture of two substances.

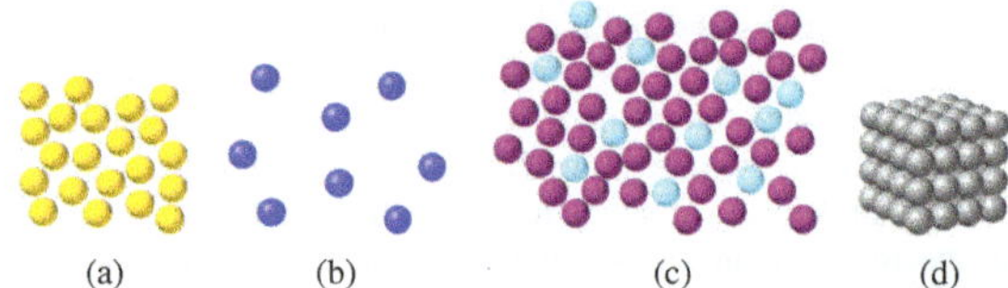

1.17 Identify each of the diagrams shown here as an element or a compound.

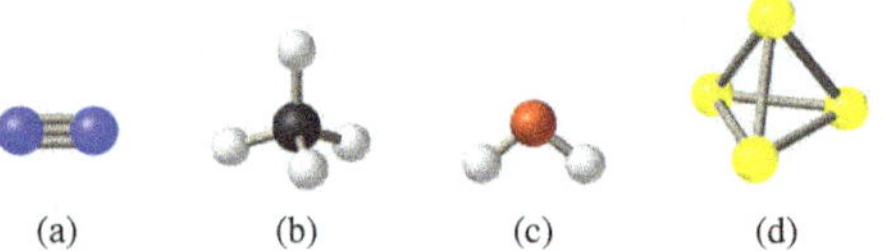

SECTION 1.3: SCIENTIFIC MEASUREMENT

Review Questions

1.18 Name the SI base units that are important in chemistry, and give the SI units for expressing the following: (a) length, (b) volume, (c) mass, (d) time, (e) temperature.

1.19 Write the numbers represented by the following prefixes: (a) mega-, (b) kilo-, (c) deci-, (d) centi-, (e) milli-, (f) micro-, (g) nano-, (h) pico-.

1.20 What units do chemists normally use for the density of liquids and solids? For the density of gas? Explain the differences.

1.21 What is the difference between mass and weight? If a person weighs 168 lb on Earth, about how much would the person weigh on the moon?

1.22 Describe the three temperature scales used in the laboratory and in everyday life: the Fahrenheit, Celsius, and Kelvin scales.

Computational Problems

1.23 Bromine is a reddish-brown liquid. Calculate its density (in g/mL) if 586 g of the substance occupies 188 mL.

1.24 The density of ethanol, a colorless liquid that is commonly known as grain alcohol, is 0.798 g/mL. Calculate the mass of 205 mL of the liquid.

1.25 Convert the following temperatures to degrees Celsius or Fahrenheit: (a) 95°F, the temperature on a hot summer day; (b) 12°F, the temperature on a cold winter day; (c) a 102°F fever; (d) a furnace operating at 1852°F; (e) −273.15°C (theoretically the lowest attainable temperature).

1.26 (a) Normally the human body can endure a temperature of 105°F for only short periods of time without permanent damage to the brain and other vital organs. What is this temperature in degrees Celsius? (b) Ethylene glycol is a liquid organic compound that is used as an antifreeze in car radiators. It freezes at −11.5°C. Calculate its freezing temperature in degrees Fahrenheit. (c) The temperature on the surface of the sun is about 6300°C. What is this temperature in degrees Fahrenheit?

1.27 The density of water at 40°C is 0.992 g/mL. What is the volume of 27.0 g of water at this temperature?

1.28 The density of platinum (Pt) is 21.5 g/cm^3 at 25°C. What is the volume of 87.6 g of Pt at this temperature?

1.29 Convert the following temperatures to kelvin: (a) 115.21°C, the melting point of sulfur; (b) 37°C, the normal body temperature; (c) 357°C, the boiling point of mercury.

1.30 Convert the following temperatures to degrees Celsius: (a) 77 K, the boiling point of liquid nitrogen, (b) 4.22 K, the boiling point of liquid helium, (c) 600.61 K, the melting point of lead.

1.31 Which of the following illustrations best represents the measurement of the temperature of boiling water using Celsius (blue) and Kelvin (red) scales? Explain.

1.32 The diagram shows balls of aluminum foil dropped into water. In one case, the ball floats on the water; in the other case, it has been squeezed into a smaller ball and sinks. What does this indicate about the relative densities of the two aluminum-foil balls? Does this mean that the density of aluminum is not constant? Explain.

SECTION 1.4: THE PROPERTIES OF MATTER

Review Questions

1.33 What is the difference between qualitative data and quantitative data?

1.34 Using examples, explain the difference between a physical property and a chemical property.

1.35 How does an intensive property differ from an extensive property?

1.36 Determine which of the following properties are intensive and which are extensive: (a) length, (b) volume, (c) temperature, (d) mass.

Computational Problems

1.37 Classify the following as qualitative or quantitative statements, giving your reasons. (a) The sun is approximately 93 million mi from Earth. (b) Leonardo da Vinci was a better painter than Michelangelo. (c) Ice is less dense than water. (d) Butter tastes better than margarine. (e) A stitch in time saves nine.

1.38 Determine whether the following statements describe chemical or physical properties: (a) Oxygen gas supports combustion. (b) Fertilizers help to increase agricultural production. (c) Water boils below 100°C on top of a mountain. (d) Lead is denser than aluminum. (e) Uranium is a radioactive element.

1.39 Determine whether each of the following describes a physical change or a chemical change: (a) The helium gas inside a balloon tends to leak out after a few hours. (b) A flashlight beam slowly gets dimmer and finally goes out. (c) Frozen orange juice is reconstituted by adding water to it. (d) The growth of plants depends on the sun's energy in a process called photosynthesis. (e) A spoonful of salt dissolves in a bowl of soup.

1.40 A student pours 44.3 g of water at 10°C into a beaker containing 115.2 g of water at 10°C. What are the final mass, temperature, and density of the combined water? The density of water at 10°C is 1.00 g/mL.

1.41 A 37.2-g sample of lead (Pb) pellets at 20°C is mixed with a 62.7 g sample of lead pellets at the same temperature. What are the final mass, temperature, and density of the combined sample? The density of Pb at 20°C is 11.35 g/cm^3.

SECTION 1.5: UNCERTAINTY IN MEASUREMENT

Review Questions

1.42 Comment on whether each of the following statements represents an exact number: (a) 50,247 tickets were sold at a sporting event, (b) 509.2 mL of water was used to make a birthday cake, (c) 3 dozen eggs were used to make a breakfast, (d) 0.41 g of oxygen was inhaled in each breath, (e) Earth orbits the sun every 365.2564 days.

1.43 What is the advantage of using scientific notation over decimal notation?

1.44 Define *significant figure*. Discuss the importance of using the proper number of significant figures in measurements and calculations.

1.45 Distinguish between the terms *accuracy* and *precision*. In general, explain why a precise measurement does not always guarantee an accurate result.

Computational Problems

1.46 Express the following numbers in scientific notation: (a) 0.000000027, (b) 356, (c) 47,764, (d) 0.096.

1.47 Express the following numbers as decimals: (a) 1.52×10^{-2}, (b) 7.78×10^{-8}, (c) 1×10^{-6}, (d) 1.6001×10^{3}.

1.48 Express the answers to the following calculations in scientific notation:
(a) $145.75 + (2.3 \times 10^{-1})$
(b) $79{,}500 \div (2.5 \times 10^{2})$
(c) $(7.0 \times 10^{-3}) - (8.0 \times 10^{-4})$
(d) $(1.0 \times 10^{4}) \times (9.9 \times 10^{6})$

1.49 Express the answers to the following calculations in scientific notation:
(a) $0.0095 + (8.5 \times 10^{-3})$
(b) $653 \div (5.75 \times 10^{-8})$
(c) $850{,}000 - (9.0 \times 10^{5})$
(d) $(3.6 \times 10^{-4}) \times (3.6 \times 10^{6})$

1.50 Determine the number of significant figures in each of the following measurements: (a) 4867 mi, (b) 56 mL, (c) 60,104 tons, (d) 2900 g, (e) 40.2 g/cm^3, (f) 0.0000003 cm, (g) 0.7 min, (h) 4.6×10^{19} atoms.

1.51 Determine the number of significant figures in each of the following measurements: (a) 0.006 L, (b) 0.0605 dm, (c) 60.5 mg, (d) 605.5 cm^2, (e) 9.60×10^{3} g, (f) 6 kg, (g) 60 m.

1.52 Carry out the following operations as if they were calculations of experimental results, and express each answer in the correct units with the correct number of significant figures:
(a) 5.6792 m + 0.6 m + 4.33 m
(b) 3.70 g − 2.9133 g
(c) 4.51 cm × 3.6666 cm

1.53 Carry out the following operations as if they were calculations of experimental results, and express each answer in the correct units with the correct number of significant figures:
(a) 7.310 km ÷ 5.70 km
(b) $(3.26 \times 10^{-3}$ mg$) - (7.88 \times 10^{-5}$ mg$)$
(c) $(4.02 \times 10^{6}$ dm$) + (7.74 \times 10^{7}$ dm$)$

1.54 Three students (A, B, and C) are asked to determine the volume of a sample of ethanol. Each student measures the volume three times with a graduated cylinder. The results in milliliters are: A (87.1, 88.2, 87.6); B (86.9, 87.1, 87.2); C (87.6, 87.8, 87.9). The true volume is 87.0 mL. Comment on the precision and the accuracy of each student's results.

1.55 Three apprentice tailors (X, Y, and Z) are assigned the task of measuring the seam of a pair of trousers. Each one makes three measurements. The results in inches are X (31.5, 31.6, 31.4); Y (32.8, 32.3, 32.7); Z (31.9, 32.2, 32.1). The true length is 32.0 in. Comment on the precision and the accuracy of each tailor's measurements.

SECTION 1.6: USING UNITS AND SOLVING PROBLEMS

Computational Problems

1.56 Carry out the following conversions: (a) 22.6 m to decimeters, (b) 25.4 mg to kilograms, (c) 556 mL to liters, (d) 10.6 kg/m^3 to g/cm^3.

1.57 Carry out the following conversions: (a) 242 lb to milligrams, (b) 68.3 cm^3 to cubic meters, (c) 7.2 m^3 to liters, (d) 28.3 μg to pounds.

1.58 The average speed of helium at 25°C is 1255 m/s. Convert this speed to miles per hour (mph).

1.59 How many minutes are there in a solar year (365.24 days)?

1.60 How many minutes does it take light from the sun to reach Earth? (The distance from the sun to Earth is 93 million mi; the speed of light is 2.99792458×10^8 m/s.)

1.61 A slow jogger runs a mile in 13 min. Calculate the speed in (a) in/s, (b) m/min, (c) km/h (1 mi = 1609 m; 1 in = 2.54 cm).

1.62 A 6.0-ft person weighs 168 lb. Express this person's height in meters and weight in kilograms (1 lb = 453.6 g; 1 m = 3.28 ft).

1.63 The highest speed limit in the United States is 85 mph on an isolated stretch of rural interstate in Texas. What is the speed limit in kilometers per hour (1 mi = 1609 m)? Report your answer as a whole number.

1.64 For a fighter jet to take off from the deck of an aircraft carrier, it must reach a speed of 62 m/s. Calculate the speed in miles per hour.

1.65 The "normal" lead content in human blood is about 0.40 part per million (i.e., 0.40 g of lead per million grams of blood). A value of 0.80 part per million (ppm) is considered to be dangerous. How many grams of lead are contained in 6.0×10^3 g of blood (the amount in an average adult) if the lead content is 0.62 ppm?

1.66 Carry out the following conversions: (a) 32.4 yd to centimeters, (b) 3.0×10^{10} cm/s to ft/s, (c) 1.42 light-years to miles (a light-year is an astronomical measure of distance—the distance traveled by light in a year, or 365 days; the speed of light is 3.00×10^8 m/s).

1.67 Carry out the following conversions: (a) 185 nm to meters, (b) 4.5 billion years (roughly the age of Earth) to seconds (assume 365 days in a year), (c) 71.2 cm^3 to cubic meters, (d) 88.6 m^3 to liters.

1.68 Aluminum is a lightweight metal (density = 2.70 g/cm^3) used in aircraft construction, high-voltage transmission lines, beverage cans, and foils. What is its density in kg/m^3?

1.69 The density of ammonia gas under certain conditions is 0.625 g/L. Calculate its density in g/cm^3.

1.70 (a) Carbon monoxide (CO) is a poisonous gas because it binds very strongly to the oxygen carrier hemoglobin in blood. A concentration of 8.00×10^2 ppm by volume of carbon monoxide is considered lethal to humans. Calculate the volume in liters occupied by carbon monoxide in a room that measures 17.6 m long, 8.80 m wide, and 2.64 m high at this concentration. (b) Prolonged exposure to mercury (Hg) vapor can cause neurological disorders and respiratory problems. For safe air quality control, the concentration of mercury vapor must be under 0.050 mg/m^3. Convert this number to g/L. (c) The general test for type II diabetes is that the blood sugar (glucose) level should be below 120 mg per deciliter (mg/dL). Convert this number to micrograms per milliliter (μg/mL).

1.71 The average time it takes for a molecule to diffuse a distance of x cm is given by

$$t = \frac{x^2}{2D}$$

where t is the time in seconds and D is the diffusion coefficient. Given that the diffusion coefficient of glucose is 5.7×10^{-7} cm^2/s, calculate the time it would take for a glucose molecule to diffuse 10 μm, which is roughly the size of a cell.

1.72 A human brain weighs about 1 kg and contains about 10^{11} cells. Assuming that each cell is completely filled with water (density = 1 g/mL), calculate the length of one side of such a cell if it were a cube. If the cells are spread out into a thin layer that is a single cell thick, what is the surface area in square meters?

ADDITIONAL PROBLEMS

1.73 Using the appropriate number of significant figures, report the length of the blue rectangle (a) using the ruler shown above the rectangle and (b) using the ruler shown below the rectangle.

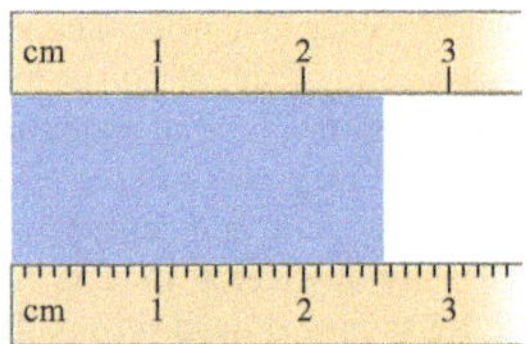

1.74 A piece of metal with a mass of 13.2 g was dropped into a graduated cylinder containing 17.00 mL of water. The graduated cylinder after the addition of the metal is shown. Determine the density of the metal to the appropriate number of significant figures.

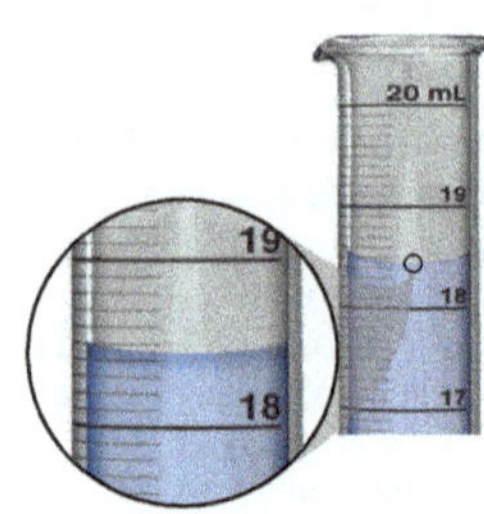

1.75 Which of the following statements describe physical properties and which describe chemical properties? (a) Iron has a tendency to rust. (b) Rainwater in industrialized regions tends to be acidic. (c) Hemoglobin molecules have a red color. (d) When a glass of water is left out in the sun, the water gradually disappears. (e) Carbon dioxide in air is converted to more complex molecules by plants during photosynthesis.

1.76 In determining the density of a rectangular metal bar, a student made the following measurements: length, 8.53 cm; width, 2.4 cm; height, 1.0 cm; mass, 52.7064 g. Calculate the density of the metal to the correct number of significant figures.

1.77 Calculate the mass of each of the following: (a) a sphere of gold with a radius of 10.0 cm (volume of a sphere with a radius r is $V = {}^4/_3\pi r^3$; density of gold = 19.3 g/cm^3), (b) a cube of platinum of edge length 0.040 mm (density = 21.4 g/cm^3), (c) 50.0 mL of ethanol (density = 0.798 g/mL).

1.78 A cylindrical glass tube 12.7 cm in length is filled with mercury (density = 13.6 g/mL). The mass of mercury needed to fill the tube is 105.5 g. Calculate the inner diameter of the tube (volume of a cylinder of radius r and length h is $V = \pi r^2 h$).

1.79 The following procedure was used to determine the volume of a flask. The flask was weighed dry and then filled with water. If the masses of the empty flask and filled flask were 56.12 g and 87.39 g, respectively, and the density of water is 0.9976 g/cm^3, calculate the volume of the flask in cubic centimeters.

1.80 The speed of sound in air at room temperature is about 343 m/s. Calculate this speed in miles per hour (1 mi = 1609 m).

1.81 A piece of silver (Ag) metal weighing 194.3 g is placed in a graduated cylinder containing 242.0 mL of water. The volume of water now reads 260.5 mL. From these data calculate the density of silver.

1.82 The experiment described in Problem 1.81 is a crude but convenient way to determine the density of some solids. Describe a similar experiment that would enable you to measure the density of ice. Specifically, what would be the requirements for the liquid used in your experiment?

1.83 A lead sphere has a mass of 1.20×10^4 g, and its volume is 1.05×10^3 cm^3. Calculate the density of lead.

1.84 Lithium is the least dense metal known (density = 0.53 g/cm^3). What is the volume occupied by 3.15×10^3 g of lithium?

1.85 At what temperature does the numerical reading on a Celsius thermometer equal that on a Fahrenheit thermometer?

1.86 Suppose that a new temperature scale has been devised on which the melting point of ethanol (−117.3°C) and the boiling point of ethanol (78.3°C) are taken as 0°S and 100°S, respectively, where S is the symbol for the new temperature scale. Derive an equation relating a reading on this scale to a reading on the Celsius scale. What would this thermometer read at 25°C?

1.87 The total volume of seawater is 1.5×10^{21} L. Assume that seawater contains 3.1 percent sodium chloride by mass and that its density is 1.03 g/mL. Calculate the total mass of sodium chloride in kilograms and in tons (1 ton = 2000 lb; 1 lb = 453.6 g).

1.88 A sheet of aluminum (Al) foil has a total area of 1.000 ft^2 and a mass of 3.636 g. What is the thickness of the foil in millimeters (density of Al = 2.699 g/cm^3)?

1.89 A student is given a crucible and asked to prove whether it is made of pure platinum. She first weighs the crucible in air and then weighs it suspended in water (density = 0.9986 g/mL). The readings are 860.2 g and 820.2 g, respectively. Based on these measurements and given that the density of platinum is 21.45 g/cm^3, what should her conclusion be? (*Hint:* An object suspended in a fluid is buoyed up by the mass of the fluid displaced by the object. Neglect the buoyancy of air.)

1.90 The surface area and average depth of the Pacific Ocean are 1.8×10^8 km^2 and 3.9×10^3 m, respectively. Calculate the volume of water in the ocean in liters.

1.91 The unit "troy ounce" is often used for precious metals such as gold (Au) and platinum (Pt) (1 troy ounce = 31.103 g). (a) A gold coin weighs 2.41 troy ounces. Calculate its mass in grams. (b) Is a troy ounce heavier or lighter than an ounce (1 lb = 16 oz; 1 lb = 453.6 g)?

1.92 Osmium (Os) is the densest element known (density = 22.57 g/cm^3). Calculate the mass in pounds and in kilograms of an Os sphere 15 cm in diameter (about the size of a grapefruit) (volume of a sphere of radius r is $\frac{4}{3}\pi r^3$).

1.93 Calculate the percent error for the following measurements: (a) The density of alcohol (ethanol) is found to be 0.802 g/mL (true value = 0.798 g/mL). (b) The mass of gold in an earring is analyzed to be 0.837 g (true value = 0.864 g).

1.94 In water conservation, chemists spread a thin film of a certain inert material over the surface of water to cut down on the rate of evaporation of water in reservoirs. This technique was pioneered by Benjamin Franklin three centuries ago. Franklin found that 0.10 mL of oil could spread over the surface of water about 40 m^2 in area. Assuming that the oil forms a *monolayer*, that is, a layer that is only one molecule thick, estimate the length of each oil molecule in nanometers (1 nm = 1×10^{-9} m).

1.95 You are given a liquid. Briefly describe the steps you would take to show whether it is a pure substance or a homogeneous mixture.

1.96 A gas company in Massachusetts charges $1.30 for 15.0 ft^3 of natural gas. (a) Convert this rate to dollars per liter of gas. (b) If it takes 0.304 ft^3 of gas to boil a liter of water, starting at room temperature (25°C), how much would it cost to boil a 2.1-L kettle of water?

1.97 A 250-mL glass bottle was filled with 242 mL of water at 20°C and tightly capped. It was then left outdoors overnight, where the average temperature was −5°C. Predict what would happen. The density of water at 20°C is 0.998 g/cm^3 and that of ice at −5°C is 0.916 g/cm^3.

1.98 A bank teller is asked to assemble \$1 sets of coins for his clients. Each set is made up of three quarters, one nickel, and two dimes. The masses of the coins are quarter, 5.645 g; nickel, 4.967 g; and dime, 2.316 g. What is the maximum number of sets that can be assembled from 33.871 kg of quarters, 10.432 kg of nickels, and 7.990 kg of dimes? What is the total mass (in grams) of the assembled sets of coins?

1.99 The men's world record for running a mile outdoors (as of 1999) is 3 min 43.13 s. At this rate, how long would it take to run a 2 km race (1 mi = 1609 m)?

1.100 Venus, the second closest planet to the sun, has a surface temperature of 7.3×10^2 K. Convert this temperature to degrees Celsius and degrees Fahrenheit.

1.101 Comment on whether each of the following is a homogeneous mixture or a heterogeneous mixture: (a) air in a closed bottle, (b) air over New York City.

1.102 It has been estimated that 8.0×10^4 tons of gold (Au) have been mined. Assume gold costs \$1350 per troy ounce. What is the total worth of this quantity of gold? (1 troy ounce = 31.103 g)

1.103 A 1.0-mL volume of seawater contains about 4.0×10^{-12} g of gold. The total volume of ocean water is 1.5×10^{21} L. Calculate the total amount of gold (in grams) that is present in seawater and the worth of the gold in dollars (see Problem 1.102). With so much gold out there, why hasn't someone become rich by mining gold from the ocean?

1.104 Measurements show that 1.0 g of iron (Fe) contains 1.1×10^{22} Fe atoms. How many Fe atoms are in 4.9 g of Fe, which is the total amount of iron in the body of an average adult?

1.105 The thin outer layer of Earth, called the crust, contains only 0.50 percent of Earth's total mass and yet is the source of almost all the elements (the atmosphere provides elements such as oxygen, nitrogen, and a few other gases). Silicon (Si) is the second most abundant element in Earth's crust (27.2 percent by mass). Calculate the mass of silicon in kilograms in Earth's crust (mass of Earth = 5.9×10^{21} tons; 1 ton = 2000 lb; 1 lb = 453.6 g).

1.106 The radius of a copper (Cu) atom is roughly 1.3×10^{-10} m. How many times can you divide evenly a 10-cm-long piece of copper wire until it is reduced to two separate copper atoms? (Assume there are appropriate tools for this procedure and that copper atoms are lined up in a straight line, in contact with each other. Round off your answer to an integer.)

1.107 A graduated cylinder is filled to the 40.00-mL mark with a mineral oil. The masses of the cylinder before and after the addition of the mineral oil are 124.966 g and 159.446 g, respectively. In a separate experiment, a metal ball bearing of mass 18.713 g is placed in the cylinder and the cylinder is again filled to the 40.00-mL mark with the mineral oil. The combined mass of the ball bearing and mineral oil is 50.952 g. Calculate the density and radius of the ball bearing (volume of a sphere of radius r is $\frac{4}{3}\pi r^3$).

1.108 A chemist mixes two liquids A and B to form a homogeneous mixture. The densities of the liquids are 2.0514 g/mL for A and 2.6678 g/mL for B. When she drops a small object into the mixture, she finds that the object becomes suspended in the liquid; that is, it neither sinks nor floats. If the mixture is made of 41.37 percent A and 58.63 percent B by volume, what is the density of the object? Can this procedure be used in general to determine the densities of solids? What assumptions must be made in applying this method?

1.109 A chemist in the nineteenth century prepared an unknown substance. In general, do you think it would be more difficult to prove that it is an element or a compound? Explain.

Industrial Problems

1.110 Chlorine is used to disinfect swimming pools. The accepted concentration for this purpose is 1 ppm chlorine, or 1 g of chlorine per million grams of water. Calculate the volume of a chlorine solution (in milliliters) a homeowner should add to her swimming pool if the solution contains 6.0 percent chlorine by mass and there are 2.0×10^4 gallons (gal) of water in the pool (1 gal = 3.79 L; density of liquids = 1.0 g/mL).

1.111 The world's total petroleum reserve is estimated at 2.0×10^{22} joules [a joule (J) is the unit of energy where $1 \text{ J} = 1 \text{ kg} \cdot \text{m}^2/\text{s}^2$]. At the present rate of consumption, 1.8×10^{20} joules per year (J/yr), how long would it take to exhaust the supply?

1.112 Bronze is an alloy made of copper (Cu) and tin (Sn). Calculate the mass of a bronze cylinder of radius 6.44 cm and length 44.37 cm. The composition of the bronze is 79.42 percent Cu and 20.58 percent Sn and the densities of Cu and Sn are 8.94 g/cm^3 and 7.31 g/cm^3, respectively. What assumption should you make in this calculation?

1.113 Chalcopyrite, the principal ore of copper (Cu), contains 34.63 percent Cu by mass. How many grams of Cu can be obtained from 7.35×10^3 kg of the ore?

Engineering Problems

1.114 Vanillin (used to flavor vanilla ice cream and other foods) is the substance whose aroma the human nose detects in the smallest amount. The threshold limit is 2.0×10^{-11} g per liter of air. If the current price of 50 g of vanillin is \$112, determine the cost to supply enough vanillin so that the aroma could be detected in a large aircraft hangar with a volume of 5.0×10^7 ft^3.

1.115 One gallon of gasoline in an automobile's engine produces on the average 9.5 kg of carbon dioxide, which is a greenhouse gas; that is, it promotes the warming of Earth's atmosphere. Calculate the annual production of carbon dioxide in kilograms if there are 40 million cars in the United States and each car covers a distance of 5000 mi at a consumption rate of 20 miles per gallon.

1.116 Magnesium (Mg) is a valuable metal used in alloys, in batteries, and in the manufacture of chemicals. It is obtained mostly from seawater, which contains about 1.3 g of Mg for every kilogram of seawater. Referring to Problem 1.87, calculate the volume of seawater (in liters) needed to extract 8.0×10^4 tons of Mg, which is roughly the annual production in the United States.

1.117 Fluoridation is the process of adding fluorine compounds to drinking water to help fight tooth decay. A concentration of 1 ppm of fluorine is sufficient for the purpose (1 ppm means one part per million, or 1 g of fluorine per 1 million g of water). The compound normally chosen for fluoridation is sodium fluoride, which is also added to some toothpastes. Calculate the quantity of sodium fluoride in kilograms needed per year for a city of 50,000 people if the daily consumption of water per person is 150 gal. What percent of the sodium fluoride is "wasted" if each person uses only 6.0 L of water a day for drinking and cooking (sodium fluoride is 45.0 percent fluorine by mass; 1 gal = 3.79 L; 1 year = 365 days; 1 ton = 2000 lb; 1 lb = 453.6 g; density of water = 1.0 g/mL)?

Biological Problems

1.118 The natural abundances of elements in the human body, expressed as percent by mass, are oxygen (O), 65 percent; carbon (C), 18 percent; hydrogen (H), 10 percent; nitrogen (N), 3 percent; calcium (Ca), 1.6 percent; phosphorus (P), 1.2 percent; all other elements, 1.2 percent. Calculate the mass in grams of each element in the body of a 62-kg person.

1.119 A resting adult requires about 240 mL of pure oxygen per minute and breathes about 12 times every minute. If inhaled air contains 20 percent oxygen by volume and exhaled air 16 percent, what is the volume of air per breath? (Assume that the volume of inhaled air is equal to that of exhaled air.)

1.120 (a) Referring to Problem 1.119, calculate the total volume (in liters) of air an adult breathes in a day. (b) In a city with heavy traffic, the air contains 2.1×10^{-6} L of carbon monoxide (a poisonous gas) per liter. Calculate the average daily intake of carbon monoxide in liters by a person.

1.121 The medicinal thermometer commonly used in homes can be read to ±0.1°F, whereas those in the doctor's office may be accurate to ±0.1°C. Percent error is often expressed as the absolute value of the difference between the true value and the experimental value, divided by the true value:

$$\text{percent error} = \frac{|\text{true value} - \text{experimental value}|}{\text{true value}} \times 100\%$$

The vertical lines indicate absolute value. In degrees Celsius, express the percent error expected from each of these thermometers in measuring a person's body temperature of 38.9°C.

1.122 TUMS is a popular remedy for acid indigestion. A typical TUMS tablet contains calcium carbonate plus some inert substances. When ingested, it reacts with the gastric juice (hydrochloric acid) in the stomach to give off carbon dioxide gas. When a 1.328-g tablet reacted with 40.00 mL of hydrochloric acid (density = 1.140 g/mL), carbon dioxide gas was given off and the resulting solution weighed 46.699 g. Calculate the number of liters of carbon dioxide gas released if its density is 1.81 g/L.

1.123 Pheromones are compounds secreted by females of many insect species to attract mates. Typically, 1.0×10^{-8} g of a pheromone is sufficient to reach all targeted males within a radius of 0.50 mi. Calculate the density of the pheromone (in grams per liter) in a cylindrical air space having a radius of 0.50 mi and a height of 40 ft (volume of a cylinder of radius r and height h is $\pi r^2 h$).

Standardized-Exam Practice Problems

Verbal Reasoning

English writer and essayist Lady Mary Wortley Montagu (1689–1762) traveled extensively and was fascinated by the customs in other countries. While in Turkey, she observed the practice of "engrafting" wherein people were inoculated against smallpox by intentional exposure to a mild form of the disease. She was so convinced of the efficacy and the safety of engrafting, that she had both of her children inoculated. She herself had survived smallpox as a child. Lady Montagu campaigned for the practice when she returned to England, and despite opposition from doctors and religious leaders, inoculation came into common use. It remained the primary defense against the scourge of smallpox for decades—until Jenner developed the practice of vaccination.

1. The main point of the passage is that
   a) Lady Montagu survived smallpox as a child.
   b) Lady Montagu brought the practice of engrafting from Turkey to England.
   c) doctors in eighteenth-century England were opposed to the practice of engrafting.
   d) Jenner developed the practice of vaccination.
2. Based on the passage, Lady Montagu was most likely
   a) a doctor.
   b) Turkish.
   c) severely scarred by smallpox.
   d) a member of a prominent British family.
3. The author refers to Lady Montagu having survived smallpox to
   a) explain why Lady Montagu was fascinated by the practice of engrafting.
   b) compare Lady Montagu to the doctors and religious leaders in England.
   c) explain why Lady Montagu herself did not undergo the engrafting procedure.
   d) emphasize Lady Montagu's fascination with other cultures.
4. Based on the passage, the author most likely thinks that Lady Montagu was
   a) educated and influential.
   b) inconsequential in the prevention of smallpox in England.
   c) trained in science and medicine.
   d) married to the British ambassador to Turkey.

Answers to In-Chapter Materials

Practice Problems

1.1A 273 K and 373 K, range = 100 K. **1.1B** −270.5°C. **1.2A** 113°F, 194°F; difference = 81°F. **1.2B** 233°C. **1.3A** (a) 13.6 g/mL, (b) 1.63×10^3 g. **1.3B** (a) 9.25 g/cm^3, (b) 3.76×10^3 g. **1.4A** (a) and (c). **1.4B** Physical: iv, chemical: ii and iii, neither: i. **1.5A** (a) 4, (b) 1, (c) 4, (d) 2, (e) 2 or 3, (f) 4. **1.5B** (a) 4, 0.0003050, 4; (b) 5, 432.00, 5; (c) 4, 0.0000008001, 4; (d) 7, 200608.0, 7; (e) 4, 0.00001503, 4; (f) 6, 60751.0, 6. **1.6A** (a) 116.2 L, (b) 80.71 m, (c) 3.813×10^{21} atoms, (d) 31 dm^2, (e) 0.504 g/mL. **1.6B** (a) 32.44 cm^3, (b) 4.2×10^2 kg/m^3, (c) 1.008×10^{10} kg, (d) 40.75 mL, (e) 227 cm^3. **1.7A** 0.8120 g/cm^3. **1.7B** 95.3 cm^3. **1.8A** 0.01 oz. **1.8B** 708.2 g. **1.9A** 1.05×10^4 kg/m^3. **1.9B** 13.6 mg/mm^3.

Checkpoints

1.3.1 c. **1.3.2** a. **1.3.3** b. **1.3.4** d. **1.4.1** b, c, e. **1.4.2** a, d, f. **1.5.1** c. **1.5.2** e. **1.5.3** c. **1.5.4** e. **1.6.1** b. **1.6.2** c. **1.6.3** a. **1.6.4** e.

CHAPTER 2

Atoms, Molecules, and Ions

Clams are a good source of dietary iron. Other iron-rich foods include meat, eggs, some vegetables, and fortified cereals. When diet alone does not provide an adequate supply, iron supplements can be taken.

Credit: © *ac_bnphotos/iStock/Getty Images.*

In This Chapter, You Will Learn

What atoms are made of and how they are arranged in molecules and ions that make up the substances that we encounter every day. You will also learn how to associate the name of a substance with its chemical formula.

Before You Begin, Review These Skills

- Significant figures [⏮ Section 1.5]
- Dimensional analysis [⏮ Section 1.6]

How Certain Atoms, Molecules, and Ions Can Affect Human Health

Atoms, molecules, and ions make up the substances we encounter every day. Some of these substances are important components of a balanced diet. An estimated 25 percent of the world's population suffers from iron deficiency, the most common nutritional deficiency in the world. Iron is necessary for the production of hemoglobin, the component in red blood cells responsible for the transport of oxygen. An inadequate supply of iron and the resulting shortage of hemoglobin can cause iron deficiency anemia (IDA). Some of the symptoms of IDA are fatigue, weakness, pale color, poor appetite, headache, and light-headedness.

Although IDA can be caused by loss of blood or by poor absorption of iron, the most common cause is insufficient iron in the diet. Dietary iron comes from such sources as meat, eggs, leafy green vegetables, dried beans, and dried fruits. Some breakfast cereals, such as Cream of Wheat, are fortified with iron in the form of iron metal, also known as *elemental* or *reduced* iron. The absorption of dietary iron can be enhanced by the intake of vitamin C (ascorbic acid). When the diet fails to provide enough iron, a nutritional supplement may be necessary to prevent a deficiency. Many supplements provide iron in the form of a compound called *ferrous sulfate.*

Student Note: Iron absorption can be diminished by certain disorders, such as Crohn's disease, and by some medications.

Elemental iron, ascorbic acid, and the iron in ferrous sulfate are examples of some of the *atoms, molecules,* and *ions* that are essential for human health.

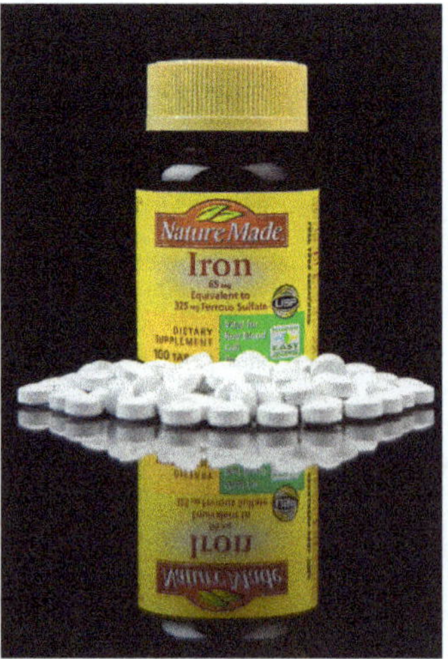

Credit: © *David A. Tietz/Editorial Image, LLC.*

At the end of this chapter, you will be able to solve a series of problems involving iron, iron sulfate, and ascorbic acid [⏭ Applying What You've Learned, page 75].

2.1 The Atomic Theory

In the fifth century B.C., the Greek philosopher Democritus proposed that all matter consists of very small, indivisible particles, which he named *atomos* (meaning uncuttable or indivisible). Although Democritus's idea was not accepted by many of his contemporaries (notably Plato and Aristotle), somehow it endured. Experimental evidence from early scientific investigations provided support for the notion of "atomism" and gradually gave rise to the modern definitions of elements and compounds. In 1808, an English scientist and schoolteacher, John Dalton[1] (Figure 2.1), formulated a precise definition of the indivisible building blocks of matter that we call atoms.

Dalton's work marked the beginning of the modern era of chemistry. The hypotheses about the nature of matter on which Dalton's atomic theory is based can be summarized as follows:

1. Elements are composed of extremely small particles called atoms. All atoms of a given element are identical, having the same size, mass, and chemical properties. The atoms of one element are different from the atoms of all other elements.
2. Compounds are composed of atoms of more than one element. In any given compound, the same types of atoms are always present in the same relative numbers.
3. A chemical reaction rearranges atoms in chemical compounds; it does not create or destroy them.

Figure 2.2 is a schematic representation of these hypotheses.

Dalton's concept of an atom was far more detailed and specific than that of Democritus. The first hypothesis states that atoms of one element are different from atoms of all other elements. Dalton made no attempt to describe the structure or composition of atoms—he had no idea what an atom was really like. He did realize, though, that the different properties shown by elements such as hydrogen and oxygen could be explained by assuming that hydrogen atoms were not the same as oxygen atoms.

The second hypothesis suggests that, to form a certain compound, we not only need atoms of the right *kinds* of elements, but specific *numbers* of these atoms as well. This idea is an extension of a law published in 1799 by Joseph Proust, a French chemist. According to Proust's ***law of definite proportions,*** different samples of a given compound always contain the same elements in the same mass *ratio*. Thus, if we were to analyze samples of carbon dioxide gas obtained from different sources, such as the exhaust from a car in Mexico City or the air above a pine forest in northern Maine, each sample would contain the same ratio by mass of oxygen to carbon. Consider the following results of the analysis of three samples of carbon dioxide, each from a different source:

Sample	Mass of O (g)	Mass of C (g)	Ratio (g O : g C)
123 g carbon dioxide	89.4	33.6	2.66:1
50.5 g carbon dioxide	36.7	13.8	2.66:1
88.6 g carbon dioxide	64.4	24.2	2.66:1

(a)

(b)

(c)

Figure 2.1 (a) John Dalton. (b) In addition to his other work, Dalton devised a system of symbols, which he used to represent the elements. (c) Dalton's only recreation is said to have been a nightly pint of ale and lawn bowling every Thursday afternoon. Note how closely the game balls resemble modern-day molecular models.

Credit: a: © SPL/The Image Works; b: © Science Source; c: © VisionsofAmerica/Joe Sohm/Getty Images.

1. John Dalton (1766–1844). English chemist, mathematician, and philosopher. In addition to the atomic theory, Dalton formulated several gas laws and gave the first detailed description of the type of color blindness, now called "Daltonism," from which he suffered. He also assigned relative weights to the elements, many of which differ considerably from those that we use today. He was described by his friends as awkward and without social grace.

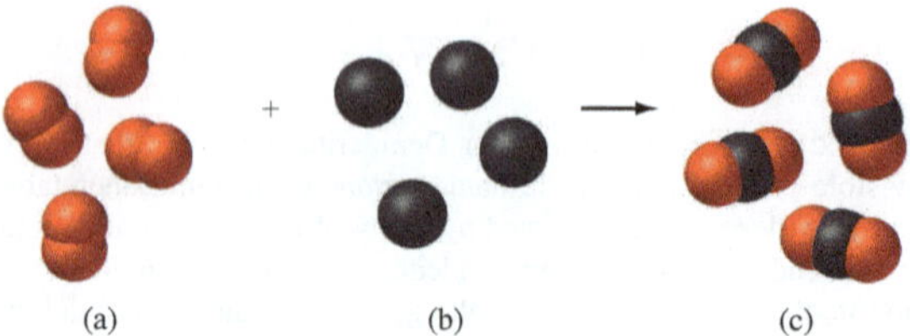

Figure 2.2 This represents a chemical reaction between the elements oxygen and carbon. (a) Oxygen does not exist as isolated atoms under ordinary conditions, but rather as *molecules,* each of which consists of *two* oxygen atoms. Note that the oxygen atoms (red spheres) appear all to be identical to one another (hypothesis 1). (b) Likewise, the carbon atoms (black spheres) all appear to be identical to one another. Carbon also exists in the form of molecules that are more varied and complex than those of oxygen. The carbon has been represented as isolated atoms to simplify the figure. (c) The compound CO_2 forms when each carbon atom combines with two oxygen atoms (hypothesis 2). Finally, the reaction results in the rearrangement of the atoms, but all the atoms present before the reaction (left of the arrow) are also present after the reaction (right of the arrow) (hypothesis 3).

In any sample of pure carbon dioxide, there are 2.66 g of oxygen for every gram of carbon present. This constant mass ratio can be explained by assuming that the elements exist in tiny particles of fixed mass (atoms), and that compounds are formed by the combination of fixed numbers of each type of particle.

Dalton's second hypothesis also supports the ***law of multiple proportions.*** According to this law, if two elements can combine to form more than one compound with each other, the masses of one element that combine with a fixed mass of the other element are in ratios of small whole numbers. That is, different compounds made up of the same elements differ in the number of atoms of each kind that combine. For example, carbon combines with oxygen to form carbon dioxide and carbon monoxide. In any sample of pure carbon monoxide, there are 1.33 g of oxygen for every gram of carbon.

Sample	Mass of O (g)	Mass of C (g)	Ratio (g O : g C)
16.3 g carbon monoxide	9.31	6.99	1.33:1
25.9 g carbon monoxide	14.8	11.1	1.33:1
88.4 g carbon monoxide	50.5	37.9	1.33:1

Thus, the ratio of oxygen to carbon in carbon *di*oxide is 2.66; and the ratio of oxygen to carbon in carbon *mon*oxide is 1.33. According to the law of multiple proportions, the ratio of two such ratios can be expressed as small whole numbers.

$$\frac{\text{ratio of O to C in carbon dioxide}}{\text{ratio of O to C in carbon monoxide}} = \frac{2.66}{1.33} = 2{:}1$$

For samples containing equal masses of carbon, the ratio of oxygen in carbon dioxide to oxygen in carbon monoxide is 2:1. Modern measurement techniques indicate that one atom of carbon combines with two atoms of oxygen in carbon dioxide and with one atom of oxygen in carbon monoxide. This result is consistent with the law of multiple proportions (Figure 2.3).

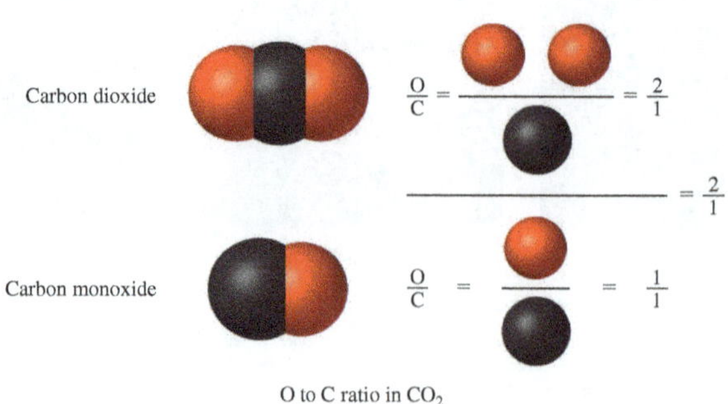

Figure 2.3 An illustration of the law of multiple proportions.

Animation
Law of conservation of mass.

Dalton's third hypothesis is another way of stating the ***law of conservation of mass,***[2] which is that matter can be neither created nor destroyed. Because matter is made up of atoms that are unchanged in a chemical reaction, it follows that mass must be conserved as well. Dalton's brilliant insight into the nature of matter was the main stimulus for the rapid progress of chemistry during the nineteenth century.

Sample Problem 2.1 shows how some common compounds obey the law of multiple proportions.

SAMPLE PROBLEM 2.1

(a) Both water (H_2O) and hydrogen peroxide (H_2O_2) are composed of hydrogen and oxygen. When water is decomposed into its constituent elements, it produces 0.125 g hydrogen for every gram of oxygen. When hydrogen peroxide is decomposed, it produces 0.063 g hydrogen for every gram of oxygen. Determine the whole number ratio of g H : 1.00 g O in water to g H : 1.00 g O in hydrogen peroxide to show how these data illustrate the law of multiple proportions. (b) Sulfur and oxygen can combine to form several compounds including sulfur dioxide (SO_2) and sulfur trioxide (SO_3). Sulfur dioxide contains 0.9978 g oxygen for every gram of sulfur. Sulfur trioxide contains 1.497 g oxygen for every gram of sulfur. Determine the whole number ratio of g O : 1.00 g S in sulfur dioxide to g O : 1.00 g S in sulfur trioxide.

Strategy For two compounds, each consisting of just two different elements, we are given the mass ratio of one element to the other. In each case, to show how the information given illustrates the law of multiple proportions, we divide the larger ratio by the smaller ratio.

Setup (a) The mass ratio of hydrogen to oxygen is higher in water than it is in hydrogen peroxide. Therefore, we divide the number of grams of hydrogen per gram of oxygen given for *water* by that given for *hydrogen peroxide*. (b) The mass ratio of oxygen to sulfur is higher in sulfur trioxide than it is in sulfur dioxide. Therefore, we divide the number of grams of oxygen per gram of sulfur given for *sulfur trioxide* by that given for *sulfur dioxide*.

Solution

(a) $$\frac{\text{g H : 1.00 g O in water}}{\text{g H : 1.00 g O in hydrogen peroxide}} = \frac{0.125}{0.063} = 1.98{:}1 \approx 2{:}1$$

(b) $$\frac{\text{g O : 1.00 g S in sulfur trioxide}}{\text{g O : 1.00 g S in sulfur dioxide}} = \frac{1.497}{0.9978} = 1.50{:}1.\ \text{Multiplying through by 2 gives 3:2.}$$

THINK ABOUT IT

When the result of such calculations is not a whole number, we must decide whether the result is close enough to *round* to a whole number as in part (a), or whether to multiply through to get whole numbers as in part (b). Only numbers that are *very* close to whole can be rounded. For example, numbers ending in approximately .25, .33, or .5 should be multiplied by 4, 3, or 2, respectively, to give a whole number.

Practice Problem ATTEMPT In each case, calculate the appropriate ratio to show that the information given is consistent with the law of multiple proportions. (a) Both ammonia (NH_3) and hydrazine (N_2H_4) are composed of nitrogen and hydrogen. Ammonia contains 0.2158 g hydrogen for every gram of nitrogen. Hydrazine contains 0.1439 g hydrogen for every gram of nitrogen. (b) Two of the compounds that consist of nitrogen and oxygen are nitric oxide, also known as nitrogen monoxide (NO) and nitrous oxide (N_2O), which is also known as dinitrogen monoxide. Nitric oxide contains 1.142 g oxygen for every gram of nitrogen. Nitrous oxide contains 0.571 g oxygen for every gram of nitrogen.

Practice Problem BUILD (a) Two of the simplest compounds containing just carbon and hydrogen are methane and ethane. Given that methane contains 0.3357 g hydrogen for every 1.00 g carbon and that the ratio of g hydrogen/1.00 g carbon in methane to g hydrogen/1.00 g carbon in ethane is 4:3, determine the number of grams of hydrogen per gram of carbon in ethane. (b) Xenon (Xe) and fluorine (F) can combine to form several different compounds, including XeF_2, which contains 0.2894 g fluorine for every gram of xenon. Use the law of multiple proportions to determine *n*, which represents the number of F atoms in another compound, XeF_n, given that it contains 0.8682 g of F for every gram of Xe.

Practice Problem CONCEPTUALIZE Which of the following diagrams illustrates the law of multiple proportions?

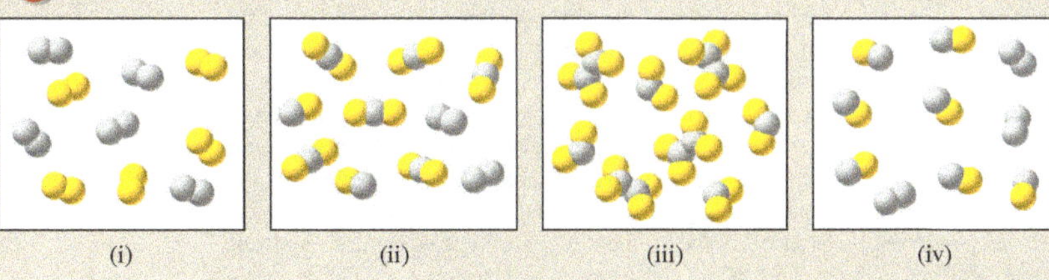

2. According to Albert Einstein, mass and energy are alternate aspects of a single entity called mass-energy. Chemical reactions usually involve a gain or loss of heat and other forms of energy. Thus, when energy is lost in a reaction, for example, mass is also lost. Except for nuclear reactions (see Chapter 20), however, changes of mass in chemical reactions are far too small to detect. Therefore, for all practical purposes mass is conserved.

CHECKPOINT – SECTION 2.1 The Atomic Theory

2.1.1  

For the two compounds pictured, evaluate the following ratio:

$$\frac{\text{g yellow : 1.00 g blue (right)}}{\text{g yellow : 1.00 g blue (left)}}$$

a) 2:1
b) 3:2
c) 3:4
d) 5:2
e) 5:4

2.1.2 For which of the following pairs of compounds is the ratio

$$\frac{\text{g yellow : 1.00 g blue (right)}}{\text{g yellow : 1.00 g blue (left)}}$$

equal to 4:1?

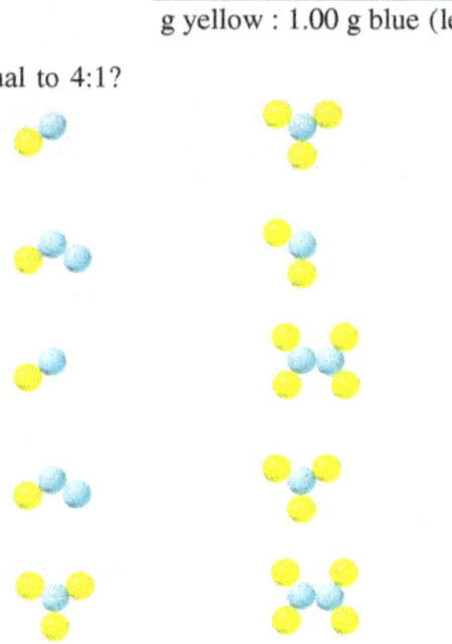

2.2 The Structure of the Atom

On the basis of Dalton's atomic theory, we can define an ***atom*** as the basic unit of an element that can enter into a chemical combination. Dalton imagined an atom that was both extremely small and indivisible. However, a series of investigations that began in the 1850s and extended into the twentieth century clearly demonstrated that atoms actually possess internal structure; that is, they are made up of even smaller particles, which are called *subatomic particles.* This research led to the discovery of electrons, protons, and neutrons.

Discovery of the Electron

Many scientists in the 1890s studied ***radiation,*** the emission and transmission of energy through space in the form of waves. Information gained from this research contributed greatly to our understanding of atomic structure. One device used to investigate this phenomenon was a cathode ray tube, the forerunner of the tubes used in older televisions and computer monitors (Figure 2.4).

Animation
Cathode ray tube experiment.

A cathode ray tube consists of two metal plates sealed inside a glass tube from which most of the air has been evacuated. When the metal plates are connected to a high-voltage source, the negatively charged plate, called the *cathode,* emits an invisible ray. The cathode ray is drawn to the positively charged plate, called the *anode,* where it passes through a hole and continues traveling to the other end of the tube. When the ray strikes the specially phosphor-coated surface, it produces a bright light.

Because consistent results are observed regardless of the composition of the cathode, cathode rays were presumed to be a component of all matter. Furthermore, because the path of the cathode rays could be deflected by magnetic and electric fields, as shown in Figure 2.4, they must be streams of charged *particles.* According to electromagnetic theory, a moving charged body behaves like a magnet and can interact with electric and magnetic fields through which it passes. Because the cathode ray is attracted by the plate bearing positive charges and repelled by the plate bearing negative charges, it must consist of negatively charged particles. We know these negatively charged particles as ***electrons.*** Figure 2.5 shows the deflection of a stream of electrons away from the south pole of a bar magnet. This effect on the path of electrons is what causes a television picture to become distorted temporarily when a magnet is brought close to the screen.

An English physicist, J. J. Thomson, used a cathode ray tube and his knowledge of electromagnetic theory to determine the ratio of electric charge to the mass of an individual electron. The number he calculated was 1.76×10^8 C/g, where C stands for *coulomb,* which is the derived

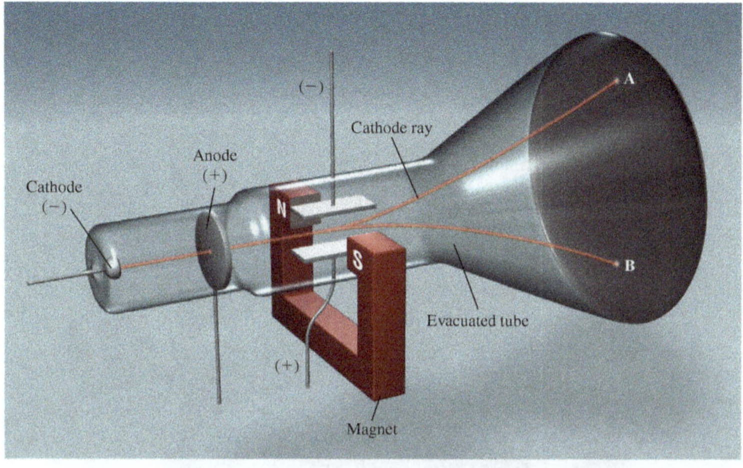

Figure 2.4 A cathode ray tube with an electric field perpendicular to the direction of the cathode rays and an external magnetic field. The symbols N and S denote the north and south poles of the magnet. The cathode rays will strike the end of the tube at point A in the presence of a magnetic field and at point B in the presence of an electric field. (In the absence of any external field—or when the effects of the electric field and magnetic field cancel each other—the cathode rays will not be deflected but will travel in a straight line and strike the middle of the circular screen.)

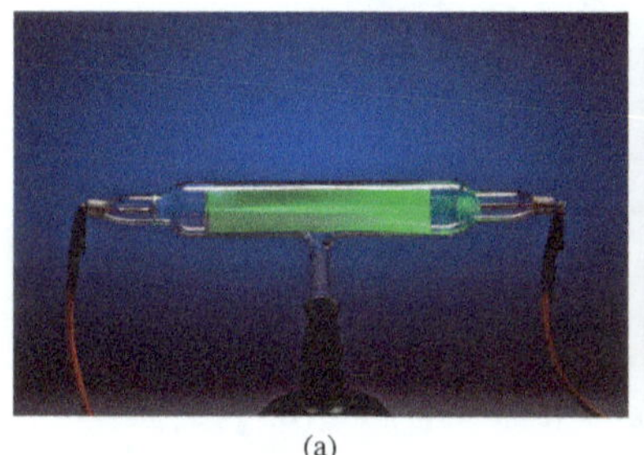
(a)

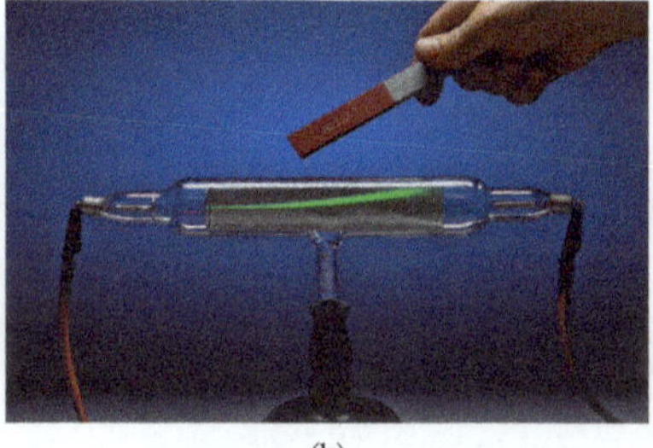
(b)

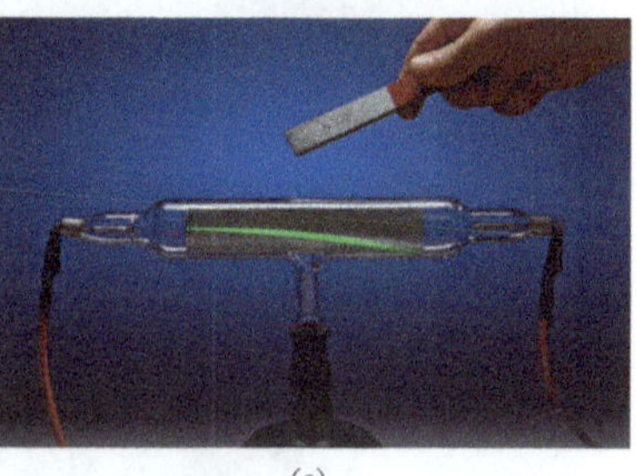
(c)

Figure 2.5 (a) A cathode ray produced in a discharge tube. The ray itself is invisible, but the fluorescence of a zinc sulfide coating on the glass causes it to appear green. (b) The cathode ray bends toward one pole of a magnet and (c) bends away from the opposite pole.

Credit: © *McGraw-Hill Education/Charles D. Winters, photographer.*

SI unit of electric charge. (In SI base units, 1 C = 1 A · s. Recall that A and s are the SI base units *ampere* and *second,* respectively [◂◂ Section 1.3, Table 1.2].) Thereafter, in a series of experiments carried out between 1908 and 1917, American R. A. Millikan succeeded in measuring the charge of the electron with great precision. In his experiment, Millikan examined the motion of single tiny drops of oil that picked up static charge from particles in the air. He suspended the charged drops in air by applying an electric field and followed their motions through a microscope (Figure 2.6). His work proved that the charge on each electron was exactly the same: -1.6022×10^{-19} C. Using Thomson's charge-to-mass ratio and the charge from his own experiments, Millikan calculated the mass of an electron as follows:

Student Note: Millikan actually determined the charge on droplets with multiple charges attached. The charges he determined were always *multiples* of -1.6022×10^{-19} C.

$$\text{mass of an electron} = \frac{\text{charge}}{\text{charge/mass}} = \frac{-1.6022 \times 10^{-19}\ \text{C}}{-1.76 \times 10^{8}\ \text{C/g}} = 9.10 \times 10^{-28}\ \text{g}$$

This is an *extremely* small mass.

Radioactivity

In 1895, the German physicist Wilhelm Röntgen noticed that cathode rays caused glass and metals to emit yet another type of ray. This highly energetic radiation penetrated matter, darkened covered photographic plates, and caused a variety of substances to *fluoresce* (give off light). These rays were *not* deflected by a magnet, however, so unlike cathode rays, they could not contain charged particles. Röntgen called them X rays because of their mysterious nature.

Not long after Röntgen's discovery, Antoine Becquerel, a professor of physics in Paris, began to study the fluorescent properties of substances. Purely by accident, he found that exposing thickly

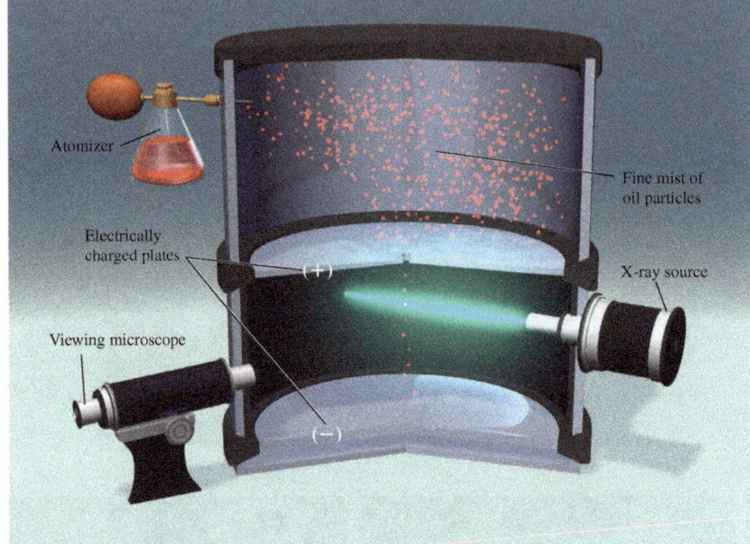

Figure 2.6 Schematic diagram of Millikan's oil-drop experiment.

Animation
Millikan oil-drop experiment.

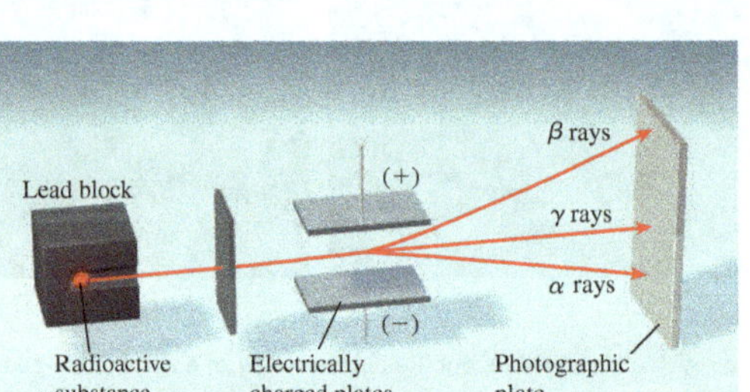

Figure 2.7 Three types of rays emitted by radioactive elements. So-called β rays actually consist of negatively charged particles (electrons) and are therefore attracted by the positively charged plate. The opposite holds true for so-called α rays—they are actually positively charged particles and are drawn to the negatively charged plate. Because γ rays are not particles and have no charge, their path is unaffected by an external electric field.

wrapped photographic plates (so no light could get in) to a uranium compound caused them to darken, even without the stimulation of cathode rays. Like X rays, the rays from the uranium compound were highly energetic and could not be deflected by a magnet, but they differed from X rays because they arose spontaneously. One of Becquerel's students, Marie Curie, suggested the name ***radioactivity*** to describe this spontaneous emission of particles and/or radiation. Today, we use the term *radioactive* to describe any element that spontaneously emits radiation.

Three types of rays are produced by the breakdown, or *decay,* of radioactive substances such as uranium. Two of the three are deflected by oppositely charged metal plates (Figure 2.7). ***Alpha (α) rays*** consist of positively charged particles, called ***α particles,*** that are deflected *away* from the positively charged plate. ***Beta (β) rays,*** or ***β particles,*** are *electrons,* so they are deflected away from the *negatively* charged plate. The third type of radioactive radiation consists of high-energy ***gamma (γ) rays.*** Like X rays, γ rays have no charge and are unaffected by external electric or magnetic fields.

The Proton and the Nucleus

By the early 1900s, scientists knew that atoms contained electrons but were electrically neutral overall. To be neutral, an atom must contain equal amounts of positive and negative charge. Thomson proposed, therefore, that an atom could be thought of as a sphere of positively charged

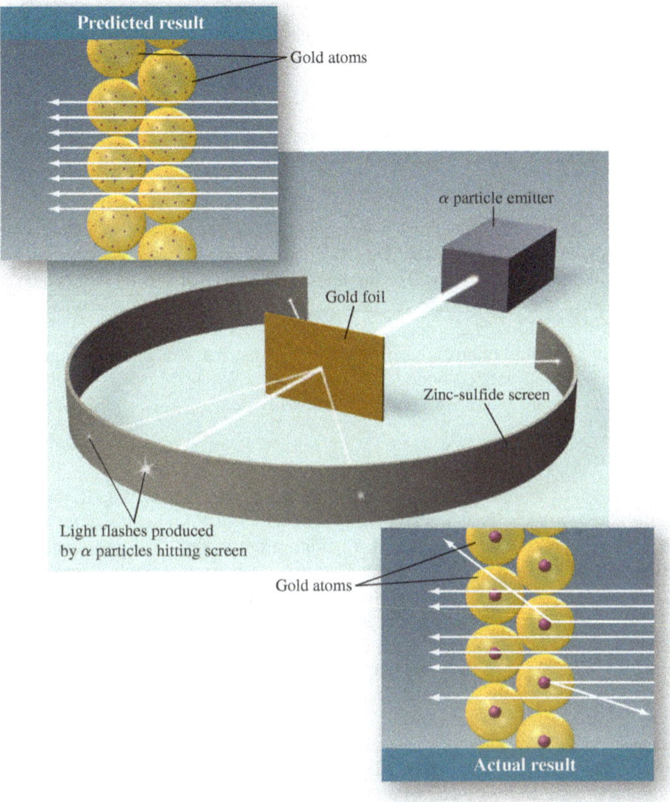

Figure 2.8 Rutherford's experimental design for measuring the scattering of α particles by a piece of gold foil. The plum-pudding model predicted that the α particles would all pass through the gold foil undeflected. The actual result: Most of the α particles do pass through the gold foil with little or no deflection, but a few are deflected at large angles. Occasionally an α particle bounces off the foil back toward the source. The nuclear model explains the results of Rutherford's experiments.

matter in which negatively charged electrons were embedded uniformly, like the chocolate chips in a scoop of mint chocolate chip ice cream. This so-called plum-pudding model was the accepted theory for a number of years.

In 1910, the New Zealand physicist Ernest Rutherford, who had studied with Thomson at Cambridge University, decided to use α particles to probe the structure of atoms. Together with his associate Hans Geiger and an undergraduate named Ernest Marsden, Rutherford carried out a series of experiments using very thin foils of gold and other metals as targets for α particles from a radioactive source. They observed that the majority of particles penetrated the foil either completely undeflected or with only a small angle of deflection. Every now and then, however, an α particle was scattered (or deflected) at a large angle. In some instances, the α particle actually bounced back in the direction from which it had come! This was an extraordinarily surprising finding. In Thomson's model, the positive charge of the atom was so diffuse that the relatively massive, positively charged particles should all have passed through the foil with little or no deflection. To quote Rutherford's initial reaction when told of this discovery: "It was as incredible as if you had fired a 15-inch shell at a piece of tissue paper and it came back and hit you." Figure 2.8 illustrates the results of Rutherford's α-scattering experiment.

Nuclear Model of the Atom

Rutherford later explained the results of the α-scattering experiment by proposing a new model for the atom. According to Rutherford, most of the atom must be empty space. This would explain why the majority of α particles passed through the gold foil with little or no deflection. The atom's positive charges, Rutherford proposed, were all concentrated in the ***nucleus,*** which is an extremely dense central *core* within the atom. Whenever an α particle came close to a nucleus in the scattering experiment, it experienced a large repulsive force and therefore a large deflection. Moreover, an α particle traveling directly toward a nucleus would be completely repelled and its direction would be reversed.

The positively charged particles in the nucleus are called ***protons***. In separate experiments, it was found that each proton carried the same *quantity* of charge as an electron (just opposite in sign) but had a mass of 1.67262×10^{-24} g. Although this is an extremely small value, it is nearly 2000 times the mass of an electron.

Based on these data, the atom was believed to consist of a nucleus that accounted for most of the mass of the atom, but which occupied only a tiny fraction of its volume. We express atomic (and molecular) dimensions using the SI unit *picometer (pm)*, where

$$1 \text{ pm} = 1 \times 10^{-12} \text{ m}$$

A typical atomic radius is about 100 pm, whereas the radius of an atomic nucleus is only about 5×10^{-3} pm. You can appreciate the relative sizes of an atom and its nucleus by imagining that if an atom were the size of the New Orleans Superdome, the volume of its nucleus would be comparable to that of a marble. While the protons are confined to the nucleus of the atom, the electrons are distributed *around* the nucleus at relatively large distances from it.

Student Note: Atomic radii are sometimes given in *angstroms*, where 1 angstrom (Å) = 1×10^{-10} m. Using angstroms, a typical atomic radius is about 1 Å and a typical nuclear radius is about 5×10^{-5} Å.

The concept of atomic radius is useful experimentally, but you should not get the impression that atoms have well-defined boundaries or surfaces. We will learn in Chapter 6 that the outer regions of atoms are relatively "fuzzy," not sharply defined.

The Neutron

Rutherford's model of atomic structure left one major problem unsolved. It was known that hydrogen, the simplest atom, contained only one proton and that the helium atom contained two protons. Therefore, the ratio of the mass of a helium atom to that of a hydrogen atom should be 2:1. (Because electrons are much lighter than protons, their contribution to atomic mass can be ignored.) In reality, however, the ratio is 4:1. Rutherford and others postulated that there must be another type of subatomic particle in the atomic nucleus, the proof of which was provided by James Chadwick, an English physicist, in 1932. When Chadwick bombarded a thin sheet of beryllium with α particles, a very high energy radiation was emitted by the metal that was not deflected by either electric or magnetic fields. Although similar to γ rays, later experiments showed that the rays actually consisted of a third type of subatomic particle, which Chadwick named ***neutrons*** because they were electrically neutral particles having a mass slightly greater than that of protons. The mystery of the mass ratio could now be explained. A typical helium nucleus consists of two protons and two neutrons, whereas a typical hydrogen nucleus contains only a proton; the mass ratio, therefore, is 4:1.

Figure 2.9 shows the location of the elementary particles (protons, neutrons, and electrons) in an atom. There are other subatomic particles, but the electron, the proton, and the neutron are the three fundamental components of the atom that are important in chemistry. Table 2.1 lists the masses and charges of these three elementary particles.

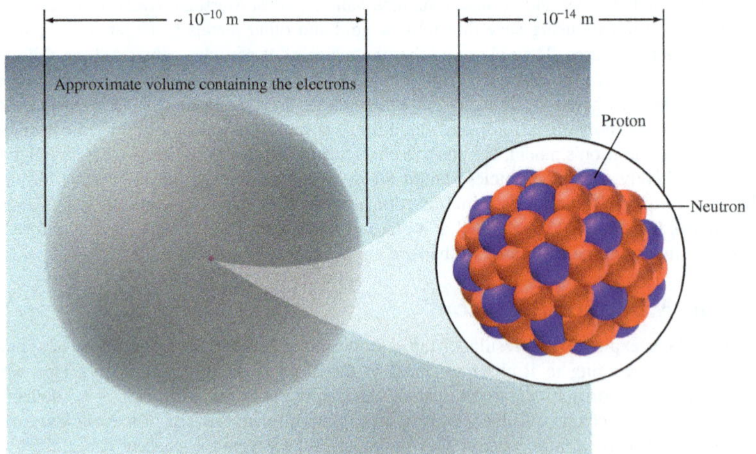

Figure 2.9 The protons and neutrons in an atom are contained in the tiny volume of the nucleus. Electrons are distributed within the sphere surrounding the nucleus.

TABLE 2.1 Masses and Charges of Subatomic Particles

Particle	Mass (g)	Charge (C)	Charge Unit
Electron*	9.10938×10^{-28}	-1.6022×10^{-19}	−1
Proton	1.67262×10^{-24}	$+1.6022 \times 10^{-19}$	+1
Neutron	1.67493×10^{-24}	0	0

*More refined measurements have resulted in a small change to Millikan's original value.

2.3 Atomic Number, Mass Number, and Isotopes

All atoms can be identified by the number of protons and neutrons they contain. The ***atomic number (Z)*** is the number of protons in the nucleus of each atom of an element. It also indicates the number of *electrons* in the atom—because atoms are *neutral* and contain the same number of protons and electrons. The chemical identity of an atom can be determined solely from its atomic number. For example, the atomic number of nitrogen is 7. Thus, each nitrogen atom has seven protons and seven electrons. Or, viewed another way, every atom in the universe that contains seven protons is a nitrogen atom.

The ***mass number (A)*** is the total number of neutrons *and* protons present in the nucleus of an atom of an element. Except for the most common form of hydrogen, which has one proton and no neutrons, all atomic nuclei contain both protons and neutrons. Collectively, protons and neutrons are called ***nucleons.*** A nucleon is a particle within the nucleus. In general, the mass number is given by

$$\text{mass number } (A) = \text{number of protons } (Z) + \text{number of neutrons}$$

The number of neutrons in an atom equals the difference between the mass number and the atomic number, or $(A - Z)$. For example, the mass number of fluorine is 19 and the atomic number is 9 (indicating 9 protons in the nucleus). Thus, the number of neutrons in an atom of fluorine is $19 - 9 = 10$. The atomic number, number of neutrons, and mass number all must be positive integers (whole numbers).

The accepted way to denote the atomic number and mass number of an atom of an element (X) is as follows:

Student Note: Because these symbols designate isotopes by specifying numbers of nucleons, they are sometimes referred to as *nuclear symbols.*

Mass number
(number of protons + neutrons)

$${}^{A}_{Z}\text{X} \leftarrow \text{Element symbol}$$

Atomic number
(number of protons)

Contrary to the first hypothesis of Dalton's atomic theory, atoms of a given element do not all have the same mass. Most elements have two or more ***isotopes,*** atoms that have the same atomic number (*Z*) but different mass numbers (*A*). For example, there are three isotopes of hydrogen, called *hydrogen* (or *protium*), *deuterium,* and *tritium.* Hydrogen has one proton and no neutrons in its nucleus, deuterium has one proton and one neutron, and tritium has one proton and two neutrons. Thus, to represent the isotopes of hydrogen, we write

${}^{1}_{1}\text{H}$	${}^{2}_{1}\text{H}$	${}^{3}_{1}\text{H}$
hydrogen	deuterium	tritium

Similarly, the two common isotopes of uranium ($Z = 92$), which have mass numbers of 235 and 238, respectively, can be represented as follows:

$${}^{235}_{92}\text{U} \qquad {}^{238}_{92}\text{U}$$

The first isotope, with $235 - 92 = 143$ neutrons in its nucleus, is used in nuclear reactors and atomic bombs, whereas the second isotope, with 146 neutrons, lacks the properties necessary for these applications. With the exception of hydrogen, which has different names for each of its isotopes, the isotopes of other elements are identified by their mass numbers. The two isotopes of uranium are called uranium-235 (pronounced "uranium two thirty-five") and uranium-238 (pronounced "uranium two thirty-eight"). Because the subscripted atomic number can be determined from the elemental symbol, it may be omitted from these representations without the loss of any information. The symbols ${}^{3}\text{H}$ and ${}^{235}\text{U}$ are sufficient to specify the isotopes tritium and uranium-235, respectively.

The chemical properties of an element are determined primarily by the protons and electrons in its atoms; neutrons do not take part in chemical changes under normal conditions. Therefore, isotopes of the same element exhibit similar chemical properties, forming the same types of compounds and displaying similar reactivities.

Sample Problem 2.2 shows how to calculate the number of protons, neutrons, and electrons using atomic numbers and mass numbers.

SAMPLE PROBLEM 2.2

Determine the numbers of protons, neutrons, and electrons in each of the following species: (a) $^{35}_{17}Cl$, (b) $^{37}_{17}Cl$, (c) ^{41}K, and (d) carbon-14.

Strategy Recall that the superscript denotes the mass number (A), and the subscript denotes the atomic number (Z). In the case where no subscript is shown, as in parts (c) and (d), the atomic number can be deduced from the elemental symbol or name. For the purpose of determining the number of electrons, remember that atoms are neutral, so the number of electrons equals the number of protons.

Setup Number of protons = Z, number of neutrons = $A - Z$, and number of electrons = number of protons. Recall that the 14 in carbon-14 is the mass number.

Solution

(a) The atomic number is 17, so there are 17 protons. The mass number is 35, so the number of neutrons is $35 - 17 = 18$. The number of electrons equals the number of protons, so there are 17 electrons.

(b) Again, the atomic number is 17, so there are 17 protons. The mass number is 37, so the number of neutrons is $37 - 17 = 20$. The number of electrons equals the number of protons, so there are 17 electrons, too.

(c) The atomic number of K (potassium) is 19, so there are 19 protons. The mass number is 41, so there are $41 - 19 = 22$ neutrons. There are 19 electrons.

(d) Carbon-14 can also be represented as ^{14}C. The atomic number of carbon is 6, so there are 6 protons and 6 electrons. There are $14 - 6 = 8$ neutrons.

THINK ABOUT IT

Verify that the number of protons and the number of neutrons for each example sum to the mass number that is given. In part (a), for example, there are 17 protons and 18 neutrons, which sum to give a mass number of 35, the value given in the problem. In part (b), 17 protons + 20 neutrons = 37. In part (c), 19 protons + 22 neutrons = 41. In part (d), 6 protons + 8 neutrons = 14.

Practice Problem ATTEMPT How many protons, neutrons, and electrons are there in an atom of (a) $^{10}_{5}B$, (b) ^{36}Ar, (c) $^{85}_{38}Sr$, and (d) carbon-11?

Practice Problem BUILD Give the correct symbols to identify an atom that contains (a) 4 protons, 4 electrons, and 5 neutrons; (b) 23 protons, 23 electrons, and 28 neutrons; (c) 54 protons, 54 electrons, and 70 neutrons; and (d) 31 protons, 31 electrons, and 38 neutrons.

Practice Problem CONCEPTUALIZE Based on the numbers of nucleons, write the nuclear symbol for each of the following diagrams:

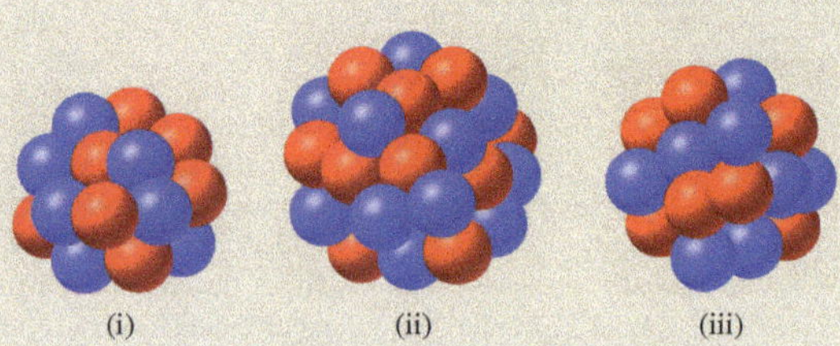

(i) (ii) (iii)

CHECKPOINT – SECTION 2.3 Atomic Number, Mass Number, and Isotopes

2.3.1 How many neutrons are there in an atom of ^{60}Ni?

a) 60
b) 30
c) 28
d) 32
e) 29

2.3.2 What is the *mass number* of an oxygen atom with nine neutrons in its nucleus?

a) 8
b) 9
c) 17
d) 16
e) 18

2.4 The Periodic Table

More than half of the elements known today were discovered between 1800 and 1900. During this period, chemists noted that the physical and chemical properties of certain groups of elements were similar to one another. These similarities, together with the need to organize the large volume of available information about the structure and properties of elemental substances, led to the development of the ***periodic table,*** a chart in which elements having similar chemical and physical properties are grouped together. Figure 2.10 shows the modern periodic table in which the elements are arranged by atomic number (shown above the element symbol) in horizontal rows called ***periods*** and in vertical columns called ***groups*** or ***families***. Elements in the same *group* tend to have similar physical and chemical properties.

The elements can be categorized as metals, nonmetals, or metalloids. A ***metal*** is a good conductor of heat and electricity, whereas a ***nonmetal*** is usually a poor conductor of heat and electricity. A ***metalloid*** has properties that are intermediate between those of metals and nonmetals. Figure 2.10 shows that the majority of known elements are metals; only 17 elements are nonmetals, and fewer than 10 elements are metalloids. Although most sources, including this text, designate the elements B, Si, Ge, As, Sb, and Te as metalloids, sources vary for the elements Po and At. In this text, we classify both Po and At as metalloids. From left to right across any period, the physical and chemical properties of the elements change gradually from metallic to nonmetallic.

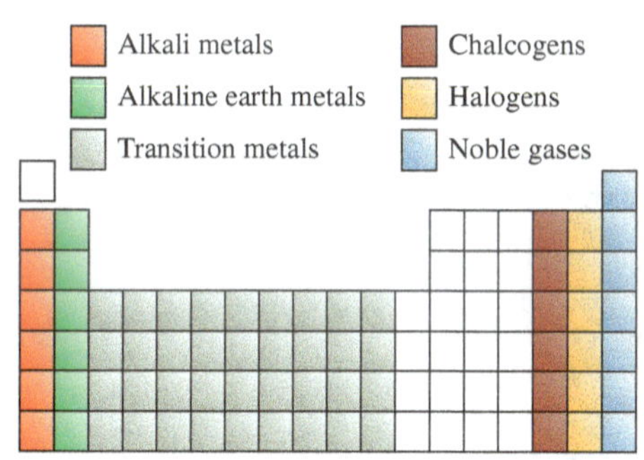

Elements are often referred to collectively by their periodic table group number (Group 1A, Group 2A, and so on). For convenience, however, some element groups have been given special names. The Group 1A elements, with the exception of H (i.e., Li, Na, K, Rb, Cs, and Fr), are called ***alkali metals,*** and the Group 2A elements (Be, Mg, Ca, Sr, Ba, and Ra) are called ***alkaline earth metals***. Elements in Group 6A (O, S, Se, Te, and Po) are sometimes referred to as the ***chalcogens***. Elements in Group 7A (F, Cl, Br, I, and At) are known as ***halogens,*** and elements in Group 8A (He, Ne, Ar, Kr, Xe, and Rn) are called ***noble gases,*** or rare gases. The elements in Group 1B and Groups 3B–8B collectively are called the ***transition elements*** or ***transition metals***.

The periodic table is a handy tool that correlates the properties of the elements in a systematic way and helps us to predict chemical behavior. At the turn of the twentieth century, the periodic table was deemed "the most predictive tool in all of science." We will take a more detailed look at this keystone of chemistry in Chapter 7.

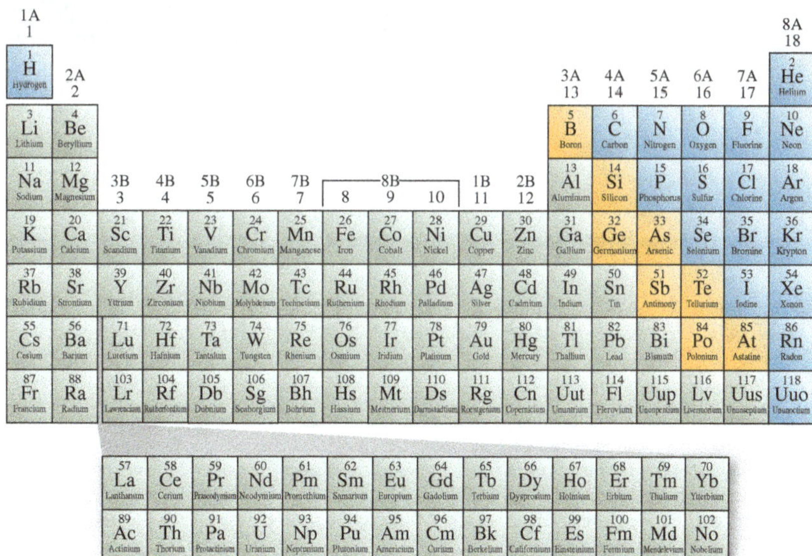

Figure 2.10 The modern periodic table. The elements are arranged according to atomic number, which is shown above each element's symbol. With the exception of hydrogen (H), nonmetals appear at the far right of the table. The two rows of metals beneath the main body of the table are set apart to keep the table from being too wide. Actually, lanthanum (57) should follow barium (56), and actinium (89) should follow radium (88). The 1–18 group designation has been recommended by the International Union of Pure and Applied Chemistry (IUPAC) but is not yet in wide use. In this text, we generally use the standard U.S. notation for group numbers (1A–8A and 1B–8B).

The following Bringing Chemistry to Life box describes the distribution of the elements in Earth's crust.

Bringing Chemistry to Life

Distribution of Elements on Earth

Earth's crust extends from the surface to a depth of about 40 km (about 25 mi). Because of technical difficulties, scientists have been unable to study the inner portions of Earth as easily and as thoroughly as the crust. Nevertheless, it is believed that there is a solid core consisting mostly of iron at the center of Earth. Surrounding the core is a layer called the *mantle,* which consists of hot fluid containing iron, carbon, silicon, and sulfur.

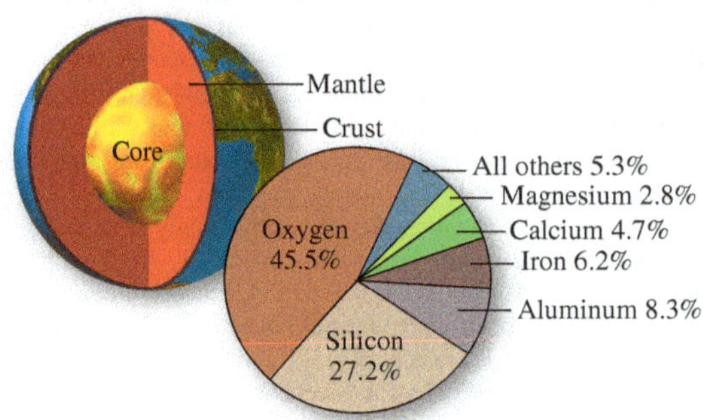

Of the 83 elements that are found in nature, 12 make up 99.7 percent of Earth's crust by mass. They are, in decreasing order of natural abundance, oxygen (O) silicon (Si), aluminum (Al), iron (Fe), calcium (Ca), magnesium (Mg), sodium (Na), potassium (K), titanium (Ti), hydrogen (H), phosphorus (P), and manganese (Mn). When discussing the natural abundance of the elements, keep in mind that the elements are unevenly distributed throughout Earth's crust, and most elements are combined chemically with other elements. Some of the elements that can be found in nature in pure form include gold, copper, and sulfur.

CHECKPOINT – SECTION 2.4 The Periodic Table

2.4.1 Which of the following series of elemental symbols lists a nonmetal, a metal, and a metalloid?

a) Ca, Cu, Si
b) K, Mg, B
c) Br, Ba, Ge
d) O, Na, S
e) Ag, Cr, As

2.4.2 Which of the following elements would you expect to have properties most similar to those of chlorine (Cl)?

a) Cu
b) F
c) Na
d) Cr
e) S

2.5 The Atomic Mass Scale and Average Atomic Mass

In some experimental work, it is important to know the masses of individual atoms, which depend on the number of protons, neutrons, and electrons they contain. However, even the smallest speck of dust that our unaided eyes can perceive consists of as many as 1×10^{16} atoms! Although we cannot weigh a single atom, it is possible to determine experimentally the mass of one atom *relative* to another. The first step is to assign a value to the mass of one atom of a given element so that it can be used as a standard.

According to international agreement, ***atomic mass*** is the mass of an atom in atomic mass units. One ***atomic mass unit (amu)*** is defined as a mass exactly equal to one-twelfth the mass of one carbon-12 atom. Carbon-12 is the carbon isotope that has six protons and six neutrons. Setting the atomic mass of carbon-12 at 12 amu provides the standard for measuring the atomic mass of the other elements. For example, experiments have shown that a hydrogen atom (^{1}H) is only 8.3985 percent as massive as the carbon-12 atom. Thus, if the mass of one carbon-12 atom is exactly 12 amu, the atomic mass of hydrogen must be 0.083985 × 12 amu, or 1.0078 amu. (Because the mass of the carbon atom is an exact number [⏮ Section 1.5], it does not limit the number of significant figures in the calculated result.) Similar calculations show that the atomic mass of fluorine-19 is 18.9984 amu and that of oxygen-16 is 15.9949 amu. Thus, although we cannot measure the mass of a single oxygen-16 atom, we know that it is approximately 16 times as massive as a hydrogen-1 atom.

When you look up the atomic mass of carbon in a table such as the one on the inside front cover of this book, you will find that its value is 12.01 amu, not 12.00 amu. The difference arises because most naturally occurring elements (including carbon) have more than one isotope. This means that when we measure the atomic mass of an element, we must generally settle for the average mass of the naturally occurring mixture of isotopes. For example, the natural abundances of carbon-12 and carbon-13 are 98.93 percent and 1.07 percent, respectively. The atomic mass of carbon-13 has been determined to be 13.003355 amu. Thus, the average atomic mass of natural carbon can be calculated as follows:

$$(0.9893)(12.00000 \text{ amu}) + (0.0107)(13.003355 \text{ amu}) = 12.01 \text{ amu}$$

In calculations involving percentages, we need to convert each percent abundance to a fractional abundance. For example, 98.93 percent becomes 98.93/100, or 0.9893. Because there are many more carbon-12 atoms than carbon-13 atoms in naturally occurring carbon, the average atomic mass is much closer to the mass of carbon-12 than to that of carbon-13.

When we say that the atomic mass of carbon is 12.01 amu, we are referring to the average value. If we could examine an individual atom of naturally occurring carbon, we would find either an atom of atomic mass of exactly 12 amu or one of 13.003355 amu, but never one of 12.01 amu. The atomic masses in the periodic table are *average atomic masses*. The term ***atomic weight*** is sometimes used to mean average atomic mass.

The atomic masses of many isotopes have been accurately determined to five or six significant figures. For most purposes, though, we will use average atomic masses, which are generally given to four significant figures (see the table of atomic masses on the inside front cover). For simplicity, we will omit the word *average* when we discuss the atomic masses of the elements.

The most direct and most accurate method for determining atomic and molecular masses is mass spectrometry. In a *mass spectrometer*, such as that depicted in Figure 2.11, a gaseous sample is bombarded by a stream of high-energy electrons. Collisions between the electrons and the gaseous atoms (or molecules) produce positively charged species, called *ions*, by dislodging an electron from the atoms or molecules. These positive ions (of mass m and charge e) are accelerated as they pass through two oppositely charged plates. The emerging ions are deflected into a circular path by a magnet. The radius of the path depends on the charge-to-mass ratio (i.e., e/m). Ions with a small e/m ratio trace a wider arc than those having a larger e/m ratio, so ions with equal charges but different masses are separated from one another. The mass of each ion (and hence its parent atom or molecule) is determined from the magnitude of its deflection.

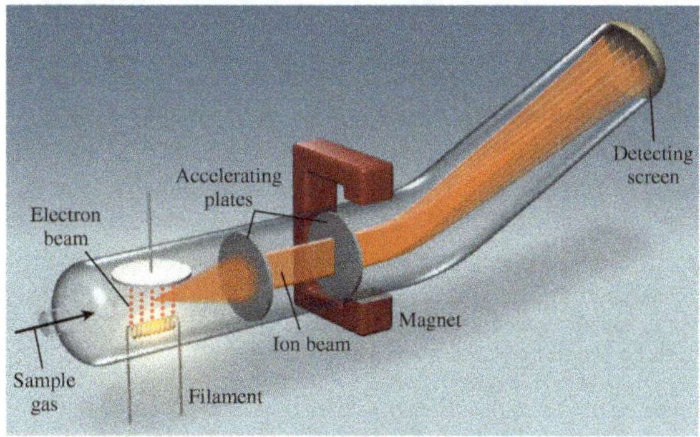

Figure 2.11 Schematic diagram of one type of mass spectrometer.

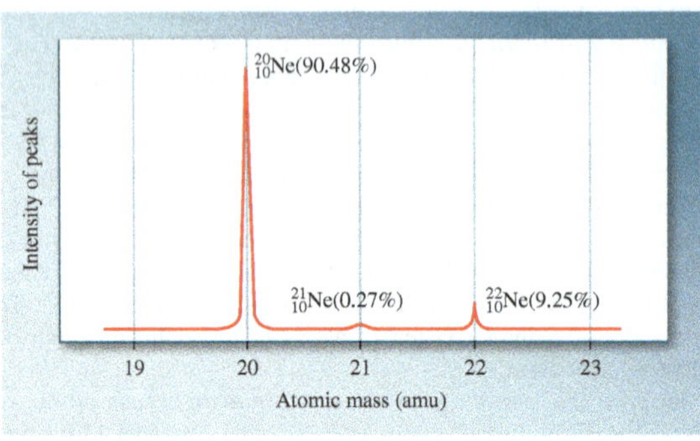

Figure 2.12 Mass spectrum of neon.

Eventually the ions arrive at the detector, which registers a current for each type of ion. The amount of current generated is directly proportional to the number of ions, so it enables us to determine the relative abundance of isotopes.

The first mass spectrometer, developed in the 1920s by the English physicist F. W. Aston, was crude by today's standards. Nevertheless, it provided indisputable evidence of the existence of isotopes, such as neon-20 (natural abundance 90.48 percent) and neon-22 (natural abundance 9.25 percent). When more sophisticated and sensitive mass spectrometers became available, scientists were surprised to discover that neon has a *third* stable isotope (neon-21) with natural abundance 0.27 percent (Figure 2.12). This example illustrates how very important experimental accuracy is to a quantitative science like chemistry. Early experiments failed to detect neon-21 because its natural abundance was so small. Only 27 in 10,000 Ne atoms are neon-21.

Sample Problem 2.3 shows how to calculate the average atomic mass of oxygen.

SAMPLE PROBLEM 2.3

Oxygen is the most abundant element in both Earth's crust and the human body. The atomic masses of its three stable isotopes, $^{16}_{8}O$ (99.757 percent), $^{17}_{8}O$ (0.038 percent), and $^{18}_{8}O$ (0.205 percent), are 15.9949, 16.9991, and 17.9992 amu, respectively. Calculate the average atomic mass of oxygen using the relative abundances given in parentheses. Report the result to four significant figures.

Strategy Each isotope contributes to the average atomic mass based on its relative abundance. Multiplying the mass of each isotope by its fractional abundance (percent value divided by 100) will give its contribution to the average atomic mass.

Setup Each percent abundance must be converted to a fractional abundance: 99.757 percent to 99.757/100 or 0.99757, 0.038 percent to 0.038/100 or 0.00038, and 0.205 percent to 0.205/100 or 0.00205. Once we find the contribution to the average atomic mass for each isotope, we can then add the contributions together to obtain the average atomic mass.

Solution

$$(0.99757)(15.9949 \text{ amu}) + (0.00038)(16.9991 \text{ amu}) + (0.00205)(17.9992) = 15.999 \text{ amu} \approx 16.00 \text{ amu}$$

THINK ABOUT IT

The average atomic mass should be closest to the atomic mass of the most abundant isotope (oxygen-16, in this case) and should, to the appropriate number of significant figures, be the same number that appears in the periodic table on the inside front cover of this book (16.00 amu, in this case)!

Practice Problem ATTEMPT The atomic masses of the two stable isotopes of copper, $^{63}_{29}Cu$ (69.17 percent) and $^{65}_{29}Cu$ (30.83 percent), are 62.929599 and 64.927793 amu, respectively. Calculate the average atomic mass of copper.

Practice Problem BUILD The average atomic mass of nitrogen is 14.0067. The atomic masses of the two stable isotopes of nitrogen, ^{14}N and ^{15}N, are 14.003074002 and 15.00010897 amu, respectively. Use this information to determine the percent abundance of each nitrogen isotope.

Practice Problem CONCEPTUALIZE The following diagrams show collections of metal spheres. Each collection consists of two or more different types of metal—represented here by different colors. The masses of metal spheres are as follows: white = 2.3575 g, black = 3.4778 g, and blue = 5.1112 g. For each diagram, determine the average mass of a single metal sphere.

(i)

(ii)

(iii)

CHECKPOINT – SECTION 2.5 The Atomic Mass Scale and Average Atomic Mass

2.5.1 Boron has two naturally occurring isotopes, ^{10}B and ^{11}B, which have masses 10.0129 and 11.0093 amu, respectively. Given the average atomic mass of boron (10.81 amu), determine the percent abundance of each isotope.

a) 50% ^{10}B, 50% ^{11}B
b) 20% ^{10}B, 80% ^{11}B
c) 98% ^{10}B, 2% ^{11}B
d) 93% ^{10}B, 7% ^{11}B
e) 22% ^{10}B, 78% ^{11}B

2.5.2 The two naturally occurring isotopes of antimony, ^{121}Sb (57.21 percent) and ^{123}Sb (42.79 percent), have masses of 120.904 and 122.904 amu, respectively. What is the average atomic mass of Sb?

a) 121.90 amu
b) 122.05 amu
c) 121.76 amu
d) 121.34 amu
e) 122.18 amu

2.6 Ions and Ionic Compounds

Out of all the elements, only the six noble gases in Group 8A of the periodic table (He, Ne, Ar, Kr, Xe, and Rn) exist in the form of isolated atoms under ordinary conditions. For this reason, they are called *monatomic* (meaning a single atom) gases. Most matter is composed of *ions,* which we discuss in this section, or *molecules,* which we will discuss in Section 2.7.

The number of positively charged protons in the nucleus of an atom remains the same during ordinary chemical reactions, but negatively charged electrons may be lost or gained—resulting in the formation of *ions.* An ***ion*** is an atom (or a *group* of atoms) that has a net positive or negative charge. The ions that make up an ionic compound are held together by strong electrostatic forces known as *ionic bonds* [⏭ Section 8.2].

Atomic Ions

An ***atomic ion*** or ***monatomic ion*** is one that consists of just *one* atom with a positive or negative charge. The *loss* of one or more electrons from an atom yields a ***cation,*** an ion with a net *positive* charge. For example, a sodium atom (Na) can readily lose an electron to become a sodium cation, which is represented by Na^+:

Na Atom	Na^+ Ion
11 protons	11 protons
11 electrons	10 electrons

An ***anion*** is an ion whose net charge is *negative* due to an *increase* in the number of electrons. A chlorine atom (Cl), for instance, can gain an electron to become a chloride ion (Cl^-):

Cl Atom	Cl^- Ion
17 protons	17 protons
17 electrons	18 electrons

Sodium chloride (NaCl), ordinary table salt, is called an ***ionic compound*** because it consists of *cations* (Na^+) and *anions* (Cl^-).

Student Note: Note that a multiple charge is denoted with the number followed by the sign; thus, 2+ not +2.

An atom can lose or gain more than one electron. Examples include Mg^{2+}, Fe^{3+}, S^{2-}, and N^{3-}. Figure 2.13 shows the charges of many more monatomic ions from across the periodic table. With very few exceptions, metals tend to form cations and nonmetals form anions. The charges on monatomic ions of elements in Groups 1A through 7A of the periodic table are fairly predictable. The cations that form from elements of Groups 1A, 2A, and 3A have charges equal to their respective group numbers. Most of the anions that form from elements of Groups 4A through 7A have charges equal to the corresponding group number minus 8. For example, the monatomic anion formed by oxygen (Group 6A) has a charge of $6 - 8 = -2$. You should be able to determine the charges on ions of elements in Groups 1A through 7A (for any element that forms only one common ion) using only a periodic table.

A monatomic cation is named simply by adding the word *ion* to the name of the element. Thus, the ion of potassium (K^+) is known as potassium ion. Similarly, the cations formed by the elements magnesium and aluminum (Mg^{2+} and Al^{3+}) are called magnesium ion and aluminum ion, respectively. It is not necessary for the name to specify the charge on these ions because their charges are equal to their group numbers.

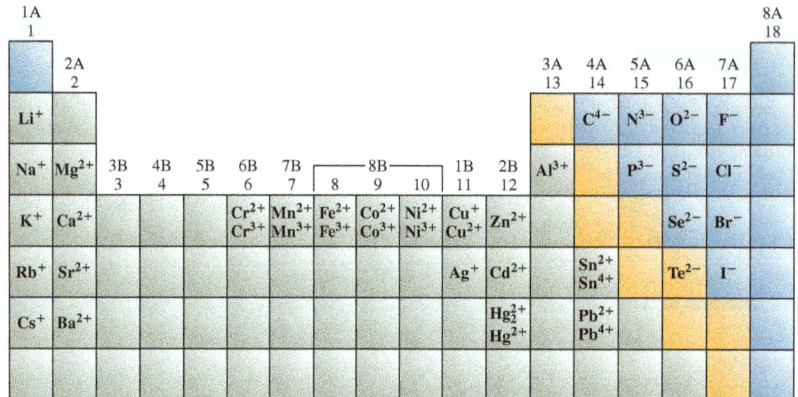

Figure 2.13 Common monatomic ions arranged by their positions in the periodic table. Note that mercury(I), Hg_2^{2+}, is actually a *polyatomic* ion.

Certain metals, especially the *transition metals,* can form cations of more than one possible charge. Iron, for example, can form Fe^{2+} and Fe^{3+}. An older nomenclature system that is still in limited use assigns the ending *–ous* to the cation with the *smaller* positive charge and the ending *–ic* to the cation with the *greater* positive charge:

Fe^{2+}: ferrous ion
Fe^{3+}: ferric ion

This method of naming ions has some distinct limitations. First, the *–ous* and *–ic* suffixes indicate the *relative* charges of the two cations involved, not the *actual* charges. Thus, Fe^{3+} is the ferric ion, but Cu^{2+} is the cupric ion. In addition, the *–ous* and *–ic* endings make it possible to name only two cations with different charges. Some metals, such as manganese (Mn), can form cations with three or more different charges.

Therefore, it has become increasingly common to designate different cations with Roman numerals, using the Stock[3] system. In this system, the Roman numeral I indicates a positive charge of one, II means a positive charge of two, and so on, as shown for manganese:

Mn^{2+}: manganese(II) ion
Mn^{3+}: manganese(III) ion
Mn^{4+}: manganese(IV) ion

These names are pronounced "manganese-two ion," "manganese-three ion," and "manganese-four ion," respectively. Using the Stock system, the ferrous and ferric ions are iron(II) and iron(III), respectively. To avoid confusion, and in keeping with modern practice, we will use the Stock system to name compounds in this textbook.

A monatomic anion is named by changing the ending of the element's name to *–ide,* and adding the word *ion.* Thus, the anion of chlorine (Cl^-), is called *chloride ion.* The anions of carbon, nitrogen, and oxygen (C^{4-}, N^{3-}, and O^{2-}) are called *carbide, nitride,* and *oxide,* respectively. Because there is only one possible charge for an ion formed from a nonmetal, it is unnecessary for the ion's name to specify its charge. Table 2.2 lists alphabetically a number of common monatomic ions.

Polyatomic Ions

Ions that consist of a combination of two or more atoms are called ***polyatomic ions.*** The atoms that make up a polyatomic ion are held together by covalent chemical bonds [⏭ Section 8.3]. Because these ions are encountered frequently throughout general chemistry, it is important that you learn and commit to memory the names, formulas, and charges of the polyatomic ions listed in Table 2.3. Although most of the common polyatomic ions are anions, a few are cations.

3. Alfred E. Stock (1876–1946). German chemist. Stock did most of his research in the synthesis and characterization of boron, beryllium, and silicon compounds. He was a pioneer in the study of mercury poisoning.

TABLE 2.2 Names and Formulas of Some Common Monatomic Ions

Name	Formula
Cations	
aluminum	Al^{3+}
barium	Ba^{2+}
cadmium	Cd^{2+}
calcium	Ca^{2+}
cesium	Cs^+
chromium(III)	Cr^{3+}
cobalt(II)	Co^{2+}
copper(I)	Cu^+
copper(II)	Cu^{2+}
hydrogen	H^+
iron(II)	Fe^{2+}
iron(III)	Fe^{3+}
lead(II)	Pb^{2+}
lithium	Li^+
magnesium	Mg^{2+}
manganese(II)	Mn^{2+}
mercury(II)	Hg^{2+}
potassium	K^+
silver	Ag^+
sodium	Na^+
strontium	Sr^{2+}
tin(II)	Sn^{2+}
zinc	Zn^{2+}
Anions	
bromide	Br^-
chloride	Cl^-
fluoride	F^-
hydride	H^-
iodide	I^-
nitride	N^{3-}
oxide	O^{2-}
sulfide	S^{2-}

TABLE 2.3 Common Polyatomic Ions

Name	Formula/Charge
Cations	
ammonium	NH_4^+
hydronium	H_3O^+
mercury(I)	Hg_2^{2+}
Anions	
acetate	$C_2H_3O_2^-$
azide	N_3^-
carbonate	CO_3^{2-}
chlorate	ClO_3^-
chlorite	ClO_2^-
chromate	CrO_4^{2-}
cyanide	CN^-
dichromate	$Cr_2O_7^{2-}$
dihydrogen phosphate	$H_2PO_4^-$
hydrogen carbonate or bicarbonate	HCO_3^-
hydrogen phosphate	HPO_4^{2-}
hydrogen sulfate or bisulfate	HSO_4^-
hydroxide	OH^-
hypochlorite	ClO^-
nitrate	NO_3^-
nitrite	NO_2^-
oxalate	$C_2O_4^{2-}$
perchlorate	ClO_4^-
permanganate	MnO_4^-
peroxide	O_2^{2-}
phosphate	PO_4^{3-}
phosphite	PO_3^{3-}
sulfate	SO_4^{2-}
sulfite	SO_3^{2-}
thiocyanate	SCN^-

Student Note: Some oxoanions occur in series of ions that contain the same central atom and have the same charge, but contain different numbers of oxygen atoms.

perchlorate	ClO_4^-
chlorate	ClO_3^-
chlorite	ClO_2^-
hypochlorite	ClO^-
nitrate	NO_3^-
nitrite	NO_2^-
phosphate	PO_4^{3-}
phosphite	PO_3^{3-}
sulfate	SO_4^{2-}
sulfite	SO_3^{2-}

Formulas of Ionic Compounds

The formulas of ionic compounds indicate the smallest whole number ratio in which the ions combine to form an electrically-neutral substance. For example, in sodium chloride, sodium ions and chloride ions combine in a 1:1 ratio, making the formula NaCl. In magnesium chloride, magnesium ions and chloride ions are combined in a 1:2 ratio, making the formula $MgCl_2$. Formulas such as these, that indicate the ratio of combination, are called ***empirical formulas.*** The word *empirical* means "from experience" or, in the context of chemical formulas, "from experiment."

An ionic compound does not exist as a collection of discrete units—there is no such thing as an NaCl *particle*. Rather, it consists of a vast, highly ordered array of interspersed cations and anions called a ***lattice.*** For example, solid sodium chloride (NaCl) consists of equal numbers of Na^+ and Cl^- ions arranged in a three-dimensional network of alternating cations and anions (Figure 2.14). As you can see in Figure 2.14, no Na^+ ion in NaCl is associated with any particular Cl^- ion. In fact, each Na^+ ion is surrounded by six Cl^- ions and vice versa. In other ionic compounds, the actual structure may be different, but the arrangement of cations and anions is

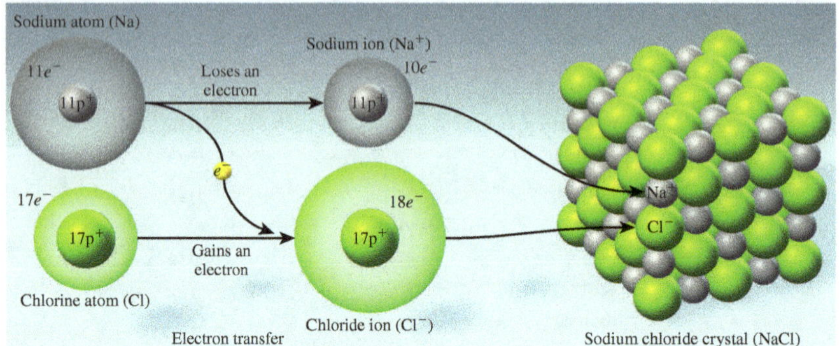

Figure 2.14 An electron is transferred from the sodium atom to the chlorine atom, giving a sodium ion and a chloride ion. The oppositely charged ions are attracted to each other electrostatically and form a solid lattice.

such that the compounds are all electrically neutral. The charges on the cation and anion are not shown in the formula for an ionic compound.

For ionic compounds to be electrically neutral, the sum of the charges on the cations and anions in each formula unit must be zero. When the charges on cation and anion are numerically equal, the ions will combine in a 1:1 ratio. If the charges on the cations and anions are numerically different, you can apply the following guideline to make the formula electrically neutral (and thus obtain the empirical formula): Write a subscript for the cation that is numerically equal to the charge on the anion and a subscript for the anion that is numerically equal to the charge on the cation. As we have seen in the cases of NaCl and $MgCl_2$, when the appropriate subscript would be 1, we do not include it in the formula.

Let's consider some examples.

Potassium Bromide The potassium ion (K^+) and the bromide ion (Br^-) combine to form the ionic compound *potassium bromide.* The sum of the charges is $1 + (-1) = 0$, so no subscripts are necessary. The formula is KBr.

Zinc Iodide The zinc ion (Zn^{2+}) and the iodide ion (I^-) combine to form *zinc iodide.* The sum of the charges of one Zn^{2+} ion and one I^- ion is $+2 + (-1) = +1$. To make the charges add up to zero, we multiply the -1 charge of the anion by 2 and add the subscript "2" to the symbol for iodine. Thus, the formula for zinc iodide is ZnI_2.

Ammonium Chloride The cation is NH_4^+ and the anion is Cl^-. The sum of the charges is $1 + (-1) = 0$, so the ions combine in a 1:1 ratio and the resulting formula is NH_4Cl.

Aluminum Oxide The cation is Al^{3+} and the anion is O^{2-}. The following diagram can be used to determine the subscripts for this compound:

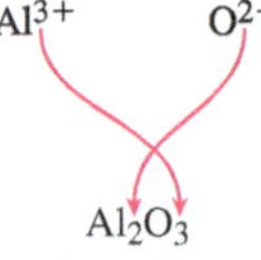

The sum of the charges for aluminum oxide is $2(+3) + 3(-2) = 0$. Thus, the formula is Al_2O_3.

Calcium Phosphate The cation is Ca^{2+} and the anion is PO_4^{3-}. The following diagram can be used to determine the subscripts:

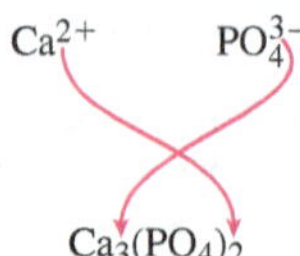

The sum of the charges is $3(+2) + 2(-3) = 0$. Thus, the formula for calcium phosphate is $Ca_3(PO_4)_2$. When we add a subscript to a polyatomic ion, we must first put parentheses around the ion's formula to indicate that the subscript applies to *all* the atoms in the polyatomic ion.

Naming Ionic Compounds

Early in the history of modern chemistry, when relatively few compounds were known, it was possible for chemists to memorize compound names—many of which were derived from physical appearance, properties, origin, or application of a compound. Examples include *milk of magnesia, laughing gas, baking soda,* and *formic acid* (*formica* is the Latin word for ant, and *formic acid* is the compound responsible for the sting of an ant bite).

Today there are many millions of known compounds, and many more being discovered or synthesized every year, so it would be impossible to memorize all of their names. Fortunately, it is unnecessary, because over the years chemists have devised a convenient system for naming chemical substances. The rules are the same worldwide, facilitating communication among scientists and providing a useful way of labeling an overwhelming variety of substances. Mastering these rules now will benefit you tremendously as you progress through your chemistry course. You must be able to name a compound, given its chemical formula, and you must be able to write the chemical formula of a compound, given its name.

An ionic compound is named simply by using the name of the cation followed by the name of the anion, eliminating the word *ion* from each. Several examples were given earlier in the Formulas of Ionic Compounds section. Other examples are sodium cyanide (NaCN), potassium permanganate ($KMnO_4$), and ammonium sulfate [$(NH_4)_2SO_4$]. Note that the names of these ionic compounds do not indicate the ratios of combination explicitly—although the chemical formulas do. There is only one neutral combination possible for each of these examples. Lithium ion has a charge of +1, always. Cyanide ion has a charge of −1, always. Their ratio of combination can only be 1:1. The same is true of potassium permanganate, in which the charges are also +1 and −1. Ammonium ion has a charge of +1 and sulfate ion has a charge of −2, always—making the only possible neutral combination of ammonium and sulfate ions 2:1. With ionic compounds, it is unnecessary to include the ratio of combination in the *name* because the charges on the ions are known.

In cases where a metal cation can have more than one possible charge, recall that the charge is indicated in the name of the ion with a Roman numeral in parentheses. Thus, the compounds $FeCl_2$ and $FeCl_3$ are named *iron(II) chloride* and *iron(III) chloride,* respectively. (These are pronounced "iron-two chloride" and "iron-three chloride.")

Sample Problems 2.4 and 2.5 illustrate how to name ionic compounds and write formulas for ionic compounds based on the information given in Tables 2.2 and 2.3.

SAMPLE PROBLEM 2.4

Name the following ionic compounds: (a) MgO, (b) $Al(OH)_3$, and (c) $Fe_2(SO_4)_3$.

Strategy Begin by identifying the cation and the anion in each compound, and then combine the names for each, eliminating the word *ion*.

Setup MgO contains Mg^{2+} and O^{2-}, the magnesium ion and the oxide ion; $Al(OH)_3$ contains Al^{3+} and OH^-, the aluminum ion and the hydroxide ion; and $Fe_2(SO_4)_3$ contains Fe^{3+} and SO_4^{2-}, the iron(III) ion and the sulfate ion. We know that the iron in $Fe_2(SO_4)_3$ is iron(III), Fe^{3+}, because it is combined with the sulfate ion in a 2:3 ratio.

Solution (a) Combining the cation and anion names, and eliminating the word *ion* from each of the individual ions' names, we get *magnesium oxide* as the name of MgO; (b) $Al(OH)_3$ is *aluminum hydroxide;* and (c) $Fe_2(SO_4)_3$ is *iron(III) sulfate.*

THINK ABOUT IT

Be careful not to confuse the subscript in a formula with the charge on the metal ion. In part (c), for example, the subscript on Fe is 2, but this is an iron(III) compound.

Practice Problem ATTEMPT Name the following ionic compounds: (a) Na_2SO_4, (b) $Cu(NO_3)_2$, (c) $Fe_2(CO_3)_3$.

Practice Problem BUILD Name the following ionic compounds: (a) $K_2Cr_2O_7$, (b) $Li_2C_2O_4$, (c) $CuNO_3$.

Practice Problem CONCEPTUALIZE The diagram represents a small sample of an ionic compound where red spheres represent nitrate ions and grey spheres represent iron ions. Deduce the correct formula and name of the compound.

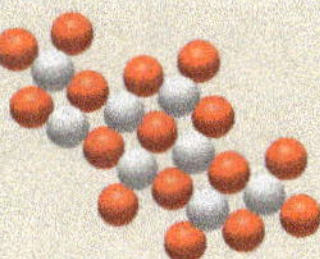

SAMPLE PROBLEM 2.5

Deduce the formulas of the following ionic compounds: (a) mercury(I) chloride, (b) lead(II) chromate, and (c) potassium hydrogen phosphate.

Strategy Identify the ions in each compound, and determine their ratios of combination using the charges on the cation and anion in each.

Setup (a) Mercury(I) chloride is a combination of Hg_2^{2+} and Cl^-. [Mercury(I) is one of the few cations listed in Table 2.3.] In order to produce a neutral compound, these two ions must combine in a 1:2 ratio. (b) Lead(II) chromate is a combination of Pb^{2+} and CrO_4^{2-}. These ions combine in a 1:1 ratio. (c) Potassium hydrogen phosphate is a combination of K^+ and HPO_4^{2-}. These ions combine in a 2:1 ratio.

Solution The formulas are (a) Hg_2Cl_2, (b) $PbCrO_4$, and (c) K_2HPO_4.

THINK ABOUT IT

Make sure that the charges sum to zero in each compound formula. In part (a), for example, $Hg_2^{2+} + Cl^- = (2+) + 2(-1) = 0$; in part (b), $(+2) + (-2) = 0$; and in part (c), $2(+1) + (-2) = 0$.

Practice Problem ATTEMPT Deduce the formulas of the following ionic compounds: (a) lead(II) chloride, (b) magnesium carbonate, and (c) ammonium phosphate.

Practice Problem BUILD Deduce the formulas of the following ionic compounds: (a) iron(III) sulfide, (b) mercury(II) nitrate, and (c) potassium sulfite.

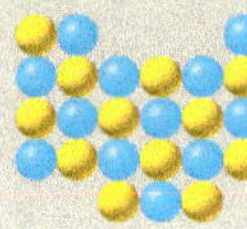

Practice Problem CONCEPTUALIZE The diagram represents a small sample of an ionic compound where yellow spheres represent sulfite ions and blue spheres represent copper ions. Deduce the correct formula and name of the compound.

Oxoanions

Oxoanions are polyatomic anions that contain one or more oxygen atoms and one atom (the "central atom") of another element. Examples include the chlorate (ClO_3^-), nitrate (NO_3^-), and sulfate (SO_4^{2-}) ions. Often, two or more oxoanions have the same central atom but different numbers of O atoms (e.g., NO_3^- and NO_2^-). Starting with the oxoanions whose names end in *–ate,* we can name these ions as follows:

1. The ion with one *more* O atom than the *–ate* ion is called the *per . . . ate* ion. Thus, ClO_3^- is the chlorate ion, so ClO_4^- is the *perchlorate ion.*
2. The ion with one *less* O atom than the *–ate* anion is called the *–ite* ion. Thus, ClO_2^- is the *chlorite ion.*
3. The ion with *two* fewer O atoms than the *–ate* ion is called the *hypo . . . ite* ion. Thus, ClO^- is the *hypochlorite ion.*

At a minimum, you must commit to memory the formulas and charges of the oxoanions whose names end in *–ate* so that you can apply these guidelines when necessary.

Sample Problem 2.6 tests your ability to name and identify oxoanions.

SAMPLE PROBLEM 2.6

Name the following species: (a) BrO_4^-, (b) HCO_3^-, and (c) SO_3^{2-}.

Strategy Each species is an oxoanion. Identify the "reference oxoanion" (the one with the *–ate* ending) for each, and apply the rules to determine appropriate names.

Setup (a) Chlorine, bromine, and iodine (members of Group 7A) all form analogous series of oxoanions with one to four oxygen atoms. Thus, the reference oxoanion is bromate (BrO_3^-), which is analogous to chlorate (ClO_3^-). In part (b), HCO_3^- has one more hydrogen than the carbonate ion (CO_3^{2-}). In part (c), the reference ion is sulfate (SO_4^{2-}).

Solution (a) BrO_4^- has one more O atom than the bromate ion (BrO_3^-), so BrO_4^- is the *perbromate* ion. (b) CO_3^{2-} is the carbonate ion. Because HCO_3^- has one ionizable hydrogen atom, it is called the *hydrogen carbonate ion.* (c) SO_3^{2-} has one less oxygen atom than the reference ion, so it is the *sulfite* ion.

(Continued on next page)

THINK ABOUT IT

Remembering all these names and formulas is greatly facilitated by memorizing the common ions that end in *–ate*.

chlorate	ClO_3^-	nitrate	NO_3^-	bromate	BrO_3^-	oxalate	$C_2O_4^{2-}$	phosphate	PO_4^{3-}
iodate	IO_3^-	carbonate	CO_3^{2-}	sulfate	SO_4^{2-}	chromate	CrO_4^{2-}	permanganate	MnO_4^-

Practice Problem ATTEMPT Name the following species: (a) BrO^-, (b) HSO_4^-, and (c) $C_2O_4^{2-}$.

Practice Problem BUILD Name the following species: (a) IO_2^-, (b) $HCrO_4^-$, and (c) $HC_2O_4^-$.

Practice Problem CONCEPTUALIZE The diagrams show models of a series of oxoanions. Which of the models represent an anion whose name ends in *–ate*?

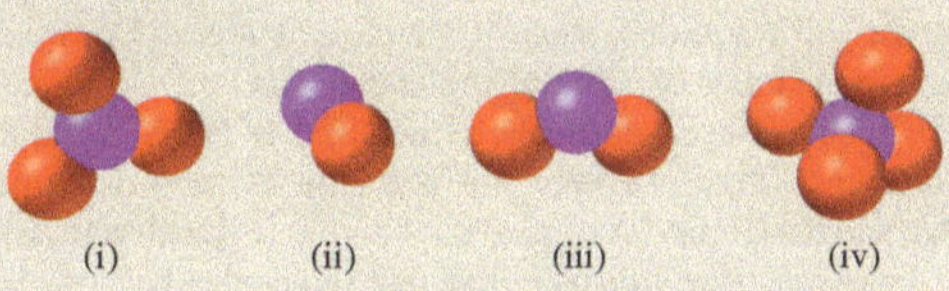

Hydrates

Figure 2.15 $CuSO_4$ is white. The pentahydrate, $CuSO_4 \cdot 5H_2O$, is blue.
Credit: © *McGraw-Hill Education/Charles D. Winters, photographer.*

Hydrates are *ionic* compounds that also contain water *molecules* within their solid structure. In its normal state, for example, each unit of copper(II) sulfate has five water molecules associated with it. The systematic name for this compound is copper(II) sulfate pentahydrate, and its formula is written as $CuSO_4 \cdot 5H_2O$. The water molecules can be driven off by heating. When this occurs, the resulting compound is $CuSO_4$, which is sometimes called anhydrous copper(II) sulfate; *anhydrous* means that the compound no longer has water molecules associated with it. Hydrates and the corresponding anhydrous compounds often have distinctly different physical and chemical properties (Figure 2.15).

Some other hydrates are

$BaCl_2 \cdot 2H_2O$	barium chloride dihydrate
$LiCl \cdot H_2O$	lithium chloride monohydrate
$MgSO_4 \cdot 7H_2O$	magnesium sulfate heptahydrate
$Sr(NO_3)_2 \cdot 4H_2O$	strontium nitrate tetrahydrate

CHECKPOINT – SECTION 2.6 Ions and Ionic Compounds

2.6.1 What is the correct name of the compound $PbSO_4$?

a) Lead sulfate
b) Lead(I) sulfate
c) Lead(II) sulfate
d) Monolead sulfate
e) Lead monosulfate

2.6.2 What is the correct formula for the compound iron(III) carbonate?

a) $FeCO_3$
b) Fe_3CO_3
c) Fe_2CO_3
d) $Fe_2(CO_3)_3$
e) $Fe_3(CO_3)_2$

2.6.3 Which of the following is the correct formula for nitrous acid?

a) HNO
b) HN_2O
c) N_2O
d) HNO_2
e) HNO_3

2.6.4 What is the formula of nickel(II) nitrate hexahydrate?

a) $NiNO_3 \cdot 6H_2O$
b) $Ni_2NO_3 \cdot 6H_2O$
c) $Ni(NO_3)_2 \cdot 6H_2O$
d) $NiNO_3 \cdot 12H_2O$
e) $Ni(NO_3)_2 \cdot 12H_2O$

2.6.5 What is the correct formula for sodium nitride?

a) NaN
b) NaN_3
c) Na_3N
d) $NaNO_3$
e) $NaNO_2$

2.6.6 What is the correct name of the compound Hg_2CrO_4?

a) Mercury(I) chromate
b) Mercury(II) chromate
c) Mercury dichromate
d) Dimercury chromate
e) Monomercury chromate

2.7 Molecules and Molecular Compounds

A ***molecule*** is an electrically neutral combination of at least two atoms in a specific arrangement held together by electrostatic forces known as *covalent chemical bonds* [⏭ Section 8.3]. A molecule may contain atoms of the same element, or it may contain atoms of two or more elements joined in a fixed ratio, in accordance with the law of definite proportions [⏮ Section 2.1]. Thus, a molecule can be an *element* or it can be a *compound,* which, by definition, is made up of two or more elements [⏮ Section 1.2]. Hydrogen gas, for example, is an element, but it consists of molecules, each of which is made up of two H atoms. Water, on the other hand, is a compound that consists of molecules, each of which contains two H atoms and one O atom.

The hydrogen molecule, symbolized as H_2, is called a ***diatomic molecule*** because it contains *two* atoms. Other elements that normally exist as diatomic molecules are nitrogen (N_2), oxygen (O_2), and the Group 7A elements—fluorine (F_2), chlorine (Cl_2), bromine (Br_2), and iodine (I_2). These are known as ***homonuclear*** diatomic molecules because both atoms in each molecule are of the same element. A diatomic molecule can also contain atoms of different elements. Examples of these ***heteronuclear*** diatomic molecules include hydrogen chloride (HCl) and carbon monoxide (CO).

Most molecules contain more than two atoms. They can all be atoms of the same element, as in ozone (O_3) and white phosphorus (P_4), or they can be combinations of two or more different elements, as in water (H_2O) and methane (CH_4). Molecules containing more than two atoms are called ***polyatomic molecules.***

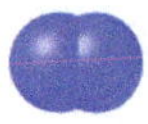
Homonuclear diatomic

Heteronuclear diatomic

Polyatomic

Molecular Formulas

A ***chemical formula*** denotes the composition of the substance. A ***molecular formula*** shows the exact number of atoms of each element in a molecule. In our discussion of molecules, each example was given with its molecular formula in parentheses. Thus, H_2 is the molecular formula for hydrogen, O_2 is that for oxygen, O_3 is that for ozone, and H_2O is that of water. The subscript numeral indicates the number of atoms of an element present in the molecule. There is no subscript for O in H_2O because there is only one atom of oxygen in a molecule of water. Just as with the formulas of ionic compounds, we do not write 1 as a subscript. Oxygen (O_2) and ozone (O_3) are allotropes of oxygen. An ***allotrope*** is one of two or more distinct forms of an element. Two of the allotropic forms of the element carbon—diamond and graphite—have dramatically different properties (and *prices*).

We can also represent molecules with *structural formulas*. The ***structural formula*** shows not only the elemental composition, but also the general arrangement of atoms within the molecule. In the case of water, each of the hydrogen atoms is connected to the oxygen atom. Figure 2.16 shows the molecular formula, structural formula, and molecular models (both ball-and-stick and space-filling variety) for water. Note that the chemical bond between two atoms can be represented with either a pair of dots or a line.

H_2O

H:O:H

H—O—H

Figure 2.16 Several ways to represent the water molecule.

In Chapters 8 and 9, you will learn how to use the molecular formula to deduce the structural formula and the three-dimensional arrangement of a molecule. We will use all of these methods for representing molecules throughout the book, so you should be familiar with each method and with the information it provides.

Sample Problem 2.7 shows how to write a molecular formula from the corresponding molecular model.

SAMPLE PROBLEM 2.7

Write the molecular formula of ethanol based on its ball-and-stick model, shown here.

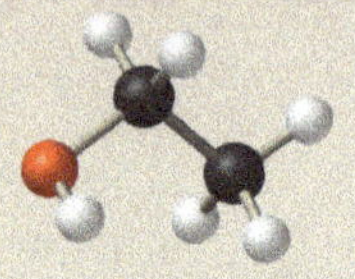
Ethanol

Strategy Refer to the labels on the atoms (or see Table 1.1).

Setup There are *two* carbon atoms, *six* hydrogen atoms, and *one* oxygen atom, so the subscript on C will be 2 and the subscript on H will be 6, and there will be no subscript on O.

Solution C_2H_6O

(Continued on next page)

THINK ABOUT IT

Often the molecular formula for an organic compound such as ethanol is written so that the formula more closely resembles the actual arrangement of atoms in the molecule. Thus, the molecular formula for ethanol is commonly written as C_2H_5OH.

Practice Problem ATTEMPT Chloroform was used as an anesthetic for childbirth and surgery during the nineteenth century. Write the molecular formula for chloroform based on the molecular model shown here.

Chloroform

Practice Problem BUILD Write the molecular formula for acetone based on the molecular model shown below.

Student Note: Note that acetone contains a *double* bond. Multiple bonds between atoms will be discussed in Chapters 8 and 9.

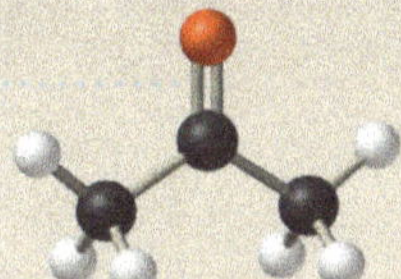

Acetone

Practice Problem CONCEPTUALIZE How many ball-and-stick models of ethanol molecules can be constructed using the collection of balls shown here? How many of each color ball will be left over?

Naming Molecular Compounds

Most molecular substances are ***binary compounds***, meaning that they are composed of atoms of two different elements. Binary molecular compounds contain two different *nonmetals* (see Figure 2.10). (The term binary can also refer to ionic compounds composed of two different elements—a metal and a nonmetal.) To name a binary molecular compound, we first name the element that appears first in the formula. For HCl that would be hydrogen. We then name the second element, changing the ending of its name to *–ide.* For HCl, the second element is chlorine, so we would change chlorine to chloride. Thus, the systematic name of HCl is *hydrogen chloride.* Similarly, HI is hydrogen iodide (iod*ine* → iod*ide*) and SiC is silicon carbide (carb*on* → carb*ide*).

Unlike the case with binary ionic compounds, which can only combine in one specific ratio, it is quite common for one pair of nonmetallic elements to form several *different* binary molecular compounds. In these cases, confusion in naming the compounds is avoided by the use of Greek prefixes to denote the number of atoms of each element present. Some of the Greek prefixes are listed in Table 2.4, and several compounds named using prefixes are listed in Table 2.5.

TABLE 2.4 Greek Prefixes

Prefix	Meaning	Prefix	Meaning
Mono–	1	Hexa–	6
Di–	2	Hepta–	7
Tri–	3	Octa–	8
Tetra–	4	Nona–	9
Penta–	5	Deca–	10

TABLE 2.5 Some Compounds Named Using Greek Prefixes

Compound	Name	Compound	Name
CO	Carbon monoxide	SO_3	Sulfur trioxide
CO_2	Carbon dioxide	NO_2	Nitrogen dioxide
SO_2	Sulfur dioxide	N_2O_5	Dinitrogen pentoxide

The prefix *mono–* is generally omitted for the first element. SO_2, for example, is named *sulfur dioxide,* not *monosulfur dioxide.* Thus, the absence of a prefix for the first element usually means there is only one atom of that element present in the molecule. In addition, for ease of pronunciation, we usually eliminate the last letter of a prefix that ends in "o" or "a" when naming an oxide. Thus, N_2O_5 is *dinitrogen pentoxide,* rather than *dinitrogen pentaoxide.*

Sample Problem 2.8 gives you some practice naming binary molecular compounds from their formulas.

SAMPLE PROBLEM 2.8

Name the following binary molecular compounds: (a) NF_3 and (b) N_2O_4.

Strategy Each compound will be named using the systematic nomenclature including, where necessary, appropriate Greek prefixes.

Setup With binary compounds, we start with the name of the element that appears *first* in the formula, and we change the ending of the *second* element's name to *–ide.* We use prefixes, where appropriate, to indicate the number of atoms of each element. In part (a) the molecule contains one nitrogen atom and three fluorine atoms. We will omit the prefix *mono–* for nitrogen because it is the first element listed in the formula, and we will use the prefix *tri–* to denote the number of fluorine atoms. In part (b) the molecule contains two nitrogen atoms and four oxygen atoms, so we will use the prefixes *di–* and *tetra–* in naming the compound. Recall that in naming an oxide, the last letter of a prefix that ends in "a" or "o" is omitted.

Solution (a) nitrogen trifluoride and (b) dinitrogen tetroxide

THINK ABOUT IT

Make sure that the prefixes match the subscripts in the molecular formulas and that the word *oxide* is not preceded immediately by an "a" or an "o."

Practice Problem ATTEMPT Name the following binary molecular compounds: (a) Cl_2O and (b) $SiCl_4$.

Practice Problem BUILD Name the following binary molecular compounds: (a) ClO_2 and (b) CBr_4.

Practice Problem CONCEPTUALIZE Name the binary molecular compound shown.

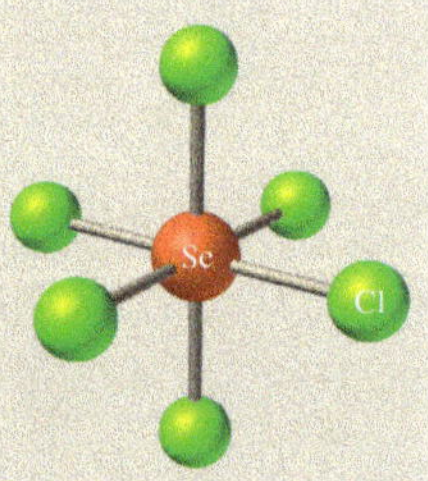

Writing the formula for a molecular compound, given its systematic name, is usually straightforward. For instance, the name *boron trichloride* indicates the presence of one boron atom (no prefix) and three chlorine atoms (*tri–*), so the corresponding molecular formula is BCl_3. Note once again that the order of the elements is the same in both the name and the formula.

Sample Problem 2.9 gives you some practice determining the formulas of binary molecular compounds from their names.

SAMPLE PROBLEM 2.9

Write the chemical formulas for the following binary molecular compounds: (a) sulfur tetrafluoride and (b) tetraphosphorus decasulfide.

Strategy The formula for each compound will be deduced using the systematic nomenclature guidelines.

Setup In part (a) there is no prefix for sulfur, so there is only one sulfur atom in a molecule of the compound. Therefore, we will use no prefix for the S in the formula. The prefix *tetra–* means that there are four fluorine atoms. In part (b) the prefixes *tetra–* and *deca–* denote four and ten, respectively.

Solution (a) SF_4 and (b) P_4S_{10}

(Continued on next page)

THINK ABOUT IT

Double-check that the subscripts in the formulas match the prefixes in the compound names: (a) 4 = *tetra* and (b) 4 = *tetra* and 10 = *deca*.

Practice Problem ATTEMPT Give the molecular formula for each of the following compounds: (a) carbon disulfide and (b) dinitrogen trioxide.

Practice Problem BUILD Give the molecular formula for each of the following compounds: (a) sulfur hexafluoride and (b) disulfur decafluoride.

Practice Problem CONCEPTUALIZE Draw a molecular model of sulfur trioxide.

Student Note: Binary compounds containing carbon and hydrogen are *organic* compounds and do not follow the same naming conventions as other molecular compounds. Organic compounds and their nomenclature are discussed in detail in Chapter 25.

The names of molecular compounds containing hydrogen do not usually conform to the systematic nomenclature guidelines. Traditionally, many of these compounds are called either by their common, nonsystematic names or by names that do not indicate explicitly the number of H atoms present:

B_2H_6	Diborane	PH_3	Phosphine
SiH_4	Silane	H_2O	Water
NH_3	Ammonia	H_2S	Hydrogen sulfide

Even the order in which the elements are written in these hydrogen-containing compounds is irregular. In water and hydrogen sulfide, H is written first, whereas it is written last in the other compounds.

Simple Acids

Student Note: In Chapter 16, we will explore acids and bases in greater detail; and we will see that there are other ways to define the terms *acid* and *base*.

Student Note: An ion is a *charged* species.

Acids make up another important class of molecular compounds. One definition of an ***acid*** is a substance that produces hydrogen ions (H^+) when dissolved in water. Several binary molecular compounds produce hydrogen ions when dissolved in water and are, therefore, acids. In these cases, two different names can be assigned to the same chemical formula. For example, HCl, *hydrogen chloride,* is a gaseous compound. When it is dissolved in water, however, we call it *hydrochloric acid.* The rules for naming simple acids of this type are as follows: remove the *–gen* ending from hydrogen (leaving *hydro–*), change the *–ide* ending on the second element to *–ic,* combine the two words, and add the word *acid.*

hydro*gen* chlor*ide* + *–ic* acid ⟶ hydrochloric acid

Likewise, hydrogen fluoride (HF) becomes *hydrofluoric acid.* Table 2.6 lists these and other examples.

For a compound to produce hydrogen ions upon dissolving, it must contain at least one ***ionizable hydrogen atom.*** An ionizable hydrogen atom is one that separates from the molecule upon dissolving and becomes a hydrogen ion (H^+).

Oxoacids

In addition to the simple acids, there is another important class of acids known as ***oxoacids,*** which ionize to produce hydrogen ions and the corresponding oxoanions. The formula of an

TABLE 2.6 Some Simple Acids

Formula	Binary Compound Name	Acid Name
HF	Hydrogen fluoride	Hydrofluoric acid
HCl	Hydrogen chloride	Hydrochloric acid
HBr	Hydrogen bromide	Hydrobromic acid
HI	Hydrogen iodide	Hydroiodic acid
HCN*	Hydrogen cyanide	Hydrocyanic acid

*Although HCN is not a *binary* compound, it is included in this table because it is similar chemically to HF, HCl, HBr, and HI.

oxoacid can be determined by adding enough H^+ ions to the corresponding oxoanion to yield a formula with no net charge. For example, the formulas of oxoacids based on the nitrate (NIO_3^-) and sulfate (SO_4^{2-}) ions are HNO_3 and H_2SO_4, respectively. The names of oxoacids are derived from the names of the corresponding oxoanions using the following guidelines:

1. An acid based on an *–ate* ion is called . . . *ic* acid. Thus, $HClO_3$ is called *chloric acid.*
2. An acid based on an *–ite* ion is called . . . *ous* acid. Thus, $HClO_2$ is called *chlorous acid.*
3. Prefixes in oxoanion names are retained in the names of the corresponding oxoacids. Thus, $HClO_4$ and HClO are called *perchloric acid* and *hypochlorous acid,* respectively.

Many oxoacids, such as H_2SO_4 and H_3PO_4, are ***polyprotic acid***—meaning that they have more than one ionizable hydrogen atom. In these cases, the names of anions in which one or more (but not all) of the hydrogen ions have been removed must indicate the number of H ions that remain, as shown for the anions derived from phosphoric acid:

H_3PO_4	phosphoric acid	HPO_4^{2-}	hydrogen phosphate ion
$H_2PO_4^-$	dihydrogen phosphate ion	PO_4^{3-}	phosphate ion

Sample Problem 2.10 lets you practice identifying and naming oxoacids.

SAMPLE PROBLEM 2.10

Determine the formula of sulfurous acid.

Strategy The *–ous* ending in the name of an acid indicates that the acid is derived from an oxoanion ending in *–ite.* Determine the formula and charge of the oxoanion, and add enough hydrogens to make a neutral formula.

Setup The sulfite ion is SO_3^{2-}.

Solution The formula of sulfurous acid is H_2SO_3.

THINK ABOUT IT

Note that none of the oxoacids' names begins with the prefix *hydro–*. The prefix *hydro–* in an acid's name indicates that the acid is *binary.*

Practice Problem ATTEMPT Determine the formula of perbromic acid. (Refer to the information in Sample Problem 2.6.)

Practice Problem BUILD Determine the formula of chromic acid.

Practice Problem CONCEPTUALIZE Referring to the diagrams in Practice Problem 2.6C, which of the ions shown would be part of an acid whose name begins with a prefix?

So far, our discussion of nomenclature has focused on ***inorganic compounds,*** which generally are defined as compounds that do not contain carbon—although some carbon-containing species such as CN^- and CO_3^{2-} are considered inorganic. Another important class of molecular substances is *organic* compounds, which have their own system of nomenclature. ***Organic compounds*** contain carbon and hydrogen, sometimes in combination with other elements such as oxygen, nitrogen, sulfur, and the halogens. The simplest organic compounds are those that contain only carbon and hydrogen and are known as ***hydrocarbons.*** Among hydrocarbons, the simplest examples are compounds known as ***alkanes.*** The name of an alkane depends on the number of carbon atoms in the molecule. Table 2.7 gives the molecular formulas, systematic names of some of the simplest alkanes, and ball-and-stick models.

Many organic compounds are derivatives of alkanes in which one of the H atoms has been replaced by a group of atoms known as a ***functional group.*** The functional group determines many of the chemical properties of a compound because it typically is where a chemical reaction *occurs.* Table 2.8 lists the names and provides ball-and-stick models of several important functional groups.

Ethanol, for example, the alcohol in alcoholic beverages, is ethane (C_2H_6) with one of the hydrogen atoms replaced by an alcohol (—OH) group. Its name is derived from that of *ethane,* indicating that it contains *two* carbon atoms.

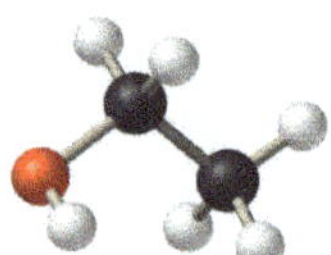

Ethanol

TABLE 2.7 Formulas, Names, and Models of Some Simple Alkanes

Formula	Name	Model
CH_4	Methane	
C_2H_6	Ethane	
C_3H_8	Propane	
C_4H_{10}	Butane	
C_5H_{12}	Pentane	
C_6H_{14}	Hexane	
C_7H_{16}	Heptane	
C_8H_{18}	Octane	
C_9H_{20}	Nonane	
$C_{10}H_{22}$	Decane	

The molecular formula of ethanol can also be written C_2H_6O, but C_2H_5OH conveys more information about the structure of the molecule. Organic compounds and several functional groups are discussed in greater detail in Chapter 25.

Empirical Formulas of Molecular Substances

In addition to the ways we have learned so far, like ionic substances, molecular substances can be represented using *empirical formulas* [rewind to Section 2.6]. The empirical formula tells what elements are present in a molecule and in what whole-number ratio they are combined. For example, the molecular formula of hydrogen peroxide is H_2O_2, but its empirical formula is

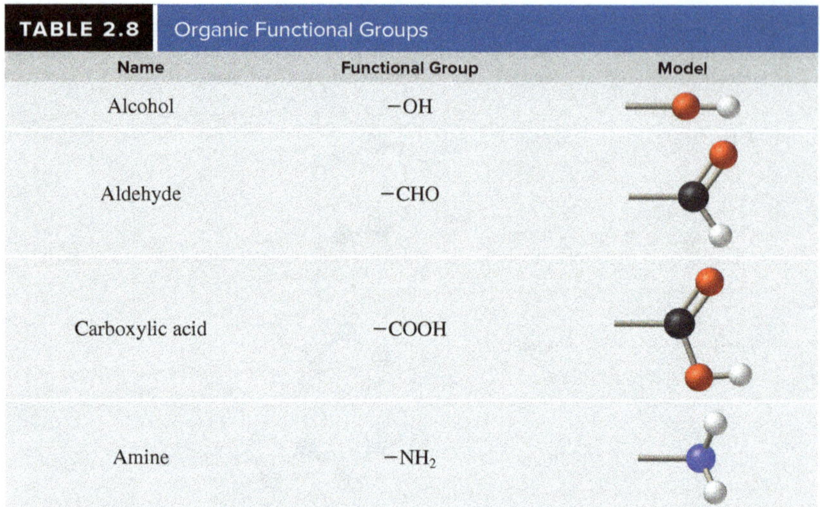

TABLE 2.8 Organic Functional Groups

Name	Functional Group	Model
Alcohol	$-OH$	
Aldehyde	$-CHO$	
Carboxylic acid	$-COOH$	
Amine	$-NH_2$	

simply HO. Hydrazine, which has been used as a rocket fuel, has the molecular formula N_2H_4, so its empirical formula is NH_2. Although the ratio of nitrogen to hydrogen is 1:2 in both the molecular formula (N_2H_4) and the empirical formula (NH_2), only the molecular formula tells us the actual number of N atoms (two) and H atoms (four) present in a hydrazine molecule.

In many cases, the empirical and molecular formulas are identical. In the case of water, for example, there is no combination of smaller whole numbers that can convey the ratio of two H atoms for every one O atom, so the empirical formula is the same as the molecular formula: H_2O. Table 2.9 lists the molecular and empirical formulas for several compounds.

TABLE 2.9 Molecular and Empirical Formulas

Compound	Molecular Formula	Model	Empirical Formula	Model
Water	H_2O		H_2O	
Hydrogen peroxide	H_2O_2		HO	
Ethane	C_2H_6		CH_3	
Propane	C_3H_8		C_3H_8	
Acetylene	C_2H_2		CH	
Benzene	C_6H_6		CH	

Empirical formulas are the *simplest* chemical formulas; they are written by reducing the subscripts in molecular formulas to the smallest possible whole numbers (without altering the relative numbers of atoms). Molecular formulas are the *true* formulas of molecules. As we will see in Chapter 3, when chemists analyze an unknown compound, the first step is usually the determination of the compound's empirical formula.

Sample Problem 2.11 lets you practice determining empirical formulas from molecular formulas.

SAMPLE PROBLEM 2.11

Write the empirical formulas for the following molecules: (a) glucose ($C_6H_{12}O_6$), a substance known as blood sugar; (b) adenine ($C_5H_5N_5$), also known as vitamin B_4; and (c) nitrous oxide (N_2O), a gas that is used as an anesthetic ("laughing gas") and as an aerosol propellant for whipped cream.

Strategy To write the empirical formula, the subscripts in the molecular formula must be reduced to the smallest possible whole numbers (without altering the relative numbers of atoms).

Setup The molecular formulas in parts (a) and (b) each contain subscripts that are divisible by common numbers. Therefore, we will be able to express the formulas with smaller whole numbers than those in the molecular formulas. In part (c), the molecule has only one O atom, so it is impossible to simplify this formula further.

Solution (a) Dividing each of the subscripts in the molecular formula for glucose by 6, we obtain the empirical formula, CH_2O. If we had divided the subscripts by 2 or 3, we would have obtained the formulas $C_3H_6O_3$ and $C_2H_4O_2$, respectively. Although the ratio of carbon to hydrogen to oxygen atoms in each of these formulas is correct (1:2:1), neither is the simplest formula because the subscripts are not in the smallest possible whole-number ratio. (b) Dividing each subscript in the molecular formula of adenine by 5, we get the empirical formula, CHN. (c) Because the subscripts in the formula for nitrous oxide are already the smallest possible whole numbers, its empirical formula is the same as its molecular formula, N_2O.

THINK ABOUT IT

Make sure that the *ratio* in each empirical formula is the same as that in the corresponding molecular formula and that the subscripts are the smallest possible whole numbers. In part (a), for example, the ratio of C:H:O in the molecular formula is 6:12:6, which is equal to 1:2:1, the ratio expressed in the empirical formula.

Practice Problem ATTEMPT Write empirical formulas for the following molecules: (a) caffeine ($C_8H_{10}N_4O_2$), a stimulant found in tea and coffee, (b) butane (C_4H_{10}), which is used in cigarette lighters, and (c) glycine ($C_2H_5NO_2$), an amino acid.

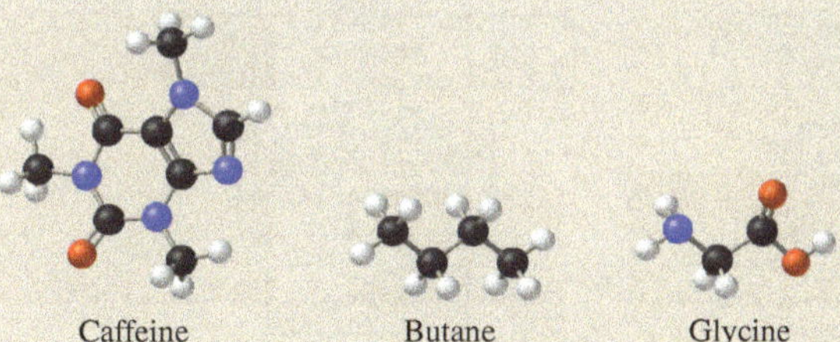

Caffeine Butane Glycine

Practice Problem BUILD For which of the following molecular formulas is the formula shown in parentheses the correct empirical formula? (a) $C_{12}H_{22}O_{11}$ ($C_{12}H_{22}O_{11}$), (b) $C_8H_{12}O_4$ ($C_4H_6O_2$), (c) H_2O_2 (H_2O)?

Practice Problem CONCEPTUALIZE Which of the following molecules has/have the same empirical formula as acetic acid ($HC_2O_2H_3$)?

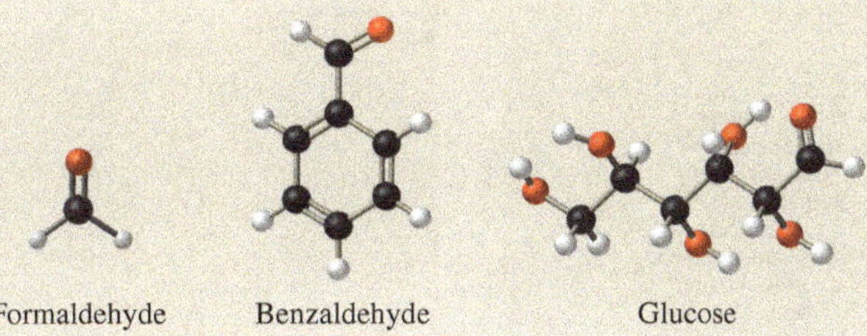

Formaldehyde Benzaldehyde Glucose

CHECKPOINT – SECTION 2.7 Molecules and Molecular Compounds

2.7.1 What is the correct systematic name of PCl_5?

a) Phosphorus chloride
b) Phosphorus pentachloride
c) Monophosphorus chloride
d) Pentachlorophosphorus
e) Pentaphosphorus chloride

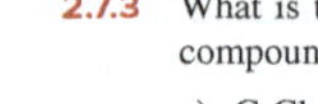

2.7.2 What is the name of the compound shown?

a) Methane
b) Carbon tetrahydrogen monoxide
c) Methanol
d) Methane monoxide
e) Tetrahydrogen carbon monoxide

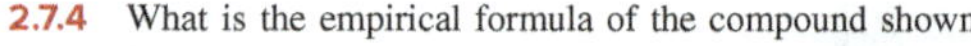

2.7.3 What is the correct formula for the compound carbon tetrachloride?

a) C_4Cl_4
b) C_4Cl
c) CCl_4
d) CCl_2
e) CCl

2.7.4 What is the empirical formula of the compound shown?

a) C_6H_6
b) C_4H_4
c) C_2H_2
d) CH_2
e) CH

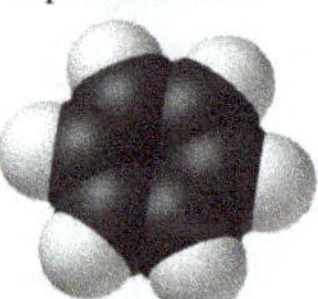

2.8 Compounds in Review

We have seen that compounds can be made up of ions, molecules. It is important that you be able to determine whether a compound is ionic or molecular from its chemical formula. In many cases, this will be possible only if you can easily recognize the common polyatomic ions. Further, we have learned a systematic approach to naming compounds. Remember that by convention, we use Greek prefixes to indicate ratios of combination in molecular compounds, but not in ionic compounds—because knowing the charges on the constituent ions means that ionic-compound names without prefixes are unambiguous. Figure 2.17 summarizes the steps for naming molecular and ionic compounds.

Student Note: In general, we determine whether a compound is molecular or ionic by asking these questions: Does the compound consist of only nonmetals? Then it is probably molecular. Does the compound contain either a metal cation or the NH_4^+ ion? Then it is probably ionic. (Note that it is possible for an ionic compound to contain only nonmetals if it contains the NH_4^+ ion.)

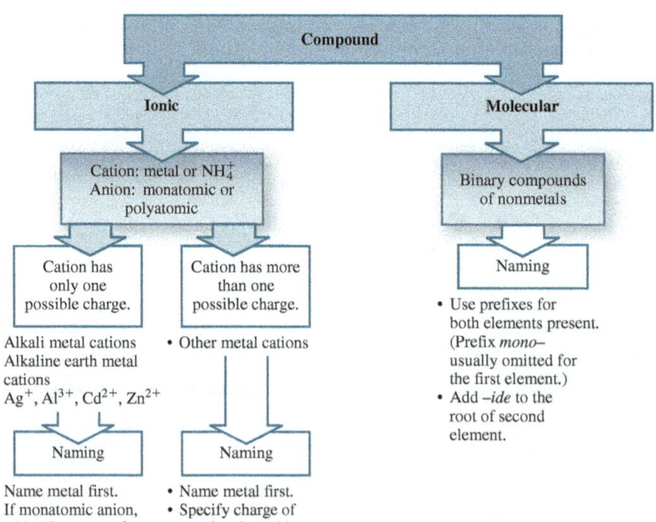

Figure 2.17 Steps for naming molecular and ionic compounds.

Some compounds are better known by their common names than by their systematic chemical names. Familiar examples are listed in Table 2.10.

TABLE 2.10 Common and Systematic Names of Some Familiar Inorganic Compounds

Formula	Common Name	Systematic Name
$CaCO_3$	Marble, chalk, limestone	Calcium carbonate
$NaHCO_3$	Baking soda	Sodium hydrogen carbonate
$Mg(OH)_2$	Milk of magnesia	Magnesium hydroxide
H_2O	Water	Dihydrogen monoxide
NH_3	Ammonia	Trihydrogen nitride
CO_2	Dry ice	Solid carbon dioxide
NaCl	Salt	Sodium chloride
N_2O	Nitrous oxide, laughing gas	Dinitrogen monoxide
$MgSO_4 \cdot 7H_2O$	Epsom salt	Magnesium sulfate heptahydrate

Chapter Summary

Section 2.1

- Dalton's atomic theory states that all matter is made up of tiny indivisible, immutable particles called *atoms*. Compounds form, moreover, when atoms of different elements combine in fixed ratios. According to the ***law of definite proportions,*** any sample of a given compound will always contain the same elements in the same mass ratio.
- The ***law of multiple proportions*** states that if two elements can form more than one compound with one another, the mass ratio of one will be related to the mass ratio of the other by a small whole number.
- The ***law of conservation of mass*** states that matter can be neither created nor destroyed.

Section 2.2

- On the basis of Dalton's atomic theory, the ***atom*** is the basic unit of an element. Studies with ***radiation*** indicated that atoms contained subatomic particles, one of which was the ***electron.***
- Experiments with ***radioactivity*** have shown that some atoms give off different types of radiation, called ***alpha (α) rays, beta (β) rays,*** and ***gamma (γ) rays.*** Alpha rays are composed of ***α particles,*** which are actually helium nuclei. Beta rays are composed of ***β particles,*** which are actually electrons. Gamma rays are high-energy radiation.
- Most of the mass of an atom resides in a tiny, dense region known as the ***nucleus.*** The nucleus contains positively charged particles called ***protons*** and electrically neutral particles called ***neutrons.*** The charge on a proton is equal in magnitude but opposite in sign to the charge on an electron. The electrons occupy the relatively large volume around the nucleus. A neutron has a slightly greater mass than a proton, but each is almost 2000 times as massive as an electron.

Section 2.3

- The ***atomic number (Z)*** is the number of protons in the nucleus of an atom. The atomic number determines the identity of the atom. The ***mass number (A)*** is the sum of the protons and neutrons in the nucleus.
- Protons and neutrons are referred to collectively as ***nucleons.***
- Atoms with the same atomic number but different mass numbers are called ***isotopes.***

Section 2.4

- The ***periodic table*** arranges the elements in rows (***periods***) and columns (***groups*** or ***families***). Elements in the same group exhibit similar properties.
- All elements fall into one of three categories: ***metal, nonmetal,*** or ***metalloid.***
- Some of the groups have special names including ***alkali metals*** (Group 1A, except hydrogen), ***alkaline earth metals*** (Group 2A), ***chalcogens*** (Group 6A), ***halogens*** (Group 7A), ***noble gases*** (Group 8A), and ***transition elements*** or ***transition metals*** (Group 1B and Groups 3B–8B).

Section 2.5

- ***Atomic mass*** is the mass of an atom in atomic mass units. One ***atomic mass unit (amu)*** is exactly one-twelfth the mass of a carbon-12 atom.
- The periodic table contains the *average* atomic mass (sometimes called the ***atomic weight***) of each element.

Section 2.6

- An ***ion*** is an atom or group of atoms with a net charge. An ***atomic ion*** or a ***monatomic ion*** consists of just one atom.
- An ion with a net positive charge is a ***cation.*** An ion with a net negative charge is an ***anion.*** An ***ionic compound*** is one that consists of cations and anions in an electrically neutral combination. A three-dimensional array of alternating cations and anions is called a ***lattice.***
- ***Polyatomic ions*** are those that contain more than one atom chemically bonded together.
- The formulas of ionic compounds are ***empirical formulas,*** indicating with subscripts the ratio of combination of a compound's constituent elements in the smallest possible whole numbers.
- Ionic compounds' names are written as the name of the cation followed by the name of the anion—each with the word "ion" removed. The *name* of an ionic compound need not indicate the ratio of combination of elements because the charges on ions are known.
- ***Oxoanions*** are polyatomic ions that contain one or more oxygen atoms.
- ***Hydrates*** are compounds whose formulas include a specific number of water molecules.

Section 2.7

- A ***molecule*** is an electrically neutral group of two or more atoms. Molecules consisting of just two atoms are called ***diatomic.*** Diatomic molecules may be ***homonuclear*** (just one kind of atom) or ***heteronuclear*** (two kinds of atoms). In general, molecules containing more than two atoms are called ***polyatomic.***
- A ***chemical formula*** denotes the composition of a substance. A ***molecular formula*** specifies the exact numbers of atoms in a molecule of a compound. A ***structural formula*** shows the arrangement of atoms in a substance.
- An ***allotrope*** is one of two or more different forms of an element.
- Molecular compounds are named according to a set of rules, including the use of Greek prefixes to specify the number of each kind of atom in the molecule.
- ***Binary*** compounds are those that consist of *two* elements. An ***acid*** is a substance that generates hydrogen ions when it dissolves in water. An ***ionizable hydrogen atom*** is one that can be removed in water to become a hydrogen ion (H^+).
- ***Oxoacids*** are acids based on oxoanions. Acids with more than one ionizable hydrogen atom are called ***polyprotic.***
- ***Inorganic compounds*** are generally those that do not contain carbon. ***Organic compounds*** contain carbon and hydrogen, sometimes in combination with other elements. ***Hydrocarbons*** contain only carbon and hydrogen. The simplest hydrocarbons are the ***alkanes.*** A ***functional group*** is a group of atoms that determines the chemical properties of an organic compound.
- The formulas of molecular compounds may be expressed as molecular formulas, or as empirical formulas. The empirical and molecular formulas of a given compound may or may not be identical.

Key Words

Naming Compounds

The process of naming binary molecular compounds follows the procedure outlined in Section 2.7. The element that appears first in the formula is named first, followed by the name of the second element—with its ending changed to *–ide*. Greek prefixes are used to indicate numbers of atoms, but the prefix *mono–* is *not* used when there is only one atom of the *first* element in the formula.

Examples:

N_2O	NO_2	Cl_2O_7	P_4O_6
dinitrogen monoxide	nitrogen dioxide	dichlorine heptoxide	tetraphosphorus hexoxide

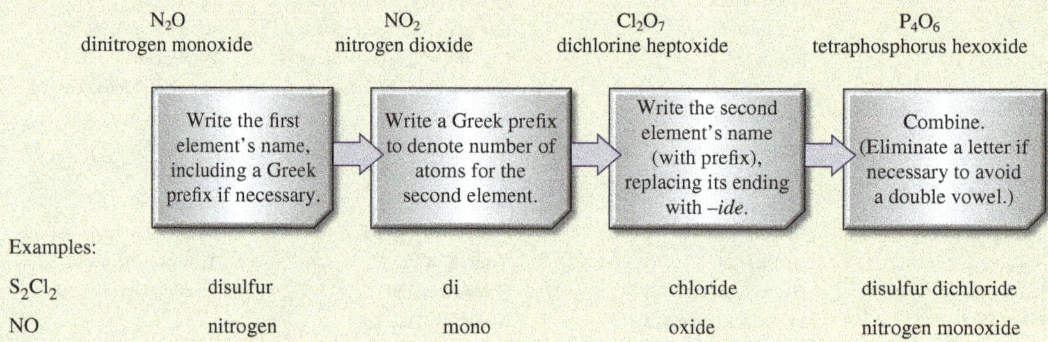

Examples:

S_2Cl_2	disulfur	di	chloride	disulfur dichloride
NO	nitrogen	mono	oxide	nitrogen monoxide

The process of naming binary ionic compounds follows the simple procedure outlined in Section 2.6. Naming compounds that contain polyatomic ions follows essentially the same procedure; but it does require you to recognize the common polyatomic ions [⏮ Table 2.3]. Because many ionic compounds contain polyatomic ions, it is important that you know their names, formulas, and charges—well enough that you can identify them readily.

In ionic compounds with ratios of combination other than 1:1, subscript numbers are used to denote the number of each ion in the formula.

Examples: $CaBr_2$, Na_2S, $AlCl_3$, Al_2O_3, FeO, Fe_2O_3

Recall that because the common ions of main group elements have predictable charges, it is unnecessary to use prefixes to denote their numbers when naming compounds that contain them. Thus, the names of the first four examples above are calcium bromide, sodium sulfide, aluminum chloride, and aluminum oxide. The last two contain transition metal ions, many of which have more than one possible charge. In these cases, in order to avoid ambiguity, the charge on the metal ion is designated with a Roman numeral in parentheses. The names of these two compounds are iron(II) oxide and iron(III) oxide, respectively.

When a subscript number is required for a polyatomic ion, the ion's formula must first be enclosed in parentheses.

Examples: $Ca(NO_3)_2$, $(NH_4)_2S$, $Ba(C_2H_3O_2)_2$, $(NH_4)_2SO_4$, $Fe_3(PO_4)_2$, $Co_2(CO_3)_3$

Names: calcium nitrate, ammonium sulfide, barium acetate, ammonium sulfate, iron(II) phosphate, cobalt(III) carbonate

The process of naming ionic compounds given their formulas can be summarized with the following flowchart:

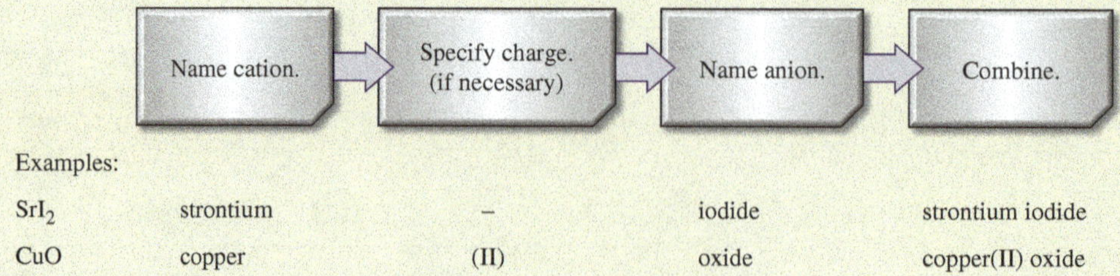

Examples:

SrI_2	strontium	–	iodide	strontium iodide
CuO	copper	(II)	oxide	copper(II) oxide

It is equally important that you be able to write the formula of an ionic compound given its name. Again, knowledge of the common polyatomic ions is critical. The process of writing an ionic compound's formula given its name is summarized as follows:

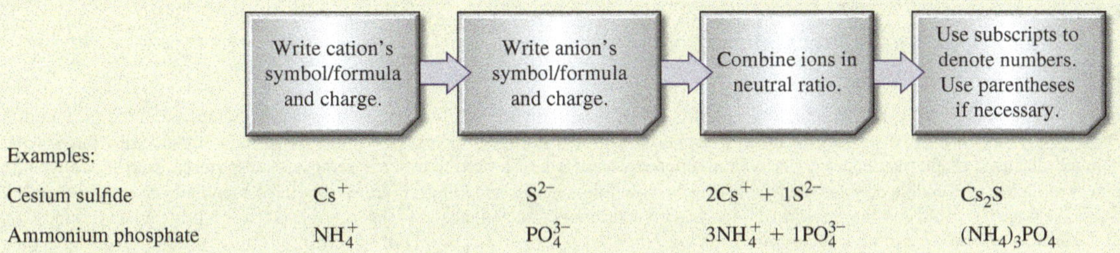

Examples:				
Cesium sulfide	Cs^+	S^{2-}	$2Cs^+ + 1S^{2-}$	Cs_2S
Ammonium phosphate	NH_4^+	PO_4^{3-}	$3NH_4^+ + 1PO_4^{3-}$	$(NH_4)_3PO_4$

The process of writing a molecular compound's formula given its name is summarized as follows:

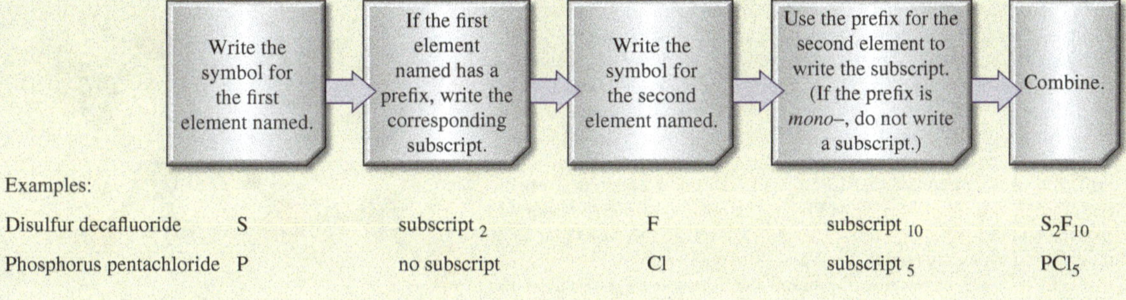

Examples:					
Disulfur decafluoride	S	subscript $_2$	F	subscript $_{10}$	S_2F_{10}
Phosphorus pentachloride	P	no subscript	Cl	subscript $_5$	PCl_5

Key Skills Problems

2.1
What is the correct name for $CaSO_4$?

(a) calcium sulfoxide (b) calcium sulfite (c) calcium sulfur oxide (d) calcium sulfate (e) calcium sulfide tetroxide

2.2
What is the correct formula for nickel(II) perchlorate?

(a) $NiClO_4$ (b) Ni_2ClO_4 (c) $Ni(ClO_4)_2$ (d) $NiClO_3$ (e) $Ni(ClO_3)_2$

2.3
What is the correct name for NCl_3?

(a) trinitrogen chloride (b) mononitrogen chloride (c) nitrogen trichloride (d) nitride trichloride (e) mononitride chloride

2.4
What is the correct formula for phosphorus pentachloride?

(a) PCl_5 (b) P_5Cl (c) $P(ClO)_5$ (d) PO_4Cl (e) $PClO$

Questions and Problems

Applying What You've Learned

Although iron is an essential element, it is also a potentially toxic substance. Hemochromatosis is one of the most common hereditary disorders, causing "iron overload" or the storage of excess iron in the tissues and organs. Individuals with hemochromatosis often must undergo periodic phlebotomy (removal of blood) in order to remove excess stored iron, which would otherwise cause irreversible damage to internal organs including the liver and kidneys. Those who have a tendency to store too much iron are advised to avoid combining iron-rich foods with substances that enhance iron absorption, such as ascorbic acid (vitamin C).

Because of iron's toxicity, iron supplements are potentially dangerous, especially to children. In fact, iron poisoning is the most common toxicological emergency in young children—due in part to the resemblance many iron supplements bear to candy. Most vitamins that contain iron are sold with childproof caps to help prevent accidental overdose. The Food and Drug Administration (FDA) recommends supplements containing more than 30 mg of iron per dose to be sold in single-dose blister packs to make it more difficult for a child to consume a dangerous amount. (a) Iron has four naturally occurring isotopes: ^{54}Fe (53.9396 amu), ^{56}Fe (55.9349 amu), ^{57}Fe (56.9354 amu), and ^{58}Fe (57.9333 amu). For each isotope, determine the number of neutrons in the nucleus [|◂◂ Sample Problem 2.2]. (b) Calculate the average atomic mass of iron given that the natural abundances of the four isotopes are 5.845, 91.754, 2.119, and 0.282 percent, respectively [|◂◂ Sample Problem 2.3]. (c) Write the formula for ferrous sulfate [iron(II) sulfate] [|◂◂ Sample Problem 2.5]. (d) Write the molecular formula for ascorbic acid (see the ball-and-stick model below) [|◂◂ Sample Problem 2.7]. (e) Determine the empirical formula of ascorbic acid [|◂◂ Sample Problem 2.11].

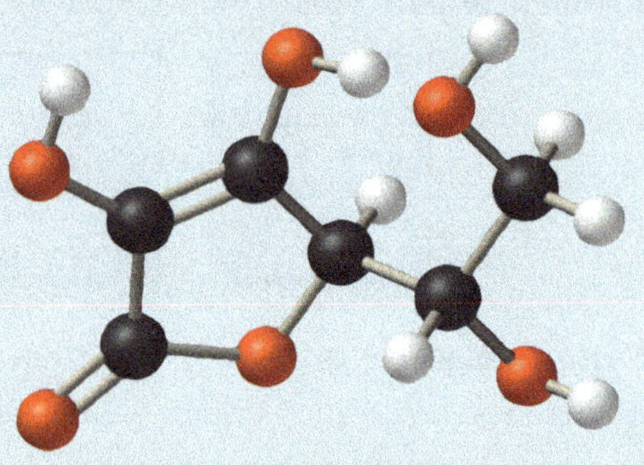

Ascorbic acid

SECTION 2.1: THE ATOMIC THEORY

Review Questions

2.1 What are the hypotheses on which Dalton's atomic theory is based?

2.2 State the laws of definite proportions and multiple proportions. Illustrate each with an example.

Computational Problems

2.3 The elements sulfur and oxygen can form a variety of different compounds, the most common of which are SO_2 and SO_3. Samples of these two compounds, were decomposed into their constituent elements. One produced 1.002 g S for every gram of O; the other produced 0.668 g S for every gram of O. Show that these results are consistent with the law of multiple proportions.

2.4 Two different compounds, each containing only phosphorus and chlorine, were decomposed into their constituent elements. One produced 0.2912 g P for every gram of Cl; the other produced 0.1747 g P for every gram of Cl. Show that these results are consistent with the law of multiple proportions.

2.5 Sulfur reacts with fluorine to produce three different compounds. The mass ratio of fluorine to sulfur for each compound is given in the following table:

Compound	mass F : mass S
S_2F_{10}	2.962
SF_4	2.370
SF_6	3.555

Show that these data are consistent with the law of multiple proportions.

2.6 Both FeO and Fe_2O_3 contain only iron and oxygen. The mass ratio of oxygen to iron for each compound is given in the following table:

Compound	mass O : mass Fe
FeO	0.2865
Fe_2O_3	0.4297

Show that these data are consistent with the law of multiple proportions.

Conceptual Problems

2.7 For the two compounds pictured, evaluate the following ratio:

$$\frac{\text{g blue : 1.00 g red(right)}}{\text{g blue : 1.00 g red(left)}}$$

2.8 For the two compounds pictured, evaluate the following ratio:

$$\frac{\text{g green : 1.00 g yellow(right)}}{\text{g green : 1.00 g yellow(left)}}$$

SECTION 2.2: THE STRUCTURE OF THE ATOM

Review Questions

2.9 Define the following terms: (a) α particle, (b) β particle, (c) γ ray, (d) X ray.

2.10 Name the types of radiation known to be emitted by radioactive elements.

2.11 Compare the properties of the following: α particles, cathode rays, protons, neutrons, and electrons.

2.12 Describe the contributions of the following scientists to our knowledge of atomic structure: J. J. Thomson, R. A. Millikan, Ernest Rutherford, and James Chadwick.

2.13 Describe the experimental basis for believing that the nucleus occupies a very small fraction of the volume of the atom.

Problems

2.14 The diameter of a neutral helium atom is about 1×10^2 pm. Suppose that we could line up helium atoms side by side in contact with one another. Approximately how many atoms would it take to make the distance 1 in from end to end?

2.15 Roughly speaking, the radius of an atom is about 10,000 times greater than that of its nucleus. If an atom were magnified so that the radius of its nucleus became 2.0 cm, about the size of a marble, what would be the radius of the atom in miles (1 mi = 1609 m)?

SECTION 2.3: ATOMIC NUMBER, MASS NUMBER, AND ISOTOPES

Review Questions

2.16 Use the helium-4 isotope to define atomic number and mass number. Why does knowledge of the atomic number enable us to deduce the number of electrons present in an atom?

2.17 Why do all atoms of an element have the same atomic number, although they may have different mass numbers?

2.18 What do we call atoms of the same elements with different mass numbers?

2.19 Explain the meaning of each term in the symbol $^{A}_{Z}X$.

Computational Problems

2.20 What is the mass number of an iron atom that has 28 neutrons?

2.21 Calculate the number of neutrons of ^{239}Pu.

2.22 For each of the following species, determine the number of protons and the number of neutrons in the nucleus: $^{3}_{2}He$, $^{4}_{2}He$, $^{24}_{12}Mg$, $^{25}_{12}Mg$, $^{48}_{22}Ti$, $^{79}_{35}Br$, $^{195}_{78}Pt$.

2.23 Indicate the number of protons, neutrons, and electrons in each of the following species: $^{15}_{7}N$, $^{33}_{16}S$, $^{63}_{29}Cu$, $^{84}_{38}Sr$, $^{130}_{56}Ba$, $^{186}_{74}W$, $^{202}_{80}Hg$.

2.24 Write the appropriate symbol for each of the following isotopes: (a) $Z = 11, A = 23$; (b) $Z = 28, A = 64$; (c) $Z = 50, A = 115$; (d) $Z = 20, A = 42$.

2.25 Write the appropriate symbol for each of the following isotopes: (a) $Z = 74, A = 186$; (b) $Z = 80, A = 201$; (c) $Z = 34, A = 76$; (d) $Z = 94, A = 239$.

2.26 Determine the mass number of (a) a boron atom with 6 neutrons, (b) a magnesium atom with 13 neutrons, (c) a bromine atom with 44 neutrons, and (d) a mercury atom with 119 neutrons.

2.27 Determine the mass number of (a) a fluorine atom with 11 neutrons, (b) a sulfur atom with 16 neutrons, (c) an arsenic atom with 45 neutrons, and (d) a platinum atom with 120 neutrons.

2.28 The following radioactive isotopes are used in medicine for imaging organs, studying blood circulation, treating cancer, and so on. Give the number of neutrons present in each isotope: ^{198}Au, ^{47}Ca, ^{60}Co, ^{18}F, ^{125}I, ^{131}I, ^{42}K, ^{43}K, ^{24}Na, ^{32}P, ^{85}Sr, ^{99}Tc.

SECTION 2.4: THE PERIODIC TABLE

Review Questions

2.29 What is the periodic table, and what is its significance in the study of chemistry?

2.30 State two differences between a metal and a nonmetal.

2.31 Write the names and symbols for four elements in each of the following categories: (a) nonmetal, (b) metal, (c) metalloid.

2.32 Give two examples of each of the following: (a) alkali metals, (b) alkaline earth metals, (c) halogens, (d) noble gases, (e) chalcogens, (f) transition metals.

2.33 The explosion of an atomic bomb in the atmosphere releases many radioactive isotopes into the environment. One of the isotopes is ^{90}Sr. Via a relatively short food chain, it can enter the human body. Considering the position of strontium in the periodic table, explain why it is particularly harmful to humans.

Computational Problems

2.34 Elements whose names end with *–ium* are usually metals; sodium is one example. Identify a nonmetal whose name also ends with *–ium*.

2.35 Describe the changes in properties (from metals to nonmetals or from nonmetals to metals) as we move (a) down a periodic group and (b) across the periodic table from left to right.

2.36 Consult webelements.com to find (a) two metals less dense than water, (b) two metals more dense than mercury, (c) the densest known solid metallic element, and (d) the densest known solid nonmetallic element.

2.37 Group the following elements in pairs that you would expect to show similar chemical properties: K, F, P, Na, Cl, and N.

2.38 Group the following elements in pairs that you would expect to show similar chemical properties: I, Ba, O, Br, S, and Ca.

2.39 Write the symbol for each of the following biologically important elements in the given periodic table: iron (present in hemoglobin for transporting oxygen), iodine (present in the thyroid gland), sodium (present in intracellular and extracellular fluids), phosphorus (present in bones and teeth), sulfur (present in proteins), and magnesium (present in chlorophyll molecules).

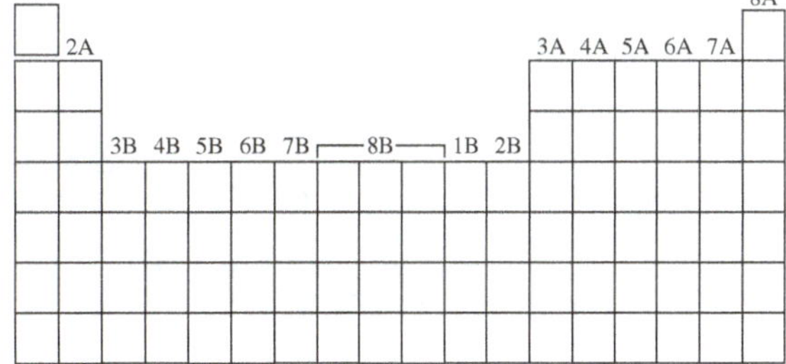

SECTION 2.5: THE ATOMIC MASS SCALE AND AVERAGE ATOMIC MASS

Review Questions

2.40 What is an atomic mass unit? Why is it necessary to introduce such a unit?

2.41 What is the mass (in amu) of a carbon-12 atom? Why is the atomic mass of carbon listed as 12.01 amu in the table on the inside front cover of this book?

2.42 Explain clearly what is meant by the statement "The atomic mass of gold is 197.0 amu."

2.43 What information would you need to calculate the average atomic mass of an element?

Problems

2.44 The atomic masses of ${}^{79}_{35}Br$ (50.69 percent) and ${}^{81}_{35}Br$ (49.31 percent) are 78.9183361 and 80.916289 amu, respectively. Calculate the average atomic mass of bromine. The percentages in parentheses denote the relative abundances.

2.45 The atomic masses of ${}^{204}Pb$ (1.4 percent), ${}^{206}Pb$ (24.1 percent), ${}^{207}Pb$ (22.1 percent), and ${}^{208}Pb$ (52.4 percent) are 203.973020, 205.974440, 206.975872, and 207.976627 amu, respectively. Calculate the average atomic mass of lead. The percentages in parentheses denote the relative abundances.

2.46 The atomic masses of ${}^{203}Tl$ and ${}^{205}Tl$ are 202.972320 and 204.974401 amu, respectively. Calculate the natural abundances of these two isotopes. The average atomic mass of thallium is 204.4 amu.

2.47 The atomic masses of ${}^{6}Li$ and ${}^{7}Li$ are 6.0151 amu and 7.0160 amu, respectively. Calculate the natural abundances of these two isotopes. The average atomic mass of Li is 6.941 amu.

2.48 What is the mass in grams of 13.2 amu?

2.49 How many atomic mass units are there in 8.4 g?

SECTION 2.6: IONS AND IONIC COMPOUNDS

Review Questions

2.50 Give an example of each of the following: (a) a monatomic cation, (b) a monatomic anion, (c) a polyatomic cation, (d) a polyatomic anion.

2.51 What is an ionic compound? How is electrical neutrality maintained in an ionic compound?

2.52 Explain why the chemical formulas of ionic compounds are usually the same as their empirical formulas.

2.53 What is the Stock system? What are its advantages over the older system of naming cations?

2.54 Give an example of an ionic compound that does not contain a metal.

Conceptual Problems

2.55 Give the number of protons and electrons in each of the following common ions: Na^+, Ca^{2+}, Al^{3+}, Fe^{2+}, I^-, F^-, S^{2-}, O^{2-}, N^{3-}.

2.56 Give the number of protons and electrons in each of the following common ions: K^+, Mg^{2+}, Fe^{3+}, Br^-, Mn^{2+}, C^{4-}, Cu^{2+}.

2.57 Write the formulas for the following ionic compounds: (a) sodium oxide, (b) iron sulfide (containing the Fe^{2+} ion), (c) cobalt sulfate (containing the Co^{3+} and SO_4^{2-} ions), (d) barium fluoride.

2.58 Write the formulas for the following ionic compounds: (a) copper bromide (containing the Cu^+ ion), (b) manganese oxide (containing the Mn^{3+} ion), (c) mercury iodide (containing the Hg_2^{2+} ion), (d) magnesium phosphate (containing the PO_4^{3-} ion).

2.59 Which of the following compounds are likely to be ionic? Which are likely to be molecular? $SiCl_4$, LiF, $BaCl_2$, B_2H_6, KCl, C_2H_4.

2.60 Which of the following compounds are likely to be ionic? Which are likely to be molecular? CH_4, NaBr, BaF_2, CCl_4, ICl, CsCl, NF_3.

2.61 Name the following compounds: (a) KH_2PO_4, (b) K_2HPO_4, (c) HBr (gas), (d) HBr (in water), (e) Li_2CO_3, (f) $K_2Cr_2O_7$, (g) NH_4NO_2, (h) HIO_3, (i) PF_5, (j) P_4O_6, (k) CdI_2, (l) $SrSO_4$, (m) $Al(OH)_3$.

2.62 Name the following compounds: (a) KClO, (b) Ag_2CO_3, (c) HNO_2, (d) $KMnO_4$, (e) $CsClO_3$, (f) KNH_4SO_4, (g) FeO, (h) Fe_2O_3, (i) $TiCl_4$, (j) NaH, (k) Li_3N, (l) Na_2O, (m) Na_2O_2.

2.63 Write the formulas for the following compounds: (a) rubidium nitrite, (b) potassium sulfide, (c) sodium hydrogen sulfide, (d) magnesium phosphate, (e) calcium hydrogen phosphate, (f) lead(II) carbonate, (g) tin(II) fluoride, (h) ammonium sulfate, (i) silver perchlorate, (j) boron trichloride.

2.64 Write the formulas for the following compounds: (a) copper(I) cyanide, (b) strontium chlorite, (c) perbromic acid, (d) hydroiodic acid, (e) disodium ammonium phosphate, (f) potassium dihydrogen phosphate, (g) iodine heptafluoride, (h) tetraphosphorus decasulfide, (i) mercury(II) oxide, (j) mercury(I) iodide, (k) selenium hexafluoride.

2.65 In the diagrams shown here, match each of the drawings with the following ionic compounds: Al_2O_3, LiH, Na_2S, $Mg(NO_3)_2$. (Green spheres represent cations and red spheres represent anions.)

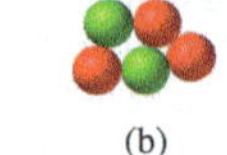

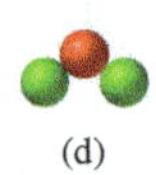

(a) (b) (c) (d)

2.66 Given the formulas for the ionic compounds, draw the correct ratio of cations to anions as shown in Problem 2.65: (a) $BaSO_4$, (b) CaF_2, (c) Mg_3N_2, (d) K_2O.

SECTION 2.7: MOLECULES AND MOLECULAR COMPOUNDS

Review Questions

2.67 What is the difference between an atom and a molecule?

2.68 What are allotropes? Give an example. How are allotropes different from isotopes?

2.69 What does a chemical formula represent? Determine the ratio of the atoms in the following molecular formulas: (a) NO, (b) NCl_3, (c) N_2O_4, (d) P_4O_6.

2.70 Define molecular formula and empirical formula. What are the similarities and differences between the empirical formula and molecular formula of a compound?

2.71 Give an example of a case in which two molecules have different molecular formulas but the same empirical formula.

2.72 What is the difference between inorganic compounds and organic compounds?

2.73 Give one example each for a binary compound and a ternary compound. (A ternary compound is one that contains three different elements.)

2.74 Explain why the formula HCl can represent two different chemical systems.

Conceptual Problems

2.75 For each of the following diagrams, determine whether it represents diatomic molecules, polyatomic molecules, molecules that are not compounds, molecules that are compounds, or an elemental form of the substance.

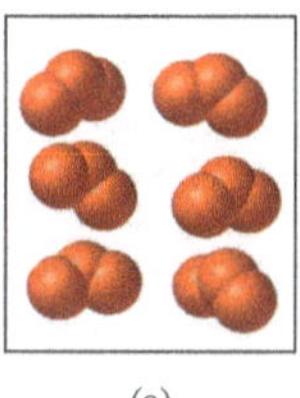
(a)

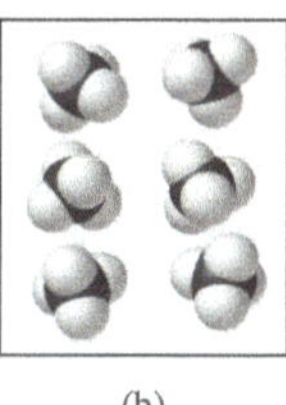
(b)

(c)

2.76 For each of the following diagrams, determine whether it represents diatomic molecules, polyatomic molecules, molecules that are not compounds, molecules that are compounds, or an elemental form of the substance.

(a)

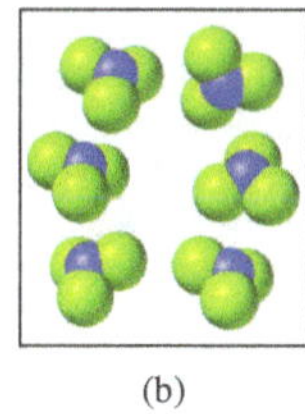
(b)

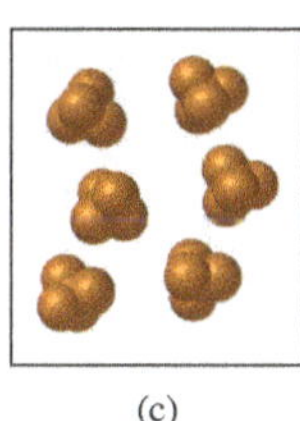
(c)

2.77 Identify the following as elements or compounds: NH_3, N_2, S_8, NO, CO, CO_2, H_2, SO_2.

2.78 Give two examples of each of the following: (a) a diatomic molecule containing atoms of the same element, (b) a diatomic molecule containing atoms of different elements, (c) a polyatomic molecule containing atoms of the same element, (d) a polyatomic molecule containing atoms of different elements.

2.79 Write the empirical formulas of the following compounds: (a) C_2N_2, (b) C_6H_6, (c) C_9H_{20}, (d) P_4O_{10}, (e) B_2H_6.

2.80 Write the empirical formulas of the following compounds: (a) Al_2Br_6, (b) $Na_2S_2O_4$, (c) N_2O_5, (d) $K_2Cr_2O_7$, (e) $H_2C_2O_4$.

2.81 Write the molecular formula of alanine, an amino acid used in protein synthesis. The color codes are black (carbon), blue (nitrogen), red (oxygen), and white (hydrogen).

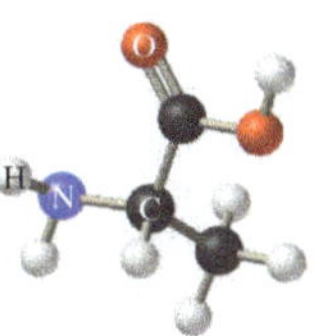

2.82 Write the molecular formula of ethanol. The color codes are black (carbon), red (oxygen), and white (hydrogen).

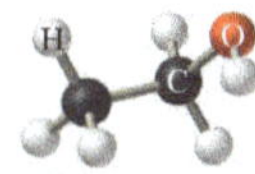

2.83 Name the following binary molecular compounds: (a) NCl_3, (b) IF_7, (c) P_4O_6, (d) S_2Cl_2.

2.84 Write chemical formulas for the following molecular compounds: (a) phosphorus tribromide, (b) dinitrogen tetrafluoride, (c) xenon tetroxide, (d) selenium trioxide.

2.85 Write the molecular formulas and names of the following compounds.

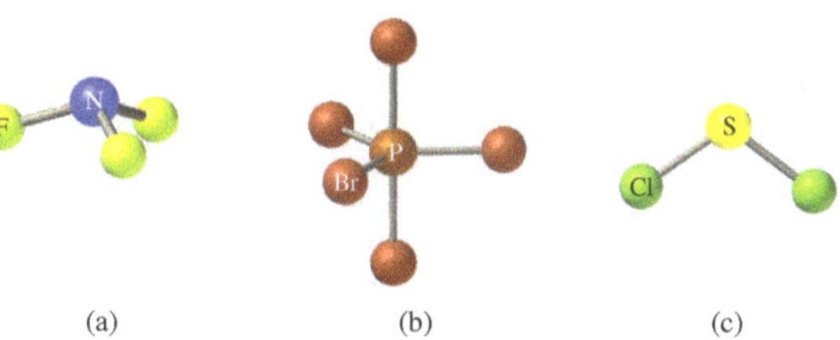

(a) (b) (c)

2.86 Write the molecular formulas and names of the following compounds.

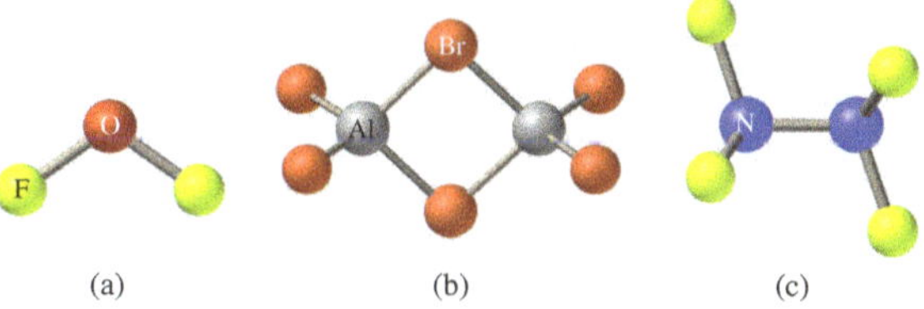

(a) (b) (c)

ADDITIONAL PROBLEMS

2.87 Define the following terms: acids, bases, oxoacids, oxoanions, and hydrates.

2.88 A sample of a uranium compound is found to be losing mass gradually. Explain what is happening to the sample.

2.89 In which one of the following pairs do the two species resemble each other most closely in chemical properties: (a) ${}^{1}_{1}H$ and ${}^{1}_{1}H^{+}$, (b) ${}^{14}_{7}N$ and ${}^{14}_{7}N^{3-}$, (c) ${}^{12}_{6}C$ and ${}^{13}_{6}C$? Explain.

2.90 One isotope of a metallic element has mass number 137 and 82 neutrons in the nucleus. The cation derived from the isotope has 54 electrons. Write the chemical symbol for this cation.

2.91 One isotope of a nonmetallic element has mass number 31 and 16 neutrons in the nucleus. The anion derived from the isotope has 18 electrons. Write the chemical symbol for this anion.

2.92 The following table gives numbers of electrons, protons, and neutrons in atoms or ions of a number of elements. Answer the following: (a) Which of the species are neutral? (b) Which are negatively charged? (c) Which are positively charged? (d) What are the conventional symbols for all the species?

	Atom or Ion of Element						
	A	B	C	D	E	F	G
Number of electrons	5	10	18	28	36	5	9
Number of protons	5	7	19	30	35	5	9
Number of neutrons	5	7	20	36	46	6	10

2.93 What is wrong with or ambiguous about the phrase "four molecules of NaCl"?

2.94 The following phosphorus sulfides are known: P_4S_3, P_4S_7, and P_4S_{10}. Do these compounds obey the law of multiple proportions?

2.95 Which of the following are elements, which are molecules but not compounds, which are compounds but not molecules, and which are both compounds and molecules? (a) SO_2, (b) S_8, (c) Cs, (d) N_2O_5, (e) O, (f) O_2, (g) O_3, (h) CH_4, (i) KBr, (j) S, (k) P_4, (l) LiF.

2.96 What is wrong with the name (given in parentheses or brackets) for each of the following compounds: (a) $BaCl_2$ (barium dichloride), (b) Fe_2O_3 [iron(II) oxide], (c) $CsNO_2$ (cesium nitrate), (d) $Mg(HCO_3)_2$ [magnesium(II) bicarbonate]?

2.97 Discuss the significance of assigning an atomic mass of exactly 12 amu to the carbon-12 isotope.

2.98 Determine what is wrong with the chemical formula and write the correct chemical formula for each of the following compounds: (a) $(NH_3)_2CO_3$ (ammonium carbonate), (b) $CaOH_2$ (calcium hydroxide), (c) $CdSO_3$ (cadmium sulfide), (d) $ZnCrO_4$ (zinc dichromate).

2.99 Fill in the blanks in the table.

Symbol		$^{54}_{26}Fe^{2+}$			
Protons	5			79	86
Neutrons	6		16	117	136
Electrons	5		18	79	
Net charge			−3		0

2.100 (a) Which elements are most likely to form ionic compounds? (b) Which metallic elements are most likely to form cations with different charges?

2.101 Write the formula of the common ion derived from each of the following: (a) Li, (b) S, (c) I, (d) N, (e) Al, (f) Cs, (g) Mg.

2.102 Which of the following symbols provides more information about the atom: ^{23}Na or $_{11}Na$? Explain.

2.103 Write the chemical formulas and names of the binary acids and oxoacids that contain Group 7A elements. Do the same for elements in Groups 3A, 4A, 5A, and 6A.

2.104 Determine the molecular and empirical formulas of the compounds shown here. (Black spheres are carbon, and white spheres are hydrogen.)

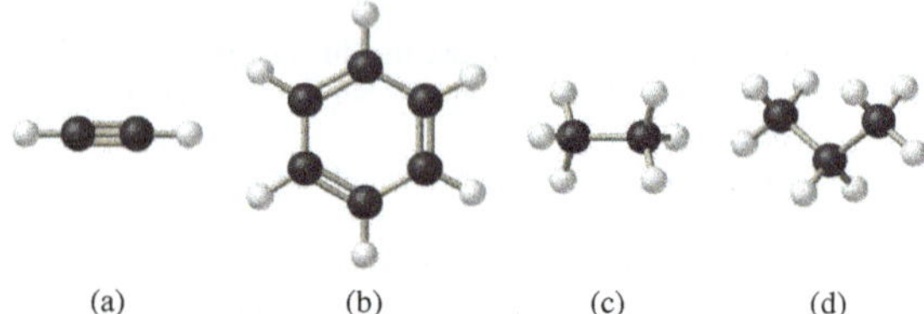

(a) (b) (c) (d)

2.105 For the noble gases (the Group 8A elements) $^{4}_{2}He$, $^{20}_{10}Ne$, $^{40}_{18}Ar$, $^{84}_{36}Kr$, and $^{132}_{54}Xe$, (a) determine the number of protons and neutrons in the nucleus of each atom, and (b) determine the ratio of neutrons to protons in the nucleus of each atom. Describe any general trend you discover in the way this ratio changes with increasing atomic number.

2.106 A monatomic ion has a charge of +2. The nucleus of the parent atom has a mass number of 55. If the number of neutrons in the nucleus is 1.2 times that of the number of protons, what is the name and symbol of the element?

2.107 The Group 1B metals, Cu, Ag, and Au, are called coinage metals. What chemical properties make them especially suitable for making coins and jewelry?

2.108 The elements in Group 8A of the periodic table are called noble gases. Can you suggest what "noble" means in this context?

2.109 The formula for calcium oxide is CaO. What are the formulas for magnesium oxide and strontium oxide?

2.110 A common mineral of barium is barytes, or barium sulfate ($BaSO_4$). Because elements in the same periodic group have similar chemical properties, we might expect to find some radium sulfate ($RaSO_4$) mixed with barytes because radium is the last member of Group 2A. However, the only source of radium compounds in nature is in uranium minerals. Explain why this is so.

2.111 Two elements form a compound that can be represented as [blue–red]. The same two elements also combine to form several other compounds, which can be represented with the diagrams shown here. For each of these, determine the ratio of g red : 1.00 g blue in this compound to g red : 1.00 g blue in [blue–red].

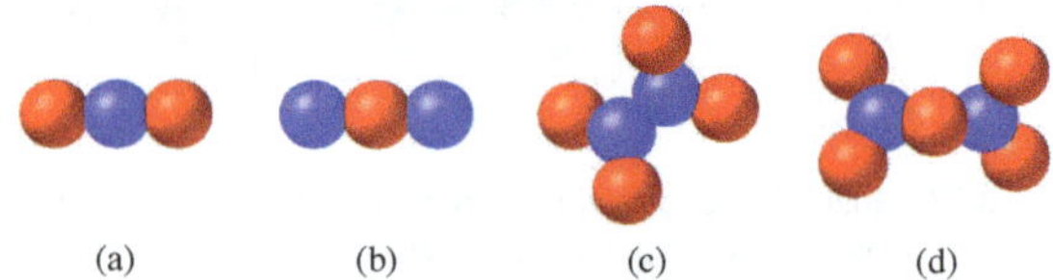

(a) (b) (c) (d)

2.112 Which of the diagrams can be used to illustrate the law of constant composition? Which can be used to illustrate the law of multiple proportions? In each case, for a diagram that cannot be used to illustrate the law, explain why.

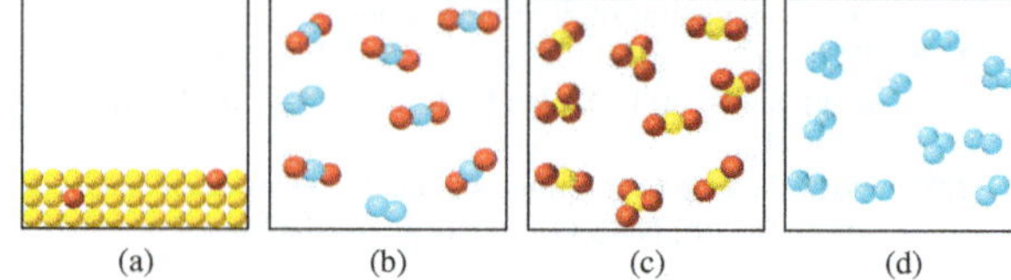

(a) (b) (c) (d)

2.113 Fluorine reacts with hydrogen (H) and deuterium (D) to form hydrogen fluoride (HF) and deuterium fluoride (DF), where deuterium (^2_1H) is an isotope of hydrogen. Would a given amount of fluorine react with different masses of the two hydrogen isotopes? Does this violate the law of definite proportion? Explain.

2.114 Predict the formula and name of a binary compound formed from the following elements: (a) Na and H, (b) B and O, (c) Na and S, (d) Al and F, (e) F and O, (f) Sr and Cl.

2.115 Identify each of the following elements: (a) a halogen whose anion contains 36 electrons, (b) a radioactive noble gas with 86 protons, (c) a Group 6A element whose anion contains 36 electrons, (d) an alkali metal cation that contains 36 electrons, (e) a Group 4A cation that contains 80 electrons.

2.116 Show the locations of (a) alkali metals, (b) alkaline earth metals, (c) the halogens, and (d) the noble gases in the given outline of a periodic table. Also draw dividing lines between metals and metalloids and between metalloids and nonmetals.

1A 8A
2A 3A 4A 5A 6A 7A
3B 4B 5B 6B 7B 8B 1B 2B

2.117 Fill in the blanks in the table.

Cation	Anion	Formula	Name
			Magnesium bicarbonate
		$SrCl_2$	
Fe^{3+}	NO_2^-		
			Manganese(II) chlorate
		$SnBr_4$	
Co^{2+}	PO_4^{3-}		
Hg_2^{2+}	I^-		
		Cu_2CO_3	
			Lithium nitride
Al^{3+}	S^{2-}		

2.118 Some compounds are better known by their common names than by their systematic chemical names. Give the chemical formulas of the following substances: (a) dry ice, (b) salt, (c) laughing gas, (d) marble (chalk, limestone), (e) baking soda, (f) ammonia, (g) water, (h) milk of magnesia, (i) epsom salt.

2.119 In the second footnote on page 42 it was pointed out that mass and energy are alternate aspects of a single entity called mass-energy. The relationship between these two physical quantities is Einstein's equation, $E = mc^2$, where E is energy, m is mass, and c is the speed of light. In a combustion experiment, it was found that 12.096 g of hydrogen molecules combined with 96.000 g of oxygen molecules to form water and released 1.715×10^3 kJ of heat. Use Einstein's equation to calculate the corresponding mass change in this process, and comment on whether or not the law of conservation of mass holds for ordinary chemical processes.

2.120 (a) Describe Rutherford's experiment and how the results revealed the nuclear structure of the atom. (b) Consider the ^{23}Na atom. Given that the radius and mass of the nucleus are 3.04×10^{-15} m and 3.82×10^{-23} g, respectively, calculate the density of the nucleus in g/cm^3. The radius of a ^{23}Na atom is 186 pm. Calculate the density of the space occupied by the electrons outside the nucleus in the sodium atom. Do your results support Rutherford's model of an atom? [The volume of a sphere of radius r is $\frac{4}{3}\pi r^3$.]

2.121 Name the given acids.

2.122 Draw two different structural formulas based on the molecular formula C_2H_6O. Is the fact that you can have more than one compound with the same molecular formula consistent with Dalton's atomic theory?

2.123 Ethane and acetylene are two gaseous hydrocarbons. Chemical analyses show that in one sample of ethane, 2.65 g of carbon are combined with 0.665 g of hydrogen, and in one sample of acetylene, 4.56 g of carbon are combined with 0.383 g of hydrogen. (a) Are these results consistent with the law of multiple proportions? (b) Write reasonable molecular formulas for these compounds.

Multiconcept Problems

2.124 A cube made of platinum (Pt) has an edge length of 1.0 cm. (a) Calculate the number of Pt atoms in the cube. (b) Atoms are spherical in shape. Therefore, the Pt atoms in the cube cannot fill all the available space. If only 74 percent of the space inside the cube is taken up by Pt atoms, calculate the radius in picometers of a Pt atom. The density Pt is 21.45 g/cm^3, and the mass of a single Pt atom is 3.240×10^{-22} g. [The volume of a sphere of radius r is $\frac{4}{3}\pi r^3$.]

2.125 (a) Assuming an atomic nucleus is spherical in shape, show that its radius r is proportional to the cube root of the mass number (A). (b) In general, the radius of a nucleus is given by $r = r_0 A^{1/3}$, where r_0 is a proportionality constant given by 1.2×10^{-15} m. Calculate the volume of the Li nucleus. (c) Given that the radius of ^7_3Li atom is 152 pm, calculate what fraction of the atom's volume is occupied by its nucleus. Does your result support Rutherford's model of the atom?

2.126 A distributor of ball bearings sells packages of "remnant" ball bearings that contain a mixture of copper bearings and titanium bearings—along with bearings made of a proprietary alloy of several different metals. Although the bearings all have the same diameter, 4.175 mm, the metals from which they are made all have different densities. The density of the proprietary alloy is 5.659 g/cm^3. Packages of the remnant bearings are, on average, 30.4 percent copper bearings and 51.2 percent titanium bearings, the remainder being proprietary alloy. Using the formula for the volume of a sphere, $V = \frac{4}{3}\pi r^3$, and the densities of the metals from webelements.com, determine the average mass of a ball bearing in one of the packages.

Standardized-Exam Practice Problems

Physical and Biological Sciences

Carbon-14, a radioactive isotope of carbon, is used to determine the ages of fossils in a technique called *carbon dating*. Carbon-14 is produced in the upper atmosphere when nitrogen-14 atoms are bombarded by neutrons from cosmic rays. ^{14}C undergoes a process called β emission in which a neutron in the nucleus decays to form a proton and an electron. The electron, or β particle, is ejected from the nucleus. Because the production and decay of ^{14}C occur simultaneously, the total amount of ^{14}C in the atmosphere is constant. Plants absorb ^{14}C in the form of CO_2 and animals consume plants and other animals. Thus, all living things contain a constant ratio of ^{12}C to ^{14}C. When a living thing dies, the ^{14}C it contains continues to decay, but because replenishment ceases, the ratio of ^{12}C to ^{14}C changes over time. Scientists use the ^{12}C to ^{14}C ratio to determine the age of material that was once living.

1. If atmospheric conditions were to change such that ^{14}C were produced at twice the current rate,
   a) the world's supply of ^{14}N would be consumed completely.
   b) the ^{12}C to ^{14}C ratio in living things would increase.
   c) the ^{12}C to ^{14}C ratio in living things would decrease.
   d) the ^{12}C to ^{14}C ratio in living things would not change.

2. When a ^{14}N nucleus is bombarded by a neutron to produce a ^{14}C nucleus, what else is produced?
   a) Nothing
   b) Another neutron
   c) An electron
   d) A proton

3. Based on the description of β emission in the passage, what nucleus results from the decay of a ^{14}C nucleus by β emission?
   a) ^{14}N
   b) ^{13}N
   c) ^{12}C
   d) ^{13}C

4. The accuracy of carbon dating depends on the assumption that
   a) ^{14}C is the only radioactive species in the material being tested.
   b) the rate of decay of ^{14}C is constant.
   c) ^{12}C and ^{14}C undergo radioactive decay at the same rate.
   d) each ^{14}C nucleus decays to give a ^{12}C nucleus.

Answers to In-Chapter Materials

Practice Problems

2.1A (a) 3:2, (b) 2:1. **2.1B** (a) 0.2518 g, (b) $n = 6$ (XeF_6). **2.2A** (a) p = 5, n = 5, e = 5. (b) p = 18, n = 18, e = 18. (c) p = 38, n = 47, e = 38. (d) p = 6, n = 5, e = 6. **2.2B** (a) $^{9}_{4}Be$, (b) $^{51}_{23}V$, (c) $^{124}_{54}Xe$, (d) $^{69}_{31}Ga$. **2.3A** 63.55 amu. **2.3B** 99.64% ^{14}N, 0.36% ^{15}N. **2.4A** (a) sodium sulfate, (b) copper(II) nitrate, (c) iron(III) carbonate. **2.4B** (a) potassium dichromate, (b) lithium oxalate, (c) copper(I) nitrate. **2.5A** (a) $PbCl_2$, (b) $MgCO_3$, (c) $(NH_4)_3PO_4$. **2.5B** (a) Fe_2S_3, (b) $Hg(NO_3)_2$, (c) K_2SO_3. **2.6A** (a) hypobromite ion, (b) hydrogen sulfate ion, (c) oxalate ion. **2.6B** (a) iodite ion, (b) hydrogen chromate ion, (c) hydrogen oxalate ion. **2.7A** $CHCl_3$. **2.7B** C_3H_6O. **2.8A** (a) dichlorine monoxide, (b) silicon tetrachloride. **2.8B** (a) chlorine dioxide, (b) carbon tetrabromide. **2.9A** (a) CS_2, (b) N_2O_3. **2.9B** (a) SF_6, (b) S_2F_{10}. **2.10A** $HBrO_4$. **2.10B** H_2CrO_4. **2.11A** (a) $C_4H_5N_2O$, (b) C_2H_5, (c) $C_2H_5NO_2$. **2.11B** (a).

Checkpoints

2.1.1 e. **2.1.2** b. **2.3.1** d. **2.3.2** c. **2.4.1** c. **2.4.2** b. **2.5.1** b. **2.5.2** c. **2.6.1** c. **2.6.2** d. **2.6.3** d. **2.6.4** c. **2.6.5** c. **2.6.6** a. **2.7.1** b. **2.7.2** c. **2.7.3** c. **2.7.4** e.

CHAPTER 5

Thermochemistry

Doughnuts, although typically high in fat and Calories, are very popular. The energy content of food such as doughnuts is determined using the principles of thermochemistry.

Credit: © *Tom Mareschal/Photographer's Choice/Getty Images.*

In This Chapter, You Will Learn

About the energy changes associated with chemical reactions and physical processes. You will also learn how calorimetry is used to measure the energy released or absorbed during a process.

Before You Begin, Review These Skills

- Tracking units [⏮ Section 1.6]
- Balancing chemical equations [⏮ Section 3.3]

How Thermochemistry Helped Uncover Deceptive Labeling of "Diet" Doughnuts

The Food and Drug Administration (FDA) and the U.S. Department of Agriculture (USDA) require "Nutrition Facts" labels on nearly all prepared foods. These labels indicate such things as serving size, grams of fat per serving, and Calories per serving. Health-conscious consumers typically are avid readers of food labels and use them to make informed dietary choices. Furthermore, because dieting to lose weight is something of a national obsession, many Americans are willing to pay premium prices for packaged foods that both taste good *and* are low in fat and Calories. For this reason, it is important that the labels on packaged foods contain accurate information.

In 1997, spurred by numerous complaints from consumers and health-care professionals, the FDA launched an investigation into the practices of a weight-loss product company with packaging facilities in Kentucky and Illinois. The company sold "low-fat, carob-coated" doughnuts that, according to package labeling, each contained just 3 g of fat and 135 Calories. According to an analysis by the FDA, the doughnuts were actually *chocolate*-coated and each contained 18 g of fat and 530 Calories! As the FDA investigation revealed, the company had simply purchased ordinary doughnuts from a Chicago bakery and repackaged them as "diet" doughnuts, marking up the price by as much as 200 percent. The investigation led to a criminal indictment.

Experimental determination of the caloric content of food involves burning a weighed sample of the food in a bomb calorimeter. The heat generated by the combustion, which is equal to the energy content of the food, is determined by measuring the resulting temperature increase in the calorimeter. Calorimetry is one of the practical applications of the principles of *thermochemistry*.

Nutrition Facts
Serving Size 3/4 Cup (27g)

Amount Per Serving	Cereal	With 1/2 Cup Skim Milk
Calories	90	130
Calories from Fat	10	10
	% Daily Value	
Total Fat 1g*	**2%**	**2%**
Saturated Fat 0g	**0%**	**0%**
Trans Fat 0g	**0%**	**0%**
Cholesterol 0mg	**0%**	**0%**
Sodium 190mg	**8%**	**11%**
Potassium 85mg	**2%**	**8%**
Total Carbohydrate 23g	**8%**	**10%**
Dietary Fiber 5g	**20%**	**20%**
Sugars 5g		
Protein 2g		
Vitamin A	0%	4%
Vitamin C	10%	15%
Calcium	0%	15%
Iron	2%	2%

Credit: © *David A. Tietz/Editorial Image, LLC.*

Student Note: Carob is a chocolate substitute made from the seed pods of the Mediterranean Carob tree.

Student Note: Cell respiration is the complex, multistep process by which living things convert the food they eat into energy, carbon dioxide, and water. Combustion occurs in a single step but is the same overall process, with the same products, as cell respiration.

At the end of this chapter, you will be able to solve a series of problems related to the fat and calorie content of doughnuts [⏭ Applying What You've Learned, page 220].

5.1 Energy and Energy Changes

You have already learned that matter can undergo physical changes and chemical changes [⏮ Section 1.4]. The melting of ice, for example, is a physical change that can be represented by the following equation:

$$H_2O(s) \longrightarrow H_2O(l)$$

The formation of water from its constituent elements, represented by the following equation, is an example of a chemical change:

$$2H_2(g) + O_2(g) \longrightarrow 2H_2O(l)$$

In each case, there is energy involved in the change. Energy (in the form of heat) must be *supplied* to melt ice, whereas energy (in the form of heat and light) is *produced* by the explosive combination of hydrogen and oxygen gases. In fact, every change that matter undergoes is accompanied by either the absorption or the release of energy. In this chapter, we will focus on the energy changes associated with physical and chemical processes.

Forms of Energy

Energy is usually defined as the capacity to do work or transfer heat. All forms of energy are either kinetic or potential. ***Kinetic energy*** is the energy that results from *motion*. It is calculated with the equation

Equation 5.1 $$E_k = \frac{1}{2}mu^2$$

Student Note: Some textbooks use the letter *v* to denote velocity.

where m is the mass of the object and u is its velocity. One form of kinetic energy of particular interest to chemists is ***thermal energy,*** which is the energy associated with the random motion of atoms and molecules. We can monitor changes in thermal energy by measuring temperature changes.

Potential energy is the energy possessed by an object by virtue of its position. The two forms of potential energy of greatest interest to chemists are chemical energy and electrostatic energy. ***Chemical energy*** is energy stored within the structural units (molecules or polyatomic ions) of chemical substances. The amount of chemical energy in a sample of matter depends on the types and arrangements of atoms in the structural units that make up the sample.

Electrostatic energy is potential energy that results from the interaction of charged particles. Oppositely charged particles *attract* each other, whereas particles of like charges *repel* each other. The magnitude of the resulting electrostatic potential energy is proportional to the product of the two charges (Q_1 and Q_2) divided by the distance (d) between them.

Equation 5.2 $$E_{el} \propto \frac{Q_1Q_2}{d}$$

If the charges Q_1 and Q_2 are opposite (i.e., one positive and one negative), the result is a negative value for E_{el}, which indicates *attraction*. Like charges (i.e., both positive or both negative) result in a positive value for E_{el}, indicating *repulsion*.

Kinetic and potential energy are interconvertible—that is, one form can be converted to the other. For example, dropping an object and allowing it to fall converts potential energy to kinetic energy. Likewise, a chemical reaction that gives off heat converts chemical energy (potential) to thermal energy (kinetic). Although energy can assume many different forms that are interconvertible, the total amount of energy in the universe is constant. When energy of one form disappears, the same amount of energy must appear in another form or forms. This principle is known as the ***law of conservation of energy.***

Energy Changes in Chemical Reactions

To analyze energy changes associated with chemical reactions, we must first define the ***system,*** the specific part of the universe that is of interest to us. For chemists, systems usually include the substances involved in chemical and physical changes. In an acid-base neutralization experiment, for example, the system may be the contents of a beaker in which HCl and NaOH react with each other. The rest of the universe outside the system, including the beaker itself, constitutes the ***surroundings.***

Many chemical reactions are carried out for the purpose of exploiting the associated energy change, rather than for the purpose of obtaining the products of the reactions. For example, combustion reactions involving fossil fuels are carried out for the thermal energy they produce, not for their products, which are carbon dioxide and water.

It is important to distinguish between thermal energy and heat. ***Heat*** is the transfer of thermal energy between two bodies that are at different temperatures. Although the term *heat* by itself implies the transfer of energy, we customarily talk of "heat flow," meaning "heat absorbed" or "heat released," when describing the energy changes that occur during a process. ***Thermochemistry*** is the study of heat (the transfer of thermal energy) in chemical reactions.

The combustion of hydrogen gas in oxygen is one of many chemical reactions that release considerable quantities of energy (Figure 5.1):

$$2H_2(g) + O_2(g) \longrightarrow 2H_2O(l) + \text{energy}$$

In this case, we label the mixture of reactants and product (hydrogen, oxygen, and water molecules) the *system*. Because energy cannot be created or destroyed, any energy released by the system must be gained by the surroundings. Thus, the heat generated by the combustion process is transferred from the system to its surroundings. This reaction is an example of an ***exothermic process,*** which is any process that gives off heat—that is, transfers thermal energy *from* the system *to* the surroundings. Figure 5.2(a) shows the energy change for the combustion of hydrogen gas.

Figure 5.1 The Hindenburg, a German airship filled with hydrogen gas, was destroyed in a horrific fire as it landed at Lakehurst, New Jersey, in 1937.
Credit: © *Bettmann/Corbis.*

Animation
Thermochemistry—heat flow in endothermic and exothermic reactions.

Next, consider the decomposition of mercury(II) oxide (HgO) at high temperatures:

$$\text{energy} + 2HgO(s) \longrightarrow 2Hg(l) + O_2(g)$$

This reaction is an ***endothermic process*** because heat has to be supplied to the system (i.e., to HgO) by the surroundings [Figure 5.2(b)] for the reaction to occur. Thus, thermal energy is transferred *from* the surroundings *to* the system in an endothermic process.

According to Figure 5.2, the energy of the products of an exothermic reaction is lower than the energy of the reactants. The difference in energy between the reactants H_2 and O_2 and the product H_2O is the heat released by the system to the surroundings. In an endothermic reaction, on the other hand, the energy of the products is higher than the energy of the reactants. Here, the difference between the energy of the reactant HgO and the products Hg and O_2 is the heat absorbed by the system from the surroundings.

Student Note: In these examples, all the energy released or absorbed by the system is in the form of heat. In Section 5.2, we will consider examples in which some of the energy is in the form of work.

Units of Energy

The SI unit of energy is the ***joule*** (J), named for the physicist James Joule.[1] The joule is a fairly small quantity of energy. It is the amount of kinetic energy possessed by a 2-kg mass moving at a speed of 1 m/s.

$$E_k = \tfrac{1}{2}mu^2 = \tfrac{1}{2}(2\text{ kg})(1\text{ m/s})^2 = 1\text{ kg}\cdot\text{m}^2/\text{s}^2 = 1\text{ J}$$

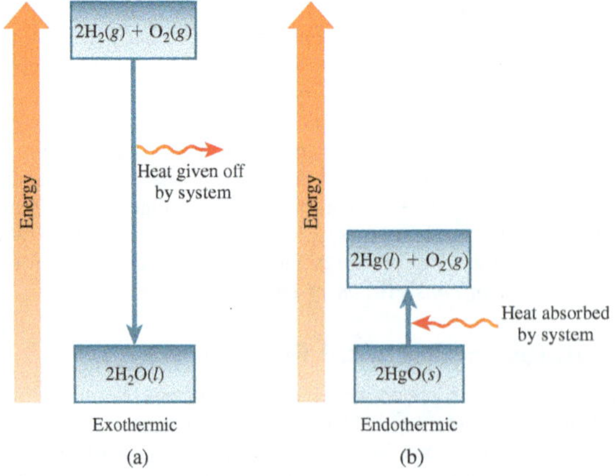

Figure 5.2 (a) An exothermic reaction. Heat is given off by the system. (b) An endothermic reaction. Heat is absorbed by the system.

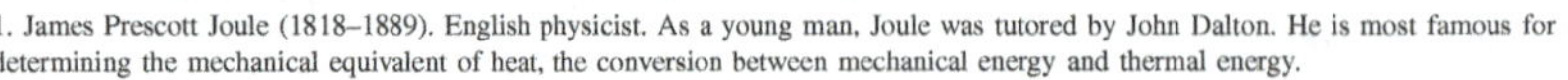

1. James Prescott Joule (1818–1889). English physicist. As a young man, Joule was tutored by John Dalton. He is most famous for determining the mechanical equivalent of heat, the conversion between mechanical energy and thermal energy.

The joule can also be defined as the amount of energy exerted when a force of 1 newton (N) is applied over a distance of 1 meter.

$$1 \text{ J} = 1 \text{ N} \cdot \text{m}$$

where

$$1 \text{ N} = 1 \text{ kg} \cdot \text{m/s}^2$$

Because the magnitude of a joule is so small, we very often express the energy changes in chemical reactions using the unit kilojoule (kJ).

$$1 \text{ kJ} = 1000 \text{ J}$$

Sample Problem 5.1 shows how to calculate kinetic and potential energies.

SAMPLE PROBLEM 5.1

(a) Calculate the kinetic energy of a helium atom moving at a speed of 125 m/s. (b) How much greater is the magnitude of electrostatic attraction between an electron and a nucleus containing three protons versus that between an electron and a nucleus containing one proton? (Assume that the distance between the nucleus and the electron is the same in each case.)

Strategy (a) Use Equation 5.1 ($E_k = \frac{1}{2}mu^2$) to calculate the kinetic energy of an atom. We will need to know the mass of the atom in kilograms. (b) Use Equation 5.2 ($E_{el} \propto Q_1Q_2/d$) to compare the electrostatic potential energy between the two charged particles in each case.

Setup

(a) The mass of a helium atom is 4.003 amu. Its mass in kilograms is

$$4.003 \text{ amu} \times \frac{1.661 \times 10^{-24} \text{ g}}{1 \text{ amu}} \times \frac{1 \text{ kg}}{1 \times 10^3 \text{ g}} = 6.649 \times 10^{-27} \text{ kg}$$

(b) The charge on a nucleus with three protons is +3; the charge on a nucleus with one proton is +1. In each case, the electron's charge is −1. Although we are not given the distance between the opposite charges in either case, we are told that the distances in both cases are equal. We can write Equation 5.2 for each case and divide one by the other to determine the relative magnitudes of the results.

Solution

(a) $E_k = \frac{1}{2}mu^2$

$= \frac{1}{2}(6.649 \times 10^{-27} \text{ kg})(125 \text{ m/s})^2$

$= 5.19 \times 10^{-23} \text{ kg} \cdot \text{m}^2/\text{s}^2 = 5.19 \times 10^{-23} \text{ J}$

Remember that the base units of the joule are kg · m²/s².

(b) $E_{el} \propto \dfrac{Q_1Q_2}{d}$

$$\frac{E_{el} \text{ where } Z = +3}{E_{el} \text{ where } Z = -1} = \frac{E_{el} \propto \dfrac{(+3)(-1)}{d}}{E_{el} \propto \dfrac{(+1)(-1)}{d}} = 3$$

The electrostatic potential energy between charges of +3 and −1 is three times that between charges of +1 and −1.

THINK ABOUT IT

We expect the energy of an atom, even a fast-moving one, to be extremely small. And we expect the attraction between charges of larger magnitude to be greater than that between charges of smaller magnitude.

Practice Problem ATTEMPT (a) Calculate the energy in joules of a 5.25-g object moving at a speed of 655 m/s, and (b) determine how much greater the electrostatic energy is between charges of +2 and −2 than it is between charges of +1 and −1 (assume that the distance between the charges is the same in each case).

Practice Problem BUILD (a) Calculate the velocity (in m/s) of a 0.340-g object that has E_k = 23.5 J, and (b) determine which of the following pairs of charged particles has the greater electrostatic energy between them: charges of +1 and −2 separated by a distance of d or charges of +2 and −2 separated by a distance of $2d$.

Practice Problem CONCEPTUALIZE For charges of +2 and −2, separated by a distance of d, the electrostatic potential energy is E. In terms of E, determine the electrostatic potential energy between each of the pairs of charges shown.

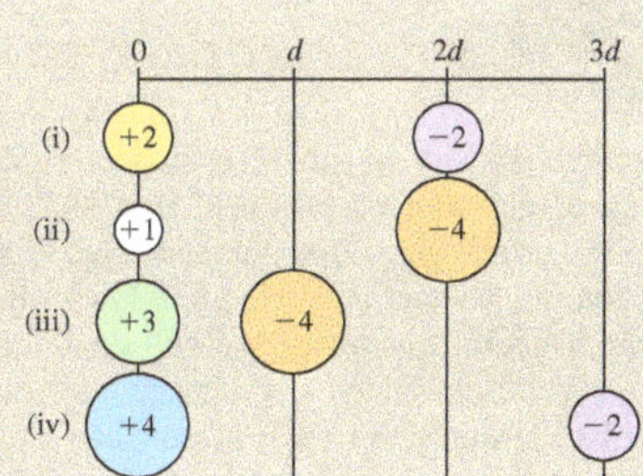

Another unit used to express energy is the calorie (cal). Although the calorie is not an SI unit, its use is still quite common. The calorie is defined in terms of the joule:

$$1 \text{ cal} \equiv 4.184 \text{ J}$$

Student Note: The triple equal sign is used to denote a definition. Recall that numbers with defined values are exact numbers [⏮ Section 1.5], which do not limit the number of significant figures in a calculation.

Because this is a definition, the number 4.184 is an exact number, which does not limit the number of significant figures in a calculation [⏮ Section 1.5]. You may be familiar with the term *calorie* from nutrition labels. In fact, the "calories" listed on food packaging are really *kilocalories*. Often the distinction is made by capitalizing the "C" in "calorie" when it refers to the energy content of food:

$$1 \text{ Cal} \equiv 1000 \text{ cal}$$

and

$$1 \text{ Cal} \equiv 4184 \text{ J}$$

CHECKPOINT – SECTION 5.1 Energy and Energy Changes

5.1.1 Calculate the kinetic energy of a 5.0-kg mass moving at 26 m/s.

a) 1.7×10^3 J
b) 3.4×10^3 J
c) 130 J
d) 65 J
e) 13×10^3 J

5.1.2 How much greater is the electrostatic potential energy between particles with charges +3 and −3 than between particles with charges +1 and −1? (Assume the same distance between particles.)

a) 3 times
b) 9 times
c) 6 times
d) 4 times
e) 30 times

5.1.3 Calculate the number of calories in 723.01 J.

a) 172.80 cal
b) 172.8 cal
c) 3025.1 cal
d) 3025 cal
e) 0.173 cal

5.1.4 The label on packaged food indicates that it contains 215 Cal per serving. Convert this amount of energy to joules.

a) 51.4 J
b) 5.14×10^4 J
c) 5.14×10^{-2} J
d) 9.00×10^2 J
e) 9.00×10^5 J

5.1.5 Arrange the following pairs of charged particles in order of increasing electrostatic potential energy.

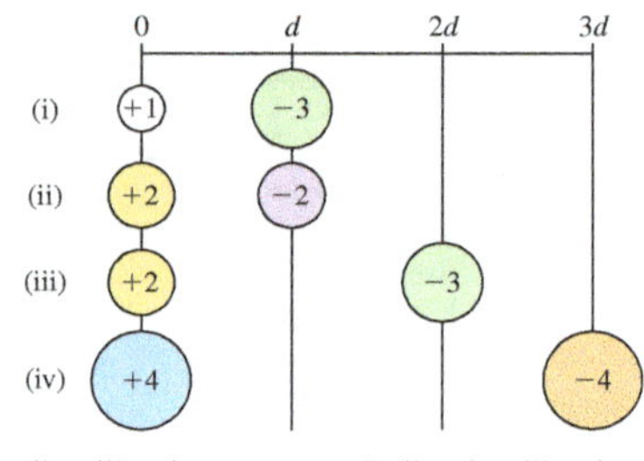

a) i < ii < iii < iv
b) iv < iii < ii < i
c) i = iii < ii < iv
d) ii < i = iii < iv
e) iv < i = ii < iii

5.1.6 Which of the following changes would double the magnitude of the electrostatic attraction between two oppositely charged particles? (Select all that apply.)

a) Doubling both charges
b) Doubling one of the charges
c) Doubling the distance between the charges
d) Reducing the distance between the charges by half
e) Doubling both charges and doubling the distance between the charges

5.2 Introduction to Thermodynamics

Thermochemistry is part of a broader subject called ***thermodynamics,*** which is the scientific study of the interconversion of heat and other kinds of energy. The laws of thermodynamics provide useful guidelines for understanding the energetics and directions of processes. In this section, we will introduce the first law of thermodynamics, which is particularly relevant to the study of thermochemistry. We will continue our discussion of thermodynamics in Chapter 18.

We have defined a system as the part of the universe we are studying. There are three types of systems. An ***open system*** can exchange mass and energy with its surroundings. For example, an open system may consist of a quantity of water in an open container, as shown in Figure 5.3(a). If we close the flask, as in Figure 5.3(b), so that no water vapor can escape from

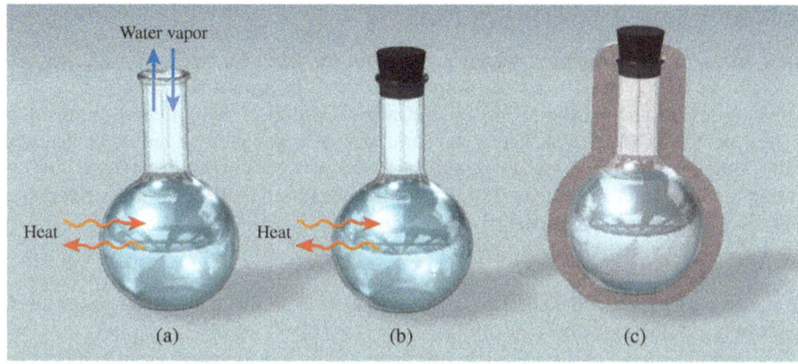

Figure 5.3 (a) An open system allows exchange of both energy and matter with the surroundings. (b) A closed system allows exchange of energy but not matter. (c) An isolated system does not allow exchange of energy or matter. (This flask is enclosed by an insulating vacuum jacket.)

Student Note: The energy exchanged between open systems or closed systems and their surroundings is usually in the form of heat.

or condense into the container, we create a ***closed system,*** which allows the transfer of *energy* but not *mass*. By placing the water in an insulated container, as shown in Figure 5.3(c), we can construct an ***isolated system,*** which does not exchange either mass or energy with its surroundings.

States and State Functions

In thermodynamics, we study changes in the ***state of a system,*** which is defined by the values of all relevant macroscopic properties, such as composition, energy, temperature, pressure, and volume. Energy, pressure, volume, and temperature are said to be ***state functions***—properties that are determined by the state of the system, regardless of how that condition was achieved. In other words, when the state of a system changes, the magnitude of change in any state function depends only on the initial and final states of the system and not on how the change is accomplished.

Consider, for example, your position in a six-story building. Your elevation depends upon which floor you are on. If you change your elevation by taking the stairs from the ground floor up to the fourth floor, the change in your elevation depends only upon your initial state (the ground floor—the floor you started on) and your final state (the fourth floor—the floor you went to). It does not depend on whether you went directly to the fourth floor or up to the sixth and then down to the fourth floor. Your overall change in elevation is the same either way because it depends only on your initial and final elevations. Thus, elevation is a state function.

The amount of effort it takes to get from the ground floor to the fourth floor, on the other hand, depends on how you get there. More effort has to be exerted to go from the ground floor to the sixth floor and back down to the fourth floor than to go from the ground floor to the fourth floor directly. The effort required for this change in elevation is *not* a state function. Furthermore, if you subsequently return to the ground floor, your overall change in elevation will be zero, because your initial and final states are the same, but the amount of effort you exerted going from the ground floor to the fourth floor and back to the ground floor is *not* zero. Even though your initial and final states are the same, you do not get back the effort that went into climbing up and down the stairs.

Energy is a state function, too. Using potential energy as an example, your net increase in gravitational potential energy is always the same, regardless of how you get from the ground floor to the fourth floor of a building (Figure 5.4).

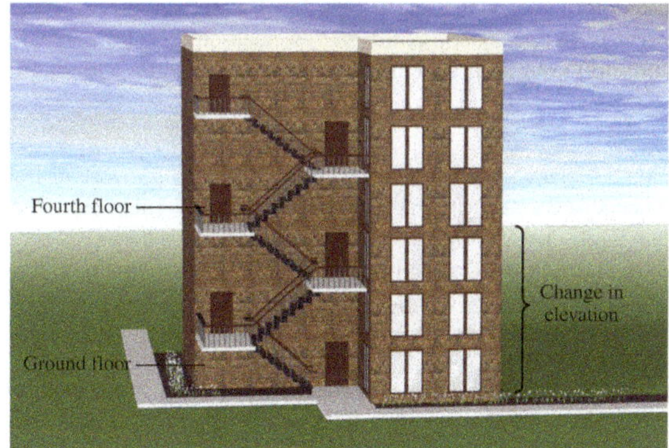

Figure 5.4 The change in elevation that occurs when a person goes from the ground floor to the fourth floor in a building does not depend on the path taken.

The First Law of Thermodynamics

The ***first law of thermodynamics,*** which is based on the law of conservation of energy, states that energy can be converted from one form to another but cannot be created or destroyed. It would be impossible to demonstrate this by measuring the total amount of energy in the universe; in fact, just determining the total energy content of a small sample of matter would be extremely difficult. Fortunately, because energy is a state function, we can demonstrate the first law by measuring the change in the energy of a system between its initial state and its final state in a process. The change in internal energy, ΔU, is given by

$$\Delta U = U_f - U_i$$

where U_i and U_f are the internal energies of the system in the initial and final states, respectively; and the symbol Δ means *final* minus *initial.*

The internal energy of a system has two components: kinetic energy and potential energy. The kinetic energy component consists of various types of molecular motion and the movement of electrons within molecules. Potential energy is determined by the attractive interactions between electrons and nuclei and by repulsive interactions between electrons and between nuclei in individual molecules, as well as by interactions between molecules. It is impossible to measure all these contributions accurately, so we cannot calculate the total energy of a system with any certainty. Changes in energy, on the other hand, can be determined experimentally.

Consider the reaction between 1 mole of sulfur and 1 mole of oxygen gas to produce 1 mole of sulfur dioxide:

$$S(s) + O_2(g) \longrightarrow SO_2(g)$$

Student Note: Elemental sulfur exists as S_8 molecules, but we typically represent it simply as S to simplify chemical equations.

In this case, our system is composed of the reactant molecules and the product molecules. We do not know the internal energy content of either the reactants or the product, but we can accurately measure the *change* in energy content, ΔU, given by

$$\Delta U = U(\text{products}) - U(\text{reactants})$$
$$= \text{energy content of 1 mol } SO_2(g) - \text{energy content of 1 mol } S(s) \text{ and 1 mol } O_2(g)$$

This reaction gives off heat. Therefore, the energy of the product is less than that of the reactants, and ΔU is negative.

The release of heat that accompanies this reaction indicates that some of the chemical energy contained in the system has been converted to thermal energy. Furthermore, the thermal energy released by the system is absorbed by the surroundings. The transfer of energy from the system to the surroundings does not change the total energy of the universe. That is, the sum of the energy changes is zero:

$$\Delta U_{sys} + \Delta U_{surr} = 0$$

where the subscripts "sys" and "surr" denote system and surroundings, respectively. Thus, if a system undergoes an energy change ΔU_{sys}, the rest of the universe, or the surroundings, must undergo a change in energy that is equal in magnitude but opposite in sign:

$$\Delta U_{sys} = -\Delta U_{surr}$$

Energy released in one place must be gained somewhere else. Furthermore, because energy can be changed from one form to another, the energy lost by one system can be gained by another system in a different form. For example, the energy released by burning coal in a power plant may ultimately turn up in our homes as electric energy, heat, light, and so on.

Work and Heat

Recall from Section 5.1 that energy is defined as the capacity to do work or transfer heat. When a system releases or absorbs heat, its internal energy changes. Likewise, when a system does work on its surroundings, or when the surroundings do work on the system, the system's internal energy also changes. The overall change in the system's internal energy is given by

Student Note: The units for heat and work are the same as those for energy: joules, kilojoules, or calories.

$$\Delta U = q + w \qquad \textbf{Equation 5.3}$$

where q is heat (released or absorbed by the system) and w is work (done *on* the system or done *by* the system). Note that it is possible for the heat and work components to cancel each other out and for there to be no change in the system's internal energy. Interestingly, although

TABLE 5.1 Sign Conventions for Heat (q) and Work (w)

Process	Sign
Heat absorbed by the system (endothermic process)	q is positive
Heat released by the system (exothermic process)	q is negative
Work done on the system by the surroundings (e.g., a volume decrease)	w is positive
Work done by the system on the surroundings (e.g., a volume increase)	w is negative

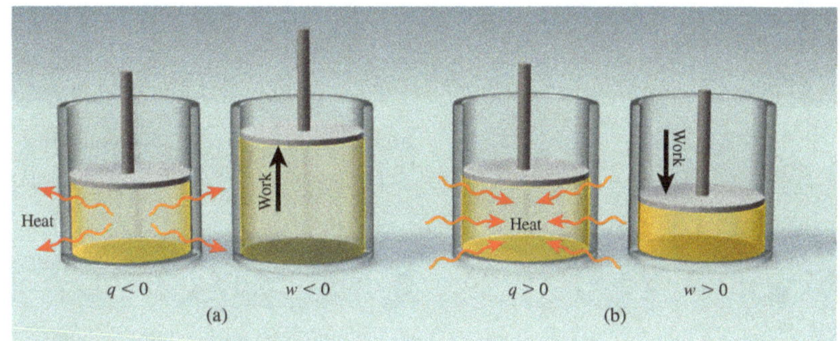

Figure 5.5 (a) When heat is released by the system (to the surroundings), q is negative. When work is done by the system (on the surroundings), w is negative. (b) When heat is absorbed by the system (from the surroundings), q is positive. When work is done on the system (by the surroundings), w is positive.

neither q nor w is a state function (each depends on the path between the initial and final states of the system), their sum, ΔU, does *not* depend on the path between initial and final states because U *is* a state function.

In chemistry, we are normally interested in the energy changes associated with the system rather than the surroundings. Therefore, unless otherwise indicated, ΔU will refer specifically to ΔU_{sys}. The sign conventions for q and w are as follows: q is positive for an endothermic process and negative for an exothermic process, and w is positive for work done on the system by the surroundings and negative for work done by the system on the surroundings. Table 5.1 summarizes the sign conventions for q and w.

The drawings in Figure 5.5 illustrate the logic behind the sign conventions for q and w. If a system releases heat to the surroundings or does work on the surroundings [Figure 5.5(a)], we would expect its internal energy to decrease because they are energy-depleting processes. For this reason, both q and w are negative. Conversely, if heat is added to the system or if work is done on the system [Figure 5.5(b)], then the internal energy of the system increases. In this case, both q and w are positive.

Sample Problem 5.2 shows how to determine the overall change in the internal energy of a system.

SAMPLE PROBLEM 5.2

Calculate the overall change in internal energy, ΔU, (in joules) for a system that absorbs 188 J of heat and does 141 J of work on its surroundings.

Strategy Combine the two contributions to internal energy using Equation 5.3 and the sign conventions for q and w.

Setup The system absorbs heat, so q is *positive*. The system does work on the surroundings, so w is negative.

Solution

$$\Delta U = q + w = 188 \text{ J} + (-141 \text{ J}) = 47 \text{ J}$$

THINK ABOUT IT

Consult Table 5.1 to make sure that you have used the proper sign conventions for q and w.

Practice Problem ATTEMPT Calculate the change in total internal energy for a system that releases 1.34×10^4 kJ of heat and does 2.98×10^4 kJ of work on the surroundings.

Practice Problem BUILD Calculate the magnitude of q for a system that does 7.05×10^5 kJ of work on its surroundings and for which the change in total internal energy is -9.55×10^3 kJ. Indicate whether heat is absorbed or released by the system.

Practice Problem CONCEPTUALIZE The diagram on the far left in the row shows a system before a process. Which of the diagrams on its right could represent the system after it undergoes a process in which the system absorbs heat and ΔU is positive?

CHECKPOINT – SECTION 5.2 Introduction to Thermodynamics

5.2.1 Calculate the overall change in internal energy for a system that releases 43 J in heat in a process in which no work is done.

a) 43 J
b) -2.3×10^{-2} J
c) 0 J
d) 2.3×10^{-2} J
e) −43 J

5.2.2 Calculate w, and determine whether work is done by the system or on the system when 928 kJ of heat is released and $\Delta U = -1.47 \times 10^3$ kJ.

a) $w = -1.36 \times 10^6$ kJ, done by the system
b) $w = 1.36 \times 10^6$ kJ, done on the system
c) $w = -5.4 \times 10^2$ kJ, done by the system
d) $w = 2.4 \times 10^3$ kJ, done on the system
e) $w = -2.4 \times 10^3$ kJ, done by the system

5.2.3 Which of the following is a closed system? (Select all that apply.)

a) Pot of boiling water on a stove
b) Pot of cold water on a countertop
c) Helium-filled Mylar balloon
d) Styrofoam cup full of hot coffee (without lid)
e) Glass of iced tea

5.2.4 Which of the following could be described as a state function? (Select all that apply.)

a) Dieter's weight
b) Calories burned by a person walking to the health club
c) Distance traveled by a person walking to the health club
d) Distance between two points on a map
e) Energy required to heat a given amount of water from 25°C to the boiling point

5.3 Enthalpy

To calculate ΔU, we must know the values and signs of both q and w. As we will see in Section 5.4, we determine q by measuring temperature changes. To determine w, we need to know whether the reaction occurs under constant-volume or constant-pressure conditions.

Reactions Carried Out at Constant Volume or at Constant Pressure

Imagine carrying out the decomposition of sodium azide (NaN_3) in two different experiments. In the first experiment, the reactant is placed in a metal cylinder with a fixed volume. When detonated, the NaN_3 reacts, generating a large quantity of N_2 gas inside the closed, fixed-volume container.

> **Student Note:** The explosive decomposition of NaN_3 is the reaction that inflates air bags in cars.

$$2NaN_3(s) \longrightarrow 2Na(s) + 3N_2(g)$$

The effect of this reaction will be an increase in the pressure inside the container, similar to what happens if you shake a bottle of soda vigorously prior to opening it. (The concept of pressure will be examined in detail in Chapter 10. However, if you have ever put air in the tire of an automobile or a bicycle, you are familiar with the concept.)

Now imagine carrying out the same reaction in a metal cylinder with a movable piston. As this explosive decomposition proceeds, the piston in the metal cylinder will move. The gas produced in the reaction pushes the cylinder upward, thereby increasing the volume of the container and preventing any increase in pressure. This is a simple example of mechanical work done by a chemical reaction. Specifically, this type of work is known as *pressure-volume*, or *PV*, work. The amount of work done by such a process is given by

Student Note: Pressure-volume work and electrical work are two important types of work done by chemical reactions. Electrical work will be discussed in detail in Chapter 19.

Equation 5.4 $$w = -P\Delta V$$

where P is the external, opposing pressure and ΔV is the change in the volume of the container as the result of the piston being pushed upward. In keeping with the sign conventions in Table 5.1, an increase in volume results in a negative value for w, whereas a decrease in volume results in a positive value for w. Figure 5.6 illustrates this reaction (a) being carried out at a constant volume, and (b) at a constant pressure.

When a chemical reaction is carried out at constant volume, then no PV work can be done because $\Delta V = 0$ in Equation 5.4. From Equation 5.3 it follows that

Equation 5.5 $$\Delta U = q - P\Delta V$$

and, because $P\Delta V = 0$ at constant volume,

Equation 5.6 $$q_V = \Delta U$$

We add the subscript "V" to indicate that this is a constant-volume process. This equality may seem strange at first. We said earlier that q is *not* a state function. However, for a process carried out under constant-volume conditions, q can have only one specific value, which is equal to ΔU. In other words, while q is *not* a state function, q_V *is* one.

Constant-volume conditions are often inconvenient and sometimes impossible to achieve. Most reactions occur in open containers, under conditions of constant pressure (usually at

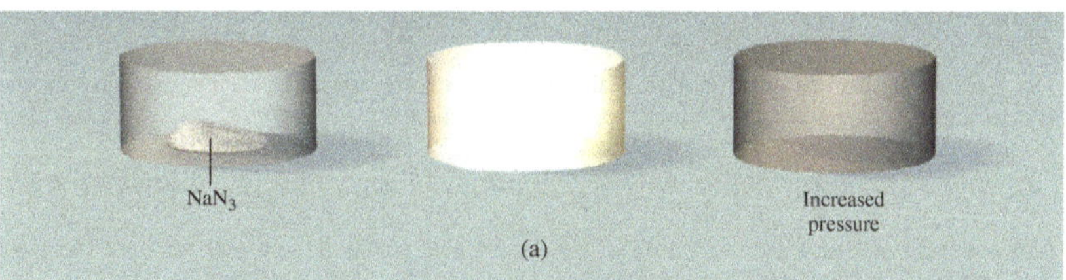

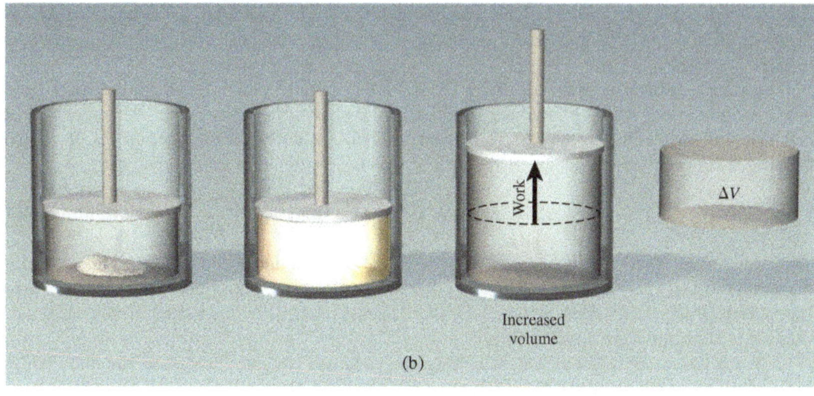

Figure 5.6 (a) The explosive decomposition of NaN_3 at constant volume results in an increase in pressure inside the vessel. (b) The decomposition at constant pressure, in a vessel with a movable piston, results in an increase in volume. The resulting change in volume, ΔV, can be used to calculate the work done by the system.

whatever the atmospheric pressure happens to be where the experiments are conducted). In general, for a constant-pressure process, we write

$$\Delta U = q + w$$
$$= q_P - P\Delta V$$

or

$$q_P = \Delta U + P\Delta V \qquad \textbf{Equation 5.7}$$

where the subscript "P" denotes constant pressure.

Enthalpy and Enthalpy Changes

There is a thermodynamic function of a system called ***enthalpy*** (H), which is defined as

$$H = U + PV \qquad \textbf{Equation 5.8}$$

where U is the internal energy of the system and P and V are the pressure and volume of the system, respectively. Because U and PV have energy units, enthalpy also has energy units. Furthermore, U, P, and V are all state functions—that is, the changes in $(U + PV)$ depend only on the initial and final states. It follows, therefore, that the change in H, or ΔH, also depends only on the initial and final states. Thus, H is a state function.

Student Note: The SI unit of pressure is the *pascal* (Pa), which in SI base units is 1 kg/(m · s^2). Volume in SI base units is cubic meters (m^3). Therefore, multiplying units of *pressure* by units of *volume* gives (1 kg/m · s^2) × m^3 = 1 kg · m^2/s^2, which is the definition of the *joule* (J). Thus, $P\Delta V$ has units of *energy.*

For any process, the *change* in enthalpy is given by

$$\Delta H = \Delta U + \Delta(PV) \qquad \textbf{Equation 5.9}$$

If the pressure is held constant, then

$$\Delta H = \Delta U + P\Delta V \qquad \textbf{Equation 5.10}$$

If we solve Equation 5.10 for ΔU,

$$\Delta U = \Delta H - P\Delta V$$

Then, substituting the result for ΔU into Equation 5.7, we obtain

$$q_P = (\Delta H - P\Delta V) + P\Delta V$$

The $P\Delta V$ terms cancel, and for a constant-pressure process, the heat exchanged between the system and the surroundings is equal to the enthalpy change:

$$q_P = \Delta H \qquad \textbf{Equation 5.11}$$

Again, q is *not* a state function, but q_P *is*; that is, the heat change at constant pressure can have only one specific value and is equal to ΔH.

We now have two quantities—ΔU and ΔH—that can be associated with a reaction. If the reaction occurs under constant-volume conditions, then the heat change, q_V, is equal to ΔU. If the reaction is carried out at constant pressure, on the other hand, the heat change, q_P, is equal to ΔH.

Because most laboratory reactions are constant-pressure processes, the heat exchanged between the system and surroundings is equal to the change in enthalpy for the process. For any reaction, we define the change in enthalpy, called the ***enthalpy of reaction*** (ΔH), as the difference between the enthalpies of the products and the enthalpies of the reactants:

Student Note: The enthalpy of reaction is often symbolized by ΔH_{rxn}. The subscript can be changed to denote a specific type of reaction or physical process: ΔH_{vap} can be used for the enthalpy of *vaporization*, for example.

$$\Delta H = H(\text{products}) - H(\text{reactants}) \qquad \textbf{Equation 5.12}$$

The enthalpy of reaction can be positive or negative, depending on the process. For an endothermic process (where heat is absorbed by the system from the surroundings), ΔH is positive (i.e., $\Delta H > 0$). For an exothermic process (where heat is released by the system to the surroundings), ΔH is negative (i.e., $\Delta H < 0$).

We will now apply the idea of enthalpy changes to two common processes, the first involving a physical change and the second involving a chemical change.

Thermochemical Equations

Student Note: Although, strictly speaking, it is unnecessary to include the sign of a positive number, we will include the sign of all positive ΔH values to emphasize the thermochemical sign convention.

Under ordinary atmospheric conditions at sea level, ice melts to form liquid water when exposed to temperatures above 0°C. Measurements show that for every mole of ice converted to liquid water under these conditions, 6.01 kJ of heat energy is absorbed by the system (the ice). Because the pressure is constant, the heat change is equal to the enthalpy change, ΔH. This is an *endothermic* process ($\Delta H > 0$), because heat is absorbed by the ice from its surroundings [Figure 5.7(a)]. The equation for this physical change is

$$H_2O(s) \longrightarrow H_2O(l) \qquad \Delta H = +6.01 \text{ kJ/mol}$$

The "per mole" in the unit for ΔH means that this is the enthalpy change *per mole of the reaction (or process) as it is written*—that is, when 1 mole of ice is converted to 1 mole of liquid water.

Now consider the combustion of methane (CH_4), the principal component of natural gas:

$$CH_4(g) + 2O_2(g) \longrightarrow CO_2(g) + 2H_2O(l) \qquad \Delta H = -890.4 \text{ kJ/mol}$$

From experience we know that burning natural gas releases heat to the surroundings, so it is an exothermic process. Under constant-pressure conditions, this heat change is equal to the enthalpy change and ΔH must have a negative sign [Figure 5.7(b)]. Again, the "per mole" in the units for ΔH means that when 1 mole of CH_4 reacts with 2 moles of O_2 to yield 1 mole of CO_2 and 2 moles of liquid H_2O, 890.4 kJ of heat is released to the surroundings. Note that when you specify that a particular amount of heat is released, it is not necessary to include a negative sign.

It is important to keep in mind that the ΔH value in kJ/mol does not mean *per mole* of a particular reactant or product. It refers to all the species in a reaction in the molar amounts specified by the coefficients in the balanced equation. Thus, for the combustion of methane, the ΔH value of −890.4 kJ/mol can be expressed in any of the following ways:

$$\frac{-890.4 \text{ kJ}}{1 \text{ mol } CH_4} \qquad \frac{-890.4 \text{ kJ}}{2 \text{ mol } O_2} \qquad \frac{-890.4 \text{ kJ}}{1 \text{ mol } CO_2} \qquad \frac{-890.4 \text{ kJ}}{2 \text{ mol } H_2O}$$

Although the importance of expressing ΔH in units of kJ/mol (rather than just kilojoules) will become apparent when we study thermodynamics in greater detail [⏭ Chapter 18], you should learn and become comfortable with this convention now.

The equations for the melting of ice and the combustion of methane are examples of ***thermochemical equations,*** which are chemical equations that show the enthalpy changes as well as the mass relationships. It is essential to specify a balanced chemical equation when quoting the enthalpy change of a reaction. The following guidelines are helpful in interpreting, writing, and manipulating thermochemical equations:

1. When writing thermochemical equations, we must always specify the physical states of all reactants and products, because they help determine the actual enthalpy changes. In the equation for the combustion of methane, for example, changing the liquid water product to water vapor changes the value of ΔH:

$$CH_4(g) + 2O_2(g) \longrightarrow CO_2(g) + 2H_2O(g) \qquad \Delta H = -802.4 \text{ kJ/mol}$$

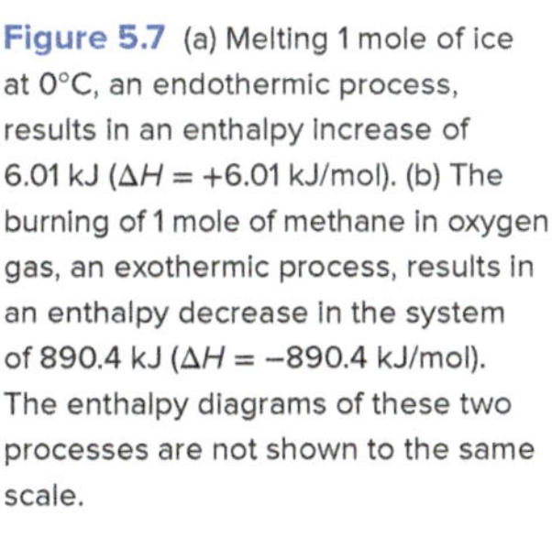

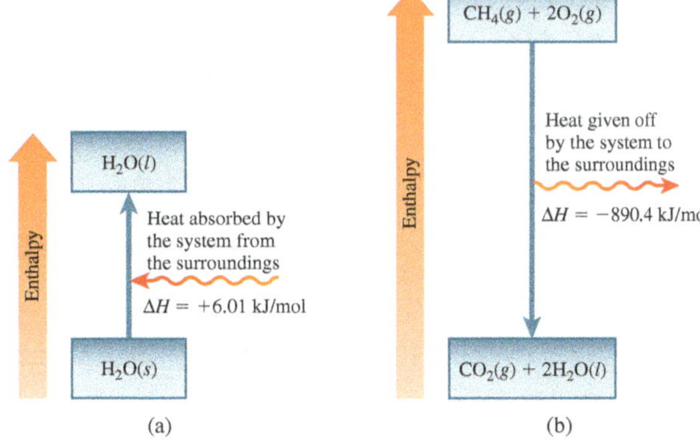

Figure 5.7 (a) Melting 1 mole of ice at 0°C, an endothermic process, results in an enthalpy increase of 6.01 kJ ($\Delta H = +6.01$ kJ/mol). (b) The burning of 1 mole of methane in oxygen gas, an exothermic process, results in an enthalpy decrease in the system of 890.4 kJ ($\Delta H = -890.4$ kJ/mol). The enthalpy diagrams of these two processes are not shown to the same scale.

The enthalpy change is −802.4 kJ rather than −890.4 kJ because 88.0 kJ is needed to convert 2 moles of liquid water to 2 moles of water vapor; that is,

$$2H_2O(l) \longrightarrow 2H_2O(g) \qquad \Delta H = +88.0 \text{ kJ/mol}$$

Student Note: Remember that the "per mole" in this context refers to per mole of reaction—not, in this case, per mole of water.

2. If we multiply both sides of a thermochemical equation by a factor n, then ΔH must also change by the same factor. Thus, for the melting of ice, if $n = 2$, we have

$$2H_2O(s) \longrightarrow 2H_2O(l) \qquad \Delta H = 2(6.01 \text{ kJ/mol}) = +12.02 \text{ kJ/mol}$$

3. When we reverse a chemical equation, we change the roles of reactants and products. Consequently, the magnitude of ΔH for the equation remains the same, but its sign changes. For example, if a reaction consumes thermal energy from its surroundings (i.e., if it is endothermic), then the reverse reaction must release thermal energy back to its surroundings (i.e., it must be exothermic) and the enthalpy change expression must also change its sign. Thus, reversing the melting of ice and the combustion of methane, the thermochemical equations become

$$H_2O(l) \longrightarrow H_2O(s) \qquad \Delta H = -6.01 \text{ kJ/mol}$$

$$CO_2(g) + 2H_2O(l) \longrightarrow CH_4(g) + 2O_2(g) \qquad \Delta H = +890.4 \text{ kJ/mol}$$

What was an endothermic process becomes an exothermic process when reversed, and vice versa.

Sample Problem 5.3 illustrates the use of a thermochemical equation to relate the mass of a product to the energy consumed in the reaction.

SAMPLE PROBLEM 5.3

Given the thermochemical equation for photosynthesis,

$$6H_2O(l) + 6CO_2(g) \longrightarrow C_6H_{12}O_6(s) + 6O_2(g) \qquad \Delta H = +2803 \text{ kJ/mol}$$

calculate the solar energy required to produce 75.0 g of $C_6H_{12}O_6$.

Strategy The thermochemical equation shows that for every mole of $C_6H_{12}O_6$ produced, 2803 kJ is absorbed. We need to find out how much energy is absorbed for the production of 75.0 g of $C_6H_{12}O_6$. We must first find out how many moles there are in 75.0 g of $C_6H_{12}O_6$.

Setup The molar mass of $C_6H_{12}O_6$ is 180.2 g/mol, so 75.0 g of $C_6H_{12}O_6$ is

$$75.0 \text{ g} \times \frac{1 \text{ mol } C_6H_{12}O_6}{180.2 \text{ g}} = 0.416 \text{ mol}$$

We will multiply the thermochemical equation, including the enthalpy change, by 0.416, in order to write the equation in terms of the appropriate amount of $C_6H_{12}O_6$.

Solution

$$(0.416 \text{ mol})[6H_2O(l) + 6CO_2(g) \longrightarrow C_6H_{12}O_6(s) + 6O_2(g)]$$

and $(0.416 \text{ mol})(\Delta H) = (0.416 \text{ mol})(2803 \text{ kJ/mol})$ gives

$$2.50\ H_2O(l) + 2.50\ CO_2(g) \longrightarrow 0.416\ C_6H_{12}O_6(s) + 2.50\ O_2(g) \qquad \Delta H = +1.17 \times 10^3 \text{ kJ}$$

Therefore, 1.17×10^3 kJ of energy in the form of sunlight is consumed in the production of 75.0 g of $C_6H_{12}O_6$. Note that the "per mole" units in ΔH are canceled when we multiply the thermochemical equation by the number of moles of $C_6H_{12}O_6$.

THINK ABOUT IT

The specified amount of $C_6H_{12}O_6$ is less than half a mole. Therefore, we should expect the associated enthalpy change to be less than half that specified in the thermochemical equation for the production of 1 mole of $C_6H_{12}O_6$.

Practice Problem ATTEMPT Calculate the solar energy required to produce 5255 g of $C_6H_{12}O_6$.

Practice Problem BUILD Calculate the mass (in grams) of O_2 that is produced by photosynthesis when 2.49×10^4 kJ of solar energy is consumed.

Practice Problem CONCEPTUALIZE The diagrams represent systems before and after reaction for two related chemical processes. ΔH for the first reaction is 1755.0 kJ/mol. Determine the value of ΔH for the second reaction.

before

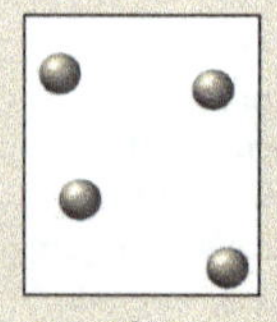
after

$\Delta H = 1755.0$ kJ/mol

before

after

$\Delta H = ?$

CHECKPOINT – SECTION 5.3 Enthalpy

5.3.1 Given the thermochemical equation: $H_2(g) + Br_2(l) \longrightarrow 2HBr(g)$, $\Delta H = -72.4$ kJ/mol, calculate the amount of heat released when a kilogram of $Br_2(l)$ is consumed in this reaction.

a) 7.24×10^4 kJ
b) 453 kJ
c) 906 kJ
d) 227 kJ
e) 724 kJ

5.3.2 Given the thermochemical equation: $2Cu_2O(s) \longrightarrow 4Cu(s) + O_2(g)$, $\Delta H = +333.8$ kJ/mol, calculate the mass of copper produced when 1.47×10^4 kJ is consumed in this reaction.

a) 11.2 kg
b) 176 kg
c) 44.0 kg
d) 334 kg
e) 782 kg

5.4 Calorimetry

In the study of thermochemistry, heat changes that accompany physical and chemical processes are measured with a *calorimeter*, a closed container designed specifically for this purpose. We begin our discussion of ***calorimetry***, the measurement of heat changes, by defining two important terms: *specific heat* and *heat capacity*.

Specific Heat and Heat Capacity

Student Note: Although heat capacity is typically given for an object rather than for a substance—the "object" may be a given quantity of a particular substance.

The ***specific heat*** (s) of a substance is the amount of heat required to raise the temperature of 1 g of the substance by 1°C. The ***heat capacity*** (C) is the amount of heat required to raise the temperature of an *object* by 1°C. We can use the specific heat of a substance to determine the heat capacity of a specified quantity of that substance. For example, we can use the specific heat of water, 4.184 J/(g · °C), to determine the heat capacity of a kilogram of water:

$$\text{heat capacity of 1 kg of water} = \frac{4.184 \text{ J}}{1 \text{ g} \cdot °\text{C}} \times 1000 \text{ g} = 4184 \quad \text{or} \quad 4.184 \times 10^3 \text{ J/°C}$$

Note that specific heat has the units J/(g · °C) and heat capacity has the units J/°C. Table 5.2 shows the specific heat values of some common substances. If we know the specific heat and the amount of a substance, then the change in the sample's temperature (ΔT) will tell us the amount of heat (q) that has been absorbed or released in a particular process. One equation for calculating the heat associated with a temperature change is given by

Equation 5.13 $q = sm\Delta T$

where s is the specific heat, m is the mass of the substance undergoing the temperature change, and ΔT is the temperature change: $\Delta T = T_{\text{final}} - T_{\text{initial}}$. Another equation for calculating the heat associated with a temperature change is given by

Student Note: Note that $C = sm$. Although specific heat is a property of substances, and heat capacity is a property of objects, we can define a specified quantity of a substance as an "object" and determine its heat capacity using its mass and its specific heat.

Equation 5.14 $q = C\Delta T$

where C is the heat capacity and ΔT is the temperature change. The sign convention for q is the same as that for an enthalpy change: q is positive for endothermic processes and negative for exothermic processes.

TABLE 5.2 Specific Heat Values of Some Common Substances

Substance	Specific Heat (J/g · °C)	Substance	Specific Heat (J/g · °C)
$Al(s)$	0.900	$Fe(s)$	0.444
$Au(s)$	0.129	$Hg(l)$	0.139
C(graphite)	0.720	$H_2O(l)$	4.184
C(diamond)	0.502	$C_2H_5OH(l)$ (ethanol)	2.46
$Cu(s)$	0.385		

Animation
Figure 5.9, Determination of $\Delta H°_{\text{rxn}}$ by Constant-Pressure Calorimetry.

Sample Problem 5.4 shows how to use the specific heat of a substance to calculate the amount of heat needed to raise the temperature of the substance by a particular amount.

SAMPLE PROBLEM 5.4

Calculate the amount of heat (in kJ) required to heat 255 g of water from 25.2°C to 90.5°C.

Strategy Use Equation 5.13 ($q = sm\Delta T$) to calculate q.

Setup $m = 255$ g, $s = 4.184$ J/g · °C, and $\Delta T = 90.5°C - 25.2°C = 65.3°C$

Solution

$$q = \frac{4.184 \text{ J}}{\text{g} \cdot °\text{C}} \times 255 \text{ g} \times 65.3°\text{C} = 6.97 \times 10^4 \text{ J} \quad \text{or} \quad 69.7 \text{ kJ}$$

THINK ABOUT IT

Look carefully at the cancellation of units and make sure that the number of kilojoules is smaller than the number of joules. It is a common error to multiply by 1000 instead of dividing in conversions of this kind.

Practice Problem ATTEMPT Calculate the amount of heat (in kJ) required to heat 1.01 kg of water from 0.05°C to 35.81°C.

Practice Problem BUILD What will be the final temperature of a 514-g sample of water, initially at 10.0°C, after 90.8 kJ have been added to it?

Practice Problem CONCEPTUALIZE Shown here are two samples of the same substance. When equal amounts of heat are added to both samples, the temperature of the sample on the left increases by 15.3°C. Determine the increase in temperature of the sample on the right.

Constant-Pressure Calorimetry

A crude constant-pressure calorimeter can be constructed from two Styrofoam coffee cups, as shown in Figure 5.8. This device, called a coffee-cup calorimeter, can be used to measure the heat exchanged between the system and surroundings for a variety of reactions, such as acid-base neutralization, heat of solution, and heat of dilution. Because the pressure is constant, the heat change for the process (q) is equal to the enthalpy change (ΔH). In such experiments, we consider the reactants and products to be the system, and the water in the calorimeter to be the surroundings. We neglect the small heat capacity of the Styrofoam cups in our calculations. In the case of an exothermic reaction, the heat released by the system is absorbed by the water (surroundings), thereby increasing its temperature. Knowing the mass of the water in the calorimeter, the specific heat of water, and the change in temperature, we can calculate q_P of the system using the equation

$$q_{\text{sys}} = -sm\Delta T \qquad \textbf{Equation 5.15}$$

Note that the minus sign makes q_{sys} a negative number if ΔT is a positive number (i.e., if the temperature goes up). This is in keeping with the sign conventions listed in Table 5.1. A negative ΔH or a negative q indicates an exothermic process, whereas a positive ΔH or a positive q indicates an endothermic process. Table 5.3 lists some of the reactions that can be studied with a constant-pressure calorimeter. Figure 5.9 shows how constant-pressure calorimetry can be used to determine ΔH for a reaction.

Constant-pressure calorimetry can also be used to determine the heat capacity of an object or the specific heat of a substance. Suppose, for example, that we have a lead pellet with a mass of 26.47 g originally at 89.98°C. We drop the pellet into a constant-pressure calorimeter

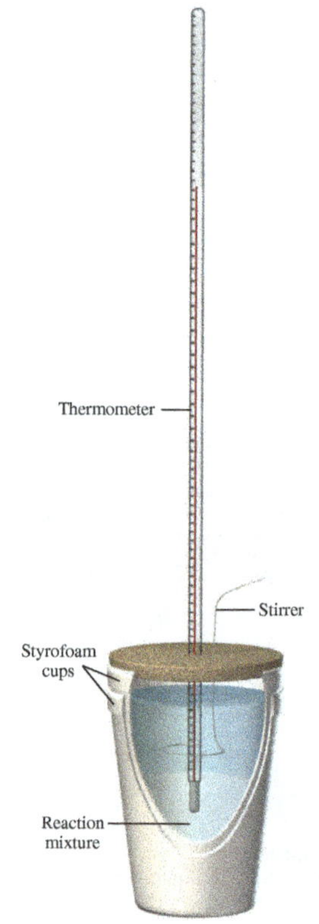

Figure 5.8 A constant-pressure coffee-cup calorimeter made of two Styrofoam cups. The nested cups help to insulate the reaction mixture from the surroundings. Two solutions of known volume containing the reactants at the same temperature are carefully mixed in the calorimeter. The heat produced or absorbed by the reaction can be determined by measuring the temperature change.

Figure 6.9

Determination of ΔH°_{rxn} by Constant-Pressure Calorimetry

One at a time, we pour the solutions into the calorimeter.

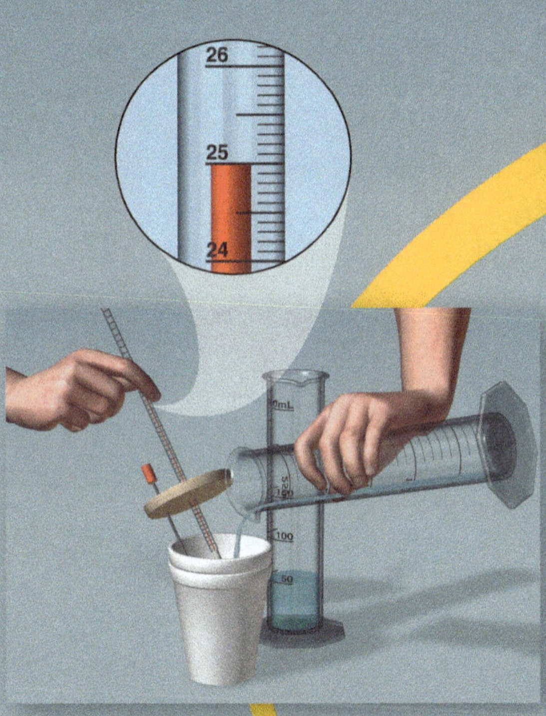

We start with 50.0 mL each of 1.00 *M* HCl and 1.00 *M* NaOH. Both solutions are at room temperature, which in this example is 25.0°C. The net ionic equation that represents the reaction is

$$H^+(aq) + OH^-(aq) \longrightarrow H_2O(l)$$

We have 0.0500 L × 1.00 *M* = 0.0500 mol of each reactant.

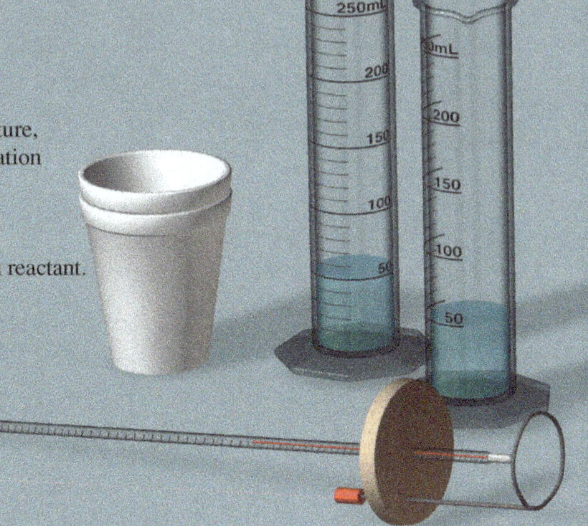

When both solutions have been added, we cap the calorimeter to prevent loss of energy to the environment and use the stirrer to ensure that the solutions are mixed thoroughly.

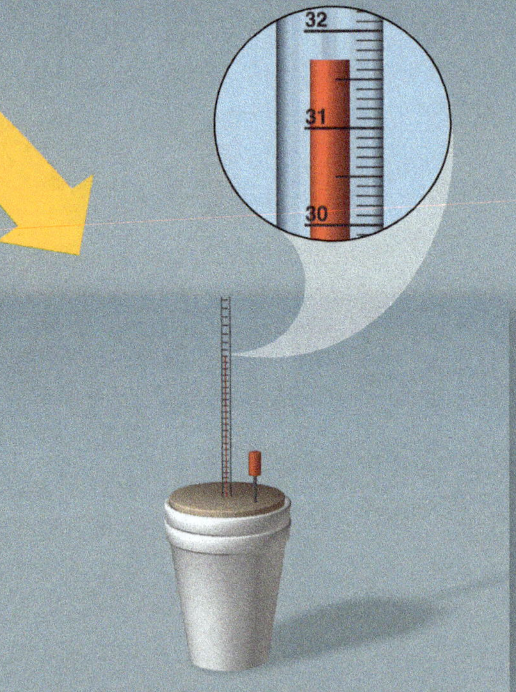

As the reaction proceeds, the temperature of the water increases as it absorbs the energy given off by the reaction. We record the maximum water temperature as 31.7°C.

Assuming the density and specific heat of the solution to be the same as those of water (1 g/mL, 4.184 J/g · °C), we calculate q_{soln} as follows:

$$q_{soln} = \text{specific heat of water} \times \text{mass of water} \times \text{temperature change}$$

$$= \frac{4.184\text{ J}}{1\text{ g} \cdot {}^\circ\text{C}} \times 100.0\text{ g} \times (31.7 - 25.0)^\circ\text{C} = 2803\text{ J}$$

Assuming that the heat capacity of the calorimeter is negligible, we know that $q_{soln} = -q_{rxn}$, and we can write

$$q_{rxn} = -2803\text{ J}$$

This is the heat of reaction when 0.0500 mol H^+ reacts with 0.0500 mol OH^-. To determine ΔH_{rxn}, we divide q_{rxn} by the number of moles. (These reactants are present in stoichiometric amounts. If there were a *limiting* reactant, we would divide by the number of moles of limiting reactant.)

$$\Delta H_{rxn} = \frac{-2803\text{ J}}{0.0500\text{ mol}} = -5.61 \times 10^4\text{ J/mol or } -56.1\text{ kJ/mol}$$

This result is very close to the number we get using Equation 5.12 and the data in Appendix 2.

(See Visualizing Chemistry questions VC 5.1–VC 5.4 on page 222.)

What's the point?

Constant-pressure calorimetry can be used to determine ΔH_{rxn}—the heat of reaction for the reactant quantities specified by the balanced equation. However, when we carry out calorimetry experiments in the laboratory, we typically use much smaller quantities of reactants than those represented in a chemical equation. By measuring the temperature change of the surroundings (a known quantity of water in which the reactants are dissolved), we can determine q_{rxn} for the reactant quantities in the experiment. We can then divide q_{rxn} by the number of moles of reactant to determine ΔH_{rxn}.

TABLE 5.3 Heats of Some Typical Reactions and Physical Processes Measured at Constant Pressure

Type of Reaction	Example	ΔH (kJ/mol)
Heat of neutralization	$HCl(aq) + NaOH(aq) \longrightarrow H_2O(l) + NaCl(aq)$	−56.2
Heat of ionization	$H_2O(l) \longrightarrow H^+(aq) + OH^-(aq)$	+56.2
Heat of fusion	$H_2O(s) \longrightarrow H_2O(l)$	+6.01
Heat of vaporization	$H_2O(l) \longrightarrow H_2O(g)$	+44.0*

*Measured at 25°C. At 100°C, the value is +40.79 kJ.

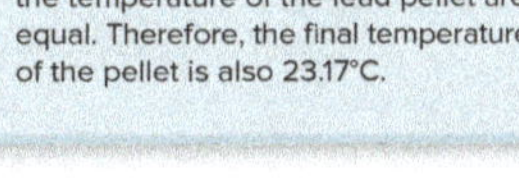

Student Note: The temperature of the water stops changing when it and the temperature of the lead pellet are equal. Therefore, the final temperature of the pellet is also 23.17°C.

containing 100.0 g of water at 22.50°C. The temperature of the water increases to 23.17°C. In this case, we consider the pellet to be the system and the water to be the surroundings. Because it is the temperature of the surroundings that we measure and because $q_{sys} = -q_{surr}$, we can rewrite Equation 5.15 as

$$q_{surr} = sm\Delta T$$

Thus, q_{water} of the water is

$$q_{water} = \frac{4.184\ J}{g \cdot °C} \times 100.0\ g \times (23.17°C - 22.50°C) = 280\ J$$

and q_{pellet} is −280 J. The negative sign indicates that heat is *released* by the pellet. Dividing q_{pellet} by the temperature change (ΔT) gives us the heat capacity of the pellet (C_{pellet}).

$$C_{pellet} = \frac{-280\ J}{23.17°C - 89.95°C} = 4.19\ J/°C$$

Furthermore, because we know the *mass* of the pellet, we can determine the specific heat of lead (s_{lead}):

$$S_{Pb} = \frac{C_{pellet}}{m_{pellet}} = \frac{4.19\ J/°C}{26.47\ g} = 0.158\ J/g \cdot °C \quad \text{or} \quad 0.16\ J/g \cdot °C$$

Animation
Figure 5.10, Determination of Specific Heat by Constant-Pressure Calorimetry.

Sample Problem 5.5 shows how to use constant-pressure calorimetry to calculate the heat capacity (C) of an object. Figure 5.10 illustrates the process of determining the specific heat of a metal using constant-pressure calorimetry.

SAMPLE PROBLEM 5.5

A metal pellet with a mass of 100.0 g, originally at 88.4°C, is dropped into 125 g of water originally at 25.1°C. The final temperature of both the pellet and the water is 31.3°C. Calculate the heat capacity C (in J/°C) of the pellet.

Strategy Use Equation 5.13 ($q = sm\Delta T$) to determine the heat absorbed by the water; then use Equation 5.14 ($q = C\Delta T$) to determine the heat capacity of the metal pellet.

Setup $m_{water} = 125$ g, $s_{water} = 4.184$ J/g · °C, and $\Delta T_{water} = 31.3°C - 25.1°C = 6.2°C$. The heat absorbed by the water must be released by the pellet: $q_{water} = -q_{pellet} \cdot m_{pellet} = 100.0$ g and $\Delta T_{pellet} = 31.3°C - 88.4°C = -57.1°C$.

Solution From Equation 5.13, we have

$$q_{water} = \frac{4.184\ J}{g \cdot °C} \times 125\ g \times 6.2°C = 3242.6\ J$$

Thus,

$$q_{pellet} = -3242.6\ J$$

From Equation 5.14, we have

$$-3242.6\ J = C_{pellet} \times (-57.1°C)$$

Thus,

$$C_{pellet} = 57\ J/°C$$

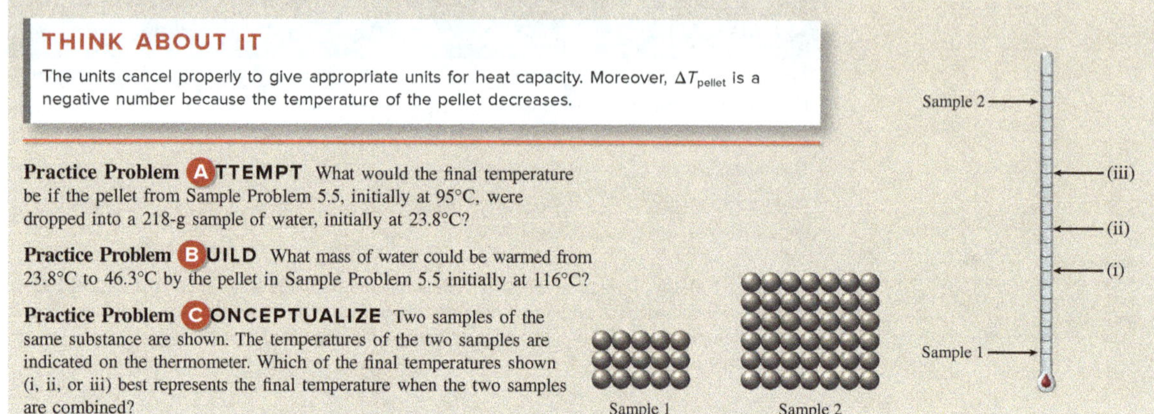

THINK ABOUT IT

The units cancel properly to give appropriate units for heat capacity. Moreover, ΔT_{pellet} is a negative number because the temperature of the pellet decreases.

Practice Problem ATTEMPT What would the final temperature be if the pellet from Sample Problem 5.5, initially at 95°C, were dropped into a 218-g sample of water, initially at 23.8°C?

Practice Problem BUILD What mass of water could be warmed from 23.8°C to 46.3°C by the pellet in Sample Problem 5.5 initially at 116°C?

Practice Problem CONCEPTUALIZE Two samples of the same substance are shown. The temperatures of the two samples are indicated on the thermometer. Which of the final temperatures shown (i, ii, or iii) best represents the final temperature when the two samples are combined?

Bringing Chemistry to Life

Heat Capacity and Hypothermia

Like a warm metal pellet, the human body loses heat when it is immersed in cold water. Because we are warm-blooded animals, our body temperature is maintained at around 37°C. The human body is about 70 percent water by mass and water has a very high specific heat, so fluctuations in body temperature normally are very small. An air temperature of 25°C (often described as "room temperature") feels warm to us because air has a small specific heat (about 1 J/g · °C) and a low density. Consequently, very little heat is lost from the body to the surrounding air. The situation is drastically different if the body is immersed in water at 25°C. The heat lost by the human body when immersed in water can be thousands of times greater than that lost to air of the same temperature.

Hypothermia occurs when the body's mechanisms for producing and conserving heat are exceeded by loss of heat to the surroundings. Although hypothermia is dangerous and potentially deadly, there are certain circumstances under which it may actually be beneficial. A colder body temperature slows down all the normal biochemical processes, reducing the brain's need for oxygen, and prolonging the time period during which resuscitation efforts can be effective. Occasionally we hear about a seemingly miraculous recovery of a near-drowning victim who was submerged for a long period of time. These victims are usually small children who were submerged in icy water. The small size and, therefore, small heat capacity of a child allows for rapid cooling and may afford some protection from hypoxia—the lack of oxygen that causes death in drowning victims.

Student Note: Hypothermia routinely is induced in patients undergoing open heart surgery, drastically reducing the body's need for oxygen. Under these conditions, the heart can be stopped for the duration of the surgery.

The lowest body temperature ever recorded for a human (who survived) was 13.7°C. Anna Bagenholm, 29, spent over an hour submerged after falling headfirst through the ice on a frozen river. Although she was clinically dead when she was pulled from the river, she has made a full recovery. There's a saying among doctors who treat hypothermia patients: "No one is dead until they're *warm* and dead."

Figure 5.10

Determination of Specific Heat by Constant-Pressure Calorimetry

We add metal shot (125.0 g at 100.0°C) to the water, and we cap the calorimeter to prevent loss of energy to the environment.

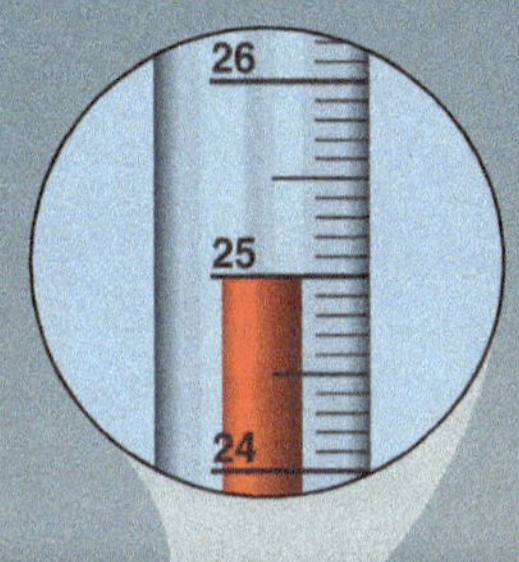

We place 100.0 mL (100.0 g) of water in the calorimeter. The temperature of this water is 25.0°C.

We place 125.0 g of metal shot in a test tube and immerse it in boiling water long enough to heat all of the metal to the boiling point of water (100.0°C).

As energy is transferred from the metal shot to the water, the temperature of the water increases and the temperature of the metal shot decreases. We use the stirrer to ensure thorough mixing. The thermometer measures the temperature of the water.

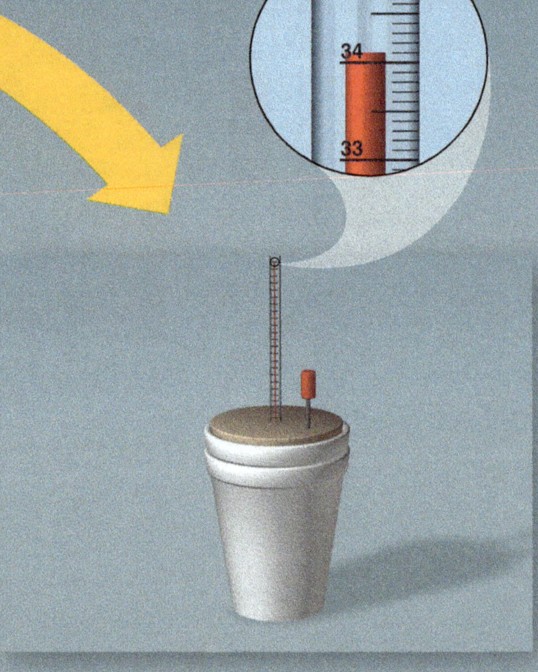

When the temperature of the metal shot and the water are equal, the temperature of the water has reached a maximum value. We record this temperature as 34.1°C.

We know that $q_{\text{water}} = -q_{\text{metal}}$.
Substituting in the information given, we write:

$$q_{\text{water}} = \text{specific heat of water} \times \text{mass of water} \times \text{temperature change}$$

$$= \frac{4.184\ \text{J}}{1\ \text{g} \cdot {}^\circ\text{C}} \times 100.0\ \text{g} \times (34.1 - 25.0){}^\circ\text{C} = 3807\ \text{J}$$

$$q_{\text{metal}} = \text{specific heat of metal} \times \text{mass of metal} \times \text{temperature change}$$

$$= x \times 125.0\ \text{g} \times (34.1 - 100.0){}^\circ\text{C} = -8238\,x\ \text{g} \cdot {}^\circ\text{C}$$

and

$$3807\ \text{J} = -(-8238\,x)\ \text{g} \cdot {}^\circ\text{C}$$

$$x = \frac{3807\ \text{J}}{8238\ \text{g} \cdot {}^\circ\text{C}} = 0.46\ \text{J/g} \cdot {}^\circ\text{C}$$

The specific heat of the metal is therefore 0.46 J/g · °C.

(See Visualizing Chemistry questions VC 5.5–VC 5.8 on page 222.)

What's the point?

We can determine the specific heat of a metal by combining a known mass of the metal at a known temperature with a known mass of water at a known temperature. Assuming the calorimeter has a negligible heat capacity, the amount of energy lost by the hotter metal is equal to the amount of energy gained by the cooler water.

Constant-Volume Calorimetry

The heat of combustion is usually measured using constant-volume calorimetry. Typically, a known mass of the compound to be analyzed is placed in a steel container called a *constant-volume bomb*, or simply a *bomb,* which is pressurized with oxygen. The closed bomb is then immersed in a known amount of water in an insulated container, as shown in Figure 5.11. (Together, the steel bomb and the water in which it is submerged constitute the *calorimeter.*) The sample is ignited electrically, and the heat released by the combustion of the sample is absorbed by the bomb and the water and can be determined by measuring the increase in temperature of the water. The special design of this type of calorimeter allows us to assume that no heat (or mass) is lost to the surroundings during the time it takes to carry out the reaction and measure the temperature change. Therefore, we can call the bomb and the water in which it is submerged an *isolated* system. Because no heat enters or leaves the system during the process, the heat change of the system overall (q_{system}) is zero and we can write

$$q_{cal} = -q_{rxn}$$

where q_{cal} and q_{rxn} are the heat changes for the calorimeter and the reaction, respectively. Thus,

$$q_{rxn} = -q_{cal}$$

To calculate q_{cal}, we need to know the heat capacity of the calorimeter (C_{cal}) and the change in temperature, that is,

Equation 5.16 $$q_{cal} = C_{cal}\Delta T$$

And, because $q_{rxn} = -q_{cal}$,

Equation 5.17 $$q_{rxn} = -C_{cal}\Delta T$$

The heat capacity of the calorimeter (C_{cal}) is determined by burning a substance with an accurately known heat of combustion. For example, it is known that the combustion of a 1.000-g sample of benzoic acid (C_6H_5COOH) releases 26.38 kJ of heat. If the measured temperature increase is 4.673°C, then the heat capacity of the calorimeter is given by

$$C_{cal} = \frac{q_{cal}}{\Delta T} = \frac{26.38 \text{ kJ}}{4.673°\text{C}} = 5.645 \text{ kJ/J°C}$$

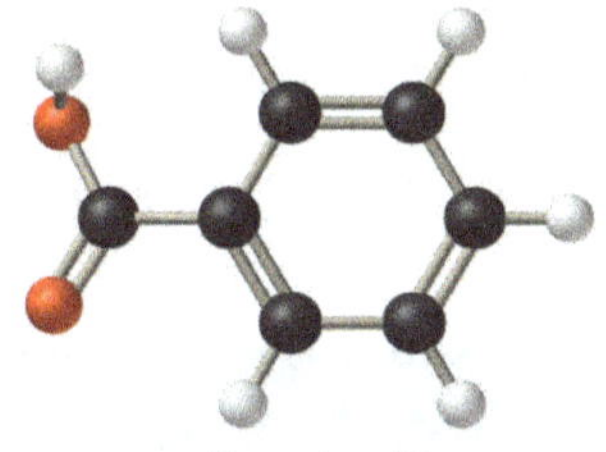

Benzoic acid

Once C_{cal} has been determined, the calorimeter can be used to measure the heat of combustion of other substances. Because a reaction in a bomb calorimeter occurs under constant-volume rather than constant-pressure conditions, the measured heat change corresponds to the *internal energy* change (ΔU) rather than to the *enthalpy* change (ΔH) (see Equations 5.6 and 5.11). It is possible to correct the measured heat changes so that they correspond to ΔH values, but the corrections usually are quite small, so we will not concern ourselves with the details here.

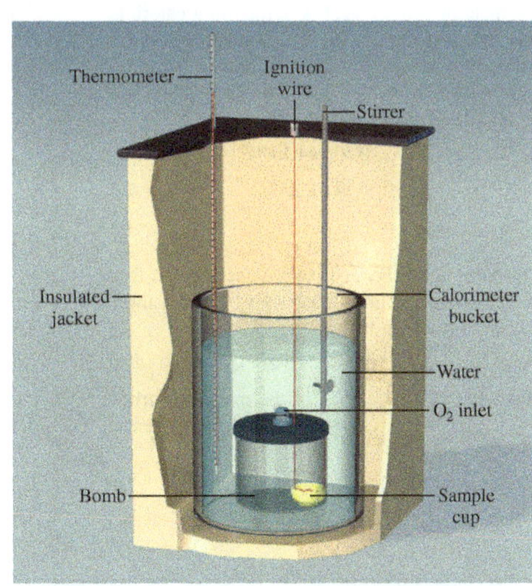

Figure 5.11 A constant-volume bomb calorimeter. The calorimeter is filled with oxygen gas at high pressure before it is placed in the bucket. The sample is ignited electrically, and the heat produced by the reaction is determined by measuring the temperature increase in the known amount of water surrounding the bomb.

Sample Problem 5.6 shows how to use constant-volume calorimetry to determine the energy content per gram of a substance.

SAMPLE PROBLEM 5.6

A Famous Amos bite-sized chocolate chip cookie weighing 7.25 g is burned in a bomb calorimeter to determine its energy content. The heat capacity of the calorimeter is 39.97 kJ/°C. During the combustion, the temperature of the water in the calorimeter increases by 3.90°C. Calculate the energy content (in kJ/g) of the cookie.

Strategy Use Equation 5.17 ($q_{rxn} = -C_{cal}\Delta T$) to calculate the heat released by the combustion of the cookie. Divide the heat released by the mass of the cookie to determine its energy content per gram.

Setup $C_{cal} = 39.97$ kJ/°C and $\Delta T = 3.90$°C

Solution From Equation 5.17 we have

$$q_{rxn} = -C_{cal}\Delta T = -(39.97 \text{ kJ/°C})(3.90\text{°C}) = -1.559 \times 10^2 \text{ kJ}$$

The negative sign in the result indicates that heat is released by the combustion. Because energy content is a positive quantity, we write

$$\text{energy content per gram} = \frac{1.559 \times 10^2 \text{ kJ}}{7.25 \text{ g}} = 21.5 \text{ kJ/g}$$

Nutrition facts label on a box of Famous Amos chocolate chip cookies.

Credit: © Brian Moeskau.

THINK ABOUT IT

According to the label on the cookie package, a serving size is four cookies, or 29 g, and each serving contains 150 Cal. Convert the energy per gram to Calories per serving to verify the result.

$$\frac{21.5 \text{ kJ}}{\text{g}} \times \frac{1 \text{ Cal}}{4.184 \text{ kJ}} \times \frac{29 \text{ g}}{\text{serving}} = 1.5 \times 10^2 \text{ Cal/serving}$$

Practice Problem ATTEMPT A serving of Grape-Nuts cereal (5.80 g) is burned in a bomb calorimeter with a heat capacity of 43.7 kJ/°C. During the combustion, the temperature of the water in the calorimeter increased by 1.92°C. Calculate the energy content (in kJ/g) of Grape-Nuts.

Practice Problem BUILD The energy content of raisin bread is 13.1 kJ/g. Calculate the temperature increase when a slice of raisin bread (32.0 g) is burned in the calorimeter in Sample Problem 5.6.

Practice Problem CONCEPTUALIZE Suppose an experiment to determine the energy content of food used a calorimeter that contained less water than it did when it was calibrated. Explain how this would affect the result of the experiment.

CHECKPOINT – SECTION 5.4 Calorimetry

5.4.1 A 1.000-g sample of benzoic acid is burned in a calorimeter to determine its heat capacity, C_{cal}. The reaction gives off 26.42 kJ of heat and the temperature of the water in the calorimeter increases from 23.40°C to 27.20°C. What is the heat capacity of the calorimeter?

a) 3.80 kJ/°C
b) 6.95 kJ/°C
c) 100 kJ/°C
d) 0.144 kJ/°C
e) 7.81 kJ/°C

5.4.2 One-gram samples of Al, Fe, and Au are each heated from 40°C to 75°C. Arrange the metals in order of increasing amount of heat absorbed in the process.

a) Al < Fe < Au
b) Al < Au < Fe
c) Fe < Au < Al
d) Au < Fe < Al
e) All of the metals absorbed the same amount of heat.

5.4.3 A reaction, carried out in a bomb calorimeter with $C_{cal} = 5.01$ kJ/°C, gives off 318 kJ of heat. The initial temperature of the water is 24.8°C. What is the final temperature of the water in the calorimeter?

a) 88.3°C
b) 63.5°C
c) 29.8°C
d) 162°C
e) 76.7°C

5.4.4 Quantities of 50.0 mL of 1.00 *M* HCl and 50.0 mL of 1.00 *M* NaOH are combined in a constant-pressure calorimeter. Both solutions are initially at 24.4°C. Calculate the final temperature of the combined solutions. (Use the data from Table 5.3. Assume that the mass of the combined solutions is 100.0 g and that the solution's specific heat is the same as that for water, 4.184 J/g · °C.) The heat capacity of the calorimeter is negligibly small.

a) 31.1°C
b) 29.0°C
c) 44.2°C
d) 91.8°C
e) 35.7°C

What if the Heat Capacity of the Calorimeter Isn't Negligible?

Although we usually assume that no heat is absorbed by the Styrofoam cups we use in the laboratory for constant-pressure calorimetry, in reality, the calorimeter generally does absorb a small portion of the heat produced by a chemical reaction. We can determine the heat capacity of a coffee-cup calorimeter by combining reactant solutions with precisely known concentrations and masses. Once we have determined the heat capacity, we can correct for the heat absorbed by the calorimeter when we carry out other experiments.

Consider an experiment in which we combine 50.0 mL of 0.250 *M* HCl(*aq*) with 50.0 mL of 0.250 *M* NaOH. The calorimeter and both solutions are initially at 23.50°C, and the density and specific heat of the combined solution are the same as that of water (1.000 g/mL and 4.184 J/g · °C, respectively). We can determine the amount of heat the reaction will generate using moles of reactants and the heat of neutralization value from Table 5.3. We have 0.0500 L × 0.250 *M* = 0.0125 mol of each reactant. According to Table 5.3, the heat of neutralization is −56.2 kJ/mol. Thus, we expect the enthalpy change of the system to be 0.0125 mol × (−56.2 kJ/mol) = −0.703 kJ. Converting this to joules and using Equation 5.15, we calculate the temperature change we expect from the combination of these reactants.

$$q_{sys} = -703 \text{ J} = -4.184 \text{ J/g} \cdot {}^\circ\text{C} \times 100.0 \text{ g} - \Delta T$$

$$\Delta T = \frac{-703 \text{ J}}{(-100.0 \text{ g})(4.184 \text{ J/g} \cdot {}^\circ\text{C})} = 1.68^\circ\text{C}$$

Thus, we expect the temperature to increase by 1.68°C. However, the measured final temperature is 25.09°C, an increase of only 1.59°C. The water temperature increased by less than expected because the calorimeter absorbed part of the heat produced by the neutralization reaction. We can determine how much heat the calorimeter absorbed by using Equation 5.15 again. This time, we use the measured temperature change to calculate the amount of heat absorbed by the water.

$$q_{surr} = 4.184 \text{ J/g} \cdot {}^\circ\text{C} \times 100.0 \text{ g} \times 1.59^\circ\text{C} = 665 \text{ J}$$

We know that the reaction produced 703 J but the water absorbed only 665 J. The remaining energy, 703 − 665 = 38 J, is $q_{calorimeter}$, the heat absorbed by the calorimeter itself. We use Equation 5.14 to calculate the heat capacity of the calorimeter.

$$38 \text{ J} = C_{calorimeter} \times 1.59^\circ\text{C}$$

$$C_{calorimeter} = 23.9 \text{ J/}^\circ\text{C}$$

Remember that the calorimeter had the same initial temperature and the same final temperature as the solutions.

5.5 Hess's Law

Because enthalpy is a state function, the change in enthalpy that occurs when reactants are converted to products in a reaction is the same whether the reaction takes place in one step or in a series of steps. This observation is called ***Hess's law.***[2] An analogy for Hess's law can be made to the floors in a building. Suppose, for example, that you take the elevator from the first floor to the sixth floor of the building. The net gain in your gravitational potential energy (which is analogous to the enthalpy change for the overall process) is the same whether you go directly there or stop at each floor on your way up (breaking the trip into a series of steps).

Recall from Section 5.3 that the enthalpy change for the combustion of a mole of methane depends on whether the product water is liquid or gas. More heat is given off by the reaction that produces liquid water. We can use this example to illustrate Hess's law by envisioning the first of these reactions happening in two steps. In step 1, methane and oxygen are converted to carbon dioxide and liquid water, releasing heat.

$$CH_4(g) + 2O_2(g) \longrightarrow CO_2(g) + 2H_2O(l) \qquad \Delta H = -890.4 \text{ kJ/mol}$$

In step 2, the liquid water is vaporized, which requires an input of heat.

$$2H_2O(l) \longrightarrow 2H_2O(g) \qquad \Delta H = +88.0 \text{ kJ/mol}$$

We can add balanced chemical equations just as we can add algebraic equalities, canceling identical items on opposite sides of the equation arrow:

$$\begin{array}{lll} CH_4(g) + 2O_2(g) \longrightarrow CO_2(g) + \cancel{2H_2O(l)} & & \Delta H = -890.4 \text{ kJ/mol} \\ + \qquad \cancel{2H_2O(l)} \longrightarrow 2H_2O(g) & & \Delta H = +88.0 \text{ kJ/mol} \\ \hline CH_4(g) + 2O_2(g) \longrightarrow CO_2(g) + 2H_2O(g) & & \Delta H = -802.4 \text{ kJ/mol} \end{array}$$

When we add thermochemical equations, we add the ΔH values as well. This gives us the overall enthalpy change for the net reaction. Using this method, we can deduce the enthalpy changes for many reactions, some of which may not be possible to carry out directly. In general, we apply

2. Germain Henri Hess (1802–1850). Swiss chemist. Hess was born in Switzerland but spent most of his life in Russia. For formulating Hess's law, he is called the father of thermochemistry.

Hess's law by arranging a series of chemical equations (corresponding to a series of steps) in such a way that they sum to the desired overall equation. Often, in applying Hess's law, we must manipulate the equations involved, multiplying by appropriate coefficients, reversing equations, or both. It is important to follow the guidelines [⏮ Section 5.3] for the manipulation of thermochemical equations and to make the corresponding change to the enthalpy change of each step.

Sample Problem 5.7 illustrates the use of this method for determining ΔH.

SAMPLE PROBLEM 5.7

Given the following thermochemical equations,

$$NO(g) + O_3(g) \longrightarrow NO_2(g) + O_2(g) \quad \Delta H = -198.9 \text{ kJ/mol}$$

$$O_3(g) \longrightarrow \tfrac{3}{2}O_2(g) \quad \Delta H = -142.3 \text{ kJ/mol}$$

$$O_2(g) \longrightarrow 2O(g) \quad \Delta H = +495 \text{ kJ/mol}$$

determine the enthalpy change for the reaction

$$NO(g) + O(g) \longrightarrow NO_2(g)$$

Strategy Arrange the given thermochemical equations so that they sum to the desired equation. Make the corresponding changes to the enthalpy changes, and add them to get the desired enthalpy change.

Setup The first equation has NO as a reactant with the correct coefficient, so we use it as is.

$$NO(g) + O_3(g) \longrightarrow NO_2(g) + O_2(g) \quad \Delta H = -198.9 \text{ kJ/mol}$$

The second equation must be reversed so that the O_3 introduced by the first equation will cancel (O_3 is not part of the overall chemical equation). We also must change the sign on the corresponding ΔH value.

$$\tfrac{3}{2}O_2(g) \longrightarrow O_3(g) \quad \Delta H = +142.3 \text{ kJ/mol}$$

These two steps sum to give the following:

$$\begin{array}{lll} & NO(g) + \cancel{O_3(g)} \longrightarrow NO_2(g) + \cancel{O_2(g)} & \Delta H = -198.9 \text{ kJ/mol} \\ + & \tfrac{3}{2}\cancel{O_2(g)} \longrightarrow \cancel{O_3(g)} & \Delta H = +142.3 \text{ kJ/mol} \\ \hline & NO(g) + \tfrac{1}{2}O_2(g) \longrightarrow NO_2(g) & \Delta H = -56.6 \text{ kJ/mol} \end{array}$$

We then replace the $\frac{1}{2}O_2$ on the left with O by incorporating the last equation. To do so, we divide the third equation by 2 and reverse its direction. As a result, we must also divide its ΔH value by 2 and change its sign.

$$O(g) \longrightarrow \tfrac{1}{2}O_2(g) \quad \Delta H = -247.5 \text{ kJ/mol}$$

Finally, we sum all the steps and add their enthalpy changes.

Solution

$$\begin{array}{ll} NO(g) + \cancel{O_3(g)} \longrightarrow NO_2(g) + \cancel{O_2(g)} & \Delta H = -198.9 \text{ kJ/mol} \\ \tfrac{3}{2}\cancel{O_2(g)} \longrightarrow \cancel{O_3(g)} & \Delta H = +142.3 \text{ kJ/mol} \\ + \; O(g) \longrightarrow \tfrac{1}{2}\cancel{O_2(g)} & \Delta H = -247.5 \text{ kJ/mol} \\ \hline NO(g) + O(g) \longrightarrow NO_2(g) & \Delta H = -304 \text{ kJ/mol} \end{array}$$

THINK ABOUT IT

Recall that when you first learned to balance chemical equations, you were advised to use the smallest possible *whole* numbers as coefficients [⏮ Section 3.3]. Using Hess's law often requires you to use *fractions* as coefficients. Remember that a balanced chemical equation can be thought of as an algebraic equality—and that you can multiply through the entire equation, including the enthalpy of reaction, by any number you choose.

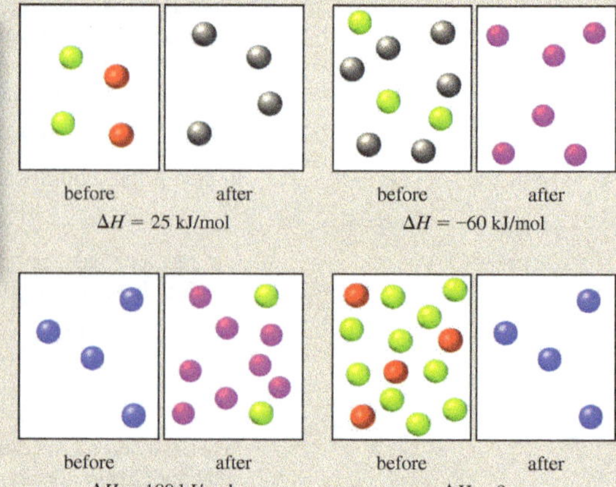

Practice Problem ATTEMPT Use the thermochemical equations provided in Sample Problem 5.7 to determine the enthalpy change for the reaction $2NO(g) + 4O(g) \longrightarrow 2NO_2(g) + O_2(g)$.

Practice Problem BUILD Use the thermochemical equations provided in Sample Problem 5.7 to determine the enthalpy change for the reaction $2NO_2(g) \longrightarrow 2NO(g) + O_2(g)$.

Practice Problem CONCEPTUALIZE The diagrams shown are representations of four systems before and after reactions involving five different chemical species—each represented by a different color sphere. The ΔH values are given for the first three. Determine ΔH for the last reaction.

CHECKPOINT – SECTION 5.5 Hess's Law

5.5.1 Given the following information:

$2H_2(g) + O_2(g) \longrightarrow 2H_2O(g) \quad \Delta H = -483.6$ kJ/mol
$3O_2(g) \longrightarrow 2O_3(g) \quad \Delta H = +284.6$ kJ/mol

what is ΔH for $3H_2(g) + O_3(g) \longrightarrow 3H_2O(g)$?

a) −199 kJ/mol
b) −1010 kJ/mol
c) −867.7 kJ/mol
d) +768.2 kJ/mol
e) −440.8 kJ/mol

5.5.2 Given the following information:

$P_4(s) + 3O_2(g) \longrightarrow P_4O_6(s) \quad \Delta H = -1640.1$ kJ/mol
$P_4(s) + 5O_2(g) \longrightarrow P_4O_{10}(s) \quad \Delta H = -2940.1$ kJ/mol

what is the value of ΔH_{rxn} for $P_4O_6(s) + 2O_2(g) \longrightarrow P_4O_{10}(s)$?

a) −1300.0 kJ/mol
b) +4580.2 kJ/mol
c) −4580.2 kJ/mol
d) +982.6 kJ/mol
e) −982.6 kJ/mol

5.5.3 Each diagram shows a system before and after a chemical reaction along with the corresponding enthalpy of reaction. Determine the enthalpy of reaction for the last reaction represented.

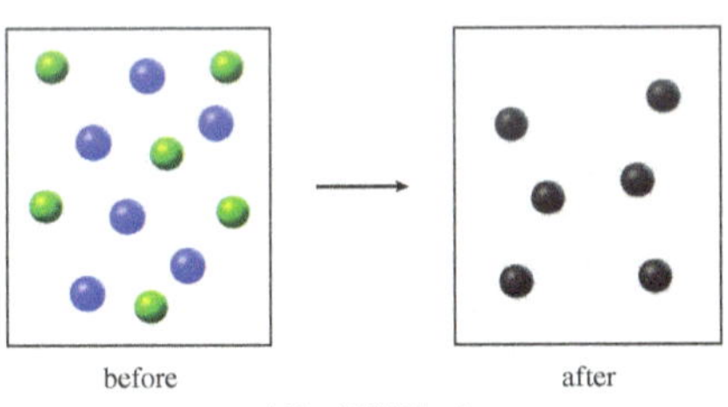

$\Delta H = 120$ kJ/mol

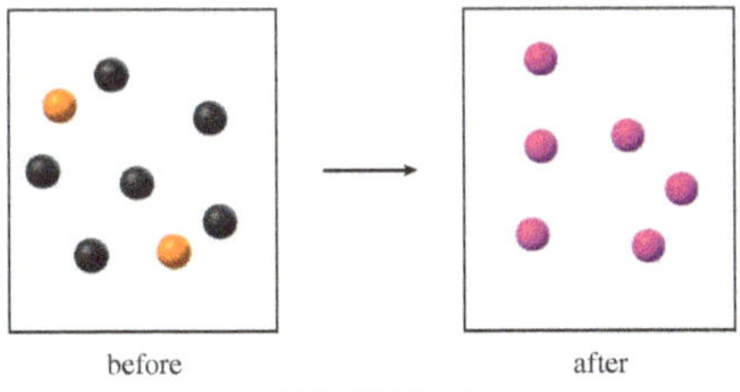

$\Delta H = 45$ kJ/mol

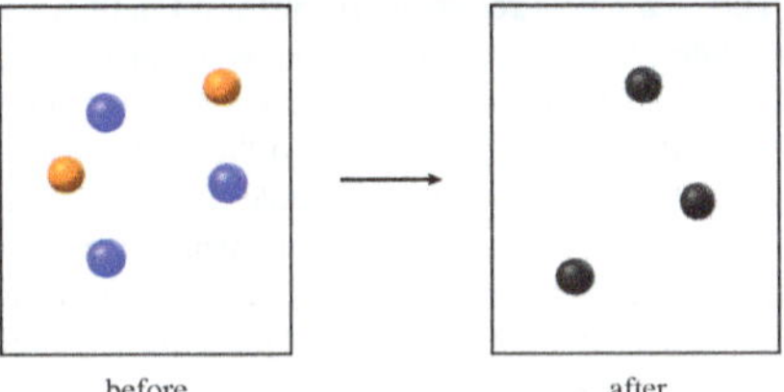

$\Delta H = -75$ kJ/mol

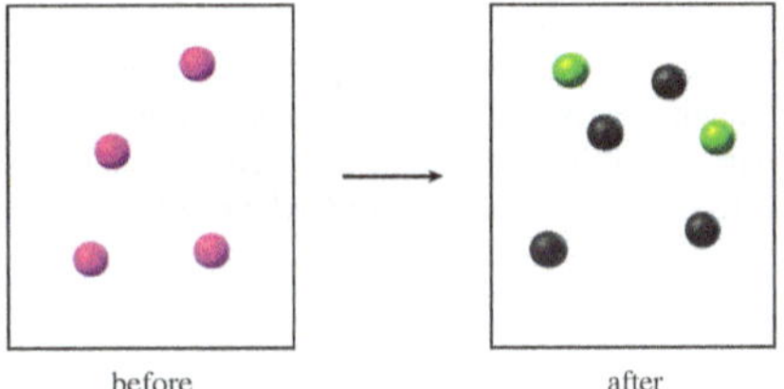

$\Delta H = ?$

a) 240 kJ/mol
b) 90 kJ/mol
c) −120 kJ/mol
d) 295 kJ/mol
e) −315 kJ/mol

5.5.4 Given the thermochemical data,

$A + B \longrightarrow C \quad \Delta H_1 = -100$ kJ/mol
$A + \frac{3}{2}B \longrightarrow D \quad \Delta H_2 = -150$ kJ/mol

determine the enthalpy of reaction for

$2C + B \longrightarrow 2D \quad \Delta H_3 = ?$

a) 500 kJ/mol
b) −500 kJ/mol
c) −50 kJ/mol
d) −100 kJ/mol
e) 200 kJ/mol

5.6 Standard Enthalpies of Formation

So far we have learned that we can determine the enthalpy change that accompanies a reaction by measuring the heat absorbed or released (at constant pressure). According to Equation 5.12, ΔH can also be calculated if we know the enthalpies of all reactants and products. However, there is no way to measure the *absolute* value of the enthalpy of a substance. Only values *relative* to an arbitrary reference can be determined. This problem is similar to the one geographers face in expressing the elevations of specific mountains or valleys. Rather than trying to devise some type of "absolute" elevation scale (perhaps based on the distance from the center of Earth), by common agreement all geographical heights and depths are expressed relative to sea level, an arbitrary reference with a defined elevation of "zero" meters or feet. Similarly, chemists have agreed on an arbitrary reference point for enthalpy.

The "sea level" reference point for all enthalpy expressions is called the ***standard enthalpy of formation*** (ΔH_f°), which is defined as the heat change that results when 1 mole of a compound is formed from its constituent elements in their standard states. The superscripted degree sign denotes standard-state conditions, and the subscripted f stands for *formation*. The phrase "in their standard states" refers to the most stable form of an element under standard conditions, meaning at ordinary atmospheric pressure. The element oxygen, for example, can exist as atomic oxygen (O), diatomic oxygen (O_2), or ozone (O_3). By far the most stable form at ordinary atmospheric pressure, though, is diatomic oxygen. Thus, the standard state of oxygen is $O_2(g)$. Although the standard state does not specify a temperature, we will always use ΔH_f° values measured at 25°C.

Appendix 2 lists the standard enthalpies of formation for a number of elements and compounds. By convention, the standard enthalpy of formation of any element in its most stable form is zero. Again, using the element oxygen as an example, we can write $\Delta H_f^\circ(O_2) = 0$, but $\Delta H_f^\circ(O_3) \neq 0$ and $\Delta H_f^\circ(O) \neq 0$. Similarly, graphite is a more stable allotropic form of carbon than diamond under standard conditions and 25°C, so we have $\Delta H_f^\circ(\text{graphite}) = 0$ and $\Delta H_f^\circ(\text{diamond}) \neq 0$.

The importance of the standard enthalpies of formation is that once we know their values, we can readily calculate the ***standard enthalpy of reaction*** (ΔH_{rxn}°), defined as the enthalpy of a reaction carried out under standard conditions. For example, consider the hypothetical reaction

$$aA + bB \longrightarrow cC + dD$$

where a, b, c, and d are stoichiometric coefficients. For this reaction, ΔH_{rxn}° is given by

$$\Delta H_{rxn}^\circ = [c\Delta H_f^\circ(C) + d\Delta H_f^\circ(D)] - [a\Delta H_f^\circ(A) + b\Delta H_f^\circ(B)] \qquad \textbf{Equation 5.18}$$

We can generalize Equation 5.18 as

$$\Delta H_{rxn}^\circ = \Sigma n\Delta H_f^\circ(\text{products}) - \Sigma m\Delta H_f^\circ(\text{reactants}) \qquad \textbf{Equation 5.19}$$

where m and n are the stoichiometric coefficients for the reactants and products, respectively, and Σ (sigma) means "the sum of." In these calculations, the stoichiometric coefficients are treated as numbers without units. Thus, the result has units of kJ/mol, where again, "per mole" means per mole of reaction as written. To use Equation 5.19 to calculate ΔH_{rxn}°, we must know the ΔH_f° values of the compounds that take part in the reaction. These values, tabulated in Appendix 2, are determined either by the direct method or the indirect method.

The *direct* method of measuring ΔH_f° works for compounds that can be synthesized from their elements easily and safely. Suppose we want to know the enthalpy of formation of carbon dioxide. We must measure the enthalpy of the reaction when carbon (graphite) and molecular oxygen in their standard states are converted to carbon dioxide in its standard state:

$$C(\text{graphite}) + O_2(g) \longrightarrow CO_2(g) \qquad \Delta H_{rxn}^\circ = -393.5 \text{ kJ/mol}$$

We know from experience that this combustion goes to completion. Thus, from Equation 5.19 we can write

$$\Delta H_{rxn}^\circ = \Delta H_f^\circ(CO_2) - [\Delta H_f^\circ(\text{graphite}) + \Delta H_f^\circ(O_2)] = -393.5 \text{ kJ/mol}$$

Because graphite and O_2 are the most stable allotropic forms of their respective elements, $\Delta H_f^\circ(\text{graphite})$ and $\Delta H_f^\circ(O_2)$ are both zero. Therefore,

$$\Delta H_{rxn}^\circ = \Delta H_f^\circ(CO_2) = -393.5 \text{ kJ/mol}$$

or

$$\Delta H_f^\circ(CO_2) = -393.5 \text{ kJ/mol}$$

Arbitrarily assigning a value of zero to ΔH_f° for each element in its standard state does *not* affect the outcome of these calculations. Remember, in thermochemistry we are interested only in enthalpy changes because they can be determined experimentally, whereas the absolute enthalpy values cannot. The choice of a zero "reference level" for enthalpy is intended to simplify the calculations. Referring again to the terrestrial altitude analogy, we find that Mt. Everest (the highest peak in the world) is 8708 ft higher than Mt. Denali (the highest peak in North America). This difference in altitude would be the same whether we had chosen sea level or the center of Earth as our reference elevation.

Other compounds that can be studied by the direct method are SF_6, P_4O_{10}, and CS_2. The equations representing their syntheses are

$$\text{S(rhombic)} + 3\text{F}_2(g) \longrightarrow \text{SF}_6(g)$$
$$\text{P}_4(\text{white}) + 5\text{O}_2(g) \longrightarrow \text{P}_4\text{O}_{10}(s)$$
$$\text{C(graphite)} + 2\text{S(rhombic)} \longrightarrow \text{CS}_2(l)$$

S(rhombic) and P(white) are the most stable allotropes of sulfur and phosphorus, respectively, at 1 atm and 25°C, so their ΔH_f° values are zero.

Sample Problem 5.8 shows how ΔH_f° values can be used to determine ΔH_{rxn}°.

SAMPLE PROBLEM 5.8

Using data from Appendix 2, calculate ΔH_{rxn}° for $Ag^+(aq) + Cl^-(aq) \longrightarrow AgCl(s)$.

Strategy Use Equation 5.19 [$\Delta H_{rxn}^\circ = \Sigma n\Delta H_f^\circ(\text{products}) - \Sigma m\Delta H_f^\circ(\text{reactants})$] and ΔH_f° values from Appendix 2 to calculate ΔH_{rxn}°.

Setup The ΔH_f° values for $Ag^+(aq)$, $Cl^-(aq)$, and $AgCl(s)$ are +105.9, −167.2, and −127.0 kJ/mol, respectively.

Solution Using Equation 5.19,

$$\begin{aligned}\Delta H_{rxn}^\circ &= \Delta H_f^\circ(\text{AgCl}) - [\Delta H_f^\circ(\text{Ag}^+) + \Delta H_f^\circ(\text{Cl}^-)]\\ &= -127.0\text{ kJ/mol} - [(+105.9\text{ kJ/mol}) + (-167.2\text{ kJ/mol})]\\ &= -127.0\text{ kJ/mol} - (-61.3\text{ kJ/mol}) = -65.7\text{ kJ/mol}\end{aligned}$$

THINK ABOUT IT

Watch out for misplaced or missing minus signs. This is an easy place to lose track of them.

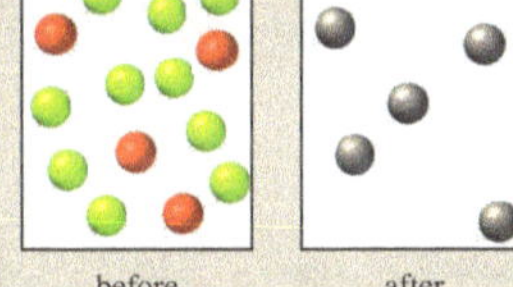

Species	ΔH_f° (kJ/mol)
gray	78.0
red	−188.5
green	106.5

Practice Problem ATTEMPT Using data from Appendix 2, calculate ΔH_{rxn}° for $CaCO_3(s) \longrightarrow CaO(s) + CO_2(g)$.

Practice Problem BUILD Using data from Appendix 2, calculate ΔH_{rxn}° for $2SO(g) + \frac{2}{3}O_3(g) \longrightarrow 2SO_2(g)$.

Practice Problem CONCEPTUALIZE The diagrams represent a system before and after a chemical reaction. Using the table of ΔH_f° values for the species involved in the reaction, determine ΔH_{rxn}° for the process represented by the diagrams.

Many compounds cannot be synthesized from their elements directly. In some cases, the reaction proceeds too slowly, or side reactions produce substances other than the desired compound. In these cases, ΔH_f° can be determined by an *indirect* approach, using Hess's law. If we know a series of reactions for which ΔH_{rxn}° can be measured, and we can arrange them in such a way as to have them sum to the equation corresponding to the formation of the compound of interest, we can calculate ΔH_f° for the compound.

Sample Problem 5.9 shows how to use Hess's law to calculate the ΔH_f° value by the indirect method for a compound that cannot be produced easily from its constituent elements.

SAMPLE PROBLEM 5.9

Given the following information, calculate the standard enthalpy of formation of acetylene (C_2H_2) from its constituent elements:

$$\text{C(graphite)} + \text{O}_2(g) \longrightarrow \text{CO}_2(g) \qquad \Delta H_{rxn}^\circ = -393.5\text{ kJ/mol} \qquad (1)$$
$$\text{H}_2(g) + \tfrac{1}{2}\text{O}_2(g) \longrightarrow \text{H}_2\text{O}(l) \qquad \Delta H_{rxn}^\circ = -285.8\text{ kJ/mol} \qquad (2)$$
$$2\text{C}_2\text{H}_2(g) + 5\text{O}_2(g) \longrightarrow 4\text{CO}_2(g) + 2\text{H}_2\text{O}(l) \qquad \Delta H_{rxn}^\circ = -2598.8\text{ kJ/mol} \qquad (3)$$

Strategy Arrange the equations that are provided so that they will sum to the desired equation. This may require reversing or multiplying one or more of the equations. For any such change, the corresponding change must also be made to the ΔH_{rxn}° value.

Setup The equation corresponding to the standard enthalpy of formation of acetylene is

$$2\text{C(graphite)} + \text{H}_2(g) \longrightarrow \text{C}_2\text{H}_2(g)$$

We multiply Equation (1) and its ΔH°_{rxn} value by 2:

$$2C(graphite) + 2O_2(g) \longrightarrow 2CO_2(g) \quad \Delta H^\circ_{rxn} = -787.0\ kJ/mol$$

We include Equation (2) and its ΔH°_{rxn} value as is:

$$H_2(g) + \tfrac{1}{2}O_2(g) \longrightarrow H_2O(l) \quad \Delta H^\circ_{rxn} = -285.8\ kJ/mol$$

We reverse Equation (3) and divide it by 2 (i.e., multiply through by $\frac{1}{2}$):

$$2CO_2(g) + H_2O(l) \longrightarrow C_2H_2(g) + \tfrac{5}{2}O_2(g) \quad \Delta H^\circ_{rxn} = +1299.4\ kJ/mol$$

The original ΔH°_{rxn} value of Equation (3) has its sign reversed and it is divided by 2.

Solution Summing the resulting equations and the corresponding ΔH°_{rxn} values:

$$\begin{aligned} 2C(graphite) + \cancel{2O_2(g)} &\longrightarrow \cancel{2CO_2(g)} & \Delta H^\circ_{rxn} &= -787.0\ kJ/mol \\ H_2(g) + \cancel{\tfrac{1}{2}O_2(g)} &\longrightarrow \cancel{H_2O(l)} & \Delta H^\circ_{rxn} &= -285.8\ kJ/mol \\ + \cancel{2CO_2(g)} + \cancel{H_2O(l)} &\longrightarrow C_2H_2(g) + \cancel{\tfrac{5}{2}O_2(g)} & \Delta H^\circ_{rxn} &= +1299.4\ kJ/mol \\ \hline 2C(graphite) + H_2(g) &\longrightarrow C_2H_2(g) & \Delta H^\circ_f &= +226.6\ kJ/mol \end{aligned}$$

THINK ABOUT IT

Watch out for misplaced or missing minus signs. This is an easy place to lose track of them.

Practice Problem ATTEMPT Use the following data to calculate ΔH°_f for $CS_2(l)$:

$$C(graphite) + O_2(g) \longrightarrow CO_2(g) \quad \Delta H^\circ_{rxn} = -393.5\ kJ/mol$$
$$S(rhombic) + O_2(g) \longrightarrow SO_2(g) \quad \Delta H^\circ_{rxn} = -296.4\ kJ/mol$$
$$CS_2(l) + 3O_2(g) \longrightarrow CO_2(g) + 2SO_2(g) \quad \Delta H^\circ_{rxn} = -1073.6\ kJ/mol$$

Practice Problem BUILD ΔH°_f of hydrogen chloride [$HCl(g)$] is −92.3 kJ/mol. Given the following data, determine the identity of the two missing products and calculate ΔH°_{rxn} for Equation 3. [*Hint:* Start by writing the chemical equation that corresponds to ΔH°_f for $HCl(g)$.]

$$N_2(g) + 4H_2(g) + Cl_2(g) \longrightarrow 2NH_4Cl(s) \quad \Delta H^\circ_{rxn} = -630.78\ kJ/mol \quad (1)$$
$$N_2(g) + 3H_2(g) \longrightarrow 2NH_3(g) \quad \Delta H^\circ_{rxn} = -92.6\ kJ/mol \quad (2)$$
$$NH_4Cl(s) \longrightarrow \quad (3)$$

Practice Problem CONCEPTUALIZE The diagrams represent a system before and after a chemical reaction for which ΔH°_{rxn} is −2624.9 kJ/mol. Use this information to complete the table of ΔH°_f values for the species involved in the reaction.

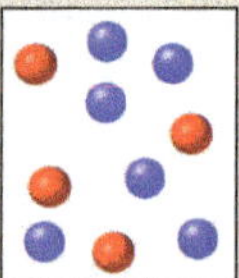
before

after

Species	ΔH°_f (kJ/mol)
blue sphere	−148.7
red sphere	?
black sphere	255.1

CHECKPOINT – SECTION 5.6 Standard Enthalpies of Formation

5.6.1 Using data from Appendix 2, calculate ΔH°_{rxn} for $H_2(g) + F_2(g) \longrightarrow 2HF(g)$.

a) −271.6 kJ/mol
b) −543.2 kJ/mol
c) +271.6 kJ/mol
d) 0 kJ/mol
e) −135.8 kJ/mol

5.6.2 Using data from Appendix 2, calculate ΔH°_{rxn} for $2NO_2(g) \longrightarrow N_2O_4(g)$.

a) −24.19 kJ/mol
b) −33.85 kJ/mol
c) +9.66 kJ/mol
d) +67.7 kJ/mol
e) −58.04 kJ/mol

5.6.3 Which of the following ΔH°_{rxn} values is a ΔH°_f value? (Select all that apply.)

a) $H_2(g) + Br_2(l) \longrightarrow 2HBr(g)$
$\Delta H^\circ_{rxn} = -72.4$ kJ/mol

b) $4Al(s) + 3O_2(g) \longrightarrow 2Al_2O_3(s)$
$\Delta H^\circ_{rxn} = -3339.6$ kJ/mol

c) $Ag(s) + \frac{1}{2}Cl_2(g) \longrightarrow AgCl(s)$
$\Delta H^\circ_{rxn} = -127.0$ kJ/mol

d) $Cu^{2+}(aq) + SO_4^{2-}(aq) \longrightarrow CuSO_4(s)$
$\Delta H^\circ_{rxn} = +73.25$ kJ/mol

e) $\frac{1}{2}H_2(g) + \frac{1}{2}N_2(g) + \frac{3}{2}O_2(g) \longrightarrow HNO_3(l)$
$\Delta H^\circ_{rxn} = -173.2$ kJ/mol

5.6.4 Using the following data, calculate ΔH°_f for $CO(g)$:

$$C(graphite) + O_2(g) \longrightarrow CO_2(g) \quad \Delta H^\circ_{rxn} = -393.5\ kJ/mol$$
$$CO(g) + \tfrac{1}{2}O_2(g) \longrightarrow CO_2(g) \quad \Delta H^\circ_{rxn} = -283.0\ kJ/mol$$

a) −393.5 kJ/mol
b) −676.5 kJ/mol
c) +676.5 kJ/mol
d) +110.5 kJ/mol
e) −110.5 kJ/mol

Chapter Summary

Section 5.1

- ***Energy*** is the capacity to do work or transfer heat. Energy may be ***kinetic energy*** (the energy associated with *motion*) or ***potential energy*** (energy possessed by virtue of *position*). ***Thermal energy*** is a form of kinetic energy. ***Chemical energy*** and ***electrostatic energy*** are forms of potential energy.
- The ***law of conservation of energy*** states that energy can neither be created nor destroyed. The SI unit of energy is the ***joule*** (J).
- The ***system*** is the particular part of the universe that we are interested in studying—such as the reactants and products in a chemical reaction. The term ***surroundings*** refers to the rest of the universe. System + surroundings = universe.
- ***Heat*** refers to the flow of thermal energy between two bodies at different temperatures. ***Thermochemistry*** is the study of the heat associated with chemical reactions and physical processes.
- In an ***exothermic process,*** heat is released to the surroundings, so the energy of the system decreases. In an ***endothermic process,*** heat is absorbed from the surroundings, so the energy of the system increases.

Section 5.2

- ***Thermodynamics*** is the study of the conversions among different types of energy. Thermochemistry is a branch of thermodynamics.
- An ***open system*** is one that can exchange both matter and energy with its surroundings. A ***closed system*** is one that can exchange energy but not matter with its surroundings. An ***isolated system*** is one that cannot exchange either energy or matter with its surroundings.
- The ***state of a system*** is defined by the values of all relevant macroscopic properties, such as temperature, volume, and pressure. A ***state function*** is one whose value depends only on the state of the system and not on how that state was achieved. State functions include energy, pressure, volume, and temperature.
- The ***first law of thermodynamics*** states that energy cannot be created or destroyed, but it can be changed from one form to another. The first law of thermodynamics is based on the law of conservation of energy.

Section 5.3

- ***Enthalpy*** (H) is the heat exchanged between the system and surroundings at constant pressure. It is a state function. ***Enthalpy of reaction*** (ΔH_{rxn}) is the heat exchanged at constant pressure for a specific reaction.
- A ***thermochemical equation*** is a balanced chemical equation for which the enthalpy change (ΔH_{rxn}) is given.

Section 5.4

- ***Calorimetry*** is the science of measuring temperature changes to determine heats associated with chemical reactions. Calorimetry may be carried out at constant pressure (in a coffee-cup calorimeter) or at constant volume (in a bomb calorimeter).
- The ***specific heat*** (s) of a substance is the amount of heat required to increase the temperature of 1 g of the substance by 1°C. The ***heat capacity*** (C) of an object is the amount of heat required to increase the temperature of the object by 1°C.

Section 5.5

- ***Hess's law*** states that the enthalpy change for a reaction that occurs in a series of steps is equal to the sum of the enthalpy changes of the individual steps. Hess's law is valid because enthalpy is a state function.

Section 5.6

- The ***standard enthalpy of formation*** (ΔH_f°) is the enthalpy change associated with the formation of 1 mole of a substance from its constituent elements, each in their standard states. The ***standard enthalpy of reaction*** ($\Delta H_{\text{rxn}}^\circ$) can be calculated for any reaction using tabulated standard enthalpies of formation (ΔH_f°) of the products and reactants.

Key Words

Calorimetry, 200
Chemical energy, 188
Closed system, 192
Electrostatic energy, 188
Endothermic process, 189
Energy, 188
Enthalpy, 197
Enthalpy of reaction, 197
Exothermic process, 189
First law of thermodynamics, 193
Heat, 189
Heat capacity, 200
Hess's law, 210
Isolated system, 192
Joule, 189
Kinetic energy, 188
Law of conservation of energy, 188
Open system, 191
Potential energy, 188
Specific heat, 200
Standard enthalpy of formation, 213
Standard enthalpy of reaction, 213
State function, 192
State of a system, 192
Surroundings, 188
System, 188
Thermal energy, 188
Thermochemical equation, 198
Thermochemistry, 189
Thermodynamics, 191

Key Equations

5.1 $E_k = \frac{1}{2}mu^2$	The kinetic energy of a moving object is calculated using the mass (m) and velocity (u) of the object.
5.2 $E_{el} \propto \frac{Q_1Q_2}{d}$	The electrostatic potential energy (E_{el}) between two charged objects is calculated using the magnitudes of charge (Q_1 and Q_2) and the distance (d) between the charges.
5.3 $\Delta U = q + w$	The change in internal energy of a system (ΔU) is the sum of heat (q) and the work (w) associated with a process. Proper sign conventions must be used for heat and work (Table 5.1).
5.4 $w = -P\Delta V$	Pressure-volume work done by (or on) a system is calculated using the external pressure (P) and the change in volume (ΔV).
5.5 $\Delta U = q - P\Delta V$	The change in internal energy of a system (ΔU) is equal to heat (q) minus pressure-volume work ($P\Delta V$).
5.6 $q_V = \Delta U$	Heat given off (or absorbed) by a system at constant volume (q_V) is equal to the change in internal energy (ΔU).
5.7 $q_P = \Delta U + P\Delta V$	Heat given off (or absorbed) by a system at constant pressure (q_P) is equal to the sum of change in internal energy (ΔU) and pressure-volume work ($P\Delta V$).
5.8 $H = U + PV$	Enthalpy (H) is equal to the sum of internal energy (U) and pressure-volume work ($P\Delta V$).
5.9 $\Delta H = \Delta U + \Delta(PV)$	The change in enthalpy (ΔH) is equal to the sum of change in internal energy (ΔU) and change in the product of pressure and volume [$\Delta(PV)$].
5.10 $\Delta H = \Delta U + P\Delta V$	The change in enthalpy (ΔH) is equal to the sum of change in internal energy (ΔU) and the product of external pressure (P) and change volume (ΔV).
5.11 $q_P = \Delta H$	Heat given off (or absorbed) by a process at constant pressure (q_P) is equal to change in enthalpy (ΔH).
5.12 $\Delta H = H(\text{products}) - H(\text{reactants})$	Enthalpy change for a reaction (ΔH) is the difference between the enthalpy of products [H(products)] and enthalpy of reactants [H(reactants)], although this is not the equation generally used to calculate enthalpy changes because the absolute values of enthalpy are not known.
5.13 $q = sm\Delta T$	Heat given off (or absorbed) by a substance (q) is equal to the product of specific heat of the substance (s), mass of the substance (m), and the change in temperature (ΔT).
5.14 $q = C\Delta T$	Heat given off (or absorbed) by an object (q) is equal to the product of specific heat of the object (C) and the change in temperature (ΔT).
5.15 $q_{sys} = -sm\Delta T$	Heat given off (or absorbed) by a system (q_{sys}) is equal in magnitude and opposite in sign to the heat given off or absorbed by the surroundings.
5.16 $q_{cal} = C_{cal}\Delta T$	Heat given off (or absorbed) by a calorimeter (q_{cal}) is equal to the product of heat capacity of the calorimeter (C_{cal}) and change in temperature (ΔT).
5.17 $q_{rxn} = -C_{cal}\Delta T$	Heat of reaction (q_{rxn}) is equal in magnitude and opposite in sign to heat of calorimeter (q_{cal}).
5.18 $\Delta H^\circ_{rxn} = [c\Delta H^\circ_f(C) + d\Delta H^\circ_f(D)] - [a\Delta H^\circ_f(A) + b\Delta H^\circ_f(B)]$	Standard enthalpy change for a reaction (ΔH°_{rxn}) can be calculated by multiplying the coefficient of each species in the reaction by the corresponding standard enthalpy of formation (ΔH°_f).
5.19 $\Delta H^\circ_{rxn} = \Sigma n\Delta H^\circ_f\,(\text{products}) - \Sigma m\Delta H^\circ_f\,(\text{reactants})$	Standard enthalpy change for a reaction (ΔH°_{rxn}) is the difference between the sum of standard enthalpies of formation of products $\Sigma\Delta H^\circ_f$ (products) and the sum of standard enthalpies of formation of reactants $\Sigma\Delta H^\circ_f$ (reactants).

KEY SKILLS Enthalpy of Reaction

Using tabulated ΔH_f° values, we can calculate the standard enthalpy of reaction (ΔH_{rxn}°) using Equation 5.19:

$$\Delta H_{rxn}^\circ = \sum n\Delta H_f^\circ(\text{products}) - \sum m\Delta H_f^\circ(\text{reactants})$$

This method of calculating thermodynamic quantities such as enthalpy of reaction is important not only in this chapter, but also in Chapters 19 and 20. The following examples illustrate the use of Equation 5.19 and data from Appendix 2. Each example provides a specific reminder of one of the important facets of this approach.

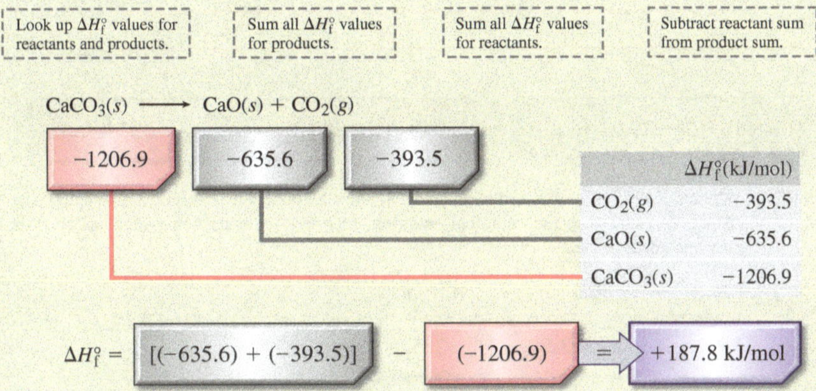

Each ΔH_f° value must be multiplied by the corresponding stoichiometric coefficient in the balanced equation.

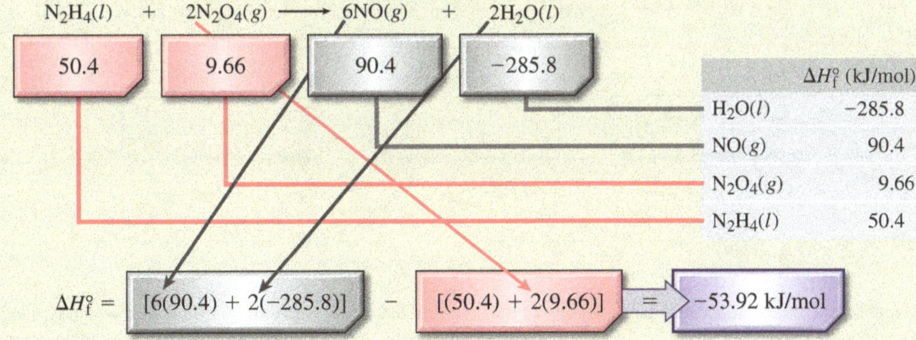

$$Ba(s) + 2H_2O(l) \longrightarrow Ba(OH)_2(aq) + H_2(g)$$

By definition, the standard enthalpy of formation for an element in its standard state is zero. In addition, many tables of thermodynamic data, including Appendix 2, do not contain values for aqueous strong electrolytes such as barium hydroxide. However, the tables do include values for the individual aqueous ions. Therefore, determination of this enthalpy of reaction is facilitated by rewriting the equation with $Ba(OH)_2$ written as separate ions:

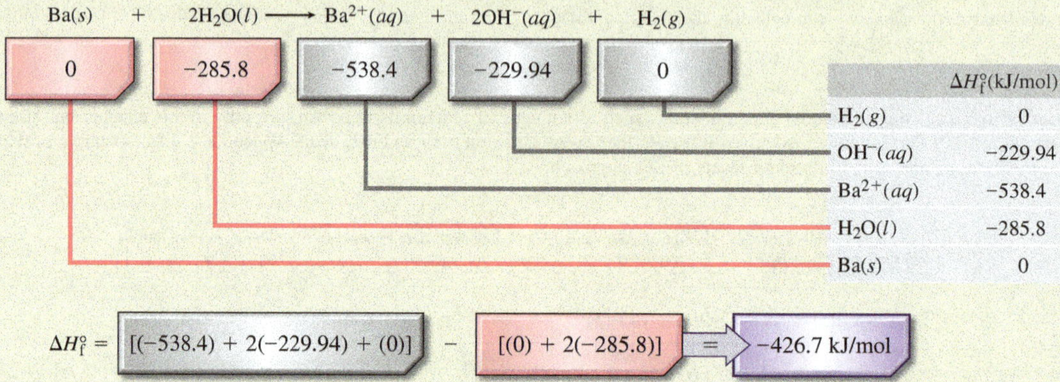

You will find more than one tabulated ΔH_f° value for some substances, such as water. It is important to select the value that corresponds to the phase of matter represented in the chemical equation. In previous examples, water has appeared in the balanced equations as a liquid. It can also appear as a gas.

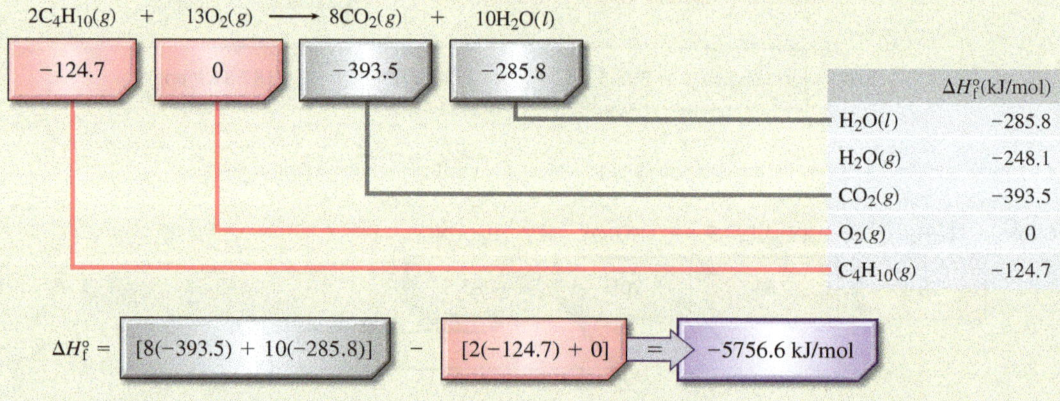

Key Skills Problems

5.1
Using data from Appendix 2, calculate the standard enthalpy of the following reaction:

$$Mg(OH)_2(s) \longrightarrow MgO(s) + H_2O(l)$$

(a) −608.7 kJ/mol (b) −81.1 kJ/mol (c) −37.1 kJ/mol
(d) +81.1 kJ/mol (e) +37.1 kJ/mol

5.2
Using data from Appendix 2, calculate the standard enthalpy of the following reaction:

$$4HBr(g) + O_2(g) \longrightarrow 2H_2O(l) + 2Br_2(l)$$

(a) −426.8 kJ/mol (b) −338.8 kJ/mol (c) −249.6 kJ/mol
(d) +426.8 kJ/mol (e) +338.8 kJ/mol

5.3
Using data from Appendix 2, calculate the standard enthalpy of the following reaction (you must first balance the equation):

$$P(\text{red}) + Cl_2(g) \longrightarrow PCl_3(g)$$

(a) −576.1 kJ/mol (b) −269.7 kJ/mol (c) −539.3 kJ/mol
(d) −602.6 kJ/mol (e) +639.4 kJ/mol

5.4
Using only whole-number coefficients, the combustion of hexane can be represented as:

$$2C_6H_{14}(l) + 19O_2(g) \longrightarrow 12CO_2(g) + 14H_2O(l) \quad \Delta H^\circ = -8388.4 \text{ kJ/mol}$$

Using this and data from Appendix 2, determine the standard enthalpy of formation of hexane.
(a) −334.8 kJ/mol (b) −167.4 kJ/mol (c) −669.6 kJ/mol
(d) +334.8 kJ/mol (e) +669.6 kJ/mol

Questions and Problems

Applying What You've Learned

One of the most popular approaches to dieting in recent years has been to reduce dietary fat. One reason many people want to avoid eating fat is its high Calorie content. Compared to carbohydrates and proteins, each of which contains an average of 4 Calories per gram (17 kJ/g), fat contains 9 Calories per gram (38 kJ/g). Tristearin, a typical fat, is metabolized (or combusted) according to the following equation:

$$C_{57}H_{110}O_6(s) + 81.5O_2(g) \longrightarrow 57CO_2(g) + 55H_2O(l) \qquad \Delta H^\circ = -37{,}760 \text{ kJ/mol}$$

Although the food industry has succeeded in producing low-fat versions of nearly everything we eat, it has thus far failed to produce a palatable low-fat doughnut. The flavor, texture, and what the industry calls "mouth feel" of a doughnut depends largely on the process of deep-fat frying. Fortunately for people in the doughnut business, though, high fat content has not diminished the popularity of doughnuts.

According to information obtained from www.krispykreme.com, a Krispy Kreme original glazed doughnut weighs 52 g and contains 200 Cal and 12 g of fat.

(a) Assuming that the fat in the doughnut is metabolized according to the given equation for tristearin, calculate the number of Calories in the reported 12 g of fat in each doughnut [⏮ Sample Problem 5.3]. (b) If all the energy contained in a Krispy Kreme doughnut (not just in the fat) were transferred to 6.00 kg of water originally at 25.5°C, what would be the final temperature of the water [⏮ Sample Problem 5.4]? (c) When a Krispy Kreme apple fritter weighing 101 g is burned in a bomb calorimeter with C_{cal} = 95.3 kJ/°C, the measured temperature increase is 16.7°C. Calculate the number of Calories in a Krispy Kreme apple fritter [⏮ Sample Problem 5.6]. (d) What would the ΔH° value be for the metabolism of 1 mole of the fat tristearin if the water produced by the reaction were gaseous instead of liquid [⏮ Sample Problem 5.7]? [*Hint:* Use data from Appendix 2 to determine the ΔH° value for the reaction $H_2O(l) \longrightarrow H_2O(g)$ [⏮ Sample Problem 5.8]].

Nutrition Facts

Serving Size 1 donut (about 52g)
Servings Per Container 12

Amount Per Serving	
Calories 200	**Calories From Fat** 100
	%Daily Value*
Total Fat 12g	**18%**
Saturated Fat 3g	**15%**
Trans Fat 4g	
Cholesterol 5mg	**1%**
Sodium 95mg	**4%**
Total Carbohydrate 22g	**7%**
Dietary Fiber <1g	**1%**
Sugars 10g	
Protein 2g	
Vitamin A 0% • Vitamin C	2%
Calcium 6% • Iron	4%

*Percent of Daily Values (DV) are based on a 2,000 calorie diet.

Nutrition facts label for Krispy Kreme original glazed doughnuts.

Credit: © *David A. Tietz/Editorial Image, LLC.*

SECTION 5.1: ENERGY AND ENERGY CHANGES

Review Questions

5.1 Define these terms: *system, surroundings, thermal energy, chemical energy, potential energy, kinetic energy, law of conservation of energy.*

5.2 What is heat? How does heat differ from thermal energy? Under what condition is heat transferred from one system to another?

5.3 What are the units for energy commonly employed in chemistry?

5.4 A truck initially traveling at 60 km/h is brought to a complete stop at a traffic light. Does this change violate the law of conservation of energy? Explain.

5.5 These are various forms of energy: chemical, heat, light, mechanical, and electrical. Suggest several ways of converting one form of energy to another.

5.6 Define these terms: *thermochemistry, exothermic process, endothermic process.*

Conceptual Problems

5.7 Stoichiometry is based on the law of conservation of mass. On what law is thermochemistry based?

5.8 Describe the interconversions of forms of energy occurring in these processes: (a) You throw a softball up into the air and catch it. (b) You switch on a flashlight. (c) You ride the ski lift to the top of the hill and then ski down. (d) You strike a match and let it burn completely.

5.9 Decomposition reactions are usually endothermic, whereas combination reactions are usually exothermic. Give a qualitative explanation for these trends.

5.10 For charges of +1 and −1, separated by a distance of d, the electrostatic potential energy is E. In terms of E, determine the electrostatic potential energy between each of the pairs of charges shown.

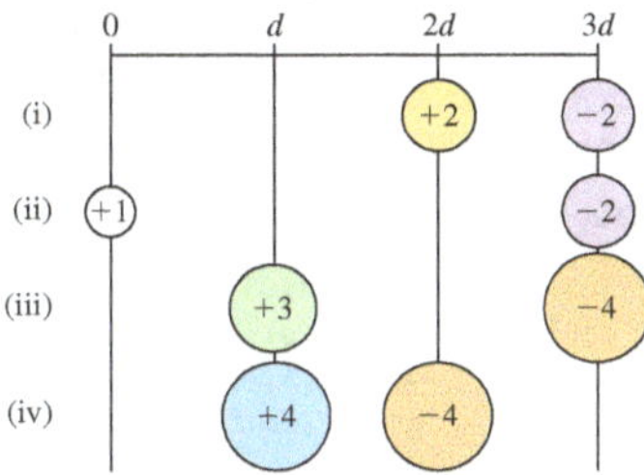

SECTION 5.2: INTRODUCTION TO THERMODYNAMICS

Review Questions

5.11 On what law is the first law of thermodynamics based? Explain the sign conventions in the equation

$$\Delta U = q + w$$

5.12 Explain what is meant by a state function. Give two examples of quantities that are state functions and two that are not state functions.

Computational Problems

5.13 The work done to compress a gas is 47 J. As a result, 93 J of heat is given off to the surroundings. Calculate the change in internal energy of the gas.

5.14 In a gas expansion, 87 J of heat is released to the surroundings and the energy of the system decreases by 128 J. Calculate the work done.

5.15 Calculate w, and determine whether work is done *by* the system or *on* the system when 415 J of heat is released and $\Delta U = 510$ J.

5.16 Calculate q, and determine whether heat is absorbed or released when a system does work on the surroundings equal to 64 J and $\Delta U = 213$ J.

Conceptual Problems

Use the following diagrams for Problems 5.17 and 5.18.

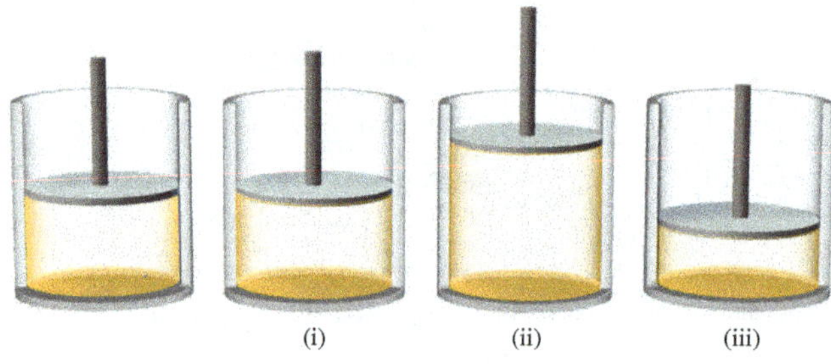

5.17 The diagram on the far left shows a system before a process. Determine which of the diagrams on the right could represent the system after it undergoes a process in which (a) the system absorbs heat and ΔU is negative; (b) the system absorbs heat and does work on the surroundings; (c) the system releases heat and does work on the surroundings.

5.18 The diagram on the far left shows a system before a process. Determine which of the diagrams on the right could represent the system after it undergoes a process in which (a) work is done on the system and ΔU is negative; (b) the system releases heat and ΔU is positive; (c) the system absorbs heat and ΔU is positive.

SECTION 5.3: ENTHALPY

Review Questions

5.19 Consider these changes.
(a) $Hg(l) \longrightarrow Hg(g)$
(b) $3O_2(g) \longrightarrow 2O_3(g)$
(c) $CuSO_4 \cdot 5H_2O(s) \longrightarrow CuSO_4(s) + 5H_2O(g)$
(d) $H_2(g) + F_2(g) \longrightarrow 2HF(g)$
At constant pressure, in which of the reactions is work done by the system on the surroundings? By the surroundings on the system? In which of them is no work done?

5.20 Define these terms: *enthalpy* and *enthalpy of reaction*. Under what condition is the heat of a reaction equal to the enthalpy change of the same reaction?

5.21 In writing thermochemical equations, why is it important to indicate the physical state (i.e., gaseous, liquid, solid, or aqueous) of each substance?

5.22 Explain the meaning of this thermochemical equation:

$$4NH_3(g) + 5O_2(g) \longrightarrow 4NO(g) + 6H_2O(g) \qquad \Delta H = -904 \text{ kJ/mol}$$

5.23 Consider this reaction:

$$2CH_3OH(l) + 3O_2(g) \longrightarrow 4H_2O(l) + 2CO_2(g) \qquad \Delta H = -1452.8 \text{ kJ/mol}$$

What is the value of ΔH if (a) the equation is multiplied throughout by 2; (b) the direction of the reaction is reversed so that the products become the reactants, and vice versa; (c) water vapor instead of liquid water is formed as the product?

Computational Problems

5.24 A sample of nitrogen gas expands in volume from 1.6 to 5.4 L at constant temperature. Calculate the work done in joules if the gas expands (a) against a vacuum, (b) against a constant pressure of 0.80 atm, and (c) against a constant pressure of 3.7 atm. (See Equation 5.4. (1 L · atm = 101.3 J)

5.25 A gas expands in volume from 26.7 to 89.3 mL at constant temperature. Calculate the work done (in joules) if the gas expands (a) against a vacuum, (b) against a constant pressure of 1.5 atm, and (c) against a constant pressure of 2.8 atm. (1 L · atm = 101.3 J)

5.26 A gas expands and does PV work on the surroundings equal to 325 J. At the same time, it absorbs 127 J of heat from the surroundings. Calculate the change in energy of the gas.

5.27 The first step in the industrial recovery of zinc from the zinc sulfide ore is roasting; that is, the conversion of ZnS to ZnO by heating:

$$2ZnS(s) + 3O_2(g) \longrightarrow 2ZnO(s) + 2SO_2(g) \qquad \Delta H = -879 \text{ kJ/mol}$$

Calculate the heat evolved (in kJ) per gram of ZnS roasted.

5.28 Determine the amount of heat (in kJ) given off when 1.26×10^4 g of NO_2 are produced according to the equation

$$2NO(g) + O_2(g) \longrightarrow 2NO_2(g) \qquad \Delta H = -114.6 \text{ kJ/mol}$$

5.29 Consider the reaction

$$2H_2O(g) \longrightarrow 2H_2(g) + O_2(g) \qquad \Delta H = +483.6 \text{ kJ/mol}$$

at a certain temperature. If the increase in volume is 32.7 L against an external pressure of 1.00 atm, calculate ΔU for this reaction. (1 L · atm = 101.3 J)

5.30 Consider the reaction

$$H_2(g) + Cl_2(g) \longrightarrow 2HCl(g) \qquad \Delta H = -184.6 \text{ kJ/mol}$$

If 3 moles of H_2 react with 3 moles of Cl_2 to form HCl, calculate the work done (in joules) against a pressure of 1.0 atm. What is ΔU for this reaction? Assume the reaction goes to completion and that $\Delta V = 0$. (1 L · atm = 101.3 J)

Conceptual Problems

5.31 The following diagrams represent systems before and after reaction for two related chemical processes. ΔH for the first reaction is -595.8 kJ/mol. Determine the value of ΔH for the second reaction.

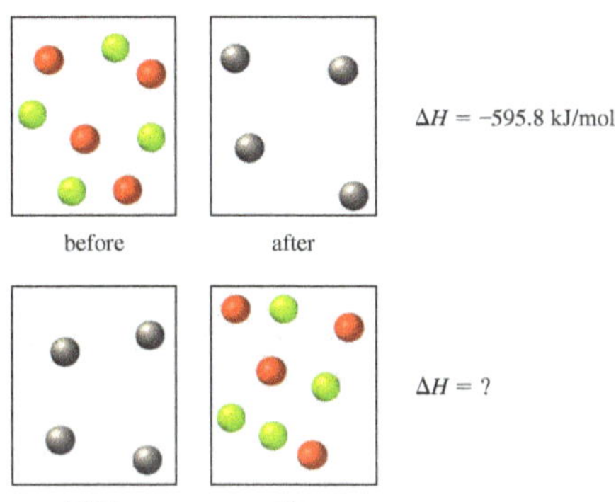

5.32 For most biological processes, the changes in internal energy are approximately equal to the changes in enthalpy. Explain.

SECTION 5.4: CALORIMETRY

Visualizing Chemistry
Figure 5.9 and Figure 5.10

VC 5.1 Referring to Figure 5.9, which of the following would result in the calculated value of ΔH_{rxn} being too high?
a) Spilling some of one of the reactant solutions before adding it to the calorimeter.
b) Reading the final temperature before it reached its maximum value.
c) Misreading the thermometer at the beginning of the experiment and recording too low an initial temperature.

VC 5.2 How would the ΔH_{rxn} calculated in Figure 5.9 be affected if the concentration of one of the reactant solutions were twice as high as it was supposed to be?
a) The calculated ΔH_{rxn} would not be affected.
b) The calculated ΔH_{rxn} would be too low.
c) The calculated ΔH_{rxn} would be too high.

VC 5.3 For an exothermic reaction like the one depicted in Figure 5.9, if the heat capacity of the calorimeter is not negligibly small, the heat absorbed by the water will be ________ the heat given off by the reaction.
a) greater than
b) less than
c) equal to

VC 5.4 Referring to Figure 5.9, how would the results of the experiment have been different if the reaction had been endothermic?
a) The results would have been the same.
b) There would have been a smaller temperature increase.
c) There would have been a temperature decrease.

VC 5.5 What would happen to the specific heat calculated in Figure 5.10 if some of the warm metal shot were lost during the transfer to the calorimeter?
a) It would not affect the calculated value of specific heat.
b) It would cause the calculated value of specific heat to be too high.
c) It would cause the calculated value of specific heat to be too low.

VC 5.6 What would happen to the specific heat calculated in Figure 5.10 if the test tube containing the metal shot were left in the boiling water for longer than the recommended time?
a) It would not affect the calculated value of specific heat.
b) It would cause the calculated value of specific heat to be too high.
c) It would cause the calculated value of specific heat to be too low.

VC 5.7 What would happen to the specific heat calculated in Figure 5.10 if some of the water were spilled prior to being added to the calorimeter?
a) It would not affect the calculated value of specific heat.
b) It would cause the calculated value of specific heat to be too high.
c) It would cause the calculated value of specific heat to be too low.

VC 5.8 Referring to the process depicted in Figure 5.10, which of the following must be known precisely for the calculated specific heat to be accurate?
a) The mass of the boiling water.
b) The temperature of the metal shot before it is immersed in the boiling water.
c) The mass of the water that is added to the calorimeter.

Review Questions

5.33 What is the difference between specific heat and heat capacity? What are the units for these two quantities? Which is the intensive property and which is the extensive property?

5.34 Define *calorimetry* and describe two commonly used calorimeters. In a calorimetric measurement, why is it important that we know the heat capacity of the calorimeter? How is this value determined?

Computational Problems

5.35 A 6.22-kg piece of copper metal is heated from 20.5°C to 324.3°C. Calculate the heat absorbed (in kJ) by the metal.

5.36 Calculate the amount of heat liberated (in kJ) from 366 g of mercury when it cools from 77.0°C to 12.0°C.

5.37 A sheet of gold weighing 10.0 g and at a temperature of 18.0°C is placed flat on a sheet of iron weighing 20.0 g and at a temperature of 55.6°C. What is the final temperature of the combined metals? Assume that no heat is lost to the surroundings. (*Hint:* The heat gained by the gold must be equal to the heat lost by the iron. The specific heats of the metals are given in Table 5.2.)

5.38 A 0.1375-g sample of solid magnesium is burned in a constant-volume bomb calorimeter that has a heat capacity of 3024 J/°C. The temperature increases by 1.126°C. Calculate the heat given off by the burning Mg, in kJ/g and in kJ/mol.

5.39 A quantity of 2.00×10^2 mL of 0.862 *M* HCl is mixed with 2.00×10^2 mL of 0.431 *M* $Ba(OH)_2$ in a constant-pressure calorimeter of negligible heat capacity. The initial temperature of the HCl and $Ba(OH)_2$ solutions is the same at 20.48°C. For the process

$$H^+(aq) + OH^-(aq) \longrightarrow H_2O(l)$$

the heat of neutralization is −56.2 kJ/mol. What is the final temperature of the mixed solution? Assume the specific heat of the solution is the same as that for pure water.

5.40 A 50.75-g sample of water at 75.6°C is added to a sample of water at 24.1°C in a constant-pressure calorimeter. If the final temperature of the combined water is 39.4°C and the heat capacity of the calorimeter is 26.3 J/°C, calculate the mass of the water originally in the calorimeter.

5.41 A 25.95-g sample of methanol at 35.6°C is added to a 38.65-g sample of ethanol at 24.7°C in a constant-pressure calorimeter. If the final temperature of the combined liquids is 28.5°C and the heat capacity of the calorimeter is 19.3 J/°C, determine the specific heat of methanol.

5.42 75.0 mL of 1.75 *M* HCl and 125.0 mL of 1.50 *M* NaOH are combined in a constant-pressure calorimeter. Both solutions are initially at 23.9°C. Calculate the final temperature of the combined solutions. (Use the data from Table 5.3. Assume that the mass of the combined solutions is 200.0 g and that the solution's specific heat is the same as that for water, 4.184 J/g · °C.) The heat capacity of the calorimeter is negligibly small.

Conceptual Problems

5.43 Consider two metals A and B, each having a mass of 100 g and an initial temperature of 20°C. The specific heat of A is larger than that of B. Under the same heating conditions, which metal would take longer to reach a temperature of 21°C?

5.44 Consider the following data:

Metal	Al	Cu
Mass (g)	10	30
Specific heat (J/g · °C)	0.900	0.385
Temperature (°C)	40	60

When these two metals are placed in contact, which of the following will take place?

(a) Heat will flow from Al to Cu because Al has a larger specific heat.

(b) Heat will flow from Cu to Al because Cu has a larger mass.

(c) Heat will flow from Cu to Al because Cu has a larger heat capacity.

(d) Heat will flow from Cu to Al because Cu is at a higher temperature.

(e) No heat will flow in either direction.

SECTION 5.5: HESS'S LAW

Review Questions

5.45 State Hess's law. Explain, with one example, the usefulness of Hess's law in thermochemistry.

5.46 Describe how chemists use Hess's law to determine the ΔH_f° of a compound by measuring its heat (enthalpy) of combustion.

Computational Problems

5.47 Given the thermochemical data,

$$A + 6B \longrightarrow 4C \qquad \Delta H_1 = -1200 \text{ kJ/mol}$$
$$C + B \longrightarrow D \qquad \Delta H_1 = -150 \text{ kJ/mol}$$

Determine the enthalpy change for each of the following:

a) $D \longrightarrow C + B$

b) $2C \longrightarrow \frac{1}{2}A + 3B$

c) $3D + \frac{1}{2}A \longrightarrow 5C$

d) $2D \longrightarrow 2C + 2B$

e) $6D + A \longrightarrow 10C$

5.48 Given the thermochemical data,

$$A + B \longrightarrow 2C \qquad \Delta H_1 = 600 \text{ kJ/mol}$$
$$2C + D \longrightarrow 2E \qquad \Delta H_1 = 210 \text{ kJ/mol}$$

Determine the enthalpy change for each of the following:

a) $4E \longrightarrow 4C + 2D$

b) $A + B + D \longrightarrow 2E$

c) $C \longrightarrow \frac{1}{2}A + \frac{1}{2}B$

d) $2C + 2E \longrightarrow 2A + 2B + D$

e) $E \longrightarrow \frac{1}{2}A + \frac{1}{2}B + \frac{1}{2}D$

5.49 From these data,

$$S(\text{rhombic}) + O_2(g) \longrightarrow SO_2(g) \qquad \Delta H^\circ_{rxn} = -296.4 \text{ kJ/mol}$$

$$S(\text{monoclinic}) + O_2(g) \longrightarrow SO_2(g) \qquad \Delta H^\circ_{rxn} = -296.7 \text{ kJ/mol}$$

calculate the enthalpy change for the transformation

$$S(\text{rhombic}) \longrightarrow S(\text{monoclinic})$$

(Monoclinic and rhombic are different allotropic forms of elemental sulfur.)

5.50 From the following data,

$$C(\text{graphite}) + O_2(g) \longrightarrow CO_2(g) \qquad \Delta H^\circ_{rxn} = -393.5 \text{ kJ/mol}$$

$$H_2(g) + \tfrac{1}{2}O_2(g) \longrightarrow H_2O(l) \qquad \Delta H^\circ_{rxn} = -285.8 \text{ kJ/mol}$$

$$2C_2H_6(g) + 7O_2(g) \longrightarrow 4CO_2(g) + 6H_2O(l) \qquad \Delta H^\circ_{rxn} = -3119.6 \text{ kJ/mol}$$

calculate the enthalpy change for the reaction

$$2C(\text{graphite}) + 3H_2(g) \longrightarrow C_2H_6(g)$$

5.51 From the following heats of combustion,

$$CH_3OH(l) + \tfrac{3}{2}O_2(g) \longrightarrow CO_2(g) + 2H_2O(l) \qquad \Delta H^\circ_{rxn} = -726.4 \text{ kJ/mol}$$

$$C(\text{graphite}) + O_2(g) \longrightarrow CO_2(g) \qquad \Delta H^\circ_{rxn} = -393.5 \text{ kJ/mol}$$

$$H_2(g) + \tfrac{1}{2}O_2(g) \longrightarrow H_2O(l) \qquad \Delta H^\circ_{rxn} = -285.8 \text{ kJ/mol}$$

calculate the enthalpy of formation of methanol (CH_3OH) from its elements:

$$C(\text{graphite}) + 2H_2(g) + \tfrac{1}{2}O_2(g) \longrightarrow CH_3OH(l)$$

5.52 Calculate the standard enthalpy change for the reaction

$$2Al(s) + Fe_2O_3(s) \longrightarrow 2Fe(s) + Al_2O_3(s)$$

given that

$$2Al(s) + \tfrac{3}{2}O_2(g) \longrightarrow Al_2O_3(s) \qquad \Delta H^\circ_{rxn} = -1669.8 \text{ kJ/mol}$$

$$2Fe(s) + \tfrac{3}{2}O_2(g) \longrightarrow Fe_2O_3(s) \qquad \Delta H^\circ_{rxn} = -822.2 \text{ kJ/mol}$$

5.53 Determine the enthalpy change for the gaseous reaction of sulfur dioxide with ozone to form sulfur trioxide given the following thermochemical data:

$$2SO(g) + O_2(g) \longrightarrow 2SO_2(g) \qquad \Delta H^\circ = -602.8 \text{ kJ/mol}$$

$$3SO(g) + 2O_3(g) \longrightarrow 3SO_3(g) \qquad \Delta H^\circ_{rxn} = -1485.03 \text{ kJ/mol}$$

$$\tfrac{3}{2}O_2(g) \longrightarrow O_3(g) \qquad \Delta H^\circ_{rxn} = 142.2 \text{ kJ/mol}$$

5.54 Determine the enthalpy change for the reaction of carbon disulfide and chlorine gases to form carbon tetrachloride and disulfur dichloride liquids given the following thermochemical data:

$$2S(\text{rhombic}) + C(\text{graphite}) \longrightarrow CS_2(g) \qquad \Delta H^\circ = 115.3 \text{ kJ/mol}$$

$$C(\text{graphite}) + 2Cl_2(g) \longrightarrow CCl_4(l) \qquad \Delta H^\circ_{rxn} = -128.2 \text{ kJ/mol}$$

$$2S(\text{rhombic}) + O_2(g) \longrightarrow SO_2(g) \qquad \Delta H^\circ_{rxn} = -296.4 \text{ kJ/mol}$$

$$2S(\text{rhombic}) + Cl_2(g) \longrightarrow S_2Cl_2(l) \qquad \Delta H^\circ_{rxn} = -58.2 \text{ kJ/mol}$$

$$3H_2(g) + S_2Cl_2(l) \longrightarrow 2H_2S(g) + 2HCl(g) \qquad \Delta H^\circ_{rxn} = -166.7 \text{ kJ/mol}$$

Conceptual Problems

5.55 Each diagram shows a system before and after a chemical reaction along with the corresponding enthalpy of reaction. Determine the enthalpy of reaction for the last reaction represented.

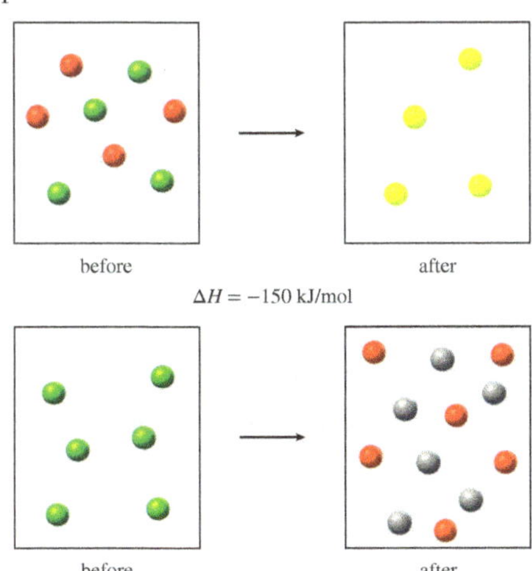

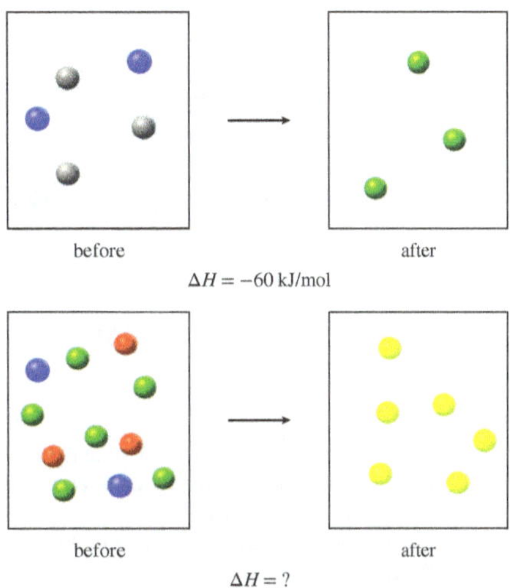

The following diagrams depict three chemical reactions involving five different chemical species—each represented by a different color sphere. Use this information to solve Problems 5.56, 5.57, and 5.58.

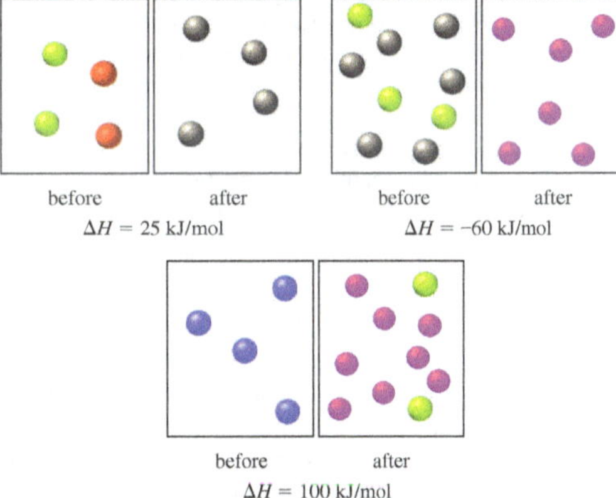

5.56 Determine the value of ΔH for the following reaction:

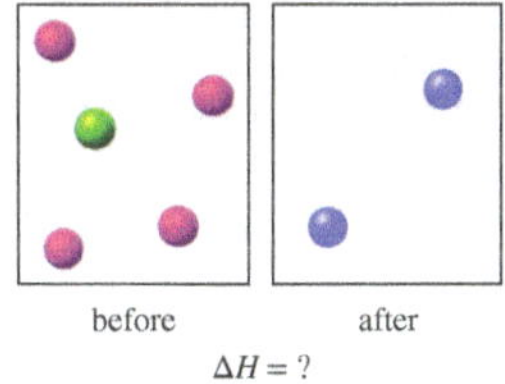

5.57 Determine the value of ΔH for the following reaction:

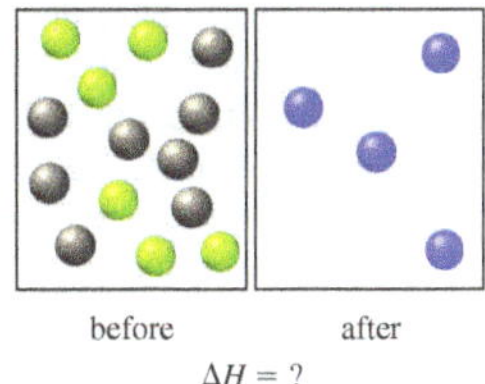

5.58 Determine the value of ΔH for the following reaction:

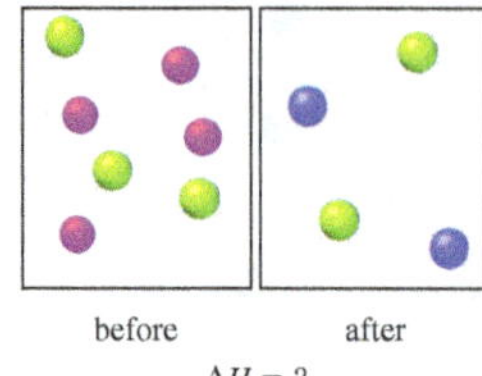

SECTION 5.6: STANDARD ENTHALPIES OF FORMATION

Review Questions

5.59 What is meant by the standard-state condition?

5.60 How are the standard enthalpies of an element and of a compound determined?

5.61 What is meant by the standard enthalpy of a reaction?

5.62 Write the equation for calculating the enthalpy of a reaction. Define all the terms.

Computational Problems

5.63 Which of the following standard enthalpy of formation values is not zero at 25°C: Na(monoclinic), Ne(g), $CH_4(g)$, S_8(monoclinic), Hg(l), H(g)?

5.64 The ΔH_f° values of the two allotropes of oxygen, O_2 and O_3, are 0 and 142.2 kJ/mol, respectively, at 25°C. Which is the more stable form at this temperature?

5.65 Which is the more negative quantity at 25°C: ΔH_f° for $H_2O(l)$ or ΔH_f° for $H_2O(g)$?

5.66 The standard enthalpies of formation of ions in aqueous solutions are obtained by arbitrarily assigning a value of zero to H^+ ions; that is, $\Delta H_f^\circ [H^+(aq)] = 0$. (a) For the following reaction

$$HCl(g) \xrightarrow{H_2O} H^+(aq) + Cl^-(aq) \qquad \Delta H^\circ = -74.9 \text{ kJ/mol}$$

calculate ΔH_f° for the Cl^- ions. (b) Given that ΔH_f° for OH^- ions is −229.6 kJ/mol, calculate the enthalpy of neutralization when 1 mole of a strong monoprotic acid (such as HCl) is titrated by 1 mole of a strong base (such as KOH) at 25°C.

5.67 Calculate the heats of combustion for the following reactions from the standard enthalpies of formation listed in Appendix 2:
(a) $2H_2(g) + O_2(g) \longrightarrow 2H_2O(l)$
(b) $2C_2H_2(g) + 5O_2(g) \longrightarrow 4CO_2(g) + 2H_2O(l)$

5.68 Calculate the heats of combustion for the following reactions from the standard enthalpies of formation listed in Appendix 2:
(a) $C_2H_4(g) + 3O_2(g) \longrightarrow 2CO_2(g) + 2H_2O(l)$
(b) $2H_2S(g) + 3O_2(g) \longrightarrow 2H_2O(l) + 2SO_2(g)$

5.69 Methanol, ethanol, and n-propanol are three common alcohols. When 1.00 g of each of these alcohols is burned in air, heat is liberated as follows: (a) methanol (CH_3OH), −22.6 kJ; (b) ethanol (C_2H_5OH), −29.7 kJ; (c) n-propanol (C_3H_7OH), −33.4 kJ. Calculate the heats of combustion of these alcohols in kJ/mol.

5.70 The standard enthalpy change for the following reaction is 436.4 kJ/mol:

$$H_2(g) \longrightarrow H(g) + H(g)$$

Calculate the standard enthalpy of formation of atomic hydrogen (H).

5.71 From the standard enthalpies of formation, calculate ΔH_{rxn}° for the reaction

$$C_6H_{12}(l) + 9O_2(g) \longrightarrow 6CO_2(g) + 6H_2O(l)$$

For $C_6H_{12}(l)$, $\Delta H_f^\circ = -151.9$ kJ/mol.

5.72 Calculate the heat of decomposition for this process at constant pressure and 25°C:

$$CaCO_3(s) \longrightarrow CaO(s) + CO_2(g)$$

(Look up the standard enthalpy of formation of the reactant and products in Appendix 2.)

5.73 Consider the reaction

$$N_2(g) + 3H_2(g) \longrightarrow 2NH_3(g) \qquad \Delta H = -92.6 \text{ kJ/mol}$$

When 2 mol of N_2 react with 6 mol of H_2 to form 4 mol of NH_3 at 1 atm and a certain temperature, there is a decrease in volume equal to 98 L. Calculate ΔU for this reaction. (The conversion factor is 1 L · atm = 101.3 J.)

5.74 Calculate the heat released when 2.00 L of $Cl_2(g)$ with a density of 1.88 g/L reacts with an excess of sodium metal at 25°C and 1 atm to form sodium chloride.

5.75 Pentaborane-9 (B_5H_9) is a colorless, highly reactive liquid that will burst into flames when exposed to oxygen. The reaction is

$$2B_5H_9(l) + 12O_2(g) \longrightarrow 5B_2O_3(s) + 9H_2O(l)$$

Calculate the kilojoules of heat released per gram of the compound reacted with oxygen. The standard enthalpy of formation of B_5H_9 is 73.2 kJ/mol.

5.76 Determine the amount of heat (in kJ) given off when 1.26×10^4 g of ammonia is produced according to the equation

$$N_2(g) + 3H_2(g) \longrightarrow 2NH_3(g) \qquad \Delta H_{rxn}^\circ = -92.6 \text{ kJ/mol}$$

Assume that the reaction takes place under standard-state conditions at 25°C.

Conceptual Problems

5.77 Predict the value of ΔH_f° (greater than, less than, or equal to zero) for these elements at 25°C: (a) $Br_2(g)$, $Br_2(l)$; (b) $I_2(g)$, $I_2(s)$.

5.78 In general, compounds with negative ΔH_f° values are more stable than those with positive ΔH_f° values. $H_2O_2(l)$ has a negative ΔH_f° (see Appendix 2). Why, then, does $H_2O_2(l)$ have a tendency to decompose to $H_2O(l)$ and $O_2(g)$?

5.79 Suggest ways (with appropriate equations) that would allow you to measure the ΔH_f° values of $Ag_2O(s)$ and $CaCl_2(s)$ from their elements. No calculations are necessary.

5.80 Using the data in Appendix 2, calculate the enthalpy change for the gaseous reaction shown here. (*Hint:* First determine the limiting reactant.)

ADDITIONAL PROBLEMS

5.81 The convention of arbitrarily assigning a zero enthalpy value for the most stable form of each element in the standard state at 25°C is a convenient way of dealing with enthalpies of reactions. Explain why this convention cannot be applied to nuclear reactions.

5.82 Consider the following two reactions:

$$A \longrightarrow 2B \qquad \Delta H^\circ_{rxn} = H_1$$
$$A \longrightarrow C \qquad \Delta H^\circ_{rxn} = H_2$$

Determine the enthalpy change for the process

$$2B \longrightarrow C$$

5.83 The standard enthalpy change ΔH° for the thermal decomposition of silver nitrate according to the following equation is +78.67 kJ:

$$AgNO_3(s) \longrightarrow AgNO_2(s) + \tfrac{1}{2}O_2(g)$$

The standard enthalpy of formation of $AgNO_3(s)$ is −123.02 kJ/mol. Calculate the standard enthalpy of formation of $AgNO_2(s)$.

5.84 Consider the reaction:

$$2Na(s) + 2H_2O(l) \longrightarrow 2NaOH(aq) + H_2(g)$$

When 2 moles of Na react with water at 25°C and 1 atm, the volume of H_2 formed is 24.5 L. Calculate the work done in joules when 0.34 g of Na reacts with water under the same conditions. (The conversion factor is 1 L · atm = 101.3 J.)

5.85 A 44.0-g sample of an unknown metal at 99.0°C was placed in a constant-pressure calorimeter containing 80.0 g of water at 24.0°C. The final temperature of the system was found to be 28.4°C. Calculate the specific heat of the metal. (The heat capacity of the calorimeter is 12.4 J/°C.)

5.86 A student mixes 88.6 g of water at 74.3°C with 57.9 g of water at 24.8°C in an insulated flask. What is the final temperature of the combined water?

5.87 You are given the following data:

$$H_2(g) \longrightarrow 2H(g) \qquad \Delta H^\circ = 436.4 \text{ kJ/mol}$$
$$Br_2(g) \longrightarrow 2Br(g) \qquad \Delta H^\circ = 192.5 \text{ kJ/mol}$$
$$H_2(g) + Br_2(g) \longrightarrow 2HBr(g) \qquad \Delta H^\circ = -72.4 \text{ kJ/mol}$$

Calculate ΔH° for the reaction

$$H(g) + Br(g) \longrightarrow HBr(g)$$

5.88 Compare the heat produced by the complete combustion of 1 mole of methane (CH_4) with a mole of water gas (0.50 mol H_2 and 0.50 mol CO) under the same conditions. On the basis of your answer, would you prefer methane over water gas as a fuel? Can you suggest two other reasons why methane is preferable to water gas as a fuel?

5.89 Ethanol (C_2H_5OH) and gasoline (assumed to be all octane, C_8H_{18}) are both used as automobile fuel. If gasoline is selling for $2.20/gal, what would the price of ethanol have to be in order to provide the same amount of heat per dollar? The density and ΔH°_f of octane are 0.7025 g/mL and −249.9 kJ/mol, respectively, and of ethanol are 0.7894 g/mL and −277.0 kJ/mol, respectively (1 gal = 3.785 L).

5.90 The combustion of how many moles of ethane (C_2H_6) would be required to heat 371 g of water from 55.0°C to 98.0°C?

5.91 The heat of vaporization of a liquid (ΔH_{vap}) is the energy required to vaporize 1.00 g of the liquid at its boiling point. In one experiment, 60.0 g of liquid nitrogen (boiling point = −196°C) is poured into a Styrofoam cup containing 2.00×10^2 g of water at 55.3°C. Calculate the molar heat of vaporization of liquid nitrogen if the final temperature of the water is 41.0°C.

5.92 Explain the cooling effect experienced when ethanol is rubbed on your skin, given that

$$C_2H_5OH(l) \longrightarrow C_2H_5OH(g) \qquad \Delta H^\circ = 42.2 \text{ kJ/mol}$$

5.93 For which of the following reactions does $\Delta H^\circ_{rxn} = \Delta H^\circ_f$?
(a) $H_2(g) + S(\text{rhombic}) \longrightarrow H_2S(g)$
(b) $C(\text{diamond}) + O_2(g) \longrightarrow CO_2(g)$
(c) $H_2(g) + CuO(s) \longrightarrow H_2O(l) + Cu(s)$
(d) $O(g) + O_2(g) \longrightarrow O_3(g)$

5.94 Calculate the work done (in joules) when 1.0 mole of water is frozen at 0°C and 1.0 atm. The volumes of 1 mole of water and ice at 0°C are 0.0180 and 0.0196 L, respectively. (The conversion factor is 1 L · atm = 101.3 J.)

5.95 A certain gas initially at 0.050 L undergoes expansion until its volume is 0.50 L. Calculate the work done (in joules) by the gas if it expands (a) against a vacuum and (b) against a constant pressure of 0.20 atm. (The conversion factor is 1 L · atm = 101.3 J.)

5.96 Calculate the standard enthalpy of formation for diamond, given that

$$C(\text{graphite}) + O_2(g) \longrightarrow CO_2(g) \qquad \Delta H^\circ = -393.5 \text{ kJ/mol}$$

$$C(\text{diamond}) + O_2(g) \longrightarrow CO_2(g) \qquad \Delta H^\circ = -395.4 \text{ kJ/mol}$$

5.97 The enthalpy of combustion of benzoic acid (C_6H_5COOH) is commonly used as the standard for calibrating constant-volume bomb calorimeters; its value has been accurately determined to be −3226.7 kJ/mol. When 1.9862 g of benzoic acid are burned in a calorimeter, the temperature rises from 21.84°C to 25.67°C. What is the heat capacity of the bomb? (Assume that the quantity of water surrounding the bomb is exactly 2000 g.)

5.98 At 25°C, the standard enthalpy of formation of $HF(aq)$ is −320.1 kJ/mol; of $OH^-(aq)$, it is −229.6 kJ/mol; of $F^-(aq)$, it is −329.1 kJ/mol; and of $H_2O(l)$, it is −285.8 kJ/mol.
(a) Calculate the standard enthalpy of neutralization of $HF(aq)$:

$$HF(aq) + OH^-(aq) \longrightarrow F^-(aq) + H_2O(l)$$

(b) Using the value of −56.2 kJ as the standard enthalpy change for the reaction

$$H^+(aq) + OH^-(aq) \longrightarrow H_2O(l)$$

calculate the standard enthalpy change for the reaction

$$HF(aq) \longrightarrow H^+(aq) + F^-(aq)$$

5.99 From the enthalpy of formation for CO_2 and the following information, calculate the standard enthalpy of formation for carbon monoxide (CO).

$$CO(g) + \frac{1}{2}O_2(g) \longrightarrow CO_2(g) \quad \Delta H° = -283.0 \text{ kJ/mol}$$

Why can't we obtain the standard enthalpy of formation directly by measuring the enthalpy of the following reaction?

$$C(\text{graphite}) + \frac{1}{2}O_2(g) \longrightarrow CO(g)$$

5.100 In the nineteenth century, two scientists named Dulong and Petit noticed that for a solid element, the product of its molar mass and its specific heat is approximately 25 J/°C. This observation, now called Dulong and Petit's law, was used to estimate the specific heat of metals. Verify the law for the metals listed in Table 5.2. The law does not apply to one of the metals. Which one is it? Why?

5.101 Determine the standard enthalpy of formation of ethanol (C_2H_5OH) from its standard enthalpy of combustion (−1367.4 kJ/mol).

5.102 Acetylene (C_2H_2) and benzene (C_6H_6) have the same empirical formula. In fact, benzene can be made from acetylene as follows:

$$3C_2H_2(g) \longrightarrow C_6H_6(l)$$

The enthalpies of combustion for C_2H_2 and C_6H_6 are −1299.4 and −3267.4 kJ/mol, respectively. Calculate the standard enthalpies of formation of C_2H_2 and C_6H_6 and hence the enthalpy change for the formation of C_6H_6 from C_2H_2.

5.103 Ice at 0°C is placed in a Styrofoam cup containing 361 g of a soft drink at 23°C. The specific heat of the drink is about the same as that of water. Some ice remains after the ice and soft drink reach an equilibrium temperature of 0°C. Determine the mass of ice that has melted. Ignore the heat capacity of the cup. (*Hint:* It takes 334 J to melt 1 g of ice at 0°C.)

5.104 A quantity of 85.0 mL of 0.600 *M* HCl is mixed with 85.0 mL of 0.600 *M* KOH in a constant-pressure calorimeter. The initial temperature of both solutions is the same at 17.35°C, and the final temperature of the mixed solution is 19.02°C. What is the heat capacity of the calorimeter? Assume that the specific heat of the solutions is the same as that of water and the molar heat of neutralization is −56.2 kJ/mol.

5.105 When 1.034 g of naphthalene ($C_{10}H_8$) is burned in a constant-volume bomb calorimeter at 298 K, 41.56 kJ of heat is evolved. Calculate ΔU and w for the reaction on a molar basis.

5.106 From a thermochemical point of view, explain why a carbon dioxide fire extinguisher or water should not be used on a magnesium fire.

5.107 A 4.117-g impure sample of glucose ($C_6H_{12}O_6$) was burned in a constant-volume calorimeter having a heat capacity of 19.65 kJ/°C. If the rise in temperature is 3.134°C, calculate the percent by mass of the glucose in the sample. Assume that the impurities are unaffected by the combustion process and that $\Delta U = \Delta H$. See Appendix 2 for thermodynamic data.

5.108 The combustion of 0.4196 g of a hydrocarbon releases 17.55 kJ of heat. The masses of the products are CO_2 = 1.419 g and H_2O = 0.290 g. (a) What is the empirical formula of the compound? (b) If the approximate molar mass of the compound is 76 g/mol, calculate its standard enthalpy of formation.

5.109 In a constant-pressure calorimetry experiment, a reaction gives off 21.8 kJ of heat. The calorimeter contains 150 g of water, initially at 23.4°C. What is the final temperature of the water? The heat capacity of the calorimeter is negligibly small.

5.110 At 850°C, $CaCO_3$ undergoes substantial decomposition to yield CaO and CO_2. Assuming that the $\Delta H_f°$ values of the reactant and products are the same at 850°C as they are at 25°C, calculate the enthalpy change (in kJ) if 66.8 g of CO_2 is produced in one reaction.

5.111 Give an example for each of the following situations: (a) adding heat to a system raises its temperature, (b) adding heat to a system does not change its temperature, and (c) a system's temperature changes despite no heat being added to it or removed from it.

5.112 Which of the constant-pressure processes given here has the smallest difference between ΔH and ΔU: (a) water ⟶ water vapor, (b) water ⟶ ice, (c) ice ⟶ water vapor? Explain.

5.113 Construct a table with the headings q, w, ΔU, and ΔH. For each of the following processes, deduce whether each of the quantities listed is positive (+), negative (−), or zero (0): (a) freezing of benzene, (b) reaction of sodium with water, (c) boiling of liquid ammonia, (d) melting of ice, (e) expansion of a gas at constant temperature.

5.114 A 3.52-g sample of ammonium nitrate (NH_4NO_3) was added to 80.0 mL of water in a constant-pressure calorimeter of negligible heat capacity. As a result, the temperature of the solution decreased from 21.6°C to 18.1°C. Calculate the heat of solution (ΔH_{soln}) in kJ/mol:

$$NH_4NO_3(s) \longrightarrow NH_4^+(aq) + NO_3^-(aq)$$

Assume the specific heat of the solution is the same as that of water.

5.115 A quantity of 50.0 mL of 0.200 *M* $Ba(OH)_2$ is mixed with 50.0 mL of 0.400 *M* HNO_3 in a constant-pressure calorimeter having a heat capacity of 496 J/°C. The initial temperature of both solutions is the same at 22.4°C. What is the final temperature of the mixed solution? Assume that the specific heat of the solutions is the same as that of water and the molar heat of neutralization is −56.2 kJ/mol.

Industrial Problems

5.116 Methanol (CH_3OH) is an organic solvent and is also used as a fuel in some automobile engines. From the following data, calculate the standard enthalpy of formation of methanol:

$$2CH_3OH(l) + 3O_2(g) \longrightarrow 2CO_2(g) + 4H_2O(l) \quad \Delta H°_{rxn} = -1452.8 \text{ kJ/mol}$$

5.117 Producer gas (carbon monoxide) is prepared by passing air over red-hot coke:

$$C(s) + \frac{1}{2}O_2(g) \longrightarrow CO(g)$$

Water gas (a mixture of carbon monoxide and hydrogen) is prepared by passing steam over red-hot coke:

$$C(s) + H_2O(g) \longrightarrow CO(g) + H_2(g)$$

For many years, both producer gas and water gas were used as fuels in industry and for domestic cooking. The large-scale preparation of these gases was carried out alternately; that is, first producer gas, then water gas, and so on. Using thermochemical reasoning, explain why this procedure was chosen.

Engineering Problems

5.118 Glauber's salt, sodium sulfate decahydrate ($Na_2SO_4 \cdot 10H_2O$), undergoes a phase transition (i.e., melting or freezing) at a convenient temperature of about 32°C:

$$Na_2SO_4 \cdot 10H_2O(s) \longrightarrow Na_2SO_4 \cdot 10H_2O(l) \qquad \Delta H° = 74.4 \text{ kJ/mol}$$

As a result, this compound is used to regulate the temperature in homes. It is placed in plastic bags in the ceiling of a room. During the day, the endothermic melting process absorbs heat from the surroundings, cooling the room. At night, it gives off heat as it freezes. Calculate the mass of Glauber's salt in kilograms needed to lower the temperature of air in a room by 8.2°C. The mass of air in the room is 605.4 kg; the specific heat of air is 1.2 J/g · °C.

5.119 An excess of zinc metal is added to 50.0 mL of a 0.100 M $AgNO_3$ solution in a constant-pressure calorimeter like the one pictured in Figure 5.8. As a result of the reaction

$$Zn(s) + 2Ag^+(aq) \longrightarrow Zn^{2+}(aq) + 2Ag(s)$$

the temperature rises from 19.25°C to 22.17°C. If the heat capacity of the calorimeter is 98.6 J/°C, calculate the enthalpy change for the given reaction on a molar basis. Assume that the density and specific heat of the solution are the same as those for water, and ignore the specific heats of the metals.

5.120 A driver's manual states that the stopping distance quadruples as the speed doubles; that is, if it takes 30 ft to stop a car moving at 25 mph, then it would take 120 ft to stop a car moving at 50 mph. Justify this statement by using mechanics and the first law of thermodynamics. (Assume that when a car is stopped, its kinetic energy ($\frac{1}{2}mu^2$) is totally converted to heat.)

5.121 A gas company in Massachusetts charges 27 cents for a mole of natural gas (CH_4). Calculate the cost of heating 200 mL of water (enough to make a cup of coffee or tea) from 20°C to 100°C. Assume that only 50 percent of the heat generated by the combustion is used to heat the water; the rest of the heat is lost to the surroundings.

5.122 Portable hot packs are available for skiers and people engaged in other outdoor activities in a cold climate. The air-permeable paper packet contains a mixture of powdered iron, sodium chloride, and other components, all moistened by a little water. The exothermic reaction that produces the heat is a very common one—the rusting of iron:

$$4Fe(s) + 3O_2(g) \longrightarrow 2Fe_2O_3(s)$$

When the outside plastic envelope is removed, O_2 molecules penetrate the paper, causing the reaction to begin. A typical packet contains 250 g of iron to warm your hands or feet for up to 4 hours. How much heat (in kJ) is produced by this reaction? (*Hint:* See Appendix 2 for $\Delta H_f°$ values.)

5.123 For reactions in condensed phases (liquids and solids), the difference between ΔH and ΔU is usually quite small. This statement holds for reactions carried out under atmospheric conditions. For certain geochemical processes, however, the external pressure may be so great that ΔH and ΔU can differ by a significant amount. A well-known example is the slow conversion of graphite to diamond under Earth's surface. Calculate $\Delta H - \Delta U$ for the conversion of 1 mole of graphite to 1 mole of diamond at a pressure of 50,000 atm. The densities of graphite and diamond are 2.25 g/cm^3 and 3.52 g/cm^3, respectively.

5.124 Consider the reaction

$$2H_2(g) + O_2(g) \longrightarrow 2H_2O(l)$$

Under atmospheric conditions (1.00 atm) it was found that the formation of water resulted in a decrease in volume equal to 73.4 L. Calculate ΔU for the process. $\Delta H = -571.6$ kJ/mol. (The conversion factor is 1 L · atm = 101.3 J.)

5.125 The total volume of the Pacific Ocean is estimated to be 7.2×10^8 km^3. A medium-sized atomic bomb produces 1.0×10^{15} J of energy upon explosion. Calculate the number of atomic bombs needed to release enough energy to raise the temperature of the water in the Pacific Ocean by 1°C.

5.126 The so-called hydrogen economy is based on hydrogen produced from water using solar energy. The gas is then burned as a fuel:

$$2H_2(g) + O_2(g) \longrightarrow 2H_2O(l)$$

A primary advantage of hydrogen as a fuel is that it is nonpolluting. A major disadvantage is that it is a gas and therefore is harder to store than liquids or solids. Calculate the number of moles of H_2 required to produce an amount of energy equivalent to that produced by the combustion of a gallon of octane (C_8H_{18}). The density of octane is 2.66 kg/gal, and its standard enthalpy of formation is −249.9 kJ/mol.

Biological Problems

5.127 Photosynthesis produces glucose ($C_6H_{12}O_6$) and oxygen from carbon dioxide and water:

$$6CO_2(g) + 6H_2O(l) \longrightarrow C_6H_{12}O_6(s) + 6O_2(g)$$

(a) How would you determine experimentally the $\Delta H°_{rxn}$ value for this reaction? (b) Solar radiation produces about 7.0×10^{14} kg of glucose a year on Earth. What is the corresponding $\Delta H°$ change?

5.128 Calculate the standard enthalpy change for the fermentation process, in which glucose ($C_6H_{12}O_6$) is converted to ethanol (C_2H_5OH) and carbon dioxide.

5.129 A 46-kg person drinks 500 g of milk, which has a "caloric" value of approximately 3.0 kJ/g. If only 17 percent of the energy in milk is converted to mechanical work, how high (in meters) can the person climb based on this energy intake? [*Hint:* The work done in ascending is given by *mgh*, where *m* is the mass (in kg), *g* is the gravitational acceleration (9.8 m/s^2), and *h* is the height (in meters).]

5.130 A man ate 0.50 pound of cheese (an energy intake of 4×10^3 kJ). Suppose that none of the energy was stored in his body. What mass (in grams) of water would he need to perspire in order to maintain his original temperature? (It takes 44.0 kJ to vaporize 1 mole of water.)

5.131 Why are cold, damp air and hot, humid air more uncomfortable than dry air at the same temperatures? [The specific heats of water vapor and air are approximately 1.9 J/(g · °C) and 1.0 J/(g · °C), respectively.]

5.132 A woman expends 95 kJ of energy walking a kilometer. The energy is supplied by the metabolic breakdown of food, which has an efficiency of 35 percent. How much energy does she save by walking the kilometer instead of driving a car that gets 8.2 km per liter of gasoline (approximately 20 mi/gal)? The density of gasoline is 0.71 g/mL, and its enthalpy of combustion is −49 kJ/g.

5.133 The carbon dioxide exhaled by sailors in a submarine is often removed by reaction with an aqueous lithium hydroxide solution. (a) Write a balanced equation for this process. (*Hint:* The products are water and a soluble salt.) (b) If every sailor consumes 1.2×10^4 kJ of energy every day and assuming that this energy is totally supplied by the metabolism of glucose ($C_6H_{12}O_6$), calculate the amounts of CO_2 produced and LiOH required to purify the air.

5.134 How much metabolic energy must a 5.2-g hummingbird expend to fly to a height of 21 m? (See the hint in Problem 5.123.)

5.135 Acetylene (C_2H_2) can be made by combining calcium carbide (CaC_2) with water. (a) Write an equation for the reaction. (b) What is the maximum amount of heat (in joules) that can be obtained from the combustion of acetylene, starting with 74.6 g of CaC_2?

5.136 (a) A person drinks four glasses of cold water (3.0°C) every day. The volume of each glass is 2.5×10^2 mL. How much heat (in kJ) does the body have to supply to raise the temperature of the water to 37°C, the body temperature? (b) How much heat would your body lose if you were to ingest 8.0×10^2 g of snow at 0°C to quench your thirst? (The amount of heat necessary to melt snow is 6.01 kJ/mol.)

5.137 Both glucose and fructose are simple sugars with the same molecular formula of $C_6H_{12}O_6$. Sucrose ($C_{12}H_{22}O_{11}$), or table sugar, consists of a glucose molecule bonded to a fructose molecule (a water molecule is eliminated in the formation of sucrose). (a) Calculate the energy released when a 2.0-g glucose tablet is burned in air. (b) To what height can a 65-kg person climb after ingesting such a tablet, assuming only 30 percent of the energy released is available for work? (See the hint for Problem 5.123.) Repeat the calculations for a 2.0-g sucrose tablet.

5.138 Metabolic activity in the human body releases approximately 1.0×10^4 kJ of heat per day. Assume that a 55-kg body has the same specific heat as water; how much would the body temperature rise if it were an isolated system? How much water must the body eliminate as perspiration to maintain the normal body temperature (98.6°F)? Comment on your results. (The heat of vaporization of water is 2.41 kJ/g.)

Environmental Problems

5.139 Calcium oxide (CaO) is used to remove sulfur dioxide generated by coal-burning power stations:

$$2CaO(s) + 2SO_2(g) + O_2(g) \longrightarrow 2CaSO_4(s)$$

Calculate the enthalpy change if 6.6×10^5 g of SO_2 is removed by this process.

5.140 About 6.0×10^{13} kg of CO_2 is fixed (converted to more complex organic molecules) by photosynthesis every year. (a) Assuming all the CO_2 ends up as glucose ($C_6H_{12}O_6$), calculate the energy (in kJ) stored by photosynthesis per year. (b) A typical coal-burning electric power station generates about 2.0×10^6 W per year. How many such stations are needed to generate the same amount of energy as that captured by photosynthesis (1 W = 1 J/s)?

5.141 The average temperature in deserts is high during the day but quite cool at night, whereas that in regions along the coastline is more moderate. Explain.

Multiconcept Problems

5.142 *Lime* is a term that includes calcium oxide (CaO, also called quicklime) and calcium hydroxide [$Ca(OH)_2$, also called slaked lime]. It is used in the steel industry to remove acidic impurities, in air-pollution control to remove acidic oxides such as SO_2, and in water treatment. Quicklime is made industrially by heating limestone ($CaCO_3$) above 2000°C:

$$CaCO_3(s) \longrightarrow CaO(s) + CO_2(g) \qquad \Delta H° = 177.8 \text{ kJ/mol}$$

Slaked lime is produced by treating quicklime with water:

$$CaO(s) + H_2O(l) \longrightarrow Ca(OH)_2(s) \qquad \Delta H° = -65.2 \text{ kJ/mol}$$

The exothermic reaction of quicklime with water and the rather small specific heats of both quicklime [0.946 J/(g · °C)] and slaked lime [1.20 J/(g · °C)] make it hazardous to store and transport lime in vessels made of wood. Wooden sailing ships carrying lime would occasionally catch fire when water leaked into the hold. (a) If a 500.0-g sample of water reacts with an equimolar amount of CaO (both at an initial temperature of 25°C), what is the final temperature of the product, $Ca(OH)_2$? Assume that the product absorbs all the heat released in the reaction. (b) Given that the standard enthalpies of formation of CaO and H_2O are −635.6 and −285.8 kJ/mol, respectively, calculate the standard enthalpy of formation of $Ca(OH)_2$.

5.143 Hydrazine (N_2H_4) decomposes to form ammonia and nitrogen gases. (a) Write a balanced chemical equation for this process. (b) Given that the standard enthalpy of formation of hydrazine is 50.42 kJ/mol, calculate ΔH°_{rxn} for its decomposition. (c) Both hydrazine and ammonia will burn in oxygen to produce $H_2O(l)$ and $N_2(g)$. Write balanced equations for these processes and determine ΔH°_{rxn} for each process. (d) If equal masses of hydrazine and ammonia were burned in separate bomb calorimeter experiments, which would cause the greater increase in temperature?

5.144 Determine the temperature change when 50.0 mL of 0.0135 *M* HBr(*aq*) and 50.0 mL of 0.00755 *M* $Ba(OH)_2(aq)$, originally both at room temperature, are combined in a coffee-cup calorimeter.

5.145 A piece of silver with a mass of 362 g has a heat capacity of 85.7 J/°C. What is the specific heat of silver?

Standardized-Exam Practice Problems

Physical and Biological Sciences

A bomb calorimeter was calibrated by burning 1.013 g of benzoic acid ($C_7H_6O_2$) ($\Delta U_{comb} = 3.221 \times 10^3$ kJ/mol). The temperature change in the calorimeter during the calibration combustion was 5.19°C. A nutritional chemist then used the calibrated calorimeter to determine the energy content of food. The chemist carefully dried a sample of food and placed 0.8996 g of the sample in the calorimeter with sufficient oxygen for the combustion to go to completion. Combustion of the food sample caused the temperature of the calorimeter to increase by 4.42°C.

1. Approximately how many moles of O_2 gas were consumed in the calibration combustion?

   a) 0.008
   b) 0.1
   c) 0.2
   d) 0.06

2. What is the heat capacity (C_V) of the calorimeter?

   a) 5.15 kJ/°C
   b) 5.08 kJ/°C
   c) 5.12 kJ/°C
   d) 4.97 kJ/°C

3. What is the energy content of the food?

   a) 22.8 kJ/g
   b) 4.97 kJ/g
   c) 25.3 kJ/g
   d) 0.201 kJ/g

4. What would be the effect on the result if the food sample were not completely dried prior to being placed in the calorimeter?

   a) The combustion of the sample would be incomplete.
   b) The calculated energy content per gram would be too low.
   c) The calculated energy content per gram would be too high.
   d) There would be no effect on the result.

Answers to In-Chapter Materials

Answers to Practice Problems

5.1A (a) 1.13×10^3 J, (b) 4. **5.1B** (a) 372 m/s, (b) neither. **5.2A** -4.32×10^4 kJ. **5.2B** 6.95×10^5 kJ, heat is absorbed. **5.3A** 8.174×10^4 kJ. **5.3B** 1.71×10^3 g. **5.4A** 151 kJ. **5.4B** 52.2°C. **5.5A** 28°C. **5.5B** 42 g. **5.6A** 14.5 kJ/g. **5.6B** 10.5°C rise. **5.7A** −1103 kJ/mol. **5.7B** 113.2 kJ/mol. **5.8A** 177.8 kJ/mol. **5.8B** −697.6 kJ/mol. **5.9A** 87.3 kJ/mol. **5.9B** $NH_3(g) + HCl(g)$, 176.8 kJ/mol.

Answers to Checkpoints

5.1.1 a. **5.1.2** b. **5.1.3** a. **5.1.4** e. **5.1.5** c. **5.1.6** b, d, e. **5.2.1** e. **5.2.2** c. **5.2.3** c. **5.2.4** a, d, e. **5.3.1** b. **5.3.2** a. **5.4.1** b. **5.4.2** d. **5.4.3** a. **5.4.4** a. **5.5.1** c. **5.5.2** a. **5.5.3** c. **5.5.4** d. **5.6.1** b. **5.6.2** e. **5.6.3** c, e. **5.6.4** e.

CHAPTER 6

Quantum Theory and the Electronic Structure of Atoms

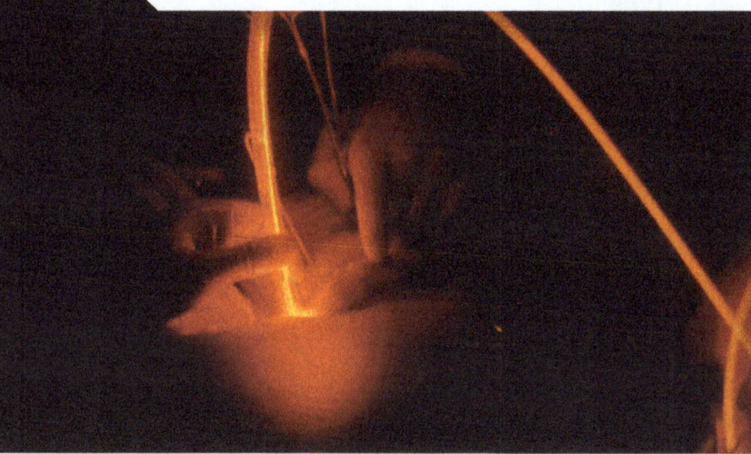

Light is used in a variety of medical procedures, including some cancer treatments. Photodynamic therapy, shown here, is done using visible light.

Credit: © *Science Source.*

In This Chapter, You Will Learn

About some of the properties of electromagnetic radiation or *light* and how these properties have been used to study and elucidate the electronic structure of atoms. You will also learn how to determine the arrangement of electrons in a particular atom.

Before You Begin, Review These Skills

- Tracking units [⏮ Section 1.6]
- The nuclear model of the atom [⏮ Section 2.2]

How Our Understanding of Electronic Structure Has Advanced Medicine

For more than a century, radiation has been used in the treatment of cancer. Traditional therapies involve the use of powerful *ionizing radiation* [⏭ Chapter 20], such as X rays or gamma rays. Ionizing radiation damages the DNA of cancer cells, preventing their continued replication—although it does the same to healthy, noncancerous cells. Over time, technicians and physicians have developed methods that minimize the damage done to healthy tissue, but there remains an inherent risk in the use of ionizing radiation, and stringent safety protocols are required for treatment.

In recent decades, a new type of radiation therapy has emerged and is becoming increasingly popular: *photodynamic therapy* (PDT). PDT works by irradiation with *visible* light, which is far less energetic than X-ray or gamma radiation, and thus safer to use. Treatment begins with the administration of a *photosensitizer* (a substance that undergoes reaction when irradiated), that is selectively taken up by cancerous tissue. Subsequent irradiation with a specific wavelength of visible light causes the photosensitizer to produce an especially reactive form of oxygen known as *singlet oxygen,* which damages the cancerous cells and prevents further replication.

The visible light used in photodynamic therapy results from electron transitions within atoms. Although the nuclear model that Rutherford proposed based on his gold-foil experiment specified the location of the protons and the neutrons in an atom, it failed to describe the location or behavior of the *electrons.* Early in the twentieth century, the application of a radical new theory in physics called *quantum theory,* and the ingenious interpretation of experimental evidence by Max Planck, Albert Einstein, and others, led to our current understanding of the *electronic structure of atoms.* This understanding of electronic structure is what makes such things as photodynamic therapy possible.

At the end of this chapter, you will be able to answer a variety of questions about the nature of light and the electronic structure of the atom **[⏭ Applying What You've Learned, page 274].**

6.1 The Nature of Light

When we say "light," we generally mean *visible* light, which is the light we can detect with our eyes. Visible light, however, is only a small part of the continuum of radiation that comprises the ***electromagnetic spectrum.*** In addition to visible light, the electromagnetic spectrum includes radio waves, microwave radiation, infrared and ultraviolet radiation, X rays, and gamma rays, as shown in Figure 6.1. Some of these terms may be familiar to you. For instance, the danger of exposure to ultraviolet radiation is why you need to use sunscreen. You may have used microwave radiation from a microwave oven to reheat food or to pop popcorn; you may have had X rays during a routine dental checkup or after breaking a bone; and you may recall from Chapter 2 that gamma rays are emitted from some radioactive materials. Although these phenomena may seem very different from each other and from visible light, they all are the transmission of energy in the form of *waves.*

Properties of Waves

The fundamental properties of waves are illustrated in Figure 6.2. Waves are characterized by their wavelength, frequency, and amplitude. ***Wavelength*** λ (lambda) is the distance between identical points on successive waves (e.g., successive peaks or successive troughs). The ***frequency*** ν (nu) is the *number* of waves that pass through a particular point in 1 second. ***Amplitude*** is the vertical distance from the midline of a wave to the top of the peak or the bottom of the trough.

The speed of a wave depends on the type of wave and the nature of the medium through which the wave is traveling (e.g., air, water, or a vacuum). The speed of light through a vacuum, c, is 2.99792458×10^8 m/s. The speed, wavelength, and frequency of a wave are related by the equation

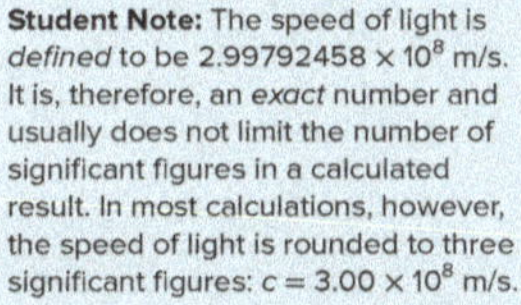

Student Note: The speed of light is *defined* to be 2.99792458×10^8 m/s. It is, therefore, an *exact* number and usually does not limit the number of significant figures in a calculated result. In most calculations, however, the speed of light is rounded to three significant figures: $c = 3.00 \times 10^8$ m/s.

Equation 6.1 $$c = \lambda\nu$$

where λ and ν are expressed in meters (m) and reciprocal seconds (s^{-1}), respectively. Although wavelength in meters is convenient for this equation, the units customarily used to express the wavelength of electromagnetic radiation depend on the type of radiation and the magnitude of the corresponding wavelength. The wavelength of visible light, for instance, is on the order of nanometers (nm, or 10^{-9} m), and that of microwave radiation is on the order of centimeters (cm, or 10^{-2} m).

Student Note: Frequency is expressed as cycles per second, or simply *reciprocal seconds* (s^{-1}), which is also known as *hertz* (Hz).

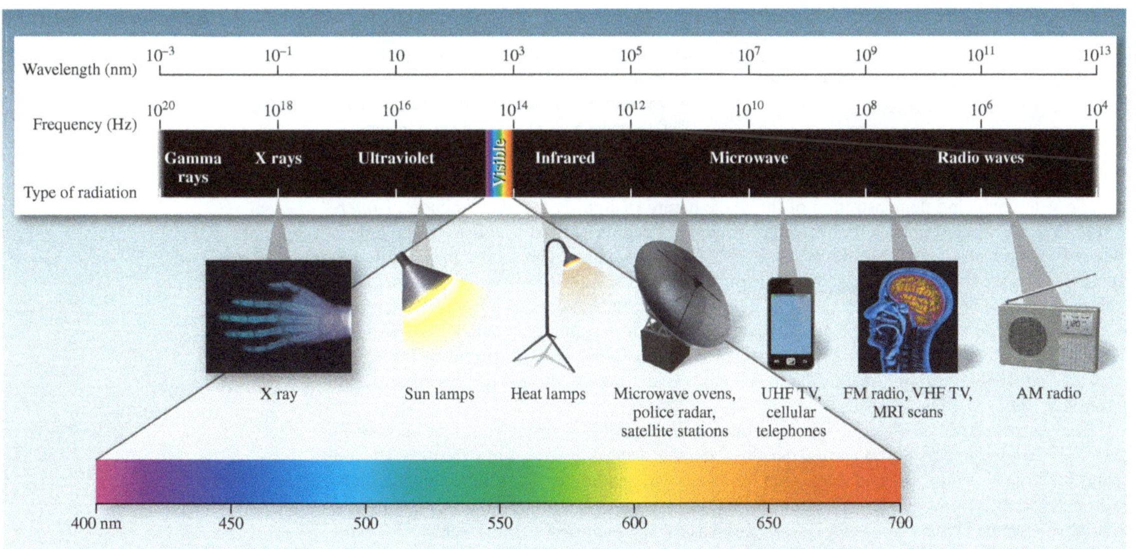

Figure 6.1 Electromagnetic spectrum. Each type of radiation is spread over a specific range of wavelengths (and frequencies). Visible light ranges from 400 nm (violet) to 700 nm (red).

Credit: (hand X ray) © Jim Wehtje/Getty Images; (MRI) © Don Farrall/Getty Images.

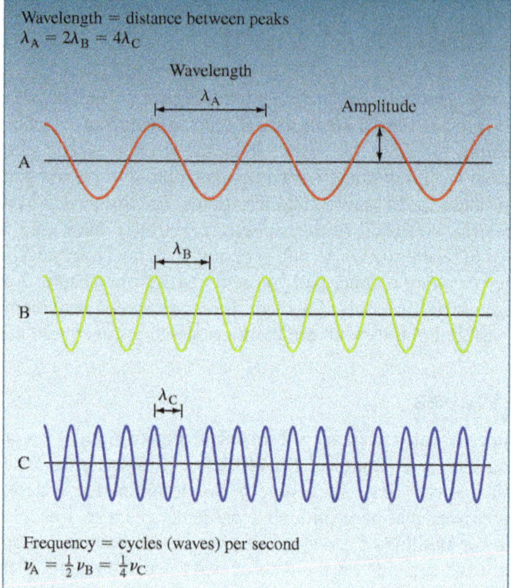

Figure 6.2 Characteristics of waves: wavelength, amplitude, and frequency.

The Electromagnetic Spectrum

In 1873, James Clerk Maxwell proposed that visible light consisted of electromagnetic waves. According to Maxwell's theory, an ***electromagnetic wave*** has an electric field component and a magnetic field component. These two components have the same wavelength and frequency, and hence the same speed, but they travel in mutually perpendicular planes (Figure 6.3). The significance of Maxwell's theory is that it provides a mathematical description of the general behavior of light. In particular, his model accurately describes how energy in the form of radiation can be propagated through space as oscillating electric and magnetic fields.

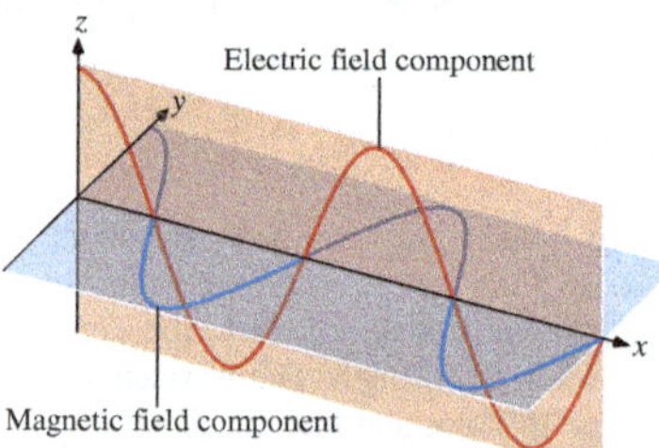

Figure 6.3 Electric field and magnetic field components of an electromagnetic wave. These two components have the same wavelength, frequency, and amplitude, but they vibrate in two mutually perpendicular planes.

The Double-Slit Experiment

A simple yet convincing demonstration of the wave nature of light is the phenomenon of *interference.* When a light source passes through a narrow opening, called a *slit,* a bright line is generated in the path of the light through the slit. When the same light source passes through two closely spaced slits, however, as shown in Figure 6.4, the result is not two bright lines, one in the path of each slit, but rather a series of light and dark lines known as an *interference pattern.* When the light sources recombine after passing through the slits, they do so *constructively* where the two waves are *in phase* (giving rise to the light lines) and *destructively* where the waves are *out of phase* (giving rise to the dark lines). Constructive interference and destructive interference are properties of waves.

The various types of electromagnetic radiation in Figure 6.1 differ from one another in wavelength and frequency. Radio waves, which have long wavelengths and low frequencies, are emitted by large antennas, such as those used by broadcasting stations. The shorter, visible light waves are produced by the motions of electrons within atoms and molecules. The shortest waves, which also have the highest frequency, are γ (gamma) rays, which result from nuclear processes [⏮ Section 2.2]. As we will see shortly, the higher the frequency, the more energetic the radiation. Thus, ultraviolet radiation, X rays, and γ rays are high-energy radiation, whereas infrared radiation, microwave radiation, and radio waves are low-energy radiation.

Sample Problem 6.1 illustrates the conversion between wavelength and frequency.

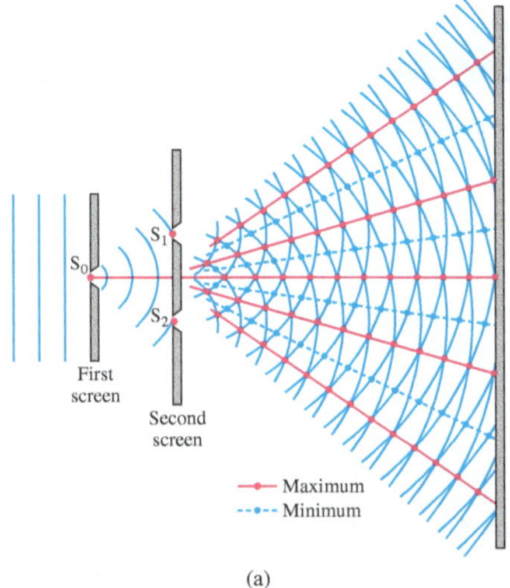

(a)

(b)

Figure 6.4 Double-slit experiment. (a) Red lines correspond to the maximum intensity resulting from constructive interference. Dashed blue lines correspond to the minimum intensity resulting from destructive interference. (b) Interference pattern with alternating bright and dark lines.

SAMPLE PROBLEM 6.1

One type of laser used in the treatment of vascular skin lesions is a neodymium-doped yttrium aluminum garnet or Nd:YAG laser. The wavelength commonly used in these treatments is 532 nm. What is the frequency of this radiation?

Strategy Wavelength and frequency are related by Equation 6.1 ($c = \lambda\nu$), so we must rearrange Equation 6.1 to solve for frequency. Because we are given the wavelength of the electromagnetic radiation in nanometers, we must convert this wavelength to meters and use $c = 3.00 \times 10^8$ m/s.

Setup Solving for frequency gives $\nu = c/\lambda$. Next we convert the wavelength to meters:

$$\lambda \text{ (in meters)} = 532 \cancel{\text{nm}} \times \frac{1 \times 10^{-9}\text{ m}}{1 \cancel{\text{nm}}} = 5.32 \times 10^{-7}\text{ m}$$

Solution

$$\nu = \frac{3.00 \times 10^8 \text{ m/s}}{5.32 \times 10^{-7}\text{ m}} = 5.64 \times 10^{14}\text{ s}^{-1}$$

THINK ABOUT IT

Make sure your units cancel properly. A common error in this type of problem is neglecting to convert wavelength to meters.

Practice Problem ATTEMPT What is the frequency (in reciprocal seconds) of electromagnetic radiation with a wavelength of 1.03 cm?

Practice Problem BUILD What is the wavelength (in meters) of an electromagnetic wave whose frequency is 1.61×10^{12} s^{-1}?

Practice Problem CONCEPTUALIZE Which of the following sets of waves best represents the relative wavelengths/frequencies of visible light of the colors shown?

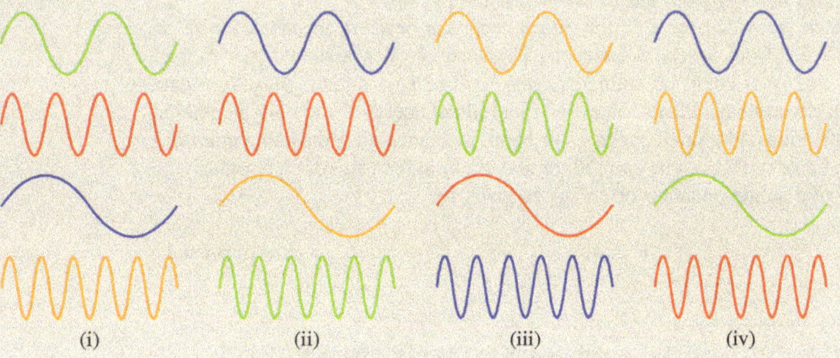

CHECKPOINT – SECTION 6.1 The Nature of Light

6.1.1 Calculate the wavelength of light with frequency $3.45 \times 10^{14}\ s^{-1}$.

a) 1.15×10^{-6} nm
b) 1.04×10^{23} nm
c) 8.70×10^{2} nm
d) 115 nm
e) 9.66×10^{-24} nm

6.1.2 Calculate the frequency of light with wavelength 126 nm.

a) $2.38 \times 10^{15}\ s^{-1}$
b) $4.20 \times 10^{-16}\ s^{-1}$
c) $37.8\ s^{-1}$
d) $2.65 \times 10^{-2}\ s^{-1}$
e) $3.51 \times 10^{19}\ s^{-1}$

6.1.3 When traveling through a translucent medium, such as glass, light travels more slowly than it travels through a vacuum. Red light with a wavelength of 684 nm travels through Pyrex glass with a frequency of $2.92 \times 10^{14}\ s^{-1}$. Calculate the speed of this light.

a) 3.00×10^{8} m/s
b) 2.00×10^{8} m/s
c) 2.92×10^{6} m/s
d) 4.23×10^{7} m/s
e) 2.23×10^{8} m/s

6.1.4 Of the waves pictured, ________ has the greatest frequency, ________ has the greatest wavelength, and ________ has the greatest amplitude.

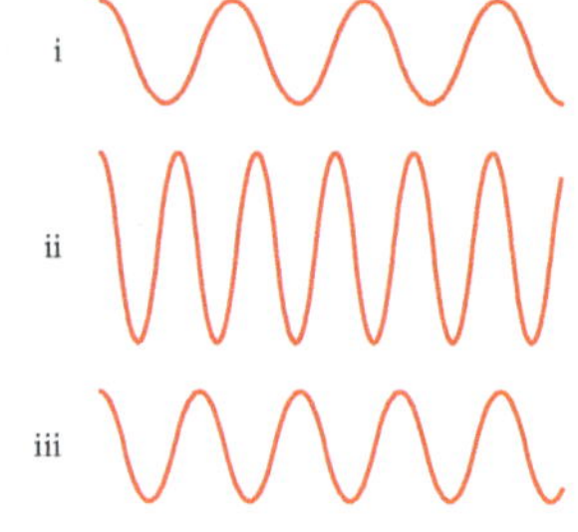

a) i, ii, iii
b) i, iii, ii
c) ii, i, ii
d) ii, i, iii
e) ii, iii, ii

6.2 Quantum Theory

Early attempts by nineteenth-century physicists to figure out the structure of the atom met with only limited success. This was largely because they were attempting to understand the behavior of subatomic particles using the laws of classical physics that govern the behavior of macroscopic objects. It took a long time to realize—and an even longer time to accept—that the properties of atoms are *not* governed by the same physical laws as larger objects.

Quantization of Energy

When a solid is heated above 800 K, it emits electromagnetic radiation, including visible wavelengths, known as ***blackbody radiation.*** As the temperature of the solid increases, the light emitted goes from red, to orange, to yellow, to white. Measurements taken in the latter part of the nineteenth century showed that the amount of energy given off by an object at a certain temperature depends on the wavelength of the emitted radiation. Attempts to account for this dependence in terms of established wave theory and thermodynamic laws were only partially successful. One theory was able to explain short-wavelength dependence but failed to account for the longer wavelengths. Another theory accounted for the longer wavelengths but failed for short wavelengths. With no *one* theory that could explain both observations, it seemed that something fundamental was missing from the laws of classical physics.

In 1900, Max Planck[1] provided the solution and launched a new era in physics with an idea that departed drastically from accepted concepts. Classical physics assumed that radiant energy was continuous; that is, it could be emitted or absorbed in any amount. Based on data from blackbody radiation experiments, Planck proposed that radiant energy could only be emitted or absorbed in discrete quantities, like small packages or bundles. Planck gave the name ***quantum*** to the smallest quantity of energy that can be emitted (or absorbed) in the form of electromagnetic radiation. The energy E of a single quantum of energy is given by

$$E = h\nu \qquad \textbf{Equation 6.2}$$

1. Max Karl Ernst Ludwig Planck (1858–1947). German physicist. Planck received the Nobel Prize in Physics in 1918 for his quantum theory. He also made significant contributions in thermodynamics and other areas of physics.

Student Note: The National Institute of Standards and Technology (NIST) gives a value of $6.6260693 \times 10^{-34}$ J · s for Planck's constant. Typically, three significant figures are sufficient for solving problems.

where h is called *Planck's constant* and ν is the *frequency* of the radiation. The value of Planck's constant is 6.63×10^{-34} J · s.

According to quantum theory, energy is always emitted in whole-number multiples of $h\nu$. At the time Planck presented his theory, he could not explain why energies should be fixed or quantized in this manner. Starting with this hypothesis, however, he had no difficulty correlating the experimental data for the emission by solids over the entire range of wavelengths; the experimental data supported his new *quantum theory*.

The idea that energy is *quantized* rather than *continuous* may seem strange, but the concept of quantization has many everyday analogies. For example, vending machines dispense cans or bottles of soft drinks only in whole numbers (you can't buy part of a can or bottle from a machine). Each can or bottle is a quantum of its soft drink. Even processes in living systems involve quantized phenomena. The eggs laid by hens are quanta (hens lay only whole eggs). Similarly, when a dog or cat gives birth to a litter, the number of offspring is always an integer. Each puppy or kitten is a quantum of that animal. Planck's quantum theory revolutionized physics. Indeed, the flurry of research that ensued altered our concept of nature forever.

Sample Problem 6.2 shows how to compare energy per photon for different wavelengths.

Bringing Chemistry to Life

Laser Pointers

The laser pointers that have become so common typically emit radiation in the red region of the visible spectrum with output wavelengths ranging from 630 to 680 nm. Low prices and availability have made the devices popular not only with instructors and other speakers, but with teenagers and even children—raising some significant safety concerns. Although the human blink reflex generally is sufficient to protect against serious or permanent injury to the eye by one of these devices, intentional prolonged exposure of the eye to the beam from a laser pointer can be dangerous. Of particular concern are the new green laser pointers that emit a wavelength of 532 nm. The lasers in these devices also produce radiation in the infrared region of the electromagnetic spectrum (1064 nm), but they are equipped with filters to prevent the emission of infrared radiation. However, some of the inexpensive imported lasers do not bear adequate safety labeling and the filters are easily removed—potentially resulting in the emission of dangerous radiation. Although the 1064-nm laser beam is less energetic than the 532-nm beam, it poses a greater danger to the eye because it is not visible and does not evoke the blink response that visible wavelengths do. Despite not being visible, a 1064-nm beam passes through the anterior structures of the eye and damages the retina. Because the beam is not visible, the damage is not immediately apparent, but it can be permanent.

SAMPLE PROBLEM 6.2

How much more energy per photon is there in green light of wavelength 532 nm than in red light of wavelength 635 nm?

Strategy Convert each wavelength to frequency using Equation 6.1, and then calculate the energy per photon using Equation 6.2.

Setup Convert the wavelengths to meters:

$$532 \text{ nm} \times \frac{1 \times 10^{-9} \text{ m}}{1 \text{ nm}} = 5.32 \times 10^{-7} \text{ m}$$

$$635 \text{ nm} \times \frac{1 \times 10^{-9} \text{ m}}{1 \text{ nm}} = 6.35 \times 10^{-7} \text{ m}$$

Planck's constant, h, is 6.63×10^{-34} J · s.

Solution For 532 nm,

$$\nu = \frac{c}{\lambda} = \frac{3.00 \times 10^8 \text{ m/s}}{5.32 \times 10^{-7} \text{ m}} = 5.64 \times 10^{14} \text{ s}^{-1}$$

and

$$E = h\nu = (6.63 \times 10^{-34} \text{ J} \cdot \text{s})(5.64 \times 10^{14} \text{ s}^{-1}) = 3.74 \times 10^{-19} \text{ J}$$

The energy of a single photon with wavelength 532 nm is 3.74×10^{-19} J.

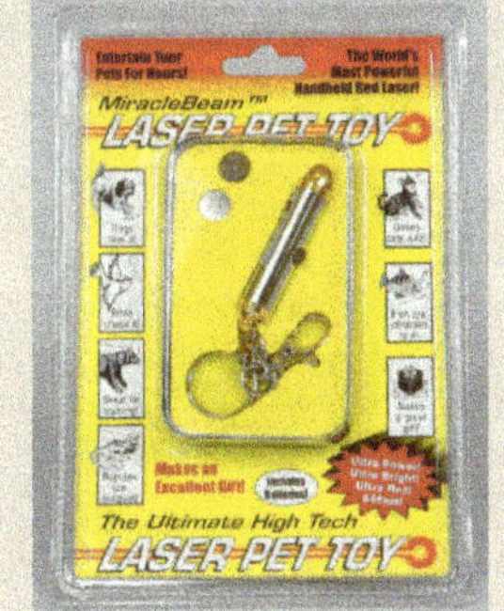

Credit: © Brian Moeskau.

Following the same procedure, the energy of a photon of wavelength 635 nm is 3.13×10^{-19} J. The difference between them is $(3.74 \times 10^{-19}\text{ J}) - (3.13 \times 10^{-19}\text{ J}) = 6.1 \times 10^{-20}$ J. Therefore, a photon of green light ($\lambda = 532$ nm) has 6.1×10^{-20} J *more* energy than a photon of red light ($\lambda = 635$ nm).

THINK ABOUT IT

As wavelength decreases, frequency increases. Energy, being directly proportional to frequency, also increases.

Practice Problem ATTEMPT Calculate the difference in energy (in joules) between a photon with $\lambda = 680$ nm (red) and a photon with $\lambda = 442$ nm (blue).

Practice Problem BUILD In what region of the electromagnetic spectrum is a photon found that possesses twice as much energy as one in the blue region ($\lambda = 442$ nm) of the visible spectrum?

Practice Problem CONCEPTUALIZE Shown here are waves of electromagnetic radiation of two different frequencies and two different amplitudes. Assume that intensity of radiation (photons/s) is directly proportional to amplitude. Which of the waves is made up of photons of greater energy? Which wave delivers more photons during a given period of time? Which wave delivers more total energy during a given time period?

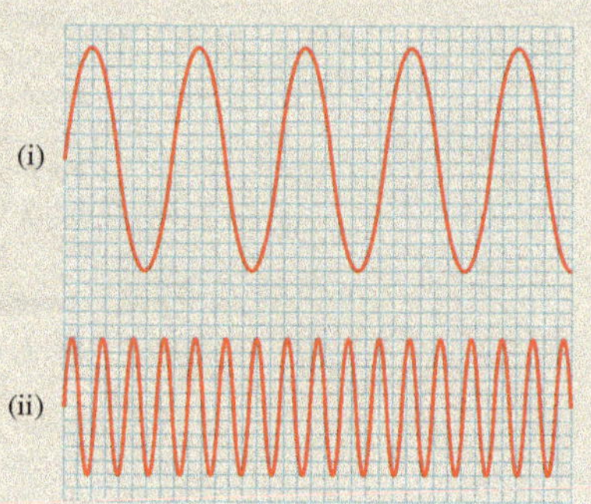

Photons and the Photoelectric Effect

In 1905, only 5 years after Planck presented his quantum theory, Albert Einstein[2] used the theory to explain another mysterious physical phenomenon, the ***photoelectric effect,*** in which electrons are ejected from the surface of a metal exposed to light of at least a certain minimum frequency, called the *threshold frequency* (Figure 6.5). The number of electrons ejected was proportional to the intensity (or brightness) of the light, but the energies of the ejected electrons were not. Below the threshold frequency no electrons were ejected no matter how intense the light.

The photoelectric effect could not be explained by the wave theory of light, which associated the energy of light with its intensity. Einstein, however, made an extraordinary assumption. He suggested that a beam of light is really a stream of particles. These *particles of light* are now called ***photons.*** Using Planck's quantum theory of radiation as a starting point, Einstein deduced that each *photon* must possess energy E given by the equation

$$E_{\text{photon}} = h\nu$$

where h is Planck's constant and ν is the frequency of the light. Electrons are held in a metal by attractive forces, and so removing them from the metal requires light of a sufficiently high frequency (which corresponds to a sufficiently high energy) to break them free. Shining a beam of light onto a metal surface can be thought of as shooting a beam of particles—*photons*—at the metal atoms. If the frequency of the photons is such that $h\nu$ exactly equals the energy that binds the electrons in the metal, then the light will have just enough energy to knock the electrons loose. If we use light of a higher frequency, then not only will the electrons be knocked loose, but they will also acquire some kinetic energy. This situation is summarized by the equation

$$h\nu = \text{KE} + W \qquad \textbf{Equation 6.3}$$

where KE is the kinetic energy of the ejected electron and W is the binding energy of the electron in the metal. [Binding energies are typically given in units of electron volts (eV), where $1\text{ eV} = 1.602 \times 10^{-19}$ J.] Rewriting Equation 6.3 as

$$\text{KE} = h\nu - W$$

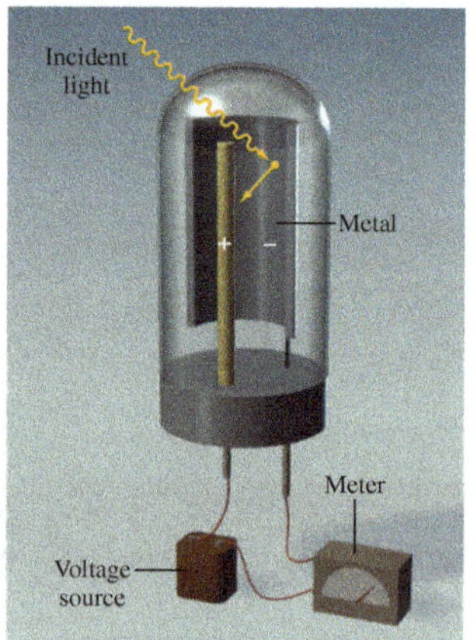

Figure 6.5 Apparatus for studying the photoelectric effect. Light of a certain frequency falls on a clean metal surface. Ejected electrons are attracted toward the positive electrode. The flow of electrons is registered by a detecting meter.

2. Albert Einstein (1879–1955). German-born American physicist. Regarded by many as one of the two greatest physicists the world has known (the other is Isaac Newton). The three papers (on special relativity, Brownian motion, and the photoelectric effect) that he published in 1905 while employed as a technical assistant in the Swiss patent office in Berne have profoundly influenced the development of physics. He received the Nobel Prize in Physics in 1921 for his explanation of the photoelectric effect.

Where Have I Encountered the Photoelectric Effect?

Chances are good that you encounter the photoelectric effect regularly. Some everyday applications include the type of device that prevents a garage door from closing when something is in the door's path and motion-detection systems used in museums and other high-security environments. These sorts of devices work simply by responding to an interruption in a beam of light. In each case, a beam of light normally shines on a *photocathode,* a surface that emits photoelectrons, and the photoelectrons, accelerated toward an anode by high voltage, constitute a current. When the beam of light is interrupted, the flow of photoelectrons stops and the current is cut off. In the case of the garage-door safety device, when the current stops, the movement of the door stops. In the case of the motion-detection systems, when the current stops, an alarm may sound or a light may turn on.

One of the more exotic uses of the photoelectric effect is in night-vision goggles. Although you may never have looked through night-vision goggles, you have probably seen night-vision images on the news or in a suspense-filled movie such as *Silence of the Lambs.* Typically, the night-vision images we see are from what are known as third-generation night-vision devices. These devices use a photocathode material (gallium arsenide) that emits photoelectrons when struck by photons in the *infrared* region of the electromagnetic spectrum. (First- and second-generation devices used different photocathode materials and relied more on the amplification of low-level visible light.) The photoelectrons emitted by the photocathode enter a microchannel plate (MCP), an array of tiny parallel tubes, where each strikes the internal surface of a tube causing many more electrons to be ejected—a process called *secondary emission.* This effectively amplifies the current generated by each photoelectron. The amplified current is then accelerated by high voltage toward a phosphorus screen, where incident electrons cause the emission of visible light, generating the familiar green glow of night vision.

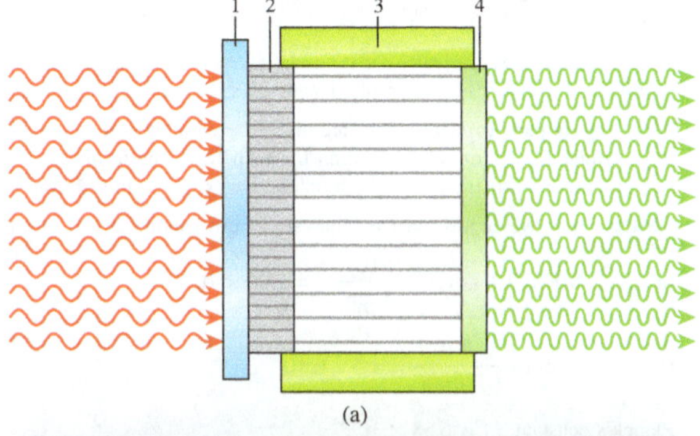

(a)

(b)

(a) Night-vision goggles schematic. 1. Photocathode (gallium arsenide). 2. Microchannel plate (MCP). 3. High-voltage source. 4. Phosphorus screen. (b) The green glow of night vision.

Credit: © *SPL/Photo Researchers.*

shows that the more energetic the photon (i.e., the higher its frequency), the greater the kinetic energy of the ejected electron. If the frequency of light is below the threshold frequency, the photon will simply bounce off the surface and no electrons will be ejected. If the frequency is equal to the threshold frequency, it will dislodge the most loosely held electron. Above the threshold frequency, it will not only dislodge the electron, but also impart some kinetic energy to the ejected electron.

Now consider two beams of light having the same frequency (greater than the threshold frequency) but different intensities. The more intense beam of light consists of a larger number of photons, so it ejects more electrons from the metal's surface than the weaker beam of light. Thus, the more intense the light, the greater the number of electrons emitted by the target metal; the higher the frequency of the light, the greater the kinetic energy of the ejected electrons.

Sample Problem 6.3 shows how to determine the energy of a single photon of light of a given wavelength and how to determine the maximum kinetic energy of an electron ejected via the photoelectric effect.

Einstein's theory of light posed a dilemma for scientists. On the one hand, it explains the photoelectric effect. On the other hand, the particle theory of light is inconsistent with the known wavelike properties of light. The only way to resolve the dilemma is to accept the idea that light possesses properties characteristic of both particles *and* waves. Depending on the experiment, light behaves either as a wave or as a stream of particles. This concept was totally alien to the way physicists had thought about matter and radiation, and it took a long time for them to accept it. We will see in Section 6.4 that possessing properties of both particles and waves is not unique to light but ultimately is characteristic of all matter, including electrons.

SAMPLE PROBLEM 6.3

Calculate the energy (in joules) of (a) a photon with a wavelength of 5.00×10^4 nm (infrared region) and (b) a photon with a wavelength of 52 nm (ultraviolet region). (c) Calculate the kinetic energy of an electron ejected by the photon in part (b) from a metal with a binding energy of 3.7 eV.

Strategy In parts (a) and (b), we are given the wavelength of light. Use Equation 6.1 to convert wavelength to frequency; then use Equation 6.2 to determine the energy of the photon for each wavelength. In part (c), we are asked to determine the kinetic energy of an ejected electron. For this we use Equation 6.3. The binding energy, given in electron volts, must be converted to joules for units to cancel.

Setup The wavelengths must be converted from nanometers to meters:

(a) $$5.00 \times 10^4 \text{ nm} \times \frac{1 \times 10^{-9}\text{ m}}{1 \text{ nm}} = 5.00 \times 10^{-5}\text{ m}$$

(b) $$52 \text{ nm} \times \frac{1 \times 10^{-9}\text{ m}}{1 \text{ nm}} = 5.2 \times 10^{-8}\text{ m}$$

Planck's constant, h, is 6.63×10^{-34} J · s.

(c) $$W = 3.7 \text{ eV} \times \frac{1.602 \times 10^{-19}\text{ J}}{1 \text{ eV}} = 5.9 \times 10^{-19}\text{ J}$$

Solution (a) $$\nu = \frac{c}{\lambda} = \frac{3.00 \times 10^8 \text{ m/s}}{5.00 \times 10^{-5}\text{ m}} = 6.00 \times 10^{12}\text{ s}^{-1}$$

and

$$E = h\nu = (6.63 \times 10^{-34}\text{ J} \cdot \text{s})(6.00 \times 10^{12}\text{ s}^{-1}) = 3.98 \times 10^{-21}\text{ J}$$

This is the energy of a single photon with wavelength 5.00×10^4 nm.

(b) Following the same procedure as in part (a), the energy of a photon of wavelength 52 nm is 3.8×10^{-18} J.

(c) $\text{KE} = h\nu - W = 3.8 \times 10^{-18}\text{ J} - 5.9 \times 10^{-19}\text{ J} = 3.2 \times 10^{-18}\text{ J}$

THINK ABOUT IT

Remember that frequency and wavelength are inversely proportional (Equation 6.1). Thus, as wavelength *decreases*, frequency and energy *increase*. Note that the binding energy becomes less significant as the energy of the incident photon increases.

Practice Problem ATTEMPT Calculate the energy (in joules) of (a) a photon with wavelength 2.11×10^2 nm, and (b) a photon with frequency $1.78 \times 10^8\text{ s}^{-1}$. (c) Calculate the kinetic energy of an electron ejected by the photon in part (a) from a metal with a binding energy of 4.66 eV.

Practice Problem BUILD (a) Calculate the wavelength (in nm) of light with energy 1.89×10^{-20} J per photon. (b) For light of wavelength 410 nm, calculate the number of photons per joule. (c) Determine the binding energy (in eV) of a metal if the kinetic energy possessed by an ejected electron [using one of the photons in part (b)] is 2.93×10^{-19} J.

Practice Problem CONCEPTUALIZE A blue billiard ball with a mass of 165 g rests in a shallow well on an otherwise flat surface. When a red billiard ball with the same mass moving at any velocity less than 1.20 m/s strikes the blue ball, the blue ball does not move (i). When the red ball strikes the blue ball moving at exactly 1.20 m/s, the blue ball is just barely dislodged from the well (ii). What will be the velocity (iii) of the blue ball when it is struck by the red ball moving at 1.75 m/s?

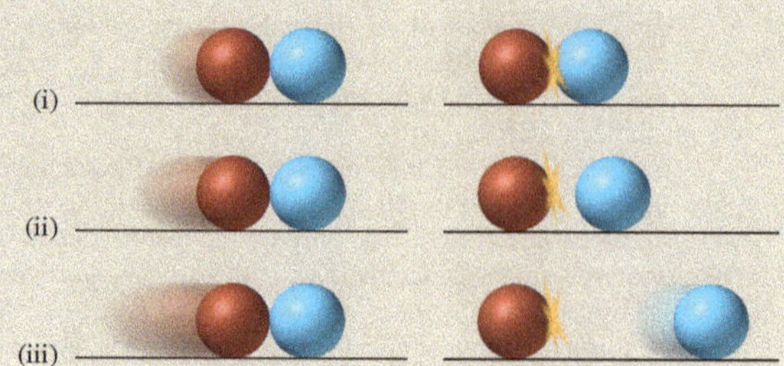

CHECKPOINT – SECTION 6.2 Quantum Theory

6.2.1 Calculate the energy per photon of light with wavelength 650 nm.

a) 1.29×10^{-31} J
b) 4.31×10^{-40} J
c) 1.02×10^{-27} J
d) 1.44×10^{-48} J
e) 3.06×10^{-19} J

6.2.2 Calculate the wavelength of light that has energy 1.32×10^{-23} J/photon.

a) 5.02×10^{-9} cm
b) 6.64×10^{3} cm
c) 2.92×10^{-63} cm
d) 1.51 cm
e) 66.4 cm

6.2.3 Which of the following is characteristic of high-energy radiation?

a) High wavelength
b) High velocity
c) High frequency
d) High amplitude
e) All of the above

6.2.4 A clean metal surface is irradiated with light of three different wavelengths: λ_1, λ_2, and λ_3. The kinetic energies of the ejected electrons are as follows: λ_1: 2.9×10^{-20} J; λ_2: approximately zero; λ_3: 4.2×10^{-19} J. Arrange the light in order of increasing wavelength.

a) $\lambda_1 < \lambda_2 < \lambda_3$
b) $\lambda_2 < \lambda_1 < \lambda_3$
c) $\lambda_3 < \lambda_2 < \lambda_1$
d) $\lambda_3 < \lambda_1 < \lambda_2$
e) $\lambda_2 < \lambda_3 < \lambda_1$

6.3 Bohr's Theory of the Hydrogen Atom

Student Note: If you have ever seen a rainbow, you are familiar with this phenomenon. The rainbow is the visible portion of the sun's emission spectrum.

In addition to explaining the photoelectric effect, Planck's quantum theory and Einstein's ideas made it possible for scientists to unravel another nineteenth-century mystery in physics: atomic line spectra.

In the seventeenth century, Newton had shown that sunlight is composed of various color components that can be recombined to produce white light. Since that time, chemists and physicists have studied the characteristics of such ***emission spectra.*** The emission spectrum of a substance can be seen by energizing a sample of material with either thermal energy or some other form of energy (such as a high-voltage electrical discharge if the substance is a gas). A "red-hot" or "white-hot" iron bar freshly removed from a fire produces a characteristic glow. The glow is the visible portion of its emission spectrum. The heat given off by the same iron bar is another portion of its emission spectrum—the infrared region. A feature common to the emission spectrum of the sun and that of a heated solid is that both are continuous; that is, all wavelengths of visible light are present in each spectrum (Figure 6.6).

(a)

(b)

Figure 6.6 The visible white light emitted by (a) the sun and (b) a white-hot iron bar. In each case, the white light is the combination of all visible wavelengths (see Figure 6.1).

Credit: a: © *Doug Menuez/Getty Images*; b: © *McGraw-Hill Education/Charles D. Winters, photographer.*

Atomic Line Spectra

Unlike those of the sun or a white-hot iron bar, the emission spectra of atoms in the gas phase do not show a continuous spread of wavelengths from red to violet; rather, the atoms produce bright lines in distinct parts of the visible spectrum. These ***line spectra*** are the emission of light only at *specific wavelengths.* Figure 6.7 is a schematic diagram of a discharge tube that is used to study emission spectra.

Every element has a unique emission spectrum, so the characteristic lines in atomic spectra can be used in chemical analysis to identify elements, much as fingerprints are used to identify people. When the lines of the emission spectrum of a known element exactly match the lines of the emission spectrum of an unknown sample, the identity of the element in the sample is established. Although the procedure of identifying elements by their line spectra had been used for many years in chemical analysis, the origin of the spectral lines was not understood until early in the twentieth century. Figure 6.8 shows the emission spectra of several elements.

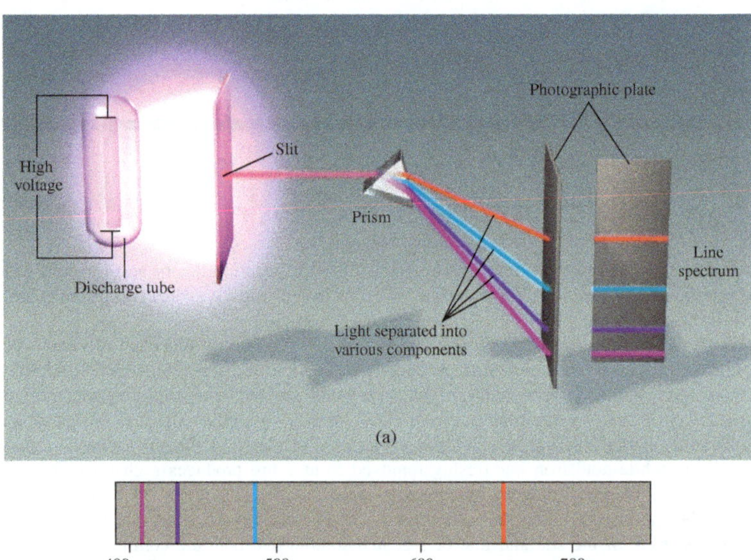

Figure 6.7 (a) Experimental arrangement for studying the emission spectra of atoms and molecules. The gas being studied is in a discharge tube containing two electrodes. As electrons flow from the negative electrode to the positive electrode, they collide with the gas particles. The collisions lead to the emission of light by the atoms (or molecules). The emitted light is separated into its components by a prism. (b) Line emission spectrum of hydrogen.

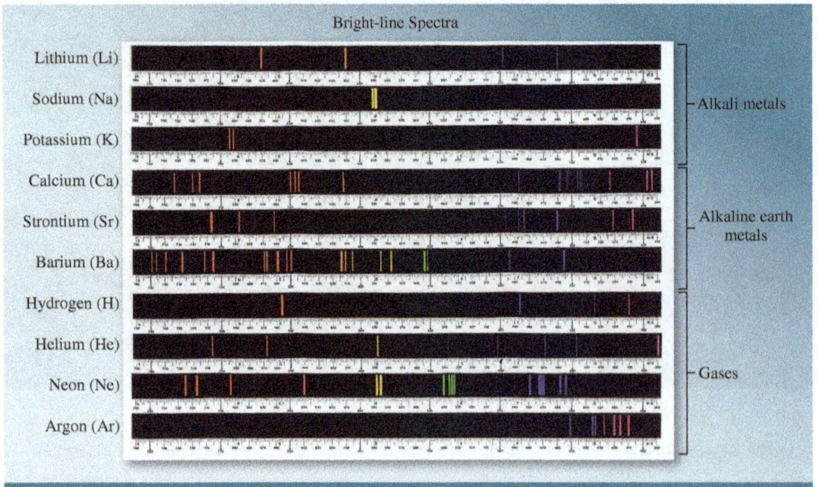

Figure 6.8 Emission spectra of several elements.

Credit: © *McGraw-Hill Education.*

In 1885, Johann Balmer[3] developed a remarkably simple equation that could be used to calculate the wavelengths of the four visible lines in the emission spectrum of hydrogen. Johannes Rydberg[4] developed Balmer's equation further, yielding an equation that could calculate not only the *visible* wavelengths, but those of *all* hydrogen's spectral lines:

Student Note: The Rydberg equation is a mathematical relationship that was derived from experimental data. Although it predates quantum theory by decades, it agrees remarkably well with it for one-electron systems such as the hydrogen atom.

Equation 6.4 $$\frac{1}{\lambda} = R_\infty\left(\frac{1}{n_1^2} - \frac{1}{n_2^2}\right)$$

In Equation 6.4, now known as the ***Rydberg equation,*** λ is the wavelength of a line in the spectrum; R_∞ is the Rydberg constant ($1.09737316 \times 10^7\ \text{m}^{-1}$); and n_1 and n_2 are positive integers, where $n_2 > n_1$.

The Line Spectrum of Hydrogen

In 1913, not long after Planck's and Einstein's discoveries, a theoretical explanation of the emission spectrum of the hydrogen atom was presented by the Danish physicist Niels Bohr.[5] Bohr's treatment is very complex and is no longer considered to be correct in all its details. We will concentrate only on his important assumptions and final results, which account for the observed spectral lines and provide an important step toward the understanding of quantum theory.

When Bohr first approached this problem, physicists already knew that the atom contains electrons and protons. They thought of an atom as an entity in which electrons whirled around the nucleus in circular orbits at high velocities. This was an appealing description because it resembled the familiar model of planetary motion around the sun. However, according to the laws of classical physics, an electron moving in an orbit of a hydrogen atom would experience an acceleration toward the nucleus by radiating away energy in the form of electromagnetic waves. Thus, such an electron would quickly spiral into the nucleus and annihilate itself with the proton. To explain why this does not happen, Bohr postulated that the electron is allowed to occupy only certain orbits of specific energies. In other words, the energies of the electron are quantized. An electron in any of the allowed orbits will not radiate energy and therefore will not spiral into the nucleus.

Animation
Figure 6.10, Emission Spectrum of Hydrogen.

Bohr attributed the emission of radiation by an energized hydrogen atom to the electron dropping from a higher-energy orbit to a lower one and giving up a quantum of energy (a photon) in the form of light (Figure 6.10). Using arguments based on electrostatic interaction and Newton's laws of motion, Bohr showed that the energies that the electron in the hydrogen atom can possess are given by

Equation 6.5 $$E_n = -2.18 \times 10^{-18}\ \text{J}\left(\frac{1}{n^2}\right)$$

where n is an integer with values $n = 1, 2, 3$, and so on. The negative sign in Equation 6.5 is an arbitrary convention, signifying that the energy of the electron in the atom is *lower* than the energy of a *free electron,* which is an electron that is infinitely far from the nucleus. The energy of a free electron is arbitrarily assigned a value of zero. Mathematically, this corresponds to setting n equal to infinity in Equation 6.5:

$$E_\infty = -2.18 \times 10^{-18}\ \text{J}\left(\frac{1}{\infty^2}\right) = 0$$

As the electron gets closer to the nucleus (as n decreases), E_n becomes larger in absolute value, but also more negative. The most negative value, then, is reached when $n = 1$, which corresponds to the most stable energy state. We call this the ***ground state,*** the *lowest* energy state of an atom. The stability of the electron diminishes as n increases. Each energy state in which $n > 1$ is

3. Johann Jakob Balmer (1825–1898). Swiss mathematician. From 1859 until his death in 1898 Balmer taught math at a secondary school for girls in Basel, Switzerland. Although physicists did not understand why his equation worked until long after his death, the visible series of lines in the spectrum of hydrogen is named for him.

4. Johannes Robert Rydberg (1854–1919). Swedish mathematician and physicist. Rydberg analyzed many atomic spectra in an effort to understand the periodic properties of elements. Although he was nominated twice for the Nobel Prize in Physics, he never received it.

5. Niels Henrik David Bohr (1885–1962). Danish physicist. One of the founders of modern physics, he received the Nobel Prize in Physics in 1922 for his theory explaining the line spectrum of hydrogen.

called an ***excited state.*** Each excited state is higher in energy than the ground state. In the hydrogen atom, an electron for which n is greater than 1 is said to be in an excited state.

The radius of each circular orbit in Bohr's model depends on n^2. Thus, as n increases from 1 to 2 to 3, the orbit radius increases very rapidly. The higher the excited state, the farther away the electron is from the nucleus (and the less tightly held it is by the nucleus).

Bohr's theory enables us to explain the line spectrum of the hydrogen atom. Radiant energy absorbed by the atom causes the electron to move from the ground state ($n = 1$) to an excited state ($n > 1$). Conversely, radiant energy (in the form of a photon) is *emitted* when the electron moves from a higher-energy excited state to a lower-energy excited state or the ground state.

The quantized movement of the electron from one energy state to another is analogous to the movement of a tennis ball either up or down a set of stairs (Figure 6.9). The ball can be on any of several steps but never between steps. The journey from a lower step to a higher one is an energy-requiring process, whereas movement from a higher step to a lower step is an energy-releasing process. The quantity of energy involved in either type of change is determined by the distance between the beginning and ending steps. Similarly, the amount of energy needed to move an electron in the Bohr atom depends on the difference in energy levels between the initial and final states.

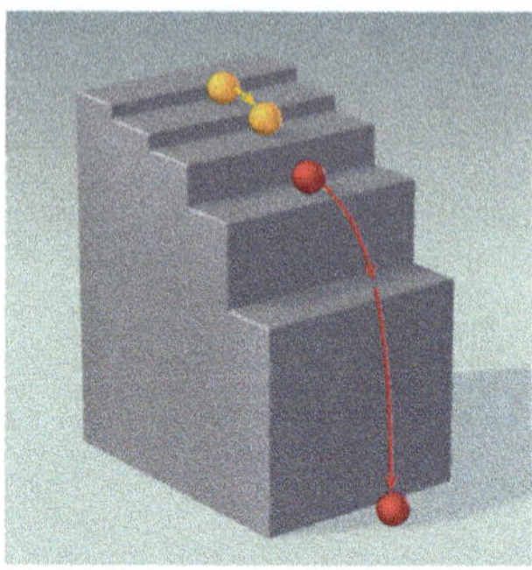

Figure 6.9 Mechanical analogy for the emission processes. The ball can rest on any step but not between steps.

To apply Equation 6.5 to the emission process in a hydrogen atom, let us suppose that the electron is initially in an excited state characterized by n_i. During emission, the electron drops to a lower-energy state characterized by n_f (the subscripts i and f denote the *initial* and *final* states, respectively). This lower-energy state may be the ground state, but it can be any state lower than the initial excited state. The difference between the energies of the initial and final states is

$$\Delta E = E_f - E_i$$

From Equation 6.5,

$$E_f = -2.18 \times 10^{-18}\,\text{J}\left(\frac{1}{n_f^2}\right)$$

and

$$E_i = -2.18 \times 10^{-18}\,\text{J}\left(\frac{1}{n_i^2}\right)$$

Therefore,

$$\Delta E = \left(\frac{-2.18 \times 10^{-18}\,\text{J}}{n_f^2}\right) - \left(\frac{-2.18 \times 10^{-18}\,\text{J}}{n_i^2}\right)$$

$$= -2.18 \times 10^{-18}\,\text{J}\left(\frac{1}{n_f^2} - \frac{1}{n_i^2}\right)$$

Student Note: When $n_i > n_f$, ΔE is *negative*, indicating energy is *emitted*. When $n_f > n_i$, ΔE is *positive*, indicating energy is *absorbed*.

Because this transition results in the emission of a photon of frequency ν and energy $h\nu$, we can write

$$\Delta E = h\nu = -2.18 \times 10^{-18}\,\text{J}\left(\frac{1}{n_f^2} - \frac{1}{n_i^2}\right) \qquad \textbf{Equation 6.6}$$

A photon is emitted when $n_i > n_f$. Consequently, the term in parentheses is *positive,* making ΔE *negative* (energy is lost to the surroundings). A photon is absorbed when $n_f > n_i$, making the term in parentheses *negative,* so ΔE is *positive.* Each spectral line in the emission spectrum of hydrogen corresponds to a particular transition in a hydrogen atom. When we study a large number of hydrogen atoms, we observe all possible transitions and hence the corresponding spectral lines. The brightness of a spectral line depends on how many photons of the same wavelength are emitted.

To calculate the wavelength of an emission line, we substitute c/λ for ν and then divide both sides of Equation 6.6 by hc. In addition, because wavelength (and frequency) can only have positive values, we take the absolute value of the right side of the equation. (In this case, we do so simply by eliminating the negative sign.)

Student Note: Because 2.18×10^{-18} J/hc = $1.096 \times 10^7\ \text{m}^{-1}$, which to three significant figures is equal to R_∞, this equation is essentially the same as the Rydberg equation (Equation 6.4).

$$\frac{1}{\lambda} = \frac{2.18 \times 10^{-18}\,\text{J}}{hc}\left(\frac{1}{n_f^2} - \frac{1}{n_i^2}\right) \qquad \textbf{Equation 6.7}$$

Figure 6.10

Emission Spectrum of Hydrogen

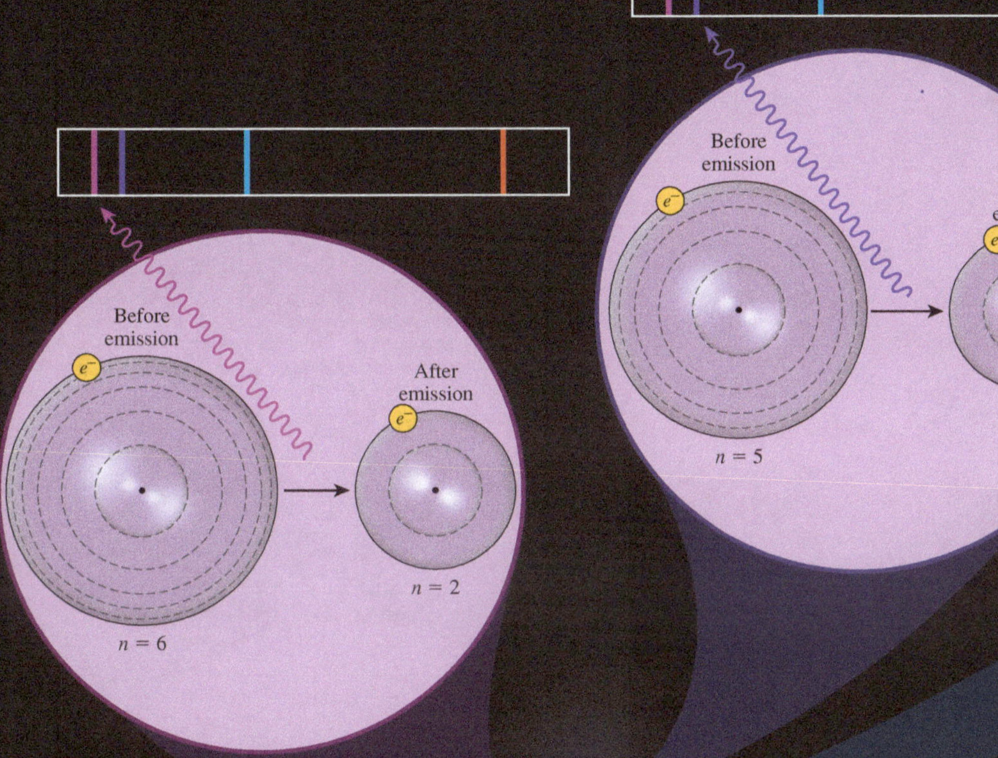

Hydrogen atoms in the ground state are excited by the addition of energy. Electrons that have been promoted to higher energy levels ($n > 1$) return to lower energy levels and emit the excess energy as electromagnetic radiation.

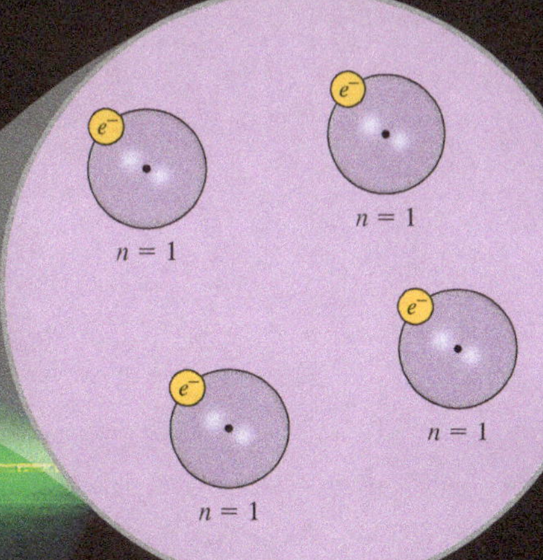

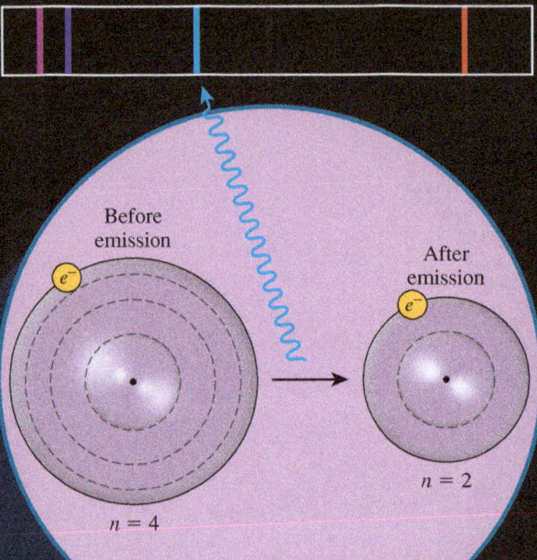

The transitions shown are as follows:

- $n = 3$ to $n = 2$
- $n = 4$ to $n = 2$
- $n = 5$ to $n = 2$
- $n = 6$ to $n = 2$

Electrons in H atoms may be promoted to excited states other than 3, 4, 5, or 6; and electrons in excited states may return to a state other than $n = 2$. However, the transitions shown are the ones that give rise to the visible lines in the hydrogen emission spectrum.

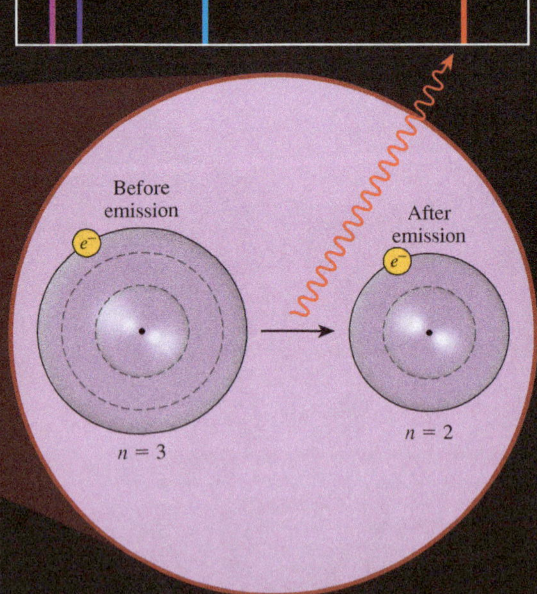

What's the point?

Each line in the visible emission spectrum of hydrogen is the result of an electronic transition from a higher excited state ($n = 3, 4, 5$, or 6) to a lower excited state ($n = 2$). The energy gap between the initial and final states determines the wavelength of the light emitted.

(See Visualizing Chemistry questions VC 6.1–VC 6.4 on page 275.)

TABLE 6.1 Emission Series in the Hydrogen Spectrum

Series	n_f	n_i	Spectrum Region
Lyman	1	2, 3, 4, . . .	Ultraviolet
Balmer	2	3, 4, 5, . . .	Visible and ultraviolet
Paschen	3	4, 5, 6, . . .	Infrared
Brackett	4	5, 6, 7, . . .	Infrared

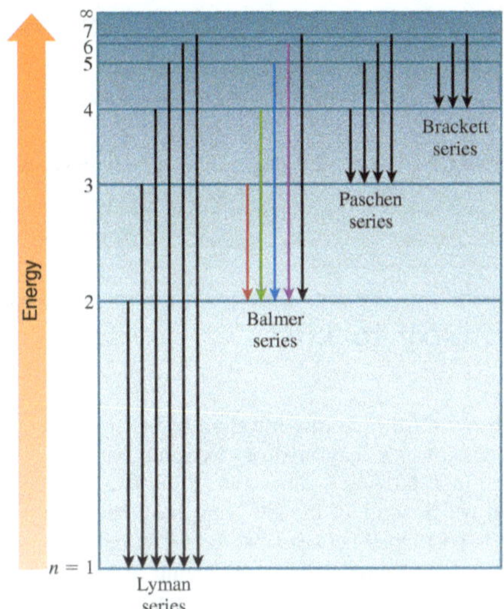

Figure 6.11 Energy levels in the hydrogen atom and the various emission series. Each series terminates at a different value of *n*.

The emission spectrum of hydrogen includes a wide range of wavelengths from the infrared to the ultraviolet. Table 6.1 lists the series of transitions in the hydrogen spectrum, each with a different value of n_f. The series are named after their discoverers (Lyman, Balmer, Paschen, and Brackett). The Balmer series was the first to be studied because some of its lines occur in the visible region.

Figure 6.11 shows transitions associated with spectral lines in each of the emission series. Each horizontal line represents one of the allowed energy levels for the electron in a hydrogen atom. The energy levels are labeled with their *n* values.

Sample Problem 6.4 illustrates the use of Equation 6.7.

SAMPLE PROBLEM 6.4

Calculate the wavelength (in nm) of the photon emitted when an electron transitions from the $n = 4$ state to the $n = 2$ state in a hydrogen atom.

Strategy Use Equation 6.7 to calculate λ.

Setup According to the problem, the transition is from $n = 4$ to $n = 2$, so $n_i = 4$ and $n_f = 2$. The required constants are $h = 6.63 \times 10^{-34}$ J · s and $c = 3.00 \times 10^8$ m/s.

Solution

$$\frac{1}{\lambda} = \frac{2.18 \times 10^{-18}\ \text{J}}{hc}\left(\frac{1}{n_f^2} - \frac{1}{n_i^2}\right)$$

$$= \frac{2.18 \times 10^{-18}\ \text{J}}{(6.63 \times 10^{-34}\ \text{J} \cdot \text{s})(3.00 \times 10^8\ \text{m/s})}\left(\frac{1}{2^2} - \frac{1}{4^2}\right)$$

$$= 2.055 \times 10^6\ \text{m}^{-1}$$

Remember to keep at least one extra digit in intermediate answers to avoid rounding error in the final result [⏮ Section 1.5]. Therefore,

$$\lambda = 4.87 \times 10^{-7} \text{ m} = 487 \text{ nm}$$

THINK ABOUT IT

Look again at the line spectrum of hydrogen in Figure 6.7 and make sure that your result matches one of them. Note that for an emission, n_i is always greater than n_f, and Equation 6.7 gives a positive result.

Practice Problem ATTEMPT What is the wavelength (in nm) of a photon emitted during a transition from the $n = 3$ state to the $n = 1$ state in the H atom?

Practice Problem BUILD What is the value of n_i for an electron that emits a photon of wavelength 93.14 nm when it returns to the ground state in the H atom?

Practice Problem CONCEPTUALIZE For each pair of transitions, determine which one results in emission of the larger amount of energy.

(a) $n = 6$ to $n = 3$ $n = 3$ to $n = 2$ (b) $n = 3$ to $n = 1$ $n = 10$ to $n = 2$ (c) $n = 2$ to $n = 1$ $n = 99$ to $n = 2$

Bringing Chemistry to Life

Lasers

The word *laser* is an acronym for *l*ight *a*mplification by *s*timulated *e*mission of *r*adiation. It is a special type of emission that may involve electronic transitions in *atoms* or *molecules*. The first laser, developed in 1960, was a ruby laser. Ruby is a deep-red mineral consisting of corundum (Al_2O_3), in which some of the Al^{3+} ions have been replaced by Cr^{3+} ions. A cylindrical ruby crystal is positioned between two perfectly parallel mirrors, one of which is only partially reflective. A light source called a flashlamp is used to excite the chromium atoms to a higher energy level. The excited atoms are unstable, so at a given instant some of them return to the ground state by emitting a photon in the red region of the spectrum ($\lambda = 694.3$ nm). Spontaneous emission of photons occurs in all directions, but photons emitted directly at either mirror will be reflected back. As the reflected photons pass back through the ruby crystal, they stimulate the emission of more photons in the same direction. These photons, in turn, are reflected back through the crystal, stimulating still more emissions, and so on. Because the light waves are *in phase*—that is, their maxima coincide and their minima coincide—the photons enhance one another, increasing their power with each passage between the mirrors. When the light reaches a certain intensity, it emerges from the partially reflective mirror as a laser beam. Laser light is characterized by three properties: It is *intense,* it has a precisely known wavelength and therefore *energy,* and it is *coherent.*

Student Note: *Coherent* means that the light waves are all in phase.

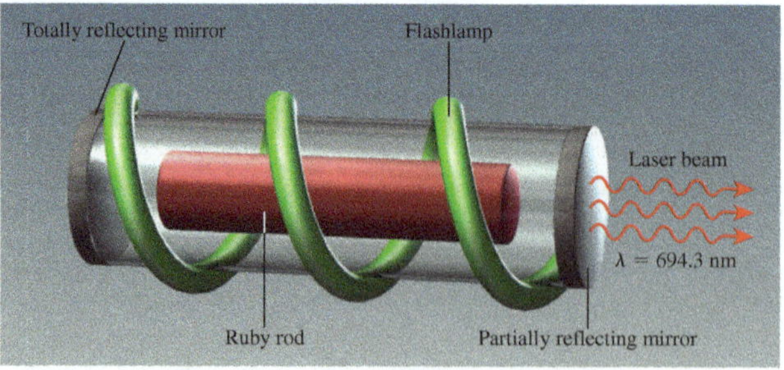

The uses of lasers are quite numerous, including many medical applications. Their high intensity and ease of focus make them suitable for drilling holes in metals, welding, and carrying out nuclear fusion. The fact that they are highly directional and have precisely known wavelengths makes them very useful for telecommunications. Lasers are also used in isotope separation, in chemical analysis, in holography (three-dimensional photography), in compact disc players, and in supermarket scanners.

CHECKPOINT – SECTION 6.3 Bohr's Theory of the Hydrogen Atom

6.3.1 Calculate the energy of an electron in the $n = 3$ state in a hydrogen atom.

a) 2.42×10^{-19} J
b) -2.42×10^{-19} J
c) 7.27×10^{-19} J
d) -7.27×10^{-19} J
e) -6.54×10^{-18} J

6.3.2 Calculate ΔE of an electron that goes from $n = 1$ to $n = 5$.

a) 8.72×10^{-20} J
b) -8.72×10^{-20} J
c) 5.45×10^{-17} J
d) 2.09×10^{-18} J
e) -2.09×10^{-18} J

6.3.3 What is the wavelength of light emitted when an electron in a hydrogen atom goes from $n = 5$ to $n = 3$?

a) 4.87×10^{-7} m
b) 6.84×10^{-7} m
c) 1.28×10^{-6} m
d) 3.65×10^{-7} m
e) 1.02×10^{-7} m

6.3.4 Which wavelength corresponds to the transition of an electron in a hydrogen atom from $n = 2$ to $n = 1$?

a) 182 nm
b) 91.2 nm
c) 724 nm
d) 812 nm
e) 122 nm

6.4 Wave Properties of Matter

Bohr's theory was both fascinating and puzzling. It fit the experimental data for hydrogen, but physicists did not understand the underlying principle. Why, for example, was an electron restricted to orbiting the nucleus at certain fixed distances? For a decade no one, not even Bohr himself, could offer a logical explanation. In 1924, Louis de Broglie[6] provided a solution to this puzzle. De Broglie reasoned that if energy (light) can, under certain circumstances, behave like a stream of particles (photons), then perhaps particles such as electrons can, under certain circumstances, exhibit wavelike properties.

The de Broglie Hypothesis

In developing his revolutionary theory, de Broglie incorporated his observations of macroscopic phenomena that exhibited quantized behavior. For example, a guitar string has certain discrete frequencies of vibration, like those shown in Figure 6.12(a). The waves generated by plucking a guitar string are *standing* or *stationary waves* because they do not travel along the string. Some points on the string, called ***nodes,*** do not move at all; that is, the amplitude of the wave at these points is *zero.* There is a node at each end, and there may be one or more nodes between the ends. The greater the frequency of vibration, the shorter the wavelength of the standing wave and the greater the number of nodes. According to de Broglie, an electron in an atom behaves like a *standing wave.* However, as Figure 6.12 shows, only certain wavelengths are possible or *allowed.*

De Broglie argued that if an electron does behave like a standing wave in the hydrogen atom, the wavelength must fit the circumference of the orbit exactly; that is, the circumference of the orbit must be an integral multiple of the wavelength, as shown in Figure 6.12(b). Otherwise, the wave would partially cancel itself by destructive interference on each successive orbit, quickly reducing its amplitude to zero.

The relationship between the circumference of an allowed orbit ($2\pi r$) and the wavelength (λ) of the electron is given by

Equation 6.8 $$2\pi r = n\lambda$$

where r is the radius of the orbit, λ is the wavelength of the electron wave, and n is a positive integer (1, 2, 3, . . .). Because n is an integer, r can have only certain values (integral multiples of λ) as n increases from 1 to 2 to 3 and so on. And, because the energy of the electron depends on the size of the orbit (or the value of r), the energy can have only certain values, too. Thus, the energy of the electron in a hydrogen atom, if it behaves like a standing wave, must be quantized.

De Broglie's reasoning led to the conclusion that waves can behave like particles and particles can exhibit wavelike properties. De Broglie deduced that the particle and wave properties are related by the following expression:

6. Louis Victor Pierre Raymond Duc de Broglie (1892–1987). French physicist. A member of an old and noble family in France, he held the title of a prince. In his doctoral dissertation, he proposed that matter and radiation have the properties of both wave and particle. For this work, de Broglie was awarded the Nobel Prize in Physics in 1929.

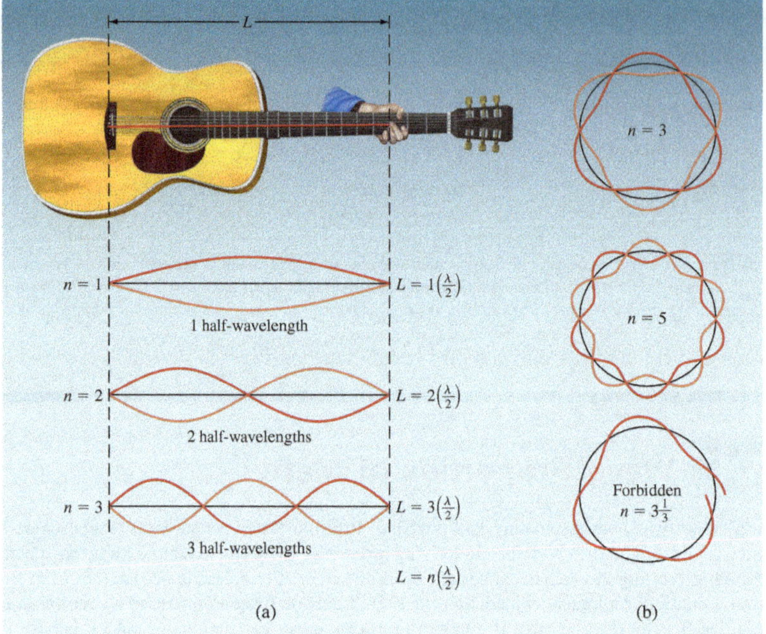

Figure 6.12 (a) Standing waves of a vibrating guitar string. The length of the string must be equal to a whole number times one-half the wavelength (λ/2). (b) In a circular orbit, only whole number multiples of wavelengths are allowed. Any fractional number of wavelengths would result in cancellation of the wave due to destructive interference.

$$\lambda = \frac{h}{mu} \qquad \textbf{Equation 6.9}$$

where λ, m, and u are the wavelength associated with a moving particle, its mass, and its velocity, respectively. Equation 6.9 implies that a particle in motion can be treated as a wave, and a wave can exhibit the properties of a particle. To help you remember this important point, notice that the left side of Equation 6.9 involves the wavelike property of wavelength, whereas the right side involves mass, a property of particles. A wavelength calculated using Equation 6.9 is usually referred to specifically as a ***de Broglie wavelength.*** Likewise, we will refer to a *mass* calculated using Equation 6.9 as a *de Broglie mass.*

Student Note: Mass (m) must be expressed in kilograms for units to cancel properly in Equation 6.9.

Sample Problem 6.5 illustrates how de Broglie's theory and Equation 6.9 can be applied.

SAMPLE PROBLEM 6.5

Calculate the de Broglie wavelength of the "particle" in the following two cases: (a) a 25-g bullet traveling at 612 m/s, and (b) an electron ($m = 9.109 \times 10^{-31}$ kg) moving at 63.0 m/s.

Strategy Use Equation 6.9 to calculate the de Broglie wavelengths. Remember that the mass in Equation 6.9 must be expressed in kilograms for the units to cancel properly.

Setup Planck's constant, h, is 6.63×10^{-34} J · s or, for the purpose of making the unit cancellation obvious, 6.63×10^{-34} kg · m²/s. Remember that 1 J = 1 kg · m²/s².

Solution

(a) $25 \text{ g} \times \frac{1 \text{ kg}}{1000 \text{ g}} = 0.025 \text{ kg}$

$$\lambda = \frac{h}{mu} = \frac{6.63 \times 10^{-34} \text{ kg} \cdot \text{m}^2/\text{s}}{(0.025 \text{ kg})(612 \text{ m/s})} = 4.3 \times 10^{-35} \text{ m}$$

(b) $\lambda = \frac{h}{mu} = \frac{6.63 \times 10^{-34} \text{ kg} \cdot \text{m}^2/\text{s}}{(9.109 \times 10^{-31} \text{ kg})(63.0 \text{ m/s})} = 1.16 \times 10^{-5} \text{ m}$

(Continued on next page)

THINK ABOUT IT

While you are new at solving these problems, always write out the units of Planck's constant (J · s) as kg · m^2/s. This will enable you to check your unit cancellations and detect common errors such as expressing mass in grams rather than kilograms. Note that the calculated wavelength of a macroscopic object, even one as small as a bullet, is extremely small. An object must be at least as small as a subatomic particle for its wavelength to be large enough for us to observe.

Practice Problem ATTEMPT Calculate the de Broglie wavelength (in nm) of a hydrogen atom ($m = 1.674 \times 10^{-27}$ kg) moving at 1500 cm/s.

Practice Problem BUILD Use Equation 6.9 to calculate the *momentum*, p (defined as mass times velocity, $m \times u$) associated with a photon of radiation of wavelength 810 nm. The velocity of a photon is the speed of light, c.

Student Note: Momentum has units of kg · m/s or N · s, where N is the *newton*, the SI unit of *force*. The newton is a derived SI unit: 1 N = 1 kg · m/s^2.

Practice Problem CONCEPTUALIZE Consider the impact of early electron diffraction experiments on scientists' understanding of the behavior of matter. Which of the following imaginary macroscopic experiments most closely corresponds to the remarkable outcome of electron diffraction?

(a) Combining one marble with another by one method yields two marbles; but combining the two marbles by another method yields four marbles.

(b) Combining one marble with another by one method yields two marbles; but combining the two marbles by another method yields zero marbles.

(c) Combining one marble with another by any method yields two marbles.

Diffraction of Electrons

Shortly after de Broglie introduced his equation and predicted that electrons should exhibit wave properties, successful electron diffraction experiments were carried out by Clinton Davisson[7] and Lester Germer[8] in the United States and G. P. Thomson[9] in England. These experiments demonstrated that electrons do indeed possess wavelike properties. By directing a beam of electrons (which are most definitely particles) through a thin piece of gold foil, Thomson obtained a set of concentric rings on a screen, similar to the diffraction pattern observed when X rays (which are most definitely waves) were used. Figure 6.13 shows X-ray and electron diffraction patterns for aluminum.

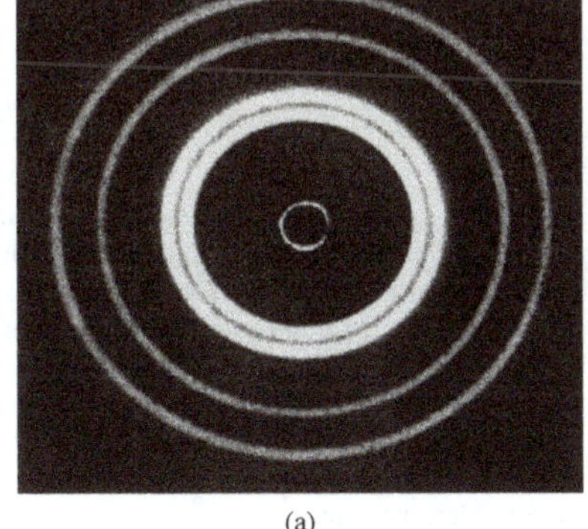

(a)

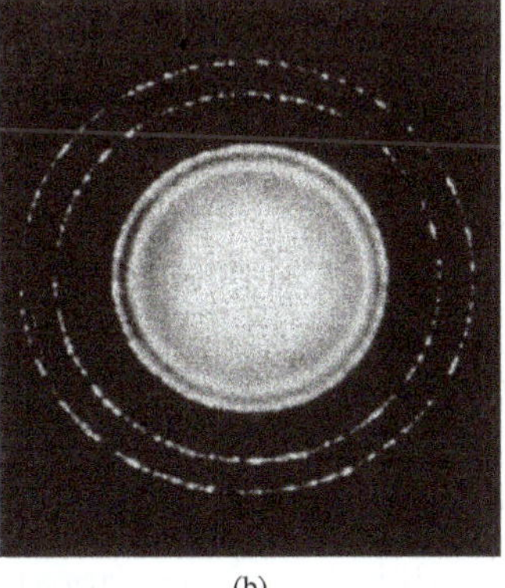

(b)

Figure 6.13 (a) X-ray diffraction pattern of aluminum foil. (b) Electron diffraction of aluminum foil. The similarity of these two patterns shows that electrons can behave like X rays and display wave properties.

Credit: *Erwin Schrödinger, Wave Mechanical Model of the Atom. 1926.*

7. Clinton Joseph Davisson (1881–1958). American physicist. He and G. P. Thomson shared the Nobel Prize in Physics in 1937 for demonstrating the wave properties of electrons.

8. Lester Halbert Germer (1896–1972). American physicist. Discoverer (with Davisson) of the wave properties of electrons.

9. George Paget Thomson (1892–1975). English physicist. Son of J. J. Thomson, he received the Nobel Prize in Physics in 1937, along with Clinton Davisson, for demonstrating the wave properties of electrons.

CHECKPOINT – SECTION 6.4 Wave Properties of Matter

6.4.1 Calculate the de Broglie wavelength associated with a helium-4 atom (4.00 amu) moving at 3.0×10^6 m/s.

a) 2.0×10^{-20} m

b) 3.3×10^{-11} m

c) 3.3×10^{-14} m

d) 1.8×10^{-19} m

e) 6.6×10^{-27} m

6.4.2 What is the momentum of a photon of light with $\lambda = 122$ nm? (The velocity of a photon is the speed of light, c.)

a) 5.43×10^{-27} kg · m/s

b) 6.63×10^{-34} kg · m/s

c) 6.00×10^{-17} kg · m/s

d) 8.64×10^{-11} kg · m/s

e) 2.71×10^{-27} kg · m/s

6.5 Quantum Mechanics

The discovery that waves could have matterlike properties and that matter could have wavelike properties was revolutionary. Although scientists had long believed that energy and matter were distinct entities, the distinction between them, at least at the atomic level, was no longer clear. Bohr's theory was tremendously successful in explaining the line spectrum of hydrogen, but it failed to explain the spectra of atoms with more than one electron. The electron appeared to behave as a particle in some circumstances and as a wave in others. Neither description could completely explain the behavior of electrons in atoms. This left scientists frustrated in their quest to understand exactly where the electrons in an atom are.

The Uncertainty Principle

To describe the problem of trying to locate a subatomic particle that behaves like a wave, Werner Heisenberg[10] formulated what is now known as the ***Heisenberg uncertainty principle:*** It is impossible to know simultaneously both the *momentum p* and the *position x* of a particle with certainty. Stated mathematically,

$$\Delta x \cdot \Delta p \geq \frac{h}{4\pi} \qquad \textbf{Equation 6.10}$$

For a particle of mass m,

$$\Delta x \cdot m\Delta u \geq \frac{h}{4\pi} \qquad \textbf{Equation 6.11}$$

Student Note: Like the de Broglie wavelength equation, Equation 6.11 requires that mass be expressed in kilograms. Unit cancellation will be more obvious if you express Planck's constant in kg · m²/s rather than J · s.

where Δx and Δu are the uncertainties in measuring the position and velocity of the particle, respectively. The $\geq$ signs have the following meaning. If the measured uncertainties of position and velocity are large (say, in a crude experiment), their product can be substantially greater than $h/4\pi$ (hence the $>$ sign). The significance of Equation 6.11 is that even in the most favorable conditions for measuring position and velocity, the product of the uncertainties can never be less than $h/4\pi$ (hence the $=$ sign). Thus, making measurement of the velocity of a particle *more* precise (i.e., making Δu a *small* quantity) means that the position must become correspondingly *less* precise (i.e., Δx will become *larger*). Similarly, if the position of the particle is known more precisely, its velocity measurement must become less precise.

If the Heisenberg uncertainty principle is applied to the hydrogen atom, we find that the electron cannot orbit the nucleus in a well-defined path, as Bohr thought. If it did, we could determine precisely both the position of the electron (from the radius of the orbit) and its speed (from its kinetic energy) at the same time. This would violate the uncertainty principle.

10. Werner Karl Heisenberg (1901–1976). German physicist. One of the founders of modern quantum theory. Heisenberg received the Nobel Prize in Physics in 1932.

Sample Problem 6.6 shows how the Heisenberg uncertainty principle can be applied.

SAMPLE PROBLEM 6.6

An electron in a hydrogen atom is known to have a velocity of 5×10^6 m/s $\pm$ 1 percent. Using the uncertainty principle, calculate the minimum uncertainty in the position of the electron and, given that the diameter of the hydrogen atom is less than 1 angstrom (Å), comment on the magnitude of this uncertainty compared to the size of the atom.

Strategy The uncertainty in the velocity, 1 percent of 5×10^6 m/s, is Δu. Using Equation 6.11, calculate Δx and compare it with the diameter of the hydrogen atom. Recall that 1 Å is equal to 1×10^{-10} m [◂◂ Section 2.2].

Setup The mass of an electron (from Table 2.1, rounded to three significant figures and converted to kilograms) is 9.11×10^{-31} kg. Planck's constant, h, is 6.63×10^{-34} kg · m²/s.

Solution

$$\Delta u = 0.01 \times 5 \times 10^6 \text{ m/s} = 5 \times 10^4 \text{ m/s}$$

$$\Delta x = \frac{h}{4\pi \cdot m\Delta u}$$

Therefore,

$$\Delta x = \frac{6.63 \times 10^{-34} \text{ kg} \cdot \text{m}^2/\text{s}}{4\pi(9.11 \times 10^{-31} \text{ kg})(5 \times 10^4 \text{ m/s})} \geq 1 \times 10^{-9} \text{ m}$$

The *minimum* uncertainty in the position x is 1×10^{-9} m = 10 Å. The uncertainty in the electron's position is 10 times larger than the atom!

THINK ABOUT IT

A common error is expressing the mass of the particle in grams instead of kilograms, but you should discover this inconsistency if you check your unit cancellation carefully. Remember that if one uncertainty is small, the other must be large. The uncertainty principle applies in a practical way only to submicroscopic particles. In the case of a macroscopic object, where the mass is much larger than that of an electron, small uncertainties, relative to the size of the object, are possible for both position and velocity.

Practice Problem ATTEMPT Calculate the minimum uncertainty in the position of the 25-g bullet from Sample Problem 6.5 if the uncertainty in its velocity is (a) ±1 percent and (b) ±0.01 percent.

Practice Problem BUILD (a) Calculate the minimum uncertainty in the momentum of an object for which the uncertainty in position is 3 Å. To what minimum uncertainty in velocity does this correspond if the particle is (b) a neutron (mass = 1.0087 amu) and (c) an electron (mass = 5.486×10^{-4} amu)?

Practice Problem CONCEPTUALIZE Using Equation 6.11, we can calculate the minimum uncertainty in the position or the velocity of any moving particle, including a macroscopic object such as a marble. Calculate the uncertainty in position of a 10-g marble moving at 2.5 m/s (±5 percent) and comment on the significance of your result.

The Schrödinger Equation

Bohr made a significant contribution to our understanding of atoms, and his suggestion that the energy of an electron in an atom is quantized remains unchallenged, but his theory did not provide a complete description of the behavior of electrons in atoms. In 1926, the Austrian physicist Erwin Schrödinger,[11] using a complex mathematical technique, formulated an equation that describes the behavior and energies of submicroscopic particles in general, an equation analogous to Newton's laws of motion for macroscopic objects. The *Schrödinger equation* requires advanced calculus to solve, and we will not discuss it here. The equation, however, incorporates both particle behavior, in terms of mass m, and wave behavior, in terms of a *wave function* ψ (psi), which depends on the location in space of the system (such as an electron in an atom).

The wave function itself has no direct physical meaning. However, the probability of finding the electron in a certain region in space is proportional to the square of the wave function, ψ^2. The idea of relating ψ^2 to probability stemmed from a wave theory analogy. According to wave theory, the intensity of light is proportional to the square of the amplitude of the wave, or ψ^2.

11. Erwin Schrödinger (1887–1961). Austrian physicist. Schrödinger formulated wave mechanics, which laid the foundation for modern quantum theory. He received the Nobel Prize in Physics in 1933.

The most likely place to find a photon is where the intensity is greatest—that is, where the value of ψ^2 is greatest. A similar argument associates ψ^2 with the likelihood of finding an electron in regions surrounding the nucleus.

Schrödinger's equation launched an entirely new field, called *quantum mechanics* (or *wave mechanics*), and began a new era in physics and chemistry. We now refer to the developments in quantum theory from 1913—when Bohr presented his model of the hydrogen atom—to 1926 as "old quantum theory."

The Quantum Mechanical Description of the Hydrogen Atom

The Schrödinger equation specifies the possible energy states the electron can occupy in a hydrogen atom and identifies the corresponding wave functions (ψ). These energy states and wave functions are characterized by a set of *quantum numbers* (to be discussed shortly), with which we can construct a comprehensive model of the hydrogen atom.

Although quantum mechanics does not allow us to specify the exact location of an electron in an atom, it does define the region where the electron is most likely to be at a given time. The concept of ***electron density*** gives the probability that an electron will be found in a particular region of an atom. The square of the wave function, ψ^2, defines the distribution of electron density in three-dimensional space around the nucleus. Regions of high electron density represent a high probability of locating the electron (Figure 6.14).

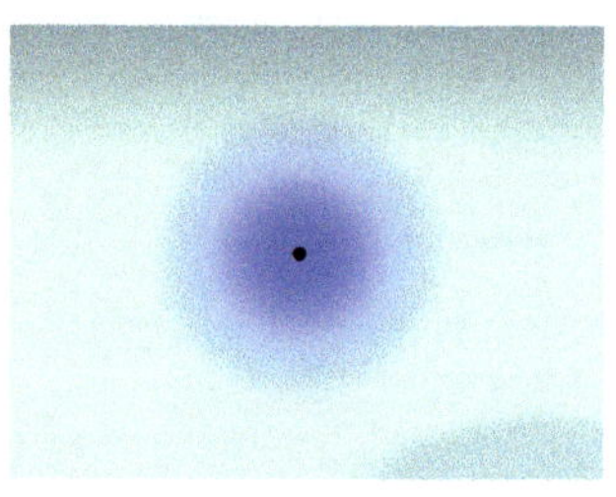

Figure 6.14 Representation of the electron density distribution surrounding the nucleus in the hydrogen atom. It shows a higher probability of finding the electron closer to the nucleus.

To distinguish the quantum mechanical description of an atom from Bohr's model, we speak of an atomic *orbital,* rather than an orbit. An ***atomic orbital*** can be thought of as the wave function of an electron in an atom. When we say that an electron is in a certain orbital, we mean that the distribution of the electron density or the probability of locating the electron in space is described by the square of the wave function associated with that orbital. An atomic orbital, therefore, has a characteristic energy, as well as a characteristic distribution of electron density.

CHECKPOINT – SECTION 6.5 Quantum Mechanics

6.5.1 What is the minimum uncertainty in the position of an electron moving at a speed of 4×10^6 m/s $\pm$ 1 percent? (The mass of an electron is 9.11×10^{-31} kg.)

a) 2×10^{-8} m
b) 1×10^{-9} m
c) 6×10^{-9} m
d) 7×10^{-8} m
e) 1×10^{-12} m

6.5.2 What is the minimum uncertainty in the position of a proton moving at a speed of 4×10^6 m/s $\pm$ 1 percent? (The mass of a proton is 1.67×10^{-27} kg.)

a) 1×10^{-13} m
b) 8×10^{-10} m
c) 4×10^{-11} m
d) 3×10^{-12} m
e) 8×10^{-13} m

6.6 Quantum Numbers

In Bohr's model of the hydrogen atom, only one number, *n,* was necessary to describe the location of the electron. In quantum mechanics, three ***quantum numbers*** are required to describe the *distribution of electron density* in an atom. These numbers are derived from the mathematical solution of Schrödinger's equation for the hydrogen atom. They are called the *principal* quantum number, the *angular momentum* quantum number, and the *magnetic* quantum number. Each atomic orbital in an atom is characterized by a unique set of these three quantum numbers.

Student Note: The three quantum numbers n, ℓ, and m_ℓ specify the *size, shape,* and *orientation* of an orbital, respectively.

Principal Quantum Number (*n*)

The ***principal quantum number (n)*** designates the *size* of the orbital. The larger *n* is, the greater the average distance of an electron in the orbital from the nucleus and therefore the larger the orbital. The principal quantum number can have integral values of 1, 2, 3, and so forth, and it corresponds to the quantum number in Bohr's model of the hydrogen atom. Recall from Equation 6.5 that in a hydrogen atom, the value of *n* alone determines the energy of an orbital. (As we will see shortly, this is *not* the case for an atom that contains more than one electron.)

Angular Momentum Quantum Number (ℓ)

The ***angular momentum quantum number*** (ℓ) describes the *shape* of the atomic orbital (see Section 6.7). The values of ℓ are integers that depend on the value of the principal quantum number, n. For a given value of n, the possible values of ℓ range from 0 to $n - 1$. If $n = 1$, there is only one possible value of ℓ; that is, 0 ($n - 1$ where $n = 1$). If $n = 2$, there are two values of ℓ: 0 and 1. If $n = 3$, there are three values of ℓ: 0, 1, and 2. The value of ℓ is designated by the letters *s, p, d,* and *f* as follows:[12]

ℓ	0	1	2	3
Orbital designation	*s*	*p*	*d*	*f*

Thus, if $\ell = 0$, we have an *s* orbital; if $\ell = 1$, we have a *p* orbital; and so on.

A collection of orbitals with the same value of *n* is frequently called a *shell.* One or more orbitals with the same *n* and ℓ values are referred to as a *subshell.* For example, the shell designated by $n = 2$ is composed of two subshells: $\ell = 0$ and $\ell = 1$ (the allowed values of ℓ for $n = 2$). These subshells are called the 2*s* and 2*p* subshells where 2 denotes the value of *n,* and *s* and *p* denote the values of ℓ.

Magnetic Quantum Number (m_ℓ)

The ***magnetic quantum number*** (m_ℓ) describes the *orientation* of the orbital in space (see Section 6.7). Within a subshell, th e value of m_ℓ depends on the value of ℓ. For a certain value of ℓ, there are $(2\ell + 1)$ integral values of m_ℓ as follows:

$$-\ell, \ldots 0, \ldots, +\ell$$

If $\ell = 0$, there is only one possible value of m_ℓ: 0. If $\ell = 1$, then there are *three* values of m_ℓ: −1, 0, and +1. If $\ell = 2$, there are *five* values of m_ℓ, namely, −2, −1, 0, +1, and +2, and so on. The number of m_ℓ values indicates the number of *orbitals* in a subshell with a particular ℓ value; that is, each m_ℓ value refers to a different orbital.

Student Note: The number of subshells in a shell is equal to *n*. The number of orbitals in a shell is equal to n^2. The number of orbitals in a subshell is equal to $2\ell + 1$.

Table 6.2 summarizes the allowed values of the three quantum numbers, *n*, ℓ, and m_ℓ, and Figure 6.15 illustrates schematically how the allowed values of quantum numbers give rise to the number of subshells and orbitals in each shell of an atom.

TABLE 6.2 Allowed Values of the Quantum Numbers *n*, ℓ, and m_ℓ

When *n* is	ℓ can be	When ℓ is	m_ℓ can be
1	only 0	0	only 0
2	0 or 1	0	only 0
		1	−1, 0, or +1
3	0, 1, or 2	0	only 0
		1	−1, 0, or +1
		2	−2, −1, 0, +1, or +2
4	0, 1, 2, or 3	0	only 0
		1	−1, 0, or +1
		2	−2, −1, 0, +1, or +2
		3	−3, −2, −1, 0, +1, +2, or +3
.	.	.	.
.	.	.	.
.	.	.	.

12. The unusual sequence of letters (*s, p, d,* and *f*) has an historical origin. Physicists who studied atomic emission spectra tried to correlate their observations of spectral lines with the energy states involved in the transitions. They described the emission lines as *s*harp, *p*rincipal, *d*iffuse, and *f*undamental.

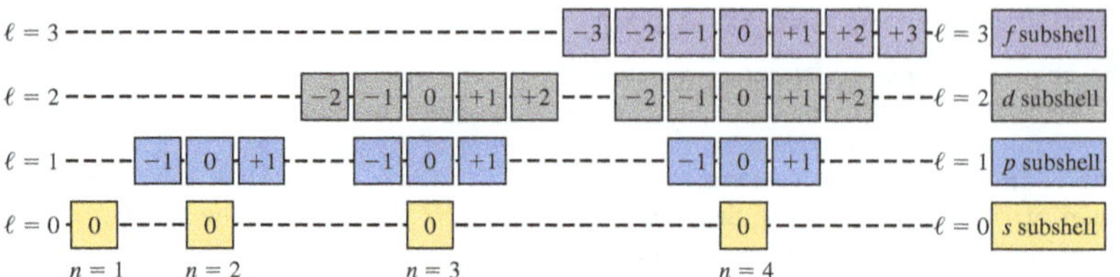

Figure 6.15 Illustration of how quantum numbers designate shells, subshells, and orbitals.

Sample Problem 6.7 gives you some practice with the allowed values of quantum numbers.

SAMPLE PROBLEM 6.7

What are the possible values for the magnetic quantum number (m_ℓ) when the principal quantum number (n) is 3 and the angular momentum quantum number (ℓ) is 1?

Strategy Use the rules governing the allowed values of m_ℓ. Recall that the possible values of m_ℓ depend on the value of ℓ, not on the value of n.

Setup The possible values of m_ℓ are $-\ell, \ldots, 0, \ldots, +\ell$.

Solution The possible values of m_ℓ are -1, 0, and $+1$.

THINK ABOUT IT

Consult Table 6.2 to make sure your answer is correct. Table 6.2 confirms that it is the value of ℓ, not the value of n, that determines the possible values of m_ℓ.

Practice Problem ATTEMPT (a) What are the possible values for m_ℓ when the principal quantum number (n) is 2 and the angular momentum quantum number (ℓ) is 0? (b) What are the possible values for m_ℓ when the principal quantum number (n) is 3 and the angular momentum quantum number (ℓ) is 2?

Practice Problem BUILD (a) What is the lowest possible value of the principal quantum number (n) when the angular momentum quantum number (ℓ) is 1? (b) What are the possible values of the angular momentum quantum number (ℓ) when the principal quantum number (n) is 4 and the magnetic quantum number (m_ℓ) is 0?

Practice Problem CONCEPTUALIZE Imagine a cobbler's business (where shoes are repaired) with trendy, V-shaped cabinets, four of which are shown here. When a pair of shoes is brought in for repair, it is kept in a shoebox in one of these cabinets. The location of each pair of shoes is recorded using a set of numbers that designate the cabinet (C), the shelf (S), and the specific box (B).

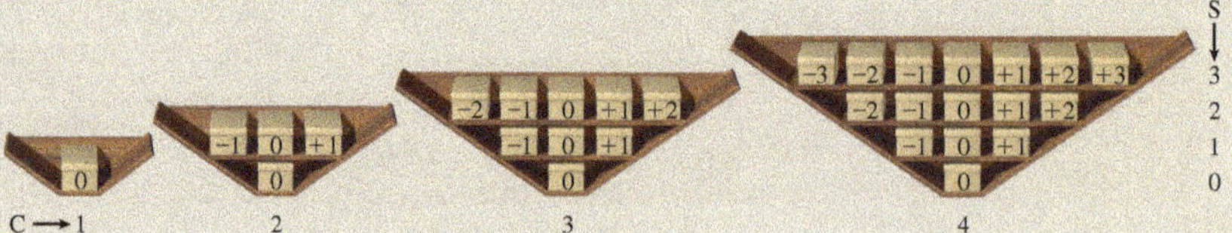

Each cabinet has a number, corresponding to the number of shelves it has. Thus, for the cabinets shown here, the value of C can be 1, 2, 3, or 4. Shelves within each cabinet are numbered sequentially from the bottom, starting with zero. For the smallest cabinet, with just one shelf, 0 is the only shelf designation. For the other cabinets, shelf designations can have integer values of 0 through C – 1. In addition, each individual box has a number on it. Boxes in the bottom row (row 0) all have the number 0 on them. Any box that resides directly above a box labeled 0, is also labeled 0. Boxes to the right or the left of the zero box on each shelf are numbered sequentially, starting with +1 (for boxes on the right), and starting with –1 (for boxes on the left). Using this numbering system, the cobbler can specify the location of a pair of shoes by designating three numbers: C, S, and B. For each of the following sets of numbers (C, S, B) determine whether or not they designate a box in one of the cabinets. For a set of numbers that does *not* designate a box in one of the cabinets, explain why.

(a) (1, 0, 0); (b) (0, 0, 0); (c) (3, 2, –2); (d) (2, 0, 0); (e) (4, 3, +1); (f) (2, 2, +2).

Electron Spin Quantum Number (m_s)

Whereas three quantum numbers are sufficient to describe an atomic orbital, an additional quantum number becomes necessary to describe an electron that occupies the orbital.

Experiments on the emission spectra of hydrogen and sodium atoms indicated that each line in the emission spectra could be split into two lines by the application of an external magnetic field. The only way physicists could explain these results was to assume that electrons act like tiny magnets. If electrons are thought of as spinning on their own axes, as Earth does, their magnetic properties can be accounted for. According to electromagnetic theory, a spinning charge generates a magnetic field, and it is this motion that causes an electron to behave like a magnet. Figure 6.16 shows the two possible spinning motions of an electron, one clockwise and the other counterclockwise. To specify the electron's spin, we use the ***electron spin quantum number*** (m_s). Because there are two possible directions of spin, opposite each other, m_s has *two* possible values: $+\frac{1}{2}$ and $-\frac{1}{2}$. Two electrons in the same orbital with opposite spins are referred to as "*paired*."

Figure 6.16 (a) Clockwise and (b) counterclockwise spins of an electron. The magnetic fields generated by these two spinning motions are analogous to those from the two magnets. The upward and downward arrows are used to denote the direction of spin.

Conclusive proof of electron spin was established by Otto Stern[13] and Walther Gerlach[14] in 1924. Figure 6.17 shows the basic experimental arrangement. A beam of gaseous atoms generated in a hot furnace passes through a nonuniform magnetic field. The interaction between an electron and the magnetic field causes the atom to be deflected from its straight-line path. Because the direction of spin is random, the electrons in *half* of the atoms will be spinning in one direction. Those atoms will be deflected in one way. The electrons in the other half of the atoms will be spinning in the *opposite* direction. Those atoms will be deflected in the other direction. Thus, two spots of equal intensity are observed on the detecting screen.

To summarize, we can designate an *orbital* in an atom with a set of *three* quantum numbers. These three quantum numbers indicate the size (n), shape (ℓ), and orientation (m_ℓ) of the orbital. A fourth quantum number (m_s) is necessary to designate the spin of an electron in the orbital.

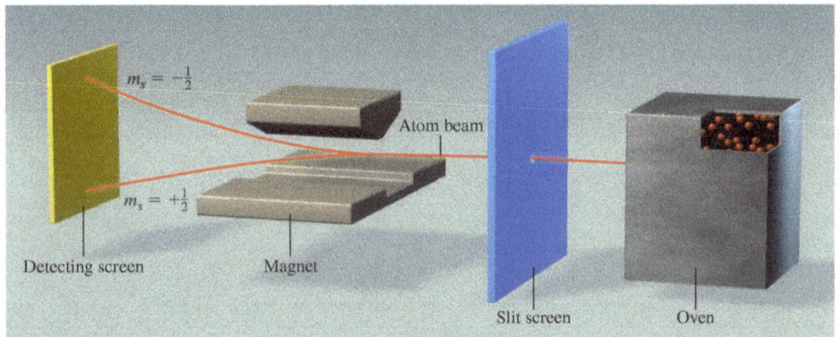

Figure 6.17 Experimental arrangement for demonstrating electron spin. A beam of atoms is directed through a magnetic field. When a hydrogen atom, with a single electron, passes through the field, it is deflected in one direction or the other, depending on the direction of the electron's spin. In a stream consisting of many atoms, there will be equal distributions of the two kinds of spins, so two spots of equal intensity are detected on the screen.

CHECKPOINT – SECTION 6.6 Quantum Numbers

6.6.1 Which of the following is a legitimate set of three quantum numbers: n, ℓ, and m_ℓ? (Select all that apply.)

a) 1, 0, 0
b) 2, 0, 0
c) 1, 0, +1
d) 2, 1, +1
e) 2, 2, −1

6.6.2 How many orbitals are there in a subshell designated by the quantum numbers $n = 3$, $\ell = 2$?

a) 2
b) 3
c) 5
d) 7
d) 10

6.6.3 How many subshells are there in the shell designated by $n = 3$?

a) 1
b) 2
c) 3
d) 6
e) 9

6.6.4 What is the total number of orbitals in the shell designated by $n = 3$?

a) 1
b) 2
c) 3
d) 6
e) 9

13. Otto Stern (1888–1969). German physicist. He made important contributions to the study of the magnetic properties of atoms and the kinetic theory of gases. Stern was awarded the Nobel Prize in Physics in 1943.

14. Walther Gerlach (1889–1979). German physicist. Gerlach's main area of research was in quantum theory.

6.7 Atomic Orbitals

Strictly speaking, an atomic orbital does not have a well-defined shape because the wave function characterizing the orbital extends from the nucleus to infinity. In that sense, it is difficult to say what an orbital looks like. On the other hand, it is certainly useful to think of orbitals as having specific shapes. Being able to visualize atomic orbitals is essential to understanding the formation of chemical bonds and molecular geometry, which are discussed in Chapters 8 and 9. In this section, we will look at each type of orbital separately.

s Orbitals

For any value of the principal quantum number (n), the value 0 is possible for the angular momentum quantum number (ℓ), corresponding to an s subshell. Furthermore, when $\ell = 0$, the magnetic quantum number (m_ℓ) has only one possible value, 0, corresponding to an s orbital. Therefore, there is an s subshell in every shell, and each s subshell contains just one orbital, an ***s* orbital.**

Figure 6.18 illustrates three ways to represent the distribution of electrons: the probability density, the spherical distribution of electron density, and the radial probability distribution (the

Student Note: The radial probability distribution can be thought of as a map of "where an electron spends most of its time."

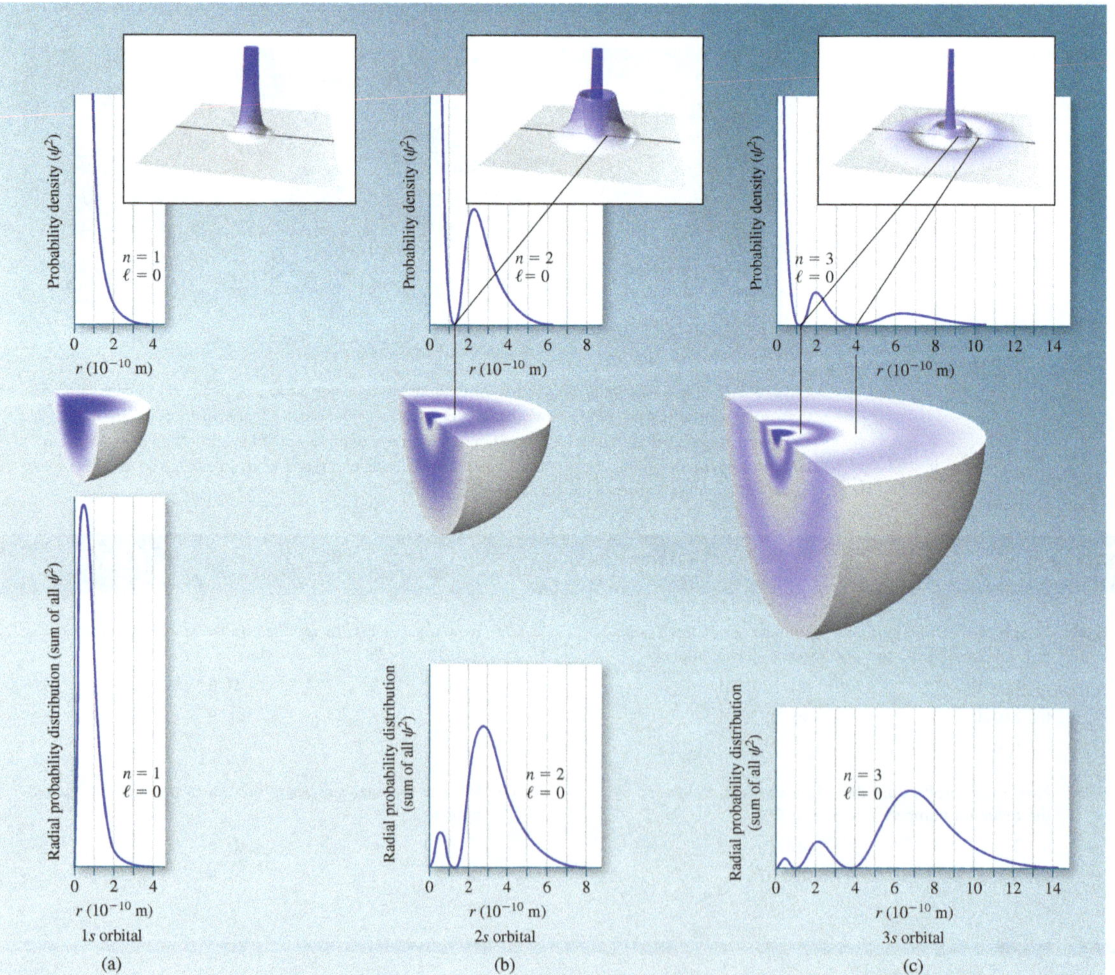

Figure 6.18 From top to bottom, the *probability density* and corresponding relief map, the distribution of electron density represented spherically with shading corresponding to the relief map, and the *radial probability distribution* for (a) the 1s, (b) the 2s, and (c) the 3s orbitals of hydrogen.

probability of finding the electron as a function of distance from the nucleus) for the 1*s*, 2*s*, and 3*s* orbitals of hydrogen. The boundary surface (the outermost surface of the spherical representation) is a common way to represent atomic orbitals, incorporating the volume in which there is about a 90 percent probability of finding the electron at any given time.

All *s* orbitals are spherical in shape but differ in size, which increases as the principal quantum number increases. The radial probability distribution for the 1*s* orbital exhibits a maximum at 52.9 pm (0.529 Å) from the nucleus. The radial probability distribution plots for the 2*s* and 3*s* orbitals exhibit two and three maxima, respectively, with the greatest probability occurring at a greater distance from the nucleus as *n* increases. Between the two maxima for the 2*s* orbital there is a point on the plot where the probability drops to zero. This corresponds to a *node* in the electron density, where the standing wave has zero amplitude. There are two such nodes in the radial probability distribution plot of the 3*s* orbital.

Student Note: Interestingly, this distance is equal to the radius of the $n = 1$ orbit in the Bohr model of the hydrogen atom and is defined as the *Bohr radius.*

Although the boundary surface diagram of an *s* orbital does not show the number of nodes, the most important features of atomic orbitals, for our purposes, are their overall shapes and *relative* sizes. These features are adequately represented by boundary surface diagrams.

p Orbitals

When the principal quantum number (n) is 2 or greater, the value 1 is possible for the angular momentum quantum number (ℓ), corresponding to a *p* subshell. And, when $\ell = 1$, the magnetic quantum number (m_ℓ) has three possible values: -1, 0, and $+1$, each corresponding to a different ***p* orbital.** Therefore, there is a *p* subshell in every shell for which $n \geq 2$, and each *p* subshell contains three *p* orbitals. These three *p* orbitals are labeled p_x, p_y, and p_z (Figure 6.19), with the subscripted letters indicating the axis along which each orbital is oriented. These three *p* orbitals are identical in size, shape, and energy; they differ from one another only in orientation. Note, however, that there is no simple relation between the values of m_ℓ and the *x, y,* and *z* directions. For our purpose, you need only remember that because there are three possible values of m_ℓ, there are three *p* orbitals with different orientations.

The boundary surface diagrams of *p* orbitals in Figure 6.19 show that each *p* orbital can be thought of as two lobes on opposite sides of the nucleus. Like *s* orbitals, *p* orbitals increase in size from 2*p* to 3*p* to 4*p* orbital and so on.

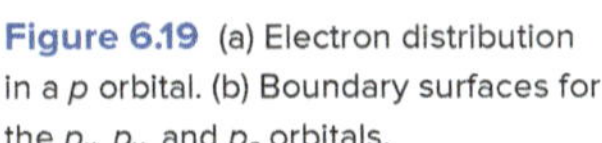

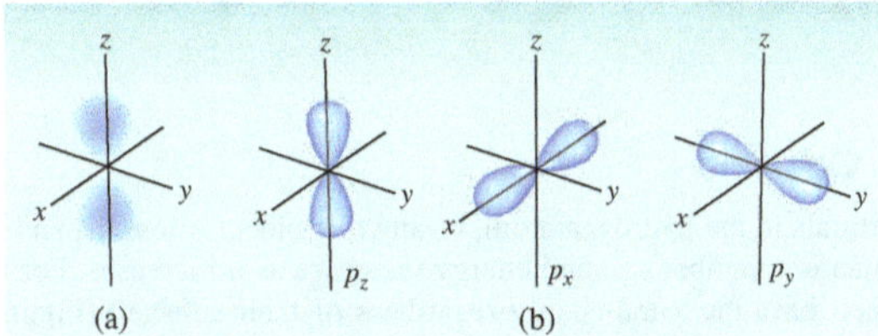

Figure 6.19 (a) Electron distribution in a *p* orbital. (b) Boundary surfaces for the p_x, p_y, and p_z orbitals.

d Orbitals and Other Higher-Energy Orbitals

When the principal quantum number (n) is 3 or greater, the value 2 is possible for the angular momentum quantum number (ℓ), corresponding to a *d* subshell. When $\ell = 2$, the magnetic quantum number (m_ℓ) has *five* possible values, -2, -1, 0, $+1$, and $+2$, each corresponding to a different ***d* orbital.** Again there is no direct correspondence between a given orientation and a particular m_ℓ value. All the 3*d* orbitals in an atom are identical in energy and are labeled with subscripts denoting their orientation with respect to the *x, y,* and *z* axes and to the planes defined by them. The *d* orbitals that have higher principal quantum numbers (4*d*, 5*d*, etc.) have shapes similar to those shown for the 3*d* orbitals in Figure 6.20.

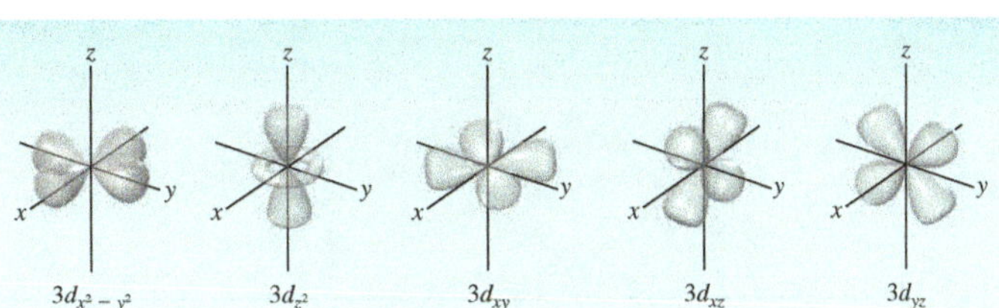

Figure 6.20 Boundary surfaces for the *d* orbitals.

The *f* ***orbitals*** are important when accounting for the behavior of elements with atomic numbers greater than 57, but their shapes are difficult to represent. In general chemistry we will not concern ourselves with the shapes of orbitals having ℓ values greater than 2.

Sample Problem 6.8 shows how to label orbitals with quantum numbers.

SAMPLE PROBLEM 6.8

List the values of n, ℓ, and m_ℓ for each of the orbitals in a 4*d* subshell.

Strategy Consider the significance of the number and the letter in the 4*d* designation and determine the values of n and ℓ. There are multiple possible values for m_ℓ, which will have to be deduced from the value of ℓ.

Setup The integer at the beginning of an orbital designation is the principal quantum number (n). The letter in an orbital designation gives the value of the angular momentum quantum number (ℓ). The magnetic quantum number (m_ℓ) can have integral values of $-\ell, \ldots, 0, \ldots, +\ell$.

Solution The values of n and ℓ are 4 and 2, respectively, so the possible values of m_ℓ are −2, −1, 0, +1, and +2.

THINK ABOUT IT

Consult Figure 6.15 to verify your answers.

Practice Problem ATTEMPT Give the values of n, ℓ, and m_ℓ for the orbitals in a 3*d* subshell.

Practice Problem BUILD Using quantum numbers, explain why there is no 2*d* subshell.

Practice Problem CONCEPTUALIZE Recall the cabinets, shelves, and shoeboxes in Practice Problem 6.7C, which are reproduced here. Write the set of three numbers that specifies each of the highlighted boxes.

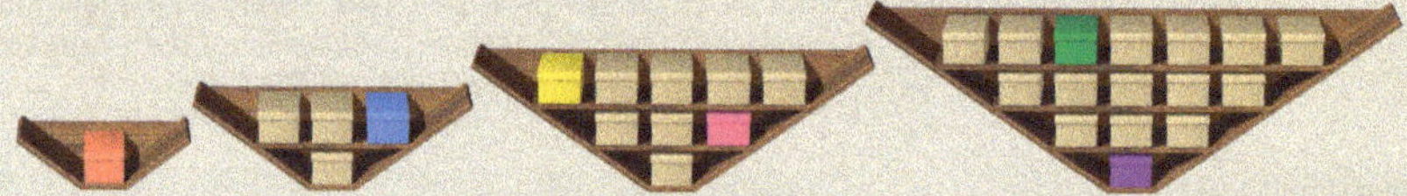

Energies of Orbitals

The energies of orbitals in the hydrogen atom, or any one-electron ion, depend only on the value of the principal quantum number (n), and energy increases as n increases. For this reason, orbitals in the same shell have the same energy regardless of their subshell (Figure 6.21).

$$1s < 2s = 2p < 3s = 3p = 3d < 4s = 4p = 4d = 4f$$

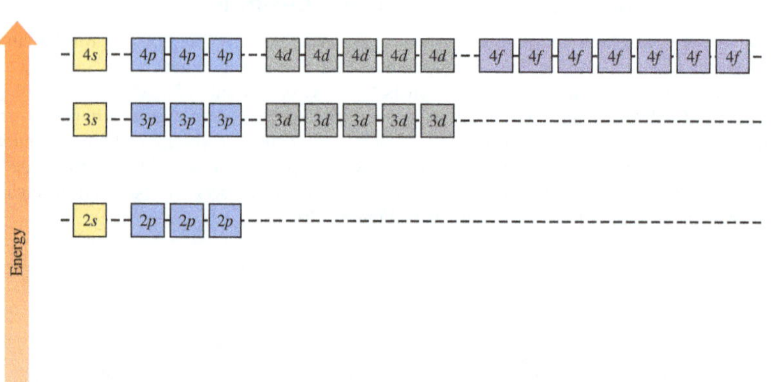

Figure 6.21 Orbital energy levels in the hydrogen atom. Each box represents one orbital. Orbitals with the same principal quantum number (n) all have the same energy.

Student Note: Recall that Bohr, whose model only *had* one quantum number, was able to calculate the energies of electrons in the hydrogen atom accurately.

Thus, all four orbitals (one 2*s* and three 2*p*) in the second shell have the same energy; all nine orbitals (one 3*s*, three 3*p*, and five 3*d*) in the third shell have the same energy; and all sixteen orbitals (one 4*s*, three 4*p*, five 4*d*, and seven 4*f*) in the fourth shell have the same energy. The energy picture is more complex for many-electron atoms than it is for hydrogen, as is discussed in Section 6.8.

CHECKPOINT – SECTION 6.7 Atomic Orbitals

6.7.1 How many orbitals are there in the 5*f* subshell?

a) 5 b) 7 c) 14 d) 16 e) 28

6.7.2 The energy of an orbital in the hydrogen atom depends on ______________.

a) n, ℓ, and m_ℓ
b) n and ℓ
c) n only
d) ℓ only
e) m_ℓ only

6.7.3 In a hydrogen atom, which orbitals are higher in energy than a 3*s* orbital? (Select all that apply.)

a) 3*p* b) 4*s* c) 2*p* d) 3*d* e) 4*p*

6.7.4 Which of the following sets of quantum numbers, n, ℓ, and m_ℓ, corresponds to a 3*p* orbital?

a) 3, 0, 0
b) 3, 1, 0
c) 3, 2, −1
d) 1, 1, −2
e) 1, 3, 1

6.8 Electron Configuration

The hydrogen atom is a particularly simple system because it contains only one electron. The electron may reside in the 1*s* orbital (the *ground state*), or it may be found in some higher-energy orbital (an *excited state*). With many-electron systems, we need to know the ground-state ***electron configuration***—that is, how the electrons are distributed in the various atomic orbitals. To do this, we need to know the relative energies of atomic orbitals in a many-electron system, which differ from those in a one-electron system such as hydrogen.

Energies of Atomic Orbitals in Many-Electron Systems

Consider the two emission spectra shown in Figure 6.22. The spectrum of helium contains more lines than that of hydrogen. This indicates that there are more possible transitions, corresponding to emission in the visible range, in a helium atom than in a hydrogen atom. This is due to the *splitting* of energy levels caused by electrostatic interactions between helium's two electrons.

Student Note: "Splitting" of energy levels refers to the splitting of a shell into subshells of different energies, as shown in Figure 6.23.

Figure 6.23 shows the general order of orbital energies in a many-electron atom. In contrast to the hydrogen atom, in which the energy of an orbital depends only on the value of n (Figure 6.21), the energy of an orbital in a many-electron system depends on both the value of n and the value of ℓ. For example, 3*p* orbitals all have the same energy, but they are higher in energy than the 3*s* orbital and lower in energy than the 3*d* orbitals. In a many-electron atom, for a given value of n, the energy of an orbital increases with increasing value of ℓ. One important consequence of the splitting of energy levels is the relative energies of *d* orbitals in one shell and the *s* orbital in the next higher shell. As Figure 6.23 shows, the 4*s* orbital is lower in energy than the 3*d* orbitals. Likewise, the 5*s* orbital is lower in energy than the 4*d* orbital, and so on. This fact becomes important when we determine how the electrons in an atom populate the atomic orbitals.

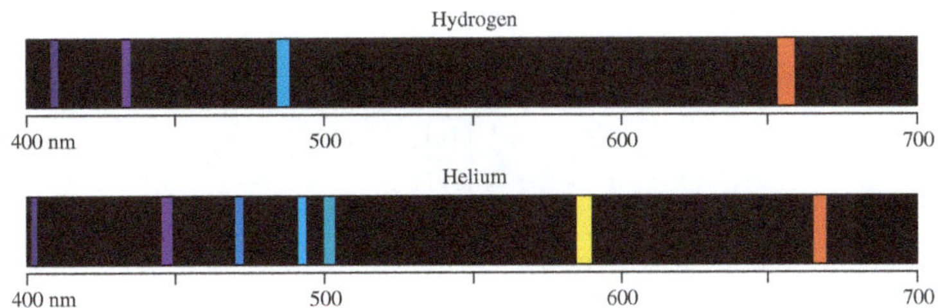

Figure 6.22 Comparison of the emission spectra of H and He.

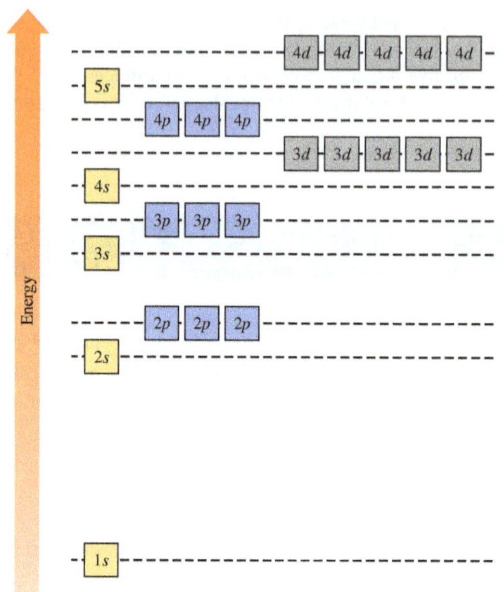

Figure 6.23 Orbital energy levels in many-electron atoms. For a given value of *n*, orbital energy increases with the value of ℓ.

The Pauli Exclusion Principle

According to the ***Pauli***[15] ***exclusion principle,*** no two electrons in the same atom can have the same four quantum numbers. If two electrons in an atom have the same n, ℓ, and m_ℓ values (meaning that they occupy the same *orbital*), then they must have different values of m_s; that is, one must have $m_s = +\frac{1}{2}$ and the other must have $m_s = -\frac{1}{2}$. Because there are only two possible values for m_s, and no two electrons in the same orbital may have the same value for m_s, a maximum of *two* electrons may occupy an atomic orbital, and these two electrons must have opposite spins. Two electrons in the same orbital with opposite spins are said to have *paired spins.*

We can indicate the arrangement of electrons in atomic orbitals with labels that identify each orbital (or subshell) and the number of electrons in it. Thus, we could describe a hydrogen atom in the ground state using $1s^1$.

Student Note: $1s^1$ is read as "one *s* one."

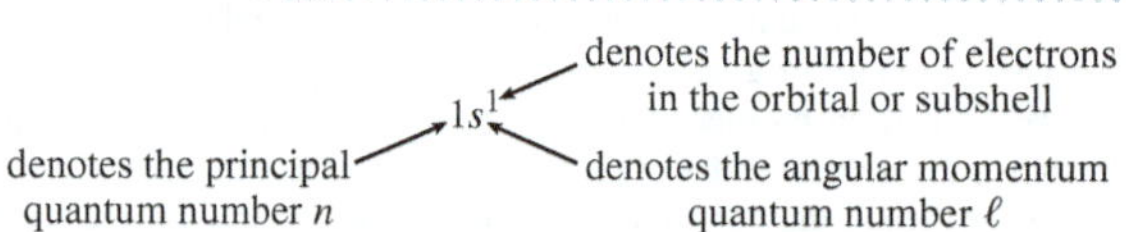

We can also represent the arrangement of electrons in an atom using *orbital diagrams,* in which each orbital is represented by a labeled box. The orbital diagram for a hydrogen atom in the ground state is

H [↑]
$1s^1$

The upward arrow denotes one of the two possible spins (one of the two possible m_s values) of the electron in the hydrogen atom (the other possible spin is indicated with a downward arrow). Under certain circumstances, as we will see shortly, it is useful to indicate the explicit locations of electrons.

The orbital diagram for a helium atom in the ground state is

Student Note: The *ground state* for a many-electron atom is the one in which all the electrons occupy orbitals of the lowest possible energy.

He [↑↓]
$1s^2$

15. Wolfgang Pauli (1900–1958). Austrian physicist. One of the founders of quantum mechanics, Pauli was awarded the Nobel Prize in Physics in 1945.

Student Note: $1s^2$ is read as "one *s* two," *not* as "one *s* squared."

The label $1s^2$ indicates there are *two* electrons in the 1*s* orbital. Note also that the arrows in the box point in opposite directions, representing opposite electron spins. Generally when an orbital diagram includes an orbital with a single electron, we represent it with an upward arrow—although we could represent it equally well with a downward arrow. The choice is arbitrary and has no effect on the energy of the electron.

The Aufbau Principle

We can continue the process of writing electron configurations for elements based on the order of orbital energies and the Pauli exclusion principle. This process is based on the ***Aufbau principle,*** which makes it possible to "build" the periodic table of the elements and determine their electron configurations by steps. Each step involves adding one proton to the nucleus and one electron to the appropriate atomic orbital. Through this process we gain a detailed knowledge of the electron configurations of the elements. As we will see in later chapters, knowledge of electron configurations helps us understand and predict the properties of the elements. It also explains why the elements fit into the periodic table the way they do.

After helium, the next element in the periodic table is lithium, which has three electrons. Because of the restrictions imposed by the Pauli exclusion principle, an orbital can accommodate no more than two electrons. Thus, the third electron cannot reside in the 1*s* orbital. Instead, it must reside in the next available orbital with the lowest possible energy. According to Figure 6.23, this is the 2*s* orbital. Therefore, the electron configuration of lithium is $1s^22s^1$, and the orbital diagram is

Li [↑↓] [↑]
$1s^2$ $2s^1$

Similarly, we can write the electron configuration of beryllium as $1s^22s^2$ and represent it with the orbital diagram

Be [↑↓] [↑↓]
$1s^2$ $2s^2$

With both the 1*s* and the 2*s* orbitals filled to capacity, the next electron, which is needed for the electron configuration of boron, must reside in the 2*p* subshell. Because all three 2*p* orbitals are of equal energy, or ***degenerate,*** the electron can occupy any one of them. By convention, we usually show the first electron to occupy the *p* subshell in the first empty box in the orbital diagram.

B [↑↓] [↑↓] [↑| |]
$1s^2$ $2s^2$ $2p^1$

Hund's Rule

Student Note: Electrons with the same spin are said to have *parallel spins.*

Will the sixth electron, which is needed to represent the electron configuration of carbon, reside in the 2*p* orbital that is already half occupied, or will it reside in one of the other, empty 2*p* orbitals? According to ***Hund's***[16] ***rule,*** the most stable arrangement of electrons in orbitals of equal energy is the one in which the number of electrons with the same spin is maximized. As we have seen, no two electrons in any orbital may have the same spin, so maximizing the number of electrons with the same spin requires putting the electrons in separate orbitals. Accordingly, in any subshell, an electron will occupy an empty orbital rather than one that already contains an electron. Because electrons are negatively charged, they repel one another. Maximizing parallel spins in separate orbitals minimizes the electron-electron repulsions in a subshell.

The electron configuration of carbon is, therefore, $1s^22s^22p^2$, and its orbital diagram is

C [↑↓] [↑↓] [↑|↑|]
$1s^2$ $2s^2$ $2p^2$

16. Frederick Hund (1896–1997). German physicist. Hund's work was mainly in quantum mechanics. He also helped to develop the molecular orbital theory of chemical bonding.

Similarly, the electron configuration of nitrogen is $1s^22s^22p^3$, and its orbital diagram is

	$1s^2$	$2s^2$	$2p^3$
N	↑↓	↑↓	↑ ↑ ↑

Once all the $2p$ orbitals are singly occupied, additional electrons will have to pair with those already in the orbitals. Thus, the electron configurations and orbital diagrams for O, F, and Ne are as follows:

		$1s^2$	$2s^2$	$2p$
O	$1s^22s^22p^4$	↑↓	↑↓	↑↓ ↑ ↑ ($2p^4$)
F	$1s^22s^22p^5$	↑↓	↑↓	↑↓ ↑↓ ↑ ($2p^5$)
Ne	$1s^22s^22p^6$	↑↓	↑↓	↑↓ ↑↓ ↑↓ ($2p^6$)

General Rules for Writing Electron Configurations

Based on the preceding examples we can formulate the following general rules for determining the electron configuration of an element in the ground state:

1. Electrons will reside in the available orbitals of the lowest possible energy.
2. Each orbital can accommodate a maximum of two electrons.
3. Electrons will not pair in degenerate orbitals if an empty orbital is available.
4. Orbitals will fill in the order indicated in Figure 6.23. Figure 6.24 provides a simple way for you to remember the proper order.

Sample Problem 6.9 illustrates the procedure for determining the ground-state electron configuration of an atom.

1s; 2s 2p; 3s 3p 3d; 4s 4p 4d 4f; 5s 5p 5d 5f; 6s 6p 6d; 7s 7p

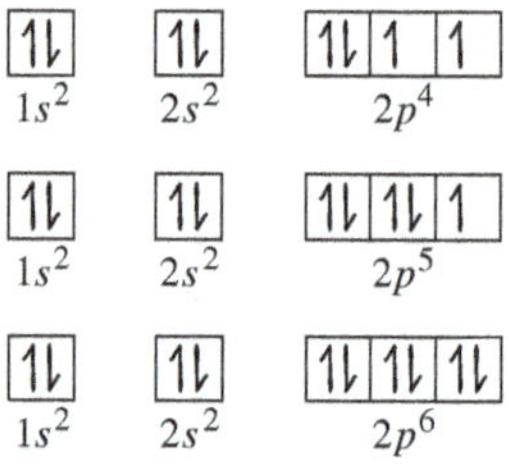

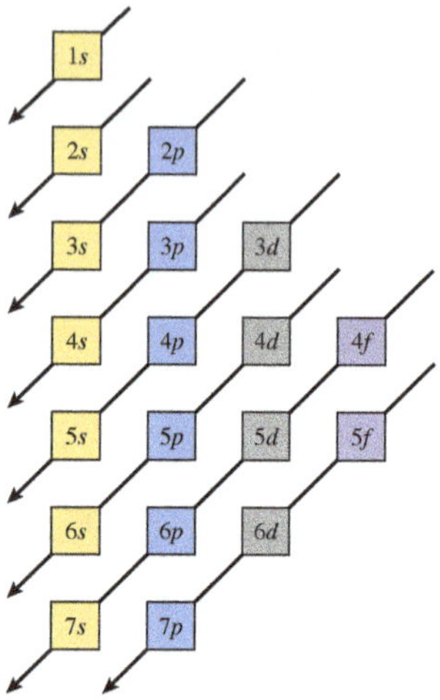

Figure 6.24 A simple way to remember the order in which orbitals fill with electrons.

Student Note: Remember that in this context, *degenerate* means "of equal energy." Orbitals in the same subshell are degenerate.

SAMPLE PROBLEM 6.9

Write the electron configuration and give the orbital diagram of a calcium (Ca) atom ($Z = 20$).

Strategy Use the general rules given and the Aufbau principle to "build" the electron configuration of a calcium atom and represent it with an orbital diagram.

Setup Because $Z = 20$, we know that a Ca atom has 20 electrons. They will fill orbitals in the order designated in Figure 6.23, obeying the Pauli exclusion principle and Hund's rule. Orbitals will fill in the following order: $1s$, $2s$, $2p$, $3s$, $3p$, $4s$. Each s subshell can contain a maximum of two electrons, whereas each p subshell can contain a maximum of six electrons.

Solution

		$1s^2$	$2s^2$	$2p^6$	$3s^2$	$3p^6$	$4s^2$
Ca	$1s^22s^22p^63s^23p^64s^2$	↑↓	↑↓	↑↓ ↑↓ ↑↓	↑↓	↑↓ ↑↓ ↑↓	↑↓

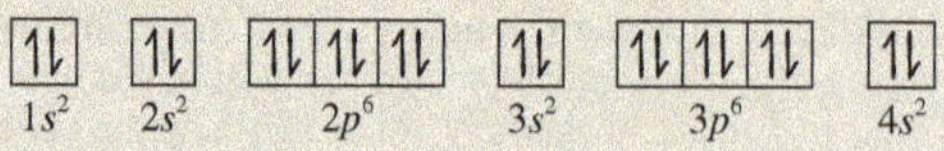

Animation
Atomic structure—electron configurations.

THINK ABOUT IT

Look at Figure 6.23 again to make sure you have filled the orbitals in the right order and that the sum of electrons is 20. Remember that the $4s$ orbital fills before the $3d$ orbitals.

Practice Problem ATTEMPT Write the electron configuration and give the orbital diagram of a rubidium (Rb) atom ($Z = 37$).

Practice Problem BUILD Write the electron configuration and give the orbital diagram of a bromine (Br) atom ($Z = 35$).

Practice Problem CONCEPTUALIZE Imagine an alternate universe in which the allowed values of the magnetic quantum number, m_ℓ, can have values of $-(\ell + 1) \ldots 0 \ldots +(\ell + 1)$. In this alternate universe, what would be the maximum number of electrons that could have the principal quantum number 3 in a given atom?

CHECKPOINT – SECTION 6.8 Electron Configuration

6.8.1 Which of the following electron configurations correctly represents the Ti atom?

a) $1s^2\ 2s^2\ 2p^6\ 3s^2\ 3p^6\ 3d^4$

b) $1s^2\ 2s^2\ 2p^6\ 3s^2\ 3p^6\ 4s^2\ 3d^2$

c) $1s^2\ 2s^2\ 2p^6\ 3s^2\ 3p^6\ 4s^2\ 3d^{10}$

d) $1s^2\ 2s^2\ 2p^6\ 3s^2\ 3p^6\ 3d^{10}$

e) $1s^2\ 2s^2\ 2p^6\ 3s^2\ 3p^6\ 4s^4$

6.8.2 What element is represented by the following electron configuration? $1s^2\ 2s^2\ 2p^6\ 3s^2\ 3p^6\ 4s^2\ 3d^{10}\ 4p^4$

a) Br

b) As

c) S

d) Se

e) Te

6.8.3 Which orbital diagram is correct for the ground-state S atom?

a) [↑↓] $1s^2$ [↑↓] $2s^2$ [↑↓|↑↓|↑↓] $2p^6$ [↑↓] $3s^2$ [↑↓|↑↓|] $3p^4$

b) [↑↓] $1s^2$ [↑↓] $2s^2$ [↑↓|↑↓|↑↓] $2p^6$ [↑↓] $3s^2$ [↑↓|↑|↑] $3p^4$

c) [↑↓] $1s^2$ [↑↓] $2s^2$ [↑↓|↑↓|] $2p^4$

d) [↑↓] $1s^2$ [↑↓] $2s^2$ [↑↓|↑|↑] $2p^4$

e) [↑↓] $1s^2$ [↑↓] $2s^2$ [↑↓|↑↓|↑↓] $2p^6$

6.9 Electron Configurations and the Periodic Table

The electron configurations of all elements except hydrogen and helium can be represented using a ***noble gas core,*** which shows in brackets the electron configuration of the noble gas element that most recently precedes the element in question, followed by the electron configuration in the outermost occupied subshells. Figure 6.25 gives the outermost ground-state electron configurations of elements from H ($Z = 1$) through Rg ($Z = 111$). Notice the similar pattern of electron configurations in the elements lithium ($Z = 3$) through neon ($Z = 10$) and those of sodium ($Z = 11$) through argon ($Z = 18$). Both Li and Na, for example, have the configuration ns^1 in their outermost occupied subshells. For Li, $n = 2$; for Na, $n = 3$. Both F and Cl have electron configuration ns^2np^5, where $n = 2$ for F and $n = 3$ for Cl, and so on.

As mentioned in Section 6.8, the 4*s* subshell is filled before the 3*d* subshell in a many-electron atom (see Figure 6.23). Thus, the electron configuration of potassium ($Z = 19$) is $1s^22s^22p^63s^23p^64s^1$. Because $1s^22s^22p^63s^23p^6$ is the electron configuration of argon, we can simplify the electron configuration of potassium by writing $[Ar]4s^1$, where [Ar] denotes the "argon core."

$$\text{K}\quad \underbrace{1s^22s^22p^63s^23p^6}_{[\text{Ar}]}4s^1 \longrightarrow [\text{Ar}]\ 4s^1$$

The placement of the outermost electron in the 4*s* orbital (rather than in the 3*d* orbital) of potassium is strongly supported by experimental evidence. The physical and chemical properties of potassium are very similar to those of lithium and sodium, the first two alkali metals. In both lithium and sodium, the outermost electron is in an *s* orbital (there is no doubt that their outermost electrons occupy *s* orbitals because there is no 1*d* or 2*d* subshell). Based on its similarities to the other alkali metals, we expect potassium to have an analogous electron configuration; that is, we expect the last electron in potassium to occupy the 4*s* rather than the 3*d* orbital.

Student Note: Although zinc and the other elements in Group 2B sometimes are included under the heading "transition metals," they neither have nor readily acquire partially filled *d* subshells. Strictly speaking, they are *not* transition metals.

The elements from Group 3B through Group 1B are *transition metals* [⏮ Section 2.4]. Transition metals either have incompletely filled *d* subshells or readily give rise to cations that have incompletely filled *d* subshells. In the first transition metal series, from scandium ($Z = 21$) through copper ($Z = 29$), additional electrons are placed in the 3*d* orbitals according to Hund's rule. However, there are two anomalies. The electron configuration of chromium ($Z = 24$) is $[Ar]4s^13d^5$ and not $[Ar]4s^23d^4$, as we might expect. A similar break in the pattern is observed

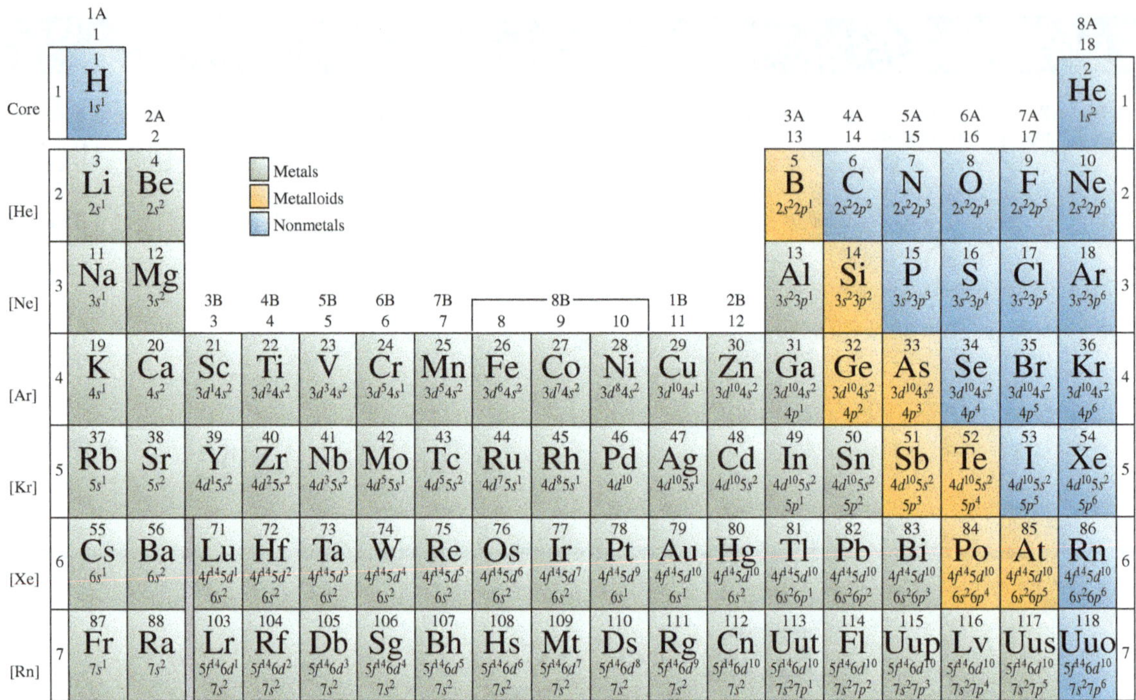

Figure 6.25 Outermost ground-state electron configurations for the known elements.

for copper, whose electron configuration is $[Ar]4s^13d^{10}$ rather than $[Ar]4s^23d^9$. The reason for these anomalies is that a slightly greater stability is associated with the half-filled ($3d^5$) and completely filled ($3d^{10}$) subshells.

> **Student Note:** Electron configurations such as these may also be written with the *d* subshell first. For example, $[Ar]4s^13d^{10}$ can also be written as $[Ar]3d^{10}4s^1$. Either way is acceptable.

Cr [Ar] [↿] (4*s*) [↿][↿][↿][↿][↿] (3*d* 3*d* 3*d* 3*d* 3*d*)

Cu [Ar] [↿] (4*s*) [↿⇂][↿⇂][↿⇂][↿⇂][↿⇂] (3*d* 3*d* 3*d* 3*d* 3*d*)

For elements Zn ($Z = 30$) through Kr ($Z = 36$), the 3*d*, 4*s*, and 4*p* subshells fill in a straightforward manner. With rubidium ($Z = 37$), electrons begin to enter the $n = 5$ energy level.

Some of the electron configurations in the second transition metal series [yttrium ($Z = 39$) through silver ($Z = 47$)] are also irregular, but the details of many of these irregularities are beyond the scope of this text and we will not be concerned with them.

The sixth period of the periodic table begins with cesium ($Z = 55$) and barium ($Z = 56$), whose electron configurations are $[Xe]6s^1$ and $[Xe]6s^2$, respectively. Following barium, there is a gap in the periodic table where the ***lanthanide (rare earth) series*** belongs. The lanthanides are a series of 14 elements that have incompletely filled 4*f* subshells or that readily give rise to *cations* that have incompletely filled 4*f* subshells. The lanthanides (and the actinides, to be discussed next) are shown at the bottom of the periodic table to keep the table from being too wide.

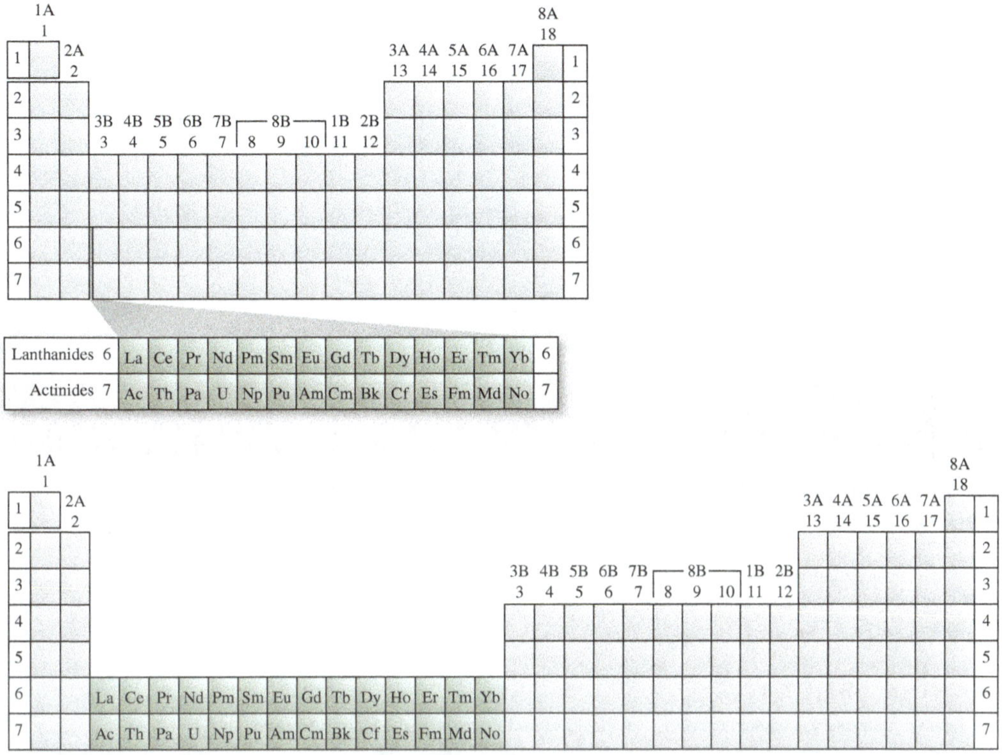

> **Student Note:** When $n = 4$, ℓ can equal 3, corresponding to an f subshell. There are seven possible values for m_ℓ when $\ell = 3$: −3, −2, −1, 0, +1, +2, and +3. Therefore, there are seven f orbitals.

In theory, the lanthanides arise from the filling of the seven degenerate 4*f* orbitals. In reality, however, the energies of the 5*d* and 4*f* orbitals are very close and the electron configurations of these elements sometimes involve 5*d* electrons. For example, in lanthanum itself ($Z = 57$) the 4*f* orbital is slightly higher in energy than the 5*d* orbital. Thus, lanthanum's electron configuration is $[Xe]6s^25d^1$ rather than $[Xe]6s^24f^1$.

After the 4*f* subshell is completely filled, the next electron enters the 5*d* subshell of lutetium ($Z = 71$). This series of elements, including lutetium and hafnium ($Z = 72$) and extending through mercury ($Z = 80$), is characterized by the filling of the 5*d* subshell. The 6*p* subshells are filled next, which takes us to radon ($Z = 86$).

The last row of elements begins with francium ($Z = 87$; electron configuration $[Rn]7s^1$) and radium ($Z = 88$; electron configuration $[Rn]7s^2$), and then continues with the ***actinide series,*** which starts at actinium ($Z = 89$) and ends with nobelium ($Z = 102$). Most of these elements are not found in nature but have been synthesized in nuclear reactions, which are the subject of Chapter 20. The actinide series has partially filled 5*f* and/or 6*d* subshells. The elements lawrencium ($Z = 103$) through darmstadtium ($Z = 110$) have a filled 5*f* subshell and are characterized by the filling of the 6*d* subshell.

With few exceptions, you should be able to write the electron configuration of any element, using Figure 6.23 (or Figure 6.24) as a guide. Elements that require particular care are the transition metals, the lanthanides, and the actinides. You may notice from looking at the electron configurations of gadolinium ($Z = 64$) and curium ($Z = 96$) that half-filled *f* subshells also appear to exhibit slightly enhanced stability. As we noted earlier, at larger values of the principal quantum number *n*, the order of subshell filling may be irregular due to the closeness of the energy levels.

Figure 6.26 groups the elements according to the type of subshell in which the outermost electrons are placed. Elements whose outermost electrons are in an *s* subshell are referred to as *s*-block elements, those whose outermost electrons are in a *p* subshell are referred to as *p*-block elements, and so on.

Sample Problem 6.10 shows how to write electron configurations.

Figure 6.26 Classification of groups of elements in the periodic table according to the type of subshell being filled with electrons.

SAMPLE PROBLEM 6.10

Without referring to Figure 6.25, write the electron configuration for an arsenic atom ($Z = 33$) in the ground state.

Strategy Use Figure 6.23 or Figure 6.24 to determine the order in which the subshells will fill, and then assign electrons to the appropriate subshells.

Setup The noble gas core for As is [Ar], where $Z = 18$ for Ar. The order of filling beyond the noble gas core is 4*s*, 3*d*, and 4*p*. Fifteen electrons must go into these subshells because there are $33 - 18 = 15$ electrons in As beyond its noble gas core.

Solution $[Ar]4s^2 3d^{10} 4p^3$

THINK ABOUT IT

Arsenic is a *p*-block element; therefore, we should expect its outermost electrons to reside in a *p* subshell.

Practice Problem ATTEMPT Without referring to Figure 6.25, write the electron configuration for a radium atom ($Z = 88$) in the ground state.

Practice Problem BUILD Without referring to Figure 6.25, determine the identity of the element with the following electron configuration:

$$[Xe]6s^2\ 4f^{14} 5d^{10} 6p^5$$

Practice Problem CONCEPTUALIZE Consider again the alternate universe and its allowed values of m_ℓ from Practice Problem 6.9C. At what atomic numbers (in the first four rows of the alternate universe's periodic table) would you expect the ground-state electron configuration to differ from that predicted by the Aufbau principle? (*Hint:* In *our* periodic table, in the first four rows, the ground-state electron configurations differ from those predicted by the Aufbau principle at atomic numbers 24 and 29.)

CHECKPOINT – SECTION 6.9 Electron Configurations and the Periodic Table

6.9.1 Which of the following electron configurations correctly represents the Ag atom?

a) $[Kr]5s^2 4d^9$
b) $[Kr]5s^2 4d^{10}$
c) $[Kr]5s^1 4d^{10}$
d) $[Xe]5s^2 4d^9$
e) $[Xe]5s^1 4d^{10}$

6.9.2 What element is represented by the following electron configuration: $[Kr]5s^2 4d^{10} 5p^5$?

a) Tc
b) Br
c) I
d) Xe
e) Te

6.9.3 Which of the following is a *d*-block element? (Select all that apply.)

a) Sb
b) Au
c) Ca
d) Zn
e) U

6.9.4 Which of the following is a *p*-block element? (Select all that apply.)

a) Pb
b) C
c) Sr
d) Xe
e) Na

Chapter Summary

Section 6.1

- What we commonly refer to as "light" is actually the visible portion of the ***electromagnetic spectrum.*** All light has certain common characteristics including wavelength, frequency, and amplitude.
- ***Wavelength*** (λ) is the distance between two crests or two troughs of a wave. ***Frequency*** (ν) is the number of waves that pass a point per unit time. ***Amplitude*** is the distance between the midpoint and crest or trough of a wave.
- ***Electromagnetic waves*** have both electric and magnetic components that are both mutually perpendicular and in phase.

Section 6.2

- ***Blackbody radiation*** is the electromagnetic radiation given off by a solid when it is heated.
- Max Planck proposed that energy, like matter, was composed of tiny, indivisible "packages" called ***quanta***. *Quanta* is the plural of *quantum.*
- Albert Einstein used Planck's revolutionary quantum theory to explain the ***photoelectric effect,*** in which electrons are emitted when light of a certain minimum frequency shines on a metal surface.
- A *quantum* of light is referred to as a ***photon.***

Section 6.3

- An ***emission spectrum*** is the light given off by an object when it is excited thermally. Emission spectra may be *continuous,* including all the wavelengths within a particular range, or they may be ***line spectra,*** consisting only of certain discrete wavelengths.
- The ***ground state*** is the lowest possible energy state for an atom. An ***excited state*** is any energy level higher than the ground state.

Section 6.4

- A ***node*** is a point at which a standing wave has zero amplitude.
- Having observed that light could exhibit particle-like behavior, de Broglie proposed that matter might also exhibit wavelike behavior. The ***de Broglie wavelength*** is the wavelength associated with a particle of very small mass. Soon after de Broglie's proposal, experiments showed that electrons could exhibit diffraction—a property of waves.

Section 6.5

- According to the ***Heisenberg uncertainty principle,*** the product of the uncertainty of the *location* and the uncertainty of the *momentum* of a very small particle must have a certain minimum value. It is thus impossible to know simultaneously both the location and momentum of an electron.
- The ***electron density*** gives the probability of finding an electron in a particular region in an atom. An ***atomic orbital*** is the region of three-dimensional space, defined by ψ^2 (the square of the wave function, ψ), where the probability of finding an electron is high. An atomic orbital can accommodate a maximum of *two* electrons.

Section 6.6

- An atomic orbital is defined by three ***quantum numbers:*** the ***principal quantum number (n),*** the ***angular momentum quantum number (ℓ),*** and the ***magnetic quantum number (m_ℓ).***
- The principal quantum number (n) indicates distance from the nucleus. Possible values of n are (1, 2, 3, . . .). The angular momentum quantum number (ℓ) indicates the shape of the orbital. Possible values of ℓ are (0, 1, . . . , $n - 1$). The magnetic quantum number (m_ℓ) indicates the orbital's orientation in space. Possible values of m_ℓ are ($-\ell$, . . . , 0, . . . , $+\ell$).
- Two electrons that occupy the same atomic orbital in the ground state must have different ***electron spin quantum numbers (m_s),*** either $+\frac{1}{2}$ or $-\frac{1}{2}$.

Section 6.7

- The value of the angular momentum quantum number (ℓ) determines the type of the atomic orbital: $\ell = 0$ corresponds to an ***s orbital,*** $\ell = 1$ corresponds to a ***p orbital,*** $\ell = 2$ corresponds to a ***d orbital,*** and $\ell = 3$ corresponds to an ***f orbital.***

Section 6.8

- The ***electron configuration*** specifies the arrangement of electrons in the atomic orbitals of an atom.
- According to the ***Pauli exclusion principle,*** no two electrons in an atom in the ground state can have the same four quantum numbers, n, ℓ, m_ℓ, and m_s.
- The ***Aufbau principle*** describes the theoretical, sequential building up of the elements in the periodic table by the stepwise addition of protons and electrons.
- Atomic orbitals that have the same energy are called ***degenerate.*** According to ***Hund's rule,*** degenerate orbitals must all contain one electron before any can contain two electrons.

Section 6.9

- The ***noble gas core*** makes it possible to abbreviate the writing of electron configurations.
- The ***lanthanide (rare earth) series*** and ***actinide series*** appear at the bottom of the periodic table. They represent the filling of *f* orbitals.

Key Words

Actinide series, 267
Amplitude, 234
Angular momentum quantum number (ℓ), 256
Atomic orbital, 255
Aufbau principle, 264
Blackbody radiation, 237
d orbital, 260
de Broglie wavelength, 251
Degenerate, 264
Electromagnetic spectrum, 234
Electromagnetic wave, 235
Electron configuration, 262
Electron density, 255
Electron spin quantum number (m_s), 258
Emission spectra, 242
Excited state, 245
f orbital, 261
Frequency (ν), 234
Ground state, 244
Heisenberg uncertainty principle, 253
Hund's rule, 264
Lanthanide (rare earth) series, 267
Line spectra, 243
Magnetic quantum number (m_ℓ), 256
Noble gas core, 266
Node, 250
p orbital, 260
Pauli exclusion principle, 263
Photoelectric effect, 239
Photons, 239
Principal quantum number (n), 255
Quantum, 237
Quantum numbers, 255
s orbital, 259
Wavelength (λ), 234

Key Equations

6.1 $c = \lambda\nu$	The wavelength (λ) and frequency (ν) of electromagnetic radiation are related to one another through the speed of light (c). If wavelength is known, frequency can be determined, and vice versa.
6.2 $E = h\nu$	The energy of a photon (E) is equal to the product of Planck's constant (h) and frequency (ν) of the photon.
6.3 $h\nu = \text{KE} + W$	The energy ($h\nu$) of a photon used to eject electrons from a metal surface via the photoelectric effect is equal to the sum of kinetic energy of the ejected electron (KE) and the binding energy (W).
6.4 $\frac{1}{\lambda} = R_\infty\left(\frac{1}{n_1^2} - \frac{1}{n_2^2}\right)$	When an electron transitions from one quantum state to another (n_i to n_f), the difference in energy between the two states is emitted (or absorbed) in the form of light. The wavelength of the emitted/absorbed light can be calculated using Equation 6.4.
6.5 $E_n = -2.18 \times 10^{-18}\text{ J}\left(\frac{1}{n^2}\right)$	The energy of an electron for a given value of n (E_n) is inversely proportional to the square of n—and is by convention a negative number.
6.6 $\Delta E = h\nu = -2.18 \times 10^{-18}\text{ J}\left(\frac{1}{n_f^2} - \frac{1}{n_i^2}\right)$	The difference in energy between two quantum states (n_i to n_f) is calculated using Equation 6.6.
6.7 $\frac{1}{\lambda} = \frac{2.18 \times 10^{-18}\text{ J}}{hc}\left(\frac{1}{n_f^2} - \frac{1}{n_i^2}\right)$	Similar to Equation 6.4, Equation 6.7 allows calculation of the wavelength of emitted/absorbed light when n_i and n_f are known.
6.8 $2\pi r = n\lambda$	This is the relationship between the allowed orbit ($2\pi r$) and wavelength (λ) of an electron behaving as a standing wave.
6.9 $\lambda = \frac{h}{mu}$	The de Broglie wavelength (λ) of a particle can be calculated using Planck's constant (h), the mass of the particle in kilograms (m), and velocity (u) of the particle.
6.10 $\Delta x \cdot \Delta p \geq \frac{h}{4\pi}$	The Heisenberg uncertainty principle states that the product of uncertainties in position (Δx) and momentum (Δp) of a particle cannot be less than Planck's constant (h) over 4π. Knowing the uncertainty in one (position or momentum) allows us to calculate the minimum uncertainty in the other.
6.11 $\Delta x \cdot m\Delta u \geq \frac{h}{4\pi}$	Similar to Equation 6.10, when mass of the particle is known, knowing the uncertainty in position (Δx) allows us to calculate the minimum uncertainty in its velocity (Δu).

KEY SKILLS

Determining Ground-State Valence Electron Configurations Using the Periodic Table

An easy way to determine the electron configuration of an element is by using the periodic table. Although the table is arranged by atomic number, it is also divided into blocks that indicate the type of orbital occupied by an element's outermost electrons. Outermost valence electrons of elements in the *s*-block (shown in yellow) reside in *s* orbitals; those of elements in the *p*-block (blue) reside in *p* orbitals; and so on.

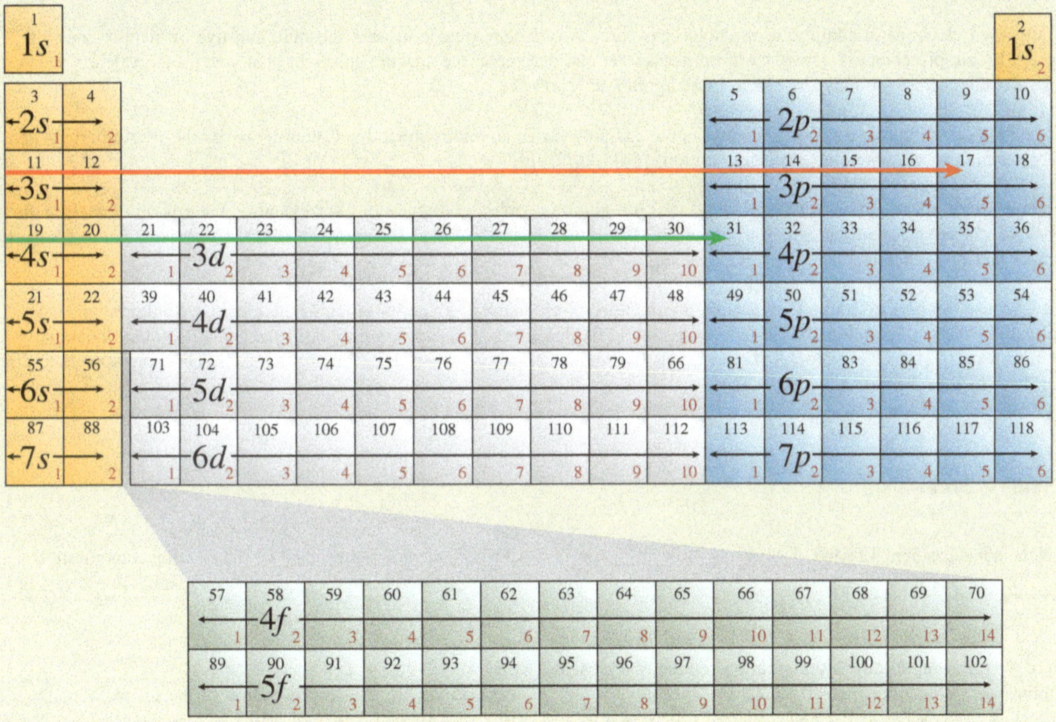

1 1s 1																	2 1s 2
3 2s 1	4 2											5 1	6 2p 2	7 3	8 4	9 5	10 6
11 3s 1	12 2											13 1	14 3p 2	15 3	16 4	17 5	18 6
19 4s 1	20 2	21 1	22 3d 2	23 3	24 4	25 5	26 6	27 7	28 8	29 9	30 10	31 1	32 4p 2	33 3	34 4	35 5	36 6
21 5s 1	22 2	39 1	40 4d 2	41 3	42 4	43 5	44 6	45 7	46 8	47 9	48 10	49 1	50 5p 2	51 3	52 4	53 5	54 6
55 6s 1	56 2	71 1	72 5d 2	73 3	74 4	75 5	76 6	77 7	78 8	79 9	66 10	81 1	6p 2	83 3	84 4	85 5	86 6
87 7s 1	88 2	103 1	104 6d 2	105 3	106 4	107 5	108 6	109 7	110 8	111 9	112 10	113 1	114 7p 2	115 3	116 4	117 5	118 6

57 1	58 4f 2	59 3	60 4	61 5	62 6	63 7	64 8	65 9	66 10	67 11	68 12	69 13	70 14
89 1	90 5f 2	91 3	92 4	93 5	94 6	95 7	96 8	97 9	98 10	99 11	100 12	101 13	102 14

To determine the ground-state electron configuration of any element, we start with the most recently completed noble gas core, and count across the following period to determine the valence electron configuration. Consider the example of Cl, which has atomic number 17. The noble gas that precedes Cl is Ne, with atomic number 10. Therefore, we begin by writing [Ne]. The noble gas symbol in square brackets represents the core electrons—with a completed *p* subshell. To complete the electron configuration, we count from the left of period 3 as shown by the red arrow, adding the last (rightmost) configuration label from each block the arrow touches:

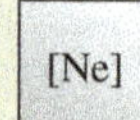

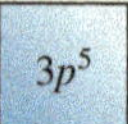

There are seven electrons in addition to the noble gas core. Two of them reside in an *s* subshell, and five of them reside in a *p* subshell. By simply counting across the third period, we can determine the specific subshells that contain the valence electrons, and arrive at the correct ground-state electron configuration: $[Ne]3s^23p^5$.

For Ga, with atomic number 31, the preceding noble gas is Ar, with atomic number 18. Counting across the fourth period (green arrow) gives the ground-state electron configuration: $[Ar]4s^23d^{10}\,4p^1$.

There are a few elements for which this method will not give the correct configuration. For example, there is no element with a ground-state valence electron configuration ending in $3d^4$ or $3d^9$. Instead, Cr and Cu are $[Ar]4s^13d^5$ and $[Ar]4s^13d^{10}$, respectively. Remember that this is the result of the unusual stability of either a *half-filled* or a *filled d* subshell [⏮ Section 6.9].

We can also use the periodic table to determine the identity of an element, given its ground-state electron configuration. For example, given the configuration $[Ne]3s^23p^4$, we focus on the last entry in the configuration: $3p^4$. This tells us that the element is in the *third* period (3), in the *p*-block (*p*), and that it has *four* electrons in its *p* subshell (superscript 4). This corresponds to atomic number 16, which is the element sulfur (S).

Key Skills Problems

6.1
What is the noble gas core for Mo?

(a) Ar (b) Kr (c) Xe (d) Ne (e) Rn

6.2
Which of the following electron configurations correctly represents the V atom?

(a) $[Ar]3d^5$ (b) $[Ar]4s^23d^2$ (c) $[Ar]4s^23d^4$ (d) $[Ar]4s^23d^3$ (e) $[Kr]4s^23d^3$

6.3
What element is represented by the electron configuration $[Kr]5s^24d^{10}5p^1$?

(a) Sn (b) Ga (c) In (d) Tl (e) Zr

6.5
What is the electron configuration of the Lu atom?

(a) $[Xe]6s^24f^{14}$ (b) $[Xe]6s^25d^1$ (c) $[Xe]6s^24f^{13}$ (d) $[Xe]6s^24f^{14}5d^1$ (e) $[Xe]4f^{14}$

Questions and Problems

Applying What You've Learned

The two most commonly used lasers in photodynamic therapy are the Ar ion laser, which emits beams at 488 nm (blue) and 514 nm (green), and a 635-nm (red) diode laser. These wavelengths do not penetrate biological tissue very deeply, and currently are useful only for tumors that are on or just beneath the surface of the skin—or of the lining of an organ that can be exposed to light, such as the bladder or esophagus. (a) Calculate the frequency and the energy per photon of the light emitted by a 635-nm diode laser [⏮ Sample Problems 6.1 and 6.2]. (b) Calculate the difference in energy per photon in the two wavelengths emitted by the Ar laser. Which of the wavelengths is more energetic [⏮ Sample Problem 6.3]? (c) Calculate the de Broglie wavelength of an argon atom moving at 1000 m/s [⏮ Sample Problem 6.5]. (d) Give the values of n, ℓ, and m_ℓ for the $3p$ orbitals in an Ar atom [⏮ Sample Problem 6.8]. (e) Write the electron configuration of Ar [⏮ Sample Problems 6.9 and 6.10].

SECTION 6.1: THE NATURE OF LIGHT

Review Questions

6.1 What is a wave? Using a diagram, define the following terms associated with waves: wavelength, frequency, amplitude.

6.2 What are the units for wavelength and frequency of electromagnetic waves? What is the speed of light in meters per second and miles per hour?

6.3 List the types of electromagnetic radiation, starting with the radiation having the longest wavelength and ending with the radiation having the shortest wavelength.

6.4 Give the high and low wavelength values that define the visible region of the electromagnetic spectrum.

Computational Problems

6.5 (a) What is the wavelength (in nm) of light having a frequency of 8.6×10^{13} Hz? (b) What is the frequency (in Hz) of light having a wavelength of 566 nm?

6.6 (a) What is the frequency of light having a wavelength of 456 nm? (b) What is the wavelength (in nm) of radiation having a frequency of 2.45×10^9 Hz? (This is the type of radiation used in microwave ovens.)

6.7 The SI unit of time is the second, which is defined as 9,192,631,770 cycles of radiation associated with a certain emission process in the cesium atom. Calculate the wavelength of this radiation (to three significant figures). In which region of the electromagnetic spectrum is this wavelength found?

Conceptual Problems

6.8 How many minutes would it take a radio wave to travel from the planet Venus to Earth? (The average distance from Venus to Earth = 28 million miles.) How long would it take an infrared wave to travel the same distance?

6.9 The average distance between Mars and Earth is about 1.3×10^8 miles. How long would it take video images transmitted from the Mars Spirit rover on Mars' surface to reach Earth (1 mile = 1.61 km)?

6.10 Four waves represent light in four different regions of the electromagnetic spectrum: visible, microwave, infrared, and ultraviolet. Determine the best match of regions to the waves shown here. Explain your choices.

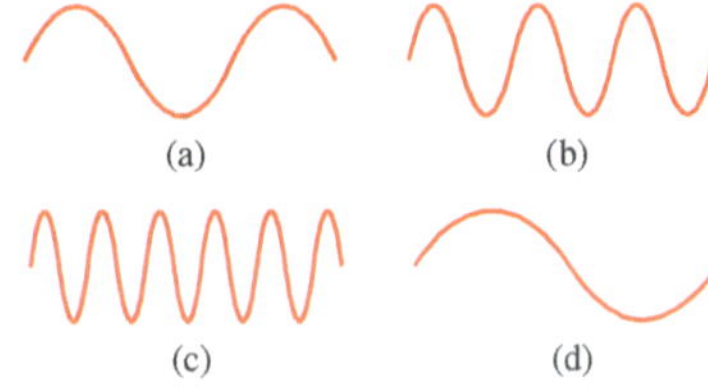

SECTION 6.2: QUANTUM THEORY

Review Questions

6.11 Briefly explain Planck's quantum theory and explain what a quantum is. What are the units for Planck's constant?

6.12 Give two everyday examples that illustrate the concept of quantization.

6.13 Explain what is meant by the photoelectric effect.

6.14 What are photons? What role did Einstein's explanation of the photoelectric effect play in the development of the particle-wave interpretation of the nature of electromagnetic radiation?

Computational Problems

6.15 A photon has a wavelength of 705 nm. Calculate the energy of the photon in joules.

6.16 The blue color of the sky results from the scattering of sunlight by molecules in the air. The blue light has a frequency of about 7.5×10^{14} Hz. (a) Calculate the wavelength (in nm) associated with this radiation, and (b) calculate the energy (in joules) of a single photon associated with this frequency.

6.17 A photon has a frequency of 6.5×10^9 Hz. (a) Convert this frequency into wavelength (nm). Does this frequency fall in the visible region? (b) Calculate the energy (in joules) of this photon. (c) Calculate the energy (in joules) of 1 mole of photons all with this frequency.

6.18 What is the wavelength (in nm) of radiation that has an energy content of 2.13×10^3 kJ/mol? In which region of the electromagnetic spectrum is this radiation found?

6.19 When copper is bombarded with high-energy electrons, X rays are emitted. Calculate the energy (in joules) associated with the photons if the wavelength of the X rays is 0.154 nm.

6.20 A particular form of electromagnetic radiation has a frequency of 9.87×10^{15} Hz. (a) What is its wavelength in nanometers? In meters? (b) To what region of the electromagnetic spectrum would you assign it? (c) What is the energy (in joules) of one quantum of this radiation?

6.21 The retina of a human eye can detect light when radiant energy incident on it is at least 4.0×10^{-17} J. For light of 585-nm wavelength, how many photons does this energy correspond to?

6.22 The radioactive ^{60}Co isotope is used in nuclear medicine to treat certain types of cancer. Calculate the wavelength and frequency of an emitted gamma particle having the energy of 1.29×10^{11} J/mol.

Conceptual Problems

6.23 Photosynthesis makes use of visible light to bring about chemical changes. Explain why heat energy in the form of infrared radiation is ineffective for photosynthesis.

6.24 A red light was shined onto a metal sample and the result shown in (i) was observed. When the light source was changed to a blue light, the result shown in (ii) was observed. Explain how these results can be interpreted with respect to the photoelectric effect.

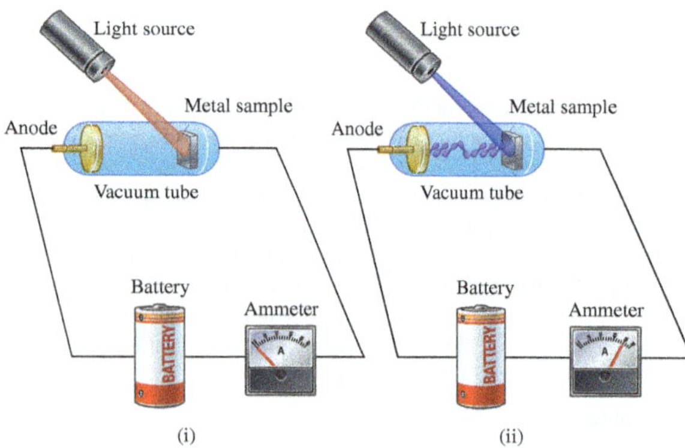

Describe the result you would expect for each of the following: (a) The intensity of red light is increased. (b) The intensity of blue light is increased. (c) Violet light is used.

6.25 A photoelectric experiment was performed by separately shining a laser at 450 nm (blue light) and a laser at 560 nm (yellow light) on a clean metal surface and measuring the number and kinetic energy of the ejected electrons. Which light would generate more electrons? Which light would eject electrons with greater kinetic energy? Assume that the same amount of energy is delivered to the metal surface by each laser and that the frequencies of the laser lights exceed the threshold frequency.

SECTION 6.3: BOHR'S THEORY OF THE HYDROGEN ATOM

Visualizing Chemistry
Figure 6.10

VC 6.1 Which of the following best explains why we see only four lines in the emission spectrum of hydrogen?
a) Hydrogen has only four different electronic transitions.
b) Only four of hydrogen's electronic transitions correspond to visible wavelengths.
c) The other lines in hydrogen's emission spectrum can't be seen easily against the black background.

VC 6.2 One way to see the emission spectrum of hydrogen is to view the hydrogen in an electric discharge tube through a *spectroscope*, a device that separates the wavelengths. Why can we not view the emission spectrum simply by pointing the spectroscope at a sample of hydrogen confined in a glass tube or flask?
a) Without the electrons being in excited states, there would be no emission of light.
b) The glass would make it impossible to see the emission spectrum.
c) Hydrogen alone does not exhibit an emission spectrum—it must be combined with oxygen.

VC 6.3 How many lines would we see in the emission spectrum of hydrogen if the downward transitions from excited states all ended at $n = 1$ and no transitions ended at $n = 2$?
a) We would still see four lines.
b) We would see five lines.
c) We would not see any lines.

VC 6.4 For a hydrogen atom in which the electron has been excited to $n = 4$, how many different transitions can occur as the electron eventually returns to the ground state?
a) 1.
b) 3.
c) 6.

Review Questions

6.26 What are emission spectra? How do line spectra differ from continuous spectra?

6.27 What is an energy level? Explain the difference between ground state and excited state.

6.28 Briefly describe Bohr's theory of the hydrogen atom and how it explains the appearance of an emission spectrum. How does Bohr's theory differ from concepts of classical physics?

Computational Problems

6.29 The first line of the Balmer series occurs at a wavelength of 656.3 nm. What is the energy difference between the two energy levels involved in the emission that results in this spectral line?

6.30 Calculate the wavelength (in nm) of a photon emitted by a hydrogen atom when its electron drops from the $n = 7$ state to the $n = 2$ state.

6.31 Calculate the frequency (Hz) and wavelength (nm) of the emitted photon when an electron drops from the $n = 4$ to the $n = 3$ level in a hydrogen atom.

6.32 Careful spectral analysis shows that the familiar yellow light of sodium lamps (such as street lamps) is made up of photons of two wavelengths, 589.0 nm and 589.6 nm. What is the difference in energy (in joules) between photons with these wavelengths?

6.33 An electron in the hydrogen atom makes a transition from an energy state of principal quantum number n_i to the $n = 1$ state. If the photon emitted has a wavelength of 94.9 nm, what is the value of n_i?

6.34 Consider the following energy levels of a hypothetical atom:

E_4: -1.0×10^{-19} J E_2: -10×10^{-19} J
E_3: -5.0×10^{-19} J E_1: -15×10^{-19} J

(a) What is the wavelength of the photon needed to excite an electron from E_1 to E_4? (b) What is the energy (in joules) a photon must have to excite an electron from E_2 to E_3? (c) When an electron drops from the E_3 level to the E_1 level, the atom is said to undergo emission. Calculate the wavelength of the photon emitted in this process.

Conceptual Problems

6.35 Some copper compounds emit green light when they are heated in a flame. How would you determine whether the light is of one wavelength or a mixture of two or more wavelengths?

6.36 Is it possible for a fluorescent material to emit radiation in the ultraviolet region after absorbing visible light? Explain your answer.

6.37 Explain how astronomers are able to tell which elements are present in distant stars by analyzing the electromagnetic radiation emitted by the stars.

SECTION 6.4: WAVE PROPERTIES OF MATTER

Review Questions

6.38 How does de Broglie's hypothesis account for the fact that the energies of the electron in a hydrogen atom are quantized?

6.39 Why is Equation 6.9 meaningful only for submicroscopic particles, such as electrons and atoms, and not for macroscopic objects?

6.40 Does a baseball in flight possess wave properties? If so, why can we not determine its wave properties?

Computational Problems

6.41 Thermal neutrons are neutrons that move at speeds comparable to those of air molecules at room temperature. These neutrons are most effective in initiating a nuclear chain reaction among ^{235}U isotopes. Calculate the wavelength (in nm) associated with a beam of neutrons moving at 7.00×10^2 m/s (mass of a neutron = 1.675×10^{-27} kg).

6.42 Protons can be accelerated to speeds near that of light in particle accelerators. Estimate the wavelength (in nm) of such a proton moving at 2.90×10^8 m/s (mass of a proton = 1.673×10^{-27} kg).

6.43 What is the de Broglie wavelength (in cm) of a 12.4-g hummingbird flying at 1.20×10^2 mph (1 mile = 1.61 km)?

6.44 What is the de Broglie wavelength (in nm) associated with a 2.5-g Ping-Pong ball traveling at 15 mph?

SECTION 6.5: QUANTUM MECHANICS

Review Questions

6.45 What are the inadequacies of Bohr's theory?

6.46 What is the Heisenberg uncertainty principle? What is the Schrödinger equation?

6.47 What is the physical significance of the wave function?

6.48 How is the concept of electron density used to describe the position of an electron in the quantum mechanical treatment of an atom?

6.49 What is an atomic orbital? How does an atomic orbital differ from an orbit?

Computational Problems

6.50 Alveoli are tiny sacs of air in the lungs. Their average diameter is 5.0×10^{-5} m. Calculate the uncertainty in the velocity of an oxygen molecule (5.3×10^{-26} kg) trapped within a sac. (*Hint:* The maximum uncertainty in the position of the molecule is given by the diameter of the sac.)

6.51 The speed of a thermal neutron (see Problem 6.41) is known to within 2.0 km/s. What is the minimum uncertainty in the position of the thermal neutron?

Conceptual Problems

6.52 In the beginning of the twentieth century, some scientists thought that a nucleus may contain both electrons and protons. Use the Heisenberg uncertainty principle to show that an electron cannot be confined within a nucleus. Repeat the calculation for a proton. Comment on your results. Assume the radius of a nucleus to be 1.0×10^{-15} m. The masses of an electron and a proton are 9.109×10^{-31} kg and 1.673×10^{-27} kg, respectively. (*Hint:* Treat the radius of the nucleus as the uncertainty in position.)

6.53 Suppose that photons of blue light (430 nm) are used to locate the position of a 2.80-g Ping-Pong ball in flight and that the uncertainty in the position is equal to one wavelength. What is the minimum uncertainty in the speed of the Ping-Pong ball? Comment on the magnitude of your result.

SECTION 6.6: QUANTUM NUMBERS

Review Questions

6.54 Describe the four quantum numbers used to characterize an electron in an atom.

6.55 Which quantum number defines a shell? Which quantum numbers define a subshell?

6.56 Which of the four quantum numbers (n, ℓ, m_ℓ, m_s) determine (a) the energy of an electron in a hydrogen atom and in a many-electron atom, (b) the size of an orbital, (c) the shape of an orbital, (d) the orientation of an orbital in space?

Conceptual Problems

6.57 An electron in a certain atom is in the $n = 2$ quantum level. List the possible values of ℓ and m_ℓ that it can have.

6.58 An electron in an atom is in the $n = 3$ quantum level. List the possible values of ℓ and m_ℓ that it can have.

6.59 List all the possible subshells and orbitals associated with the principal quantum number n, if $n = 4$.

6.60 List all the possible subshells and orbitals associated with the principal quantum number n, if $n = 5$.

SECTION 6.7: ATOMIC ORBITALS

Review Questions

6.61 Describe the shapes of *s*, *p*, and *d* orbitals. How are these orbitals related to the quantum numbers n, ℓ, and m_ℓ?

6.62 List the hydrogen orbitals in increasing order of energy.

6.63 Describe the characteristics of an *s* orbital, *p* orbital, and *d* orbital. Which of the following orbitals do not exist: 1*p*, 2*s*, 2*d*, 3*p*, 3*d*, 3*f*, 4*g*?

6.64 Why is a boundary surface diagram useful in representing an atomic orbital?

Conceptual Problems

6.65 Give the values of the quantum numbers associated with the following orbitals: (a) 2*p*, (b) 3*s*, (c) 5*d*.

6.66 Give the values of the four quantum numbers of an electron in the following orbitals: (a) 3*s*, (b) 4*p*, (c) 3*d*.

6.67 Discuss the similarities and differences between a 1*s* and a 2*s* orbital.

6.68 What is the difference between a $2p_x$ and a $2p_y$ orbital?

6.69 Why do the 3*s*, 3*p*, and 3*d* orbitals have the same energy in a hydrogen atom but different energies in a many-electron atom?

6.70 Make a chart of all allowable orbitals in the first four principal energy levels of the hydrogen atom. Designate each by type (e.g., *s*, *p*), and indicate how many orbitals of each type there are.

6.71 For each of the following pairs of hydrogen orbitals, indicate which is higher in energy: (a) 1*s*, 2*s*; (b) 2*p*, 3*p*; (c) $3d_{xy}$, $3d_{yz}$; (d) 3*s*, 3*d*; (e) 4*f*, 5*s*.

6.72 Which orbital in each of the following pairs is lower in energy in a many-electron atom: (a) 2*s*, 2*p*; (b) 3*p*, 3*d*; (c) 3*s*, 4*s*; (d) 4*d*, 5*f*?

6.73 A 3*s* orbital is illustrated here. Using this as a reference to show the relative size of the other four orbitals, answer the following questions.

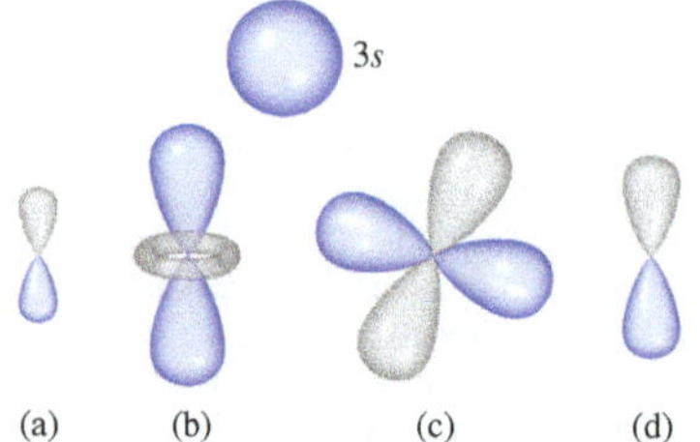

(a) Which orbital has the greatest value of n? (b) How many orbitals have a value of $\ell = 1$? (c) How many other orbitals with the same value of n would have the same general shape as orbital (b)?

SECTION 6.8: ELECTRON CONFIGURATION

Review Questions

6.74 What is electron configuration? Describe the roles that the Pauli exclusion principle and Hund's rule play in writing the electron configuration of elements.

6.75 Explain the meaning of the symbol $4d^6$.

6.76 State the Aufbau principle, and explain the role it plays in classifying the elements in the periodic table.

Computational Problems

6.77 Calculate the total number of electrons that can occupy (a) one *s* orbital, (b) three *p* orbitals, (c) five *d* orbitals, (d) seven *f* orbitals.

6.78 What is the total number of electrons that can be held in all orbitals having the same principal quantum number n?

6.79 Determine the maximum number of electrons that can be found in each of the following subshells: 3*s*, 3*d*, 4*p*, 4*f*, 5*f*.

6.80 Indicate the total number of (a) *p* electrons in N ($Z = 7$), (b) *s* electrons in Si ($Z = 14$), and (c) 3*d* electrons in S ($Z = 16$).

Conceptual Problems

6.81 Indicate which of the following sets of quantum numbers in an atom are unacceptable and explain why: (a) $(1, 1, +\frac{1}{2}, -\frac{1}{2})$, (b) $(3, 0, -1, +\frac{1}{2})$, (c) $(2, 0, +1, +\frac{1}{2})$, (d) $(4, 3, -2, +\frac{1}{2})$, (e) $(3, 2, +1, 1)$.

6.82 The ground-state electron configurations listed here are incorrect. Explain what mistakes have been made in each and write the correct electron configurations.
Al: $1s^22s^22p^43s^23p^3$
B: $1s^22s^22p^5$
F: $1s^22s^22p^6$

6.83 Indicate the number of unpaired electrons present in each of the following atoms: B, Ne, P, Sc, Mn, Se, Kr, Fe, Cd, I, Pb.

6.84 The electron configuration of a neutral atom is $1s^22s^22p^63s^2$. Write a complete set of quantum numbers for each of the electrons. Name the element.

6.85 Which of the following species has the greatest number of unpaired electrons: S^+, S, or S^-?

6.86 Portions of orbital diagrams representing the ground-state electron configurations of certain elements are shown here. Which of them violate the Pauli exclusion principle? Which violate Hund's rule?

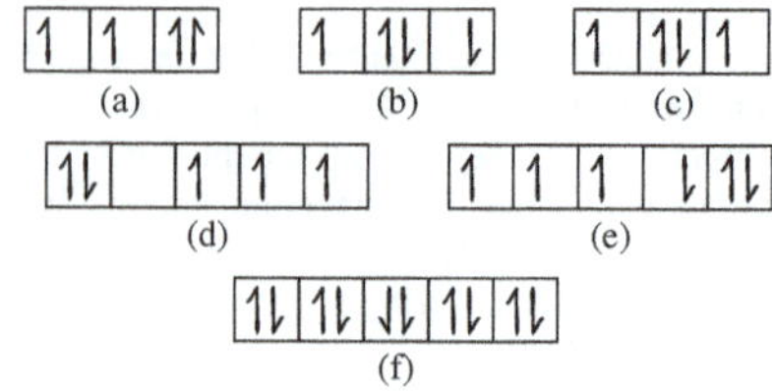

SECTION 6.9: ELECTRON CONFIGURATIONS AND THE PERIODIC TABLE

Review Questions

6.87 Describe the characteristics of transition metals.

6.88 What is the noble gas core? How does it simplify the writing of electron configurations?

6.89 What are the group and period of the element osmium?

6.90 Define the following terms and give an example of each: lanthanides, actinides.

6.91 Explain why the ground-state electron configurations of Cr and Cu are different from what we might expect.

6.92 Write the electron configuration of a xenon core.

6.93 Comment on the correctness of the following statement: The probability of finding two electrons with the same four quantum numbers in an atom is zero.

Conceptual Problems

6.94 Use the Aufbau principle to obtain the ground-state electron configuration of selenium.

6.95 Use the Aufbau principle to obtain the ground-state electron configuration of technetium.

6.96 Write the ground-state electron configurations for the following elements: B, V, C, As, I, Au.

6.97 Write the ground-state electron configurations for the following elements: Ge, Fe, Zn, Ni, W, Tl.

ADDITIONAL PROBLEMS

6.98 Spectral lines of the Lyman and Balmer series do not overlap. Verify this statement by calculating the longest wavelength associated with the Lyman series and the shortest wavelength associated with the Balmer series (in nm).

6.99 Discuss the current view of the correctness of the following statements. (a) The electron in the hydrogen atom is in an orbit that never brings it closer than 100 pm to the nucleus. (b) Atomic absorption spectra result from transitions of electrons from lower to higher energy levels. (c) A many-electron atom behaves somewhat like a solar system that has a number of planets.

6.100 Distinguish carefully between the following terms: (a) wavelength and frequency, (b) wave properties and particle properties, (c) quantization of energy and continuous variation in energy.

6.101 What is the maximum number of electrons in an atom that can have the following quantum numbers? Specify the orbitals in which the electrons would be found. (a) $n = 2$, $m_s = +\frac{1}{2}$; (b) $n = 4$, $m_\ell = +1$; (c) $n = 3$, $\ell = 2$; (d) $n = 2$, $\ell = 0$, $m_s = -\frac{1}{2}$; (e) $n = 4$, $\ell = 3$, $m_\ell = -2$.

6.102 Identify the following individuals and their contributions to the development of quantum theory: Bohr, de Broglie, Einstein, Planck, Heisenberg, Schrödinger.

6.103 Consider the graph here.

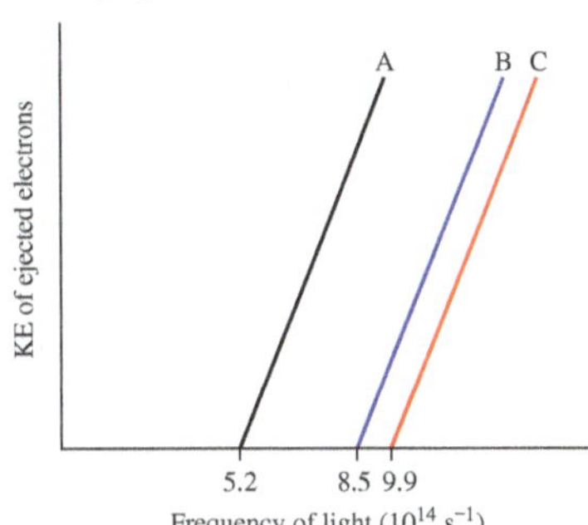

(a) Calculate the binding energy (W) of each metal. Which metal has the highest binding energy? (b) A photon with a wavelength of 333 nm is fired at the three metals. Which, if any, of the metals will eject an electron?

6.104 A baseball pitcher's fastballs have been clocked at about 100 mph. (a) Calculate the wavelength of a 0.141-kg baseball (in nm) at this speed. (b) What is the wavelength of a hydrogen atom at the same speed (1 mile = 1609 m)?

6.105 The He ion contains only one electron and is therefore a hydrogen-like ion. Calculate the wavelengths, in increasing order, of the first four transitions in the Balmer series of the He^+ ion. Compare these wavelengths with the same transitions in an H atom. Comment on the differences. (The Rydberg constant for He is 4.39×10^7 m^{-1}.)

6.106 Draw the shapes (boundary surfaces) of the following orbitals: (a) $2p_y$, (b) $3d_{z^2}$, (c) $3d_{x^2-y^2}$. (Show coordinate axes in your sketches.)

6.107 Draw orbital diagrams for atoms with the following electron configurations:
(a) $1s^22s^22p^5$
(b) $1s^22s^22p^63s^23p^3$
(c) $1s^22s^22p^63s^23p^64s^23d^7$

6.108 Ionization energy is the minimum energy required to remove an electron from an atom. It is usually expressed in units of kJ/mol, that is, the energy in kilojoules required to remove one mole of electrons from one mole of atoms. (a) Calculate the ionization energy for the hydrogen atom. (b) Repeat the calculation, assuming in this second case that the electrons are removed from the $n = 2$ state, instead of from the ground state.

6.109 An electron in a hydrogen atom is excited from the ground state to the $n = 4$ state. Comment on the correctness of the following statements (true or false).
(a) $n = 4$ is the first excited state.
(b) It takes more energy to ionize (remove) the electron from $n = 4$ than from the ground state.
(c) The electron is farther from the nucleus (on average) in $n = 4$ than in the ground state.
(d) The wavelength of light emitted when the electron drops from $n = 4$ to $n = 1$ is longer than that from $n = 4$ to $n = 2$.
(e) The wavelength the atom absorbs in going from $n = 1$ to $n = 4$ is the same as that emitted as it goes from $n = 4$ to $n = 1$.

6.110 The ionization energy of a certain element is 412 kJ/mol (see Problem 6.108). However, when the atoms of this element are in the first excited state, the ionization energy is only 126 kJ/mol. Based on this information, calculate the wavelength of light emitted in a transition from the first excited state to the ground state.

6.111 The electron configurations described in this chapter all refer to gaseous atoms in their ground states. An atom may absorb a quantum of energy and promote one of its electrons to a higher-energy orbital. When this happens, we say that the atom is in an excited state. The electron configurations of some excited atoms are given. Identify these atoms and write their ground-state configurations:
(a) $1s^1 2s^1$
(b) $1s^2 2s^2 2p^2 3d^1$
(c) $1s^2 2s^2 2p^6 4s^1$
(d) $[Ar]4s^1 3d^{10} 4p^4$
(e) $[Ne]3s^2 3p^4 3d^1$

6.112 All molecules undergo vibrational motions. Quantum mechanical treatment shows that the vibrational energy E_{vib} of a diatomic molecule such as HCl is given by

$$E_{vib} = \left(n + \frac{1}{2}\right)h\nu$$

where n is a quantum number ($n = 0, 1, 2, 3, \ldots$) and ν is the fundamental frequency of vibration. (a) Sketch the first three vibrational energy levels for HCl. (b) Calculate the energy required to excite an HCl molecule from the ground level to the first excited level. The fundamental frequency of vibration for HCl is $8.66 \times 10^{13}\ s^{-1}$. (c) The fact that the lowest vibrational energy in the ground level is not zero but equal to $\frac{1}{2}h\nu$ means that molecules will vibrate at all temperatures, including absolute zero. Use the Heisenberg uncertainty principle to justify this prediction. (*Hint:* Consider a molecule that is not vibrating and start by predicting the uncertainty in its momentum.)

6.113 When an electron makes a transition between energy levels of a hydrogen atom, there are no restrictions on the initial and final values of the principal quantum number n. However, there is a quantum mechanical rule that restricts the initial and final values of the orbital angular momentum ℓ. This is the *selection rule,* which states that $\Delta\ell = \pm 1$; that is, in a transition, the value of ℓ can only increase or decrease by 1. According to this rule, which of the following transitions are allowed: (a) $1s \longrightarrow 2s$, (b) $2p \longrightarrow 1s$, (c) $1s \longrightarrow 3d$, (d) $3d \longrightarrow 4f$, (e) $4d \longrightarrow 3s$? In view of this selection rule, explain why it is possible to observe the various emission series shown in Figure 6.11.

6.114 In 1996, physicists created an anti-atom of hydrogen. In such an atom, which is the antimatter equivalent of an ordinary atom, the electric charges of all the component particles are reversed. Thus the nucleus of an anti-atom is made of an antiproton, which has the same mass as a proton but bears a negative charge, while the electron is replaced by an anti-electron (also called a positron) with the same mass as an electron, but bearing a positive charge. Would you expect the energy levels, emission spectra, and atomic orbitals of an antihydrogen atom to be different from those of a hydrogen atom? What would happen if an anti-atom of hydrogen collided with a hydrogen atom?

6.115 An electron in an excited state in a hydrogen atom can return to the ground state in two different ways: (a) via a direct transition in which a photon of wavelength λ_1 is emitted and (b) via an intermediate excited state reached by the emission of a photon of wavelength λ_2. This intermediate excited state then decays to the ground state by emitting another photon of wavelength λ_3. Derive an equation that relates λ_1 to λ_2 and λ_3.

6.116 (a) An electron in the ground state of the hydrogen atom moves at an average speed of 5×10^6 m/s. If the speed is known to an uncertainty of 20 percent, what is the minimum uncertainty in its position? Given that the radius of the hydrogen atom in the ground state is 5.29×10^{-11} m, comment on your result. The mass of an electron is 9.1094×10^{-31} kg. (b) A 0.15-kg baseball thrown at 100 mph has a momentum of 6.7 kg · m/s. If the uncertainty in measuring the momentum is 1.0×10^{-7} of the momentum, calculate the uncertainty in the baseball's position.

6.117 The wave function for the 2s orbital in the hydrogen atom is

$$\psi_{2s} = \frac{1}{\sqrt{2a_0^3}}\left(1 - \frac{\rho}{2}\right)e^{-\rho/2}$$

where a_0 is the value of the radius of the first Bohr orbit, equal to 0.529 nm; ρ is $Z(r/a_0)$; and r is the distance from the nucleus in meters. Calculate the distance from the nucleus (in nm) of the node of the 2s wave function.

6.118 Calculate the energies needed to remove an electron from the $n = 1$ state and the $n = 5$ state in the Li^{2+} ion. What is the wavelength (in nm) of the emitted photon in a transition from $n = 5$ to $n = 1$? Solving Equation 6.4 for energy gives $\Delta E = R_\infty hc\left(\frac{1}{n_1^2} - \frac{1}{n_2^2}\right)$, where the Rydberg constant R_∞ for hydrogen-like atoms is $1.097 \times 10^7\ m^{-1} \cdot Z^2$, and Z is the atomic number.

6.119 According to Einstein's special theory of relativity, the mass of a moving particle, m_{moving}, is related to its mass at m_{rest}, by the following equation

$$m_{moving} = \frac{m_{rest}}{\sqrt{1 - (u/c)^2}}$$

where u and c are the speeds of the particle and light, respectively. (a) In particle accelerators, protons, electrons, and other charged particles are often accelerated to speeds close to the speed of light. Calculate the wavelength (in nm) of a proton moving at 50.0 percent the speed of light. The mass of a proton is 1.67×10^{-27} kg. (b) Calculate the mass of a 6.0×10^{-2} kg tennis ball moving at 63 m/s. Comment on your results.

6.120 The mathematical equation for studying the photoelectric effect is

$$h\nu = W + \frac{1}{2}m_e u^2$$

where ν is the frequency of light shining on the metal; W is the energy needed to remove an electron from the metal; and m_e and u are the mass and speed of the ejected electron, respectively. In an experiment, a student found that a maximum wavelength of 351 nm is needed to just dislodge electrons from a zinc metal surface. Calculate the velocity (in m/s) of an ejected electron when the student employed light with a wavelength of 313 nm.

6.121 Calculate the wavelength and frequency of an emitted gamma particle having the energy of 3.14×10^{11} J/mol.

6.122 The figure illustrates a series of transitions that occur in a hydrogen atom.

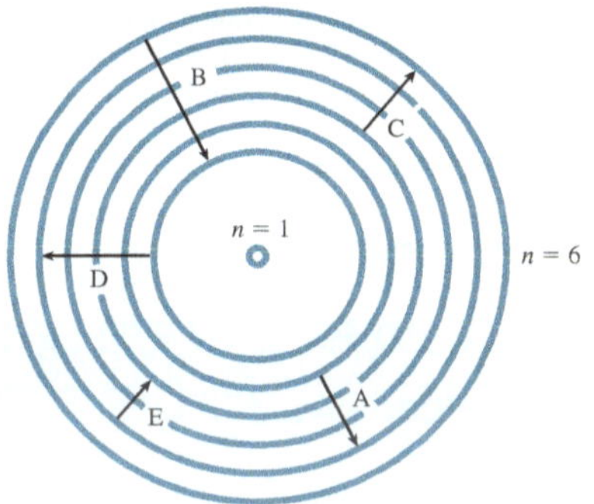

(a) Which transitions are absorptions and which are emissions?
(b) Rank the emissions in order of increasing energy.
(c) Rank the emissions in order of increasing wavelength of light emitted.

Engineering Problems

6.123 In a photoelectric experiment a student uses a light source whose frequency is greater than that needed to eject electrons from a certain metal. However, after continuously shining the light on the same area of the metal for a long period of time the student notices that the maximum kinetic energy of ejected electrons begins to decrease, even though the frequency of the light is held constant. How would you account for this behavior?

6.124 When a compound containing cesium ion is heated in a Bunsen burner flame, photons with an energy of 4.30×10^{-19} J are emitted. What color is the cesium flame?

6.125 In an electron microscope, electrons are accelerated by passing them through a voltage difference. The kinetic energy thus acquired by the electrons is equal to the voltage times the charge on the electron. Thus a voltage difference of 1 volt imparts a kinetic energy of 1.602×10^{-19} volt-coulomb or 1.602×10^{-19} J. Calculate the wavelength associated with electrons accelerated by 5.00×10^3 volts.

6.126 The sun is surrounded by a white circle of gaseous material called the corona, which becomes visible during a total eclipse of the sun. The temperature of the corona is in the millions of degrees Celsius, which is high enough to break up molecules and remove some or all of the electrons from atoms. One way astronomers have been able to estimate the temperature of the corona is by studying the emission lines of ions of certain elements. For example, the emission spectrum of Fe^{14+} ions has been recorded and analyzed. Knowing that it takes 3.5×10^4 kJ/mol to convert Fe^{13+} to Fe^{14+}, estimate the temperature of the sun's corona. (*Hint:* The average kinetic energy of one mole of a gas is $\frac{3}{2}RT$.)

6.127 Scientists have found interstellar hydrogen atoms with quantum number n in the hundreds. Calculate the wavelength of light emitted when a hydrogen atom undergoes a transition from $n = 236$ to $n = 235$. In what region of the electromagnetic spectrum does this wavelength fall?

6.128 Only a fraction of the electric energy supplied to a tungsten lightbulb is converted to visible light. The rest of the energy shows up as infrared radiation (i.e., heat). A 75-W lightbulb converts 15.0 percent of the energy supplied to it into visible light (assume the wavelength to be 550 nm). How many photons are emitted by the lightbulb per second (1 W = 1 J/s)?

6.129 Certain sunglasses have small crystals of silver chloride (AgCl) incorporated in the lenses. When the lenses are exposed to light of the appropriate wavelength, the following reaction occurs:

$$AgCl \longrightarrow Ag + Cl$$

The Ag atoms formed produce a uniform grey color that reduces the glare. If ΔH for the preceding reaction is 248 kJ/mol, calculate the maximum wavelength of light that can induce this process.

6.130 A ruby laser produces radiation of wavelength 633 nm in pulses whose duration is 1.00×10^{-9} s. (a) If the laser produces 0.376 J of energy per pulse, how many photons are produced in each pulse? (b) Calculate the power (in watts) delivered by the laser per pulse (1 W = 1 J/s).

Biological Problems

6.131 The UV light that is responsible for tanning the skin falls in the 320- to 400-nm region. Calculate the total energy (in joules) absorbed by a person exposed to this radiation for 2.5 h, given that there are 2.0×10^{16} photons hitting Earth's surface per square centimeter per second over a 80-nm (320 to 400 nm) range and that the exposed body area is 0.45 m^2. Assume that only half of the radiation is absorbed and the other half is reflected by the body. (*Hint:* Use an average wavelength of 360 nm in calculating the energy of a photon.)

6.132 The retina of a human eye can detect light when radiant energy incident on it is at least 4.0×10^{-17} J. For light of 575-nm wavelength, how many photons does this correspond to?

6.133 Blackbody radiation is the term used to describe the dependence of the radiation energy emitted by an object on wavelength at a certain temperature. Planck proposed the quantum theory to account for the dependence. Shown in the figure is a plot of the radiation energy emitted by our sun versus wavelength. This curve is characteristic of objects at about 6000 K, which is the temperature at the surface of the sun. At a higher temperature, the curve has a similar shape but the maximum will shift to a shorter wavelength. (a) What does this curve reveal about two consequences of great biological significance on Earth? (b) How are astronomers able to determine the temperature at the surface of stars in general?

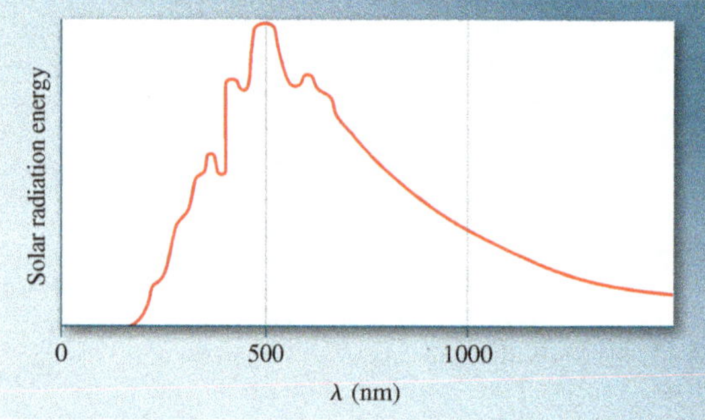

Multiconcept Problems

6.134 Photodissociation of water

$$H_2O(l) + h\nu \longrightarrow H_2(g) + \tfrac{1}{2}O_2(g)$$

has been suggested as a source of hydrogen. The ΔH°_{rxn} for the reaction, calculated from thermochemical data, is 285.8 kJ per mole of water decomposed. Calculate the maximum wavelength (in nm) that would provide the necessary energy. In principle, is it feasible to use sunlight as a source of energy for this process?

6.135 A microwave oven operating at 1.22×10^8 nm is used to heat 150 mL of water (roughly the volume of a tea cup) from 20°C to 100°C. Calculate the number of photons needed if 92.0 percent of microwave energy is converted to the thermal energy of water.

6.136 How many photons at 586 nm must be absorbed to melt 5.0×10^2 g of ice? On average, how many H_2O molecules does one photon convert from ice to water? (*Hint:* It takes 334 J to melt 1 g of ice at 0°C.)

6.137 A 368-g sample of water absorbs infrared radiation at 1.06×10^4 nm from a carbon dioxide laser. Suppose all the absorbed radiation is converted to heat. Calculate the number of photons at this wavelength required to raise the temperature of the water by 5.00°C.

Standardized-Exam Practice Problems

Physical and Biological Sciences

According to Wien's law, the wavelength of maximum intensity in blackbody radiation, λ_{max}, is inversely proportional to the temperature of the radiating body. Mathematically,

$$\lambda_{max} = \frac{b}{T}$$

where b is Wien's displacement constant (2.898×10^6 nm · K) and T is the temperature of the radiating body in kelvins. The sun, composed primarily of hydrogen, emits a continuous spectrum from the region known as the photosphere, with the most intense emission occurring at approximately 500 nm.

1. What is the approximate surface temperature of the sun?

   a) 60,000 K b) 6000 K c) 1500 K d) 500 K

2. What is the frequency of the sun's λ_{max}?

   a) 6.0×10^{14} s^{-1} c) 2.5×10^2 s^{-1}
   b) 5.0×10^{14} s^{-1} d) 5.9×10^{-14} s^{-1}

3. What is the energy of a photon with wavelength λ_{max} for a blackbody at 5200 K?

   a) 5.4×10^{-14} J c) 3.6×10^{-19} J
   b) 5.4×10^{14} J d) 5.6×10^{-7} J

4. The visible region of the electromagnetic spectrum includes wavelengths from about 400 nm to about 700 nm. What is the minimum approximate temperature that would be required for the maximum intensity wavelength emitted by a blackbody to occur in the ultraviolet region of the spectrum?

   a) 71,000 K b) 9100 K c) 7300 K d) 64,000 K

Answers to In-Chapter Materials

Answers to Practice Problems

6.1A 2.91×10^{10} s^{-1}. **6.1B** 1.86×10^{-4} m. **6.2A** 1.58×10^{-19} J. **6.2B** UV. **6.3A** (a) 9.43×10^{-19} J, (b) 1.18×10^{-25} J, (c) 1.96×10^{-19} J. **6.3B** (a) 1.05×10^4 nm, (b) 2.1×10^{18}, (c) 1.2 eV. **6.4A** 103 nm. **6.4B** 7. **6.5A** 26 nm. **6.5B** 8.2×10^{-28} kg · m/s. **6.6A** (a) 3×10^{-34} m, (b) 3×10^{-32} m. **6.6B** (a) $\pm 2 \times 10^{-25}$ kg · m/s, (b) $\pm 1 \times 10^2$ m/s, (c) $\pm 2 \times 10^5$ m/s. **6.7A** Only 0, (b) −2, −1, 0, +1, +2. **6.7B** (a) 2, (b) 0, 1, 2, 3. **6.8A** $n = 3$, $\ell = 2$, $m_\ell = -2, -1, 0, +1, +2$. **6.8B** For a d orbital, $\ell = 2$, but when $n = 2$, ℓ cannot be 2. **6.9A** $1s^22s^22p^63s^23p^64s^23d^{10}4p^65s^1$. **6.9B** $1s^22s^22p^63s^23p^64s^23d^{10}4p^5$. **6.10A** [Rn]$7s^2$. **6.10B** At.

Answers to Checkpoints

6.1.1 c. **6.1.2** a. **6.1.3** b. **6.1.4** c. **6.2.1** e. **6.2.2** d. **6.2.3** c. **6.2.4** d. **6.3.1** b. **6.3.2** d. **6.3.3** c. **6.3.4** e. **6.4.1** c. **6.4.2** a. **6.5.1** b. **6.5.2** e. **6.6.1** a, b, d. **6.6.2** c. **6.6.3** c. **6.6.4** e. **6.7.1** b. **6.7.2** c. **6.7.3** b, e. **6.7.4** b. **6.8.1** b. **6.8.2** d. **6.8.3** b. **6.9.1** c. **6.9.2** c. **6.9.3** b, d. **6.9.4** a, b, d.

CHAPTER 7

Electron Configuration and the Periodic Table

Table salt, one of the most common seasonings, contains a Group 1A metal.

Credit: © *PM Images/Getty Images.*

In This Chapter, You Will Learn

Some of the chemical and physical properties of the elements and how these properties are related to each element's position in the periodic table.

Before You Begin, Review These Skills

- Atomic ions [⏮ Section 2.6]
- Electron configurations and the periodic table [⏮ Section 6.9]

What an Element's Position in the Periodic Table Can Tell Us

Elements in the same group of the periodic table tend to exhibit similar physical and chemical properties. The alkali metals, for example, Li, Na, and K, all form chloride salts that have similar properties. Because of their similarities to sodium chloride (NaCl), lithium chloride (LiCl) and potassium chloride (KCl) have both been used as salt substitutes for people on low-sodium diets. Despite their similarities, only one of these salt substitutes proved to be safe. In the 1940s, a group of cardiac patients on low-sodium diets were given lithium chloride to use in place of sodium chloride. Lithium chloride looks just like sodium chloride and has a very similar taste. The resulting unregulated intake of lithium had disastrous results. Many of the patients developed symptoms of severe lithium toxicity and several of them died. Today, salt substitutes such as Morton Salt Substitute and Nu-Salt contain *potassium* chloride, which also looks and tastes like sodium chloride. The similarities in appearance and flavor of these compounds are largely due to the chemical similarity of the Group 1A elements they contain.

Credit: © *David A. Tietz/Editorial Image, LLC.*

Credit: © *David A. Tietz/Editorial Image, LLC.*

	1A 1	2A 2	3B 3	4B 4	5B 5	6B 6	7B 7	8B 8	8B 9	8B 10	1B 11	2B 12	3A 13	4A 14	5A 15	6A 16	7A 17	8A 18	
1																			1
2	Li																		2
3	Na																		3
4	K																		4
5																			5
6																			6
7																			7

Although some of their biochemical properties vary, the physical and chemical properties of the alkali metals are quite similar. This is why they were placed in the same group in the periodic table in the first place. It turns out they belong in the same group because they have the same valence electron configuration.

At the end of this chapter, you will be able to answer a series of questions about the alkali metals and the ions that they form [⏭ Applying What You've Learned, page 318].

7.1 Development of the Periodic Table

In the nineteenth century, chemists had only a vague idea of atoms and molecules and did not yet know about electrons and protons. Nevertheless, they devised the periodic table using their knowledge of atomic masses. Accurate measurements of the atomic masses of many elements had already been made. Arranging elements according to their atomic masses in a periodic table seemed logical to those chemists, who believed that chemical behavior should somehow be related to atomic mass.

In 1864, the English chemist John Newlands[1] noticed that when the elements were arranged in order of atomic mass, every eighth element had similar properties. Newlands referred to this peculiar relationship as the *law of octaves*. However, this "law" turned out to be inadequate for elements beyond calcium, and Newlands's work was not accepted by the scientific community.

In 1869, the Russian chemist Dmitri Mendeleev[2] and the German chemist Lothar Meyer[3] independently proposed a much more extensive tabulation of the elements based on the regular, periodic recurrence of properties—a phenomenon known as *periodicity*.

Mendeleev's classification system was a great improvement over Newlands's for two reasons. First, it grouped the elements together more accurately, according to their properties. Second, and equally important, it made it possible to predict the properties of several elements that had not yet been discovered. For example, Mendeleev proposed the existence of an unknown element that he called eka-aluminum and predicted a number of its properties. (*Eka* is a Sanskrit word meaning "first"; thus, eka-aluminum would be the first element under aluminum in the same group.) When gallium was discovered 4 years later, its properties matched the predicted properties of eka-aluminum remarkably well:

	Eka-Aluminum (Ea)	**Gallium (Ga)**
Atomic mass	68 amu	69.9 amu
Melting point	Low	30.15°C
Density	5.9 g/cm^3	5.94 g/cm^3
Formula of oxide	Ea_2O_3	Ga_2O_3

Mendeleev's periodic table included 66 known elements. By 1900, some 30 more had been added to the list, filling in some of the empty spaces. Figure 7.1 gives the time period during which each element was discovered.

Although this periodic table was remarkably successful, the early versions had some inconsistencies that were impossible to overlook. For example, the atomic mass of argon (39.95 amu) is greater than that of potassium (39.10 amu), but argon comes before potassium in the periodic table. If elements were arranged solely according to increasing atomic mass, argon would appear in the position occupied by potassium in our modern periodic table (see the inside front cover). No chemist would place argon, a gas with no tendency to react, in the same group as lithium and sodium, two highly reactive metals! This and other discrepancies suggested that some fundamental property other than atomic mass must be the basis of periodicity. The fundamental property turned out to be the number of protons in an atom's nucleus, something that could not have been known by Mendeleev and his contemporaries.

Figure 7.1 Periodic table of elements classified by dates of discovery.

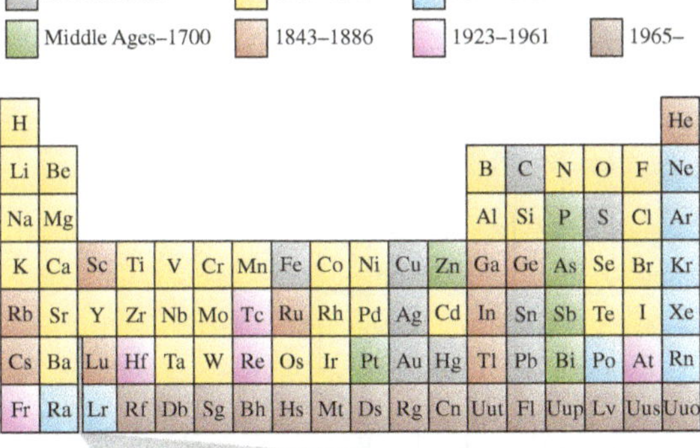

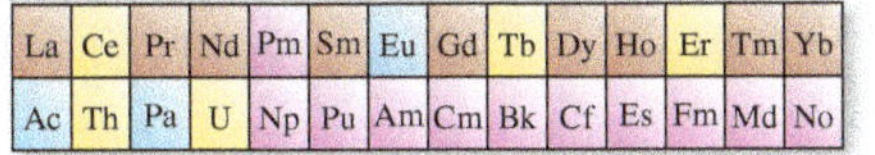

1. John Alexander Reina Newlands (1838–1898). English chemist. Newlands's work was a step in the right direction in the classification of the elements. Unfortunately, because of its shortcomings, he was subjected to much criticism, and even ridicule. At one meeting he was asked if he had ever examined the elements according to the order of their initial letters! Nevertheless, in 1887 Newlands was honored by the Royal Society of London for his contribution.

2. Dmitri Ivanovich Mendeleev (1836–1907). Russian chemist. His work on the periodic classification of elements is regarded by many as the most significant achievement in chemistry in the nineteenth century.

3. Julius Lothar Meyer (1830–1895). German chemist. In addition to his contribution to the periodic table, Meyer also discovered the chemical affinity of hemoglobin for oxygen.

In 1913, a young English physicist, Henry Moseley,[4] discovered a correlation between what he called *atomic number* and the frequency of X rays generated by bombarding an element with high-energy electrons. Moseley noticed that, in general, the frequencies of X rays emitted from the elements increased with increasing atomic mass. Among the few exceptions he found were argon and potassium. Although argon has a greater atomic mass than potassium, the X-ray emission from potassium indicated that it has the greater atomic number. Ordering the periodic table using atomic number enabled scientists to make sense out of the discrepancies that had puzzled them earlier. Moseley concluded that the atomic number was equal to the number of protons in the nucleus and to the number of electrons in an atom.

Entries in modern periodic tables usually include an element's atomic number along with its symbol. Electron configurations of elements help to explain the periodic recurrence of physical and chemical properties. The importance and usefulness of the periodic table lie in the fact that we can use our understanding of the general properties and trends within a group or a period to predict with considerable accuracy the properties of any element, even though that element may be unfamiliar to us.

Sample Problem 7.1 shows how the periodic table can be used to predict similarities and differences in the properties of elements.

SAMPLE PROBLEM 7.1

What elements would you expect to exhibit properties most similar to those of chlorine?

Strategy Because elements in the same group tend to have similar properties, you should identify elements in the same group as chlorine.

Setup Chlorine is a member of Group 7A.

Solution Fluorine, bromine, and iodine, the other nonmetals in Group 7A, should have properties most similar to those of chlorine.

THINK ABOUT IT

Astatine (At) is also in Group 7A. Astatine, though, is classified as a metalloid, and we have to be careful comparing nonmetals to *metalloids* (or to *metals*). As a metalloid, the properties of astatine should be less similar to those of chlorine than the other members of Group 7A. (Actually, astatine is radioactive and very little is known about its properties.)

Practice Problem ATTEMPT What element(s) would you expect to exhibit properties most similar to those of silicon (Si)?

Practice Problem BUILD Arrange the following Group 5A elements in order of increasing similarity of properties to N: As, Bi, and P.

Practice Problem CONCEPTUALIZE Three different groups are highlighted in the periodic table shown here. Which of the three highlighted groups contains the largest number of elements with similar properties? Explain.

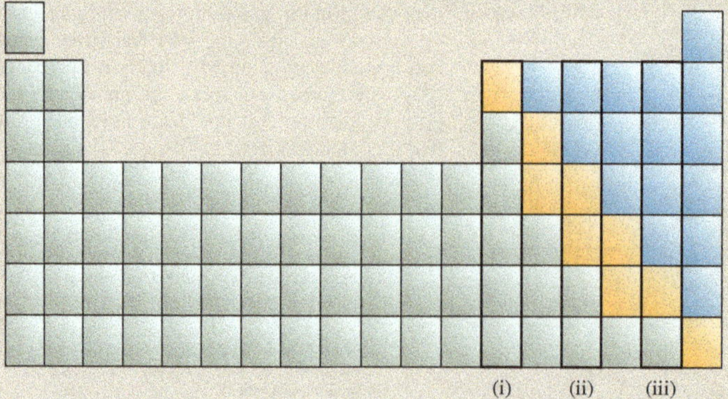

4. Henry Gwyn-Jeffreys Moseley (1887–1915). English physicist. Moseley discovered the relationship between X-ray spectra and atomic number. A lieutenant in the Royal Engineers, he was killed in action at the age of 28 during the British campaign in Gallipoli, Turkey.

Bringing Chemistry to Life

The Chemical Elements of Life

Of the 117 elements currently known, relatively few are essential to living systems. Actually, only six nonmetallic elements form the building blocks of cells: C, H, O, N, P, and S. They are the constituent elements of proteins, nucleic acids, and carbohydrates. Although the natural abundance of carbon is quite low (about 0.1 percent by mass of Earth's crust), it is present in nearly all biological molecules. Carbon is perhaps the most versatile element because it has the ability to form various types of chemical bonds. Carbon atoms can form bonds to each other, linking up to form an enormous variety of chains and ring structures.

The metals play several different roles in living systems. As cations (Na^+, K^+, Ca^{2+}, and Mg^{2+}), they serve to maintain the balance between intracellular and extracellular fluids, nerve transmissions, and other activities. They are also needed for protein functions. For example, the Fe^{2+} ion binds oxygen in hemoglobin molecules, and Cu^{2+}, Zn^{2+}, and Mg^{2+} ions are essential for enzyme activity. In addition, calcium in the form of $Ca_5(PO_4)_3(OH)$ and $Ca_3(PO_4)_2$ is an essential component of teeth and bones.

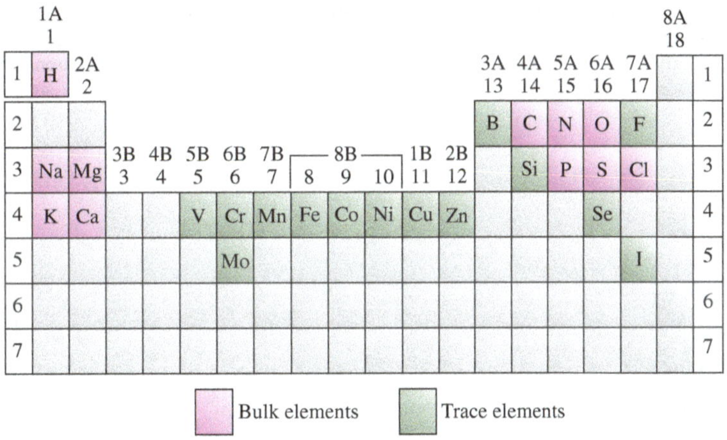

	1A 1	2A 2	3B 3	4B 4	5B 5	6B 6	7B 7	8B 8	8B 9	8B 10	1B 11	2B 12	3A 13	4A 14	5A 15	6A 16	7A 17	8A 18	
1	H																		1
2													B	C	N	O	F		2
3	Na	Mg												Si	P	S	Cl		3
4	K	Ca			V	Cr	Mn	Fe	Co	Ni	Cu	Zn				Se			4
5						Mo											I		5
6																			6
7																			7

Bulk elements Trace elements

The periodic table shown here highlights the essential elements in living systems. Of special interest are the *trace elements,* such as iron (Fe), copper (Cu), zinc (Zn), iodine (I), cobalt (Co), selenium (Se), and fluorine (F), which together make up about 0.1 percent of the human body's mass. Although the trace elements are present in very small amounts, they are crucial for our health. In many cases, however, their exact biological role is still not fully understood.

These elements are necessary for biological functions such as growth, the transport of oxygen for metabolism, and defense against disease. There is a delicate balance in the amounts of these elements in our bodies. Too much or too little over an extended period of time can lead to serious illness, retardation, or even death.

CHECKPOINT – SECTION 7.1 Development of the Periodic Table

7.1.1 Which of the following elements would you expect to have chemical properties most similar to those of S?

a) P b) Cl c) Se d) Na e) Sr

7.1.2 Which of the following elements would you expect to have properties similar to those of Ba?

a) Sr b) Rb c) Na d) K e) B

7.1.3 The first synthesis of element 117 was reported in 2010. Based on its position in the periodic table, which of the following properties would you expect it to exhibit?

(i) gaseous at room temperature
(ii) unreactive
(iii) properties similar to At
(iv) properties similar to Ra

a) i and iv
b) i, ii, and iv
c) iii only
d) iv only
e) i and iii

7.2 The Modern Periodic Table

Figure 7.2 shows the modern periodic table together with the outermost ground-state electron configurations of the elements. (The electron configurations of the elements are also given in Figure 6.25.) Starting with hydrogen, the electronic subshells are filled in the order shown in Figure 6.23 [⏮ Section 6.8].

Classification of Elements

Based on the type of subshell containing the outermost electrons, the elements can be divided into categories—the main group elements, the noble gases, the transition elements (or transition metals), the lanthanides, and the actinides. The ***main group elements*** (also called the *representative elements*) are the elements in Groups 1A through 7A. With the exception of helium, each of the *noble gases* (the Group 8A elements) has a completely filled *p* subshell. The outermost electron configurations are $1s^2$ for helium and ns^2np^6 for the other noble gases, where n is the principal quantum number for the outermost shell.

Student Note: In this context, *outermost* electrons refers to those that are placed in orbitals *last* using the Aufbau principle [⏮ Section 6.8].

The transition metals are the elements in Groups 1B and 3B through 8B. Transition metals either have incompletely filled *d* subshells or readily produce cations with incompletely filled *d* subshells. According to this definition, the elements of Group 2B are *not* transition metals. They typically form +2 ions, although they can also form +1 ions. In either case, the electron configuration includes a completed *d* subshell [⏭ Section 7.6]. Zinc, cadmium, and mercury are *d*-block elements, though, so they generally are included in the discussion of transition metals.

	1A 1	2A 2	3B 3	4B 4	5B 5	6B 6	7B 7	8B 8	8B 9	8B 10	1B 11	2B 12	3A 13	4A 14	5A 15	6A 16	7A 17	8A 18	
1	1 **H** $1s^1$																	2 **He** $1s^2$	1
2	3 **Li** $2s^1$	4 **Be** $2s^2$											5 **B** $2s^22p^1$	6 **C** $2s^22p^2$	7 **N** $2s^22p^3$	8 **O** $2s^22p^4$	9 **F** $2s^22p^5$	10 **Ne** $2s^22p^6$	2
3	11 **Na** $3s^1$	12 **Mg** $3s^2$											13 **Al** $3s^23p^1$	14 **Si** $3s^23p^2$	15 **P** $3s^23p^3$	16 **S** $3s^23p^4$	17 **Cl** $3s^23p^5$	18 **Ar** $3s^23p^6$	3
4	19 **K** $4s^1$	20 **Ca** $4s^2$	21 **Sc** $4s^23d^1$	22 **Ti** $4s^23d^2$	23 **V** $4s^23d^3$	24 **Cr** $4s^13d^5$	25 **Mn** $4s^23d^5$	26 **Fe** $4s^23d^6$	27 **Co** $4s^23d^7$	28 **Ni** $4s^23d^8$	29 **Cu** $4s^13d^{10}$	30 **Zn** $4s^23d^{10}$	31 **Ga** $4s^24p^1$	32 **Ge** $4s^24p^2$	33 **As** $4s^24p^3$	34 **Se** $4s^24p^4$	35 **Br** $4s^24p^5$	36 **Kr** $4s^24p^6$	4
5	37 **Rb** $5s^1$	38 **Sr** $5s^2$	39 **Y** $5s^24d^1$	40 **Zr** $5s^24d^2$	41 **Nb** $5s^14d^4$	42 **Mo** $5s^14d^5$	43 **Tc** $5s^24d^5$	44 **Ru** $5s^14d^7$	45 **Rh** $5s^14d^8$	46 **Pd** $4d^{10}$	47 **Ag** $5s^14d^{10}$	48 **Cd** $5s^24d^{10}$	49 **In** $5s^25p^1$	50 **Sn** $5s^25p^2$	51 **Sb** $5s^25p^3$	52 **Te** $5s^25p^4$	53 **I** $5s^25p^5$	54 **Xe** $5s^25p^6$	5
6	55 **Cs** $6s^1$	56 **Ba** $6s^2$	71 **Lu** $6s^24f^{14}5d^1$	72 **Hf** $6s^25d^2$	73 **Ta** $6s^25d^3$	74 **W** $6s^25d^4$	75 **Re** $6s^25d^5$	76 **Os** $6s^25d^6$	77 **Ir** $6s^25d^7$	78 **Pt** $6s^15d^9$	79 **Au** $6s^15d^{10}$	80 **Hg** $6s^25d^{10}$	81 **Tl** $6s^26p^1$	82 **Pb** $6s^26p^2$	83 **Bi** $6s^26p^3$	84 **Po** $6s^26p^4$	85 **At** $6s^26p^5$	86 **Rn** $6s^26p^6$	6
7	87 **Fr** $7s^1$	88 **Ra** $7s^2$	103 **Lr** $7s^25f^{14}6d^1$	104 **Rf** $7s^26d^2$	105 **Db** $7s^26d^3$	106 **Sg** $7s^26d^4$	107 **Bh** $7s^26d^5$	108 **Hs** $7s^26d^6$	109 **Mt** $7s^26d^7$	110 **Ds** $7s^26d^8$	111 **Rg** $7s^26d^9$	112 **Cn** $7s^26d^{10}$	113 **Uut** $7s^27p^1$	114 **Fl** $7s^27p^2$	115 **Uup** $7s^27p^3$	116 **Lv** $7s^27p^4$	117 **Uus** $7s^27p^5$	118 **Uuo** $7s^27p^6$	7

57 **La** $6s^25d^1$	58 **Ce** $6s^24f^15d^1$	59 **Pr** $6s^24f^3$	60 **Nd** $6s^24f^4$	61 **Pm** $6s^24f^5$	62 **Sm** $6s^24f^6$	63 **Eu** $6s^24f^7$	64 **Gd** $6s^24f^75d^1$	65 **Tb** $6s^24f^9$	66 **Dy** $6s^24f^{10}$	67 **Ho** $6s^24f^{11}$	68 **Er** $6s^24f^{12}$	69 **Tm** $6s^24f^{13}$	70 **Yb** $6s^24f^{14}$
89 **Ac** $7s^26d^1$	90 **Th** $7s^26d^2$	91 **Pa** $7s^25f^26d^1$	92 **U** $7s^25f^36d^1$	93 **Np** $7s^25f^46d^1$	94 **Pu** $7s^25f^6$	95 **Am** $7s^25f^7$	96 **Cm** $7s^25f^76d^1$	97 **Bk** $7s^25f^9$	98 **Cf** $7s^25f^{10}$	99 **Es** $7s^25f^{11}$	100 **Fm** $7s^25f^{12}$	101 **Md** $7s^25f^{13}$	102 **No** $7s^25f^{14}$

Figure 7.2 Valence electron configurations of the elements. For simplicity, the filled *f* subshells are not shown in elements 72 through 86 and 104 through 118.

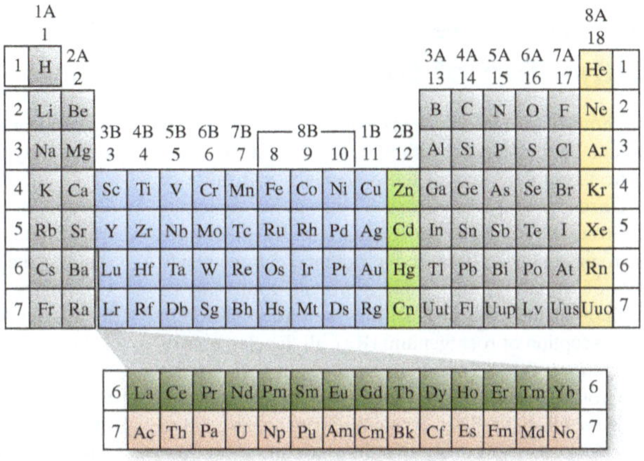

	1A 1	2A 2	3B 3	4B 4	5B 5	6B 6	7B 7	8B 8	8B 9	8B 10	1B 11	2B 12	3A 13	4A 14	5A 15	6A 16	7A 17	8A 18	
1	H																	He	1
2	Li	Be											B	C	N	O	F	Ne	2
3	Na	Mg											Al	Si	P	S	Cl	Ar	3
4	K	Ca	Sc	Ti	V	Cr	Mn	Fe	Co	Ni	Cu	Zn	Ga	Ge	As	Se	Br	Kr	4
5	Rb	Sr	Y	Zr	Nb	Mo	Tc	Ru	Rh	Pd	Ag	Cd	In	Sn	Sb	Te	I	Xe	5
6	Cs	Ba	Lu	Hf	Ta	W	Re	Os	Ir	Pt	Au	Hg	Tl	Pb	Bi	Po	At	Rn	6
7	Fr	Ra	Lr	Rf	Db	Sg	Bh	Hs	Mt	Ds	Rg	Cn	Uut	Fl	Uup	Lv	Uus	Uuo	7

6	La	Ce	Pr	Nd	Pm	Sm	Eu	Gd	Tb	Dy	Ho	Er	Tm	Yb	6
7	Ac	Th	Pa	U	Np	Pu	Am	Cm	Bk	Cf	Es	Fm	Md	No	7

Figure 7.3 Periodic table with color-coding of main group elements, noble gases, transition metals, group 2B metals, lanthanides, and actinides.

The lanthanides and actinides are sometimes called *f*-block transition elements because they have incompletely filled *f* subshells. Figure 7.3 distinguishes the groups of elements discussed here.

There is a distinct pattern to the electron configurations of the elements in a particular group. See, for example, the electron configurations of Groups 1A and 2A in Table 7.1. Each member of Group 1A has a noble gas core plus one additional electron, giving each alkali metal the general electron configuration of [noble gas]ns^1. Similarly, the Group 2A alkaline earth metals have a noble gas core and an outer electron configuration of ns^2.

The outermost electrons of an atom are called ***valence electrons,*** which are the ones involved in the formation of chemical bonds between atoms. The similarity of the valence electron configurations (i.e., they have the same number and type of valence electrons) is what makes the elements in the same group resemble one another chemically. This observation holds true for the other main group elements as well. For instance, the halogens (Group 7A) all have outer electron configurations of ns^2np^5, and they have similar properties.

In predicting properties for Groups 3A through 7A, we must take into account that each of these groups contains elements on both sides of the line that divides metals and nonmetals. For example, the elements in Group 4A all have the same outer electron configuration, ns^2np^2, but there is considerable variation in chemical properties among these elements because carbon is a nonmetal, silicon and germanium are metalloids, and tin and lead are metals.

As a group, the noble gases behave very similarly. The noble gases are generally unreactive because they all have completely filled outer *ns* and *np* subshells, a condition that imparts unusual stability.

Although the outer electron configuration of the transition metals is not always the same within a group and there is often no regular pattern in the way the electron configuration changes from one metal to the next in the same period, all transition metals share many characteristics (multiple oxidation states, richly colored compounds, magnetic properties, and so on) that set them apart from other elements. These properties are similar because all these metals have incompletely filled *d* subshells. Likewise, the lanthanide and actinide elements resemble one another because they have incompletely filled *f* subshells.

Sample Problem 7.2 shows how to determine the electron configuration from the number of electrons in an atom.

Student Note: Although hydrogen's electron configuration is $1s^1$ [⏮ Section 6.9, Figure 6.25], it is a nonmetal and is not really a member of Group 1A.

TABLE 7.1 Electron Configurations of Group 1A and Group 2A Elements

Group 1A		Group 2A	
Li	[He]$2s^1$	Be	[He]$2s^2$
Na	[Ne]$3s^1$	Mg	[Ne]$3s^2$
K	[Ar]$4s^1$	Ca	[Ar]$4s^2$
Rb	[Kr]$5s^1$	Sr	[Kr]$5s^2$
Cs	[Xe]$6s^1$	Ba	[Xe]$6s^2$
Fr	[Rn]$7s^1$	Ra	[Rn]$7s^2$

Why Are There Two Different Sets of Numbers at the Top of the Periodic Table?

The numbering of the transition metal groups 3B through 7B indicates the similarity between the outer electron configurations of these elements and those of the corresponding main group elements. For example, scandium (Sc; Group 3B) and gallium (Ga; Group 3A) each have three outer electrons. However, because their outer electrons reside in different types of atomic orbitals (*s* and *d* orbitals in the case of Sc; *s* and *p* orbitals in the case of Ga), they belong in different groups. With the exception of roentgenium (Rg), all the elements of Groups 1B and 2B have filled *d* subshells [ Section 6.9]. Unlike the elements of Group 2B, elements of Group 1B form cations with incompletely filled *d* subshells. Their group numbers correspond to the one and two electrons, respectively, that they have in *s* orbitals—just like the main group elements in Groups 1A and 2A. The elements iron (Fe), cobalt (Co), and nickel (Ni), and the elements that appear beneath them in the periodic table, cannot be classified in this way and are all placed in Group 8B.

The designation of A and B groups is not universal. In Europe, the practice is to use B for main group elements and A for transition or *d*-block elements, which is just the opposite of the American convention. The International Union of Pure and Applied Chemistry (IUPAC) has recommended eliminating ambiguity by numbering the columns sequentially with Arabic numerals 1 through 18 (see Figure 7.2). The proposal has not been accepted universally, and many modern periodic tables retain the traditional group designations. Periodic tables in this text display both the IUPAC-recommended Arabic numerals and the traditional American numbering system. Discussions in the text will refer to the traditional American group numbers.

SAMPLE PROBLEM 7.2

Without using a periodic table, give the ground-state electron configuration and block designation (*s*-, *p*-, *d*-, or *f*-block) of an atom with (a) 17 electrons, (b) 37 electrons, and (c) 22 electrons. Classify each atom as a main group element or transition metal.

Strategy Use the Aufbau principle discussed in Section 6.8. Start writing each electron configuration with principal quantum number $n = 1$, and then continue to assign electrons to orbitals in the order presented in Figure 6.23 until all the electrons have been accounted for.

Setup According to Figure 6.23, orbitals fill in the following order: 1*s*, 2*s*, 2*p*, 3*s*, 3*p*, 4*s*, 3*d*, 4*p*, 5*s*, 4*d*, 5*p*, 6*s*, and so on. Recall that an *s* subshell contains one orbital, a *p* subshell contains three orbitals, and a *d* subshell contains five orbitals. Remember, too, that each orbital can accommodate a maximum of two electrons. The block designation of an element corresponds to the type of subshell occupied by the last electrons added to the configuration according to the Aufbau principle.

Solution

(a) $1s^22s^22p^63s^23p^5$, *p*-block, main group

(b) $1s^22s^22p^63s^23p^64s^23d^{10}4p^65s^1$, *s*-block, main group

(c) $1s^22s^22p^63s^23p^64s^23d^2$, *d*-block, transition metal

THINK ABOUT IT

Consult Figure 6.25 on page 267 to confirm your answers.

Practice Problem ATTEMPT Without using a periodic table, give the ground-state electron configuration and block designation (*s*-, *p*-, *d*-, or *f*-block) of an atom with (a) 15 electrons, (b) 20 electrons, and (c) 35 electrons.

Practice Problem BUILD Identify the elements represented by (a) $1s^22s^22p^63s^23p^1$, (b) $1s^22s^22p^63s^23p^64s^23d^{10}$, and (c) $1s^22s^22p^63s^23p^64s^23d^{10}4p^65s^2$.

Practice Problem CONCEPTUALIZE Determine the *total* number of electrons and the number of *valence* electrons for each of the indicated elements.

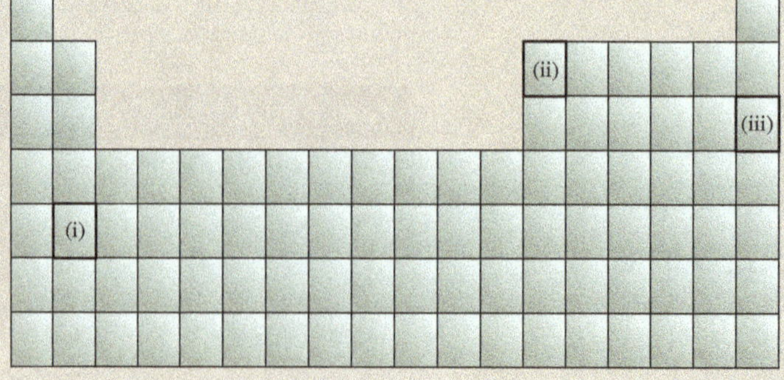

Representing Free Elements in Chemical Equations

Having classified the elements according to their ground-state electron configurations, we can now learn how chemists represent elements in chemical equations.

Metals Because metals typically do not exist in discrete molecular units but rather in complex, three-dimensional networks of atoms, we always use their empirical formulas in chemical equations. The empirical formulas are the same as the symbols that represent the elements. For example, the empirical formula for iron is Fe, the same as the symbol for the element.

Student Note: Recall that allotropes are different forms of the same element [⏮ Section 2.6].

Nonmetals There is no single rule regarding the representation of nonmetals in chemical equations. Carbon, for example, exists in several allotropic forms. Regardless of the allotrope, we use its empirical formula C to represent elemental carbon in chemical equations. Often the symbol C will be followed by the specific allotrope in parentheses as in the equation representing the conversion of graphite to diamond, two of carbon's allotropic forms:

$$\text{C(graphite)} \longrightarrow \text{C(diamond)}$$

For nonmetals that exist as polyatomic molecules, we generally use the molecular formula in equations: H_2, N_2, O_2, F_2, Cl_2, Br_2, I_2, and P_4, for example. In the case of sulfur, however, we usually use the empirical formula S rather than the molecular formula S_8. Thus, instead of writing the equation for the combustion of sulfur as

$$S_8(s) + 8O_2(g) \longrightarrow 8SO_2(g)$$

we usually write

$$S(s) + O_2(g) \longrightarrow SO_2(g)$$

although, technically, both ways are correct.

Noble Gases All the noble gases exist as isolated atoms, so we use their symbols: He, Ne, Ar, Kr, Xe, and Rn.

Metalloids The metalloids, like the metals, all have complex three-dimensional networks, so we also represent them with their empirical formulas—that is, their symbols: B, Si, Ge, and so on.

CHECKPOINT – SECTION 7.2 The Modern Periodic Table

7.2.1 Which electron configuration is correct for a germanium (Ge) atom in the ground state?

a) $1s^2 2s^2 2p^6 3s^2 3p^6 4s^2 4p^2$

b) $1s^2 2s^2 2p^6 3s^2 3p^6 4s^2 3d^{10} 4p^2$

c) $1s^2 2s^2 2p^6 3s^2 3p^2$

d) $1s^2 2s^2 2p^6 3s^2 3p^6 4s^2 4p^2 4d^{10}$

e) $1s^2 2s^2 2p^6 3s^2 3p^2 3d^{10}$

7.2.2 Which of the following equations correctly represent the chemical reaction in which graphite combines with sulfur to form carbon disulfide gas [$CS_2(g)$]? (Select all that apply.)

a) $\text{C(graphite)} + 2S(s) \longrightarrow CS_2(g)$

b) $\text{C(graphite)} + S_2(s) \longrightarrow CS_2(g)$

c) $\text{C(graphite)} + S_8(s) \longrightarrow CS_8(g)$

d) $\text{C(graphite)} + \frac{1}{4}S_8(s) \longrightarrow CS_2(g)$

e) $4\text{C(graphite)} + S_8(s) \longrightarrow 4CS_2(g)$

7.3 Effective Nuclear Charge

As we have seen, the electron configurations of the elements show a periodic variation with increasing atomic number. In this and the next few sections, we will examine how electron configuration explains the periodic variation of physical and chemical properties of the elements. We begin by introducing the concept of *effective nuclear charge.*

Nuclear charge (Z) is simply the number of protons in the nucleus of an atom. ***Effective nuclear charge*** **(Z_{eff})** is the actual magnitude of positive charge that is "experienced" by an electron in the atom. The only atom in which the nuclear charge and effective nuclear charge are the same is hydrogen, which has only one electron. In all other atoms, the electrons are simultaneously attracted to the nucleus and repelled by one another. This results in a phenomenon

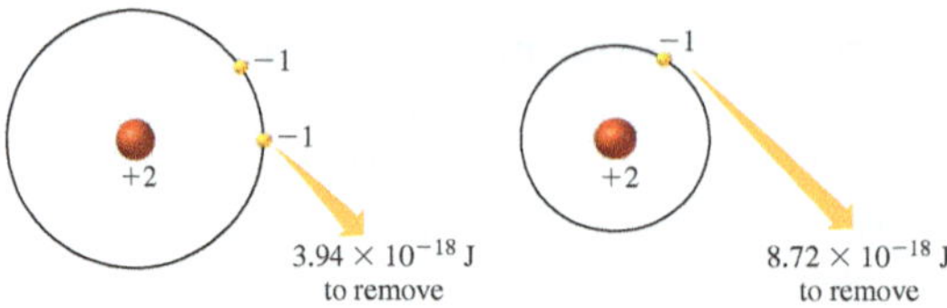

Figure 7.4 Removal of the first electron in He requires less energy than removal of the second electron because of shielding.

known as ***shielding.*** An electron in a many-electron atom is partially shielded from the positive charge of the nucleus by the other electrons in the atom.

Student Note: *Shielding* is also known as *screening*.

One way to illustrate how electrons in an atom shield one another is to consider the amounts of energy required to remove the two electrons from a helium atom, shown in Figure 7.4. Experiments show that it takes 3.94×10^{-18} J to remove the first electron but 8.72×10^{-18} J to remove the second one. There is no shielding once the first electron is removed, so the second electron feels the full effect of the +2 nuclear charge and is more difficult to remove.

Although all the electrons in an atom shield one another to some extent, those that are most effective at shielding are the *core* electrons. As a result, the value of Z_{eff} increases steadily from left to right across a period of the periodic table because the number of core electrons remains the same (only the number of protons, Z, and the number of *valence* electrons increases).

Student Note: Core electrons are those in the completed inner shells.

As we move to the right across period 2, the nuclear charge increases by 1 with each new element, but the *effective* nuclear charge increases only by an average of 0.64. (If the valence electrons did *not* shield one another, the effective nuclear charge would also increase by 1 each time a proton was added to the nucleus.)

	Li	Be	B	C	N	O	F
Z	3	4	5	6	7	8	9
Z_{eff} (felt by valence electrons)	1.28	1.91	2.42	3.14	3.83	4.45	5.10

In general, the effective nuclear charge is given by

$$Z_{eff} = Z - \sigma \qquad \textbf{Equation 7.1}$$

where σ is the shielding constant. The shielding constant is greater than zero but smaller than Z.

The change in Z_{eff} as we move from the top of a group to the bottom is generally less significant than the change as we move across a period. Although each step down a group represents a large increase in the nuclear charge, there is also an additional shell of core electrons to shield the valence electrons from the nucleus. Consequently, the *effective* nuclear charge changes less than the nuclear charge as we move down a column of the periodic table.

7.4 Periodic Trends in Properties of Elements

Several physical and chemical properties of the elements depend on effective nuclear charge. To understand the trends in these properties, it is helpful to visualize the electrons of an atom in *shells.* Recall that the value of the principal quantum number (n) increases as the distance from the nucleus increases [◀◀ Section 6.7]. If we take this statement literally, and picture all the electrons in a shell at the same distance from the nucleus, the result is a sphere of uniformly distributed negative charge, with its distance from the nucleus depending on the value of n. With this as a starting point, we will examine the periodic trends in atomic radius, ionization energy, and electron affinity.

Atomic Radius

Intuitively, we think of the ***atomic radius*** as the distance between the nucleus of an atom and its valence shell (i.e., the outermost shell that is occupied by one or more electrons), because we usually envision atoms as spheres with discrete boundaries. According to the quantum mechanical model of the atom, though, there is no specific distance from the nucleus beyond which an electron may not be found [◀◀ Section 6.7]. Therefore, the atomic radius requires a specific definition.

There are two ways in which the atomic radius is commonly defined. One is the ***metallic radius,*** which is half the distance between the nuclei of two adjacent, identical metal atoms [Figure 7.5(a)]. The other is the ***covalent radius,*** which is half the distance between adjacent, identical nuclei in a molecule [Figure 7.5(b)].

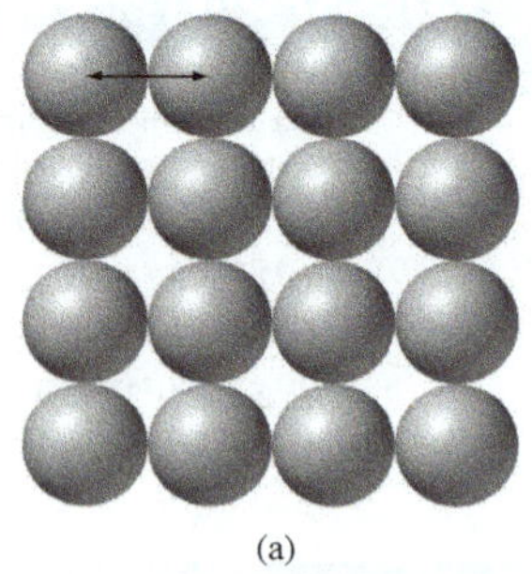

(a)

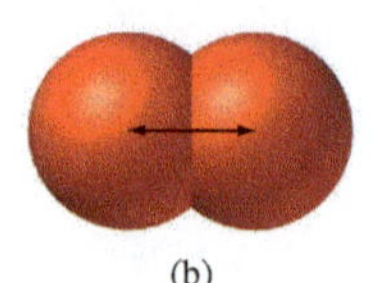

(b)

Figure 7.5 (a) Atomic radius in metals is defined as half the distance between adjacent metal atoms. (b) Atomic radius in nonmetals is defined as half the distance between bonded identical atoms in a molecule.

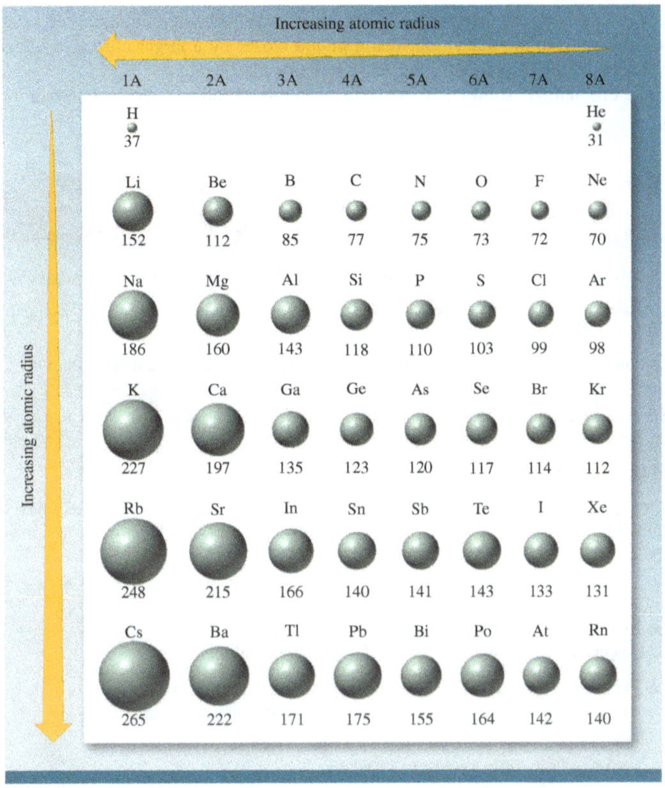

Figure 7.6 Atomic radii of main group elements.

Increasing atomic radius (← across a period); Increasing atomic radius (↓ down a group)

1A	2A	3A	4A	5A	6A	7A	8A
H 37							He 31
Li 152	Be 112	B 85	C 77	N 75	O 73	F 72	Ne 70
Na 186	Mg 160	Al 143	Si 118	P 110	S 103	Cl 99	Ar 98
K 227	Ca 197	Ga 135	Ge 123	As 120	Se 117	Br 114	Kr 112
Rb 248	Sr 215	In 166	Sn 140	Sb 141	Te 143	I 133	Xe 131
Cs 265	Ba 222	Tl 171	Pb 175	Bi 155	Po 164	At 142	Rn 140

Animation
Periodic table—atomic radius.

Figure 7.6 shows the atomic radii of the main group elements according to their positions in the periodic table. There are two distinct trends. The atomic radius *decreases* as we move from left to right across a period and *increases* from top to bottom as we move down within a group. The increase down a group is fairly easily explained. As we step down a column, the outermost occupied shell has an ever-increasing value of n, so it lies farther from the nucleus, making the radius bigger.

Now let's try to understand the decrease in radius from left to right across a period. Although this trend may at first seem counterintuitive, given that the number of valence electrons is increasing with each new element, consider the shell model in which all the electrons in a shell form a uniform sphere of negative charge around the nucleus at a distance specified by the value of n. As we move from left to right across a period, the effective nuclear charge increases and each step to the right adds another electron to the valence shell. Coulomb's law dictates that there will be a more powerful attraction between the nucleus and the valence shell when the magnitudes of both charges increase. The result is that as we step across a period the valence shell is drawn closer to the nucleus, making the atomic radius smaller. Figure 7.7 shows how the effective nuclear charge, charge on the valence shell, and atomic radius vary across

Student Note: Although the overall trend in atomic size for transition elements is also to decrease from left to right and increase from top to bottom, the observed radii do not vary in as regular a way as do the main group elements.

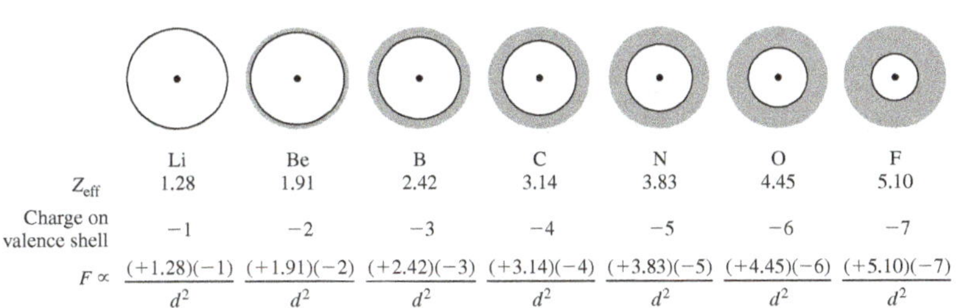

Figure 7.7 Atomic radius decreases from left to right across a period because of the increased electrostatic attraction between the effective nuclear charge and the charge on the valence shell. The white circle shows the atomic size in each case. The comparison of attractive forces between the nuclei and valence shells is done using Coulomb's law [⏭ Page 298].

period 2. We can picture the valence shells in all the atoms as being initially at the same distance (determined by n) from the nuclei, but being pulled closer by a larger attractive force resulting from increases in both Z_{eff} and the number of valence electrons.

Sample Problem 7.3 shows how to use these trends to compare the atomic radii of different elements.

SAMPLE PROBLEM 7.3

Referring only to a periodic table, arrange the elements P, S, and O in order of increasing atomic radius.

Strategy Use the left-to-right (decreasing) and top-to-bottom (increasing) trends to compare the atomic radii of two of the three elements at a time.

Setup Sulfur is to the right of phosphorus in the third row, so sulfur should be smaller than phosphorus. Oxygen is above sulfur in Group 6A, so oxygen should be smaller than sulfur.

Solution: O < S < P

THINK ABOUT IT

Consult Figure 7.6 to confirm the order. Note that there are circumstances under which the trends alone will be insufficient to compare the radii of two elements. Using only a periodic table, for example, it would not be possible to determine that chlorine (r = 99 pm) has a larger radius than oxygen (r = 73 pm).

Practice Problem ATTEMPT Referring only to a periodic table, arrange the elements Ge, Se, and F in order of increasing atomic radius.

Practice Problem BUILD For which of the following pairs of elements can the atomic radii *not* be compared using the periodic table alone: P and Se, Se and Cl, or P and O?

Practice Problem CONCEPTUALIZE Based on size and using only a periodic table, identify the colored spheres as Al, B, Mg, and Sr.

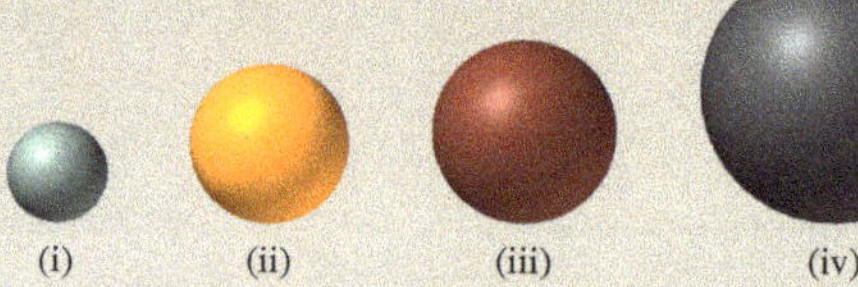

Ionization Energy

Ionization energy (IE) is the minimum energy required to remove an electron from an atom in the gas phase. Typically, we express ionization energy in kJ/mol, the number of kilojoules required to remove a mole of electrons from a mole of gaseous atoms. Sodium, for example, has an ionization energy of 495.8 kJ/mol, meaning that the energy input required to drive the process

$$Na(g) \longrightarrow Na^{+}(g) + e^{-}$$

is 495.8 kJ/mol. Specifically, this is the *first* ionization energy of sodium, IE_1(Na), which corresponds to the removal of the most loosely held electron. Figure 7.8(a) shows the first ionization energies of the main group elements according to their positions in the periodic table. Figure 7.8(b) shows a graph of IE_1 as a function of atomic number.

Student Note: As with atomic radius, ionization energy changes in a similar but somewhat less regular way among the transition elements.

In general, as effective nuclear charge increases, ionization energy also increases. Thus, IE_1 increases from left to right across a period. Despite this trend, the graph in Figure 7.8(b) shows that IE_1 for a Group 3A element is smaller than that for the corresponding Group 2A element. Likewise, IE_1 for a Group 6A element is smaller than that for the corresponding Group 5A element. Both of these *interruptions* of the upward trend in IE_1 can be explained by using electron configuration.

Recall that the energy of an electron in a many-electron system depends not only on the principal quantum number (n), but also on the angular momentum quantum number (ℓ) [⏮ Section 6.8, Figure 6.23]. Within a given shell, electrons with the higher value of ℓ have a higher energy (are less tightly held by the nucleus) and are therefore *easier* to remove. Figure 7.9(a) shows the relative energies of an s subshell ($\ell = 0$) and a p subshell ($\ell = 1$). Ionization of an element in Group 2A requires the removal of an electron from an s orbital, whereas ionization of an element in Group 3A requires the removal of an electron from a p orbital; therefore, the element in Group 3A has a lower ionization energy than the element in Group 2A.

As for the decrease in ionization energy in elements of Group 6A compared to those in Group 5A, both ionizations involve the removal of a p electron, but the ionization of an atom

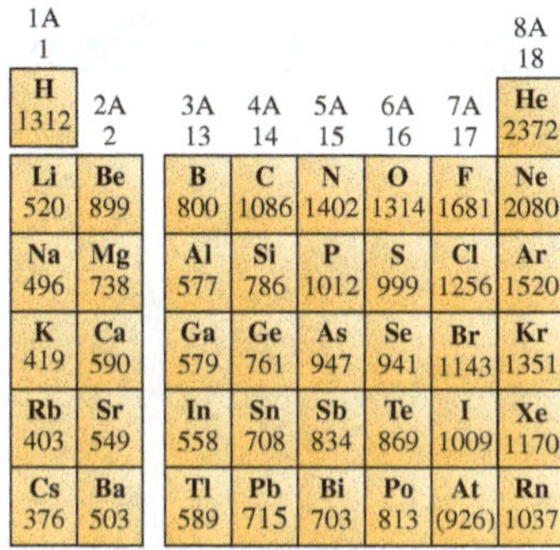

1A 1	2A 2	3A 13	4A 14	5A 15	6A 16	7A 17	8A 18
H 1312							He 2372
Li 520	Be 899	B 800	C 1086	N 1402	O 1314	F 1681	Ne 2080
Na 496	Mg 738	Al 577	Si 786	P 1012	S 999	Cl 1256	Ar 1520
K 419	Ca 590	Ga 579	Ge 761	As 947	Se 941	Br 1143	Kr 1351
Rb 403	Sr 549	In 558	Sn 708	Sb 834	Te 869	I 1009	Xe 1170
Cs 376	Ba 503	Tl 589	Pb 715	Bi 703	Po 813	At (926)	Rn 1037

IE_1 values for main group elements (kJ/mol)

(a)

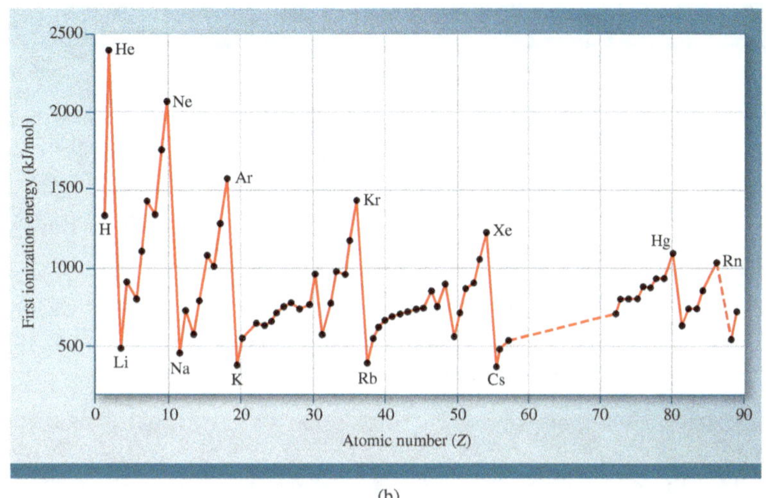

(b)

Figure 7.8 (a) First ionization energies (in kJ/mol) of the main group elements. (b) First ionization energy as a function of atomic number.

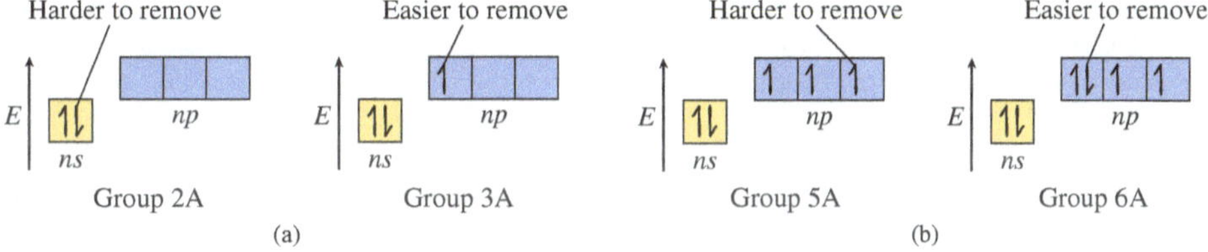

Figure 7.9 (a) It is harder to remove an electron from an *s* orbital than it is to remove an electron from a *p* orbital with the same principal quantum number. (b) Within a *p* subshell, it is easier to remove an electron from a doubly occupied orbital than from a singly occupied orbital.

in Group 6A involves the removal of a *paired* electron. The repulsive force between two electrons in the same orbital makes it easier to remove one of them, making the ionization energy for the Group 6A element actually lower than that for the Group 5A element. [See Figure 7.9(b).]

The first ionization energy IE_1 decreases as we move from top to bottom within a group due to the increasing atomic radius. Although the effective nuclear charge does not change significantly as we step down a group, the atomic radius increases because the value of n for the valence shell increases. According to Coulomb's law, the attractive force between a valence electron and the effective nuclear charge gets *weaker* as the distance between them increases. This makes it easier to remove an electron, and so IE_1 decreases.

It is possible to remove additional electrons in subsequent ionizations, giving IE_2, IE_3, and so on. The second and third ionizations of sodium, for example, can be represented, respectively, as

$$Na^+(g) \longrightarrow Na^{2+}(g) + e^- \quad \text{and} \quad Na^{2+}(g) \longrightarrow Na^{3+}(g) + e^-$$

However, the removal of successive electrons requires ever-increasing amounts of energy because it is harder to remove an electron from a cation than from an atom (and it gets even harder as the charge on the cation increases). Table 7.2 lists the ionization energies of the elements in period 2 and of sodium. These data show that it takes much more energy to remove core electrons than to remove valence electrons. There are two reasons for this. First, core electrons are closer to the nucleus, and second, core electrons experience a greater effective nuclear charge because there are fewer filled shells shielding them from the nucleus. Both of these factors contribute to a greater attractive force between the electrons and the nucleus, which must be overcome to remove the electrons.

Sample Problem 7.4 shows how to use these trends to compare first ionization energies, and subsequent ionization energies, of specific atoms.

TABLE 7.2 Ionization Energies (in kJ/mol) for Elements 3 Through 11*

	Z	IE_1	IE_2	IE_3	IE_4	IE_5	IE_6	IE_7	IE_8	IE_9	IE_{10}
Li	3	520	7,298	11,815							
Be	4	899	1,757	14,848	21,007						
B	5	800	2,427	3,660	25,026	32,827					
C	6	1,086	2,353	4,621	6,223	37,831	47,277				
N	7	1,402	2,856	4,578	7,475	9,445	53,267	64,360			
O	8	1,314	3,388	5,301	7,469	10,990	13,327	71,330	84,078		
F	9	1,681	3,374	6,050	8,408	11,023	15,164	17,868	92,038	106,434	
Ne	10	2,080	3,952	6,122	9,371	12,177	15,238	19,999	23,069	115,380	131,432
Na	11	496	4,562	6,910	9,543	13,354	16,613	20,117	25,496	28,932	141,362

*Cells shaded with blue represent the removal of core electrons.

SAMPLE PROBLEM 7.4

Would you expect Na or Mg to have the greater first ionization energy (IE_1)? Which should have the greater second ionization energy (IE_2)?

Strategy Consider effective nuclear charge and electron configuration to compare the ionization energies. Effective nuclear charge increases from left to right in a period (thus increasing *IE*), and it is more difficult to remove a paired core electron than an unpaired valence electron.

Setup Na is in Group 1A, and Mg is beside it in Group 2A. Na has one valence electron, and Mg has two valence electrons.

Solution IE_1(Mg) > IE_1(Na) because Mg is to the right of Na in the periodic table (i.e., Mg has the greater effective nuclear charge, so it is more difficult to remove its electron). IE_2(Na) > IE_2(Mg) because the second ionization of Mg removes a valence electron, whereas the second ionization of Na removes a core electron.

THINK ABOUT IT

The first ionization energies of Na and Mg are 496 and 738 kJ/mol, respectively. The second ionization energies of Na and Mg are 4562 and 1451 kJ/mol, respectively.

Practice Problem ATTEMPT Which element, Mg or Al, will have the higher first ionization energy and which will have the higher third ionization energy?

Practice Problem BUILD Explain why Rb has a lower IE_1 than Sr, but Sr has a lower IE_2 than Rb.

Practice Problem CONCEPTUALIZE Imagine an arrangement of atomic orbitals in an alternate universe, in which the *s* subshell contains *two* orbitals instead of one, and the *p* subshell contains *four* orbitals rather than three. Under these circumstances, in which groups would you expect the anomalously low first ionization energies to occur?

Electron Affinity

Electron affinity (EA) is the energy released (the negative of the enthalpy change ΔH) when an atom in the gas phase accepts an electron. Consider the process in which a gaseous chlorine atom accepts an electron:

$$Cl(g) + e^- \longrightarrow Cl^-(g) \qquad \Delta H = -349.0 \text{ kJ/mol}$$

A negative value of ΔH indicates an exothermic process [◀◀ Section 5.3], so 349.0 kJ/mol of energy is released (the definition of electron affinity) when a mole of gaseous chlorine atoms accepts a mole of electrons. A positive electron affinity indicates a process that is energetically favorable. In general, the larger and more positive the *EA* value, the more favorable the process and the more apt it is to occur. Figure 7.10 shows electron affinities for the main group elements.

Like ionization energy, electron affinity increases from left to right across a period. This trend in *EA* is due to the increase in effective nuclear charge from left to right (i.e., it becomes progressively easier to add a negatively charged electron as the positive charge of the element's nucleus increases). There are also periodic interruptions of the upward trend of *EA* from left to right, similar to those observed for IE_1, although they do *not* occur for the same elements. For example, the *EA*

Student Note: Some books define electron affinity as ΔH, rather than the negative of ΔH for the process of adding an electron. This simply changes the sign of *EA* relative to what we show here.

Student Note: Although *IE* and *EA* both increase from left to right across a period, an increase in *IE* means that it is *less* likely that an electron will be *removed* from an atom. An increase in *EA*, on the other hand, means that it is *more* likely that an electron will be *accepted* by an atom.

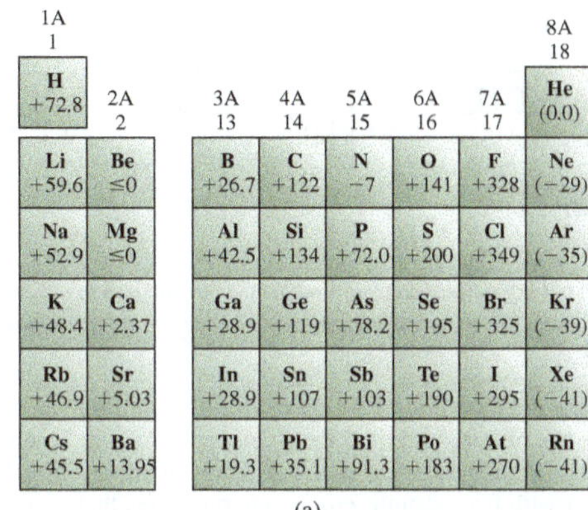

(a)

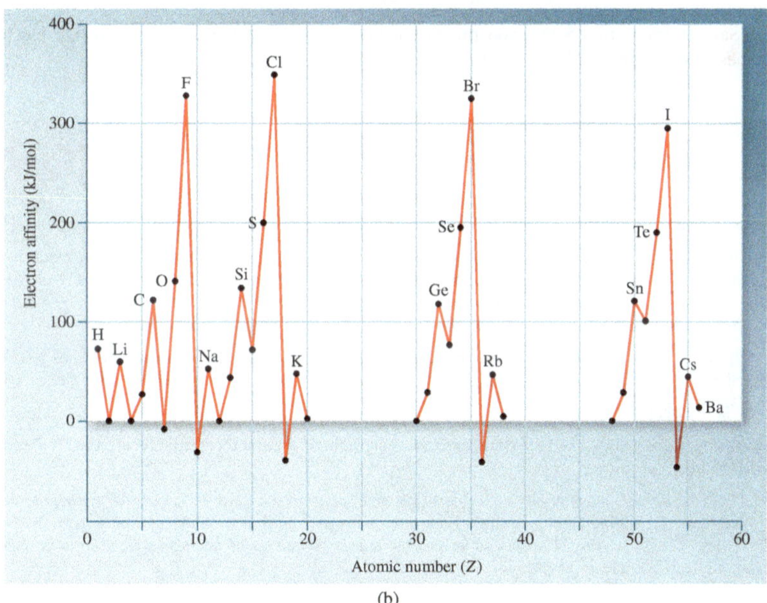

(b)

Figure 7.10 (a) Electron affinities (kJ/mol) of the main group elements. (b) Electron affinity as a function of atomic number.

of a Group 2A element is lower than that for the corresponding Group 1A element, and the *EA* of a Group 5A element is lower than that for the corresponding Group 4A element. These exceptions to the trend are due to the electron configurations of the elements involved.

It is harder to add an electron to a Group 2A element (ns^2) than to the Group 1A element (ns^1) in the same period because the electron added to the Group 2A element is placed in an orbital of higher energy (a *p* orbital versus an *s* orbital). Likewise, it is harder to add an electron to a Group 5A element (ns^2np^3) than to the corresponding Group 4A element (ns^2np^2) because the electron added to the Group 5A element must be placed in an orbital that already contains an electron. Figure 7.11 illustrates these points. Note that there is a much less significant and less regular variation in electron affinities from top to bottom within a group [Figure 7.10(a)].

Just as more than one electron can be removed from an atom, more than one electron can also be added to an atom. While many first electron affinities are positive, subsequent electron affinities are always negative. Considerable energy is required to overcome the repulsive forces between the electron and the negatively charged ion. The addition of two electrons to a gaseous oxygen atom can be represented as:

Process	ΔH (kJ/mol)	Electron Affinity
$O(g) + e^- \longrightarrow O^-(g)$	−141	$EA_1 = 141$ kJ/mol
$O^-(g) + e^- \longrightarrow O^{2-}(g)$	744	$EA_2 = -744$ kJ/mol

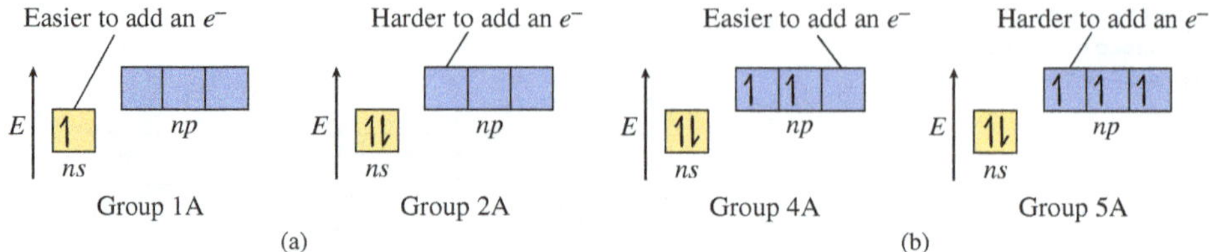

Figure 7.11 (a) It is easier to add an electron to an *s* orbital than to add one to a *p* orbital with the same principal quantum number. (b) Within a *p* subshell, it is easier to add an electron to an empty orbital than to add one to an orbital that already contains an electron.

The term *second electron affinity* may seem like something of a misnomer, because an anion in the gas phase has no real "affinity" for an electron. As will be discussed in Chapter 8, a significantly *endothermic* process such as the addition of an electron to a gaseous O^- ion happens only in concert with one or more *exothermic* processes that more than compensate for the required energy input.

Sample Problem 7.5 lets you practice using the periodic table to compare the electron affinities of elements.

SAMPLE PROBLEM 7.5

For each pair of elements, indicate which one you would expect to have the greater first electron affinity, EA_1: (a) Al or Si, (b) Si or P.

Strategy Consider the effective nuclear charge and electron configuration to compare the electron affinities. The effective nuclear charge increases from left to right in a period (thus generally increasing *EA*), and it is more difficult to add an electron to a partially occupied orbital than to an empty one. Writing out orbital diagrams for the valence electrons is helpful for this type of problem.

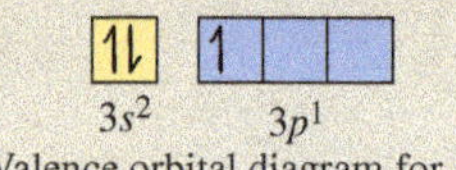

Valence orbital diagram for Al

Setup (a) Al is in Group 3A and Si is beside it in Group 4A. Al has three valence electrons ([Ne]$3s^23p^1$), and Si has four valence electrons ([Ne]$3s^23p^2$).

(b) P is in Group 5A (to the right of Si), so it has five valence electrons ([Ne]$3s^23p^3$).

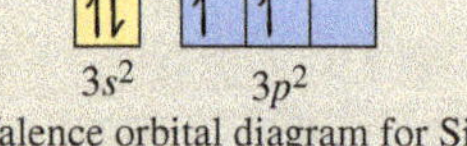

Valence orbital diagram for Si

Solution (a) EA_1(Si) > EA_1(Al) because Si is to the right of Al and therefore has a greater effective nuclear charge.

(b) EA_1(Si) > EA_1(P) because although P is to the right of Si in the third period of the periodic table (giving P the larger Z_{eff}), adding an electron to a P atom requires placing it in a 3*p* orbital that is partially occupied. The energy cost of *pairing* electrons outweighs the energy advantage of adding an electron to an atom with a larger effective nuclear charge.

$3s^2$ $3p^3$

Valence orbital diagram for P

THINK ABOUT IT

The first electron affinities of Al, Si, and P are 42.5, 134, and 72.0 kJ/mol, respectively.

Practice Problem ATTEMPT Would you expect Mg or Al to have the higher EA_1?

Practice Problem BUILD Why is the EA_1 for Ge greater than the EA_1 for As?

Practice Problem CONCEPTUALIZE In the same hypothetical arrangement described in Practice Problem 7.4C (page 295), in which groups would you expect the anomalously low electron affinities to occur?

Metallic Character

Metals tend to

- Be shiny, lustrous, and malleable
- Be good conductors of both heat and electricity
- Have low ionization energies (so they commonly form *cations*)
- Form ionic compounds with chlorine (metal chlorides)
- Form basic, ionic compounds with oxygen (metal oxides)

Student Note: *Malleability* is the property that allows metals to be pounded into thin sheets. *Ductility*, the capacity to be drawn out into wires, is another characteristic of metals.

Nonmetals, on the other hand, tend to

- Vary in color and lack the shiny appearance associated with metals
- Be brittle, rather than malleable
- Be poor conductors of both heat and electricity
- Form acidic, molecular compounds with oxygen
- Have high electron affinities (so they commonly form *anions*)

Metallic character increases from top to bottom in a group and decreases from left to right within a period.

Metalloids are elements with properties intermediate between those of metals and nonmetals. Because the definition of metallic character depends on a combination of properties, there may be some variation in the elements identified as metalloids in different sources. Astatine (At), for example, is listed as a metalloid in some sources and a nonmetal in others.

Explaining Periodic Trends

Many of the periodic trends in properties of the elements can be explained using ***Coulomb's law,*** which states that the force (F) between two charged objects (Q_1 and Q_2) is directly proportional to the product of the two charges and *inversely* proportional to the distance (d) between the objects squared. Recall that the *energy* between two oppositely charged particles (E_{el}) is inversely proportional to d [⏮ Section 5.1, Equation 5.2]. The SI unit of force is the newton (1 N = 1 kg · m/s^2), and the SI unit of energy is the joule (1 J = 1 kg · m^2/s^2).

$$F \propto \frac{Q_1 \times Q_2}{d^2}$$

When the charges have opposite signs, F is negative—indicating an *attractive* force between the objects. When the charges have the same sign, F is positive—indicating a *repulsive* force. Table 7.3 shows how the magnitude of the attractive force between two oppositely charged objects at a fixed distance from each other varies with changes in the magnitudes of the charges.

Sample Problem 7.6 illustrates how Coulomb's law can be used to compare the magnitudes of attractive forces between charged objects.

TABLE 7.3 Attractive Force Between Oppositely Charged Objects at a Fixed Distance (d = 1) from Each Other

Q_1	Q_2	Attractive force is proportional to
+1	−1	1
+2	−2	4
+3	−3	9

SAMPLE PROBLEM 7.6

For carbon and nitrogen, use the effective nuclear charges given in Figure 7.7 (page 292) and the atomic radii given in Figure 7.6 (page 292) to compare the attractive force between the nucleus in each atom and the valence electron that would be removed by the first ionization.

Strategy Use Coulomb's law to calculate a number to which the attractive force will be proportional in each case.

Setup From Figure 7.7, the effective nuclear charges of C and N are 3.14 and 3.83, respectively, and the radii of C and N are 77 pm and 75 pm, respectively. The first ionization energies are 1086 kJ/mol (C) and 1402 kJ/mol (N). The charge on the valence electron in each case is −1.

Solution For C: $F \propto \frac{3.14 \times (-1)}{(77\text{ pm})^2} = -5.3 \times 10^{-4}$

For N: $F \propto \frac{3.83 \times (-1)}{(75\text{ pm})^2} = -6.8 \times 10^{-4}$

Note that in this type of comparison, it doesn't matter what units we use for the distance between the charges. We are not trying to calculate a particular attractive force, only to compare the magnitudes of these two attractive forces.

THINK ABOUT IT

Remember that the negative sign simply indicates that the force is attractive rather than repulsive. The calculated number for nitrogen is about 28 percent larger than that for carbon.

Practice Problem ATTEMPT Between which two charges is the attractive force larger: +3.26 and −1.15 separated by a distance of 1.5 pm, or +2.84 and −3.63 separated by a distance of 2.5 pm?

Practice Problem BUILD What must the distance be between charges of +2.25 and −1.86 for the attractive force between them to be the same as that between charges of +4.06 and −2.11 separated by a distance of 2.16 pm?

Practice Problem CONCEPTUALIZE Rank these pairs of charged objects in order of increasing magnitude of the attractive force between them.

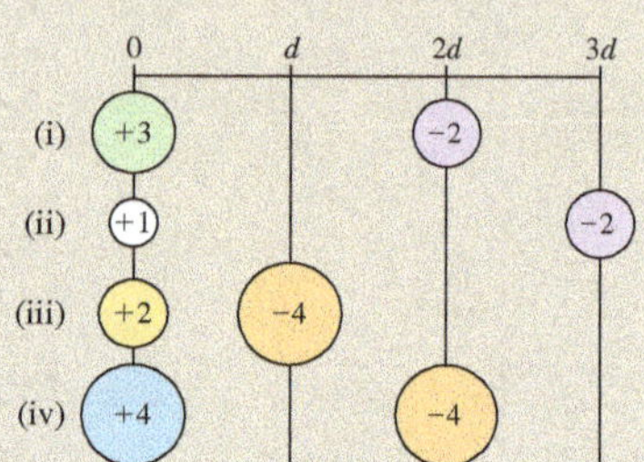

CHECKPOINT – SECTION 7.4 Periodic Trends in Properties of Elements

7.4.1 Arrange the elements Ca, Sr, and Ba in order of increasing IE_1.

a) Ca < Sr < Ba
b) Ba < Sr < Ca
c) Ba < Ca < Sr
d) Sr < Ba < Ca
e) Sr < Ca < Ba

7.4.2 Arrange the elements Li, Be, and B in order of increasing IE_2.

a) Li < Be < B
b) Li < B < Be
c) Be < B < Li
d) Be < Li < B
e) B < Be < Li

7.4.3 For each of the following pairs of elements, indicate which will have the greater EA_1: Rb or Sr, C or N, O or F.

a) Rb, C, O
b) Sr, N, F
c) Sr, C, F
d) Sr, N, O
e) Rb, C, F

7.4.4 Which element, K or Ca, will have the greater IE_1, which will have the greater IE_2, and which will have the greater EA_1?

a) Ca, K, K
b) K, K, Ca
c) K, Ca, K
d) Ca, Ca, K
e) Ca, Ca, Ca

7.5 Electron Configuration of Ions

Because many ionic compounds are made up of monatomic anions and cations, it is helpful to know how to write the electron configurations of these ionic species. Just as for atoms, we use the Pauli exclusion principle and Hund's rule to write the ground-state electron configurations of cations and anions.

Recall from Chapter 2 that we can use the periodic table to predict the charges on many of the ions formed by main group elements. Elements in Groups 1A and 2A, for example, form ions with charges of +1 and +2, respectively. Elements in Groups 6A and 7A form ions with charges of −2 and −1, respectively. Knowing something about electron configurations enables us to explain these charges.

Ions of Main Group Elements

In Section 7.4, we learned about the tendencies of atoms to lose or gain electrons. In every period of the periodic table, the element with the highest IE_1 is the Group 8A element, the noble gas. [See Figure 7.8(b).] Also, Group 8A is the only group in which none of the members has *any* tendency to accept an electron; that is, they all have negative EA values. [See Figure 7.10(b).] High ionization energies and low electron affinities make the noble gases almost completely unreactive. Ultimately, the cause of this lack of reactivity is electron configuration. The $1s^2$ configuration of He and the ns^2np^6 ($n \geq 2$) valence electron configurations of the other noble

Student Note: It is a common error to mistake species with the same valence electron configuration for isoelectronic species. For example, F^- and Ne are isoelectronic. F^- and Cl^- are not.

gases are extraordinarily stable. Other main group elements tend to either lose or gain the number of electrons needed to achieve the same number of electrons as the nearest noble gas. Species with identical electron configurations are called ***isoelectronic.***

To write the electron configuration of an ion formed by a main group element, we first write the configuration for the atom and either add or remove the appropriate number of electrons. Electron configurations for the sodium and chloride ions are

$$\text{Na: } 1s^22s^22p^63s^1 \longrightarrow \text{Na}^+\text{: } 1s^22s^22p^6 \quad \text{(10 electrons total, isoelectronic with Ne)}$$

$$\text{Cl: } 1s^22s^22p^63s^23p^5 \longrightarrow \text{Cl}^-\text{: } 1s^22s^22p^63s^23p^6 \quad \text{(18 electrons total, isoelectronic with Ar)}$$

We can also write electron configurations for ions using the noble gas core.

$$\text{Na: [Ne]}3s^1 \longrightarrow \text{Na}^+\text{: [Ne]}$$

$$\text{Cl: [Ne]}3s^23p^5 \longrightarrow \text{Cl}^-\text{: [Ne]}3s^23p^6 \quad \text{or} \quad \text{[Ar]}$$

Sample Problem 7.7 gives you some practice writing electron configurations for the ions of main group elements.

SAMPLE PROBLEM 7.7

Write electron configurations for the following ions of main group elements: (a) N^{3-}, (b) Ba^{2+}, and (c) Be^{2+}.

Strategy First write electron configurations for the atoms. Then add electrons (for anions) or remove electrons (for cations) to account for the charge.

Setup (a) N^{3-} forms when N ($1s^22s^22p^3$ or [He]$2s^22p^3$), a main group nonmetal, gains three electrons.

(b) Ba^{2+} forms when Ba ($1s^22s^22p^63s^23p^64s^23d^{10}4p^65s^24d^{10}5p^66s^2$ or [Xe]$6s^2$) loses two electrons.

(c) Be^{2+} forms when Be ($1s^22s^2$ or [He]$2s^2$) loses two electrons.

Solution (a) [He]$2s^22p^6$ or [Ne]

(b) [Kr]$5s^24d^{10}5p^6$ or [Xe]

(c) $1s^2$ or [He]

THINK ABOUT IT

Be sure to add electrons to form an anion, and remove electrons to form a cation.

Practice Problem ATTEMPT Write electron configurations for (a) O^{2-}, (b) Ca^{2+}, and (c) Se^{2-}.

Practice Problem BUILD List all the species (atoms and/or ions) that are likely to have the following electron configuration: $1s^22s^22p^6$.

Practice Problem CONCEPTUALIZE Select the correct valence orbital diagram for the Mg^{2+} ion and for the S^{2-} ion.

Mg^{2+} S^{2-}

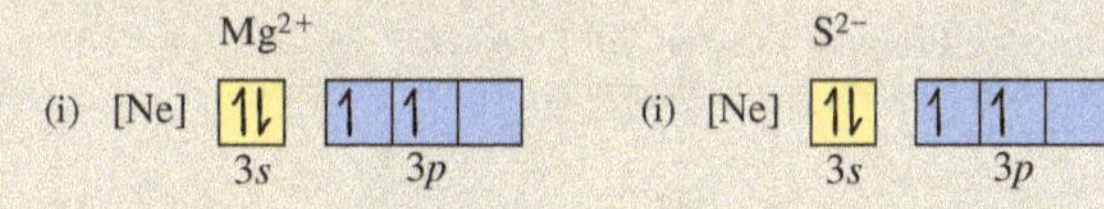

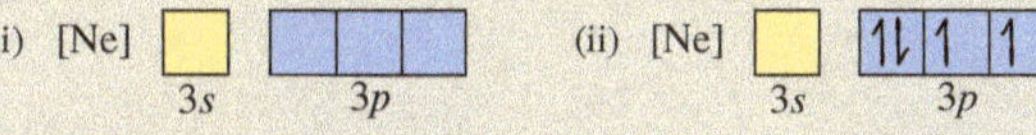

(iii) [Ne] 3s 3p (iii) [Ne] 3s 3p

(iv) [Ne] 3s 3p (iv) [Ne] 3s 3p

Ions of *d*-Block Elements

Recall from Section 6.8 that the 4*s* orbital fills before the 3*d* orbitals for the elements in the first row of the *d*-block (Sc to Zn) [⏮ Section 6.8]. Following the pattern for writing electron configurations for main group ions, then, we might expect the two electrons lost in the formation of the Fe^{2+} ion to come from the 3*d* subshell. It turns out, though, that an atom always loses electrons first from the shell with the *highest* value of *n*. In the case of Fe, that would be the 4*s* subshell.

$$\text{Fe: [Ar]}4s^23d^6 \longrightarrow \text{Fe}^{2+}\text{: [Ar]}3d^6$$

Iron can also form the Fe^{3+} ion, in which case the third electron is removed from the 3*d* subshell.

$$\text{Fe: [Ar]}4s^23d^6 \longrightarrow \text{Fe}^{3+}\text{: [Ar]}3d^5$$

In general, when a *d*-block element becomes an ion, it loses electrons first from the *ns* subshell and then from the $(n-1)d$ subshell. This explains, in part, why many of the transition metals can form ions with a +2 charge.

Sample Problem 7.8 gives you some practice writing electron configurations for the ions of *d*-block elements.

SAMPLE PROBLEM 7.8

Write electron configurations for the following ions of *d*-block elements: (a) Zn^{2+}, (b) Mn^{2+}, and (c) Cr^{3+}.

Strategy First write electron configurations for the atoms. Then add electrons (for anions) or remove electrons (for cations) to account for the charge. The electrons removed from a *d*-block element must come first from the outermost *s* subshell, not the partially filled *d* subshell.

Setup (a) Zn^{2+} forms when Zn ($1s^22s^22p^63s^23p^64s^23d^{10}$ or $[Ar]4s^23d^{10}$) loses two electrons.

(b) Mn^{2+} forms when Mn ($1s^22s^22p^63s^23p^64s^23d^5$ or $[Ar]4s^23d^5$) loses two electrons.

(c) Cr^{3+} forms when Cr ($1s^22s^22p^63s^23p^64s^13d^5$ or $[Ar]4s^13d^5$) loses three electrons—one from the 4*s* subshell and two from the 3*d* subshell. Remember that the electron configuration of Cr is anomalous in that it has only one 4*s* electron, making its *d* subshell half filled [◀◀ Section 6.9].

Solution (a) $[Ar]3d^{10}$ (b) $[Ar]3d^5$ (c) $[Ar]3d^3$

THINK ABOUT IT

Be sure to add electrons to form an anion and remove electrons to form a cation. Also, double-check to make sure that electrons removed from a *d*-block element come first from the *ns* subshell and then, if necessary, from the $(n-1)d$ subshell.

Practice Problem ATTEMPT Write electron configurations for (a) Co^{3+}, (b) Cu^{2+}, and (c) Ag^+.

Practice Problem BUILD What common *d*-block ion (see Figure 2.14) is isoelectronic with Zn^{2+}?

Practice Problem CONCEPTUALIZE Select the correct valence orbital diagram for the Fe^{2+} ion and for the Fe^{3+} ion.

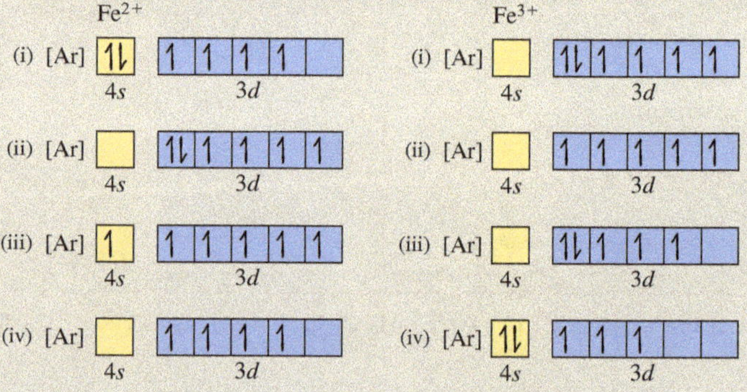

CHECKPOINT – SECTION 7.5 Electron Configuration of Ions

7.5.1 Which of the following ions are isoelectronic with a noble gas? (Select all that apply.)

a) Mn^{2+}
b) Ca^{2+}
c) Br^-
d) O^{2+}
e) F^-

7.5.2 Which of the following pairs are isoelectronic with each other? (Select all that apply.)

a) Ca^{2+} and Sr^{2+}
b) O^{2-} and Mg^{2+}
c) I^- and Kr
d) S^{2-} and Cl^-
e) He and H^+

7.5.3 Select the correct ground-state electron configuration for Ti^{2+}.

a) $[Ar]4s^23d^2$
b) $[Ar]4s^23d^4$
c) $[Ar]4s^2$
d) $[Ar]3d^2$
e) $[Ar]4s^13d^1$

7.5.4 Select the correct ground-state electron configuration for S^{2-}.

a) $[Ne]3p^4$
b) $[Ne]3s^23p^6$
c) $[Ne]3s^23p^2$
d) $[Ne]3p^6$
e) $[Ne]$

7.6 Ionic Radius

When an atom gains or loses one or more electrons to become an ion, its radius changes. The ***ionic radius,*** the radius of a cation or an anion, affects the physical and chemical properties of an ionic compound. The three-dimensional structure of an ionic compound, for example, depends on the relative sizes of its cations and anions.

Comparing Ionic Radius with Atomic Radius

When an atom loses an electron and becomes a cation, its radius decreases due in part to a reduction in electron-electron repulsions (and consequently a reduction in shielding) in the valence shell. A significant decrease in radius occurs when *all* of an atom's valence electrons are removed. This is the case with ions of most main group elements, which are isoelectronic with the noble gases preceding them. Consider Na, which loses its 3*s* electron to become Na^+:

Valence orbitals in Na and in Na^+

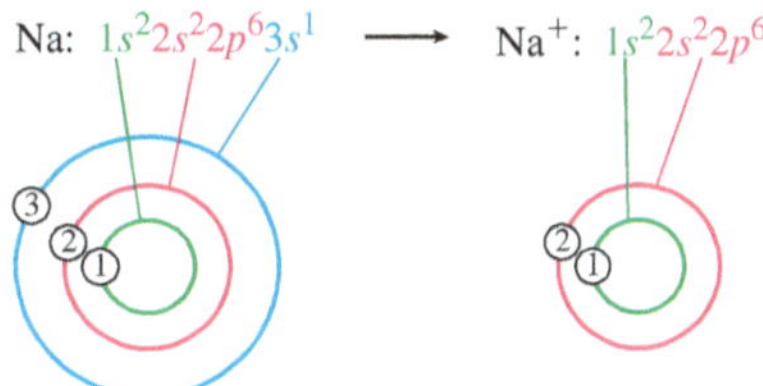

The valence electron of Na has a principal quantum number of $n = 3$. When it has been removed, the resulting Na^+ ion no longer has any electrons in the $n = 3$ shell. The outermost electrons of the Na^+ ion have a principal quantum number of $n = 2$. Because the value of n determines the distance from the nucleus, this corresponds to a smaller radius.

When an atom gains one or more electrons and becomes an anion, its radius increases due to increased electron-electron repulsions. Adding an electron causes the rest of the electrons in the valence shell to spread out and take up more space to maximize the distance between them.

Figure 7.12 shows the ionic radii for those ions of main group elements that are isoelectronic with noble gases and compares them to the radii of the parent atoms. Note that the ionic radius, like the atomic radius, increases from top to bottom in a group.

Isoelectronic Series

An ***isoelectronic series*** is a series of two or more species that have identical electron configurations, but different nuclear charges. For example, O^{2-}, F^-, and Ne constitute an isoelectronic series. Although these three species have identical electron configurations, they have different radii. In an isoelectronic series, the species with the smallest nuclear charge (i.e., the smallest atomic number Z) will have the largest radius. The species with the largest nuclear charge (i.e., the largest Z) will have the smallest radius.

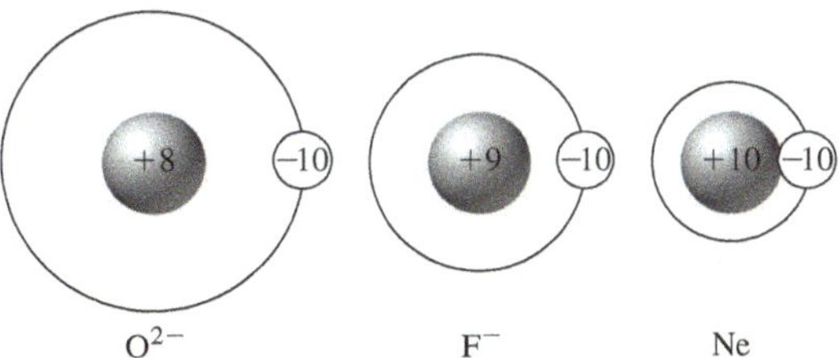

Sample Problem 7.9 shows how to identify members of an isoelectronic series and how to arrange them according to radius.

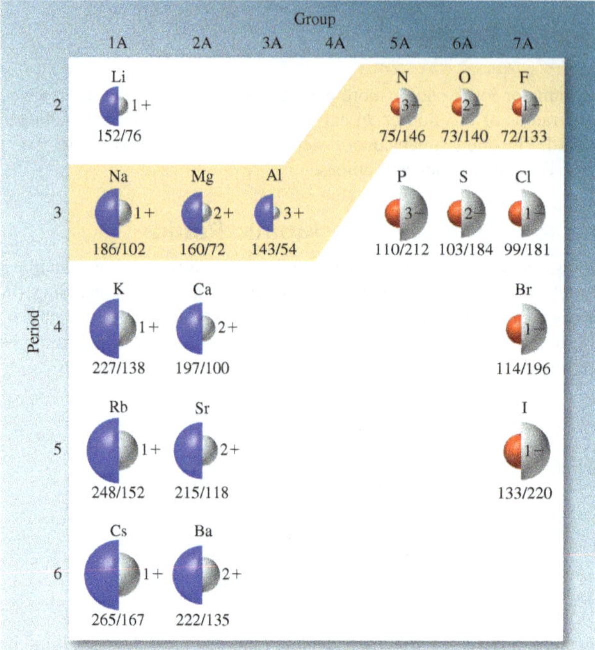

Figure 7.12 A comparison of atomic and ionic radii (in picometers) for main group elements and their common ions (those that are isoelectronic with noble gases).

SAMPLE PROBLEM 7.9

Identify the isoelectronic series in the following group of species, and arrange the ions in order of increasing radius: K^+, Ne, Ar, Kr, P^{3-}, S^{2-}, and Cl^-.

Strategy Isoelectronic series are species with identical electron configurations but different nuclear charges. Determine the number of electrons in each species. The radii of isoelectronic ions within a series decrease with increasing nuclear charge.

Setup The number of electrons in each species is as follows: 18 (K^+), 10 (Ne), 18 (Ar), 36 (Kr), 18 (P^{3-}), 18 (S^{2-}), and 18 (Cl^-). The species with 18 electrons constitute the isoelectronic series. The nuclear charges of the ions with 18 electrons are +19 (K^+), +15 (P^{3-}), +16 (S^{2-}), and +17 (Cl^-).

Solution The isoelectronic series includes K^+, Ar, P^{3-}, S^{2-}, and Cl^-. The ions, in order of increasing radius, are: $K^+ < Cl^- < S^{2-} < P^{3-}$.

THINK ABOUT IT

Consult Figure 7.12 to check your result. With identical electron configurations, the attractive force between the valence electrons and the nucleus will be strongest for the largest nuclear charge. Thus, the larger the nuclear charge, the closer in the valence electrons will be pulled and the smaller the radius of an ion will be.

Practice Problem ATTEMPT Arrange the following isoelectronic series in order of increasing radius: Se^{2-}, Br^-, and Rb^+.

Practice Problem BUILD List all the common ions that are isoelectronic with Ne.

Practice Problem CONCEPTUALIZE Which periodic table's highlighted portion includes elements that can form an isoelectronic series? (Select all that apply.)

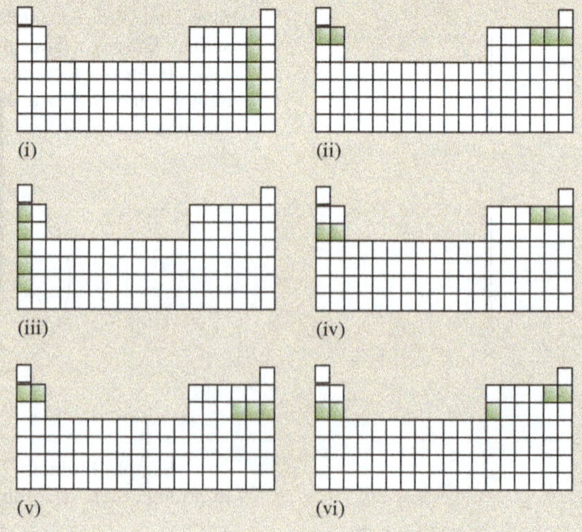

CHECKPOINT – SECTION 7.6 Ionic Radius

7.6.1 Which of the following species are isoelectronic with Kr? (Select all that apply.)

a) He

b) Ne

c) Ar

d) Br^-

e) Rb^+

7.6.2 Which of the following are arranged correctly in order of increasing radius? (Select all that apply.)

a) $F^- < Cl^- < Br^- < I^-$

b) $O^{2-} < F^- < Na^+ < Mg^{2+}$

c) $Ca^{2+} < K^+ < S^{2-} < Cl^-$

d) $Rb^+ < K^+ < Na^+ < Li^+$

e) $Sr^{2+} < Ca^{2+} < Mg^{2+} < Be^{2+}$

7.6.3 Which of the following is the most realistic representation of an atom from Group 6A becoming an ion?

a) 

b) 

c)

d)

e)

7.6.4 Which of the following is the most realistic representation of an atom from Group 1A becoming an ion?

a)

b)

c)

d)

e)

7.7 Periodic Trends in Chemical Properties of the Main Group Elements

Ionization energy and electron affinity enable us to understand the types of reactions that elements undergo and the types of compounds they form. These two parameters actually measure similar things. Ionization energy is a measure of how powerfully an atom attracts its own electrons, while electron affinity is a measure of how powerfully an atom can attract electrons from another source. As a very simple example of how this helps us understand a chemical reaction, consider the combination of a sodium atom and a chlorine atom, shown in Figure 7.13.

Sodium, with its low ionization energy, has a relatively weak attraction for its one valence electron. Chlorine, with its energetically favorable electron affinity, has the ability to attract electrons from another source. In this case, the electron that is loosely held by the Na atom,

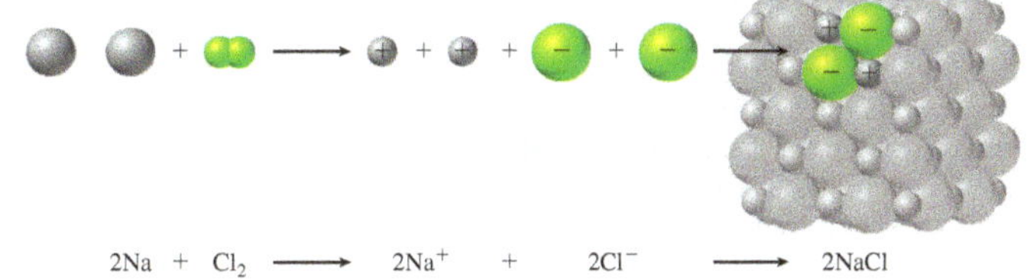

Figure 7.13 Formation of NaCl from its constituent elements. Note that although the charges are not all shown, the solid consists of a three-dimensional array of alternating oppositely-charged ions.

and powerfully attracted by the Cl atom, is transferred from Na to Cl, thus producing a sodium ion (Na^+) and a chloride ion (Cl^-). According to Coulomb's law, oppositely charged objects attract each other. The positively charged sodium ion and the negatively charged chloride ion are drawn together by electrostatic attraction, and the result is the formation of the solid ionic compound sodium chloride (NaCl).

Animation
Periodic table—properties of alkali and alkaline earth metals.

General Trends in Chemical Properties

Before we examine the elements in individual groups, let's identify some overall trends. We have said that elements in the same group resemble one another in chemical behavior because they have similar valence electron configurations. This statement, although correct in the general sense, must be applied with caution. Chemists have long known that the properties of the first member of each group (Li, Be, B, C, N, O, and F) are different from those of the rest of the members of the same group. Lithium, for example, exhibits many, but not all, of the properties characteristic of the Group 1A (alkali) metals. For example, unlike the other Group 1A elements, Li reacts with the O_2 and N_2 in air to form a simple oxide (Li_2O) and nitride (Li_3N), respectively. Similarly, beryllium is a somewhat atypical member of Group 2A (alkaline earth metals) in that it forms covalent compounds, and so on. The differences can be attributed to the unusually small size of the first element in each group (see Figure 7.6).

Another trend in the chemical behavior of main group elements is the diagonal relationship. ***Diagonal relationships*** refer to similarities between pairs of elements in different groups and periods of the periodic table. Specifically, the first three members of the second period (Li, Be, and B) exhibit many similarities to the elements located diagonally below them in the periodic table (Mg, Al, and Si). The reason for this phenomenon is the similarity of charge densities of their cations. (Charge density is the charge on an ion divided by its volume.) Cations with comparable charge densities react similarly with anions and therefore form the same types of compounds. Thus, the chemistry of lithium resembles that of magnesium in some ways; the same holds for beryllium and aluminum and for boron and silicon. Each of these pairs is said to exhibit a diagonal relationship.

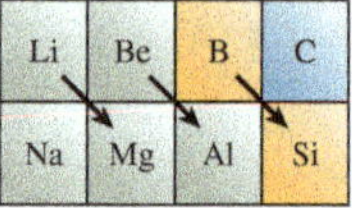

Diagonal relationships

A comparison of the properties of elements in the same group is most valid if the elements in question have a similar metallic (or nonmetallic) character. The elements in Groups 1A and 2A, for example, are all metals, whereas those in Groups 7A and 8A are all nonmetals. We have to be more careful when comparing the elements of Groups 3A through 6A, though, because a single group may contain metals, metalloids, and nonmetals. In these groups, we should expect a greater variation in chemical properties even though all group members have similar valence electron configurations.

Hydrogen ($1s^1$)

There is no completely suitable position for hydrogen in the periodic table (it really belongs in a group by itself). Traditionally hydrogen is shown at the top of Group 1A, because, like the alkali metals, it has a single *s* valence electron and forms a cation with a charge of +1 (H^+), which is hydrated in solution. On the other hand, hydrogen also forms the *hydride* ion (H^-) in ionic compounds such as NaH and CaH_2. In this respect, hydrogen resembles the members of Group 7A (halogens), all of which form −1 anions (F^-, Cl^-, Br^-, and I^-) in ionic compounds. Ionic hydrides react with water to produce hydrogen gas and the corresponding metal hydroxides:

$$2NaH(s) + 2H_2O(l) \longrightarrow 2NaOH(aq) + 2H_2(g)$$

$$CaH_2(s) + 2H_2O(l) \longrightarrow Ca(OH)_2(aq) + 2H_2(g)$$

The most important compound of hydrogen is water, which forms when hydrogen burns in air:

$$2H_2(g) + O_2(g) \longrightarrow 2H_2O(l)$$

Properties of the Active Metals

Group 1A Elements (ns^1, $n \geq 2$)

Figure 7.14 shows samples of the Group 1A elements. These elements all have low ionization energies, making it easy for them to become M^+ ions. In fact, these metals are so reactive that they are never found in nature in the pure elemental state. They react with water to produce hydrogen gas and the corresponding metal hydroxide:

$$2M(s) + 2H_2O(l) \longrightarrow 2MOH(aq) + H_2(g)$$

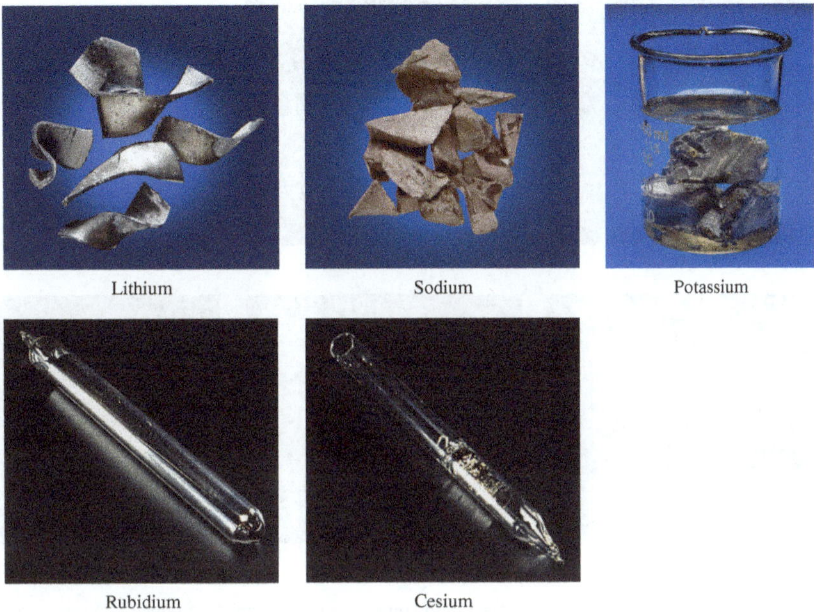

Lithium Sodium Potassium

Rubidium Cesium

Figure 7.14 Group 1A elements.

Credit: (lithium, sodium, potassium): © *McGraw-Hill Education/Charles D. Winters, photographer*; (rubidium, cesium): © *McGraw-Hill Education/Stephen Frisch, photographer.*

where M denotes an alkali metal. When exposed to air, they gradually lose their shiny appearance as they react with oxygen to form metal oxides. Lithium forms lithium oxide (containing the oxide ion, O^{2-}):

$$4Li(s) + O_2(g) \longrightarrow 2Li_2O(s)$$

The other alkali metals all form oxides or *peroxides* (containing the peroxide ion, O_2^{2-}):

$$2Na(s) + O_2(g) \longrightarrow Na_2O_2(s)$$

Potassium, rubidium, and cesium also form *superoxides* (containing the superoxide ion, O_2^-):

$$K(s) + O_2(g) \longrightarrow KO_2(s)$$

The type of oxide that forms when an alkali metal reacts with oxygen has to do with the stability of the various oxides. Because these oxides are all ionic compounds, their stability depends on how strongly the cations and anions attract one another. Lithium tends to form predominantly the oxide because lithium oxide is more stable than lithium peroxide.

Group 2A Elements (ns^2, $n \geq 2$)

Figure 7.15 shows samples of the Group 2A elements. As a group, the alkaline earth metals are somewhat less reactive than the alkali metals. Both the first and the second ionization energies decrease (and metallic character increases) from beryllium to barium. Group 2A elements tend to form M^{2+} ions, where M denotes an alkaline earth metal atom.

Student Note: Because they have less metallic character than the other Group 2A elements, beryllium and magnesium form some molecular compounds such as BeH_2 and MgH_2.

The reactions of alkaline earth metals with water vary considerably. Beryllium does not react with water; magnesium reacts slowly with steam; and calcium, strontium, and barium react vigorously with cold water.

$$Ca(s) + 2H_2O(l) \longrightarrow Ca(OH)_2(s) + H_2(g)$$

$$Sr(s) + 2H_2O(l) \longrightarrow Sr(OH)_2(s) + H_2(g)$$

$$Ba(s) + 2H_2O(l) \longrightarrow Ba(OH)_2(aq) + H_2(g)$$

The reactivity of the alkaline earth metals toward oxygen also increases from Be to Ba. Beryllium and magnesium form oxides (BeO and MgO) only at elevated temperatures, whereas CaO, SrO, and BaO form at room temperature.

Magnesium reacts with aqueous acid to produce hydrogen gas:

$$Mg(s) + 2H^+(aq) \longrightarrow Mg^{2+}(aq) + H_2(g)$$

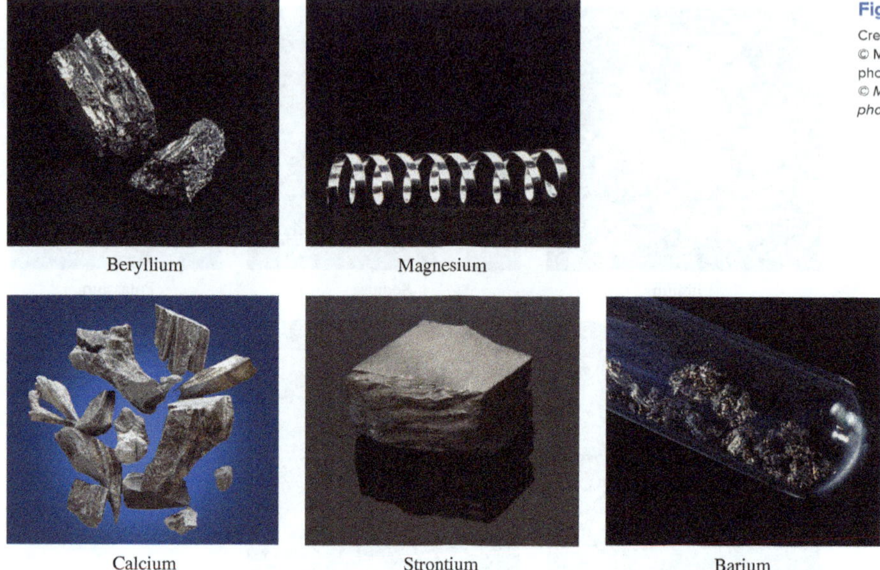

Figure 7.15 Group 2A elements.

Credit: (beryllium, magnesium, barium): © McGraw-Hill Education/Stephen Frisch, photographer; (calcium, strontium): © *McGraw-Hill Education/Charles D. Winters, photographer.*

Calcium, strontium, and barium also react with aqueous acid solutions to produce hydrogen gas. However, because the metals also react with water, the two different reactions (with H^+ and with H_2O) occur simultaneously.

Properties of Other Main Group Elements

Group 3A Elements (ns^2np^1, $n \geq 2$)

Figure 7.16 shows samples of the first four Group 3A elements. Boron, the first member of the group, is a metalloid; the others (Al, Ga, In, and Tl) are metals. Boron does not form binary ionic compounds and is unreactive toward both oxygen and water. Aluminum, the next element in the group, readily forms aluminum oxide when exposed to air:

$$4Al(s) + 3O_2(g) \longrightarrow 2Al_2O_3(s)$$

The aluminum oxide forms a protective coating, preventing the underlying metal from reacting further. This fact makes it possible to use aluminum for structural materials, such as aluminum siding and the shells of airplanes. Without the protective coating, layer after layer of Al atoms would become oxidized, and the structure would eventually crumble.

Aluminum forms the Al^{3+} ion. It reacts with hydrochloric acid according to the equation:

$$2Al(s) + 6H^+(aq) \longrightarrow 2Al^{3+}(aq) + 3H_2(g)$$

The other Group 3A metals (Ga, In, and Tl) can form both M^+ and M^{3+} ions. As we move down the group, the M^+ ion becomes the more stable of the two.

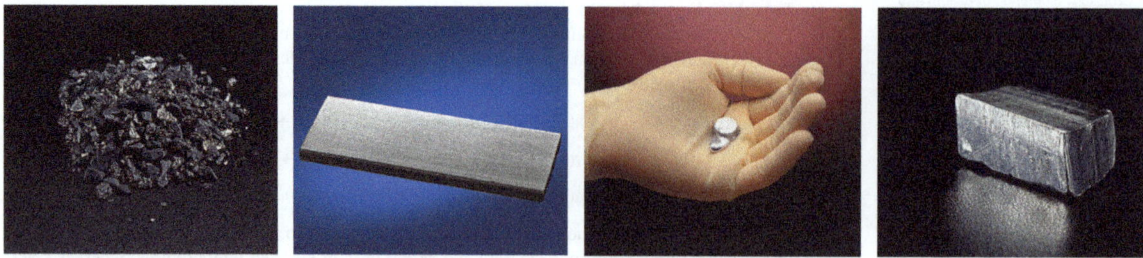

Figure 7.16 Group 3A elements.

Credit: (boron, gallium, indium): © *McGraw-Hill Education/Stephen Frisch, photographer;* (aluminum): © *McGraw-Hill Education/Charles D. Winters, photographer.*

The metallic elements in Group 3A also form many molecular compounds. For example, aluminum reacts with hydrogen to form AlH_3, which has properties similar to those of BeH_2. The progression of properties across the second row of the periodic table illustrates the gradual shift from metallic to nonmetallic character in the main group elements.

Group 4A Elements (ns^2np^2, $n \geq 2$)

Figure 7.17 shows samples of the Group 4A elements. Carbon, the first member of the group, is a nonmetal, whereas silicon and germanium, the next two members, are metalloids. Tin and lead, the last two members of the group, are metals. They do not react with water, but they do react with aqueous acid to produce hydrogen gas:

$$Sn(s) + 2H^+(aq) \longrightarrow Sn^{2+}(aq) + H_2(g)$$

$$Pb(s) + 2H^+(aq) \longrightarrow Pb^{2+}(aq) + H_2(g)$$

Student Note: Carbon, being a nonmetal, achieves its +4 oxidation state without actually losing four electrons, which would be very "expensive" in terms of the energy required. For the same reason, compounds containing metals in very high oxidation states also tend to be molecular rather than ionic.

The Group 4A elements form compounds in both the +2 and +4 oxidation states. For carbon and silicon, the +4 oxidation state is the more stable one. For example, CO_2 is more stable than CO, and SiO_2 is a stable compound, but SiO does not exist under ordinary conditions. As we move down the group, however, the relative stability of the two oxidation states is reversed. In tin compounds the +4 oxidation state is only slightly more stable than the +2 oxidation state. In lead compounds the +2 oxidation state is the more stable one. The outer electron configuration of lead is $6s^26p^2$, and lead tends to lose only the $6p$ electrons to form Pb^{2+} rather than both the $6p$ and $6s$ electrons to form Pb^{4+}.

Group 5A Elements (ns^2np^3, $n \geq 2$)

Figure 7.18 shows samples of the Group 5A elements. Nitrogen and phosphorus are nonmetals, arsenic and antimony are metalloids, and bismuth is a metal. Because Group 5A contains elements in all three categories, we expect greater variation in their chemical properties.

Student Note: Elemental bismuth appears colorful because of a thin surface layer of bismuth oxide (Bi_2O_3).

Credit: © McGraw-Hill Education/Charles D. Winters, photographer.

Elemental nitrogen is a diatomic gas (N_2). It forms a variety of oxides (NO, N_2O, NO_2, N_2O_4, and N_2O_5), all of which are gases except for N_2O_5, which is a solid at room temperature. Nitrogen has a tendency to accept three electrons to form the nitride ion (N^{3-}). Most metal nitrides, such as Li_3N and Mg_3N_2, are ionic compounds. Phosphorus exists as individual P_4 molecules (white phosphorus) or chains of P_4 molecules (red phosphorus). It forms two solid oxides with the formulas P_4O_6 and P_4O_{10}. The industrially important oxoacids nitric acid and phosphoric acid form when N_2O_5 and P_4O_{10}, respectively, react with water:

$$N_2O_5(s) + H_2O(l) \longrightarrow 2HNO_3(aq)$$

$$P_4O_{10}(s) + 6H_2O(l) \longrightarrow 4H_3PO_4(aq)$$

Arsenic, antimony, and bismuth have extensive three-dimensional structures. Bismuth is far less reactive than metals in the preceding groups.

Carbon (graphite) | Carbon (diamond) | Silicon | Germanium | Tin | Lead

Figure 7.17 Group 4A elements.

Credit: (graphite, diamond, silicon, germanium, tin, lead): © *McGraw-Hill Education/Charles D. Winters, photographer.*

Figure 7.18 Group 5A elements.

Credit: (nitrogen, white phosphorus, red phosphorus, antimony): © *McGraw-Hill Education/Charles D. Winters, photographer;* (arsenic, bismuth): © *McGraw-Hill Education/Stephen Frisch, photographer.*

Group 6A Elements (ns^2np^4, $n \geq 2$)

Figure 7.19 shows samples of the first three Group 6A elements. The first three members of the group (oxygen, sulfur, and selenium) are nonmetals, whereas the last two (tellurium and polonium) are metalloids. Oxygen is a colorless, odorless, diatomic gas; elemental sulfur and selenium exist as the molecules S_8 and Se_8, respectively; and tellurium and polonium have more extensive three-dimensional structures. (Polonium is a radioactive element that is difficult to study in the laboratory.) Oxygen has a tendency to accept two electrons to form the oxide ion (O^{2-}) in many compounds. Sulfur, selenium, and tellurium also form ions by accepting two electrons: S^{2-}, Se^{2-}, and Te^{2-}. The elements in Group 6A (especially oxygen) form a large number of molecular compounds with nonmetals. Some of the important compounds of sulfur are SO_2, SO_3, and H_2S. Sulfuric acid, an oxoacid, forms when sulfur trioxide reacts with water:

$$SO_3(g) + H_2O(l) \longrightarrow H_2SO_4(aq)$$

Figure 7.19 Group 6A elements.

Credit: (oxygen): © *David A. Tietz/Editorial Image, LLC;* (sulfur): © *Richard Treptow/Science Source;* (selenium): © *McGraw-Hill Education/Charles D. Winters, photographer;* (tellurium, fluorine): © *McGraw-Hill Education/Stephen Frisch, photographer.*

Figure 7.20 Group 7A elements.

Credit: (fluorine, bromine): © *McGraw-Hill Education/Stephen Frisch, photographer;* (chlorine): © *Charles D. Winters/Science Source;* (iodine): © *McGraw-Hill Education/ Charles D. Winters, photographer.*

Group 7A Elements (ns^2np^5, $n \geq 2$)

Figure 7.20 shows samples of the first four Group 7A elements. All the halogens are nonmetals with the general formula X_2, where X denotes a halogen element. Like the Group 1A metals, the Group 7A nonmetals are too reactive to be found in nature in the elemental form. (Astatine, the last member of Group 7A, is radioactive. Very little is known about its properties.)

The halogens have high ionization energies and large, energetically favorable electron affinities. Anions derived from the halogens (F^-, Cl^-, Br^-, and I^-) are called *halides.* The vast majority of alkali metal halides are ionic compounds. The halogens also form many molecular compounds among themselves, such as ICl and BrF_3, and with nonmetals in other groups, such as NF_3, PCl_5, and SF_6. The halogens react with hydrogen to form hydrogen halides:

$$H_2(g) + X_2(g) \longrightarrow 2HX(g)$$

This reaction is explosive when it involves fluorine, but it becomes less and less violent as we substitute chlorine, bromine, and iodine. The hydrogen halides dissolve in water to form hydrohalic acids. Hydrofluoric acid (HF) is a weak acid (meaning it is a weak electrolyte), but the other hydrohalic acids (HCl, HBr, and HI) are all strong acids (strong electrolytes).

Group 8A Elements (ns^2np^6, $n \geq 2$)

Figure 7.21 shows samples of the Group 8A elements. All the noble gases exist as monatomic species. With the exception of helium, which has the electron configuration $1s^2$, their atoms have completely filled outer *ns* and *np* subshells. Their electron configurations give the noble gases their great stability. The Group 8A ionization energies are among the highest of all the elements (see Figure 7.8). Their electron affinities are all less than zero (Figure 7.10), so they have no tendency to accept extra electrons.

Student Note: Note that the common ions formed by the other main group elements are those that make them isoelectronic with a noble gas. In Group 1A elements, for example, each atom *loses* one electron to become isoelectronic with the noble gas that immediately precedes it; in Group 7A elements, each atom *gains* one electron to become isoelectronic with the noble gas that immediately follows it; and so on.

For years the noble gases were called *inert gases* because they were not known to react with anything. Beginning in 1963, however, compounds were prepared from the heavier members of the group by exposing them to very strong oxidizing agents such as fluorine and oxygen. Some of the compounds that have been prepared are XeF_4, XeO_3, $XeOF_4$, KrF_2, and most recently, HArF. Although the chemistry of the noble gases is interesting, their compounds are not involved in any natural biological processes and they currently have no major commercial applications.

Helium

Neon

Argon

Krypton

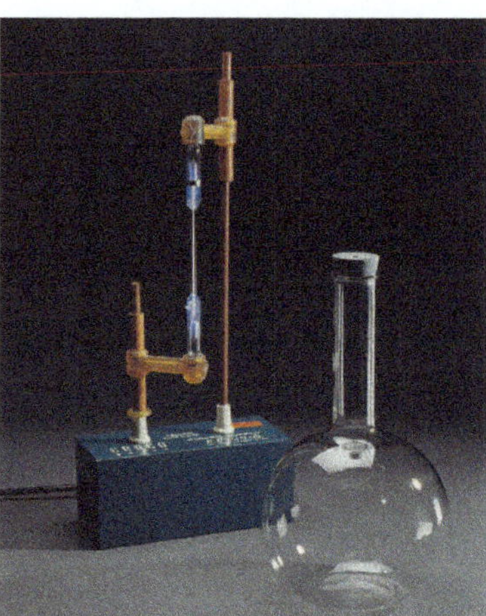

Xenon

Figure 7.21 Discharge tubes containing Group 8A elements.

Credit: (helium, neon, argon, krypton, xenon): © *McGraw-Hill Education/Stephen Frisch, photographer.*

Comparison of Group 1A and Group 1B Elements

Although the outer electron configurations of Groups 1A and 1B are similar (members of both groups have a single valence electron in an *s* orbital), their chemical properties are very different.

The first ionization energies of Cu, Ag, and Au are 745, 731, and 890 kJ/mol, respectively. Because these values are considerably larger than those of the alkali metals, the Group 1B elements are much less reactive. The higher ionization energies of the Group 1B elements result from incomplete shielding of the nucleus by the inner *d* electrons (compared with the more effective shielding by the completely filled noble gas cores). Consequently, the outer *s* electrons of the Group 1B elements are more strongly attracted by the nucleus. In fact, copper, silver, and gold are so unreactive that they are usually found in the uncombined state in nature. The inertness, rarity, and attractive appearance of these metals make them valuable in the manufacture of coins and jewelry. For this reason, these metals are also known as "coinage metals." The differences in the chemistry of the elements in Group 2A from that of the elements in Group 2B can be explained in a similar way.

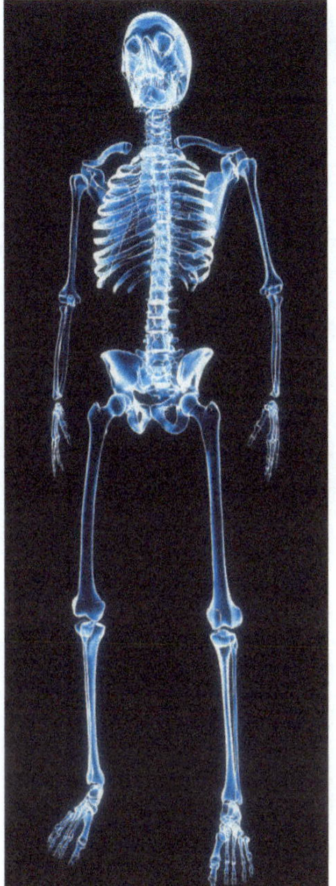
X-ray image of a human skeleton
Credit: © 3D4Medical.com/Getty Images.

Bringing Chemistry to Life

Radioactive Bone

One of the consequences of two chemical species having similar properties is that human physiology sometimes mistakes one species for another. Healthy bones require constant replenishment of calcium. Our dietary calcium comes primarily from dairy products but can also be found in some vegetables—especially dark, leafy greens such as spinach and kale. Strontium-90, a radioactive isotope, is found in the fallout of atomic bomb explosions and is a component in the waste generated by nuclear power facilities. Strontium-90 released into the atmosphere will eventually settle on land and water, and it can reach our bodies through ingestion—especially of vegetation—and through the inhalation of airborne particles. Because calcium and strontium are chemically similar, Sr^{2+} ions are mistaken by the body for Ca^{2+} ions and are incorporated into the bones. Constant exposure to the radiation emitted by strontium-90 affects not only the bone and surrounding soft tissue, but also the bone marrow, damaging and destroying stem cells vital to the immune system. Such long-term exposure leads to an increased risk of leukemia and other cancers.

	1A 1	2A 2	3B 3	4B 4	5B 5	6B 6	7B 7	8B 8	8B 9	8B 10	1B 11	2B 12	3A 13	4A 14	5A 15	6A 16	7A 17	8A 18	
1																			1
2																			2
3																			3
4		Ca																	4
5		Sr																	5
6																			6
7																			7

Variation in Properties of Oxides Within a Period

One way to compare the properties of the main group elements across a period is to examine the properties of a series of similar compounds. Because oxygen combines with almost all elements, we will compare the properties of oxides of the third-period elements to see how metals differ from metalloids and nonmetals. Some elements in the third period (P, S, and Cl) form several types of oxides, but for simplicity we will consider only those oxides in which the elements have the highest oxidation number. Table 7.4 lists a few general characteristics of these oxides and some specific physical properties of the oxides of third-period elements.

The tendency of oxygen to form the oxide ion is greatly favored when oxygen combines with metals that have low ionization energies, such as those in Groups 1A and 2A and aluminum. Thus, Na_2O, MgO, and Al_2O_3 are ionic compounds, as evidenced by their high melting points and boiling points. They have extensive three-dimensional structures in which each cation is surrounded by a specific number of anions, and vice versa. As the ionization energies of the elements increase from left to right, so does the molecular nature of the oxides that form. Silicon is a metalloid; its oxide (SiO_2) also has a huge three-dimensional network, although it is not an ionic compound. The oxides of phosphorus, sulfur, and chlorine are molecular compounds composed of small discrete units. The weak attractions among these molecules result in relatively low melting points and boiling points.

Most oxides can be classified as acidic or basic depending on whether they produce acidic or basic solutions when dissolved in water (or whether they *react* as acids or bases). Some oxides are ***amphoteric,*** which means that they display both acidic and basic properties. The first

TABLE 7.4 Some Properties of Oxides of the Third-Period Elements

	Na_2O	MgO	Al_2O_3	SiO_2	P_4O_{10}	SO_3	Cl_2O_7
Type of compound	←	Ionic	→	←	Molecular		→
Structure	←	Extensive three-dimensional		→	← Discrete molecular units		→
Melting point (°C)	1275	2800	2045	1610	580	16.8	−91.5
Boiling point (°C)	?	3600	2980	2230	?	44.8	82
Acid-base nature	Basic	Basic	Amphoteric	←	Acidic		→

two oxides of the third period, Na_2O and MgO, are basic oxides. For example, Na_2O reacts with water to form the base sodium hydroxide:

$$Na_2O(s) + H_2O(l) \longrightarrow 2NaOH(aq)$$

Magnesium oxide is quite insoluble; it does not react with water to any appreciable extent. However, it does react with acids in a manner that resembles an acid-base reaction:

$$MgO(s) + 2HCl(aq) \longrightarrow MgCl_2(aq) + H_2O(l)$$

The products of this reaction are a salt ($MgCl_2$) and water, the same kind of products that are obtained in an acid-base neutralization.

Aluminum oxide is even less soluble than magnesium oxide. It, too, does not react with water, but it exhibits the properties of a base by reacting with acids:

$$Al_2O_3(s) + 6HCl(aq) \longrightarrow 2AlCl_3(aq) + 3H_2O(l)$$

It also exhibits acidic properties by reacting with bases:

$$Al_2O_3(s) + 2NaOH(aq) + 3H_2O(l) \longrightarrow 2NaAl(OH)_4(aq)$$

Thus, Al_2O_3 is classified as an amphoteric oxide because it has properties of both acids and bases. Other amphoteric oxides are ZnO, BeO, and Bi_2O_3.

Silicon dioxide is insoluble and does not react with water. It has acidic properties, however, because it reacts with a very concentrated aqueous base:

$$SiO_2(s) + 2OH^-(aq) \longrightarrow SiO_3^{2-}(aq) + H_2O(l)$$

For this reason, concentrated aqueous, strong bases such as sodium hydroxide (NaOH) should *not* be stored in Pyrex glassware, which is made of SiO_2.

The remaining third-period oxides (P_4O_{10}, SO_3, and Cl_2O_7) are acidic. They react with water to form phosphoric acid, sulfuric acid, and perchloric acid, respectively:

$$P_4O_{10}(s) + 6H_2O(aq) \longrightarrow 4H_3PO_4(aq)$$

$$SO_3(g) + H_2O(l) \longrightarrow 2H_2SO_4(aq)$$

$$Cl_2O_7(l) + H_2O(l) \longrightarrow 2HClO_4(aq)$$

Student Note: Certain oxides such as CO and NO are neutral; that is, they do not react with water to produce acidic or basic solutions. In general, oxides of nonmetals are either acidic or neutral.

This brief examination of oxides of the third-period elements shows that as the metallic character of the elements decreases from left to right across the period, their oxides change from basic to amphoteric to acidic. Metal oxides are usually basic, and most oxides of nonmetals are acidic. The intermediate properties of the oxides (as demonstrated by the amphoteric oxides) are exhibited by elements whose positions are intermediate within the period. Because the metallic character of the elements increases from top to bottom within a group of main group elements, the oxides of elements with higher atomic numbers are more basic than the lighter elements.

Chapter Summary

Section 7.1

- The modern periodic table was devised independently by Dmitri Mendeleev and Lothar Meyer in the nineteenth century. The elements that were known at the time were grouped based on their physical and chemical properties. Using his arrangements of the elements, Mendeleev successfully predicted the existence of elements that had not yet been discovered.
- Early in the twentieth century, Henry Moseley refined the periodic table with the concept of the *atomic number,* thus resolving a few inconsistencies in the tables proposed by Mendeleev and Meyer.
- Elements in the same group of the periodic table tend to have similar physical and chemical properties.

Section 7.2

- The periodic table can be divided into the ***main group elements*** (also known as the *representative elements*) and the *transition metals.* It is further divided into smaller groups or *columns* of elements that all have the same configuration of ***valence electrons.***
- The 18 columns of the periodic table are labeled 1A through 8A (*s*- and *p*-block elements) and 1B through 8B (*d*-block elements), or by the numbers 1 through 18.

Section 7.3

- ***Effective nuclear charge*** *(Z_{eff})* is the nuclear charge that is "felt" by the valence electrons. It is usually lower than the nuclear charge due to ***shielding*** by the core electrons.

Section 7.4

- ***Atomic radius*** is the distance between an atom's nucleus and its valence shell. The atomic radius of a metal atom is defined as the ***metallic radius,*** which is one-half the distance between adjacent, identical nuclei in a metal solid. The atomic radius of a nonmetal is defined as the ***covalent radius,*** which is one-half the distance between adjacent, identical nuclei in a molecule. In general, atomic radii *decrease* from left to right across a *period* of the periodic table and *increase* from top to bottom down a *group.*
- ***Ionization energy (IE)*** is the energy required to remove an electron from a gaseous atom. The first ionization energy (IE_1) is smaller than subsequent ionization energies [e.g., second (IE_2), third (IE_3), and so on]. The first ionization of any atom removes a valence electron. Ionization energies increase dramatically when core electrons are being removed.
- First ionization energies (IE_1 values) tend to increase across a period and decrease down a group. Exceptions to this trend can be explained based upon the electron configuration of the element.
- ***Electron affinity (EA)*** is the energy released when an atom in the gas phase accepts an electron. *EA* is equal to $-\Delta H$ for the process $A(g) + e^- \longrightarrow A^-(g)$
- Electron affinities tend to increase across a period. As with first ionization energies, exceptions to the trend can be explained based on the electron configuration of the element.
- Metals tend to be shiny, lustrous, malleable, ductile, and conducting (for both heat and electricity). Metals typically lose electrons to form cations, and they tend to form ionic compounds (including basic oxides).
- Nonmetals tend to be brittle and not good conductors (for either heat or electricity). They can gain electrons to form anions but they commonly form molecular compounds (including acidic oxides).
- In general, metallic character decreases across a period and increases down a group of the periodic table. ***Metalloids*** are elements with properties intermediate between metals and nonmetals.
- According to ***Coulomb's law,*** the attractive force (F) between two oppositely charged particles (Q_1 and Q_2) is directly proportional to the product of the charges and inversely proportional to the distance (d) between the objects squared: ($F \propto Q_1 \cdot Q_2/d^2$).

Section 7.5

- The common ions of main group elements are ***isoelectronic*** with noble gases. When a *d*-block element loses one or more electrons, it loses them first from the shell with the highest principal quantum number (e.g., electrons in the 4*s* subshell are lost before electrons in the 3*d* subshell).

Section 7.6

- ***Ionic radius*** is the distance between the nucleus and valence shell of a cation or an anion. A cation is smaller than its parent atom. An anion is larger than its parent atom.
- An ***isoelectronic series*** consists of one or more ions and sometimes an atom, all of which have identical electron configurations. Within an isoelectronic series of ions, the greater the nuclear charge, the smaller the radius.

Section 7.7

- A ***diagonal relationship*** describes similarities in the chemical properties of elements that are in different groups, but that are positioned diagonally from each other in the periodic table.
- Although members of a group in the periodic table exhibit similar chemical and physical properties, the first member of each group tends to be significantly different from the other members. Hydrogen is essentially a group unto itself.
- The alkali metals (Group 1A) tend to be highly reactive toward oxygen, water, and acid. Group 2A metals are less reactive than Group 1A metals, but the heavier members all react with water to produce metal hydroxides and hydrogen gas. Groups that contain both metals and nonmetals (e.g., Groups 4A, 5A, and 6A) tend to show greater variability in their physical and chemical properties.
- ***Amphoteric*** oxides, such as Al_2O_3, are those that exhibit both acidic and basic behavior.

Key Words

Key Equation

7.1 $Z_{eff} = Z - \sigma$	Effective nuclear charge (Z_{eff}) is equal to *nuclear* charge (Z) minus the shielding constant (σ).

KEY SKILLS

Periodic Trends in Atomic Radius, Ionization Energy, and Electron Affinity

The properties of an element are determined in large part by the *size* (radius) and the valence-shell *electron configuration* of its atoms and ions. Together, principal quantum number (n), effective nuclear charge (Z_{eff}), and charge on the valence shell (number of valence electrons) determine atomic radius.

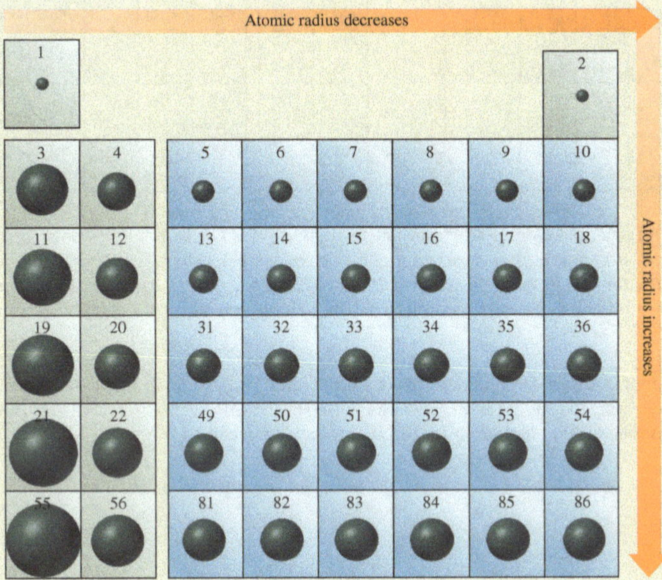

From left to right across a period, Z_{eff} and charge on the valence shell both *increase.* Each step to the right involves the addition of a proton, which increases Z, and the addition of an electron. Each additional electron resides in the same shell. Remember that electrons in the same shell do not shield one another well. The result is that Z_{eff} increases by nearly as much as Z. As the magnitude of opposite charges increases, coulombic attraction between them increases, and they are drawn closer together, thus reducing the atomic radius.

From top to bottom within a group, valence electrons reside in shells with increasingly larger values of n, putting them farther away from the nucleus—thereby increasing the atomic radius. Valence-shell electron configuration and, to a large extent, Z_{eff} remain the same from top to bottom in a group. Variations in atomic radii of the d-block elements vary less regularly than the main group elements.

From left to right across a period, Z_{eff} *increases* and radius *decreases.* Both factors contribute to stronger coulombic attraction between the positively charged nucleus and the valence electrons—making it harder to remove an electron; therefore, ionization energy (IE_1) *increases.* From top to bottom within a group, although Z_{eff} remains fairly constant, radius increases. The increased distance between the nucleus and the valence electrons corresponds to weaker coulombic attraction, making it easier to remove an electron; therefore, IE_1 *decreases.*

Because Z_{eff} *increases* and radius *decreases* from left to right across a period, the coulombic attraction between the positively charged nucleus and an added electron gets stronger, making it easier to add an electron; therefore, electron affinity (EA) *increases.* And because Z_{eff} remains fairly constant and radius *increases* from top to bottom within a group, there is a smaller coulombic attraction between the nucleus and an added electron—making it harder to add an electron; therefore, EA *decreases.*

Specific anomalies in the general trends of IE_1 and EA are determined by valence-electron configuration.

Ionization energy (IE_1) increases →

Ionization energy (IE_1) decreases ↓

1 1312							2 2372
3 520	4 899	5 800	6 1086	7 1402	8 1314	9 1681	10 2080
11 496	12 738	13 577	14 786	15 1012	16 999	17 1256	18 1520
19 419	20 590	31 579	32 761	33 947	34 941	35 1143	36 1351
21 403	22 549	49 558	50 708	51 834	52 869	53 1009	54 1170
55 376	56 503	81 589	82 715	83 703	84 813	85 (926)	86 1037

Electron affinity (EA) increases →

Electron affinity (EA) decreases ↓

1 +72.8							2 (0.0)
3 +59.6	4 ≤0	5 +26.7	6 +122	7 −7	8 +141	9 +328	10 (−29)
11 +52.9	12 ≤0	13 +42.5	14 +134	15 +72.0	16 +200	17 +349	18 (−35)
19 +48.4	20 +2.37	31 +28.9	32 +119	33 +78.2	34 +195	35 +325	36 (−39)
21 +46.9	22 +5.03	49 +28.9	50 +107	51 +103	52 +190	53 +295	54 (−41)
55 +45.5	56 +13.95	81 +19.3	82 +35.1	83 +91.3	84 +183	85 +270	86 (−41)

For example, although the trend is an increase in IE_1 from left to right, IE_1 for Group 3A is lower than IE_1 for Group 2A—within a period. This is because the elec tron removed by ionization comes from the p subshell, which is higher in energy than the s subshell [⏮ Figure 7.9(a)]. Likewise, IE_1 for Group 6A is lower than IE_1 for Group 5A. In this case, the electron removed by ionization is one of a pair of electrons in a p orbital. Because paired electrons in a single orbital repel one another, removing one of them is relatively easy [⏮ Figure 7.9(b)].

Similarly, despite the general trend, EA for Group 2A is lower than EA for Group 1A. In this case, the electron added by electron affinity goes into the p subshell, which is higher in energy than the s subshell [⏮ Figure 7.11(a)]. And EA for Group 5A is lower than that for Group 4A because the added electron must go into an already occupied p orbital [⏮ Figure 7.11(b)].

Each of these periodic properties can be explained and understood using Coulomb's law: $F \propto \frac{Q_1 \times Q_2}{d^2}$.

Key Skills Problems

7.1
Often we can compare properties of two elements based solely on periodic trends. For which pair of elements is the periodic trend not sufficient to determine which has the higher first ionization energy?

(a) C and Si (b) Al and Ga (c) Ga and Si (d) Tl and Sn (e) B and Si

7.2
The colored spheres represent the ions Ca^{2+}, Cl^-, K^+, P^{3-}, and S^{2-}. Based on size and using only a periodic table, determine which of the following has the ions in the same order as the diagram.

(a) Ca^{2+}, Cl^-, K^+, P^{3-}, S^{2-}
(b) Ca^{2+}, K^+, P^{3-}, S^{2-}, Cl^-
(c) P^{3-}, S^{2-}, Cl^-, K^+, Ca^{2+}
(d) Ca^{2+}, K^+, Cl^-, S^{2-}, P^{3-}
(e) P^{3-}, S^{2-}, Cl^-, Ca^{2+}, K^+

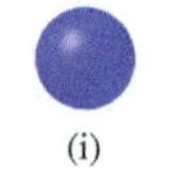
(i)

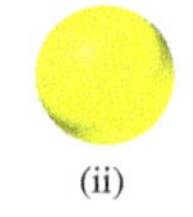
(ii)

(iii)

(iv)

(v)

7.3
Group 8A exhibits the highest first ionization energy. Which group do you expect to exhibit the highest *second* ionization energy (EA_2)?

(a) 7A (b) 6A (c) 5A (d) 2A (e) 1A

7.4
Which of the following best describes why Z_{eff} does not change significantly from top to bottom within a group?

(a) There are fewer completed electron shells.
(b) There are fewer valence electrons.
(c) There are more valence electrons.
(d) There are more completed electron shells.
(e) The number of protons in the nucleus does not change significantly.

Questions and Problems

Applying What You've Learned

In 1949, the Australian psychiatrist John Cade published the results of his studies showing that lithium was useful in the treatment of "manic episodes," one of the phases of what is known today as bipolar disorder. Although the research had shown real promise, its publication coincided with news of the lithium poisoning and resulting deaths of a group of cardiac patients that had used lithium chloride as a dietary salt substitute. Reports of this disaster prompted drug manufacturers to withdraw all lithium salts from the market, and for a time, any medical use of lithium was viewed as too dangerous even to consider. Additional research in Europe and the United States resulted in the gradual acceptance of lithium as a potentially valuable psychiatric therapy. The FDA approved lithium carbonate (Li_2CO_3) in 1970 for the treatment of manic illness, and in 1974 for the treatment of bipolar disorder. (a) Without referring to a periodic table, write the electron configuration of lithium (Z = 3) [⏮ Sample Problem 7.2]. (b) Referring only to a periodic table, arrange Li and the other alkali metals (not including Fr) in order of increasing atomic radius [⏮ Sample Problem 7.3]. (c) Again referring only to the periodic table, arrange the members of Group 1A (not including Fr) in order of increasing ionization energy (IE_1) [⏮ Sample Problem 7.4]. (d) Write the electron configuration for each of the alkali metal cations [⏮ Sample Problem 7.7]. (e) For each alkali metal cation in part (d), identify an isoelectronic series consisting of a noble gas and, where appropriate, one or more common ions (see Figure 2.14) [⏮ Sample Problem 7.8]. List each series in order of increasing radius [⏮ Sample Problem 7.9].

SECTION 7.1: DEVELOPMENT OF THE PERIODIC TABLE

Review Questions

7.1 Briefly describe the significance of Mendeleev's periodic table.

7.2 What is Moseley's contribution to the modern periodic table?

7.3 Describe the general layout of a modern periodic table.

7.4 What is the most important relationship among elements in the same group in the periodic table?

SECTION 7.2: THE MODERN PERIODIC TABLE

Review Questions

7.5 Classify each of the following elements as a metal, a nonmetal, or a metalloid: As, Xe, Fe, Li, B, Cl, Ba, P, I, Si.

7.6 Compare the physical and chemical properties of metals and nonmetals.

7.7 Draw a rough sketch of a periodic table (no details are required). Indicate regions where metals, nonmetals, and metalloids are located.

7.8 What is a main group element? Give names and symbols of four main group elements.

7.9 Without referring to a periodic table, write the name and give the symbol for one element in each of the following groups: 1A, 2A, 3A, 4A, 5A, 6A, 7A, 8A, transition metals.

7.10 Indicate whether the following elements exist as atomic species, molecular species, or extensive three-dimensional structures in their most stable states at room temperature, and write the molecular or empirical formula for each one: phosphorus, iodine, magnesium, neon, carbon, sulfur, cesium, and oxygen.

7.11 You are given a sample of a dark, shiny solid and asked to determine whether it is the nonmetal *iodine* or a *metallic* element. What test could you do that would enable you to answer the question without destroying the sample?

7.12 What are valence electrons? For main group elements, the number of valence electrons of an element is equal to its group number. Show that this is true for the following elements: Al, Sr, K, Br, P, S, C.

7.13 Write the outer electron configurations for the (a) alkali metals, (b) alkaline earth metals, (c) halogens, (d) noble gases.

7.14 Use the first-row transition metals (Sc to Cu) as an example to illustrate the characteristics of the electron configurations of transition metals.

7.15 Arsenic is not an essential element for the human body. Based on its position in the periodic table, suggest a reason for its toxicity.

Conceptual Problems

7.16 In the periodic table, the element hydrogen is sometimes grouped with the alkali metals and sometimes with the halogens. Explain why hydrogen can resemble the Group 1A and the Group 7A elements.

7.17 A neutral atom of a certain element has 34 electrons. Consulting only the periodic table, identify the element and write its ground-state electron configuration.

7.18 Group the following electron configurations in pairs that would represent elements with similar chemical properties:
(a) $1s^22s^22p^63s^2$
(b) $1s^22s^22p^3$
(c) $1s^22s^22p^63s^23p^64s^23d^{10}4p^6$
(d) $1s^22s^2$
(e) $1s^22s^22p^6$
(f) $1s^22s^22p^63s^23p^3$

7.19 Group the following electron configurations in pairs that would represent elements with similar chemical properties:
(a) $1s^22s^22p^5$
(b) $1s^22s^1$
(c) $1s^22s^22p^6$
(d) $1s^22s^22p^63s^23p^5$
(e) $1s^22s^22p^63s^23p^64s^1$
(f) $1s^22s^22p^63s^23p^64s^23d^{10}4p^6$

7.20 Without referring to a periodic table, write the electron configuration of elements with the following atomic numbers: (a) 9, (b) 20, (c) 26, (d) 33.

7.21 Specify the group of the periodic table in which each of the following elements is found: (a) $[Ne]3s^1$, (b) $[Ne]3s^23p^3$, (c) $[Ne]3s^23p^6$, (d) $[Ar]4s^23d^8$.

SECTION 7.3: EFFECTIVE NUCLEAR CHARGE

Review Questions

7.22 Explain the term *effective nuclear charge.*

7.23 Explain why the atomic radius of Be is smaller than that of Li.

Computational Problems

7.24 The electron configuration of B is $1s^22s^22p^1$. (a) If each core electron (i.e., the $1s$ electrons) were totally effective in shielding the valence electrons (i.e., the $2s$ and $2p$ electrons) from the nucleus and the valence electrons did not shield one another, what would be the shielding constant (σ) and the effective nuclear charge (Z_{eff}) for the $2s$ and $2p$ electrons? (b) In reality, the shielding constants for the $2s$ and $2p$ electrons in B are slightly different. They are 2.42 and 2.58, respectively. Calculate Z_{eff} for these electrons, and explain the differences from the values you determined in part (a).

7.25 The electron configuration of C is $1s^22s^22p^2$. (a) If each core electron (i.e., the $1s$ electrons) were totally effective in shielding the valence electrons (i.e., the $2s$ and $2p$ electrons) from the nucleus and the valence electrons did not shield one another, what would be the shielding constant (σ) and the effective nuclear charge (Z_{eff}) for the $2s$ and $2p$ electrons? (b) In reality, the shielding constants for the $2s$ and $2p$ electrons in C are slightly different. They are 2.78 and 2.86, respectively. Calculate Z_{eff} for these electrons, and explain the differences from the values you determined in part (a).

SECTION 7.4: PERIODIC TRENDS IN PROPERTIES OF ELEMENTS

Review Questions

7.26 Define *atomic radius.* Does the size of an atom have a precise meaning? Explain.

7.27 How does atomic radius change (a) from left to right across a period and (b) from top to bottom in a group?

7.28 Define *ionization energy.* Explain why ionization energy measurements are usually made when atoms are in the gaseous state. Why is the second ionization energy always greater than the first ionization energy for any element?

7.29 Sketch the outline of the periodic table, and show group and period trends in the first ionization energy of the elements. What types of elements have the highest ionization energies and what types have the lowest ionization energies?

7.30 (a) Define *electron affinity.* (b) Explain why electron affinity measurements are made with gaseous atoms. (c) Ionization energy is always a positive quantity, whereas electron affinity may be either positive or negative. Explain.

7.31 Explain the trends in electron affinity from aluminum to chlorine (see Figure 7.10).

Computational Problems

7.32 A hydrogen-like ion is an ion containing only one electron. The energies of the electron in a hydrogen-like ion are given by

$$E_n = -(2.18 \times 10^{-18}\text{ J})Z^2\left(\frac{1}{n^2}\right)$$

where n is the principal quantum number and Z is the atomic number of the element. Calculate the ionization energy (in kJ/mol) of the He^+ ion.

7.33 Plasma is a state of matter consisting of positive gaseous ions and electrons. In the plasma state, a mercury atom could be stripped of its 80 electrons and therefore would exist as Hg^{80+}. Use the equation in Problem 7.32 to calculate the energy required for the last ionization step, that is,

$$Hg^{79+}(g) \longrightarrow Hg^{80+}(g) + e^-$$

Conceptual Problems

7.34 On the basis of their positions in the periodic table, select the atom with the larger atomic radius in each of the following pairs: (a) Na, Si; (b) Ba, Be; (c) N, F; (d) Br, Cl; (e) Ne, Kr.

7.35 Arrange the following atoms in order of increasing atomic radius: Na, Al, P, Cl, Mg.

7.36 Which is the largest atom in the third period of the periodic table?

7.37 Which is the smallest atom in Group 7A?

7.38 Based on size, identify the spheres shown as Na, Mg, O, and S.

7.39 Based on size, identify the spheres shown as K, Ca, S, and Se.

7.40 Why is the radius of the lithium atom considerably larger than the radius of the hydrogen atom?

7.41 Use the second period of the periodic table as an example to show that the size of atoms decreases as we move from left to right. Explain the trend.

7.42 Arrange the following in order of increasing first ionization energy: Na, Cl, Al, S, and Cs.

7.43 Arrange the following in order of increasing first ionization energy: F, K, P, Ca, and Ne.

7.44 Use the third period of the periodic table as an example to illustrate the change in first ionization energies of the elements as we move from left to right. Explain the trend.

7.45 In general, the first ionization energy increases from left to right across a given period. Aluminum, however, has a lower first ionization energy than magnesium. Explain.

7.46 The first and second ionization energies of K are 419 and 3052 kJ/mol, and those of Ca are 590 and 1145 kJ/mol, respectively. Compare their values and comment on the differences.

7.47 Two atoms have the electron configurations $1s^22s^22p^6$ and $1s^22s^22p^63s^1$. The first ionization energy of one is 2080 kJ/mol, and that of the other is 496 kJ/mol. Match each ionization energy with one of the given electron configurations. Justify your choice.

7.48 Arrange the elements in each of the following groups in order of increasing electron affinity: (a) Li, Na, K; (b) F, Cl, Br, I.

7.49 Specify which of the following elements you would expect to have the greatest electron affinity: He, K, Co, S, Cl.

7.50 Considering their electron affinities, do you think it is possible for the alkali metals to form an anion like M^-, where M represents an alkali metal?

7.51 Explain why alkali metals have a greater affinity for electrons than alkaline earth metals.

SECTION 7.5: ELECTRON CONFIGURATION OF IONS

Review Questions

7.52 How does the electron configuration of ions derived from main group elements give them stability?

7.53 What do we mean when we say that two ions or an atom and an ion are *isoelectronic?*

7.54 Is it possible for the atoms of one element to be isoelectronic with the atoms of another element? Explain.

7.55 Give three examples of first-row transition metal (Se to Cu) ions that are isoelectronic with argon.

Conceptual Problems

7.56 A M^{2+} ion derived from a metal in the first transition metal series has four electrons in the $3d$ subshell. What element might M be?

7.57 A metal ion with a net +3 charge has five electrons in the $3d$ subshell. Identify the metal.

7.58 Write the ground-state electron configurations of the following ions: (a) Li^+, (b) H^-, (c) N^{3-}, (d) F^-, (e) S^{2-}, (f) Al^{3+}, (g) Se^{2-}, (h) Br^-, (i) Rb^+, (j) Sr^{2+}, (k) Sn^{2+}, (l) Te^{2-}, (m) Ba^{2+}, (n) Pb^{2+}, (o) In^{3+}, (p) Tl^+, (q) Tl^{3+}.

7.59 Group the species that are isoelectronic: Be^{2+}, F^-, Fe^{2+}, N^{3-}, He, S^{2-}, Co^{3+}, Ar.

7.60 Write the ground-state electron configurations of the following transition metal ions: (a) Sc^{3+}, (b) Ti^{4+}, (c) V^{5+}, (d) Cr^{3+}, (e) Mn^{2+}, (f) Fe^{2+}, (g) Fe^{3+}, (h) Co^{2+}, (i) Ni^{2+}, (j) Cu^+, (k) Cu^{2+}, (l) Ag^+, (m) Au^+, (n) Au^{3+}, (o) Pt^{2+}.

7.61 Name the ions with three charges that have the following electron configurations: (a) $[Ar]3d^3$, (b) [Ar], (c) $[Kr]4d^6$, (d) $[Xe]4f^{14}5d^6$.

7.62 Which of the following species are isoelectronic with each other: C, Cl^-, Mn^{2+}, B^-, Ar, Zn, Fe^{3+}, Ge^{2+}?

SECTION 7.6: IONIC RADIUS

Review Questions

7.63 Define *ionic radius.* How does the size of an atom change when it is converted to (a) an anion and (b) a cation?

7.64 Explain why, for isoelectronic ions, the anions are larger than the cations.

Conceptual Problems

7.65 Indicate which one of the two species in each of the following pairs is smaller: (a) Cl or Cl^-, (b) Na or Na^+, (c) O^{2-} or S^{2-}, (d) Mg^{2+} or Al^{3+}, (e) Au^+ or Au^{3+}.

7.66 List the following ions in order of increasing ionic radius: N^{3-}, Na^+, F^-, Mg^{2+}, O^{2-}.

7.67 Explain which of the following cations is larger, and why: Cu^+ or Cu^{2+}.

7.68 Explain which of the following anions is larger, and why: Se^{2-} or Te^{2-}.

SECTION 7.7: PERIODIC TRENDS IN CHEMICAL PROPERTIES OF THE MAIN GROUP ELEMENTS

Review Questions

7.69 Why do members of a group exhibit similar chemical properties?

7.70 Explain why hydrogen belongs in its own group.

7.71 Which elements are more likely to form acidic oxides? Basic oxides? Amphoteric oxides?

Conceptual Problems

7.72 Give the physical states (gas, liquid, or solid) of the main group elements in the fourth period (K, Ca, Ga, Ge, As, Se, Br) at room temperature.

7.73 The boiling points of neon and krypton are −246.1°C and −153.2°C, respectively. Using these data, estimate the boiling point of argon.

7.74 Use the alkali metals and alkaline earth metals as examples to show how we can predict the chemical properties of elements simply from their electron configurations.

7.75 Based on your knowledge of the chemistry of the alkali metals, predict some of the chemical properties of francium, the last member of the group.

7.76 As a group, the noble gases are very stable chemically. Why?

7.77 Why are Group 1B elements more stable than Group 1A elements even though they seem to have the same outer electron configuration, ns^1, where n is the principal quantum number of the outermost shell?

7.78 How do the chemical properties of oxides change from left to right across a period? How do they change from top to bottom within a particular group?

7.79 Write balanced equations for the reactions between each of the following oxides and water: (a) Li_2O, (b) CaO, (c) SO_3.

7.80 Write formulas for and name the binary hydrogen compounds of the second-period elements (Li to F). Describe how the physical and chemical properties of these compounds change from left to right across the period.

7.81 Which oxide is more basic, MgO or BaO? Why?

ADDITIONAL PROBLEMS

7.82 State whether each of the following properties of the main group elements generally increases or decreases (a) from left to right across a period and (b) from top to bottom within a group: metallic character, atomic size, ionization energy, acidity of oxides.

7.83 Referring to the periodic table, name (a) the halogen in the fourth period, (b) an element similar to phosphorus in chemical properties, (c) the most reactive metal in the fifth period, (d) an element that has an atomic number smaller than 20 and is similar to strontium.

7.84 Write equations representing the following processes:
(a) The electron affinity of S^-
(b) The third ionization energy of titanium
(c) The electron affinity of Mg^{2+}
(d) The ionization energy of O^{2-}

7.85 Arrange the following isoelectronic species in order of increasing ionization energy: O^{2-}, F^-, Na^+, Mg^{2+}.

7.86 Write the empirical (or molecular) formulas of compounds that the elements in the third period (sodium to chlorine) should form with (a) molecular oxygen and (b) molecular chlorine. In each case indicate whether you would expect the compound to be ionic or molecular in character.

7.87 Arrange the following species in isoelectronic pairs: O^+, Ar, S^{2-}, Ne, Zn, Cs^+, N^{3-}, As^{3+}, N, Xe.

7.88 In which of the following are the species written in decreasing order by size of radius: (a) Be, Mg, Ba, (b) N^{3-}, O^{2-}, F^-, (c) Tl^{3+}, Tl^{2+}, Tl^+?

7.89 Which of the following properties show a clear periodic variation: (a) first ionization energy, (b) molar mass of the elements, (c) number of isotopes of an element, (d) atomic radius?

7.90 When carbon dioxide is bubbled through a clear calcium hydroxide solution, the solution appears milky. Write an equation for the reaction, and explain how this reaction illustrates that CO_2 is an acidic oxide.

7.91 You are given four substances: a fuming red liquid, a dark metallic-looking solid, a pale-yellow gas, and a yellow-green gas that attacks glass. You are told that these substances are the first four members of Group 7A, the halogens. Name each one.

7.92 For each pair of elements listed, give three properties that show their chemical similarity: (a) sodium and potassium and (b) chlorine and bromine.

7.93 What is the most reactive element on the periodic table?

7.94 Explain why the first electron affinity of sulfur is 200 kJ/mol but the second electron affinity is −649 kJ/mol.

7.95 The H^- ion and the He atom have two 1*s* electrons each. Which of the two species is larger? Explain.

7.96 Predict the products of the following oxides with water: Na_2O, BaO, CO_2, N_2O_5, P_4O_{10}, SO_3. Write an equation for each of the reactions. Specify whether the oxides are acidic, basic, or amphoteric.

7.97 Write the formulas and names of the oxides of the second-period elements (Li to N). Identify the oxides as acidic, basic, or amphoteric. Use the highest oxidation state of each element.

7.98 State whether each of the following elements is a gas, liquid, or solid under atmospheric conditions. Also state whether it exists in the elemental form as atoms, molecules, or a three-dimensional network: Mg, Cl, Si, Kr, O, I, Hg, Br.

7.99 The formula for calculating the energies of an electron in a hydrogen-like ion is given in Problem 7.32. This equation can be applied only to one-electron atoms or ions. One way to modify it for more complex species is to replace Z with $Z - \sigma$ or Z_{eff}. Calculate the value of σ if the first ionization energy of helium is 3.94×10^{-18} J per atom. (Disregard the minus sign in the given equation in your calculation.)

7.100 Why do noble gases have negative electron affinity values?

7.101 The atomic radius of K is 227 pm and that of K^+ is 138 pm. Calculate the percent decrease in volume that occurs when K(g) is converted to $K^+(g)$. (The volume of a sphere is $\frac{4}{3}\pi r^3$, where r is the radius of the sphere.)

7.102 The atomic radius of F is 72 pm and that of F^- is 133 pm. Calculate the percent increase in volume that occurs when F(g) is converted to $F^-(g)$. (See Problem 7.101 for the volume of a sphere.)

7.103 Match each of the elements on the right with its description on the left:

(a) A dark-red liquid	Calcium (Ca)
(b) A colorless gas that burns in oxygen gas	Gold (Au)
(c) A metal that reacts violently with water	Hydrogen (H_2)
(d) A shiny metal that is used in jewelry	Argon (Ar)
(e) An inert gas	Bromine (Br_2)

7.104 The energy needed for the following process is 1.96×10^4 kJ/mol:

$$\text{Li}(g) \longrightarrow \text{Li}^{3+}(g) + 3e^-$$

If the first ionization energy of lithium is 520 kJ/mol, calculate the second ionization energy of lithium, that is, the energy required for the process

$$\text{Li}^+(g) \longrightarrow \text{Li}^{2+}(g) + e^-$$

(*Hint:* You need the equation in Problem 7.32.)

7.105 A student is given samples of three elements, X, Y, and Z, which could be an alkali metal, a member of Group 4A, or a member of Group 5A. She makes the following observations: Element X has a metallic luster and conducts electricity. It reacts slowly with hydrochloric acid to produce hydrogen gas. Element Y is a light yellow solid that does not conduct electricity. Element Z has a metallic luster and conducts electricity. When exposed to air, it slowly forms a white powder. A solution of the white powder in water is basic. What can you conclude about the elements from these observations?

7.106 What is the electron affinity (in kJ/mol) of the Na^+ ion?

7.107 The ionization energies of sodium (in kJ/mol), starting with the first and ending with the eleventh, are 496, 4562, 6910, 9543, 13,354, 16,613, 20,117, 25,496, 28,932, 141,362, 159,075. Plot the log of ionization energy (y axis) versus the number of ionization (x axis); for example, log 496 is plotted versus 1 (labeled IE_1, the first ionization energy), log 4562 is plotted versus 2 (labeled IE_2, the second ionization energy), and so on. (a) Label IE_1 through IE_{11} with the electrons in orbitals such as 1*s*, 2*s*, 2*p*, and 3*s*. (b) What can you deduce about electron shells from the breaks in the curve?

7.108 Explain, in terms of their electron configurations, why Fe^{2+} is more easily oxidized to Fe^{3+} than Mn^{2+} is to Mn^{3+}.

7.109 Write the formulas and names of the hydrides of the following second-period elements: Li, C, N, O, F. Predict their reactions with water.

7.110 In halogen displacement reactions a halogen element can be generated by oxidizing its anions with a halogen element that lies above it in the periodic table. This means that there is no way to prepare elemental fluorine, because it is the first member of Group 7A. Indeed, for years the only way to prepare elemental fluorine was to oxidize F^- ions by electrolytic means. Then, in 1986, a chemist reported that by combining potassium hexafluoromanganate(IV) (K_2MnF_6) with antimony pentafluoride (SbF_5) at 150°C, he had generated elemental fluorine. Balance the following equation representing the reaction:

$$K_2MnF_6 + SbF_5 \longrightarrow KSbF_6 + MnF_3 + F_2$$

7.111 Write a balanced equation for the preparation of (a) molecular oxygen, (b) ammonia, (c) carbon dioxide, (d) molecular hydrogen, (e) calcium oxide. Indicate the physical state of the reactants and products in each equation.

7.112 Write chemical formulas for oxides of nitrogen with the following oxidation numbers: +1, +2, +3, +4, +5. (*Hint:* There are two oxides of nitrogen with a +4 oxidation number.)

7.113 Most transition metal ions are colored. For example, a solution of $CuSO_4$ is blue. How would you show that the blue color is due to the hydrated Cu^{2+} ions and not the SO_4^{2-} ions?

7.114 In general, atomic radius and ionization energy have opposite periodic trends. Why?

7.115 Explain why the electron affinity of nitrogen is approximately zero, while the elements on either side, carbon and oxygen, have substantial positive electron affinities.

7.116 Consider the halogens chlorine, bromine, and iodine. The melting point and boiling point of chlorine are −101.5°C and −34.0°C and those of iodine are 113.7°C and 184.3°C, respectively. Thus chlorine is a gas and iodine is a solid under room conditions. Estimate the melting point and boiling point of bromine. Compare your values with those from the webelements.com website.

7.117 Although it is possible to determine the second, third, and higher ionization energies of an element, the same cannot usually be done with the electron affinities of an element. Explain.

7.118 Why do elements that have high ionization energies also have more positive electron affinities? Which group of elements would be an exception to this generalization?

7.119 The first four ionization energies of an element are approximately 738, 1450, 7.7×10^3, and 1.1×10^4 kJ/mol. To which periodic group does this element belong? Explain your answer.

7.120 Predict the atomic number and ground-state electron configuration of the next member of the alkali metals after francium.

7.121 (a) The formula of the simplest hydrocarbon is CH_4 (methane). Predict the formulas of the simplest compounds formed between hydrogen and the following elements: silicon, germanium, tin, and lead. (b) Sodium hydride (NaH) is an ionic compound. Would you expect rubidium hydride (RbH) to be more or less ionic than NaH? (c) Predict the reaction between radium (Ra) and water. (d) When exposed to air, aluminum forms a tenacious oxide (Al_2O_3) coating that protects the metal from corrosion. Which metal in Group 2A would you expect to exhibit similar properties? (See page 305.)

7.122 Match each of the elements on the right with its description on the left:

(a) A pale yellow gas that reacts with water	Nitrogen (N_2)
(b) A soft metal that reacts with water to produce hydrogen	Boron (B)
	Fluorine (F_2)
(c) A metalloid that is hard and has a high melting point	Aluminum (Al)
	Sodium (Na)
(d) A colorless, odorless gas	
(e) A metal that is more reactive than iron, but does not corrode in air	

7.123 One way to estimate the effective charge (Z_{eff}) of a many-electron atom is to use the equation $IE_1 =$ (1312 kJ/mol) (Z_{eff}^2/n^2), where IE_1 is the first ionization energy and n is the principal quantum number of the shell in which the electron resides. Use this equation to calculate the effective nuclear charges of Li, Na, and K. Also calculate Z_{eff}/n for each metal. Comment on your results.

7.124 Use your knowledge of thermochemistry to calculate the ΔH for the following processes: (a) $Cl^-(g) \longrightarrow Cl^+(g) + 2e^-$ and (b) $K^+(g) + 2e^- \longrightarrow K^-(g)$.

7.125 To prevent the formation of oxides, peroxides, and superoxides, alkali metals are sometimes stored in an inert atmosphere. Which of the following gases should not be used for lithium: Ne, Ar, N_2, Kr? Explain. (*Hint:* As mentioned in the chapter, Li and Mg exhibit a diagonal relationship. Compare the common compounds of these two elements.)

7.126 On one graph, plot the effective nuclear charge (shown in parentheses) and atomic radius (see Figure 7.6) versus atomic number for the second-period elements: Li(1.28), Be(1.91), B(2.42), C(3.14), N(3.83), O(4.45), F(5.10), Ne(5.76). Comment on the trends.

7.127 One allotropic form of an element X is a colorless crystalline solid. The reaction of X with an excess amount of oxygen produces a colorless gas. This gas dissolves in water to yield an acidic solution. Choose one of the following elements that matches X: (a) sulfur, (b) phosphorus, (c) carbon, (d) boron, (e) silicon.

Engineering Problems

7.128 Calculate the maximum wavelength of light (in nm) required to ionize a single sodium atom.

7.129 A technique called photoelectron spectroscopy is used to measure the ionization energy of atoms. A gaseous sample is irradiated with UV light, and electrons are ejected from the valence shell. The kinetic energies of the ejected electrons are measured. Because the energy of the UV photon and the kinetic energy of the ejected electron are known, we can write

$$h\nu = IE + \tfrac{1}{2}mu^2$$

where ν is the frequency of the UV light, and m and u are the mass and velocity of the electron, respectively. In one experiment the kinetic energy of the ejected electron from potassium is found to be 5.34×10^{-19} J using a UV source of wavelength 162 nm. Calculate the ionization energy of potassium. How can you be sure that this ionization energy corresponds to the electron in the valence shell (i.e., the most loosely held electron)?

7.130 Element M is a shiny and highly reactive metal (melting point 63°C), and element X is a highly reactive nonmetal (melting point −7.2°C). They react to form a compound with the empirical formula MX, a colorless, brittle white solid that melts at 734°C. When dissolved in water or when in the molten state, the substance conducts electricity. When chlorine gas is bubbled through an aqueous solution containing MX, a reddish-brown liquid appears and Cl^- ions are formed. From these observations, identify M and X. (You may need to consult a handbook of chemistry for the melting-point values.)

Biological Problems

7.131 Write the ground-state electron configurations of the following ions, which play important roles in biochemical processes in our bodies: (a) Na^+, (b) Mg^{2+}, (c) Cl^-, (d) K^+, (e) Ca^{2+}, (f) Fe^{2+}, (g) Cu^{2+}, (h) Zn^{2+}.

7.132 Thallium (Tl) is a neurotoxin and exists mostly in the Tl(I) oxidation state in its compounds. Aluminum (Al), which causes anemia and dementia, is only stable in the Al(III) form. The first, second, and third ionization energies of Tl are 589, 1971, and 2878 kJ/mol, respectively. The first, second, and third ionization energies of Al are 577.5, 1817, and 2745 kJ/mol, respectively. Plot the ionization energies of Al and Tl versus the number of electrons removed and explain the trends.

7.133 Both Mg^{2+} and Ca^{2+} are important biological ions. One of their functions is to bind to the phosphate group of ATP molecules or amino acids of proteins. For Group 2A metals in general, the tendency for binding to the anions increases in the order $Ba^{2+} < Sr^{2+} < Ca^{2+} < Mg^{2+}$. Explain this trend.

7.134 The air in a manned spacecraft or submarine needs to be purified of exhaled carbon dioxide. Write equations for the reactions between carbon dioxide and (a) lithium oxide (Li_2O), (b) sodium peroxide (Na_2O_2), and (c) potassium superoxide (KO_2).

Multiconcept Problems

7.135 As discussed in the chapter, the atomic mass of argon is greater than that of potassium. This observation created a problem in the early development of the periodic table because it meant that argon should be placed after potassium. (a) How was this difficulty resolved? (b) From the following data, calculate the average atomic masses of argon and potassium: Ar-36 (35.9675 amu, 0.337 percent), Ar-38 (37.9627 amu, 0.063 percent), Ar-40 (39.9624 amu, 99.60 percent), K-39 (38.9637 amu, 93.258 percent), K-40 (39.9640 amu, 0.0117 percent), K-41 (40.9618 amu, 6.730 percent).

7.136 Little is known of the chemistry of astatine, the last member of Group 7A. Describe the physical characteristics that you would expect this halogen to have. Predict the products of the reaction between sodium astatide (NaAt) and sulfuric acid. (*Hint:* Sulfuric acid is an oxidizing agent.)

7.137 Based on knowledge of the electronic configuration of titanium, state which of the following compounds of titanium is unlikely to exist: K_3TiF_6, $K_2Ti_2O_5$, $TiCl_3$, K_2TiO_4, K_2TiF_6.

7.138 The ionization energy of a certain element is 412 kJ/mol. When the atoms of this element are in the first excited state, however, the ionization energy is only 126 kJ/mol. Based on this information, calculate the wavelength of light emitted in a transition from the first excited state to the ground state.

7.139 Experimentally, the electron affinity of an element can be determined by using a laser light to ionize the anion of the element in the gas phase:

$$X^-(g) + h\nu \longrightarrow X(g) + e^-$$

Referring to Figure 7.10, calculate the photon wavelength (in nm) corresponding to the electron affinity for chlorine. In what region of the electromagnetic spectrum does this wavelength fall?

Standardized-Exam Practice Problems

Physical and Biological Sciences

These questions are not based on a descriptive passage.

1. A halogen has valence electrons in which orbitals?

   a) s b) s and p c) p d) s, p, and d

2. How many subshells does a shell with principal quantum number n contain?

   a) n b) n^2 c) $n - 1$ d) $2n - 1$

3. In a shell that contains an f subshell, what is the ratio of f orbitals to s orbitals?

   a) 14:1 b) 7:1 c) 7:3 d) 7:5

4. What is the maximum number of electrons that can be in the $n = 3$ shell?

   a) 2 b) 6 c) 8 d) 18

Answers to In-Chapter Materials

Answers to Practice Problems

7.1A Ge. **7.1B** Bi < As < P. **7.2A** (a) $1s^22s^22p^63s^23p^3$, p-block, (b) $1s^22s^22p^63s^23p^64s^2$, s-block, (c) $1s^22s^22p^63s^23p^64s^23d^{10}4p^5$, p-block. **7.2B** (a) Al, (b) Zn, (c) Sr. **7.3A** F < Se < Ge. **7.3B** P and Se. **7.4A** Mg, Mg. **7.4B** Rb has a smaller Z_{eff}, IE_2 for Rb corresponds to the removal of a core electron. **7.5A** Al. **7.5B** Adding an electron to As involves pairing. **7.6A** The attractive force is slightly larger between +3.26 and −1.15 separated by 1.5 pm. **7.6B** 1.51 pm. **7.7A** (a) [Ne], (b) [Ar], (c) [Kr]. **7.7B** N^{3-}, O^{2-}, F^-, Ne, Na^+, Mg^{2+}, Al^{3+}. **7.8A** (a) [Ar]$3d^6$, (b) [Ar]$3d^9$, (c) [Kr]$4d^{10}$. **7.8B** Cu^+. **7.9A** $Rb^+ < Br^- < Se^{2-}$. **7.9B** F^-, O^{2-}, N^{3-}, Na^+, Mg^{2+}, Al^{3+}.

Answers to Checkpoints

7.1.1 c. **7.1.2** a. **7.1.3** c. **7.2.1** b. **7.2.2** a, d, e. **7.4.1** b. **7.4.2** c. **7.4.3** e. **7.4.4** a. **7.5.1** b, c, e. **7.5.2** b, d. **7.5.3** d. **7.5.4** b. **7.6.1** d, e. **7.6.2** a. **7.6.3** c. **7.6.4** d.

CHAPTER 8

Chemical Bonding I: Basic Concepts

Dynamite, a stabilized form of the explosive nitroglycerin, is used to blast through solid rock.

Credit: © *Dinodia Photos/Alamy.*

In This Chapter, You Will Learn

About two different types of chemical bonds, called *ionic* and *covalent*. You will also learn how to represent atoms and atomic ions using Lewis dot symbols and how to represent molecules and polyatomic ions using Lewis *structures*.

Before You Begin, Review These Skills

- Atomic radius [⏮ Section 7.4]
- Ions of main group elements [⏮ Section 7.5]

What Chemical Bonds Have to Do with Explosives

When the Swedish chemist Alfred Nobel died in 1896, his will specified that the bulk of his considerable fortune was to be used to establish the prizes that bear his name. The prizes, given annually in five categories (Chemistry, Physics, Physiology or Medicine, Literature, and Peace), are intended to recognize significant contributions to the betterment of humankind. In life, Nobel had been a prolific scientist and entrepreneur. He held more than 300 patents, including the one for dynamite, a stabilized form of the explosive nitroglycerin. His extensive work on the development and manufacture of explosives earned him the title "merchant of death" and was the cause of personal tragedy when his younger brother was killed in an explosion at one of the family's factories.

Student Note: A sixth prize, the *Nobel Memorial Prize in Economics*, is awarded along with the others but was not specified in Nobel's will, and it is not really a Nobel Prize.

Ironically, toward the end of his life, Nobel developed heart disease–related chest pain (angina pectoris) and was directed by his physician to take nitroglycerin orally, which Nobel refused to do. Glyceryl trinitrate, the name used by the medical community—perhaps to avoid the impression that doctors are prescribing explosives to patients—still is widely used to treat the symptoms of heart disease and other medical conditions. One of the more recently developed and intriguing uses of nitroglycerin is its placement in the tip of a condom to stimulate an erection.

Both the explosive nature of nitroglycerin and its effectiveness in treating such conditions as heart disease and erectile dysfunction can be illuminated by an understanding of the basic concepts of *chemical bonding*.

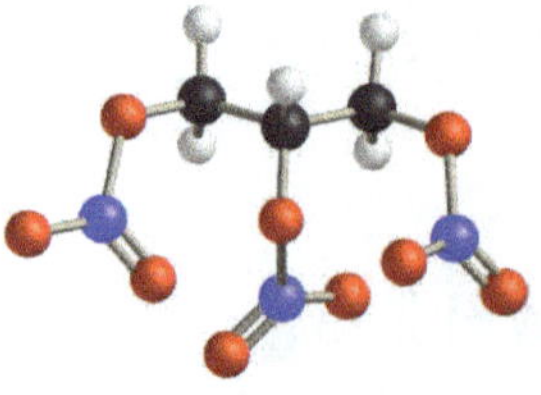

Nitroglycerin

At the end of this chapter, you will be able to answer a series of questions about nitric oxide and nitroglycerin [⏭ Applying What You've Learned, page 362].

8.1 Lewis Dot Symbols

Student Note: Remember that for two species to be isoelectronic they must have exactly the same electron configuration [⏮ Section 7.5].

The development of the periodic table and the concept of electron configuration gave chemists a way to explain the formation of compounds. The explanation, formulated by Gilbert Lewis,[1] is that atoms combine to achieve a more stable electron configuration. Maximum stability results when an atom is *isoelectronic* with a noble gas.

When atoms interact to form compounds, it is their *valence electrons* that actually interact. Therefore, it is helpful to have a method for depicting the valence electrons of the atoms involved. This can be done using Lewis dot symbols. A ***Lewis dot symbol*** consists of the element's symbol surrounded by dots, where each dot represents a valence electron. For the main group elements, the number of dots in the Lewis dot symbol is the same as the traditional group number (1A–8A), as shown in Figure 8.1. (Because they have incompletely filled inner shells, transition metals typically are not represented with Lewis dot symbols.)

Figure 8.1 shows that dots are placed above and below as well as to the left and right of the symbol. The exact order in which the dots are placed around the element symbol is not important, but the number of dots is. Thus, any of the following would be correct for the Lewis dot symbol for boron:

When writing Lewis dot symbols, though, we do not "pair" dots until absolutely necessary. Thus, we would *not* represent boron with a pair of dots on one side and a single dot on another.

For main group metals such as Na or Mg, the number of dots in the Lewis dot symbol is the number of electrons that are lost when the atom forms a cation that is isoelectronic with the preceding noble gas. For nonmetals of the second period (B through F), the number of unpaired dots is the number of bonds the atom can form. (As we will see shortly [⏭ Section 8.8], a larger nonmetal atom [one in the third period or beyond] can actually form as many bonds as the total number of dots in its Lewis dot symbol.)

In addition to atoms, we can also represent atomic *ions* with Lewis dot symbols. To do so, we simply add (for anions) or subtract (for cations) the appropriate number of dots from the Lewis dot symbol of the atom and include the ion's charge.

Sample Problem 8.1 shows how to use Lewis dot symbols to represent atomic ions.

SAMPLE PROBLEM 8.1

Write Lewis dot symbols for (a) fluoride ion (F^-), (b) potassium ion (K^+), and (c) sulfide ion (S^{2-}).

Strategy Starting with the Lewis dot symbol for each element, add dots (for anions) or remove dots (for cations) as needed to achieve the correct charge on each ion. Don't forget to include the appropriate charge on the Lewis dot symbol.

Setup The Lewis dot symbols for F, K, and S are :Ḟ̤:, K·, and :Ṡ̤:, respectively.

Solution (a) $[:\ddot{\underset{..}{F}}:]^-$

(b) K^+

(c) $[:\ddot{\underset{..}{S}}:]^{2-}$

THINK ABOUT IT

For ions that are isoelectronic with noble gases, cations should have no dots remaining around the element symbol, whereas anions should have eight dots around the element symbol. Note, too, that we put square brackets around the Lewis dot symbol for an anion and place the negative charge outside the brackets.

1. Gilbert Newton Lewis (1875–1946). American chemist. Lewis made many important contributions in the areas of chemical bonding, thermodynamics, acids and bases, and spectroscopy. Despite the significance of Lewis's work, he was never awarded a Nobel Prize.

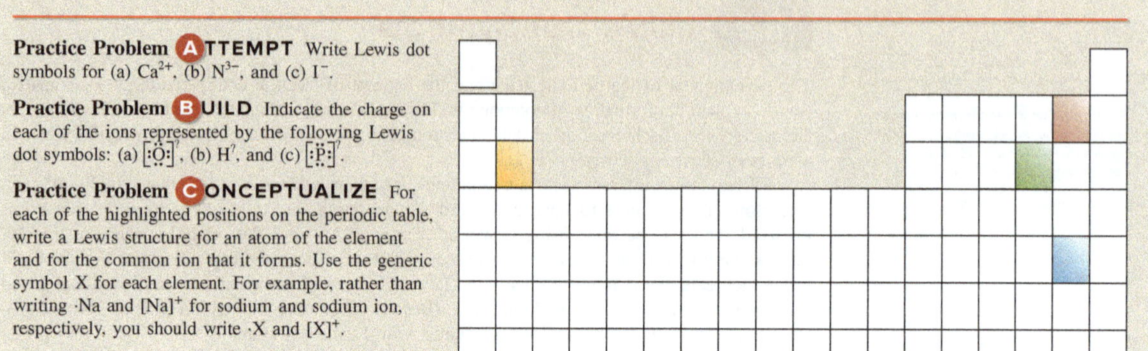

Practice Problem ATTEMPT Write Lewis dot symbols for (a) Ca^{2+}, (b) N^{3-}, and (c) I^-.

Practice Problem BUILD Indicate the charge on each of the ions represented by the following Lewis dot symbols: (a) $[:\ddot{O}:]^?$, (b) $H^?$, and (c) $[:\ddot{P}:]^?$.

Practice Problem CONCEPTUALIZE For each of the highlighted positions on the periodic table, write a Lewis structure for an atom of the element and for the common ion that it forms. Use the generic symbol X for each element. For example, rather than writing ·Na and $[Na]^+$ for sodium and sodium ion, respectively, you should write ·X and $[X]^+$.

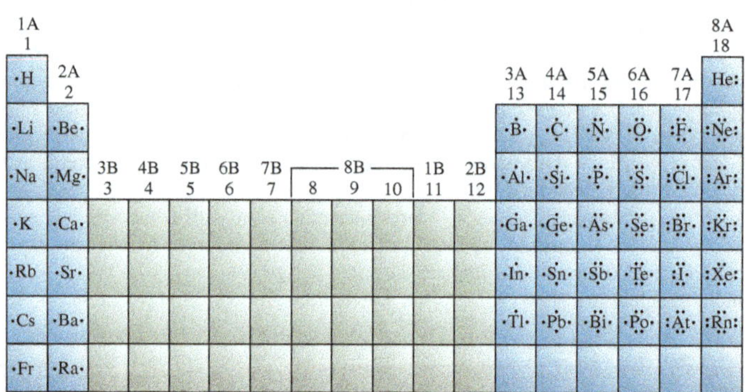

Figure 8.1 Lewis dot symbols of the main group elements.

CHECKPOINT – SECTION 8.1 Lewis Dot Symbols

8.1.1 Using only a periodic table, determine the correct Lewis dot symbol for a silicon (Si) atom.

a) :Si:
b) ·$\dot{\text{Si}}$·
c) :$\dot{\text{Si}}$·
d) ·Si·
e) :Si

8.1.2 Using only a periodic table, determine the correct Lewis dot symbol for the bromide ion (Br^-).

a) $[:\ddot{Br}:]^-$
b) $[:\ddot{Br}]^-$
c) $[\cdot\ddot{Br}\cdot]^-$
d) $[Br\cdot]^-$
e) $[:\ddot{Br}:]^-$

8.1.3 To which group does element X belong if its Lewis symbol is ·$\ddot{X}$·?

a) 1A
b) 2A
c) 3A
d) 4A
e) 5A
f) 6A
g) 7A
h) It is not possible to tell.

8.1.4 To which group does element Y belong if the Lewis symbol for its anion is $[:\ddot{Y}:]^{m-}$, where m represents the charge?

a) 4A
b) 5A
c) 6A
d) 7A
e) It is not possible to tell.

8.2 Ionic Bonding

Student Note: Iodized salt is used to prevent the devastating effects of iodine deficiency disorders, which include stillbirths, birth defects, and mental retardation.

Recall from Chapter 7 that atoms of elements with low ionization energies tend to form cations, while those with high positive electron affinities tend to form anions. ***Ionic bonding*** refers to the electrostatic attraction that holds these oppositely charged ions together in an ionic compound, such as K^+ and I^- in potassium iodide (KI), the "iodizing" ingredient in iodized salt. The electron configuration of potassium is $[Ar]4s^1$, and that of iodine is $[Kr]5s^24d^{10}5p^5$. When potassium and iodine atoms come into contact with each other, the valence electron of potassium is transferred to the iodine atom. We can imagine these processes taking place separately and represent each process using Lewis dot symbols.

$$\cdot K \longrightarrow K^+ + e^-$$

$$:\ddot{\underset{..}{I}}\cdot + e^- \longrightarrow [:\ddot{\underset{..}{I}}:]^-$$

The sum of these two equations is

$$K\cdot + :\ddot{\underset{..}{I}}\cdot \longrightarrow K^+ + [:\ddot{\underset{..}{I}}:]^-$$

The electrostatic attraction between the resulting cation and anion draws them together to form the electrically neutral compound KI.

Student Note: Remember that the electron affinity, *EA*, is the energy *released* when a gaseous atom accepts an electron. A *positive EA* corresponds to a *negative* ΔH for the process [⏮ Section 7.4]. Hess's law says that we can add the ΔH values for the individual steps to determine the overall ΔH [⏮ Section 5.5].

The net energy change associated with the formation of K^+ and I^- ions is *endothermic;* that is, energy must be supplied for the overall transfer of an electron from K to I to take place. The ionization of potassium requires the input of 419 kJ/mol, and the electron affinity of iodine is only 295 kJ/mol [⏮ Section 7.4]. If a mole of KI were to form as we have described it, an input of [419 + (−295)] = 124 kJ of energy would be required. Ionic compounds do form, though, and often in vigorous reactions. In fact, the formation of ionic bonds is highly *exothermic* and more than compensates for the energy input required to transfer electrons from metal atoms to nonmetal atoms. We can quantify the energy change associated with the formation of ionic bonds with *lattice energy*.

Lattice Energy

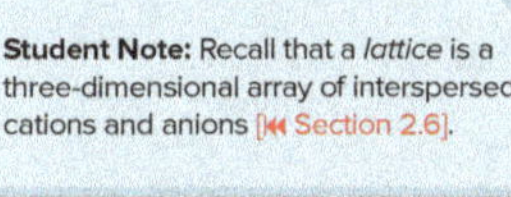

Student Note: Recall that a *lattice* is a three-dimensional array of interspersed cations and anions [⏮ Section 2.6].

Lattice energy is the amount of energy required to convert a mole of ionic solid to its constituent ions in the gas phase. For example, the lattice energy of potassium iodide is 632 kJ/mol. Thus, it takes 632 kJ of energy to convert 1 mole of KI(*s*) to 1 mole each of $K^+(g)$ and $I^-(g)$.

$$KI(s) \longrightarrow K^+(g) + I^-(g) \qquad \Delta H = 632 \text{ kJ/mol}$$

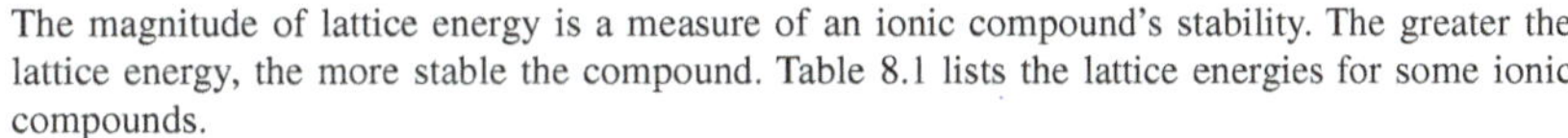

The magnitude of lattice energy is a measure of an ionic compound's stability. The greater the lattice energy, the more stable the compound. Table 8.1 lists the lattice energies for some ionic compounds.

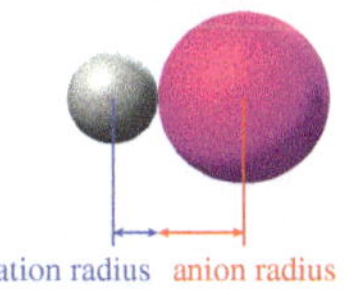

Lattice energy depends on the magnitudes of the charges and on the distance between them.

For example, LiI, NaI, and KI all have the same anion (I^-) and all have cations with the same charge (+1). The trend in their lattice energies (LiI > NaI > KI) can be explained on the

TABLE 8.1 Lattice Energies of Selected Ionic Compounds

Compound	Lattice Energy (kJ/mol)	Melting Point (°C)	Compound	Lattice Energy (kJ/mol)	Melting Point (°C)
LiF	1017	845	KCl	699	772
LiCl	860	610	KBr	689	735
LiBr	787	550	KI	632	680
LiI	732	450	$MgCl_2$	2527	714
NaCl	787	801	Na_2O	2570	Sub*
NaBr	736	750	MgO	3890	2800
NaI	686	662			

*Na_2O sublimes at 1275°C.

basis of ionic radius of the cations (Figure 8.2). The radii of alkali metal ions increase as we move down a group in the periodic table ($r_{Li^+} < r_{Na^+} < r_{K^+}$) [⏮ Section 7.6]. Knowing the radius of each ion, we can use Coulomb's law to compare the attractive forces between the ions in these three compounds (Figure 8.3).

LiI, with the smallest distance between ions, has the strongest attractive forces. It should therefore have the largest lattice energy. KI, with the largest distance between ions, has the weakest attractive forces, and should have the smallest lattice energy. NaI has an intermediate distance between ions and should have an intermediate lattice energy. Thus, Coulomb's law correctly predicts the relative magnitudes of the lattice energies of LiI, NaI, and KI.

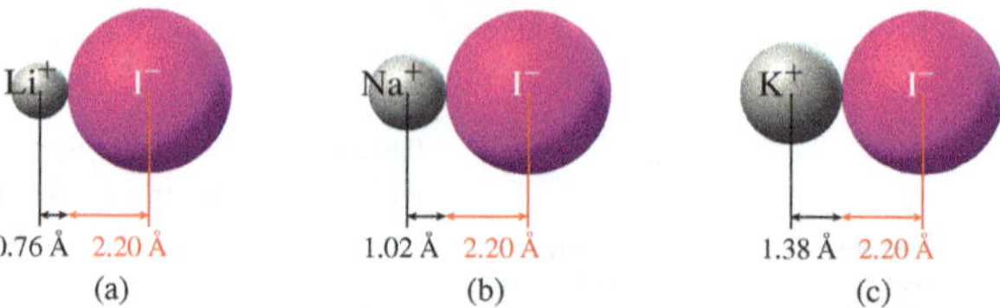

Figure 8.2 Ionic compounds LiI, NaI, and KI all contain the same anion, I^-. In this series of compounds, the relative lattice energies depend on cation radii.

Now consider the compounds LiF and MgO. With the distances between ions roughly equal (0.76 + 1.33 = 2.09 Å for LiF versus 0.72 + 1.40 = 2.12 Å for MgO) and the magnitude of each charge increased by a factor of 2 in MgO as compared to LiF, the lattice energy of MgO is roughly four times as large as that of LiF (Figure 8.4).

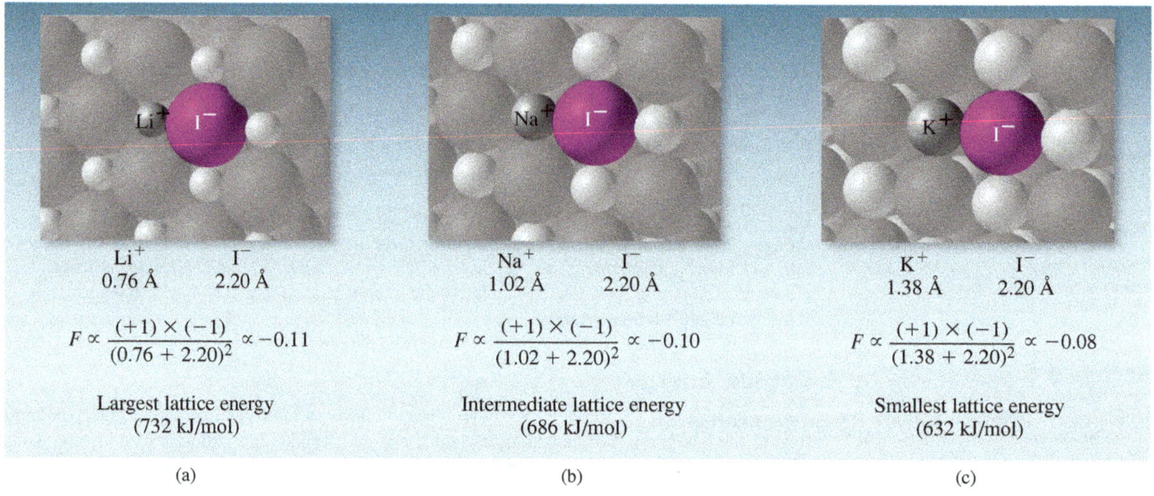

(a) (b) (c)

Figure 8.3 Relative magnitudes of lattice energy of alkali iodides calculated with Coulomb's law. (a) Li^+ has the smallest ionic radius of the alkali metal cations, making its lattice energy the highest of these three. (b) The lattice energy of NaI is slightly lower because of the larger ionic radius of Na^+. (c) KI has the smallest lattice energy of these three because it contains the cation with the largest ionic radius.

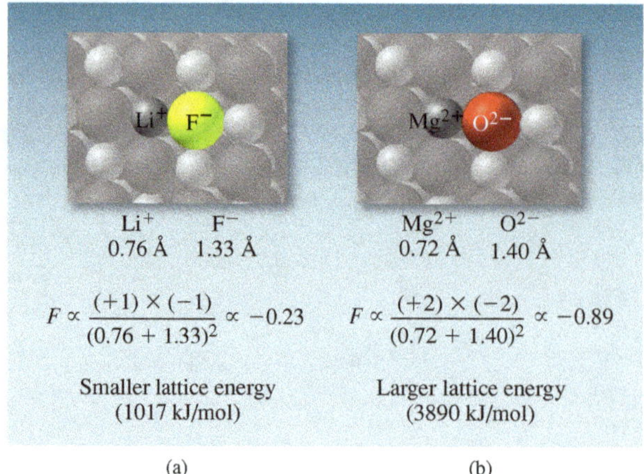

(a) (b)

Figure 8.4 (a) Distance between charges is the sum of cation and anion radii. (b) Relative lattice energies of LiF and MgO, calculated with Coulomb's law.

Sample Problem 8.2 shows how to use ionic radii and Coulomb's law to compare lattice energies for ionic compounds.

SAMPLE PROBLEM 8.2

Arrange MgO, CaO, and SrO in order of increasing lattice energy.

Strategy Consider the charges on the ions and the distances between them. Apply Coulomb's law to determine the relative lattice energies.

Setup MgO, CaO, and SrO all contain the same anion (O^{2-}), and all contain cations with the same charge (+2). In this case, then, the distance between ions will determine the relative lattice energies. Recall that lattice energy increases as the distance between ions decreases (because the force between oppositely charged particles increases as the distance between them decreases). Because all three compounds contain the same anion, we need only consider the radii of the cations when determining the distance between ions. From Figure 7.12, the ionic radii are 0.72 Å (Mg^{2+}), 1.00 Å (Ca^{2+}), and 1.18 Å (Sr^{2+}).

Solution MgO has the smallest distance between ions, whereas SrO has the largest distance between ions. Therefore, in order of increasing lattice energy: SrO < CaO < MgO.

THINK ABOUT IT

Mg, Ca, and Sr are all Group 2A metals, so we could have predicted this result without knowing their radii. Recall that ionic radii increase as we move down a column in the periodic table, and charges that are farther apart are more easily separated (meaning the lattice energy will be smaller). The lattice energies of SrO, CaO, and MgO are 3217, 3414, and 3890 kJ/mol, respectively.

Practice Problem ATTEMPT Determine which compound has the larger lattice energy: $MgCl_2$ or $SrCl_2$.

Practice Problem BUILD Arrange the compounds NaF, MgO, and AlN in order of increasing lattice energy.

Practice Problem CONCEPTUALIZE Common ions of four hypothetical elements are shown along with their radii in nanometers. Arrange all the binary ionic compounds that could form from these ions in order of increasing lattice energy.

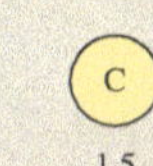

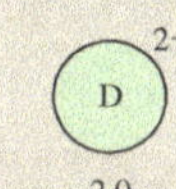

Although lattice energy is a useful measure of an ionic compound's stability, it is *not* a quantity that we can measure directly. Instead, we use various thermodynamic quantities that can be measured, and calculate lattice energy using Hess's law [⏮ Section 5.5].

The Born-Haber Cycle

Animation
Figure 8.6, Born-Haber Cycle.

We have described the formation of an ionic compound as though it happens when gaseous ions coalesce into a solid. In fact, the reactions that produce ionic solids generally do not occur this way. Figure 8.5 illustrates the formation of sodium chloride (NaCl) from its constituent elements.

We can imagine the reaction of Na(*s*) and $Cl_2(g)$ to form NaCl(*s*) as taking place in a series of steps for which most of the energy changes can be measured directly (see Table 8.2). Using these energy changes and the enthalpy of formation for NaCl, we can calculate the lattice energy using Hess's law. This method of determining the lattice energy is known as the ***Born-Haber cycle.***

The net reaction resulting from the series of steps in Table 8.2 is

$$Na(s) + \tfrac{1}{2}Cl_2(g) \longrightarrow Na^+(g) + Cl^-(g)$$

The final step in the formation of NaCl(*g*) would be the coalescence of $Na^+(g) + Cl^-(g)$. This is the step for which we cannot measure the energy change directly. However, we *can* measure the standard heat of formation of NaCl(*s*). (It is tabulated in Appendix 2 as −410.9 kJ/mol.) Although the formation of NaCl(*s*) from its constituent elements is not actually a step in our imaginary process, knowing its value enables us to calculate the lattice energy of NaCl. Figure 8.6 illustrates how this is done using all of these thermodynamic data and Hess's law.

Sample Problem 8.3 shows how to use the Born-Haber cycle to calculate the lattice energy.

TABLE 8.2 Hypothetical Steps in the Formation of $Na^+(g)$ and $Cl^-(g)$ from Na(*s*) and $Cl_2(g)$

Chemical Equation	Energy Change (kJ/mol)
$Na(s) \longrightarrow Na(g)$	107.7*
$\tfrac{1}{2}Cl_2(g) \longrightarrow Cl(g)$	121.7†
$Na(g) \longrightarrow Na^+(g) + e^-$	495.9‡
$Cl(g) + e^- \longrightarrow Cl^-(g)$	−349§

*Standard heat of formation (ΔH_f°) of Na(*g*) from Appendix 2.
†Standard heat of formation (ΔH_f°) of Cl(*g*) from Appendix 2.
‡First ionization energy (IE_1) of Na from Figure 7.8.
§ΔH° for this process is negative. (Recall that by definition, *EA* is the amount of energy *released* [⏮ Section 7.4]. This ΔH is equal to $-EA$.)

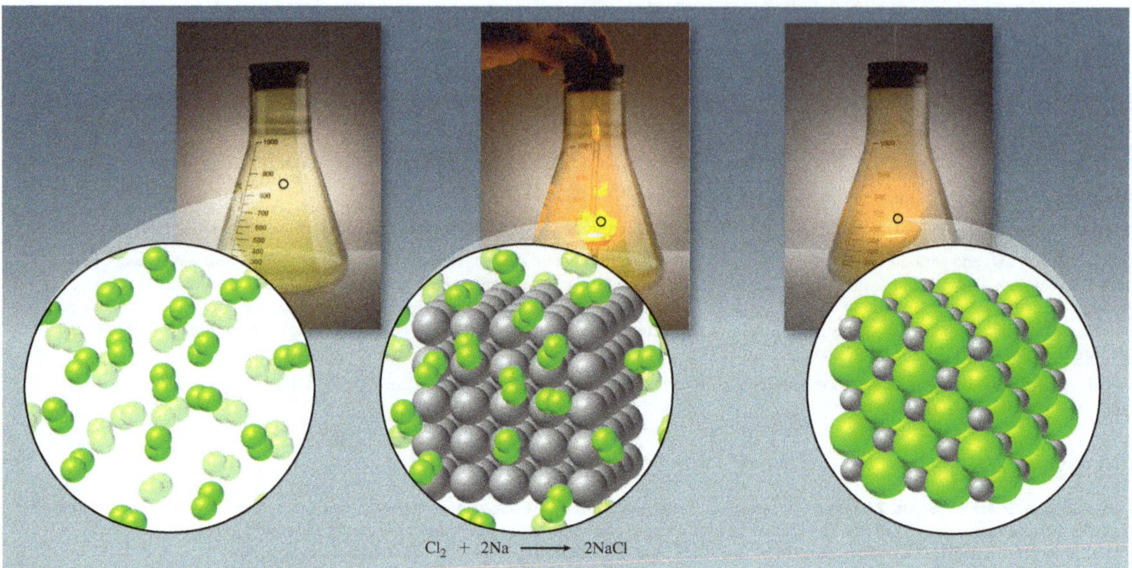

Figure 8.5 Sodium metal and chlorine gas combine to produce sodium chloride in a highly exothermic reaction.

SAMPLE PROBLEM 8.3

Using data from Figures 7.8 and 7.10 and Appendix 2, calculate the lattice energy of cesium chloride (CsCl).

Strategy Using Figure 8.6 as a guide, combine the pertinent thermodynamic data and use Hess's law to calculate the lattice energy.

Setup From Figure 7.8, $IE_1(\text{Cs}) = 376$ kJ/mol. From Figure 7.10, $EA_1(\text{Cl}) = 349.0$ kJ/mol. From Appendix 2, $\Delta H_f^\circ[\text{Cs}(g)] = 76.50$ kJ/mol, $\Delta H_f^\circ[\text{Cl}(g)] = 121.7$ kJ/mol, and $\Delta H_f^\circ[\text{CsCl}(s)] = -442.8$ kJ/mol. Because we are interested in magnitudes only, we can use the absolute values of the thermodynamic data. And, because only the standard heat of formation of CsCl(s) is a negative number, it is the only one for which the sign changes.

Solution

$$\{\Delta H_f^\circ[\text{Cs}(g)] + \Delta H_f^\circ[\text{Cl}(g)] + IE_1(\text{Cs}) + |\Delta H_f^\circ[\text{CsCl}(s)]|\} - EA_1(\text{Cl}) = \text{lattice energy}$$

$$= (76.50 \text{ kJ/mol} + 121.7 \text{ kJ/mol} + 376 \text{ kJ/mol} + 442.8 \text{ kJ/mol}) - 349.0 \text{ kJ/mol}$$

$$= 668 \text{ kJ/mol}$$

THINK ABOUT IT

Compare this value to that for NaCl in Figure 8.6 (787 kJ/mol). Both compounds contain the same anion (Cl^-) and both have cations with the same charge (+1), so the relative sizes of the cations will determine the relative strengths of their lattice energies. Because Cs^+ is larger than Na^+, the lattice energy of CsCl is smaller than that of NaCl.

Practice Problem ATTEMPT Using data from Figures 7.8 and 7.10 and Appendix 2, calculate the lattice energy of rubidium iodide (RbI).

Practice Problem BUILD The lattice energy of MgO is 3890 kJ/mol, and the second ionization energy (IE_2) of Mg is 1450.6 kJ/mol. Using these data, as well as data from Figures 7.8 and 7.10 and Appendix 2, determine the second electron affinity for oxygen, EA_2(O).

Practice Problem CONCEPTUALIZE Five points (A through E) lie along a line. The known distances between points are given. Determine the distance between points A and C.

A — B — C — D — E

AB = 5.05 cm

DE = 4.65 cm

BD = 7.65 cm

CE = 6.27 cm

Figure 8.6
Born-Haber Cycle

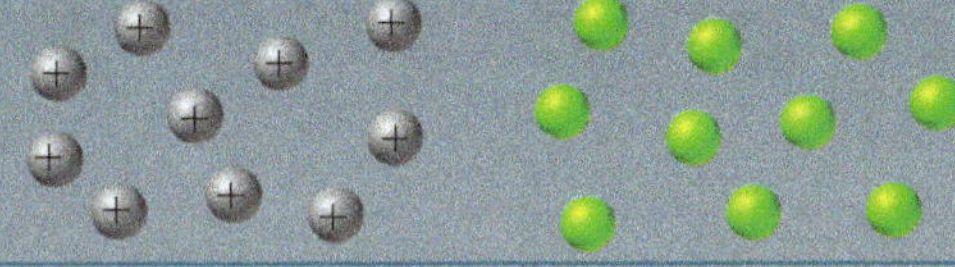

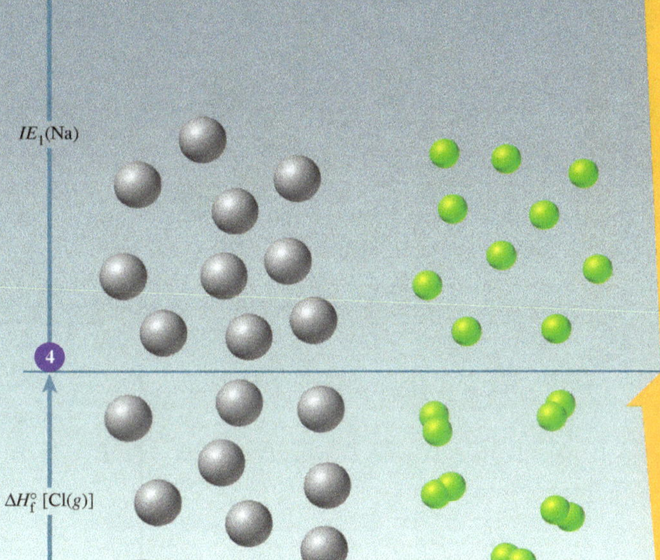

Step 4

The first ionization energy of sodium, IE_1 (Na), gives the amount of energy required to convert 1 mole of Na(g) to 1 mole of Na^+(g):

$$Na(g) \longrightarrow Na^+(g) + e^-$$

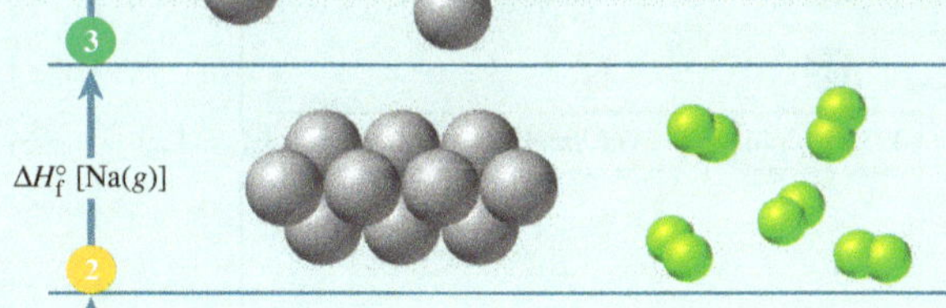

Step 3

The tabulated value of ΔH_f° for Cl(g) gives the amount of energy needed to convert $\frac{1}{2}$ mole of Cl_2(g) to 1 mole of Cl(g):

$$\tfrac{1}{2}Cl_2(g) \longrightarrow Cl(g)$$

Step 2

The tabulated value of ΔH_f° for Na(g) gives the amount of energy needed to convert 1 mole of Na(s) to 1 mole of Na(g):

$$Na(s) \longrightarrow Na(g)$$

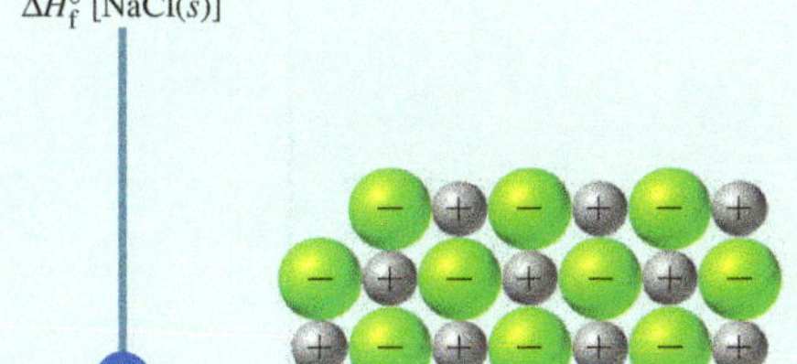

Step 1

The tabulated value of ΔH_f° for NaCl(s) gives us the energy produced when 1 mole of Na and $\frac{1}{2}$ mole of Cl_2 combine to form 1 mole of NaCl:

$$Na(s) + \tfrac{1}{2}Cl_2(g) \longrightarrow NaCl(s)$$

Step 1 in the Born-Haber cycle involves converting 1 mole of NaCl into 1 mole of Na and $\frac{1}{2}$ mole of Cl_2 (the reverse of the ΔH_f° reaction):

$$NaCl(s) \longrightarrow Na(s) + \tfrac{1}{2}Cl_2(g)$$

Therefore, ΔH for step 1 is $-\Delta H_f^\circ$ [NaCl(s)]

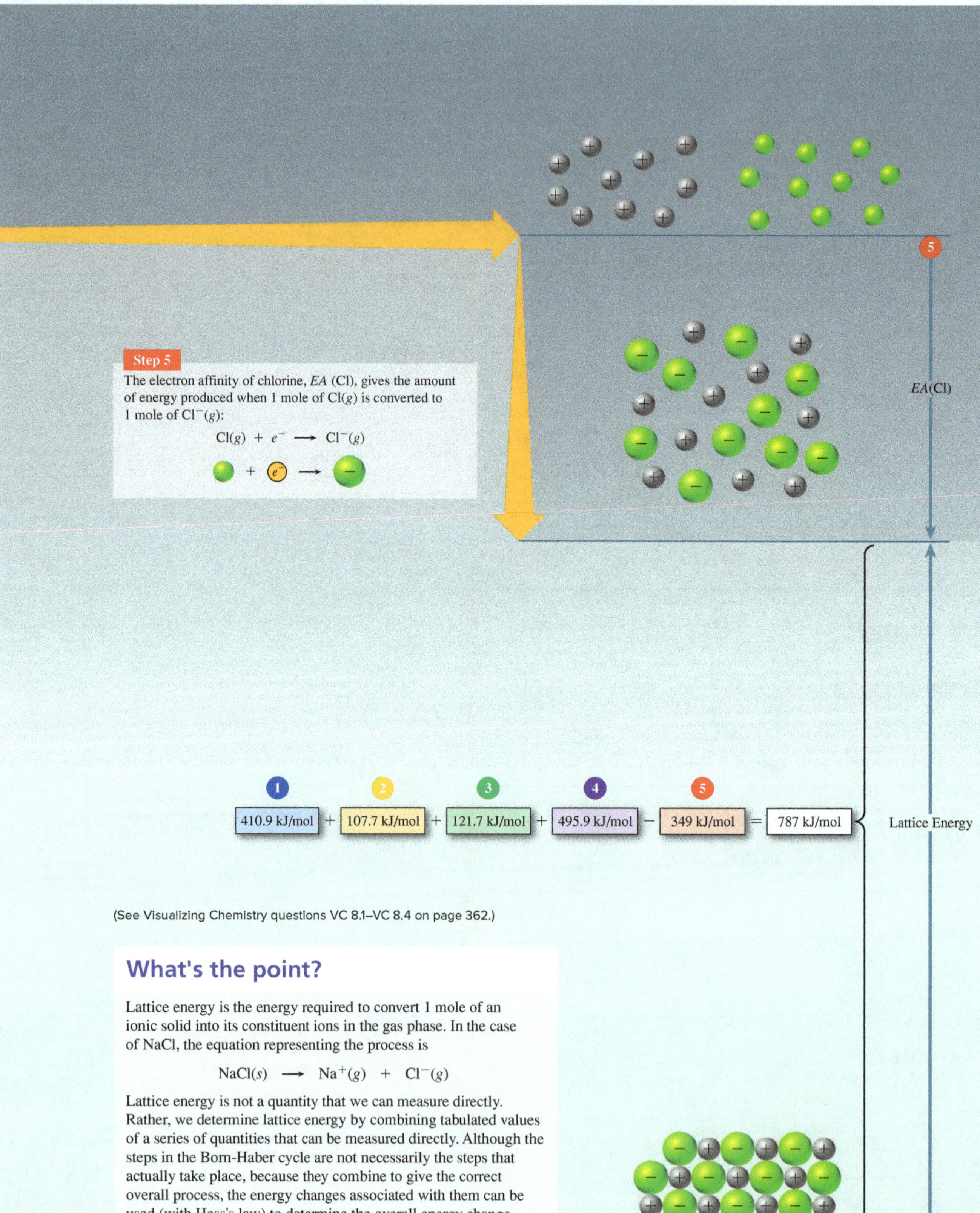

(See Visualizing Chemistry questions VC 8.1–VC 8.4 on page 362.)

What's the point?

Lattice energy is the energy required to convert 1 mole of an ionic solid into its constituent ions in the gas phase. In the case of NaCl, the equation representing the process is

$$NaCl(s) \longrightarrow Na^+(g) + Cl^-(g)$$

Lattice energy is not a quantity that we can measure directly. Rather, we determine lattice energy by combining tabulated values of a series of quantities that can be measured directly. Although the steps in the Born-Haber cycle are not necessarily the steps that actually take place, because they combine to give the correct overall process, the energy changes associated with them can be used (with Hess's law) to determine the overall energy change.

CHECKPOINT – SECTION 8.2 Ionic Bonding

8.2.1 Will the lattice energy of KF be larger or smaller than that of LiF, larger or smaller than that of KCl, and larger or smaller than that of KI?

a) larger, larger, and smaller

b) smaller, larger, and smaller

c) smaller, larger, and larger

d) smaller, smaller, and smaller

e) larger, smaller, and larger

8.2.2 Using the following data, calculate the lattice energy of KF: $\Delta H_f^\circ[K(g)] = 89.99$ kJ/mol, $\Delta H_f^\circ[F(g)] = 80.0$ kJ/mol, $IE_1(K) = 418.8$ kJ/mol, $\Delta H_f^\circ[KF(s)] = -547$ kJ/mol, and $EA_1(F) = 328$ kJ/mol.

a) 808 kJ/mol

b) −286 kJ/mol

c) 261 kJ/mol

d) 1355 kJ/mol

e) −261 kJ/mol

8.2.3 Lattice energies are graphed for three series of compounds in which the ion charges are +2, −2; +2, −1; and +1, −1. The ions in each series of compounds are separated by different distances. Identify the series.

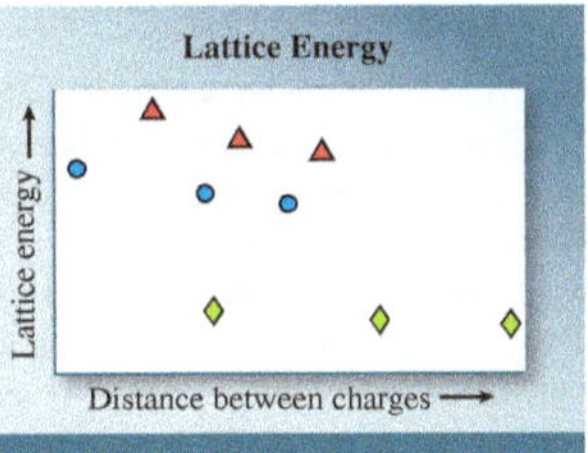

a) red +2, −2; blue +1, −1; green +2, −1

b) red +2, −2; blue +2, −1; green +1, −1

c) red +2, −1; blue +2, −2; green +1, −1

d) red +1, −1; blue +2, −1; green +2, −2

e) red +2, −1; blue +1, −1; green +2, −2

8.3 Covalent Bonding

We learned in Section 8.2 that ionic compounds tend to form between metals and nonmetals when electrons are transferred from an element with a low ionization energy (the metal) to one with a high electron affinity (the nonmetal). When compounds form between elements with more similar properties, electrons are not transferred from one element to another but instead are shared to give each atom a noble gas electron configuration. It was Gilbert Lewis who first suggested that a chemical bond involves atoms sharing electrons, and this approach is known as the ***Lewis theory of bonding.***

Lewis theory depicts the formation of the bond in H_2 as

$$\mathrm{H\cdot + \cdot H \longrightarrow H{:}H}$$

In essence, two H atoms move close enough to each other to *share* the electron pair. Although there are still two atoms and just two electrons, this arrangement allows each H atom to "count" both electrons as its own and to "feel" as though it has the noble gas electron configuration of helium. This type of arrangement, where two atoms share a pair of electrons, is known as ***covalent bonding,*** and the shared pair of electrons constitutes the ***covalent bond.*** (For the sake of simplicity, the shared pair of electrons can be represented by a dash, rather than by two dots: H—H.) In a covalent bond, each electron in a shared pair is attracted to the nuclei of both atoms. It is this attraction that holds the two atoms together.

Student Note: For nearly all elements, achieving a noble gas electron configuration results in eight electrons around each atom—hence the name *octet* rule. For H, the octet rule dictates that it have two electrons, giving it the electron configuration of the noble gas He.

Lewis summarized much of his theory of chemical bonding with the octet rule. According to the ***octet rule,*** atoms will lose, gain, or share electrons to achieve a noble gas electron configuration. This rule enables us to predict many of the formulas for compounds consisting of specific elements. The octet rule holds for nearly all the compounds made up of second-period elements and is therefore especially important in the study of organic compounds, which contain mostly C, N, and O atoms.

As with ionic bonding, covalent bonding of many-electron atoms involves only the valence electrons. Consider the fluorine molecule (F_2). The electron configuration of F is $1s^22s^22p^5$. The $1s$ electrons are low in energy and stay near the nucleus most of the time, so they do not participate in bond formation. Thus, each F atom has seven valence electrons (the two $2s$ and five $2p$ electrons). According to Figure 8.1, there is only one unpaired electron on F, so the formation of the F_2 molecule can be represented as follows:

$$\mathrm{:\ddot{\underset{..}{F}}\cdot + \cdot\ddot{\underset{..}{F}}: \longrightarrow :\ddot{\underset{..}{F}}{:}\ddot{\underset{..}{F}}: \text{ or } :\ddot{\underset{..}{F}}{-}\ddot{\underset{..}{F}}:}$$

Only two valence electrons participate in the bond that forms F_2. The other, nonbonding electrons, are called ***lone pairs***—pairs of valence electrons that are not involved in covalent bond formation. Thus, each F in F_2 has three lone pairs of electrons.

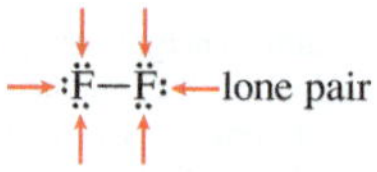

:F̤̈—F̤̈: ← lone pair

Lewis Structures

The structures used to represent molecules held together by covalent bonds, such as H_2 and F_2, are called *Lewis structures.* A ***Lewis structure*** is a representation of covalent bonding in which shared electron pairs are shown either as dashes or as pairs of dots between two atoms, and lone pairs are shown as pairs of dots on individual atoms. Only valence electrons are shown in a Lewis structure.

Student Note: Lewis structures are also referred to as Lewis dot structures. In this book, we will use the term *Lewis structure* to avoid confusion with the term *Lewis dot symbol.*

To draw the Lewis structure of the water molecule, recall that the Lewis dot symbol for oxygen has two unpaired dots (Figure 8.1), meaning that it has two unpaired electrons and can form two bonds. Because hydrogen has only one electron, it can form only one covalent bond. Thus, the Lewis structure for water is

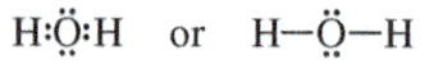

H:Ö̤:H or H—Ö̤—H

In this case, the O atom has two lone pairs. The hydrogen atom has no lone pairs because its only valence electron is used to form a covalent bond.

In the F_2 and H_2O molecules, the F, H, and O atoms each achieve a stable noble gas configuration by sharing electrons, thus illustrating the octet rule:

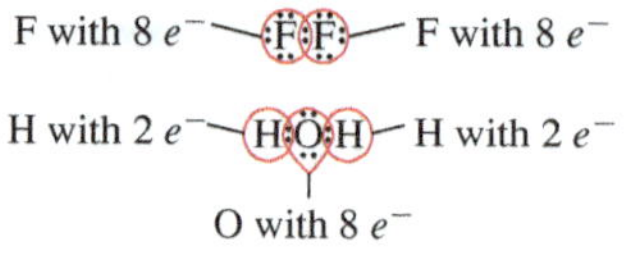

F with 8 e^- — :F̤̈:F̤̈: — F with 8 e^-

H with 2 e^- — H:Ö̤:H — H with 2 e^-

O with 8 e^-

The octet rule works best for elements in the second period of the periodic table. These elements have only $2s$ and $2p$ valence subshells, which can hold a total of eight electrons. When an atom of one of these elements forms a covalent compound, it can attain the noble gas electron configuration [Ne] by sharing electrons with other atoms in the same compound. In Section 8.8, we will discuss some important exceptions to the octet rule.

Multiple Bonds

Atoms can form several different types of covalent bonds, such as single bonds and multiple bonds. In a ***single bond,*** two atoms are held together by one electron pair. ***Multiple bonds*** form, on the other hand, when two atoms share two or more pairs of electrons. A multiple bond in which the atoms share two pairs of electrons is called a ***double bond.*** Double bonds are found in molecules such as carbon dioxide (CO_2) and ethylene (C_2H_4):

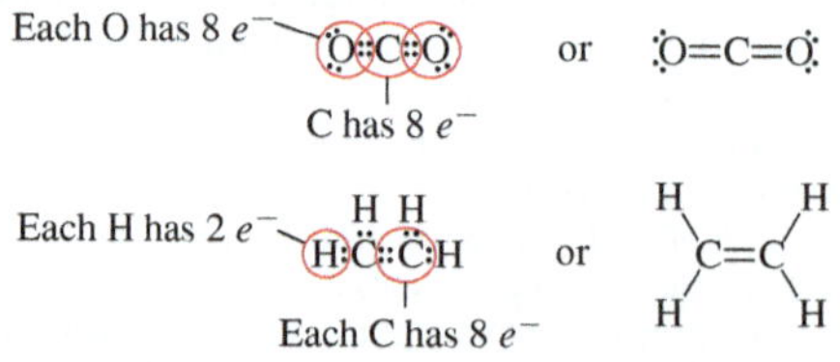

Each O has 8 e^- — Ö::C::Ö or Ö=C=Ö

C has 8 e^-

Each H has 2 e^- — H:C̈::C̈:H (with H H above) or H₂C=CH₂

Each C has 8 e^-

A ***triple bond*** arises when two atoms share three pairs of electrons, as in molecules such as nitrogen (N_2) and acetylene (C_2H_2):

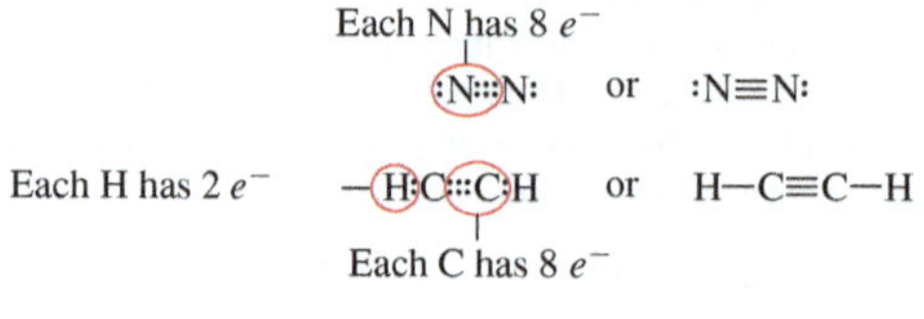

Each N has 8 e^-

:N⋮⋮N: or :N≡N:

Each H has 2 e^- — H:C⋮⋮C:H or H—C≡C—H

Each C has 8 e^-

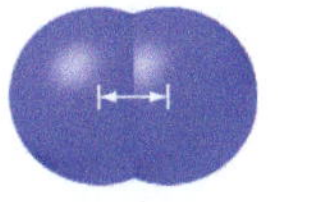

N_2
Bond length 1.10 Å

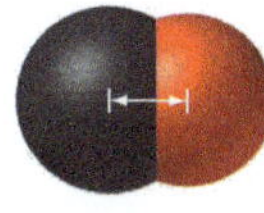

CO
Bond length 1.13 Å

Figure 8.7 Bond length is the distance between the nuclei of two bonded atoms.

TABLE 8.3 Average Bond Lengths of Some Common Single, Double, and Triple Bonds

Bond Type	Bond Length (pm)
C–H	107
O–H	96
C–O	143
C=O	121
C≡O	113
C–C	154
C=C	133
C≡C	120
C–N	143
C=N	138
C≡N	116
N–N	147
N=N	124
N≡N	110
N–O	136
N=O	122
O–O	148
O=O	121

TABLE 8.4 Comparison of Some Properties of an Ionic Compound (NaCl) and a Covalent Compound (CCl_4)

Property	NaCl	CCl_4
Appearance	White solid	Colorless liquid
Melting point (°C)	801	−23
Molar heat of fusion* (kJ/mol)	30.2	2.5
Boiling point (°C)	1413	76.5
Molar heat of vaporization* (kJ/mol)	600	30
Density (g/cm^3)	2.17	1.59
Solubility in water	High	Very low
Electrical conductivity		
Solid	Poor	Poor
Liquid	Good	Poor
Aqueous	Good	Poor

*The *molar heat of fusion* and *molar heat of vaporization* are the amounts of heat needed to melt 1 mole of the solid and to vaporize 1 mole of the liquid, respectively.

In ethylene and acetylene, all the valence electrons are used in bonding; there are no lone pairs on the carbon atoms. In fact, with the important exception of carbon monoxide (CO), most stable molecules containing carbon do not have lone pairs on the carbon atoms.

Multiple bonds are shorter than single bonds. *Bond length* is defined as the distance between the nuclei of two covalently bonded atoms in a molecule (Figure 8.7). Table 8.3 lists some experimentally determined bond lengths. For a given pair of atoms, such as carbon and nitrogen, triple bonds are shorter than double bonds, and double bonds are shorter than single bonds. The shorter multiple bonds are also stronger than single bonds, as we will see in Section 8.9.

Comparison of Ionic and Covalent Compounds

Ionic and covalent compounds differ markedly in their general physical properties because of differences in the nature of their bonds. There are two types of attractive forces in covalent compounds, the *intramolecular* bonding force that holds the atoms together in a molecule, and the *intermolecular* forces between molecules. Bond enthalpy, discussed in Section 8.9, can be used to quantify the intramolecular bonding force. Intermolecular forces are usually quite weak compared to the forces holding atoms together within a molecule, so molecules of a covalent compound are not held together tightly. As a result, covalent compounds are usually gases, liquids, or low-melting solids.

On the other hand, the electrostatic forces holding ions together in an ionic compound are usually very strong, so ionic compounds are solids at room temperature and have high melting points. Many ionic compounds are soluble in water, and the resulting aqueous solutions conduct electricity because the compounds are strong electrolytes. Most covalent compounds are insoluble in water, or if they do dissolve, their aqueous solutions generally do not conduct electricity, because the compounds are nonelectrolytes. Molten ionic compounds conduct electricity because they contain mobile cations and anions; liquid or molten covalent compounds do *not* conduct electricity because no ions are present. Table 8.4 compares some of the properties of a typical ionic compound, sodium chloride (NaCl), with those of a covalent compound, carbon tetrachloride (CCl_4).

8.4 Electronegativity and Polarity

So far, we have described chemical bonds as either *ionic*, when they occur between a metal and a nonmetal, or *covalent*, when they occur between nonmetals. In fact, ionic and covalent bonds are simply the extremes in a spectrum of bonding. Bonds that fall between these two extremes are ***polar***, meaning that electrons are shared but are not shared equally. Such bonds are referred to as ***polar covalent bonds.*** The following shows a comparison of the different types of bonds, where M and X represent two different elements:

M:X	$M^{\delta+}X^{\delta-}$	$M^{+}X^{-}$
Pure covalent bond	**Polar covalent bond**	**Ionic bond**
Neutral atoms held together by *equally* shared electrons	Partially charged atoms held together by *unequally* shared electrons	Oppositely charged ions held together by electrostatic attraction

To illustrate the spectrum of bonding, let's consider three substances: H_2, HF, and NaF. In the H_2 molecule, where the two bonding atoms are identical, the electrons are shared equally. That is, the electrons in the covalent bond spend roughly the same amount of time in the vicinity of each H atom. In the HF molecule, on the other hand, where the two bonding atoms are different, the electrons are *not* shared equally. They spend more time in the vicinity of the F atom than in the vicinity of the H atom. (The δ symbol is used to denote partial charges on the atoms.) In NaF, the electrons are not shared at all but rather are transferred from sodium to fluorine.

Student Note: In fact, the transfer of electrons in NaF is *nearly* complete. Even in an ionic bond, the electrons in question spend a small amount of time near the cation.

One way to visualize the distribution of electrons in species such as H_2, HF, and NaF is to use electrostatic potential models (Figure 8.8). These models show regions where electrons spend a lot of time in red, and regions where electrons spend very little time in blue. (Regions where electrons spend a moderate amount of time appear in green.)

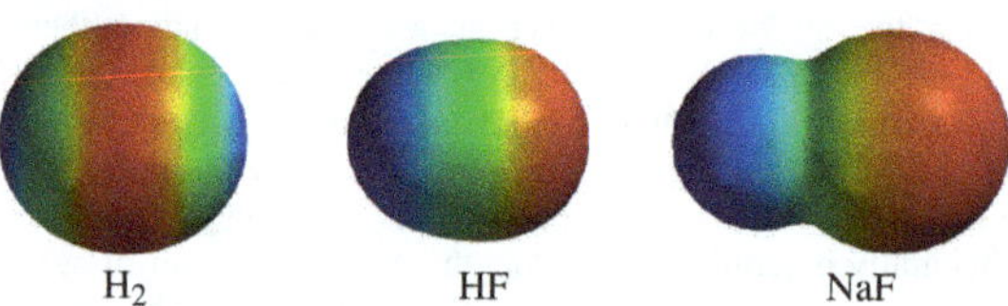

Figure 8.8 Electron density maps show the distribution of charge in a covalent species (H_2), a polar covalent species (HF), and an ionic species (NaF). The most electron-rich regions are red; the most electron-poor regions are blue.

Electronegativity

Electronegativity is the ability of an atom in a compound to draw electrons to itself. It determines where electrons in a compound spend most of their time. Elements with high electronegativity have a greater tendency to attract electrons than do elements with low electronegativity. Electronegativity is related to electron affinity and ionization energy. An atom such as fluorine, which has a high electron affinity (tends to accept electrons) and a high ionization energy (does not lose electrons easily), has a high electronegativity. Sodium, on the other hand, has a low electron affinity, a low ionization energy, and therefore a low electronegativity.

Animation
Periodic table—electronegativity.

Electronegativity is a relative concept, meaning that an element's electronegativity can be measured only in relation to the electronegativity of other elements. Linus Pauling[2] devised a method for calculating the relative electronegativities of most elements. These values are shown in Figure 8.9. In general, electronegativity increases from left to right across a period in the periodic table, as the metallic character of the elements decreases. Within each group, electronegativity decreases with increasing atomic number and increasing metallic character. The transition metals do not follow these trends. The most electronegative elements (the halogens, oxygen, nitrogen, and sulfur) are found in the upper right-hand corner of the periodic table, and the least electronegative elements (the alkali and alkaline earth metals) are clustered near the lower left-hand corner. These trends are readily apparent in the graph in Figure 8.10.

Electronegativity and electron affinity are related but distinct concepts. Both indicate the tendency of an atom to attract electrons. Electron affinity, however, refers to an isolated atom's ability to attract an additional electron in the gas phase, whereas electronegativity refers to the ability of an atom in a chemical bond (with another atom) to attract the shared electrons. Electron affinity, moreover, is an experimentally measurable quantity, whereas electronegativity is an estimated number that cannot be measured directly.

2. Linus Carl Pauling (1901–1994). American chemist. Regarded by many as the most influential chemist of the twentieth century. Pauling received the Nobel Prize in Chemistry in 1954 for his work on protein structure, and the Nobel Peace Prize in 1962 for his tireless campaign against the testing and proliferation of nuclear arms. He is the only person ever to have received two unshared Nobel Prizes.

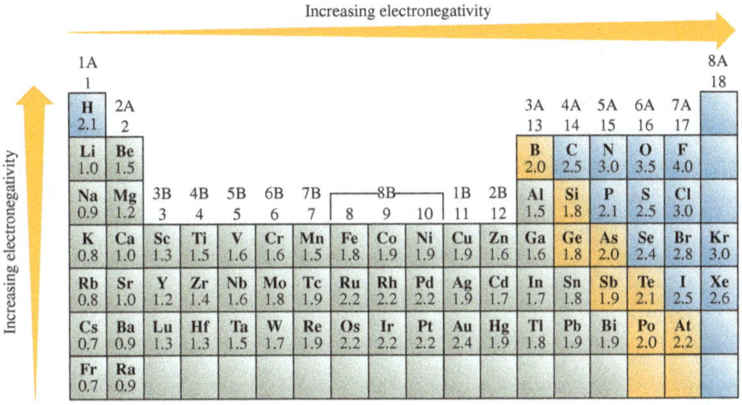

Figure 8.9 Electronegativities of common elements.

Increasing electronegativity (→ across; ↑ up)

1A 1	2A 2	3B 3	4B 4	5B 5	6B 6	7B 7	8B 8	8B 9	8B 10	1B 11	2B 12	3A 13	4A 14	5A 15	6A 16	7A 17	8A 18
H 2.1																	
Li 1.0	Be 1.5											B 2.0	C 2.5	N 3.0	O 3.5	F 4.0	
Na 0.9	Mg 1.2											Al 1.5	Si 1.8	P 2.1	S 2.5	Cl 3.0	
K 0.8	Ca 1.0	Sc 1.3	Ti 1.5	V 1.6	Cr 1.6	Mn 1.5	Fe 1.8	Co 1.9	Ni 1.9	Cu 1.9	Zn 1.6	Ga 1.6	Ge 1.8	As 2.0	Se 2.4	Br 2.8	Kr 3.0
Rb 0.8	Sr 1.0	Y 1.2	Zr 1.4	Nb 1.6	Mo 1.8	Tc 1.9	Ru 2.2	Rh 2.2	Pd 2.2	Ag 1.9	Cd 1.7	In 1.7	Sn 1.8	Sb 1.9	Te 2.1	I 2.5	Xe 2.6
Cs 0.7	Ba 0.9	Lu 1.3	Hf 1.3	Ta 1.5	W 1.7	Re 1.9	Os 2.2	Ir 2.2	Pt 2.2	Au 2.4	Hg 1.9	Tl 1.8	Pb 1.9	Bi 1.9	Po 2.0	At 2.2	
Fr 0.7	Ra 0.9																

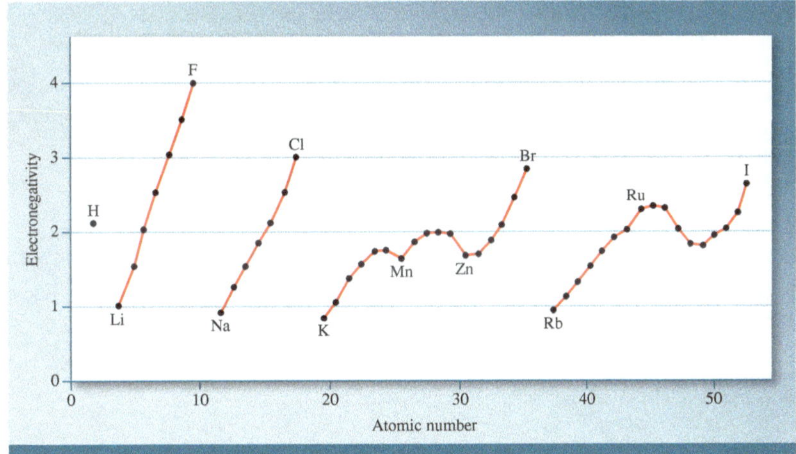

Figure 8.10 Variation of electronegativity with atomic number.

Atoms of elements with widely different electronegativities tend to form ionic compounds with each other because the atom of the less electronegative element gives up its electrons to the atom of the more electronegative element. Atoms of elements with comparable electronegativities tend to form polar covalent bonds, or simply polar bonds, with each other because the shift in electron density from one atom to the other is usually small. Only atoms of the same element, which have the same electronegativity, can be joined by a pure covalent bond.

There is no sharp distinction between *nonpolar covalent* and *polar covalent* or between *polar covalent* and *ionic*, but the following guidelines can help distinguish among them:

- A bond between atoms whose electronegativities differ by less than 0.5 is generally considered purely covalent or ***nonpolar.***
- A bond between atoms whose electronegativities differ by the range of 0.5 to 2.0 is generally considered *polar covalent.*
- A bond between atoms whose electronegativities differ by 2.0 or more is generally considered *ionic.*

Sometimes chemists describe bonds using the term *percent ionic character* [⏭ Page 341]. A purely ionic bond would have 100 percent ionic character (although no such bond is known). A purely covalent, nonpolar bond has 0 percent ionic character.

Sample Problem 8.4 shows how to use electronegativities to identify a chemical bond as nonpolar, polar, or ionic.

SAMPLE PROBLEM 8.4

Classify the following bonds as nonpolar, polar, or ionic: (a) the bond in ClF, (b) the bond in CsBr, and (c) the carbon-carbon double bond in C_2H_4.

Strategy Using the information in Figure 8.9, determine which bonds have identical, similar, and widely different electronegativities.

Setup Electronegativity values from Figure 8.9 are: Cl (3.0), F (4.0), Cs (0.7), Br (2.8), C (2.5).

Solution (a) The difference between the electronegativities of F and Cl is $4.0 - 3.0 = 1.0$, making the bond in ClF polar.

(b) In CsBr, the difference is $2.8 - 0.7 = 2.1$, making the bond ionic.

(c) In C_2H_4, the two atoms are identical. (Not only are they the same element, but each C atom is bonded to two H atoms.) The carbon-carbon double bond in C_2H_4 is nonpolar.

THINK ABOUT IT

By convention, the difference in electronegativity is always calculated by subtracting the smaller number from the larger one, so the result is always positive.

Practice Problem ATTEMPT Classify the following bonds as nonpolar, polar, or ionic: (a) the bonds in H_2S, (b) the H—O bonds in H_2O_2, and (c) the O—O bond in H_2O_2.

Practice Problem BUILD In order of increasing polarity, list the bonds between carbon and each of the Group 6A elements.

Practice Problem CONCEPTUALIZE Electrostatic potential maps are shown for HCl and LiH. Determine which diagram is which. (The H atom is shown on the left in both.)

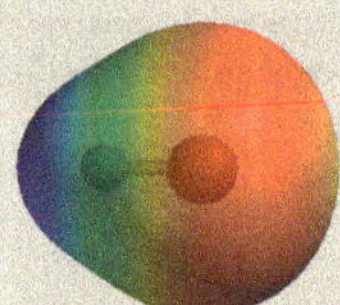

Dipole Moment, Partial Charges, and Percent Ionic Character

The shift of electron density in a polar bond is symbolized by placing a crossed arrow (a dipole arrow) above the Lewis structure to indicate the direction of the shift. For example,

$$\overset{\longmapsto}{\text{H—}\ddot{\text{F}}\text{:}}$$

The consequent charge separation can be represented as

$$\overset{\delta+}{\text{H}}\text{—}\overset{\delta-}{\ddot{\text{F}}\text{:}}$$

A quantitative measure of the polarity of a bond is its ***dipole moment (μ),*** which is calculated as the product of the charge (Q) and the distance (r) between the charges:

$$\mu = Q \times r \qquad \textbf{Equation 8.1}$$

The distance, r, between partial charges in a polar diatomic molecule is the bond length expressed in meters. Bond lengths are usually given in angstroms (Å) or picometers (pm), so it is generally necessary to convert to meters. For a diatomic molecule containing a polar bond to be electrically neutral, the partial positive and partial negative charges must have the same magnitude. Therefore, the Q term in Equation 8.1 refers to the magnitude of the partial charges and the calculated value of μ is always positive. Dipole moments are usually expressed in debye units (D), named for Peter Debye.[3] In terms of more familiar SI units,

$$1\ \text{D} = 3.336 \times 10^{-30}\ \text{C} \cdot \text{m}$$

where C is coulombs and m is meters. Table 8.5 lists several polar diatomic molecules, their bond lengths, and their experimentally measured dipole moments.

Sample Problem 8.5 shows how to use bond lengths and dipole moments to determine the magnitude of the partial charges in a polar diatomic molecule.

Student Note: We usually express the charge on an electron as −1. This refers to *units of electronic charge*. However, remember that the charge on an electron can also be expressed in *coulombs* [⏮ Section 2.2]. The conversion factor between the two is necessary to calculate dipole moments: $1e^- = 1.6022 \times 10^{-19}$ C.

3. Peter Joseph William Debye (1884–1966). American chemist and physicist of Dutch origin. Debye made many significant contributions to the study of molecular structure, polymer chemistry, X-ray analysis, and electrolyte solutions. He was awarded the Nobel Prize in Chemistry in 1936.

TABLE 8.5 Bond Lengths and Dipole Moments of the Hydrogen Halides

Molecule	Bond Length (Å)	Dipole Moment (D)
HF	0.92	1.82
HCl	1.27	1.08
HBr	1.41	0.82
HI	1.61	0.44

SAMPLE PROBLEM 8.5

Hydrofluoric acid [HF(*aq*)] has several important industrial applications, including the etching of glass and the manufacture of electronic components. Burns caused by hydrofluoric acid are unlike any other acid burns and present unique medical complications. HF solutions typically penetrate the skin and damage internal tissues, including bone, often with minimal surface damage. Less concentrated solutions actually can cause greater injury than more concentrated ones by penetrating more deeply before causing injury, thus delaying the onset of symptoms and preventing timely treatment. Determine the magnitude of the partial positive and partial negative charges in the HF molecule.

Strategy Rearrange Equation 8.1 to solve for Q. Convert the resulting charge in coulombs to charge in units of electronic charge.

Setup According to Table 8.5, $\mu = 1.82$ D and $r = 0.92$ Å for HF. The dipole moment must be converted from debye to C · m and the distance between the ions must be converted to meters.

$$\mu = 1.82\text{ D} \times \frac{3.336 \times 10^{-30}\text{ C}\cdot\text{m}}{1\text{ D}} = 6.07 \times 10^{-30}\text{ C}\cdot\text{m}$$

$$r = 0.92\text{ Å} \times \frac{1 \times 10^{-10}\text{ m}}{1\text{ Å}} = 9.2 \times 10^{-11}\text{ m}$$

Solution In coulombs:

$$Q = \frac{\mu}{r} = \frac{6.07 \times 10^{-30}\text{ C}\cdot\text{m}}{9.2 \times 10^{-11}\text{ m}} = 6.598 \times 10^{-20}\text{ C}$$

In units of electronic charge:

$$6.598 \times 10^{-20}\text{ C} \times \frac{1e^-}{1.6022 \times 10^{-19}\text{ C}} = 0.41e^-$$

Therefore, the partial charges in HF are +0.41 and −0.41 on H and F, respectively.

$$^{+0.41}\text{H—}\ddot{\underset{\cdot\cdot}{\text{F}}}\text{:}^{-0.41}$$

THINK ABOUT IT

Calculated partial charges should always be less than 1. If a "partial" charge were 1 or greater, it would indicate that at least one electron had been transferred from one atom to the other. Remember that polar bonds involve unequal sharing of electrons, not a complete transfer of electrons.

Practice Problem ATTEMPT Using data from Table 8.5, determine the magnitude of the partial charges in HBr.

Practice Problem BUILD Given that the partial charges on C and O in carbon monoxide are +0.020 and −0.020, respectively, calculate the dipole moment of CO. (The distance between the partial charges, r, is 113 pm.)

Practice Problem CONCEPTUALIZE Based on the answers to Sample Problem 8.5 and Practice Problem 8.5 A, estimate the magnitude of the partial charges in HCl.

Although the designations "covalent," "polar covalent," and "ionic" can be useful, sometimes chemists wish to describe and compare chemical bonds with more precision. For this purpose, we can use Equation 8.1 to calculate the dipole moment we would expect if the charges

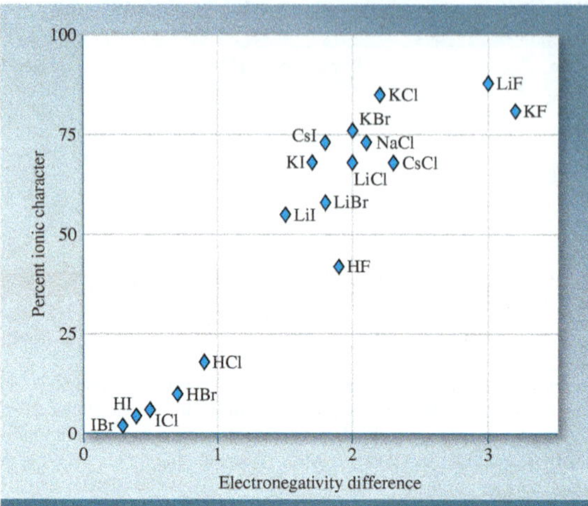

Figure 8.11 Relationship between percent ionic character and electronegativity difference.

on the atoms were discrete instead of partial; that is, if an electron had actually been transferred from one atom to the other. Comparing this calculated dipole moment with the measured value gives us a quantitative way to describe the nature of a bond using the term ***percent ionic character,*** which is defined as the ratio of observed μ to calculated μ, multiplied by 100.

$$\text{percent ionic character} = \frac{\mu(\text{observed})}{\mu(\text{calculated assuming discrete charges})} \times 100\% \qquad \textbf{Equation 8.2}$$

Figure 8.11 illustrates the relationship between percent ionic character and the electronegativity difference in a heteronuclear diatomic molecule.

Sample Problem 8.6 shows how to calculate percent ionic character using Equation 8.2.

SAMPLE PROBLEM 8.6

Using data from Table 8.5, calculate the percent ionic character of the bond in HI.

Strategy Use Equation 8.1 to calculate the dipole moment in HI assuming that the charges on H and I are +1 and −1, respectively; and use Equation 8.2 to calculate percent ionic character. The magnitude of the charges must be expressed as coulombs ($1\ e^- = 1.6022 \times 10^{-19}$ C); the bond length (r) must be expressed as meters ($1\ \text{Å} = 1 \times 10^{-10}$ m); and the calculated dipole moment should be expressed as debyes ($1\ \text{D} = 3.336 \times 10^{-30}\ \text{C} \cdot \text{m}$).

Setup From Table 8.5, the bond length in HI is 1.61Å (1.61×10^{-10} m) and the measured dipole moment of HI is 0.44 D.

Solution The dipole moment we would expect if the magnitude of charges were 1.6022×10^{-19} C is

$$\mu = Q \times r = (1.6022 \times 10^{-19}\ \text{C}) \times (1.61 \times 10^{-10}\ \text{m}) = 2.58 \times 10^{-29}\ \text{C} \cdot \text{m}$$

Converting to debyes gives

$$2.58 \times 10^{-29}\ \cancel{\text{C} \cdot \text{m}} \times \frac{1\ \text{D}}{3.336 \times 10^{-30}\ \cancel{\text{C} \cdot \text{m}}} = 7.73\ \text{D}$$

The percent ionic character of the H—I bond is

$$\frac{0.44\ \text{D}}{7.73\ \text{D}} \times 100\% = 5.7\%$$

(Continued on next page)

THINK ABOUT IT

A purely covalent bond (in a homonuclear diatomic molecule such as H_2) would have 0 percent ionic character. In theory, a purely ionic bond would be expected to have 100 percent ionic character, although no such bond is known.

Practice Problem ATTEMPT Using data from Table 8.5, calculate the percent ionic character of the bond in HF.

Practice Problem BUILD Using information from Figure 7.12, and given that the NaI bond has 59.7 percent ionic character, determine the measured dipole moment of NaI.

Practice Problem CONCEPTUALIZE One metal and three nonmetals are highlighted on the periodic table shown here. List the nonmetal elements in order of increasing ionic character of the bond each might form with the highlighted metal.

CHECKPOINT – SECTION 8.4 Electronegativity and Polarity

8.4.1 In which of the following molecules are the bonds *most* polar?

a) H_2Se
b) H_2O
c) CO_2
d) BCl_3
e) PCl_5

8.4.2 Using data from Table 8.5, calculate the magnitude of the partial charges in HI.

a) 0.39
b) 1.8
c) 0.057
d) 0.60
e) 0.15

8.4.3 Arrange molecules A through E in order of increasing percent ionic character.

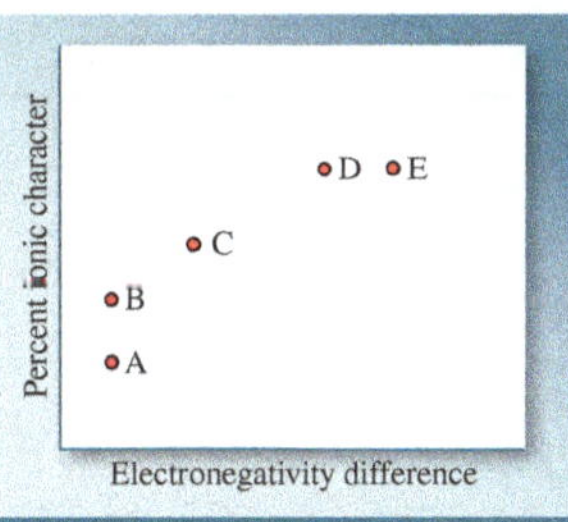

a) A < B < C < D < E
b) A = B < C < D < E
c) A = B < C < D = E
d) A < B < C < D = E
e) A < B = C = D < E

8.4.4 Using data from Table 8.5, calculate the percent ionic character of HCl.

a) 85.0 percent
b) 6.10 percent
c) 17.7 percent
d) 2.03 percent
e) 20.3 percent

8.5 Drawing Lewis Structures

Although the octet rule and Lewis structures alone do not present a complete picture of covalent bonding, they do help us account for some of the properties of molecules. In addition, Lewis structures provide a starting point for the bonding theories that we will examine in Chapter 9. It is crucial, therefore, that you learn a system for drawing correct Lewis structures for molecules and polyatomic ions. The basic steps, illustrated in Figure 8.12, are as follows:

1. From the molecular formula, draw the skeletal structure of the compound, using chemical symbols and placing bonded atoms next to one another. For simple compounds, this step is fairly easy. Often there will be a unique central atom surrounded by a group of other identical atoms. In general, the *least* electronegative atom will be the central atom. (H cannot be a central atom because it only forms *one* covalent bond.) Draw a single covalent bond (dash) between the central atom and each of the surrounding atoms. (For more complex compounds whose structures might not be obvious, you may need to be given information in addition to the molecular formula.)
2. Count the total number of valence electrons present. Remember that an element's *group number* (1A–8A) gives the number of valence electrons it contributes to the total number. For polyatomic ions, add electrons to the total number to account for negative charges; subtract electrons from the total number to account for positive charges.
3. For each bond (dash) in the skeletal structure, subtract two electrons from the total valence electrons (determined in step 2) to determine the number of remaining electrons.
4. Use the remaining electrons to complete the octets of the terminal atoms (those bonded to the central atom) by placing pairs of electrons on each atom. (Remember that an H atom only requires two electrons to complete its valence shell.) If there is more than one type of terminal atom, complete the octets of the most electronegative atoms first.
5. If any electrons remain after step 4, place them in pairs on the central atom.
6. If the central atom has fewer than eight electrons after completing steps 1 to 5, move one or more pairs from the terminal atoms to form multiple bonds between the central atom and the terminal atoms. (Unless the central atom is a Group 3A element.) Like Lewis dot symbols for atomic anions, Lewis structures for polyatomic anions are enclosed by square brackets.

Steps for Drawing Lewis Structures

Step	CH_4	CCl_4	H_2O	O_2	CN^-
1	H H—C—H H	Cl Cl—C—Cl Cl	H—O—H	O—O	C—N
2	8	32	8	12	10
3	8 − 8 = 0	32 − 8 = 24	8 − 4 = 4	12 − 2 = 10	10 − 2 = 8
4	H H—C—H H	:C̈l: :C̤̈l—C—C̤̈l: :C̤l:	H—O—H	:Ö̤—Ö̤:	:C—N̤̈:
5	—	—	H—Ö̤—H	—	—
6	—	—	—	Ö̤=Ö̤	[:C≡N:]$^-$

Student Note: To summarize the process of drawing Lewis structures:

1. Draw the skeletal structure.
2. Count the valence electrons.
3. Subtract two electrons for each bond.
4. Distribute the remaining electrons.
5. Complete the octets of all atoms.
6. Use multiple bonds if necessary.

Figure 8.12 Drawing Lewis structures.

Sample Problem 8.7 shows how to draw a Lewis structure.

SAMPLE PROBLEM 8.7

Draw the Lewis structure for carbon disulfide (CS_2).

Strategy Use the procedure described in steps 1 through 6 on page 343 for drawing Lewis structures.

Setup

Step 1: C and S have identical electronegativities. We will draw the skeletal structure with the unique atom, C, at the center.

S—C—S

Step 2: The total number of valence electrons is 16, six from each S atom and four from the C atom [2(6) + 4 = 16].
Step 3: Subtract four electrons to account for the bonds in the skeletal structure, leaving us 12 electrons to distribute.
Step 4: Distribute the 12 remaining electrons as three lone pairs on each S atom.

:S̤̈—C—S̤̈:

Step 5: There are no electrons remaining after step 4, so step 5 does not apply.
Step 6: To complete carbon's octet, use one lone pair from each S atom to make a double bond to the C atom.

Solution

:S̈=C=S̈:

THINK ABOUT IT

Counting the total number of valence electrons should be relatively simple to do, but it is often done hastily and is therefore a potential source of error in this type of problem. Remember that the number of valence electrons for each element is equal to the group number of that element.

Practice Problem ATTEMPT Draw the Lewis structure for NF_3.

Practice Problem BUILD Draw the Lewis structure for ClO_3^-.

Practice Problem CONCEPTUALIZE Of the three Lewis structures shown here, identify any that are not correct and specify what is wrong.

[:C=N:]⁻ (i) :O=C̤̈=O: (ii)

:Ö̤—S̈—Ö: (iii)

CHECKPOINT – SECTION 8.5 Drawing Lewis Structures

8.5.1 Identify the correct Lewis structure for formic acid (HCOOH).

a) H—C̤̈—Ö̤—Ö̤—H

b) H—Ö=C=Ö—H

c) H—Ö̤—C̤̈—Ö̤—H

d) H—C(=Ö:)—Ö̤—H

e) :Ö̤—C(—Ö:)(—H)—H

8.5.2 Identify the correct Lewis structure for hydrogen peroxide (H_2O_2).

a) H—Ö=Ö—H

b) H—O≡O—H

c) H—Ö̤—Ö̤—H

d) H=O=O=H

e) H—H—Ö̤—Ö̤:

8.6 Lewis Structures and Formal Charge

So far you have learned two different methods of electron "bookkeeping." In Chapter 4, you learned about oxidation numbers [⏮ Section 4.4], and in Section 8.4, you learned how to calculate partial charges. There is one additional commonly used method of electron bookkeeping—namely, ***formal charge,*** which can be used to determine the most plausible Lewis structures when more than one possibility exists for a compound. Formal charge is determined by comparing the number of electrons associated with an atom in a Lewis structure with the number of electrons that would be associated with the isolated atom. In an isolated atom, the number of electrons associated with the atom is simply the number of valence electrons. (As usual, we need not be concerned with the core electrons.)

Animation
Chemical Bonding—formal charge calculations.

To determine the number of electrons associated with an atom in a Lewis structure, keep in mind the following:

- All the atom's nonbonding electrons are associated with the atom.
- Half of the atom's bonding electrons are associated with the atom.

formal charge = valence electrons − associated electrons **Equation 8.3**

An atom's formal charge is calculated as follows: We can illustrate the concept of formal charge using the ozone molecule (O_3). Use the step-by-step method for drawing Lewis structures to draw the Lewis structure for ozone, and then determine the formal charge on each O atom by subtracting the number of associated electrons from the number of valence electrons.

$$2 \text{ unshared} + \frac{6 \text{ shared}}{2} = 5\,e^-$$

$$\ddot{\text{O}}{=}\ddot{\text{O}}{-}\ddot{\text{O}}\text{:}$$

$$4 \text{ unshared} + \frac{4 \text{ shared}}{2} = 6\,e^- \qquad 6 \text{ unshared} + \frac{2 \text{ shared}}{2} = 7\,e^-$$

Valence e^-	6	6	6
e^- associated with atom	6	5	7
Difference (formal charge)	0	+1	−1

$$\ddot{\text{O}}{=}\ddot{\text{O}}{-}\ddot{\text{O}}\text{:}$$

As with oxidation numbers, the sum of the formal charges must equal the overall charge on the species [⏮ Section 4.4]. Because O_3 is a molecule, its formal charges must sum to zero. For ions, the formal charges must sum to the overall charge on the ion.

Formal charges do *not* represent actual charges on atoms in a molecule. In the O_3 molecule, for example, there is no evidence that the central atom bears a net +1 charge or that one of the terminal atoms bears a −1 charge. Assigning formal charges to the atoms in the Lewis structure merely helps us keep track of the electrons involved in bonding in the molecule.

Sample Problem 8.8 on page 346 lets you practice determining formal charges.

Student Note: While you are new at determining formal charges, it may be helpful to draw Lewis structures with all dots, rather than dashes. This can make it easier to see how many electrons are associated with each atom.

Remember that for the purpose of counting associated electrons, those shared by two atoms are evenly split between them.

SAMPLE PROBLEM 8.8

The widespread use of fertilizers has resulted in the contamination of some groundwater with nitrates, which are potentially harmful. Nitrate toxicity is due primarily to its conversion in the body to nitrite (NO_2^-), which interferes with the ability of hemoglobin to transport oxygen. Determine the formal charges on each atom in the nitrate ion (NO_3^-).

Strategy Use steps 1 through 6 on page 343 for drawing Lewis stuctures to draw the Lewis structure of NO_3^-. For each atom, subtract the associated electrons from the valence electrons.

Setup

$$\left[:\ddot{O}-\overset{\overset{:\ddot{O}:}{|}}{N}=\ddot{O}: \right]^-$$

The N atom has five valence electrons and four associated electrons (one from each single bond and two from the double bond). Each singly bonded O atom has six valence electrons and seven associated electrons (six in three lone pairs and one from the single bond). The doubly bonded O atom has six valence electrons and six associated electrons (four in two lone pairs and two from the double bond).

Solution The formal charges are as follows: +1 (N atom), −1 (singly bonded O atoms), and 0 (doubly bonded O atom).

THINK ABOUT IT

The sum of formal charges (+1) + (−1) + (−1) + (0) = −1 is equal to the overall charge on the nitrate ion.

Practice Problem ATTEMPT Determine the formal charges on each atom in the carbonate ion (CO_3^{2-}).

Practice Problem BUILD Determine the formal charges and use them to determine the overall charge, if any, on the species represented by the following Lewis structure:

$$\left[:\ddot{O}-\overset{\overset{:\ddot{O}:}{|}}{\underset{\cdot\cdot}{S}}-\ddot{O}: \right]^?$$

Practice Problem CONCEPTUALIZE The hypothetical element A is shown here in three different partial Lewis structures. For each structure, determine what the formal charge is on A (a) if it is a member of Group 7A, (b) if it is a member of Group 5A, and (c) if it is a member of Group 3A.

$$-\ddot{\underset{\cdot\cdot}{A}}: \qquad -\ddot{\underset{\cdot\cdot}{A}}- \qquad -\overset{\cdot\cdot}{\underset{|}{A}}-$$

(i) (ii) (iii)

Sometimes, there is more than one possible skeletal arrangement of atoms for the Lewis structure for a given species. In such cases, we often can select the best skeletal arrangement by using formal charges and the following guidelines:

- For molecules, a Lewis structure in which all formal charges are zero is preferred to one in which there are nonzero formal charges.
- Lewis structures with small formal charges (0 and ±1) are preferred to those with large formal charges (±2, ±3, and so on).
- The best skeletal arrangement of atoms will give rise to Lewis structures in which the formal charges are consistent with electronegativities. For example, the more electronegative atoms should have the more negative formal charges.

Sample Problem 8.9 shows how formal charge can be used to determine the best skeletal arrangement of atoms for the Lewis structure of a molecule or polyatomic ion.

SAMPLE PROBLEM 8.9

Formaldehyde (CH_2O), which can be used to preserve biological specimens, is commonly sold as a 37% aqueous solution. Use formal charges to determine which skeletal arrangement of atoms shown here is the best choice for the Lewis structure of CH_2O.

H—C—O—H   H—C(—O)—H

Strategy Complete the Lewis structures for each of the CH_2O skeletons shown and determine the formal charges on the atoms in each one.

Setup The completed Lewis structures for the skeletons shown are:

H—C̈=Ö—H   H—C(=Ö)—H

In the structure on the left, the formal charges are as follows:

Both H atoms: 1 valence e^- − 1 associated e^- (from single bond) = 0

C atom: 4 valence e^- − 5 associated e^- (two in the lone pair, one from the single bond, and two from the double bond) = −1

O atom: 6 valence e^- − 5 associated e^- (two from the lone pair, one from the single bond, and two from the double bond) = +1

H—C̈=Ö—H
Formal charges 0 −1 +1 0

In the structure on the right, the formal charges are as follows:

Both H atoms: 1 valence e^- − 1 associated e^- (from single bond) = 0

C atom: 4 valence e^- − 4 associated e^- (one from each single bond, and two from the double bond) = 0

O atom: 6 valence e^- − 6 associated e^- (four from the two lone pairs and two from the double bond) = 0

H—C(=Ö)—H
Formal charges all zero

Solution Of the two possible arrangements, the structure on the left has an O atom with a positive formal charge, which is inconsistent with oxygen's high electronegativity. Therefore, the structure on the right, in which both H atoms are attached directly to the C atom and all atoms have a formal charge of zero, is the better choice for the Lewis structure of CH_2O.

THINK ABOUT IT

For a molecule, formal charges of zero are preferred. When there are nonzero formal charges, they should be consistent with the electronegativities of the atoms in the molecule. A positive formal charge on oxygen, for example, is inconsistent with oxygen's high electronegativity.

Practice Problem ATTEMPT Two possible arrangements are shown for the Lewis structure of a carboxyl group, —COOH. Use formal charges to determine which of the two arrangements is better.

—C(=Ö)—Ö—H   —C̤—Ö=Ö—H

Practice Problem BUILD Use Lewis structures and formal charges to determine the best skeletal arrangement of atoms in NCl_2^-.

Practice Problem CONCEPTUALIZE For each partial Lewis structure shown here, determine what group element A must belong to in order for its formal charge to be zero.

(i)	(ii)	(iii)	(iv)
—A(=)—	—Ä— (with lone pairs)	—Ä— (with single bond below)	=A=

CHECKPOINT – SECTION 8.6 Lewis Structures and Formal Charge

8.6.1 Determine the formal charges on H, C, and N, respectively, in HCN.

a) 0, +1, and −1

b) −1, +1, and 0

c) 0, −1, and +1

d) 0, +1, and +1

e) 0, 0, and 0

8.6.2 Which of the Lewis structures shown is most likely preferred for NCO^-?

a) $[:\ddot{N}=C=\ddot{O}:]^-$

b) $[:\ddot{\underset{..}{N}}-C\equiv O:]^-$

c) $[:N\equiv C-\ddot{\underset{..}{O}}:]^-$

d) $[:\ddot{\underset{..}{N}}-C=\ddot{O}:]^-$

e) $[:\ddot{N}=C-\ddot{\underset{..}{O}}:]^-$

8.7 Resonance

Our drawing of the Lewis structure for ozone (O_3) satisfied the octet rule for the central O atom because we placed a double bond between it and one of the two terminal O atoms. In fact, we can put the double bond at either end of the molecule, as shown by the following two equivalent Lewis structures:

$$\ddot{O}=\ddot{O}-\ddot{\underset{..}{O}}: \longleftrightarrow :\ddot{\underset{..}{O}}-\ddot{O}=\ddot{O}$$

A single bond between O atoms should be longer than a double bond between O atoms, but experimental evidence indicates that both of the bonds in O_3 are equal in length (128 pm). Because neither one of these two Lewis structures accounts for the known bond lengths in O_3, we use both Lewis structures to represent the ozone molecule.

Each of the Lewis structures is called a resonance structure. A ***resonance structure*** is one of two or more Lewis structures for a single molecule that cannot be represented accurately by only one Lewis structure. The double-headed arrow indicates that the structures shown are resonance structures. Like the medieval European traveler to Africa who described a rhinoceros as a cross between a griffin and a unicorn (two familiar but imaginary animals), we describe ozone, a real molecule, in terms of two familiar but nonexistent structures.

A common misconception about resonance is that a molecule such as ozone somehow shifts quickly back and forth from one resonance structure to the other. Neither resonance structure, though, adequately represents the actual molecule, which has its own unique, stable structure. "Resonance" is a human invention, designed to address the limitations of a simple bonding model. To extend the animal analogy, a rhinoceros is a distinct, real creature, not some oscillation between the mythical griffin and unicorn!

The carbonate ion provides another example of resonance:

$$\left[\begin{array}{c}:\ddot{O}:\\ |\\ \ddot{O}=C-\ddot{\underset{..}{O}}:\end{array}\right]^{2-} \longleftrightarrow \left[\begin{array}{c}\ddot{O}\\ \|\\ :\ddot{\underset{..}{O}}-C-\ddot{\underset{..}{O}}:\end{array}\right]^{2-} \longleftrightarrow \left[\begin{array}{c}:\ddot{O}:\\ |\\ :\ddot{\underset{..}{O}}-C=\ddot{O}\end{array}\right]^{2-}$$

According to experimental evidence, all three carbon-oxygen bonds in CO_3^{2-} are equivalent. Therefore, the properties of the carbonate ion are best explained by considering its resonance structures together.

The concept of resonance applies equally well to organic systems. A good example is the benzene molecule (C_6H_6):

[Two Kekulé resonance structures of benzene, each C bonded to one H, connected by a double-headed arrow]

If one of these resonance structures corresponded to the actual structure of benzene, there would be two different bond lengths between adjacent C atoms, one with the properties of a single bond and the other with the properties of a double bond. In fact, the distance between all adjacent C atoms in benzene is 140 pm, which is shorter than a C–C bond (154 pm) and longer than a C=C bond (133 pm).

Resonance structures differ only in the positions of their *electrons*—not in the positions of their atoms. Thus, :N=N=O: and :N≡N–O: are resonance structures of each other, whereas :N=N=O: and :N=O=N: are not.

Sample Problem 8.10 shows how to draw resonance structures.

SAMPLE PROBLEM 8.10

High oil and gasoline prices have renewed interest in alternative methods of producing energy, including the "clean" burning of coal. Part of what makes "dirty" coal *dirty* is its high sulfur content. Burning dirty coal produces sulfur dioxide (SO_2), among other pollutants. Sulfur dioxide is oxidized in the atmosphere to form sulfur trioxide (SO_3), which subsequently combines with water to produce sulfuric acid—a major component of acid rain. Draw all possible resonance structures of sulfur trioxide.

Strategy Draw two or more Lewis structures for SO_3 in which the atoms are arranged the same way but the electrons are arranged differently.

Setup Following the steps for drawing Lewis structures, we determine that a correct Lewis structure for SO_3 contains two sulfur-oxygen single bonds and one sulfur-oxygen double bond.

:O:
|
:O=S–O:

But the double bond can be put in any one of three positions in the molecule.

Solution

:O: ⟷ O ⟷ :O:
| ‖ |
:O=S–O: ⟷ :O–S–O: ⟷ :O–S=O:

THINK ABOUT IT

Always make sure that resonance structures differ only in the positions of the electrons, *not* in the positions of the atoms.

Practice Problem ATTEMPT Draw all possible resonance structures for the nitrate ion (NO_3^-).

Practice Problem BUILD Draw three resonance structures for the thiocyanate ion (NCS^-), and determine the formal charges in each resonance structure. Based on formal charges, list the resonance structures in order of increasing relative importance.

Practice Problem CONCEPTUALIZE The Lewis structure of a molecule consisting of the hypothetical elements A, B, and C is shown here. Of the four other structures, identify any that is *not* a resonance structure of the original and explain why it is not a resonance structure.

:B: / :C–A=C: :A: / :C=B–C: (i)   B / :C–A–C: (ii)   :B: / :C–A–C: (iii)   :B: / :C–A–C: (iv)

CHECKPOINT – SECTION 8.7 Resonance

8.7.1 Indicate which of the following are resonance structures of :Cl–Be–Cl: (select all that apply).

a) :Cl=Be=Cl:
b) :Cl≡Be–Cl:
c) :Be≡Cl–Cl:
d) :Cl–Be≡Cl:
e) :Be=Cl=Cl:

8.7.2 How many resonance structures can be drawn for the nitrite ion (NO_2^-)? (N and O must obey the octet rule.)

a) 1
b) 2
c) 3
d) 4
e) 5

8.8 Exceptions to the Octet Rule

The octet rule almost always holds for second-period elements. Exceptions to the octet rule fall into three categories:

1. The central atom has fewer than eight electrons due to a shortage of electrons.
2. The central atom has fewer than eight electrons due to an odd number of electrons.
3. The central atom has more than eight electrons.

Incomplete Octets

In some compounds, the number of electrons surrounding the central atom in a stable molecule is fewer than eight. Beryllium, for example, which is the Group 2A element in the second period, has the electron configuration $[He]2s^2$. Thus, it has two valence electrons in the $2s$ orbital. In the gas phase, beryllium hydride (BeH_2) exists as discrete molecules. The Lewis structure of BeH_2 is

H—Be—H

Only four electrons surround the Be atom, so there is no way to satisfy the octet rule for beryllium in this molecule.

Elements in Group 3A also tend to form compounds in which they are surrounded by fewer than eight electrons. Boron, for example, has the electron configuration $[He]2s^22p^1$, so it has only three valence electrons. Boron reacts with the halogens to form a class of compounds having the general formula BX_3, where X is a halogen atom. Thus, there are only six electrons around the boron atom in boron trifluoride:

:F̤:
|
:F̤—B—F̤:

We actually *can* satisfy the octet rule for boron in BF_3 by using a lone pair on one of the F atoms to form a double bond between the F atom and boron. This gives rise to three additional resonance structures:

:F̤: | F=B—F̤: ⟷ F‖ :F̤—B—F̤: ⟷ :F̤: | :F̤—B=F

Although these resonance structures result in boron carrying a negative formal charge while fluorine carries a positive formal charge, a situation that is inconsistent with the electronegativities of the atoms involved, the experimentally determined bond length in BF_3 (130.9 pm) is shorter than a single bond (137.3 pm). The shorter bond length would appear to support the idea behind the three resonance structures.

On the other hand, boron trifluoride combines with ammonia in a reaction that is better represented using the Lewis structure in which boron has only six valence electrons around it:

$$BF_3 + :NH_3 \longrightarrow F_3B{-}NH_3$$

It seems, then, that the properties of BF_3 are best explained by all four resonance structures.

The B—N bond in $F_3B{-}NH_3$ is different from the covalent bonds discussed so far in the sense that both electrons are contributed by the N atom. This type of bond is called a ***coordinate covalent bond*** (also referred to as a ***dative bond***), which is defined as a covalent bond in which one of the atoms donates both electrons. Although the properties of a coordinate covalent bond do not differ from those of a normal covalent bond (i.e., the electrons are shared in both cases), the distinction is useful for keeping track of valence electrons and assigning formal charges.

Odd Numbers of Electrons

Some molecules, such as nitrogen dioxide (NO_2), contain an odd number of electrons.

$$:\ddot{O}{=}\dot{N}{-}\ddot{\underset{..}{O}}:$$

Because we need an even number of electrons for every atom in a molecule to have a complete octet, the octet rule cannot be obeyed for all the atoms in these molecules. Molecules with an odd number of electrons are sometimes referred to as ***free radicals*** (or just *radicals*). Many radicals are highly reactive, because there is a tendency for the unpaired electron to form a covalent bond with an unpaired electron on another molecule. When two nitrogen dioxide molecules collide, for example, they form dinitrogen tetroxide, a molecule in which the octet rule *is* satisfied for both the N and O atoms.

$$O{=}\dot{N}{-}O \;+\; O{-}\dot{N}{=}O \longrightarrow (O{=})(O{-})N{-}N({-}O)({=}O)$$

Bringing Chemistry to Life

The Power of Radicals

Beginning about a week after the September 11, 2001, attacks, letters containing anthrax bacteria were mailed to several news media offices and to two U.S. senators. Of the 22 people who subsequently contracted anthrax, five died. Anthrax is a spore-forming bacterium (*Bacillus anthracis*) and, like smallpox, is classified by the CDC as a *Category A bioterrorism agent*. Spore-forming bacteria are notoriously difficult to kill, making the cleanup of the buildings contaminated by anthrax costly and time-consuming. The American Media Inc. (AMI) building in Boca Raton, Florida, was not deemed safe to enter until July 2004, after it had been treated with chlorine dioxide (ClO_2), the only structural fumigant approved by the Environmental Protection Agency (EPA) for anthrax decontamination. The effectiveness of ClO_2 in killing anthrax and other hardy biological agents stems in part from its being a *radical,* meaning that it contains an odd number of electrons.

The American Media Inc. building in Boca Raton, Florida.

Credit: © *Eliot J. Schechte/Stringer/AFP/Getty Images.*

Sample Problem 8.11 lets you practice drawing Lewis structures for species with odd numbers of electrons.

SAMPLE PROBLEM 8.11

Draw the Lewis structure of chlorine dioxide (ClO_2).

Strategy The skeletal structure is

$$O{-}Cl{-}O$$

This puts the unique atom, Cl, in the center and puts the more electronegative O atoms in terminal positions.

Setup There are a total of 19 valence electrons (seven from the Cl and six from each of the two O atoms). We subtract four electrons to account for the two bonds in the skeleton, leaving us with 15 electrons to distribute as follows: three lone pairs on each O atom, one lone pair on the Cl atom, and the last remaining electron also on the Cl atom.

Solution

$$:\ddot{\underset{..}{O}}{-}\ddot{\underset{.}{Cl}}{-}\ddot{\underset{..}{O}}:$$

THINK ABOUT IT

ClO_2 is used primarily to bleach wood pulp in the manufacture of paper, but it is also used to bleach flour, disinfect drinking water, and deodorize certain industrial facilities. In recent years, it has become popular for use in the disinfection of produce—particularly sprouts, which have been implicated in numerous bacterial-infection outbreaks.

(Continued on next page)

Practice Problem ATTEMPT Draw the Lewis structure for the OH species. [The OH species is a *radical*, not to be confused with the hydroxide *ion* (OH^-).]

Practice Problem BUILD Draw the Lewis structure for the NS_2 molecule.

Practice Problem CONCEPTUALIZE Hypothetical elements A and B combine to form a number of molecules and polyatomic ions. (Element A is a member of Group 5A; element B is a member of Group 6A.) Using the chemical formulas, determine which of the following species must be represented by a Lewis structure with an unpaired electron.

AB AB_2 A_2B AB_3 AB_3^- AB_2^-

Expanded Octets

Atoms of the second-period elements cannot have more than eight valence electrons around them, but atoms of elements in and beyond the third period of the periodic table can. In addition to the 3*s* and 3*p* orbitals, elements in the third period also have 3*d* orbitals that can be used in bonding. These orbitals enable an atom to form an *expanded octet*. One compound in which there is an expanded octet is sulfur hexafluoride, a very stable compound. The electron configuration of sulfur is $[Ne]3s^23p^4$. In SF_6, each of sulfur's 6 valence electrons forms a covalent bond with a fluorine atom, so there are 12 electrons around the central sulfur atom:

SF_6 (S bonded to six :F: atoms)

In Chapter 9, we will see that these 12 electrons, or six bonding pairs, are accommodated in six orbitals that originate from the one 3*s*, the three 3*p*, and two of the five 3*d* orbitals. Sulfur also forms many compounds in which it does obey the octet rule. In sulfur dichloride, for instance, S is surrounded by only eight electrons:

:C̈l—S̈—C̈l:

When drawing Lewis structures of compounds containing a central atom from the third period and beyond, the octet rule may be satisfied for all the atoms before all the valence electrons have been used up. When this happens, the extra electrons should be placed as lone pairs on the central atom.

Which Is More Important: Formal Charge or the Octet Rule?

We have learned that although central atoms from the third period (and beyond) often obey the octet rule, it is not necessary for them to do so. Rather, they can have what we call "expanded octets," meaning they are surrounded by more than eight electrons.

The sulfate ion (SO_4^{2-}), for example, can be represented by a resonance structure that obeys the octet rule (structure I), or by a structure that does not (structure II):

Structure I   Structure II

In structure I, although the octet rule is obeyed for the central atom, there are nonzero formal charges (shown in blue) on all the atoms. In structure II, by relocating a lone pair from each of two oxygen atoms and creating two double bonds, we change three of the formal charges to zero.

In some species, including the sulfate ion, it is possible to incorporate too many double bonds. Structures with *three* and *four* double bonds to sulfur would give formal charges on S and O that are inconsistent with the electronegativities of these elements. In general, if you are trying to minimize formal charges by expanding the central atom's octet, only add enough double bonds to make the formal charge on the central atom zero.

Whether structure I or structure II is better (or "more important") has been the subject of some debate among educators over the past two decades. Although some chemists may have a strong preference for one or the other, it is important for you to understand that *both* are valid Lewis structures and you should be able to draw both types of structures.

(See end-of-chapter problems 8.90, 8.102, and 8.103.)

Sample Problem 8.12 involves compounds that do not obey the octet rule.

SAMPLE PROBLEM 8.12

Draw the Lewis structures of (a) boron triiodide (BI_3), (b) arsenic pentafluoride (AsF_5), and (c) xenon tetrafluoride (XeF_4).

Strategy Follow the step-by-step procedure for drawing Lewis structures. The skeletal structures are

(a) I—B(—I)—I (b) As bonded to five F (c) F—Xe—F with F above and below Xe

Note that the skeletal structure already has more than an octet around the As atom.

Setup (a) There are a total of 24 valence electrons in BI_3 (three from the B and seven from each of the three I atoms). We subtract six electrons to account for the three bonds in the skeleton, leaving 18 electrons to distribute as three lone pairs on each I atom. (b) There are 40 total valence electrons [five from As (Group 5A) and seven from each of the five F atoms (Group 7A)]. We subtract 10 electrons to account for the five bonds in the skeleton, leaving 30 to be distributed. Next, place three lone pairs on each F atom, thereby completing all their octets and using up all the electrons. (c) There are 36 total valence electrons (eight from Xe and seven from each of the four F atoms). We subtract eight electrons to account for the bonds in the skeleton, leaving 28 to distribute. We first complete the octets of all four F atoms. When this is done, four electrons remain, so we place two lone pairs on the Xe atom.

Solution

(a) :Ï—B(—Ï:)—Ï: (b) As bonded to five :F̤: (c) :F̤—Xe(with two lone pairs)—F̤: with :F̤: above and below Xe

THINK ABOUT IT

Boron is one of the elements that does not always follow the octet rule. Like BF_3, however, BI_3 can be drawn with a double bond to satisfy the octet of boron. This gives rise to a total of four resonance structures:

:Ï—B(—Ï:)—Ï: ⟷ Ï=B(—Ï:)—Ï: ⟷ :Ï—B(=I)—Ï: ⟷ :Ï—B(—Ï:)=I

Atoms beyond the second period can accommodate more than an octet of electrons, whether those electrons are used in bonds or reside on the central atom as lone pairs.

Practice Problem ATTEMPT Draw the Lewis structures of (a) beryllium fluoride (BeF_2), (b) phosphorus pentachloride (PCl_5), and (c) the iodine tetrachloride ion (ICl_4^-).

Practice Problem BUILD Draw the Lewis structures of (a) boron trichloride (BCl_3), (b) antimony pentafluoride (SbF_5), and (c) krypton difluoride (KrF_2).

Practice Problem CONCEPTUALIZE Elements in the same group exhibit similar chemistry and sometimes form analogous species. For example, nitrogen and phosphorus (both members of Group 5A) can combine with chlorine in a 1:3 ratio to form NCl_3 and PCl_3, respectively. Phosphorus can also combine with chlorine in a 1:5 ratio. Explain why nitrogen cannot.

CHECKPOINT – SECTION 8.8 Exceptions to the Octet Rule

8.8.1 In which of the following species does the central atom *not* obey the octet rule?

a) ClO_2^-
b) CO_2
c) BrO_3^-
d) HCN
e) ICl_4^-

8.8.2 Which elements cannot have more than an octet of electrons? (Select all that apply.)

a) N
b) C
c) S
d) Br
e) O

8.8.3 In which species does the central atom *obey* the octet rule? (Select all that apply.)

a) I_3^-
b) BH_3
c) AsF_6^-
d) NO_2
e) ClO_2^-

8.8.4 How many lone pairs are there on the central atom in the Lewis structure of ICl_2^-?

a) 0
b) 1
c) 2
d) 3
e) 4

8.9 Bond Enthalpy

One measure of the stability of a molecule is its ***bond enthalpy***, which is the enthalpy change associated with breaking a particular bond in 1 mole of gaseous molecules. (Bond enthalpies in solids and liquids are affected by neighboring molecules.) The experimentally determined bond enthalpy of the diatomic hydrogen molecule, for example, is

$$H_2(g) \longrightarrow H(g) + H(g) \qquad \Delta H° = 436.4 \text{ kJ/mol}$$

According to this equation, breaking the covalent bonds in 1 mole of gaseous H_2 molecules requires 436.4 kJ of energy. For the less stable chlorine molecule,

$$Cl_2(g) \longrightarrow Cl(g) + Cl(g) \qquad \Delta H° = 242.7 \text{ kJ/mol}$$

Bond enthalpies can also be directly measured for heteronuclear diatomic molecules, such as HCl,

$$HCl(g) \longrightarrow H(g) + Cl(g) \qquad \Delta H° = 431.9 \text{ kJ/mol}$$

as well as for molecules containing multiple bonds:

$$O_2(g) \longrightarrow O(g) + O(g) \qquad \Delta H° = 498.7 \text{ kJ/mol}$$

$$N_2(g) \longrightarrow N(g) + N(g) \qquad \Delta H° = 941.4 \text{ kJ/mol}$$

Measuring the strength of covalent bonds in polyatomic molecules is more complicated. For example, measurements show that the energy needed to break the first O—H bond in H_2O is different from that needed to break the second O—H bond:

$$H_2O(g) \longrightarrow H(g) + OH(g) \qquad \Delta H° = 502 \text{ kJ/mol}$$

$$OH(g) \longrightarrow H(g) + O(g) \qquad \Delta H° = 427 \text{ kJ/mol}$$

In each case, an O—H bond is broken, but the first step requires the input of more energy than the second. The difference between the two $\Delta H°$ values suggests that the second O—H bond itself undergoes change, because of the changes in its chemical environment.

We can now understand why the bond enthalpy of the same O—H bond in two different molecules, such as methanol (CH_3OH) and water (H_2O), will not be the same: their environments are different. For polyatomic molecules, therefore, we speak of the *average* bond enthalpy of a particular bond. For example, we can measure the enthalpy of the O—H bond in 10 different polyatomic molecules and obtain the average O—H bond enthalpy by dividing the sum of the bond enthalpies by 10. Table 8.6 lists the average bond enthalpies of a number of diatomic and polyatomic molecules. As we noted earlier, triple bonds are stronger than double bonds, and double bonds are stronger than single bonds.

A comparison of the thermochemical changes that take place during a number of reactions reveals a strikingly wide variation in the enthalpies of different reactions. For example, the combustion of hydrogen gas in oxygen gas is fairly *exothermic:*

$$H_2(g) + \tfrac{1}{2}O_2(g) \longrightarrow H_2O(l) \qquad \Delta H° = -285.8 \text{ kJ/mol}$$

The formation of glucose from carbon dioxide and water, on the other hand, best achieved by photosynthesis, is highly *endothermic:*

$$6CO_2(g) + 6H_2O(l) \longrightarrow C_6H_{12}O_6(s) + 6O_2(g) \qquad \Delta H° = 2801 \text{ kJ/mol}$$

TABLE 8.6 Bond Enthalpies

Bond	Bond Enthalpy (kJ/mol)	Bond	Bond Enthalpy (kJ/mol)	Bond	Bond Enthalpy (kJ/mol)
H−H*	436.4	C≡O	1070	O−O	142
H−N	393	C−P	263	O=O	498.7
H−O	460	C−S	255	O−P	502
H−S	368	C=S	477	O=S	469
H−P	326	C−F	453	O−F	190
H−F	568.2	C−Cl	339	O−Cl	203
H−Cl	431.9	C−Br	276	O−Br	234
H−Br	366.1	C−I	216	O−I	234
H−I	298.3	N−N	193	P−P	197
C−H	414	N=N	418	P=P	489
C−C	347	N≡N	941.4	S−S	268
C=C	620	N−O	176	S=S	352
C≡C	812	N=O	607	F−F	156.9
C−N	276	N−F	272	Cl−Cl	242.7
C=N	615	N−Cl	200	Cl−F	193
C≡N	891	N−Br	243	Br−Br	192.5
C−O	351	N−I	159	I−I	151.0
C=O†	745				

*Bond enthalpies shown in red are for diatomic molecules.

†The C=O bond enthalpy in CO_2 is 799 kJ/mol.

Student Note: Bond enthalpies for diatomic molecules have more significant figures than those for polyatomic molecules. Those for polyatomic molecules are average values based on the bonds in more than one compound.

We can account for such variations by looking at the stability of individual reactant and product molecules. After all, most chemical reactions involve the making and breaking of bonds. Therefore, knowing the bond enthalpies and hence the stability of molecules reveals something about the thermochemical nature of the reactions that molecules undergo.

In many cases, it is possible to predict the approximate enthalpy of a reaction by using the average bond enthalpies. Because energy is always required to break chemical bonds and chemical bond formation is always accompanied by a release of energy, we can estimate the enthalpy of a reaction by counting the total number of bonds broken and formed in the reaction and recording all the corresponding enthalpy changes. The enthalpy of reaction in the gas phase is given by

$$\Delta H^\circ = \Sigma \text{BE(reactants)} - \Sigma \text{BE(products)} \qquad \textbf{Equation 8.4}$$

$$= \text{total energy } \textit{input} \text{ (to } \textit{break} \text{ bonds)} - \text{total energy } \textit{released} \text{ (by bond } \textit{formation}\text{)}$$

where BE stands for average bond enthalpy and Σ is the summation sign. As written, Equation 8.4 takes care of the sign convention for ΔH°. Thus, if the total energy input needed to break bonds in the reactants is less than the total energy released when bonds are formed in the products, then ΔH° is negative and the reaction is exothermic [Figure 8.13(a)]. On the other hand, if less energy is released (bond making) than absorbed (bond breaking), ΔH° is positive and the reaction is endothermic [Figure 8.13(b)].

If all the reactants and products are diatomic molecules, then the equation for the enthalpy of reaction will yield accurate results because the bond enthalpies of diatomic molecules are accurately known. If some or all of the reactants and products are polyatomic molecules, the equation will yield only approximate results because the bond enthalpies used will be averages.

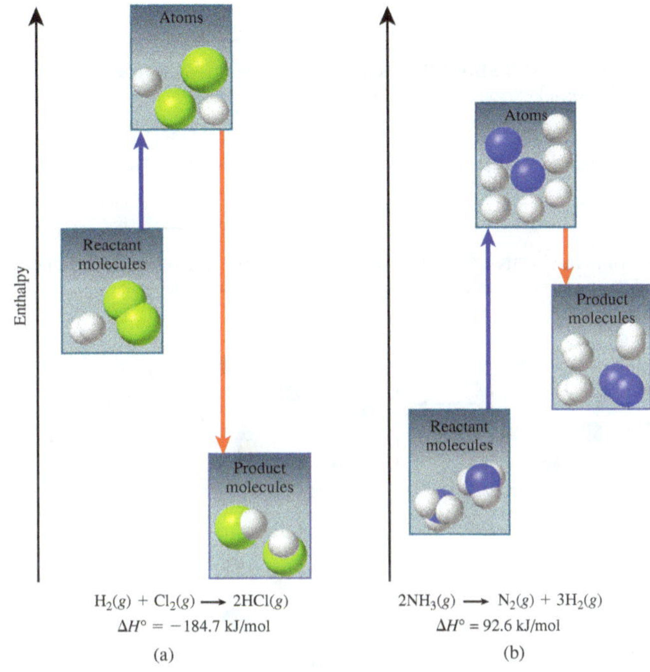

Figure 8.13 Enthalpy changes in (a) an exothermic reaction and (b) an endothermic reaction. The $\Delta H°$ values are calculated using Equation 5.19 and tabulated $\Delta H_f°$ values from Appendix 2.

Sample Problem 8.13 shows how to estimate enthalpies of reaction using bond enthalpies.

SAMPLE PROBLEM 8.13

Use bond enthalpies from Table 8.6 to estimate the enthalpy of reaction for the combustion of methane:

$$CH_4(g) + 2O_2(g) \longrightarrow CO_2(g) + 2H_2O(l)$$

Strategy Draw Lewis structures to determine what bonds are to be broken and what bonds are to be formed. (Don't skip the step of drawing Lewis structures. This is the only way to know for certain what types and numbers of bonds must be broken and formed.)

Setup

$$\text{H}-\underset{\text{H}}{\overset{\text{H}}{\underset{|}{\overset{|}{\text{C}}}}}-\text{H} + \begin{matrix}\ddot{\text{O}}=\ddot{\text{O}} \\ \ddot{\text{O}}=\ddot{\text{O}}\end{matrix} \longrightarrow \ddot{\text{O}}=\text{C}=\ddot{\text{O}} + \begin{matrix}\text{H}-\ddot{\text{O}}-\text{H} \\ \text{H}-\ddot{\text{O}}-\text{H}\end{matrix}$$

Bonds to break: 4 C—H and 2 O=O.

Bonds to form: 2 C=O and 4 H—O.

Bond enthalpies from Table 8.6: 414 kJ/mol (C—H), 498.7 kJ/mol (O=O), 799 kJ/mol (C=O in CO_2), and 460 kJ/mol (H—O).

Solution

$$[4(414 \text{ kJ/mol}) + 2(498.7 \text{ kJ/mol})] - [2(799 \text{ kJ/mol}) + 4(460 \text{ kJ/mol})] = -785 \text{ kJ/mol}$$

Remember that heats of reaction are expressed in kJ/mol, where the "per mole" refers to *per mole of reaction as written* [⏮ Section 5.3].

THINK ABOUT IT

Use Equation 5.19 [⏮ Section 5.6] and data from Appendix 2 to calculate this enthalpy of reaction again; then compare your results using the two approaches. The difference in this case is due to two things: Most tabulated bond enthalpies are averages and, by convention, we show the product of combustion as liquid water—but average bond enthalpies apply to species in the gas phase, where there is little or no influence exerted by neighboring molecules.

Practice Problem ATTEMPT Use bond enthalpies from Table 8.6 to estimate the enthalpy of reaction for the combination of carbon monoxide and oxygen to produce carbon dioxide:

$$2CO(g) + O_2(g) \longrightarrow 2CO_2(g)$$

Practice Problem BUILD Using the following chemical equation, data from Table 8.6, and data from Appendix 2, determine the P—Cl bond enthalpy:

$$PH_3(g) + 3HCl(g) \longrightarrow PCl_3(g) + 3H_2(g)$$

Practice Problem CONCEPTUALIZE Four different chemical reactions are represented here. For each reaction, indicate whether it is endothermic or exothermic—or if there is not enough information to determine.

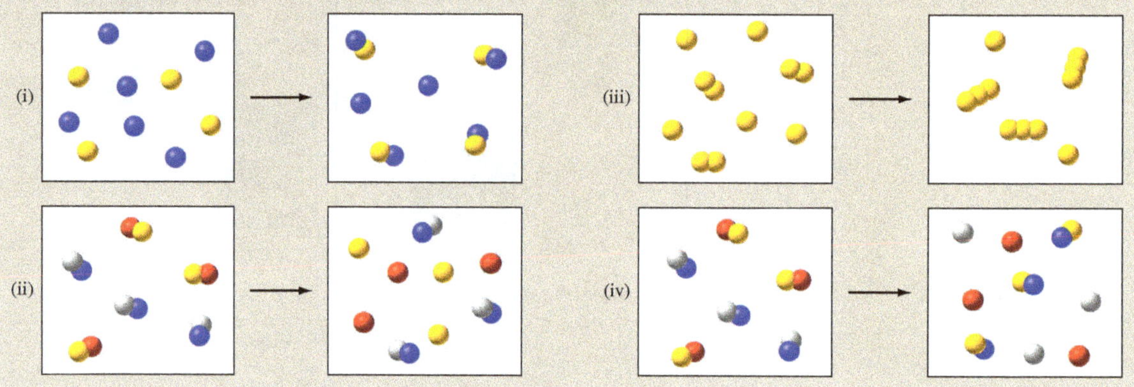

CHECKPOINT – SECTION 8.9 Bond Enthalpy

8.9.1 Use data from Table 8.6 to estimate ΔH_{rxn} for the decomposition of hydrogen peroxide to form water and oxygen.

$$H_2O_2(l) \longrightarrow H_2O(l) + \tfrac{1}{2}O_2(g)$$

a) −71 kJ/mol
b) +357 kJ/mol
c) −357 kJ/mol
d) −107 kJ/mol
e) +71 kJ/mol

8.9.2 Use data from Table 8.6 to estimate ΔH_{rxn} for the reaction of fluorine and chlorine to produce ClF.

$$F_2(g) + Cl_2(g) \longrightarrow 2ClF(g)$$

a) −77.5 kJ/mol
b) −206.6 kJ/mol
c) 206.6 kJ/mol
d) −13.6 kJ/mol
e) 13.6 kJ/mol

8.9.3 Use bond enthalpies to determine ΔH_{rxn} for the reaction shown.

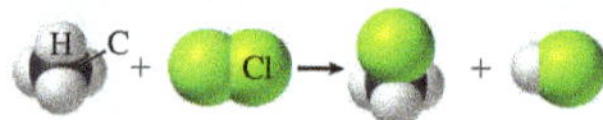

a) −1139 kJ/mol
b) +1139 kJ/mol
c) −346 kJ/mol
d) +114 kJ/mol
e) −114 kJ/mol

8.9.4 Use bond enthalpies to determine ΔH_{rxn} for the reaction shown.

a) −1028 kJ/mol
b) −200 kJ/mol
c) −392 kJ/mol
d) +392 kJ/mol
e) +200 kJ/mol

Chapter Summary

Section 8.1

- A ***Lewis dot symbol*** depicts an atom or an atomic ion of a main group element with dots (representing the valence electrons) arranged around the element's symbol. Main group atoms lose or gain one or more electrons to become isoelectronic with noble gases.

Section 8.2

- The electrostatic attraction that holds ions together in an ionic compound is referred to as ***ionic bonding.***
- ***Lattice energy*** is the amount of energy required to convert a mole of ionic solid to its constituent ions in the gas phase. Lattice energy cannot be measured directly, but is determined using the ***Born-Haber cycle*** and thermodynamic quantities that can be measured directly.

Section 8.3

- According to the ***Lewis theory of bonding, covalent bonding*** results when atoms *share* valence electrons. The atoms in molecules and those in polyatomic ions are held together by ***covalent bonds.***
- According to the ***octet rule,*** atoms will lose, gain, or share electrons to achieve a noble gas configuration. Pairs of valence electrons that are *not* involved in the covalent bonding in a molecule or polyatomic ion (i.e., valence electrons that are not shared) are called ***lone pairs.***
- ***Lewis structures*** are drawn to represent molecules and polyatomic ions, showing the arrangement of atoms and the positions of all valence electrons. Lewis structures represent the shared pairs of valence electrons either as two dots, ··, or as a single dash, —. Any *unshared* electrons are represented as dots.
- One shared pair of electrons between atoms constitutes a ***single bond. Multiple bonds*** form between atoms that share more than one pair of electrons. Two shared pairs constitute a ***double bond,*** and three shared pairs constitute a ***triple bond.***

Section 8.4

- Bonds in which electrons are not shared equally are ***polar*** and are referred to as ***polar covalent bonds.***
- ***Electronegativity*** is an atom's ability to draw shared electrons toward itself. Bonds between elements of widely different electronegativities ($\Delta \geq 2.0$) are *ionic.* Covalent bonds between atoms with significantly different electronegativities ($0.5 \leq \Delta < 2.0$) are *polar.* Bonds between atoms with very similar electronegativities ($\Delta < 0.5$) are ***nonpolar.***
- ***Percent ionic character*** quantifies the polarity of a bond, and is determined by comparing the measured dipole moment to the one predicted by assuming that the bonded atoms have discrete charges.
- The ***dipole moment (μ)*** is a quantitative measure of the polarity of a bond.

Section 8.5

- Lewis structures of molecules or polyatomic ions can be drawn using the following step-by-step procedure:
  1. Use the molecular formula to draw the skeletal structure.
  2. Count the total number of valence electrons, adding electrons to account for a negative charge and subtracting electrons to account for a positive charge.
  3. Subtract two electrons for each bond in the skeletal structure.
  4. Distribute the remaining valence electrons to complete octets, completing the octets of the more electronegative atoms first.
  5. Place any remaining electrons on the central atom.
  6. Include double or triple bonds, if necessary, to complete the octets of all atoms.

Section 8.6

- ***Formal charge*** is a way of keeping track of the valence electrons in a species. Formal charges should be consistent with electronegativities and can be used to determine the best arrangement of atoms and electrons for a Lewis structure.

Section 8.7

- ***Resonance structures*** are two or more equally correct Lewis structures that differ in the positions of the electrons but *not* in the positions of the atoms. Different resonance structures of a compound can be separated by a resonance arrow, $\longleftrightarrow$.

Section 8.8

- In an ordinary covalent bond, each atom contributes one electron to the shared pair of electrons. In cases where just one of the atoms contributes *both* of the electrons, the bond is called a ***coordinate covalent bond*** or a ***dative bond.***
- A species that contains an odd number of electrons is called a ***free radical.***

Section 8.9

- ***Bond enthalpy*** is the energy required to break 1 mole of a particular type of bond. Bond enthalpies are a measure of the stability of covalent bonds and can be used to estimate the enthalpy change for a reaction.

Key Words

Key Equations

8.1 $\mu = Q \times r$	Dipole moment (μ) is calculated as the product of charge magnitude (Q) and distance between the charges (bond length, r) in a diatomic molecule. Because molecules are neutral, the partial charges in a heteronuclear diatomic molecule are equal in magnitude and opposite in sign. Equation 8.1 can be used to calculate the dipole moment when the magnitude of partial charges is known—or it can be used to determine the magnitude of partial charges when the experimentally determined dipole moment is known.
8.2 $\text{percent ionic character} = \dfrac{\mu(\text{observed})}{\mu(\text{calculated assuming discrete charges})} = 100\%$	Percent ionic character of a bond is equal to the ratio of the observed dipole moment to the dipole moment calculated, assuming discrete charges on the atoms.
8.3 formal charge = valence electrons − associated electrons	Formal charge on an atom in a Lewis structure is equal to the number of valence electrons (group number) minus half of the electrons it shares with other atoms in the structure.
8.4 $\Delta H^\circ = \Sigma\text{BE(reactants)} - \Sigma\text{BE(products)}$	The enthalpy change of a reaction can be estimated by subtracting the sum of bond enthalpies in products from the sum of bond enthalpies in reactants.

KEY SKILLS Drawing Lewis Structures

The first step in solving many problems is drawing a correct Lewis structure. The process of drawing a Lewis structure was first described on page 343. The steps, which are summarized in the flowchart below, are:

1. Count the total valence electrons. Recall that each atom contributes a number equal to its group number; and remember to add or subtract valence electrons to account for charge on a polyatomic ion.
2. Use the chemical formula to draw a skeletal structure. Usually, the central atom is less electronegative than the terminal atoms; although hydrogen cannot be a central atom because it can form only one bond.
3. For each bond in the skeletal formula, subtract two from the total number of valence electrons.
4. Distribute the remaining electrons, satisfying first the octets of the more electronegative (usually terminal) atoms.
5. If all terminal atoms have complete octets, and there are valence electrons still to be distributed, place them on the central atom as lone pairs.
6. If the valence electrons run out before all octets are satisfied, use multiple bonds to complete the octets of all atoms.
7. For a charged species, enclose the Lewis structure in square brackets and add a superscript charge.

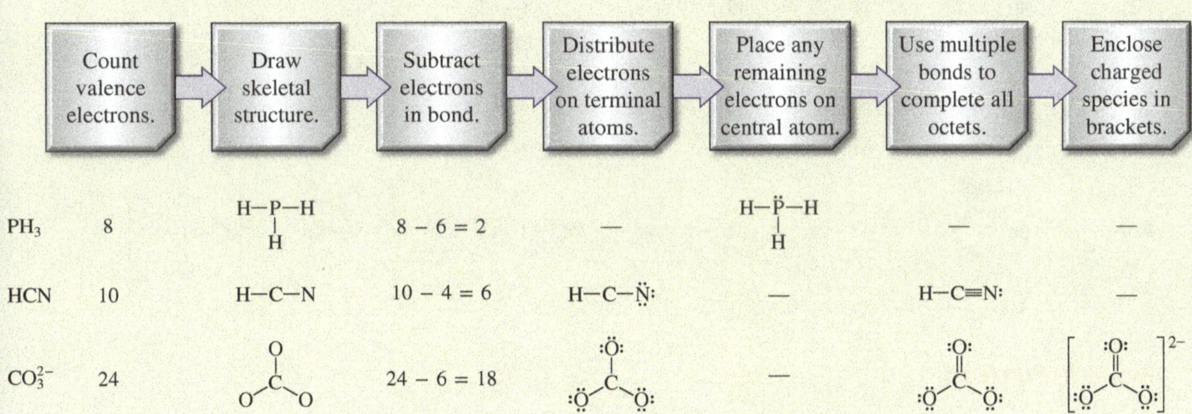

There are exceptions to the octet rule.

- Be and B, small atoms with low electronegativity, need not obey the octet rule.
- Species with an odd number of valence electrons cannot obey the octet rule.
- Elements in the third period and beyond need not obey the octet rule.
- A larger central atom (from the third period or beyond) can accommodate more than eight electrons and can have an "expanded" octet.

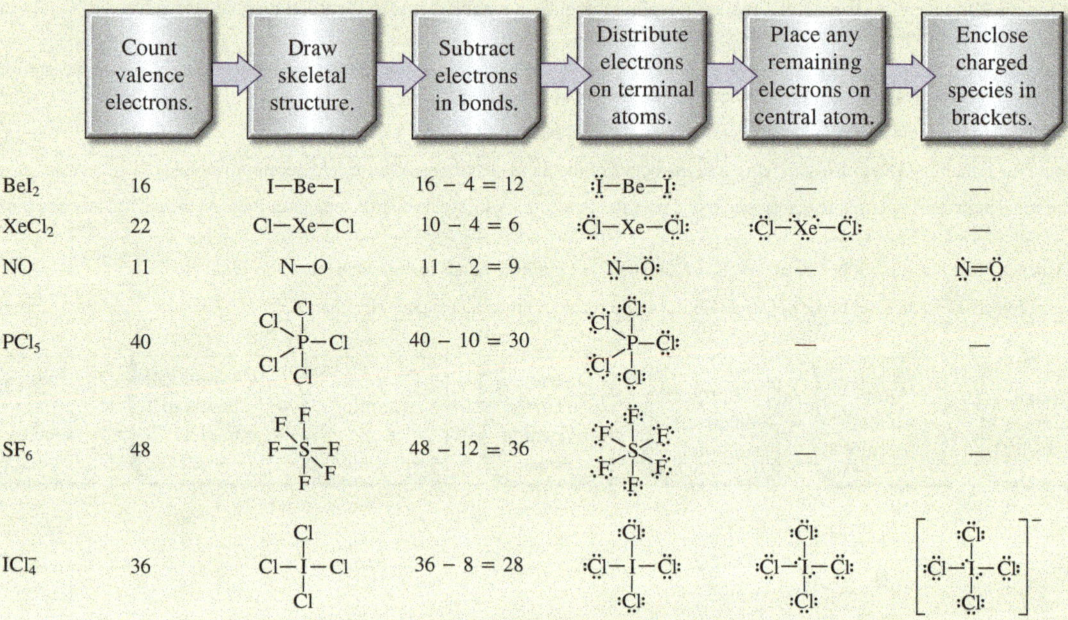

Key Skills Problems

8.1
Which of the following atoms must always obey the octet rule? (Select all that apply.)

(a) C (b) N (c) S (d) Br (e) Xe

8.2
Which of the following species has an odd number of electrons? (Select all that apply.)

(a) N_2O (b) NO_2 (c) NO_2^- (d) NO_3^- (e) NS

8.3
How many lone pairs are on the central atom in $XeOF_2$?

(a) 0 (b) 1 (c) 2 (d) 3 (e) 4

8.4
How many lone pairs are on the central atom in the perchlorate ion?

(a) 0 (b) 1 (c) 2 (d) 3 (e) 4

Questions and Problems

Applying What You've Learned

Researchers in the early 1990s made the sensational announcement that nitric oxide (NO), which had long been thought of only as a component of air pollution, turns out to play an important role in human physiology. They found that NO serves as a *signal* molecule, being produced *in vivo* and regulating a wide variety of cell functions in the body including in the cardiovascular, nervous, and immune systems.

The discovery of the biological role of nitric oxide has shed light on how nitroglycerin works as a drug. For many years, nitroglycerin tablets have been prescribed for cardiac patients to relieve the pain caused by brief interruptions in the flow of blood to the heart, although how it worked was not understood. We now know that nitroglycerin produces nitric oxide in the body, which causes muscles to relax and allows the arteries to dilate.

Research continues to uncover the role nitric oxide plays in biological processes, and medicine continues to find new uses for this molecule.

Problems:

(a) Without consulting Figure 8.1, give the Lewis dot symbols for N and O [⏮ Sample Problem 8.1]. (b) Classify the bond in NO as nonpolar, polar, or ionic [⏮ Sample Problem 8.4]. (c) Given the experimentally determined dipole moment (0.16 D) and the bond length (1.15 Å), determine the magnitude of the partial charges in the NO molecule [⏮ Sample Problem 8.5]. (d) Draw the Lewis structures for NO and for nitroglycerin ($C_3H_5N_3O_9$) [⏮ Sample Problem 8.7]. (e) Determine the formal charges on each atom in NO and in nitroglycerin [⏮ Sample Problem 8.8]. (f) Nitroglycerin decomposes explosively to give carbon dioxide, water, nitrogen, and oxygen. Given the balanced equation for this reaction,

$$C_3H_5(NO_3)_3 \longrightarrow 3CO_2 + 2.5H_2O + 1.5N_2 + 0.25O_2$$

use bond enthalpies to estimate $\Delta H°$ for the reaction [⏮ Sample Problem 8.13].

SECTION 8.1: LEWIS DOT SYMBOLS

Review Questions

8.1 What is a Lewis dot symbol? What elements do we generally represent with Lewis symbols?

8.2 Use the second member of each group from Group 1A to Group 7A to show that the number of valence electrons on an atom of the element is the same as its group number.

Conceptual Problems

8.3 Without referring to Figure 8.1, write Lewis dot symbols for atoms of the following elements: (a) Be, (b) K, (c) Ca, (d) Ga, (e) O, (f) Br, (g) N, (h) I, (i) As, (j) F.

8.4 Write Lewis dot symbols for the following ions: (a) Li^+, (b) Cl^-, (c) S^{2-}, (d) Sr^{2+}, (e) N^{3-}.

8.5 Write Lewis dot symbols for the following atoms and ions: (a) I, (b) I^-, (c) S, (d) S^{2-}, (e) P, (f) P^{3-}, (g) Na, (h) Na^+, (i) Mg, (j) Mg^{2+}, (k) Al, (l) As^{3+}, (m) Pb, (n) Pb^{2+}.

SECTION 8.2: IONIC BONDING

Visualizing Chemistry
Figure 8.6

VC 8.1 What additional information would you need to calculate the lattice energy for a compound if the charges on the cation and anion were +2 and −1, respectively, rather than +1 and −1?
a) No additional information is needed.
b) IE_2 of the cation.
c) IE_2 of the cation and EA_2 of the anion.

VC 8.2 What additional information would you need to calculate the lattice energy for a compound if the charges on the cation and anion were +2 and −2, respectively, rather than +1 and −1?
a) No additional information is needed.
b) IE_2 of the cation.
c) IE_2 of the cation and EA_2 of the anion.

VC 8.3 How would the magnitude of the lattice energy calculated using the Born-Haber cycle change if the charges on the cation and anion were +2 and −2, respectively, rather than +1 and −1?
a) Lattice energy would increase.
b) Lattice energy would decrease.
c) Whether lattice energy would increase or decrease depends on the relative magnitudes of IE_2 of the cation and EA_2 of the anion.

VC 8.4 What law enables us to use the Born-Haber cycle to calculate lattice energy?
a) Coulomb's law.
b) Hess's law.
c) Law of multiple proportions.

Review Questions

8.6 Explain what *ionic bonding* is.

8.7 Explain how ionization energy and electron affinity determine whether atoms of elements will combine to form ionic compounds.

8.8 Name five metals and five nonmetals that are very likely to form ionic compounds. Write formulas for compounds that might result from the combination of these metals and nonmetals. Name these compounds.

8.9 Name one ionic compound that contains only nonmetallic elements.

8.10 Name one ionic compound that contains a polyatomic cation and a polyatomic anion (see Table 2.3).

8.11 Explain why ions with charges greater than ± 3 are seldom found in ionic compounds.

8.12 The term *molar mass* was introduced in Chapter 3. Molar mass is numerically equivalent to molecular mass, although the units are different, for a covalent compound. What is the advantage of using the term molar mass when we discuss ionic compounds?

8.13 In which of the following states would NaCl be electrically conducting: (a) solid, (b) molten (i.e., melted), (c) dissolved in water? Explain.

8.14 Beryllium forms a compound with chlorine that has the empirical formula $BeCl_2$. How would you determine whether it is an ionic compound? (The compound is not soluble in water.)

8.15 What is *lattice energy*, and what does it indicate about the stability of an ionic compound?

8.16 Explain how the lattice energy of an ionic compound such as KCl can be determined using the Born-Haber cycle. On what law is this procedure based?

8.17 Specify which compound in each of the following pairs of ionic compounds should have the higher lattice energy: (a) KCl or MgO, (b) LiF or LiBr, (c) Mg_3N_2 or NaCl. Explain your choice.

8.18 Specify which compound in each of the following pairs of ionic compounds should have the higher lattice energy: (a) AlN or CaO, (b) NaF or CsF, (c) $MgCl_2$ or MgF_2. Explain your choice.

Computational Problems

8.19 Use the Born-Haber cycle outlined in Section 8.2 for NaCl to calculate the lattice energy of LiCl. Use data from Figures 7.8 and 7.10 and Appendix 2.

8.20 Calculate the lattice energy of $CaCl_2$. Use data from Figures 7.8 and 7.10 and Appendix 2. (The second ionization energy of Ca, IE_2, is 1145 kJ/mol.)

Conceptual Problems

8.21 An ionic bond is formed between a cation A^+ and an anion B^-. Based on Coulomb's law

$$E \propto \frac{Q_1 \times Q_2}{d}$$

how would the energy of the ionic bond be affected by the following changes: (a) doubling the radius of A^+, (b) tripling the charge on A^+, (c) doubling the charges on A^+ and B^-, (d) decreasing the radii of A^+ and B^- to half their original values?

8.22 Give the empirical formulas and names of the compounds formed from the following pairs of ions: (a) Rb^+ and I^-, (b) Cs^+ and SO_4^{2-}, (c) Sr^{2+} and N^{3-}, (d) Al^{3+} and S^{2-}.

8.23 Use Lewis dot symbols to show the transfer of electrons between the following atoms to form cations and anions: (a) Na and F, (b) K and S, (c) Ba and O, (d) Al and N.

8.24 Write the Lewis dot symbols of the reactants and products in the following reactions. (First balance the equations.)

(a) $Sr + Se \longrightarrow SrSe$ (c) $Li + N_2 \longrightarrow Li_3N$

(b) $Ca + H_2 \longrightarrow CaH_2$ (d) $Al + S \longrightarrow Al_2S_3$

SECTION 8.3: COVALENT BONDING

Review Questions

8.25 Describe Lewis's contribution to our understanding of the covalent bond.

8.26 Use an example to illustrate each of the following terms: *lone pair, Lewis structure, octet rule, bond length.*

8.27 What is the difference between a Lewis *symbol* and a Lewis *structure?*

8.28 How many lone pairs are on the underlined atoms in these compounds: H<u>Br</u>, H_2<u>S</u>, <u>C</u>H_4?

8.29 Compare single, double, and triple bonds in a molecule, and give an example of each. For the same bonding atoms, how does the bond length change from single bond to triple bond?

8.30 Compare the properties of ionic compounds and covalent compounds.

8.31 Summarize the essential features of the Lewis octet rule. The octet rule applies mainly to the second-period elements. Explain.

Conceptual Problems

8.32 For each of the following pairs of elements, state whether the binary compound they form is likely to be ionic or covalent. Write the empirical formula and name of the compound: (a) I and Cl, (b) Mg and F.

8.33 For each of the following pairs of elements, state whether the binary compound they form is likely to be ionic or covalent. Write the empirical formula and name of the compound: (a) B and F, (b) K and Br.

SECTION 8.4: ELECTRONEGATIVITY AND POLARITY

Review Questions

8.34 Define *electronegativity*, and explain the difference between electronegativity and electron affinity. Describe in general how the electronegativities of the elements change according to their position in the periodic table.

8.35 What is a *polar covalent bond?* Name two compounds that contain one or more polar covalent bonds.

Computational Problems

8.36 Four atoms are arbitrarily labeled D, E, F, and G. Their electronegativities are as follows: D = 3.8, E = 3.3, F = 2.8, and G = 1.3. If the atoms of these elements form the molecules DE, DG, EG, and DF, how would you arrange these molecules in order of increasing covalent bond character?

8.37 Using information in Table 8.5, calculate the magnitude of the partial charges on the atoms in HI.

8.38 Using information in Table 8.5, calculate the percent ionic character of the bond in HBr.

Conceptual Problems

8.39 List the following bonds in order of increasing ionic character: cesium to fluorine, chlorine to chlorine, bromine to chlorine, silicon to carbon.

8.40 Classify the following bonds as covalent, polar covalent, or ionic, and explain: (a) the CC bond in H_3CCH_3, (b) the KI bond in KI, (c) the NB bond in H_3NBCl_3, (d) the CF bond in CF_4.

8.41 Classify the following bonds as covalent, polar covalent, or ionic, and explain: (a) the SiSi bond in $Cl_3SiSiCl_3$, (b) the SiCl bond in $Cl_3SiSiCl_3$, (c) the CaF bond in CaF_2, (d) the NH bond in NH_3.

8.42 List the following bonds in order of increasing ionic character: the lithium-to-fluorine bond in LiF, the potassium-to-oxygen bond in K_2O, the nitrogen-to-nitrogen bond in N_2, the sulfur-to-oxygen bond in SO_2, the chlorine-to-fluorine bond in ClF_3.

8.43 Arrange the following bonds in order of increasing ionic character: carbon to hydrogen, fluorine to hydrogen, bromine to hydrogen, sodium to chlorine, potassium to fluorine, lithium to chlorine.

8.44 Two pairs of elements are highlighted in the periodic table shown here. Consider two binary compounds, one consisting of the two elements highlighted in yellow and one consisting of the two elements highlighted in blue. For which pair of elements will the partial charges be largest? Explain.

8.45 Considering all four highlighted elements in the periodic table shown here, which pair of elements would produce a binary compound with bonds of the greatest percent ionic character? (The binary compound may consist of two yellow elements, two blue elements, or one of each.) Explain.

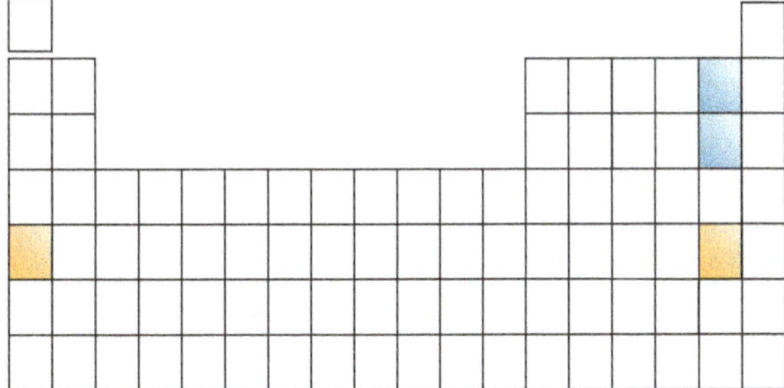

SECTION 8.5: DRAWING LEWIS STRUCTURES

Conceptual Problems

8.46 Draw Lewis structures for the following molecules and ions: (a) NCl_3, (b) OCS, (c) H_2O_2, (d) CH_3COO^-, (e) CN^-, (f) $CH_3CH_2NH_3^+$.

8.47 Draw Lewis structures for the following molecules and ions: (a) OF_2, (b) N_2F_2, (c) Si_2H_6, (d) OH^-, (e) CH_2ClCOO^-, (f) $CH_3NH_3^+$.

8.48 Draw Lewis structures for the following molecules: (a) ICl, (b) PH_3, (c) P_4 (each P is bonded to three other P atoms), (d) H_2S, (e) N_2H_4, (f) $HClO_3$.

8.49 Draw Lewis structures for the following molecules: (a) $COBr_2$ (C is bonded to O and Br atoms), (b) H_2Se, (c) NH_2OH, (d) CH_3NH_2, (e) CH_3CH_2Br, (f) NCl_3.

SECTION 8.6: LEWIS STRUCTURES AND FORMAL CHARGE

Review Question

8.50 Explain the concept of *formal charge*. Do formal charges represent an actual separation of charges?

Conceptual Problems

8.51 Draw Lewis structures for the following ions: (a) NO_2^+, (b) SCN^-, (c) S_2^{2-}, (d) ClF_2^+. Show formal charges.

8.52 Draw Lewis structures for the following ions: (a) O_2^{2-}, (b) C_2^{2-}, (c) NO^+, (d) NH_4^+. Show formal charges.

8.53 The skeletal structure of acetic acid shown here is correct, but some of the bonds are wrong. (a) Identify the incorrect bonds and explain what is wrong with them. (b) Write the correct Lewis structure for acetic acid.

```
  H  :O:
  |   |
H=C—C—O—H
  |
  H
```

8.54 The following Lewis structures are incorrect. Explain what is wrong with each one, and give a correct Lewis structure for the molecule. (Relative positions of atoms are shown correctly.)

(a) H—C̈=N̈

(b) H=C=C=H

(c) Ö—Sn—Ö

(d) :F̈—B̈—F̈: with :F̈: below B

(e) H—Ö=F̈:

(f) H—C̈(—F̈:)—Ȯ:

(g) :F̈—N—F̈: with :F̈: below N

SECTION 8.7: RESONANCE

Review Questions

8.55 What is a *resonance structure?* Is it possible to isolate one resonance structure of a compound for analysis? Explain.

8.56 What are the rules for writing resonance structures?

Conceptual Problems

8.57 Draw Lewis structures for the following species, including all resonance forms, and show formal charges: (a) HCO_2^-, (b) $CH_2NO_2^-$, The relative positions of the atoms are as follows:

```
    O            H     O
H   C              C   N
    O            H     O
```

8.58 Draw three resonance structures for the chlorate ion (ClO_3^-). Show formal charges.

8.59 Draw three resonance structures for hydrazoic acid (HN_3). The atomic arrangement is HNNN. Show formal charges.

8.60 Draw two resonance structures for diazomethane (CH_2N_2). Show formal charges. The skeletal structure of the molecule is

```
H
  C  N  N
H
```

8.61 Draw three reasonable resonance structures for the OCN^- ion. Show formal charges and rank the importance of the structures.

8.62 Draw three resonance structures for the molecule N_2O in which the atoms are arranged in the order NNO. Indicate formal charges and arrange the resonance structures in order of increasing relative importance.

8.63 Draw a resonance structure of the adenine molecule shown here, which is part of the DNA structure. Show all the lone pairs and label the formal charges.

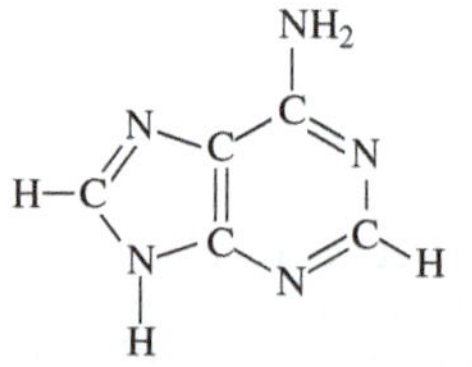

SECTION 8.8: EXCEPTIONS TO THE OCTET RULE

Review Questions

8.64 Why does the octet rule not hold for many compounds containing elements in the third period of the periodic table and beyond?

8.65 Give three examples of compounds that do not satisfy the octet rule. Write a Lewis structure for each.

8.66 Because fluorine has seven valence electrons ($2s^22p^5$), seven covalent bonds in principle could form around the atom. Such a compound might be FH_7 or FCl_7. These compounds have never been prepared. Why?

8.67 What is a *coordinate covalent bond?* Is it different from an ordinary covalent bond?

8.68 Because the central atom in each case is from the same group of the periodic table, the Lewis structures we draw for SO_2 and O_3 are essentially the same. Explain why we can draw a resonance structure for SO_2 in which the formal charge on the central atom is zero, but we cannot do this for O_3.

8.69 What is the advantage of drawing the Lewis structures of oxoanions and oxoacids using expanded octets?

Conceptual Problems

8.70 The AlI_3 molecule has an incomplete octet around Al. Draw three resonance structures of the molecule in which the octet rule is satisfied for both the Al and the I atoms. Show formal charges.

8.71 In the vapor phase, beryllium chloride consists of discrete $BeCl_2$ molecules. Is the octet rule satisfied for Be in this compound? If not, can you form an octet around Be by drawing another resonance structure? How plausible is this structure?

8.72 Of the noble gases, only Kr, Xe, and Rn are known to form a few compounds with O and/or F. Write Lewis structures for the following molecules: (a) XeF_2, (b) XeF_4, (c) XeF_6, (d) $XeOF_4$, (e) XeO_2F_2. In each case Xe is the central atom.

8.73 Write a Lewis structure for $SbCl_5$. Does this molecule obey the octet rule?

8.74 Write Lewis structures for SeF_4 and SeF_6. Is the octet rule satisfied for Se?

8.75 Write Lewis structures for the reaction

$$AlCl_3 + Cl^- \longrightarrow AlCl_4^-$$

What kind of bond joins Al and Cl in the product?

8.76 Draw two resonance structures for the bromate ion (BrO_3^-), one that obeys the octet rule and one in which the formal charge on the central atom is zero.

8.77 Draw two resonance structures for the sulfite ion (SO_3^{2-}), one that obeys the octet rule and one in which the formal charge on the central atom is zero.

SECTION 8.9: BOND ENTHALPY

Review Questions

8.78 What is *bond enthalpy?* Bond enthalpies of polyatomic molecules are average values, whereas those of diatomic molecules can be accurately determined. Why?

8.79 Explain why the bond enthalpy of a molecule is usually defined in terms of a gas-phase reaction. Why are bond-breaking processes always endothermic and bond-forming processes always exothermic?

Computational Problems

8.80 From the following data, calculate the average bond enthalpy for the NH bond:

$$NH_3(g) \longrightarrow NH_2(g) + H(g) \quad \Delta H° = 435 \text{ kJ/mol}$$
$$NH_2(g) \longrightarrow NH(g) + H(g) \quad \Delta H° = 381 \text{ kJ/mol}$$
$$NH(g) \longrightarrow N(g) + H(g) \quad \Delta H° = 360 \text{ kJ/mol}$$

8.81 For the reaction

$$O(g) + O_2(g) \longrightarrow O_3(g) \qquad \Delta H° = -107.2 \text{ kJ/mol}$$

Calculate the average bond enthalpy in O_3.

8.82 The bond enthalpy of $F_2(g)$ is 156.9 kJ/mol. Calculate ΔH_f° for F(*g*).

8.83 For the reaction

$$2C_2H_6(g) + 7O_2(g) \longrightarrow 4CO_2(g) + 6H_2O(g)$$

(a) Predict the enthalpy of reaction from the average bond enthalpies in Table 8.6. (b) Calculate the enthalpy of reaction from the standard enthalpies of formation (see Appendix 2) of the reactant and product molecules, and compare the result with your answer for part (a).

Conceptual Problems

8.84 Use average bond enthalpies from Table 8.6 to estimate ΔH_{rxn} for the following reaction.

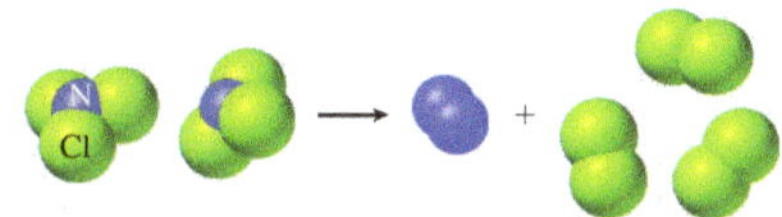

8.85 Use average bond enthalpies from Table 8.6 to estimate ΔH_{rxn} for the following reaction.

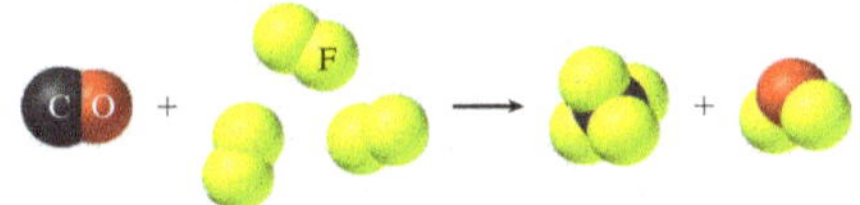

ADDITIONAL PROBLEMS

8.86 Classify the following substances as ionic compounds or covalent compounds containing discrete molecules: CH_4, KF, CO, $SiCl_4$, $BaCl_2$.

8.87 Which of the following are ionic compounds and which are covalent compounds: RbCl, PF_5, BrF_3, KO_2, CI_4?

8.88 Match each of the following energy changes with one of the processes given: ionization energy, electron affinity, bond enthalpy, and standard enthalpy of formation.
(a) $F(g) + e^- \longrightarrow F^-(g)$
(b) $F_2(g) \longrightarrow 2F(g)$
(c) $Na(g) \longrightarrow Na^+(g) + e^-$
(d) $Na(s) + \frac{1}{2}F_2(g) \longrightarrow NaF(s)$

8.89 The formulas for the fluorides of the third-period elements are NaF, MgF_2, AlF_3, SiF_4, PF_5, SF_6, and ClF_3. Classify these compounds as covalent or ionic.

8.90 Use ionization energy (see Figure 7.8) and electron affinity (see Figure 7.10) values to calculate the energy change (in kJ/mol) for the following reactions:
(a) $Li(g) + I(g) \longrightarrow Li^+(g) + I^-(g)$
(b) $Na(g) + F(g) \longrightarrow Na^+(g) + F^-(g)$
(c) $K(g) + Cl(g) \longrightarrow K^+(g) + Cl^-(g)$

8.91 Describe some characteristics of an ionic compound such as KF that would distinguish it from a covalent compound such as benzene (C_6H_6).

8.92 Write Lewis structures for BrF_3, ClF_5, and IF_7. Identify those in which the octet rule is not obeyed.

8.93 Write three reasonable resonance structures for the azide ion N_3^- in which the atoms are arranged as NNN. Show formal charges.

8.94 Draw two resonance structures for sulfurous acid (H_2SO_3): one that obeys the octet rule for the central atom, and one that minimizes the formal charges. Determine the formal charge on each atom in both structures.

8.95 Give an example of an ion or molecule containing Al that (a) obeys the octet rule, (b) has an expanded octet, and (c) has an incomplete octet.

8.96 Draw four reasonable resonance structures for the PO_3F^{2-} ion. The central P atom is bonded to the three O atoms and to the F atom. Show formal charges.

8.97 Attempts to prepare the compounds CF_2, LiO_2, $CsCl_2$, and PI_5 as stable species under atmospheric conditions have failed. Suggest possible reasons for the failure.

8.98 Draw reasonable resonance structures for the following ions: (a) HSO_4^-, (b) PO_4^{3-}, (c) HPO_3^{2-}, (d) IO_3^-.

8.99 Are the following statements true or false? (a) Formal charges represent an actual separation of charges. (b) ΔH°_{rxn} can be estimated from the bond enthalpies of reactants and products. (c) All second-period elements obey the octet rule in their compounds. (d) The resonance structures of a molecule can be separated from one another in the laboratory.

8.100 A rule for drawing plausible Lewis structures is that the central atom is generally less electronegative than the surrounding atoms. Explain why this is so.

8.101 Using the following information and the fact that the average C—H bond enthalpy is 414 kJ/mol, estimate the standard enthalpy of formation of methane (CH_4).

$$C(s) \longrightarrow C(g) \quad \Delta H^\circ_{rxn} = 716 \text{ kJ/mol}$$
$$2H_2(g) \longrightarrow 4H(g) \quad \Delta H^\circ_{rxn} = 872.8 \text{ kJ/mol}$$

8.102 Based on changes in enthalpy, which of the following reactions will occur more readily?
(a) $Cl(g) + CH_4(g) \longrightarrow CH_3Cl(g) + H(g)$
(b) $Cl(g) + CH_4(g) \longrightarrow CH_3(g) + HCl(g)$

8.103 Which of the following molecules has the shortest nitrogen-to-nitrogen bond: N_2H_4, N_2O, N_2, N_2O_4? Explain.

8.104 Most organic acids can be represented as RCOOH, where COOH is the carboxyl group and R is the rest of the molecule. [For example, R is CH_3 in acetic acid (CH_3COOH).] (a) Draw a Lewis structure for the carboxyl group. (b) Upon ionization, the carboxyl group is converted to the carboxylate group (COO^-). Draw resonance structures for the carboxylate group.

8.105 Which of the following species are isoelectronic: NH_4^+, C_6H_6, CO, CH_4, N_2, $B_3N_3H_6$?

8.106 Draw three resonance structures for the hydrogen sulfite ion (HSO_3^-)—one that obeys the octet rule for the central atom, and two that expand the octet of the central atom. Calculate the formal charges on all atoms in each structure and determine which, if any, of the resonance structures has formal charges that are inconsistent with the elements' electronegativities.

8.107 Draw two resonance structures for each species—one that obeys the octet rule, and one in which the formal charge on the central atom is zero: PO_4^{3-}, $HClO_3$, SO_3, SO_2.

8.108 The following species have been detected in interstellar space: (a) CH, (b) OH, (c) C_2, (d) HNC, (e) HCO. Draw Lewis structures for these species.

8.109 The amide ion (NH_2^-) is a Brønsted base. Use Lewis structures to represent the reaction between the amide ion and water.

8.110 Draw Lewis structures for the following organic molecules: (a) tetrafluoroethylene (C_2F_4), (b) propane (C_3H_8), (c) butadiene ($CH_2CHCHCH_2$), (d) propyne (CH_3CCH), (e) benzoic acid (C_6H_5COOH). (To draw C_6H_5COOH, replace an H atom in benzene with a COOH group.)

8.111 The triiodide ion (I_3^-) in which the I atoms are arranged in a straight line is stable, but the corresponding F_3^- ion does not exist. Explain.

8.112 Compare the bond enthalpy of F_2 with the overall energy change for the following process:

$$F_2(g) \longrightarrow F^+(g) + F^-(g)$$

Which is the preferred dissociation for F_2, energetically speaking?

8.113 In 1999, an unusual cation containing only nitrogen (N_5^+) was prepared. Draw three resonance structures of the ion, showing formal charges. (*Hint:* The N atoms are joined in a linear fashion.)

8.114 Write the formulas of the binary hydrides for the second-period elements from Li to F. Identify the bonding in each as covalent, polar covalent, or ionic.

8.115 Several resonance structures for the molecule CO_2 are shown here. Explain why some of them are likely to be of little importance in describing the bonding in this molecule.

(a) $:\ddot{O}=C=\ddot{O}:$ (c) $:O\equiv\ddot{C}-\ddot{O}:$
(b) $:O\equiv C-\ddot{O}:$ (d) $:\ddot{O}-C-\ddot{O}:$

8.116 For each of the following organic molecules draw a Lewis structure in which the carbon atoms are bonded to each other by single bonds: (a) C_2H_6, (b) C_4H_{10}, (c) C_5H_{12}. For parts (b) and (c), show only structures in which each C atom is bonded to no more than two other C atoms.

8.117 In the gas phase, aluminum chloride exists as a dimer (a unit of two) with the formula Al_2Cl_6. Its skeletal structure is given by

```
Cl     Cl     Cl
  \   /  \   /
   Al     Al
  /   \  /   \
Cl     Cl     Cl
```

Complete the Lewis structure and indicate the coordinate covalent bonds in the molecule. Does this dimer possess a dipole moment? Explain.

8.118 Draw Lewis structures for the following organic molecules: C_2H_3F, C_3H_6, C_4H_8. In each there is one C=C bond, and the rest of the carbon atoms are joined by C—C bonds.

8.119 Calculate $\Delta H°$ for the reaction

$$H_2(g) + I_2(g) \longrightarrow 2HI(g)$$

using (a) Equation 8.4 and (b) Equation 5.19, given that ΔH_f° for $I_2(g)$ is 61.0 kJ/mol.

8.120 Draw Lewis structures for the following organic molecules: (a) methanol (CH_3OH); (b) ethanol (CH_3CH_2OH); (c) tetraethyl lead [$Pb(CH_2CH_3)_4$], which is used in "leaded gasoline"; (d) methylamine (CH_3NH_2), which is used in tanning; (e) mustard gas ($ClCH_2CH_2SCH_2CH_2Cl$), a poisonous gas used in World War I; (f) urea [$(NH_2)_2CO$], a fertilizer; and (g) glycine (NH_2CH_2COOH), an amino acid.

8.121 Write Lewis structures for the following four isoelectronic species: (a) CO, (b) NO^+, (c) CN^-, (d) N_2. Show formal charges.

8.122 Oxygen forms three types of ionic compounds in which the anions are oxide (O^{2-}), peroxide (O_2^{2-}), and superoxide (O_2^-). Draw Lewis structures of these ions.

8.123 Comment on the correctness of the statement, "All compounds containing a noble gas atom violate the octet rule."

8.124 Write three resonance structures for (a) the cyanate ion (NCO^-) and (b) the isocyanate ion (CNO^-). In each case, rank the resonance structures in order of increasing importance.

8.125 (a) From the following data calculate the bond enthalpy of the F_2^- ion.

$$F_2(g) \longrightarrow 2F(g) \qquad \Delta H^\circ_{rxn} = 156.9 \text{ kJ/mol}$$
$$F^-(g) \longrightarrow F(g) + e^- \qquad \Delta H^\circ_{rxn} = 333 \text{ kJ/mol}$$
$$F_2^-(g) \longrightarrow F_2(g) + e^- \qquad \Delta H^\circ_{rxn} = 290 \text{ kJ/mol}$$

(b) Explain the difference between the bond enthalpies of F_2 and F_2^-.

8.126 The resonance concept is sometimes described by analogy to a mule, which is a cross between a horse and a donkey. Compare this analogy with the one used in this chapter, that is, the description of a rhinoceros as a cross between a griffin and a unicorn. Which description is more appropriate? Why?

8.127 The N—O bond distance in nitric oxide is 115 pm, which is intermediate between a triple bond (106 pm) and a double bond (120 pm). (a) Draw two resonance structures for NO, and comment on their relative importance. (b) Is it possible to draw a resonance structure having a triple bond between the atoms?

8.128 Vinyl chloride (C_2H_3Cl) differs from ethylene (C_2H_4) in that one of the H atoms is replaced with a Cl atom. Vinyl chloride is used to prepare poly(vinyl chloride), which is an important polymer used in pipes. (a) Draw the Lewis structure of vinyl chloride. (b) The repeating unit in poly(vinyl chloride) is $-CH_2-CHCl-$. Draw a portion of the molecule showing three such repeating units. (c) Calculate the enthalpy change when 1.0×10^3 kg of vinyl chloride forms poly(vinyl chloride).

8.129 Experiments show that it takes 1656 kJ/mol to break all the bonds in methane (CH_4) and 4006 kJ/mol to break all the bonds in propane (C_3H_8). Based on these data, calculate the average bond enthalpy of the C—C bond.

8.130 Draw a Lewis structure for nitrogen pentoxide (N_2O_5) in which each N is bonded to three O atoms.

8.131 The American chemist Robert S. Mulliken suggested a different definition for the electronegativity (EN) of an element, given by

$$EN = \frac{IE_1 + EA}{2}$$

where IE_1 is the first ionization energy and EA is the electron affinity of the element. Calculate the electronegativities of O, F, and Cl using the preceding equation. Compare the electronegativities of these elements on the Mulliken and Pauling scales. (To convert to the Pauling scale, divide each EN value by 230 kJ/mol.)

8.132 Among the common inhaled anesthetics are:

Halothane ($CF_3CHClBr$)
Isoflurane ($CF_3CHClOCHF_2$)
Enflurane ($CHFClCF_2OCHF_2$)
Methoxyflurane ($CHCl_2CF_2OCH_3$)

Draw Lewis structures of these molecules.

8.133 Using Table 8.6, compare the following bond enthalpies: C−C in C_2H_6, N−N in N_2H_4, and O−O in H_2O_2. What effect do lone pairs on adjacent atoms appear to have on bond enthalpy?

8.134 From the lattice energy of KCl in Table 8.6, and the ionization energy of K and electron affinity on Cl in Figures 7.8 and 7.10, respectively, calculate the $\Delta H°$ for the reaction

$$K(g) + Cl(g) \longrightarrow KCl(s)$$

Engineering Problems

8.135 In 1998, scientists using a special type of electron microscope were able to measure the force needed to break a *single* chemical bond. If 2.0×10^{-9} N was needed to break a C−Si bond, estimate the bond enthalpy in kJ/mol. Assume that the bond has to be stretched by a distance of 2 Å (2×10^{-10} m) before it is broken.

8.136 ΔH_f° [$MgF_2(g)$] = −1123 kJ/mol. Using this and data from Appendix 2, Figures 7.8 and 7.10, and Practice Problem 8.3B, calculate the lattice energy of MgF_2.

Biological Problems

8.137 A student in your class claims that magnesium oxide actually consists of Mg^+ and O^- ions, not Mg^{2+} and O^{2-} ions. Suggest some experiments one could do to show that your classmate is wrong.

8.138 The following is a simplified (skeletal) structure of the amino acid histidine. Draw a complete Lewis structure of the molecule.

```
        N
      /   \
     C     C
      \   /
       C—N
      /   \
  H—C—H    H
    |
  H—C—C—O
    |  |
  H—N  O
    |  |
    H  H
```

8.139 The following is a simplified (skeletal) structure of the amino acid tryptophan. Draw a complete Lewis structure of the molecule.

```
       H
       |                  H  H
 H    C                   |  |
   \ /  \                 |  |
    C    C——————C——C——C——————C—O—H
    |    |      |  |  |
    C    C      C  H  N—H  O
   / \  / \    / \    |
 H    C    N      H   H
      |    |
      H    H
```

8.140 Do a Web search of the following ionic compounds and give brief descriptions of their medical uses: $AgNO_3$, $BaSO_4$, $CaSO_4$, KI, Li_2CO_3, $Mg(OH)_2$, $NaHCO_3$, NaF, TiO_2, ZnO.

8.141 Methyl isocyanate (CH_3NCO) is used to make certain pesticides. In December 1984, water leaked into a tank containing this substance at a chemical plant, producing a toxic cloud that killed thousands of people in Bhopal, India. Draw Lewis structures for CH_3NCO, showing formal charges.

8.142 The amide group plays an important role in determining the structure of proteins:

```
       :O:
        ||
   —N̈—C—
    |
    H
```

Draw another resonance structure for this group. Show formal charges.

Environmental Problems

8.143 Draw Lewis structures for the following chlorofluorocarbons (CFCs), which are partly responsible for the depletion of ozone in the stratosphere: (a) $CFCl_3$, (b) CF_2Cl_2, (c) CHF_2Cl, (d) CF_3CHF_2.

8.144 Although nitrogen dioxide (NO_2) is a stable compound, there is a tendency for two such molecules to combine to form dinitrogen tetroxide (N_2O_4). Why? Draw four resonance structures of N_2O_4, showing formal charges.

8.145 The chlorine nitrate ($ClONO_2$) molecule is believed to be involved in the destruction of ozone in the Antarctic stratosphere. Draw a plausible Lewis structure for this molecule.

Multiconcept Problems

8.146 The hydroxyl radical (OH) plays an important role in atmospheric chemistry. It is highly reactive and has a tendency to combine with an H atom from other compounds, causing them to break up. Thus OH is sometimes called a "detergent" radical because it helps to clean up the atmosphere. (a) Draw the Lewis structure for the radical. (b) Refer to Table 8.6 and explain why the radical has a high affinity for H atoms. (c) Estimate the enthalpy change for the following reaction:

$$OH(g) + CH_4(g) \longrightarrow CH_3(g) + H_2O(g)$$

(d) The radical is generated when sunlight hits water vapor. Calculate the maximum wavelength (in nm) required to break an O−H bond in H_2O.

8.147 The species H_3^+ is the simplest polyatomic ion. The geometry of the ion is that of an equilateral triangle. (a) Draw three resonance structures to represent the ion. (b) Given the following information

$$2H + H^+ \longrightarrow H_3^+ \qquad \Delta H° = -849 \text{ kJ/mol}$$
$$H_2 \longrightarrow 2H \qquad \Delta H° = 436.4 \text{ kJ/mol}$$

calculate $\Delta H°$ for the reaction

$$H^+ + H_2 \longrightarrow H_3^+$$

8.148 The bond enthalpy of the C−N bond in the amide group of proteins (see Problem 8.142) can be treated as an average of C−N and C=N bonds. Calculate the maximum wavelength of light needed to break the bond.

Standardized-Exam Practice Problems

Physical and Biological Sciences

Nitrous oxide (N_2O) is an anesthetic commonly used for dental procedures. Because of the euphoria caused by inhaling it, N_2O is commonly known as "laughing gas." It is licensed for use as a food additive and as an aerosol propellant. It is used to displace air from potato chip bags to extend shelf life and as the propellant in whipped cream canisters. In recent years N_2O has become popular as a recreational drug, due in part to its ready availability to consumers. Although N_2O is legal, it is regulated by the FDA; its sale and distribution for the purpose of human consumption are not permitted.

Bond	Bond Enthalpy (kJ/mol)
N−O	176
N=O	607
N−N	193
N=N	418
N≡N	941.4
O=O	498.7

1. Which of the following Lewis structures are possible for N_2O?

   :N≡N−Ö: (I)   :N=N=O: (II)   :N=O=N: (III)   :N̈−O−N̈: (IV)

   a) I only b) I and II c) I, II, and III d) I, II, III, and IV
2. Use formal charges to choose the best of the resonance structures shown.

   a) I b) II c) III d) IV
3. Using the best resonance structure and the average bond enthalpies given, determine the ΔH_f° for N_2O.

   a) 73 kJ/mol b) −73 kJ/mol c) 166 kJ/mol d) −166 kJ/mol
4. Why does the calculated ΔH_f° value differ from the tabulated value of 81.56 kJ/mol?

   a) The tabulated value is wrong.
   b) None of the resonance structures depicts the bonds realistically.
   c) To a reasonable number of significant figures, the calculated and tabulated values are the same.
   d) Using bond enthalpies gives only an estimate of ΔH°_{rxn}.

Answers to In-Chapter Materials

Practice Problems

8.1A (a) Ca^{2+}, (b) $[:\ddot{N}:]^{3-}$, (c) $[:\ddot{I}:]^{-}$. **8.1B** (a) 2−, (b) +, (c) 3−. **8.2A** $MgCl_2$. **8.2B** NaF < MgO < AlN. **8.3A** 629 kJ/mol. **8.3B** −841 kJ/mol. **8.4A** (a) nonpolar, (b) polar, (c) nonpolar. **8.4B** C−S < C−Se < C−Te < C−Po < C−O. **8.5A** 0.12. **8.5B** 0.11 D. **8.6A** 41 percent. **8.6B** 9.23 D. **8.7A** :F̈−N̈(−:F̈:)−F̈: **8.7B** [:Ö−Cl(−:Ö:)−Ö:]$^-$ **8.8A** C atom = 0, double-bonded O atom = 0, single-bonded O atoms = −1. **8.8B** S atom = +1, O atoms = −1, overall charge = −2. **8.9A** −C(=Ö)−Ö−H **8.9B** Cl−N−Cl.

8.10A [:Ö=N(−:Ö:)−Ö:]$^-$ ⇄ [:Ö−N(=Ö)−Ö:]$^-$ ⇄ [:Ö−N(−:Ö:)=Ö:]$^-$

8.10B [:N̈−C≡S:]$^-$ (−2, 0, +1) ⇄ [:N=C=S:]$^-$ (−1, 0, 0) ⇄ [:N≡C−S̈:]$^-$ (0, 0, −1)

8.11A ·Ö−H **8.11B** :S=N̈−S̈: **8.12A** (a) :F̈−Be−F̈: (b) P bonded to five :Cl̈: atoms (c) [:Cl̈−I(−:Cl̈:)(−:Cl̈:)−Cl̈:]$^-$ **8.12B** (a) :Cl̈−B(−:Cl̈:)−Cl̈: (b) Sb bonded to four :F̈: atoms (c) :F̈−K̈r−F̈: **8.13A** −557 kJ/mol. **8.13B** 328 kJ/mol.

Answers to Checkpoints

8.1.1 b. **8.1.2** e. **8.1.3** e. **8.1.4** e. **8.2.1** c. **8.2.2** a. **8.2.3** b. **8.4.1** b. **8.4.2** c. **8.4.3** d. **8.4.4** c. **8.5.1** d. **8.5.2** c. **8.6.1** e. **8.6.2** c. **8.7.1** a, b, d. **8.7.2** b. **8.8.1** e. **8.8.2** a, b, e. **8.8.3** e. **8.8.4** d. **8.9.1** a. **8.9.2** e. **8.9.3** e. **8.9.4** b.

CHAPTER 9

Chemical Bonding II: Molecular Geometry and Bonding Theories

Molecules responsible for the aroma of a substance, such as freshly brewed coffee, have distinctive shapes. Their shapes enable molecules to stimulate our sense of smell.

Credit: © *Jamie Grill/Getty Images.*

In This Chapter, You Will Learn

How to determine the three-dimensional shape of a molecule and how the interactions of atomic orbitals give rise to chemical bonds.

Before You Begin, Review These Skills

- Shapes of atomic orbitals [⏮ Section 6.7]
- Electron configurations of atoms [⏮ Section 6.8]
- Drawing Lewis structures [⏮ Section 8.5]

How Molecular Shape Affects Our Perception of the World

Research has revealed that the geometry or *shape* of a molecule determines some of its important properties. The 2004 Nobel Prize in Physiology or Medicine was awarded to Richard Axel and Linda Buck for their discoveries of odorant receptors and the organization of the olfactory system. Their work shed new light on how our sense of smell works. Although it is somewhat more complicated than can be explained in detail here, the odor of a molecule depends in large part on which of the olfactory receptors it stimulates. *Olfactory receptors* are specialized proteins, located on hairlike cilia in the back of the nose. Stimulation of a receptor triggers an electrical signal that travels along nerve fibers to the brain, where the scent is identified. Which olfactory receptors a molecule stimulates depends on its shape.

We are thought to have receptors for at least seven different, fundamental scents: floral, pepperminty, musky, pungent, camphorlike, ethereal, and putrid. Different parts of a larger molecule may have distinctly different shapes, enabling it to stimulate more than one type of olfactory receptor. This results in the perception of a combination of odors. For example, parts of the benzaldehyde molecule are shaped such that they stimulate the camphorlike, floral, and pepperminty receptors. This is a combination that we perceive as the smell of almonds. Such combinations enable us to recognize thousands of different smells.

This discussion illustrates the importance of *molecular geometry* to the most enigmatic of our senses—smell. Many biochemical processes are specific in that they depend on the shapes of the molecules involved. *Chemical bonding theories* help us to predict and/or explain these shapes.

Benzaldehyde

At the end of this chapter, you will be able to answer several questions about the geometry and structure of a molecule with a distinctive, musky odor [⏭ Applying What You've Learned, page 414].

9.1 Molecular Geometry

Many familiar chemical and biochemical processes depend heavily on the three-dimensional shapes of the molecules and/or ions involved. Our sense of smell is one example; the effectiveness of a particular drug is another. Although the shape of a molecule or polyatomic ion must be determined experimentally, we can predict their shapes reasonably well using Lewis structures [⏮ Section 8.5] and the ***valence-shell electron-pair repulsion (VSEPR)*** model. In this section, we will focus primarily on determining the shapes of molecules of the general type AB_x, where A is a central atom surrounded by x B atoms and x can have integer values of 2 to 6. (Any atom that is bonded to two or more other atoms can be considered a "central" atom.) For example, NH_3 is an AB_3 molecule in which A is nitrogen, B is hydrogen, and $x = 3$. Its Lewis structure is shown in the left margin.

H—N̈—H
  |
  H

Lewis structure of NH_3

Table 9.1 lists examples of each type of AB_x molecule and polyatomic ion that we will consider. Throughout this chapter, we will discuss concepts that apply both to molecules and to polyatomic ions, but we will usually refer to them collectively as "molecules."

Having the molecular formula alone is insufficient to predict the shape of a molecule. For instance, AB_2 molecules may be linear or bent:

Animation
Chemical bonding—valence-shell electron-pair repulsion theory.

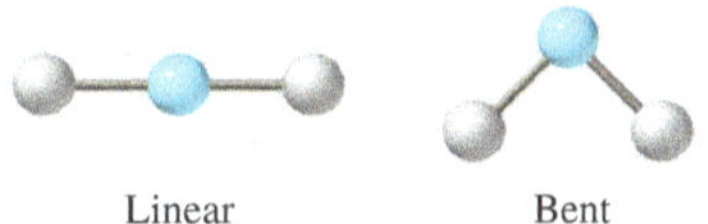

Moreover, AB_3 molecules may be planar, pyramidal, or T-shaped:

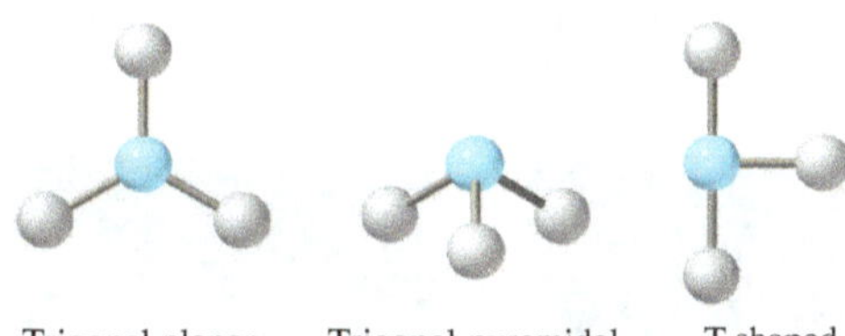

To determine shape, we must start with a correct Lewis structure and apply the VSEPR model.

The VSEPR Model

Student Note:

Electron Domains
Lone pair
Single bond
Double bond
Triple bond

Recall that the electrons in the valence shell are the ones involved in chemical bonding [⏮ Section 8.1]. The basis of the VSEPR model is that electron pairs in the valence shell of an atom *repel* one another. As we learned in Chapter 8, there are two types of electron pairs: bonding pairs and nonbonding pairs (also known as lone pairs). Furthermore, bonding pairs may be found in single bonds or in multiple bonds. For clarity, we will refer to electron *domains* instead of electron pairs when we use the VSEPR model. An ***electron domain*** in this context is a lone pair or a bond, regardless of whether the bond is single, double, or triple. Consider the following examples:

	CO_2	O_3	NH_3	PCl_5	XeF_4
	:Ö=C=Ö:	:Ö=Ö—Ö:	H—N̈—H (with H below N)	P bonded to five :C̈l:	:F̈—Xe—F̈: with :F̈: above and below Xe
	2 double bonds	1 single bond 1 double bond + 1 lone pair	3 single bonds + 1 lone pair	5 single bonds	4 single bonds + 2 lone pairs
Total number of electron domains on central atom	2 electron domains	3 electron domains	4 electron domains	5 electron domains	6 electron domains

TABLE 9.1 Examples of AB_x Molecules and Polyatomic Ions

AB_2	$BeCl_2$, SO_2, H_2O, NO_2^-
AB_3	BF_3, NH_3, ClF_3, SO_3^{2-}
AB_4	CCl_4, NH_4^+, SF_4, XeF_4, ClO_4^-
AB_5	PCl_5, IF_5, SbF_5, BrF_5
AB_6	SF_6, UF_6, $TiCl_6^{3-}$

Note the number of electron domains on the central atom in each molecule. The VSEPR model predicts that because these electron domains repel one another, they will arrange themselves to be as far apart as possible, thus minimizing the repulsive interactions between them. It is important to understand that you cannot tell the shape of a molecule or ion simply from its formula—you must apply VSEPR theory.

Student Note: It is essential that you are able to draw the correct Lewis structure and that you count the electron domains on the central atom carefully. This will be your "blueprint" for determining the shape of a molecule or polyatomic ion.

We can visualize the arrangement of electron domains using balloons, as shown in Figure 9.1. Like the B atoms in our AB_x molecules, the balloons are all connected to a central, fixed point, which represents the central atom (A). When they are as far apart as possible, they adopt the five geometries shown in the figure. When there are only two balloons, they orient themselves to point in opposite directions [Figure 9.1(a)]. With three balloons the arrangement is a trigonal plane [Figure 9.1(b)]. With four balloons the arrangement adopted is a tetrahedron [Figure 9.1(c)]. With five balloons, three of them adopt positions in a trigonal plane whereas the other two point opposite to each other, forming an axis that is perpendicular to the trigonal plane [Figure 9.1(d)]. This geometry is called a trigonal bipyramid. Finally, with six balloons the arrangement is an octahedron, which is essentially a square bipyramid [Figure 9.1(e)]. Each of the AB_x molecules we consider will have one of these five electron-domain geometries: linear, trigonal planar, tetrahedral, trigonal bipyramidal, or octahedral.

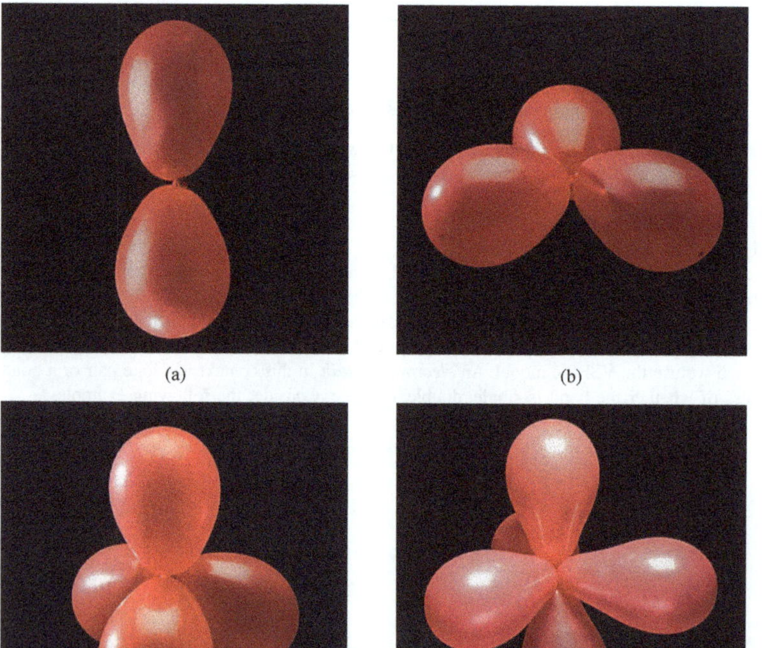

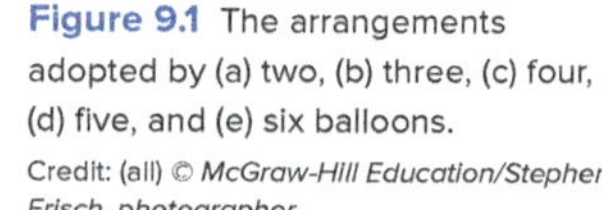

Figure 9.1 The arrangements adopted by (a) two, (b) three, (c) four, (d) five, and (e) six balloons.

Credit: (all) © *McGraw-Hill Education/Stephen Frisch, photographer.*

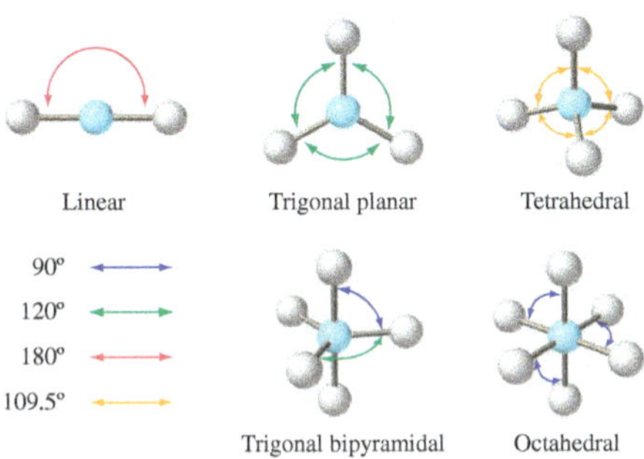

Figure 9.2 Five geometries of AB_x molecules in which all the electron domains are bonds.

Electron-Domain Geometry and Molecular Geometry

It is important to distinguish between the ***electron-domain geometry,*** which is the arrangement of electron domains (bonds and lone pairs) around the central atom, and the ***molecular geometry,*** which is the arrangement of bonded *atoms.* Figure 9.2 illustrates the molecular geometries of AB_x molecules in which all the electron domains are bonds—that is, there are no lone pairs on any of the central atoms. In these cases, the molecular geometry is the same as the electron-domain geometry.

In an AB_x molecule, a ***bond angle*** is the angle between two adjacent A−B bonds. In an AB_2 molecule there are only two bonds and therefore only one bond angle, and, provided that there are no lone pairs on the central atom, the bond angle is 180°. AB_3 and AB_4 molecules have three and four bonds, respectively. However, in each case there is only one bond angle possible between any two A−B bonds. In an AB_3 molecule, the bond angle is 120°, and in an AB_4 molecule, the bond angle is 109.5°—again, provided that there are no lone pairs on the central atoms. Similarly, in an AB_6 molecule the bond angles between adjacent bonds are all 90°. (The angle between any two A−B bonds that point in opposite directions is 180°.)

AB_5 molecules contain two different bond angles between adjacent bonds. The reason for this is that, unlike those in the other AB_x molecules, the positions occupied by bonds in a trigonal bipyramid are not all equivalent. The three bonds that are arranged in a trigonal plane are referred to as ***equatorial.*** The bond angle between any two of the three equatorial bonds is 120°. The two bonds that form an axis perpendicular to the trigonal plane are referred to as ***axial.*** The bond angle between either of the axial bonds and any one of the equatorial bonds is 90°. (As in the case of the AB_6 molecule, the angle between any two A−B bonds that point in opposite directions is 180°.) Figure 9.2 illustrates all these bond angles. The angles shown in the figure are the bond angles that are observed when all the electron domains on the central atom are identical. As we will see later in this section, the bond angles in many molecules will differ slightly from these *ideal* values.

Animation
Chemical bonding—valence-shell electron-pair repulsion theory.

Student Note: It would be impossible to overstate the importance of being able to draw Lewis structures correctly, especially for students who will go on to study organic chemistry.

When the central atom in an AB_x molecule bears one or more lone pairs, the electron-domain geometry and the molecular geometry are no longer the same. However, we still use the electron-domain geometry as a first step in determining the molecular geometry. The first step in determining the molecular geometry of O_3 (or any species), for example, is to draw its Lewis structure. Two different resonance structures can be drawn for O_3:

$$:\ddot{O}=\ddot{O}-\ddot{\underset{..}{O}}: \longleftrightarrow :\ddot{\underset{..}{O}}-\ddot{O}=\ddot{O}:$$

Either one can be used to determine its geometry.

The next step is to count the electron domains on the central atom. In this case, there are three: one single bond, one double bond, and one lone pair. Using the VSEPR model, we first determine the electron-domain geometry. According to the information in Figure 9.2, three electron domains on the central atom will be arranged in a trigonal plane. Molecular geometry, however, is dictated by the arrangement of *atoms.* If we consider only the positions of the three atoms in this molecule, the molecular geometry (the molecule's shape) is *bent.*

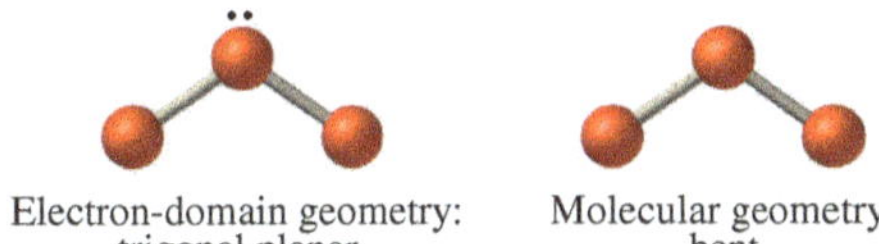

Electron-domain geometry: trigonal planar — Molecular geometry: bent

In addition to the five basic geometries depicted in Figure 9.2, you must be familiar with how molecular geometry can differ from electron-domain geometry. Table 9.2 shows the common molecular geometries where there are one or more lone pairs on the central atom. Note the positions occupied by the lone pairs in the trigonal bipyramidal electron-domain geometry. When there are lone pairs on the central atom in a trigonal bipyramid, the lone pairs preferentially occupy equatorial positions because repulsion is greater when the angle between electron domains is 90° or less. Placing a lone pair in an *axial* position would put it at 90° to three other electron domains. Placing it in an *equatorial* position puts it at 90° to only two other domains, thus minimizing the number of strong repulsive interactions.

All positions are equivalent in the octahedral geometry, so one lone pair on the central atom can occupy any of the positions. If there is a second lone pair in this geometry, though, it must occupy the position opposite the first. This arrangement minimizes the repulsive forces between the two lone pairs (they are 180° apart instead of 90° apart).

TABLE 9.2 Electron-Domain and Molecular Geometries of Molecules with Lone Pairs on the Central Atom

Total Number of Electron Domains	Type of Molecule	Electron-Domain Geometry	Number of Lone Pairs	Placement of Lone Pairs	Molecular Geometry	Example
3	AB_2	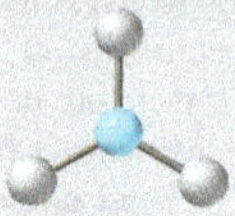 Trigonal planar	1	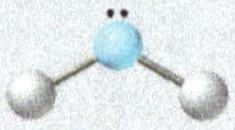	 Bent	SO_2
4	AB_3	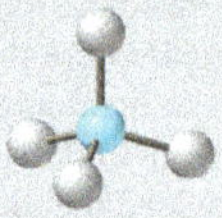 Tetrahedral	1		 Trigonal pyramidal	NH_3
4	AB_2	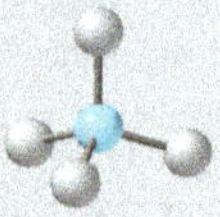 Tetrahedral	2		 Bent	H_2O
5	AB_4	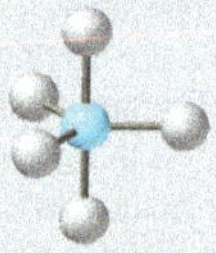 Trigonal bipyramidal	1		 Seesaw-shaped	SF_4
5	AB_3	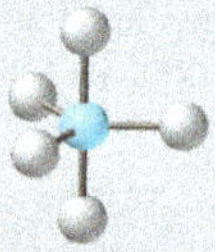 Trigonal bipyramidal	2	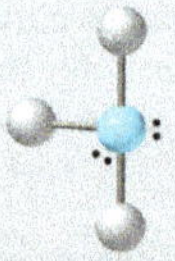	 T-shaped	ClF_3
5	AB_2	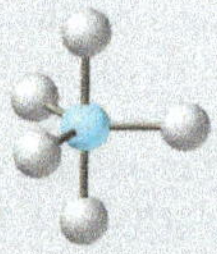 Trigonal bipyramidal	3		 Linear	IF_2^-
6	AB_5	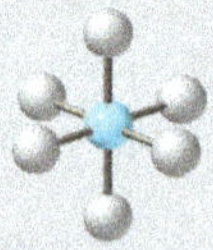 Octahedral	1		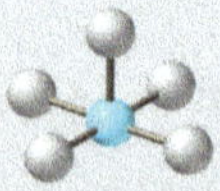 Square pyramidal	BrF_5
6	AB_4	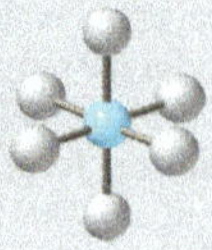 Octahedral	2	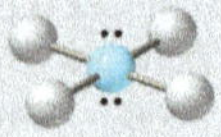	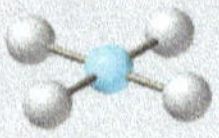 Square planar	XeF_4

In summary, the steps to determine the electron-domain and molecular geometries are as follows:

1. Draw the Lewis structure of the molecule or polyatomic ion.
2. Count the number of electron domains on the central atom.
3. Determine the electron-domain geometry by applying the VSEPR model.
4. Determine the molecular geometry by considering the positions of the atoms only.

Sample Problem 9.1 shows how to determine the shape of a molecule or polyatomic ion.

SAMPLE PROBLEM 9.1

Determine the shapes of (a) SO_3 and (b) ICl_4^-.

Strategy Use Lewis structures and the VSEPR model to determine first the electron-domain geometry and then the molecular geometry (shape).

Setup (a) The Lewis structure of SO_3 is:

:Ö:
|
:O=S—Ö:

There are three electron domains on the central atom: one double bond and two single bonds.

(b) The Lewis structure of ICl_4^- is:

[:Cl: | :Cl—I—Cl: | :Cl:]$^-$

There are six electron domains on the central atom in ICl_4^-: four single bonds and two lone pairs.

Solution

(a) According to the VSEPR model, three electron domains will be arranged in a trigonal plane. Since there are no lone pairs on the central atom in SO_3, the molecular geometry is the same as the electron-domain geometry. Therefore, the shape of SO_3 is trigonal planar.

Electron-domain geometry: trigonal planar ⟶ Molecular geometry: trigonal planar

(b) Six electron domains will be arranged in an octahedron. Two lone pairs on an octahedron will be located on opposite sides of the central atom, making the shape of ICl_4^- square planar.

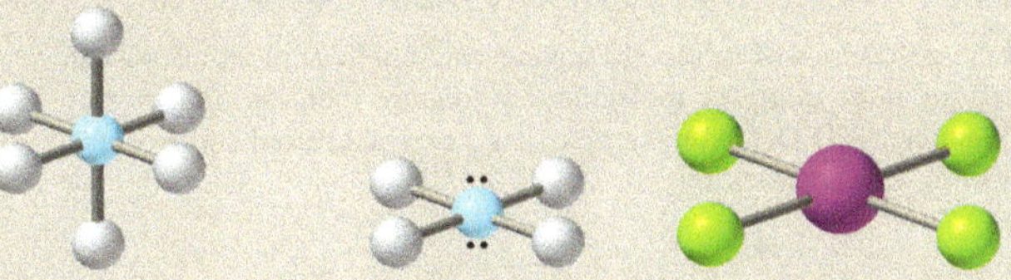

Electron-domain geometry: octahedral ⟶ Molecular geometry: square planar

THINK ABOUT IT

Compare these results with the information in Figure 9.2 and Table 9.2. Make sure that you can draw Lewis structures correctly. Without a correct Lewis structure, you will be unable to determine the shape of a molecule.

Practice Problem ATTEMPT Determine the shapes of (a) CO_2 and (b) SCl_2.

Practice Problem BUILD (a) From what group must the terminal atoms come in an AB_x molecule where the central atom is from Group 6A, for the electron-domain geometry and the molecular geometry both to be trigonal planar? (b) From what group must the terminal atoms come in an AB_x molecule where the central atom is from Group 7A, for the electron-domain geometry to be octahedral and the molecular geometry to be square pyramidal?

Practice Problem CONCEPTUALIZE These four models may represent molecules or polyatomic ions. Lone pairs on the central atom, if any, are not shown. Which of these could represent a species in which there are lone pairs on the central atom? Which could represent a species in which there are no lone pairs on the central atom?

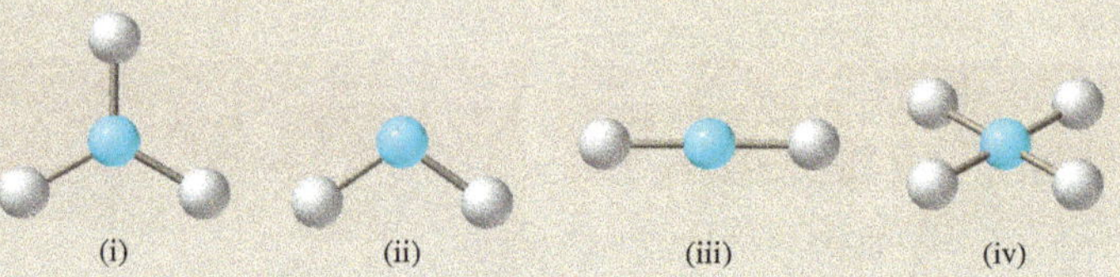

Deviation from Ideal Bond Angles

Some electron domains are better than others at repelling neighboring domains. As a result, the bond angles may be slightly different from those shown in Figure 9.2. For example, the electron-domain geometry of ammonia (NH_3) is tetrahedral, so we might predict the H—N—H bond angles to be 109.5°. In fact, the bond angles are about 107°, slightly smaller than predicted. The lone pair on the nitrogen atom repels the N—H bonds more strongly than the bonds repel one another. It therefore "squeezes" them closer together than the ideal tetrahedral angle of 109.5°.

In effect, a lone pair takes up more *space* than the bonding pairs. This can be understood by considering the attractive forces involved in determining the location of the electron pairs. A lone pair on a central atom is attracted only to the nucleus of that atom. A bonding pair of electrons, on the other hand, is simultaneously attracted by the nuclei of both of the bonding atoms. As a result, the lone pair has more freedom to spread out and greater capacity to repel other electron domains. Also, because they contain more electron density, multiple bonds repel more strongly than single bonds. Consider the bond angles in each of the following examples:

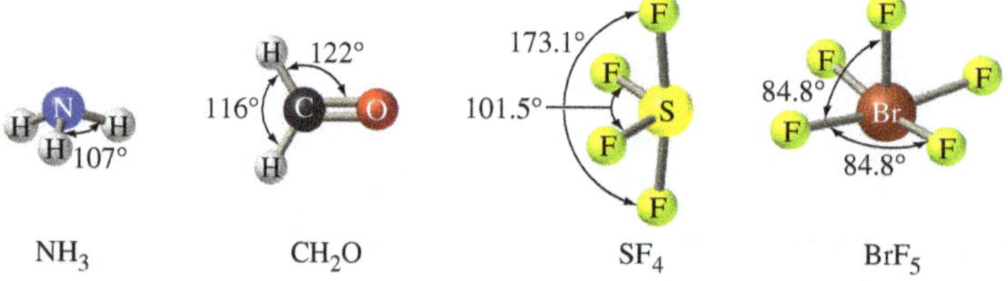

Geometry of Molecules with More than One Central Atom

Thus far, we have considered the geometries of molecules having only one central atom. We can determine the overall geometry of more complex molecules by treating them as though they have multiple central atoms. Methanol (CH_3OH), for example, has a central C atom and a central O atom, as shown in the following Lewis structure:

```
    H
    |
H—C—Ö—H
    |  ..
    H
```

Both the C and the O atoms are surrounded by four electron domains. In the case of C, they are three C—H bonds and one C—O bond. In the case of O, they are one O—C bond, one O—H bond, and two lone pairs. In each case, the electron-domain geometry is tetrahedral. However, the molecular geometry of the C part of the molecule is *tetrahedral,* whereas the molecular geometry of the O part of the molecule is *bent.* Note that although the Lewis structure makes it appear as though there is a 180° angle between the O—C and O—H bonds, the angle is actually approximately 109.5°, the angle in a tetrahedral arrangement of electron domains.

Student Note: When we specify the geometry of a particular portion of a molecule, we refer to it as the geometry "about" a particular atom. In methanol, for example, we say that the geometry is *tetrahedral about the C atom* and *bent about the O atom.*

Sample Problem 9.2 shows how to determine when bond angles differ from ideal values.

SAMPLE PROBLEM 9.2

Acetic acid, the substance that gives vinegar its characteristic smell and sour taste, is sometimes used in combination with corticosteroids to treat certain types of ear infections. Its Lewis structure is:

```
   H  :Ö:
   |  ||
H—C—C—Ö—H
   |
   H
```

Determine the molecular geometry about each of the central atoms, and determine the approximate value of each of the bond angles in the molecule. Which if any of the bond angles would you expect to be smaller than the ideal values?

Strategy Identify the central atoms and count the number of electron domains around each of them. Use the VSEPR model to determine each electron-domain geometry, and the information in Table 9.2 to determine the molecular geometry about each central atom.

Setup The leftmost C atom is surrounded by four electron domains: one C—C bond and three C—H bonds. The middle C atom is surrounded by three electron domains: one C—C bond, one C—O bond, and one C=O (double) bond. The O atom is surrounded by four electron domains: one O—C bond, one O—H bond, and two lone pairs.

Solution The electron-domain geometry of the leftmost C is tetrahedral. Because all four electron domains are bonds, the molecular geometry of this part of the molecule is also tetrahedral. The electron-domain geometry of the middle C is trigonal planar. Again, because all the domains are bonds, the molecular geometry is also trigonal planar. The electron-domain geometry of the O atom is tetrahedral. Because two of the domains are lone pairs, the molecular geometry about the O atom is bent.

Bond angles are determined using electron-domain geometry. Therefore, the approximate bond angles about the leftmost C are 109.5°, those about the middle C are 120°, and those about the O are 109.5°. The angle between the two single bonds on the middle carbon will be *less* than 120° because the double bond repels the single bonds more strongly than they repel each other. Likewise, the bond angle between the two bonds on the O will be less than 109.5° because the lone pairs on O repel the single bonds more strongly than they repel each other and push the two bonding pairs closer together. The angles are labeled as follows:

```
~109.5°         >120°
   H  :Ö:
   |  ||
H—C—C—Ö—H
   |
   H
      <120°   <109.5°
```

THINK ABOUT IT

Compare these answers with the information in Figure 9.2 and Table 9.2.

Practice Problem ATTEMPT Ethanolamine ($HOCH_2CH_2NH_2$) has a smell similar to ammonia and is commonly found in biological tissues. Its Lewis structure is:

```
      H  H
      |  |
H—Ö—C—C—N̈—H
      |  |  |
      H  H  H
```

Determine the molecular geometry about each central atom and label all the bond angles. Cite any expected deviations from ideal bond angles.

Practice Problem BUILD The bond angle in NH_3 is significantly smaller than the ideal bond angle of 109.5° because of the lone pair on the central atom. Explain why the bond angle in SO_2 is very close to 120° despite there being a lone pair on the central atom.

Practice Problem CONCEPTUALIZE Which of these models represents a species in which there is deviation from ideal bond angles?

(i) (ii) (iii) (iv)

How Are Larger, More Complex Molecules Represented?

Organic molecules, which can be quite large and complex, commonly are represented using *bond-line* structures. As the name suggests, each line in a bond-line structure represents a bond. Because organic compounds all contain carbon,most of the lines represent bonds between carbon atoms. In fact, with only a few exceptions, the carbon atoms in an organic molecule are not shown explicitly in bond-line structures—rather they are understood to be at the end of each line, unless an atom *other* than carbon is shown. Also not shown in bond-line structures are the hydrogen atoms that are bonded to carbon. They are understood to be attached as appropriate to complete the octet of each carbon atom. For example, the bond-line structure of propane (C_3H_8) is:

Each line represents a carbon-carbon bond; there are carbon atoms at the ends of each line; and each carbon atom is bonded to the number of hydrogen atoms necessary to satisfy its octet. Thus, the bond-line structure shown here corresponds to the molecular model

Atoms other than carbon and hydrogen are shown explicitly in bond-line structures. For example, the bond-line structure of ethanol (C_2H_5OH) is

OH

Note how similar this structures appears to the structure of propane. In this case, though, one of the lines represents a carbon-carbon bond and the other represents a carbon-oxygen bond. This structure corresponds to the molecular model

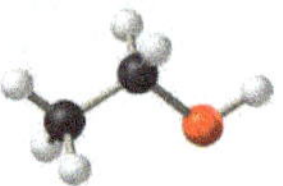

Bond-line structures make is possible to represent large, complex molecules, such as those found in biological systems. You may have noticed that the benzaldehyde molecule shown at the beginning of this chapter is represented with a bond-line structure. Some other examples are shown here along with their molecular formulas.

2-ethyl-3,5-dimethypyrazine, $C_8H_{12}N_2$

3-methylbutanal, $C_5H_{10}O$

(Two of the many volatile organic compounds responsible for the aroma of coffee.)

CHECKPOINT – SECTION 9.1 Molecular Geometry

9.1.1 What are the electron-domain geometry and molecular geometry of CO_3^{2-}?

a) tetrahedral, trigonal planar
b) tetrahedral, trigonal pyramidal
c) trigonal pyramidal, trigonal pyramidal
d) trigonal planar, trigonal planar
e) tetrahedral, tetrahedral

9.1.2 What are the electron-domain geometry and molecular geometry of ClO_3^-?

a) tetrahedral, trigonal planar
b) tetrahedral, trigonal pyramidal
c) trigonal pyramidal, trigonal pyramidal
d) trigonal planar, trigonal planar
e) tetrahedral, tetrahedral

9.1.3 What is the approximate value of the bond angle indicated?

H H
H–C=C–H

a) < 90°
b) < 109.5°
c) > 109.5°
d) > 120°
e) < 120°

9.1.4 What is the approximate value of the bond angle indicated?

H
H–C–Ö–H
H

a) < 180°
b) > 180°
c) < 109.5°
d) > 109.5°
e) < 90°

9.1.5 Which of the following shows a deviation from ideal bond angles that is not possible for an AB_x molecule?

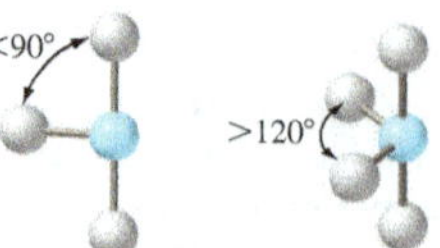
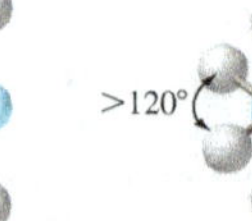
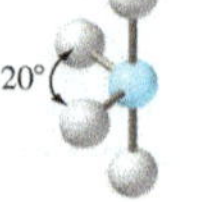
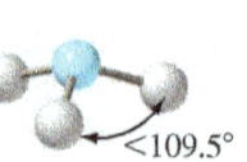
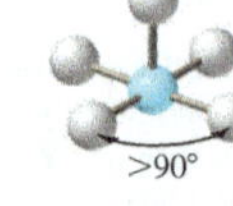
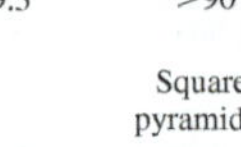

T-shaped (i)
Seesaw-shaped (ii)
Trigonal pyramidal (iii)
Square pyramidal (iv)

a) ii only
b) i and iii
c) i, ii, and iv
d) ii and iv
e) All these deviations from ideal bond angles are possible.

9.2 Molecular Geometry and Polarity

Animation
Chemical bonding—molecular geometry and polarity.

Molecular geometry is tremendously important in understanding the physical and chemical behavior of a substance. *Molecular polarity,* for example, is one of the most important consequences of molecular geometry, because molecular polarity influences physical, chemical, and biological properties. Recall from Section 8.4 that a bond between two atoms of different electronegativities is polar and that a diatomic molecule containing a polar bond is a *polar molecule.* Whether a molecule made up of three or more atoms is polar depends not only on the polarity of the individual bonds, but also on its molecular geometry.

Each of the CO_2 and H_2O molecules contains two identical atoms bonded to a central atom and two polar bonds. However, only one of these molecules is polar. To understand why, think of each individual bond dipole as a vector. The overall dipole moment of the molecule is determined by vector addition of the individual bond dipoles.

Student Note: Recall that we can represent an individual bond dipole using a crossed arrow that points toward the more electronegative atom [Section 8.4].

In the case of CO_2, we have two identical vectors pointing in opposite directions. When the vectors are placed on a Cartesian coordinate system, they have no y component and their x components are equal in magnitude but opposite in sign. The sum of these two vectors is zero in both the x and y directions. Thus, although the *bonds* in CO_2 are polar, the *molecule* is nonpolar.

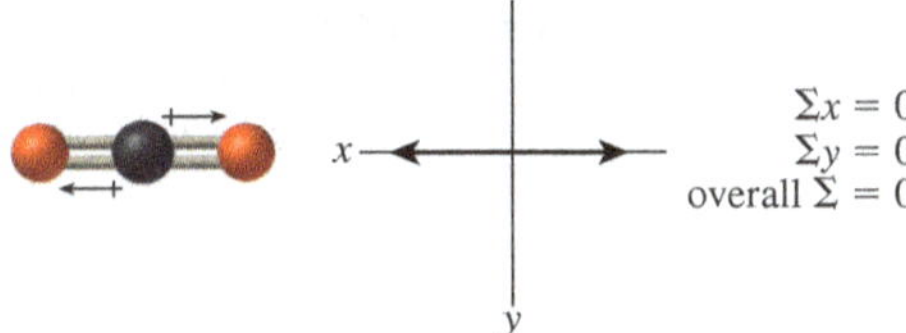

The vectors representing the bond dipoles in water, although equal in magnitude and opposite in the x direction, are not opposite in the y direction. Therefore, although their x components sum to zero, their y components do not. This means that there is a net resultant dipole and H_2O is *polar.*

$$\Sigma x = 0 \quad \Sigma y \neq 0 \quad \text{overall } \Sigma \neq 0$$

Dipole moments can be used to distinguish between molecules that have the same chemical formula but different arrangements of atoms. Such compounds are called ***structural isomers.*** For example, there are two structural isomers of dichloroethylene ($C_2H_2Cl_2$). Because the individual bond dipoles sum to zero in *trans*-dichloroethylene, the *trans* isomer is nonpolar:

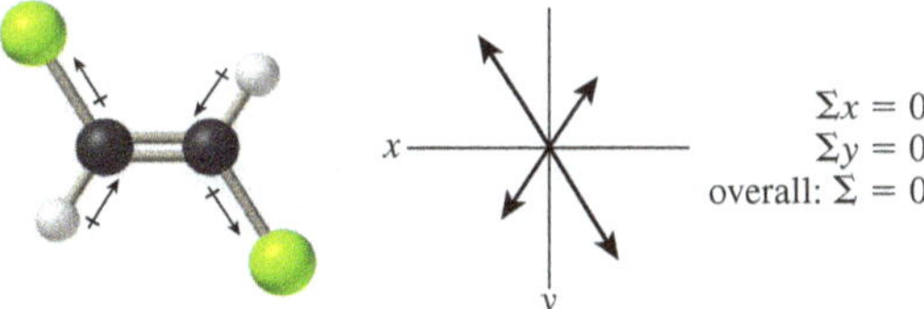

The bond dipoles in the *cis* isomer do not cancel one another, so *cis*-dichloroethylene is polar:

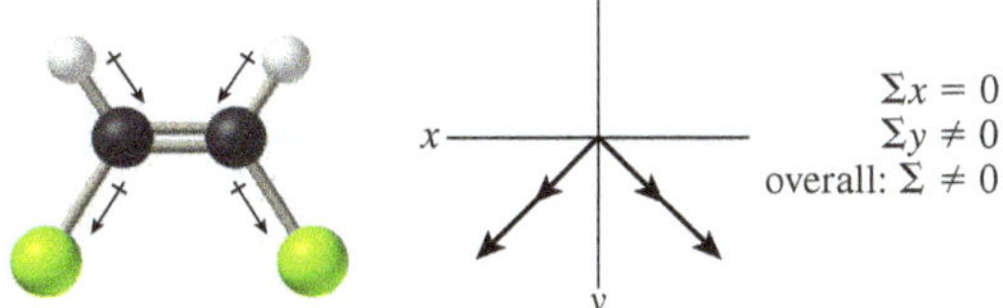

Because of the difference in polarity, these two isomers can be distinguished experimentally by measuring the dipole moment.

Can More Complex Molecules Contain Polar Bonds and Still Be Nonpolar?

In AB_x molecules where $x \geq 3$, it may be less obvious whether the individual bond dipoles cancel one another. Consider the molecule BF_3, for example, which has a trigonal planar geometry:

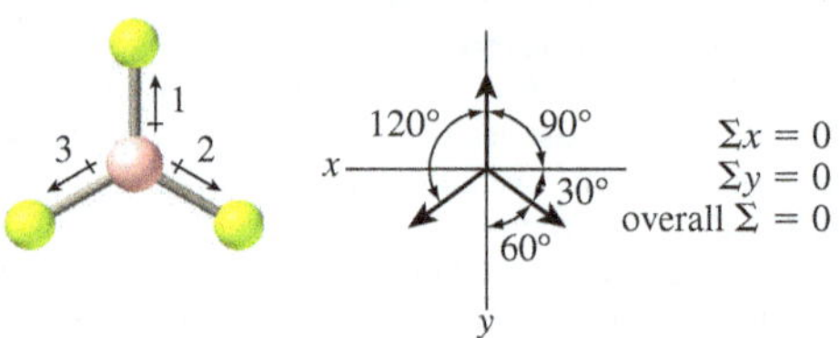

We will simplify the math in this analysis by assigning the vectors representing the three identical B—F bonds an arbitrary magnitude of 1.00. The *x, y* coordinates for the end of arrow 1 are (0, 1.00). Determining the coordinates for the ends of arrows 2 and 3 requires the use of trigonometric functions. You may have learned the mnemonic SOH CAH TOA, where the letters stand for

Sin = Opposite over Hypotenuse

Cos = Adjacent over Hypotenuse

Tan = Opposite over Adjacent

The *x* coordinate for the end of arrow 2 corresponds to the length of the line *opposite* the 60° angle. The hypotenuse of the triangle has a length of 1.00 (the arbitrarily assigned value). Therefore, using SOH,

$$\sin 60° = 0.866 = \frac{\text{opposite}}{\text{hypotenuse}} = \frac{\text{opposite}}{1}$$

so the *x* coordinate for the end of arrow 2 is 0.866.

The magnitude of the *y* coordinate corresponds to the length of the line *adjacent* to the 60° angle. Using TOA,

$$\tan 60° = 1.73 = \frac{\text{opposite}}{\text{adjacent}} = \frac{0.866}{\text{adjacent}}$$

$$\text{adjacent} = \frac{0.866}{1.73} = 0.500$$

so the *y* coordinate for the end of arrow 2 is −0.500. (The trigonometric formula gives us the length of the side. We know from the diagram that the sign of this *y* component is negative.)

Arrow 3 is similar to arrow 2. Its *x* component is equal in magnitude but opposite in sign, and its *y* component is the same magnitude and sign as that for arrow 2. Therefore, the *x* and *y* coordinates for all three vectors are

	x	*y*
Arrow 1	0	1
Arrow 2	0.866	−0.500
Arrow 3	−0.866	−0.500
Sum =	0	0

Because the individual bond dipoles (represented here as the vectors) sum to zero, the molecule is nonpolar overall.

Although it is somewhat more complicated, a similar analysis can be done to show that all *x, y,* and *z* coordinates sum to zero when there are four identical polar bonds arranged in a tetrahedron about a central atom. In fact, any time there are identical bonds symmetrically distributed around a central atom, with no lone pairs on the central atom, the molecule will be nonpolar overall, even if the bonds themselves are polar.

In cases where the bonds are distributed symmetrically around the central atom, the nature of the atoms surrounding the central atom determines whether the molecule is polar overall. For example, CCl_4 and $CHCl_3$ have the same molecular geometry (tetrahedral), but CCl_4 is nonpolar because the bond dipoles cancel one another. In $CHCl_3$, however, the bonds are not all identical, and therefore the bond dipoles do not sum to zero. The $CHCl_3$ molecule is polar.

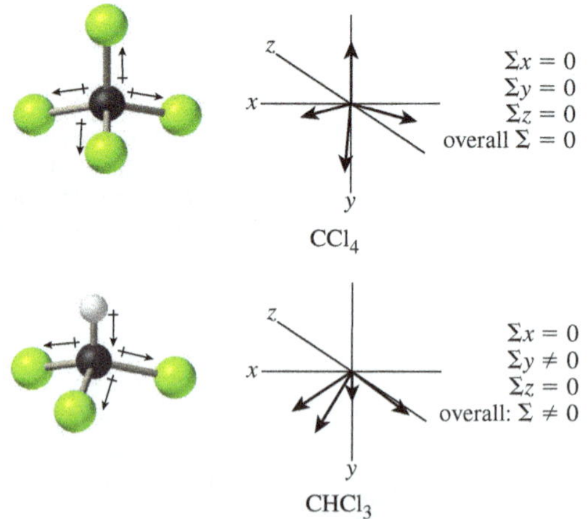

(See end-of-chapter problems 9.79, 9.85, and 9.87.)

CHECKPOINT – SECTION 9.2 Molecular Geometry and Polarity

9.2.1 Identify the polar molecules in the following group: HBr, CH_4, CS_2.

a) HBr only
b) HBr and CS_2
c) HBr, CH_4, and CS_2
d) CH_4 and CS_2
e) CH_4 only

9.2.2 Identify the nonpolar molecules in the following group: SO_2, NH_3, XeF_2.

a) SO_2, NH_3, and XeF_2
b) SO_2 only
c) XeF_2 only
d) SO_2 and XeF_2
e) SO_2 and NH_3

H—H

:F̤—F̤:

H—F̤:

Lewis dot structures of H_2, F_2, and HF

H_2

F_2

HF

Ball-and-stick models

9.3 Valence Bond Theory

The Lewis theory of chemical bonding provides a relatively simple way for us to visualize the arrangement of electrons in molecules. It is insufficient, however, to explain the differences between the covalent bonds in compounds such as H_2, F_2, and HF, shown in the left margin. Although Lewis theory describes the bonds in these three molecules in exactly the same way, they really are quite different from one another, as evidenced by their bond lengths and bond enthalpies listed in Table 9.3. Understanding these differences and why covalent bonds form in the first place requires a bonding model that combines Lewis's notion of atoms sharing electron pairs and the quantum mechanical descriptions of atomic orbitals.

According to ***valence bond theory,*** atoms share electrons when an atomic orbital on one atom overlaps with an atomic orbital on the other. Each of the overlapping atomic orbitals must contain a single, unpaired electron. Furthermore, the two electrons shared by the bonded atoms must have opposite spins [⏮ Section 6.6]. The nuclei of both atoms are attracted to the shared pair of electrons. It is this mutual attraction for the shared electrons that holds the atoms together.

Representing Electrons in Atomic Orbitals

In Chapter 6, we learned that although an electron is a particle with a known mass, it exhibits wavelike properties. The quantum mechanical model of the atom, which gives rise to the familiar shapes of *s* and *p* atomic orbitals, treats electrons in atoms as waves, rather than particles. Therefore, rather than use arrows to denote the locations and spins of electrons, we will adopt a convention whereby a singly occupied orbital will appear as a light color and a doubly occupied orbital will appear as a darker version of the same color (Figure 9.3). In the representations of orbitals that follow, atomic *s* orbitals will be represented as yellow, and atomic *p* orbitals will be represented as blue. (Empty *p* orbitals will appear white.) When two electrons occupy the same atomic orbital in the ground state, their spins are paired—meaning that they have opposite spins [⏮ Section 6.6].

The H—H bond in H_2 forms when the singly occupied 1*s* orbitals of the two H atoms overlap:

Student Note: Keep in mind that although there are still just two electrons, each atom "thinks" it owns them both, so when the singly occupied orbitals overlap, both orbitals end up doubly occupied.

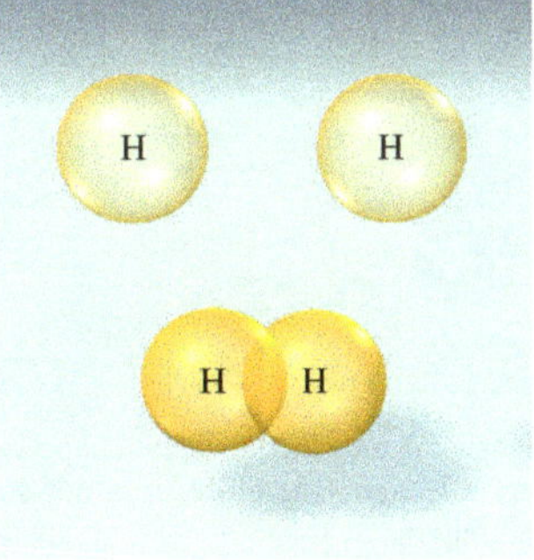

TABLE 9.3 Bond Lengths and Bond Enthalpies of H_2, F_2, and HF

	Bond Length (Å)	Bond Enthalpy (kJ/mol)
H_2	0.74	436.4
F_2	1.42	150.6
HF	0.92	568.2

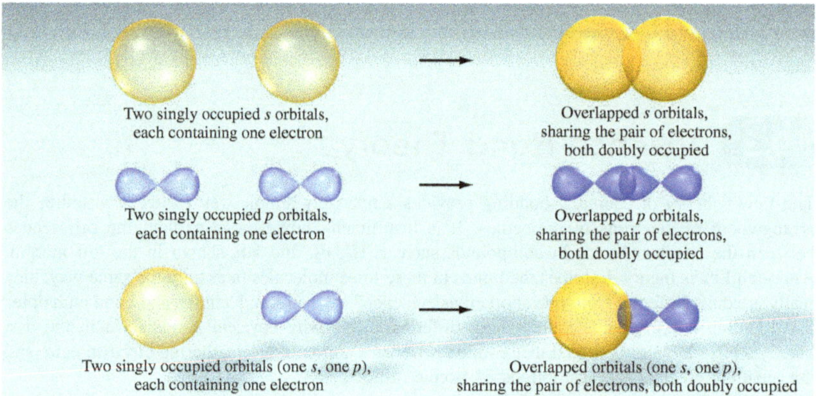

Figure 9.3 Representations of atomic orbitals. Singly occupied *s* orbitals appear light yellow. Doubly occupied *s* orbitals appear darker yellow. Singly and doubly occupied *p* orbitals appear lighter blue and darker blue, respectively.

Similarly, the F—F bond in F_2 forms when the singly occupied 2*p* orbitals of the two F atoms overlap:

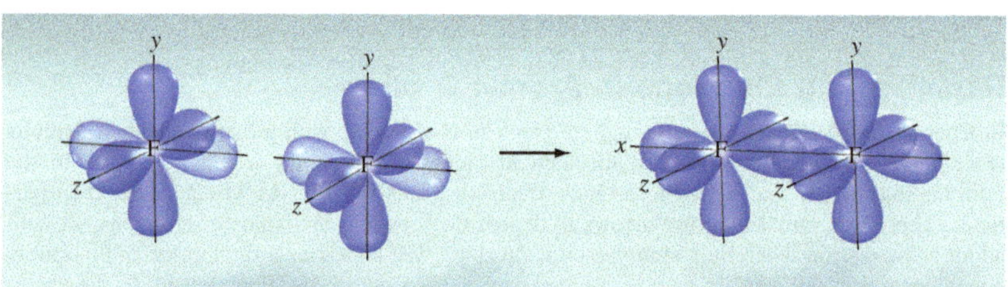

Recall that the ground-state electron configuration of the F atom is $[He]2s^22p^5$ [⏮ Section 6.8]. (The ground-state orbital diagram of F is shown in the margin.)

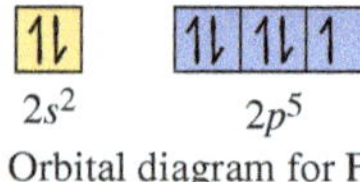

Orbital diagram for F

We can also depict the formation of an H—F bond using the valence bond model. In this case, the singly occupied 1*s* orbital of the H atom overlaps with the singly occupied 2*p* orbital of the F atom:

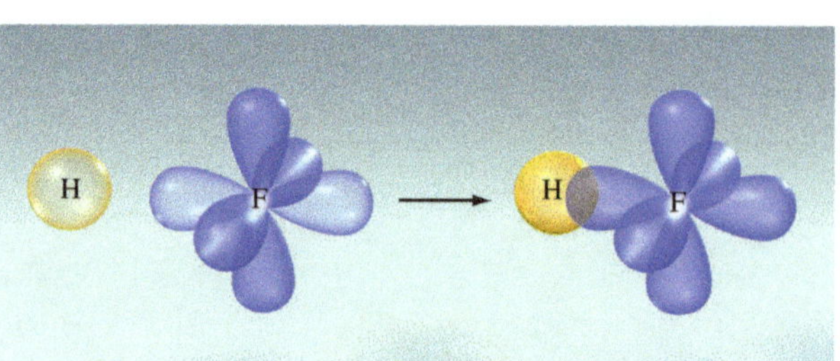

According to the quantum mechanical model, the sizes, shapes, and energies of the 1*s* orbital of H and the 2*p* orbital of F are different. Therefore, it is not surprising that the bonds in H_2, F_2, and HF vary in strength and length.

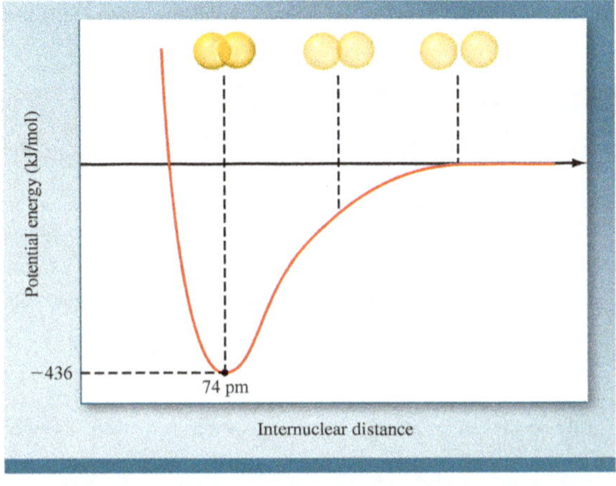

Figure 9.4 The change in potential energy of two hydrogen atoms as a function of internuclear distance. The minimum potential energy (−436 kJ/mol) occurs when the distance between the nuclei is 74 pm. The yellow spheres represent the 1s orbitals of hydrogen.

Energetics and Directionality of Bonding

Why do covalent bonds form? According to valence bond theory, a covalent bond will form between two atoms if the potential energy of the resulting molecule is lower than that of the isolated atoms. Simply put, this means that the formation of covalent bonds is exothermic. While this fact may not seem intuitively obvious, you know that energy must be supplied to a molecule to *break* covalent bonds [◀◀ Section 8.9]. Because the formation of a bond is the reverse process, we should expect energy to be given off. (Recall that the enthalpy change for a forward process and that for the reverse process differ only in sign: $\Delta H_{\text{forward}} = -\Delta H_{\text{reverse}}$ [◀◀ Section 5.3].) Figure 9.4 illustrates how the potential energy of two hydrogen atoms varies with distance between the nuclei.

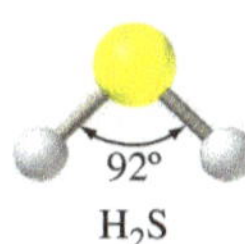

Valence bond theory also introduces the concept of *directionality* to chemical bonds. For example, we expect the bond formed by the overlap of a *p* orbital to coincide with the axis along which the *p* orbital lies. Consider the molecule H_2S, shown in the left margin. Unlike the other molecules that we have encountered, H_2S does not have the bond angle that Lewis theory and the VSEPR model would lead us to predict. (With four electron domains on the central atom, we would expect the bond angle to be on the order of 109.5°.) In fact, the H−S−H bond angle is 92°. Looking at this in terms of valence bond theory, the central atom (S) has two unpaired electrons, each of which resides in a 3*p* orbital. The orbital diagram for the ground-state electron configuration of the S atom is

Student Note: For you to understand the material in this section and Section 9.4, you must be able to draw orbital diagrams for ground-state electron configurations [◀◀ Section 6.8].

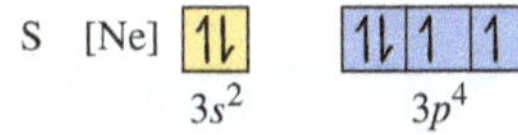

Remember that *p* orbitals are mutually perpendicular, lying along the *x*, *y*, and *z* axes [◀◀ Section 6.7]. We can rationalize the observed bond angle by envisioning the overlap of each of the singly occupied 3*p* orbitals with the 1*s* orbital of a hydrogen atom:

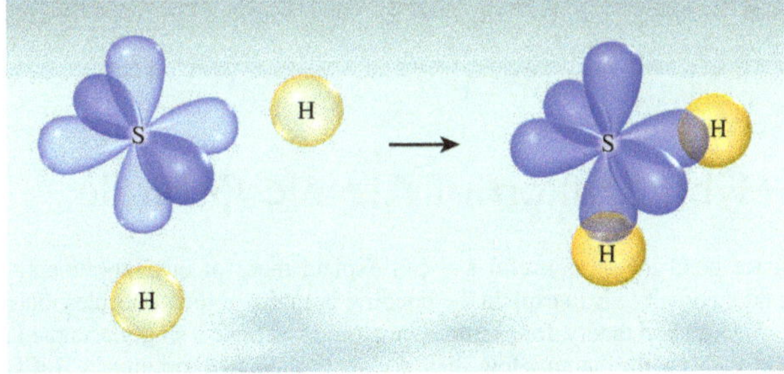

In summary, the important features of valence bond theory are as follows:

- A bond forms when singly occupied atomic orbitals on two atoms overlap.
- The two electrons shared in the region of orbital overlap must be of opposite spin.
- Formation of a bond results in a lower potential energy for the system.

Sample Problem 9.3 shows how to use valence bond theory to explain the bonding in a molecule.

SAMPLE PROBLEM 9.3

Hydrogen selenide (H_2Se) is a foul-smelling gas that can cause eye and respiratory tract inflammation. The H—Se—H bond angle in H_2Se is approximately 92°. Use valence bond theory to describe the bonding in this molecule.

Strategy Consider the central atom's ground-state electron configuration, and determine what orbitals are available for bond formation.

Setup The ground-state electron configuration of Se is $[Ar]4s^23d^{10}4p^4$. Its orbital diagram (showing only the 4*p* orbitals) is

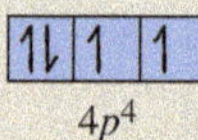

Solution Two of the 4*p* orbitals are singly occupied and therefore available for bonding. The bonds in H_2Se form as the result of the overlap of a hydrogen 1*s* orbital with each of these orbitals on the Se atom.

THINK ABOUT IT

Because the 4*p* orbitals on the Se atom are all mutually perpendicular, we should expect the angles between bonds formed by their overlap to be approximately 90°.

Practice Problem ATTEMPT Use valence bond theory to describe the bonding in phosphine (PH_3), which has H—P—H bond angles of approximately 94°.

Practice Problem BUILD For which molecule(s) can we *not* use valence bond theory to explain the bonding: SO_2 (O—S—O bond angle ~120°), CH_4 (H—C—H bond angles = 109.5°), AsH_3 (H—As—H bond angles = 92°)? Explain.

Practice Problem CONCEPTUALIZE Which of these models could represent a species for which valence bond theory is sufficient to explain the observed bond angle? Explain.

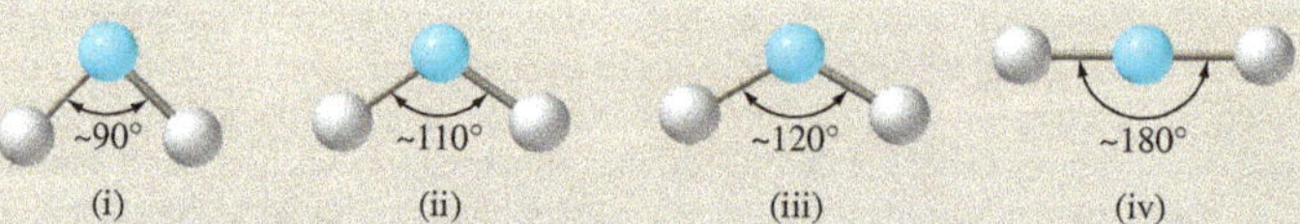

CHECKPOINT – SECTION 9.3 Valence Bond Theory

9.3.1 Which of the following atoms, in its ground state, does not have unpaired electrons? (Select all that apply.)

a) O b) Be c) B d) F e) Ne

9.3.2 According to valence bond theory, how many bonds would you expect a nitrogen atom (in its ground state) to form?

a) 2 b) 3 c) 4 d) 5 e) 6

9.4 Hybridization of Atomic Orbitals

Although valence bond theory is useful and can explain more of our experimental observations than Lewis bond theory, it fails to explain the bonding in many of the molecules that we encounter. According to valence bond theory, for example, an atom must have a singly occupied atomic orbital to form a bond with another atom. How then do we explain the bonding in $BeCl_2$? The central atom, Be, has a ground-state electron configuration of $[He]2s^2$, so it has no unpaired electrons. With no singly occupied atomic orbitals in its ground state, how does Be form two bonds?

:C̈l—Be—C̈l:
$BeCl_2$

Furthermore, in cases where the ground-state electron configuration of the central atom *does* have the required number of unpaired electrons, how do we explain the observed bond

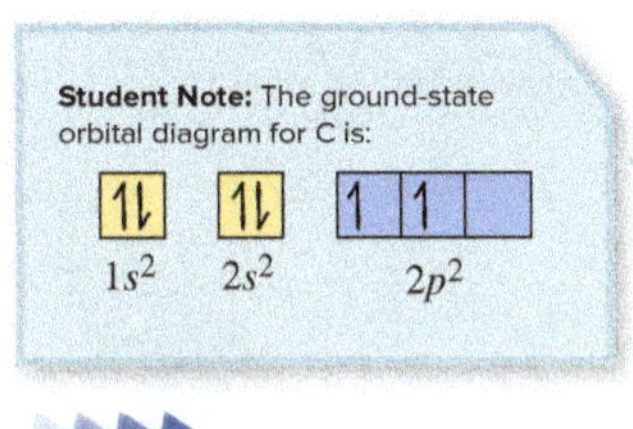

Animation
Hybrid orbitals—orbital hybridization and valence bond theory.

angles? Carbon, like sulfur, has two unpaired electrons in its ground state. Using valence bond theory as our guide, we might envision the formation of two covalent bonds with oxygen, as in CO_2. If the two unpaired electrons on C (each residing in a $2p$ orbital) were to form bonds, however, the O—C—O bond angle should be on the order of 90°, like the bond angle in H_2S. In fact, the bond angle in CO_2 is 180°:

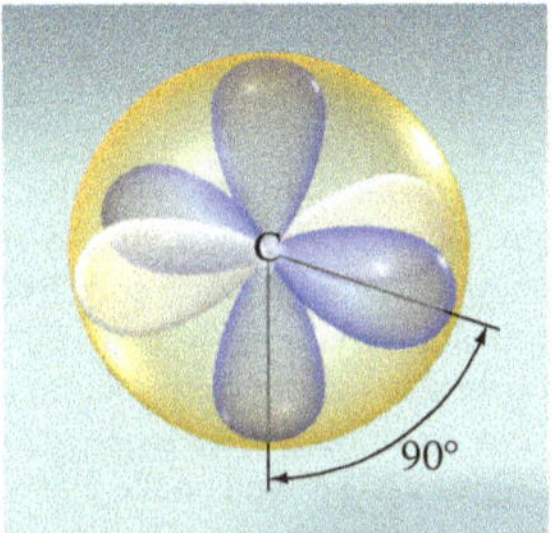

Bond angle should be 90°.

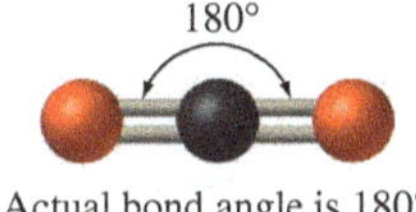

Actual bond angle is 180°.

To explain these and other observations, we need to extend our discussion of orbital overlap to include the concept of ***hybridization*** or *mixing* of atomic orbitals.

The idea of hybridization of atomic orbitals begins with the molecular geometry and works backward to explain the bonds and the observed bond angles in a molecule. To extend our discussion of orbital overlap and introduce the concept of hybridization of atomic orbitals, we first consider beryllium chloride ($BeCl_2$), which has two electron domains on the central atom. Using its Lewis structure (shown in the margin) and the VSEPR model, we predict that $BeCl_2$ will have a Cl—Be—Cl bond angle of 180°. If this is true, though, how does Be form two bonds with no unpaired electrons, and why is the angle between the two bonds 180°?

Student Note: Remember that Be is one of the atoms that does not obey the octet rule [⏮ Section 8.1]. The Lewis structure of $BeCl_2$ is

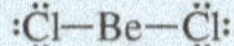

To answer the first part of the question, we envision the *promotion* of one of the electrons in the $2s$ orbital to an empty $2p$ orbital. Recall that electrons can be promoted from a lower atomic orbital to a higher one [⏮ Section 6.3]. The ground-state electron configuration is the one in which all the electrons occupy orbitals of the lowest possible energy. A configuration in which one or more electrons occupy a *higher* energy orbital is called an *excited* state. An excited state generally is denoted with an asterisk (e.g., Be* for an excited-state Be atom). Showing only the valence orbitals, we can represent the promotion of one of the valence electrons of beryllium as

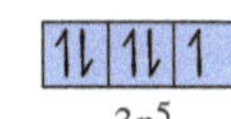

$3s^2$ $3p^5$

Orbital diagram for Cl

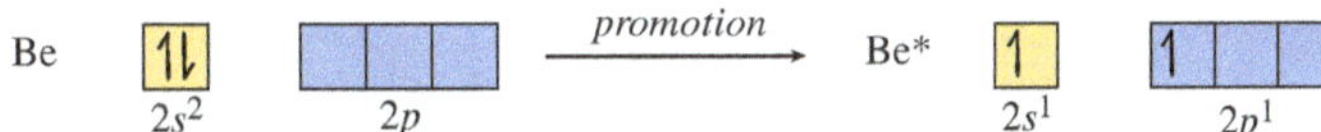

With one of its valence electrons promoted to the $2p$ subshell, the Be atom now has two unpaired electrons and therefore can form two bonds. However, the orbitals in which the two unpaired electrons reside are different from each other, so we would expect bonds formed as a result of the overlap of these two orbitals (each with a $3p$ orbital on a Cl atom) to be different:

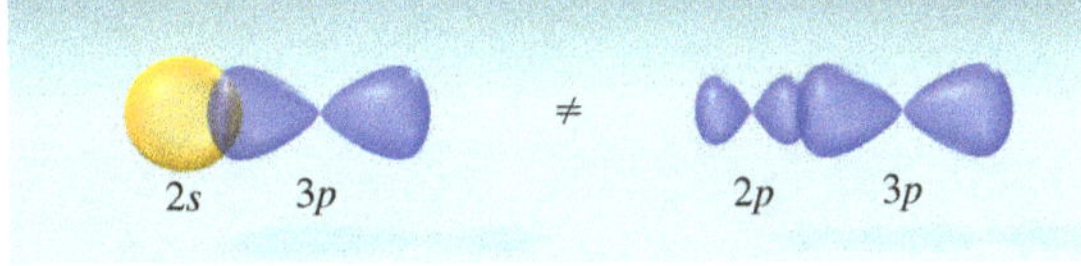

Experimentally, though, the bonds in $BeCl_2$ are identical in length and strength.

Hybridization of *s* and *p* Orbitals

To explain how beryllium forms two identical bonds, we must mix the orbitals in which the unpaired electrons reside, thus yielding two equivalent orbitals. The mixing of beryllium's $2s$ orbital with one of its $2p$ orbitals, a process known as *hybridization*, yields two *hybrid orbitals* that are neither *s* nor *p*, but have some character of each. The hybrid orbitals are designated 2*sp* or simply *sp*.

Student Note: Hybrid orbitals are another type of *electron domain*.

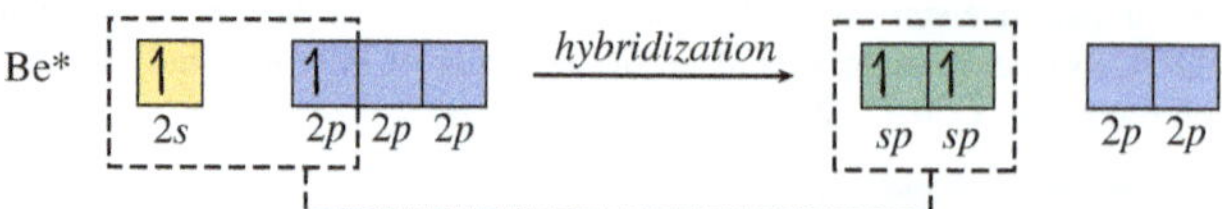

Mixing of one *s* orbital and one *p* orbital to yield two *sp* orbitals

The mathematical combination of the quantum mechanical wave functions for an *s* orbital and a *p* orbital gives rise to two new, equivalent wave functions. As shown in Figure 9.5(a), each *sp* hybrid orbital has one small lobe and one large lobe and, like any two electron domains on an atom, they are oriented in opposite directions with a 180° angle between them. The figure shows the atomic and hybrid orbitals separately for clarity. Note that the hybrid orbitals are shown in two ways: the first is a more realistic shape, whereas the second is a simplified shape that we use to keep the figures clear and make the visualization of orbitals easier. Note also that the representations of hybrid orbitals are green. Figure 9.5(b) shows the locations of the atomic orbitals and the hybrid orbitals, respectively, relative to the beryllium nucleus.

Student Note: In picturing the shapes of the resulting hybrid orbitals it may help to remember that orbitals, like waves, can combine constructively or destructively. We can think of the large lobe of each *sp* hybrid orbital as the result of a *constructive* combination and the small lobe of each as the result of a *destructive* combination.

With two *sp* hybrid orbitals, each containing a single unpaired electron, we can see how the Be atom is able to form two identical bonds with two Cl atoms [Figure 9.5(c)]. Each of the singly occupied *sp* hybrid orbitals on the Be atom overlaps with the singly occupied 3*p* atomic orbital on a Cl atom. The energy required to promote an electron in an atom is more than compensated for by the energy given off when a bond forms.

We can do a similar analysis of the bonds and the trigonal-planar geometry of boron trifluoride (BF_3). The ground-state electron configuration of the B atom is $[He]2s^22p^1$, containing just one unpaired electron. Promotion of one of the 2*s* electrons to an empty 2*p* orbital gives

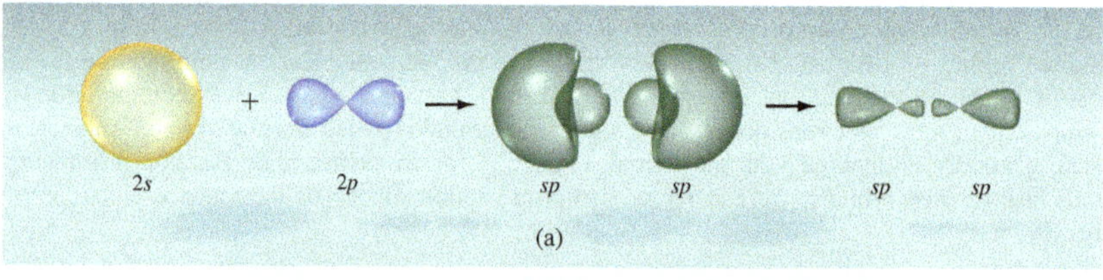

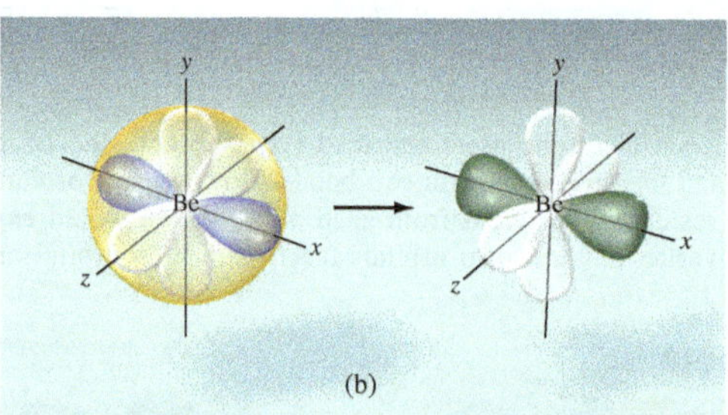

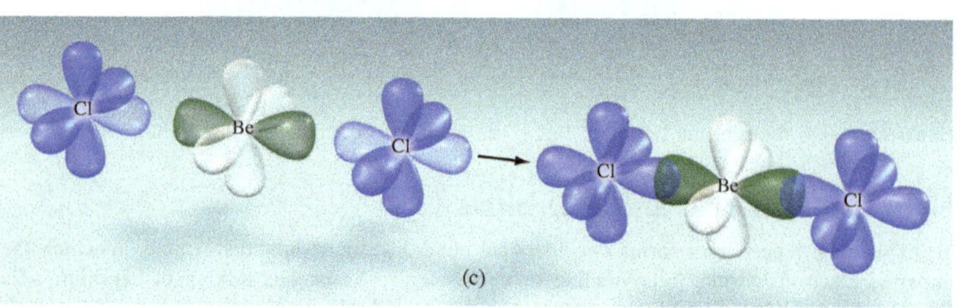

Figure 9.5 (a) An atomic *s* orbital (yellow) and one atomic *p* orbital (blue) combine to form two *sp* hybrid orbitals (green). The realistic hybrid orbital shapes are shown first. The thinner representations are used to keep diagrams clear. (b) The 2*s* orbital and one of the 2*p* orbitals on Be combine to form two *sp* hybrid orbitals. Unoccupied orbitals are shown in white. Like any two electron domains, the hybrid orbitals on Be are 180° apart. (c) The hybrid orbitals on Be each overlap with a singly occupied 3*p* orbital on a Cl atom.

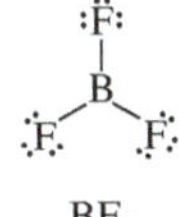

BF_3

Student Note: A hybrid orbital analysis starts with a known molecular geometry and known bond angles. It is not used to predict geometries.

the three unpaired electrons needed to explain the formation of *three* bonds. The ground-state and excited-state electron configurations can be represented by

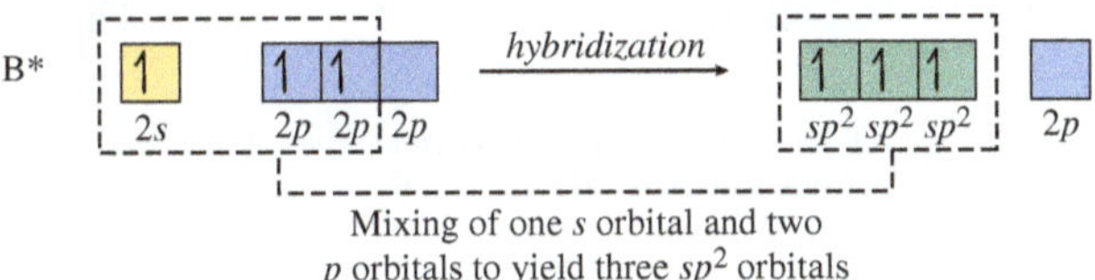

Because the three bonds in BF_3 are identical, we must hybridize the three singly occupied atomic orbitals (the one *s* and two *p* orbitals) to give three singly occupied hybrid orbitals:

Figure 9.6 illustrates the hybridization and bond formation in BF_3.

In both cases (i.e., for $BeCl_2$ and BF_3), some but not *all* of the *p* orbitals are hybridized. When the remaining unhybridized atomic *p* orbitals do not contain electrons, as in the case of BF_3, they will not be part of the discussion of bonding in this chapter. As we will see in Section 9.5, though, unhybridized atomic orbitals that *do* contain electrons are important in our description of the bonding in a molecule.

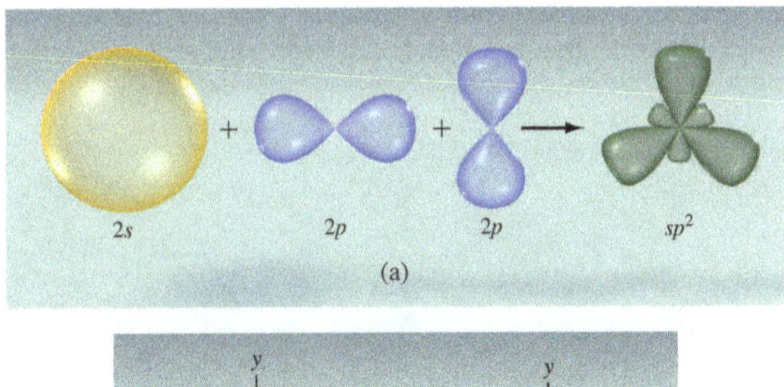

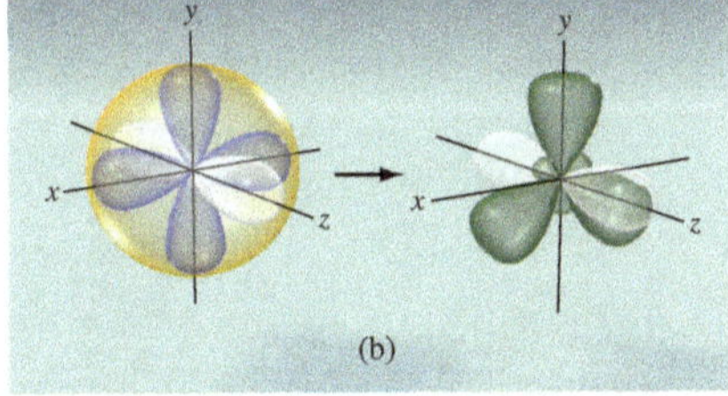

Figure 9.6 (a) An *s* atomic orbital and two *p* atomic orbitals combine to form three sp^2 hybrid orbitals. (b) The three sp^2 hybrid orbitals on B are arranged in a trigonal plane. (Empty atomic orbitals are shown in white.) (c) Hybrid orbitals on B overlap with 2*p* orbitals on F.

We can now apply the same kind of analysis to the methane molecule (CH_4). The Lewis structure of CH_4 has four electron domains around the central carbon atom. This means that we need four hybrid orbitals, which in turn means that four atomic orbitals must be hybridized. The ground-state electron configuration of the C atom contains two unpaired electrons. Promotion of one electron from the 2*s* orbital to the empty 2*p* orbital yields the four unpaired electrons needed for the formation of four bonds:

H
H—C—H
H
CH_4

C [↑↓] $2s^2$ [↑|↑|] $2p^2$ —*promotion*→ C* [↑] $2s^1$ [↑|↑|↑] $2p^3$

Hybridization of the *s* orbital and the three *p* orbitals yields four hybrid orbitals designated sp^3. We can then place the electrons that were originally in the *s* and *p* atomic orbitals into the sp^3 hybrid orbitals:

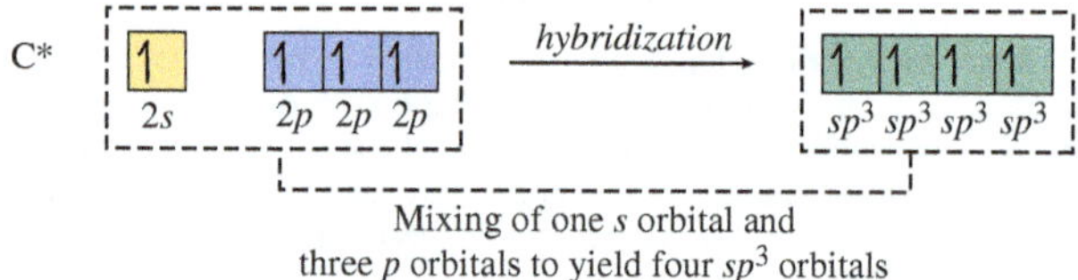

The set of four sp^3 hybrid orbitals on carbon, like any four electron domains on a central atom, assumes a tetrahedral arrangement. Figure 9.7 illustrates how the hybridization of the C atom results in the formation of the four bonds and the 109.5° bond angles observed in CH_4.

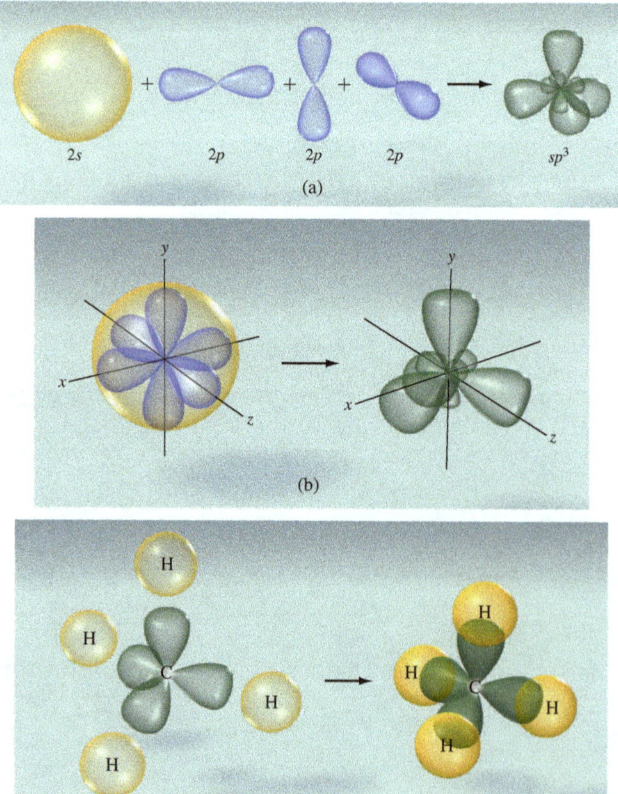

Figure 9.7 (a) An *s* atomic orbital and three *p* atomic orbitals combine to form four sp^3 hybrid orbitals. (b) The four sp^3 hybrid orbitals on C are arranged in a tetrahedron. (c) Hybrid orbitals on C overlap with 1*s* orbitals on H. For clarity, the small lobes of the hybrid orbitals are not shown.

Hybridization of *s*, *p*, and *d* Orbitals

PCl_5

Recall that elements in the third period of the periodic table and beyond do not necessarily obey the octet rule because they have *d* orbitals that can hold additional electrons [⏮ Section 8.5]. To explain the bonding and geometry of molecules in which there are more than four electron domains on the central atom, we must include *d* orbitals in our hybridization scheme. PCl_5, for example, has five electron domains around the P atom. To explain the five bonds in this molecule, we will need five singly occupied hybrid orbitals. The ground-state electron configuration of the P atom is $[Ne]3s^2 3p^3$, which contains three unpaired electrons. In this case, though, because all three of the *p* orbitals are occupied, promotion of an electron from the 3*s* orbital to a 3*p* orbital would *not* result in additional unpaired electrons. However, we can promote an electron from the 3*s* orbital to an empty 3*d* orbital, thus forming the five unpaired electrons needed:

Student Note: Note that the superscript numbers in hybrid orbital notation are used to designate the number of atomic orbitals that have undergone hybridization. When the superscript is 1, it is not shown (analogous to the subscripts in chemical formulas).

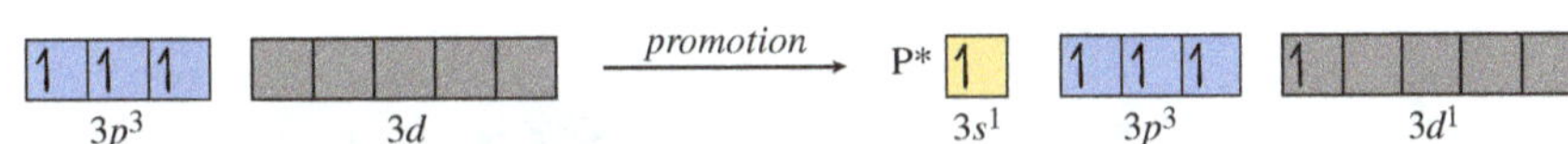

Hybridization of the *s* orbital, the three *p* orbitals, and one of the *d* orbitals yields hybrid orbitals that are designated sp^3d. After placing the five electrons in the five hybrid orbitals, we can rationalize the formation of five bonds in the molecule:

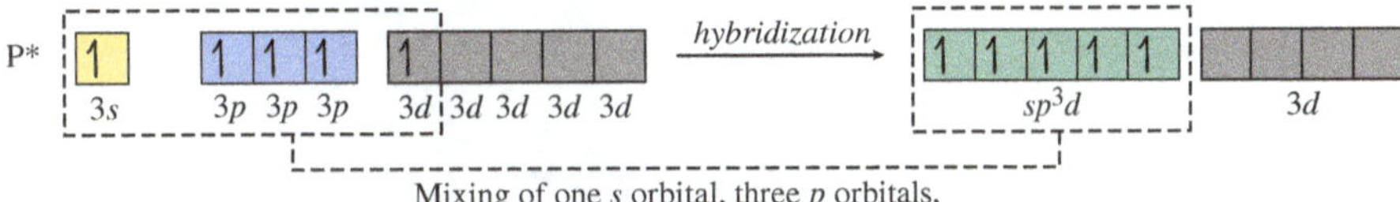

The sp^3d orbitals have shapes similar to those we have seen for the *sp*, sp^2, and sp^3 hybrid orbitals; that is, one large lobe and one small lobe. In addition, the five hybrid orbitals adopt a trigonal bipyramidal arrangement, enabling us to explain the geometry and bond angles in PCl_5:

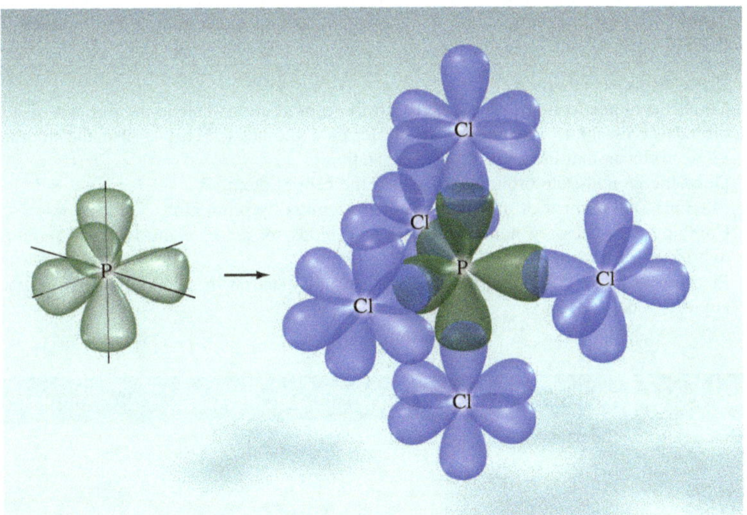

A similar analysis can be done with the SF_6 molecule. The ground-state electron configuration of the S atom is $[Ne]3s^2 3p^4$, giving it only two unpaired electrons. To obtain the six unpaired electrons needed to form six S—F bonds, we must promote *two* electrons to empty *d* orbitals: one from the 3*s* orbital and one from the doubly occupied 3*p* orbital. The resulting hybrid orbitals are designated sp^3d^2.

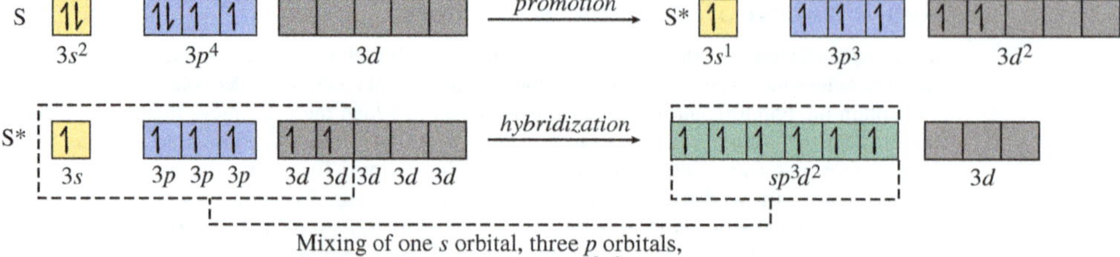

The six bonds in SF_6 form, therefore, when each sp^3d^2 hybrid orbital on the S atom overlaps with a singly occupied $2p$ orbital on an F atom:

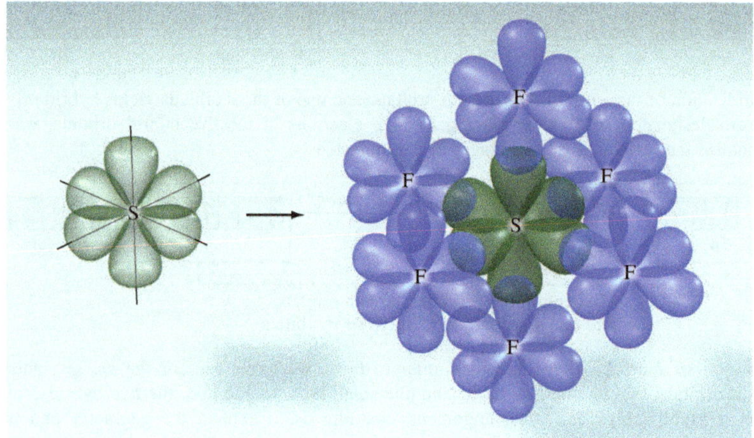

Table 9.4 shows how the number of electron domains on a central atom corresponds to a set of hybrid orbitals. In general, the hybridized bonding in a molecule can be described using the following steps:

1. Draw the Lewis structure.
2. Count the number of electron domains on the central atom. This is the number of hybrid orbitals necessary to account for the molecule's geometry. (This is also the number of *atomic* orbitals that must undergo hybridization.)
3. Draw the ground-state orbital diagram for the central atom.
4. Maximize the number of unpaired valence electrons by promotion.
5. Combine the necessary number of atomic orbitals to generate the required number of hybrid orbitals.
6. Place electrons in the hybrid orbitals, putting one electron in each orbital before pairing any electrons.

Student Note: The electrons that you place in hybrid orbitals are those that originally resided in the atomic orbitals that have undergone hybridization.

TABLE 9.4 Number of Electron Domains and Hybrid Orbitals on Central Atom

Number of Electron Domains on Central Atom	2	3	4	5	6
Hybrid Orbitals	sp	sp^2	sp^3	sp^3d	sp^3d^2
Geometry	Linear	Trigonal planar	Tetrahedral	Trigonal bipyramidal	Octahedral

It is important to recognize that we do not use hybrid orbitals to *predict* molecular geometries, but rather to *explain* geometries that are already known. As we saw in Section 9.3, the bonding in many molecules can be explained without the use of hybrid orbitals. Hydrogen sulfide (H_2S), for example, has a bond angle of 92°. This bond angle is best explained without the use of hybrid orbitals.

Sample Problem 9.4 shows how to use hybridization to explain the bonding and geometry in a molecule.

SAMPLE PROBLEM 9.4

Ammonia (NH_3) is a trigonal pyramidal molecule with H—N—H bond angles of about 107°. Describe the formation of three equivalent N—H bonds, and explain the angles between them.

Strategy Starting with a Lewis structure, determine the number and type of hybrid orbitals necessary to rationalize the bonding in NH_3.

Setup The Lewis structure of NH_3 is

H—N̈—H
H 107°

The ground-state electron configuration of the N atom is $[He]2s^22p^3$. Its valence orbital diagram is

N [⇅] [↑|↑|↑]
$2s^2$ $2p^3$

Solution Although the N atom has the three unpaired electrons needed to form three N—H bonds, we would expect bond angles of ~90° (not 107°) to form from the overlap of the three mutually perpendicular $2p$ orbitals. Hybridization, therefore, is necessary to explain the bonding in NH_3. (Recall from Section 9.3 that valence bond theory alone could be used to explain the bonding in H_2S, where the bond angles are ~90°—hybridization was unnecessary.) Although we often need to promote an electron to maximize the number of unpaired electrons, no promotion is necessary for the nitrogen in NH_3. We already have the three unpaired electrons necessary, and the promotion of an electron from the $2s$ orbital to one of the $2p$ orbitals would not result in any additional unpaired electrons. Furthermore, there are no empty d orbitals in the second shell. According to the Lewis structure, there are four electron domains on the central atom (three bonds and a lone pair of electrons). Four electron domains on the central atom require four hybrid orbitals, and four hybrid orbitals require the hybridization of four atomic orbitals: one s and three p. This corresponds to sp^3 hybridization. Because the atomic orbitals involved in the hybridization contain a total of five electrons, we place five electrons in the resulting hybrid orbitals. This means that one of the hybrid orbitals will contain a lone pair of electrons:

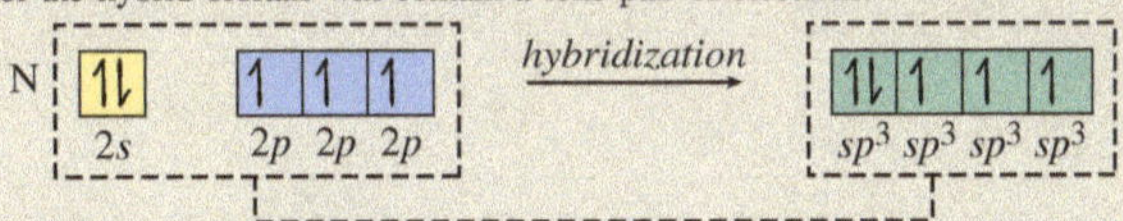

Each N—H bond is formed by the overlap between an sp^3 hybrid orbital on the N atom and the $1s$ atomic orbital on an H atom. Because there are four electron domains on the central atom, we expect them to be arranged in a tetrahedron. In addition, because one of the electron domains is a lone pair, we expect the H—N—H bond angles to be slightly smaller than the ideal tetrahedral bond angle of 109.5°:

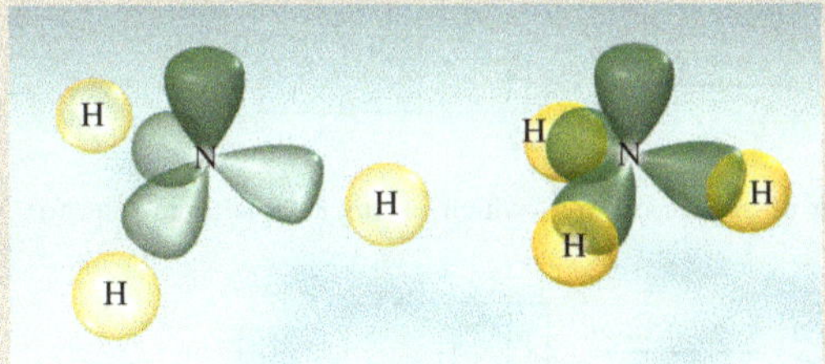

THINK ABOUT IT

This analysis agrees with the experimentally observed geometry and bond angles of 107° in NH_3.

Practice Problem ATTEMPT Use hybrid orbital theory to describe the bonding and explain the bond angles in bromine pentafluoride (BrF_5).

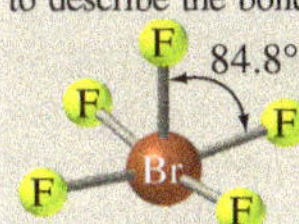

Practice Problem BUILD Use hybrid orbital theory to describe the bonding and explain the bond angles in BeF_2:

180°

F—Be—F

Practice Problem CONCEPTUALIZE Indicate for which of the following species hybrid orbitals must be used to explain the geometry: CCl_4, Cl_2, SO_3^{2-}, ClF.

CHECKPOINT – SECTION 9.4 Hybridization of Atomic Orbitals

9.4.1 How many orbitals does a set of sp^2 hybrid orbitals contain?

a) 2 b) 3 c) 4 d) 5 e) 6

9.4.2 How many p atomic orbitals are required to generate a set of sp^3 hybrid orbitals?

a) 0 b) 1 c) 2 d) 3 e) 4

9.5 Hybridization in Molecules Containing Multiple Bonds

The concept of valence bond theory and hybridization can also be used to describe the bonding in molecules containing double and triple bonds, such as ethylene (C_2H_4) and acetylene (C_2H_2). The Lewis structure of ethylene is

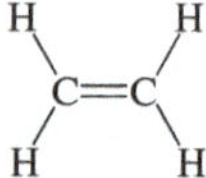

Each carbon atom is surrounded by three electron domains (two single bonds and one double bond). Thus, we expect the hybridization about each C atom to be sp^2, just like the B atom in BF_3. Applying the procedure described in Section 9.4, we first maximize the number of unpaired electrons by promoting an electron from the $2s$ orbital to the empty $2p$ orbital:

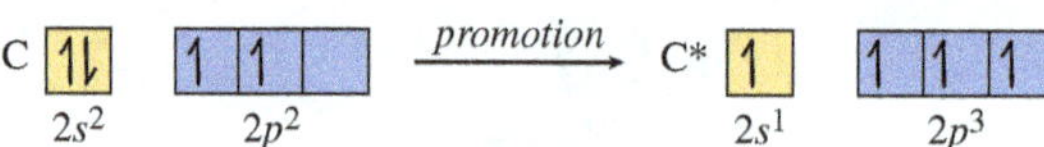

We then hybridize the required number of atomic orbitals, which in this case is three (one for each electron domain on the C atom):

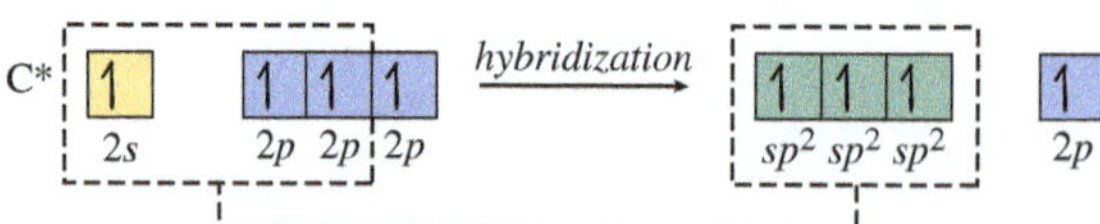

The three equivalent sp^2 hybrid orbitals, arranged in a trigonal plane, enable us to explain the three bonds about each C atom. In this case, however, each C atom is left with a singly occupied, *unhybridized* atomic orbital. As we will see, it is the singly occupied p orbitals *not* involved in hybridization that give rise to multiple bonds in molecules.

In the bonding schemes that we have described thus far, the overlap of atomic orbitals or hybrid orbitals occurs directly between the two nuclei involved in bonding. Such bonds, in which the shared electron density is concentrated directly along the internuclear axis, are called ***sigma* (σ) *bonds*.** The ethylene molecule (also known as *ethene*) contains five sigma bonds: one between

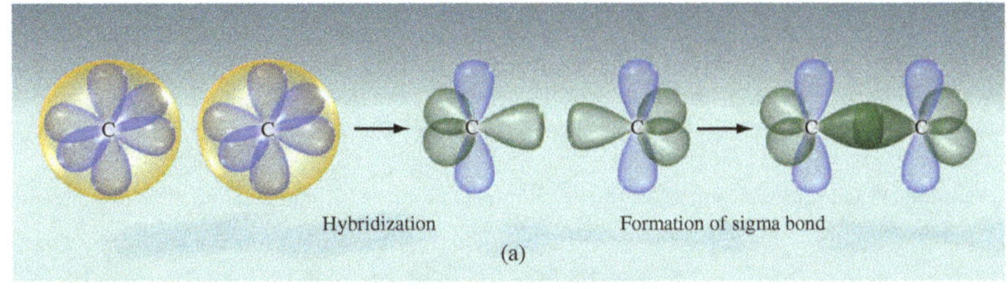

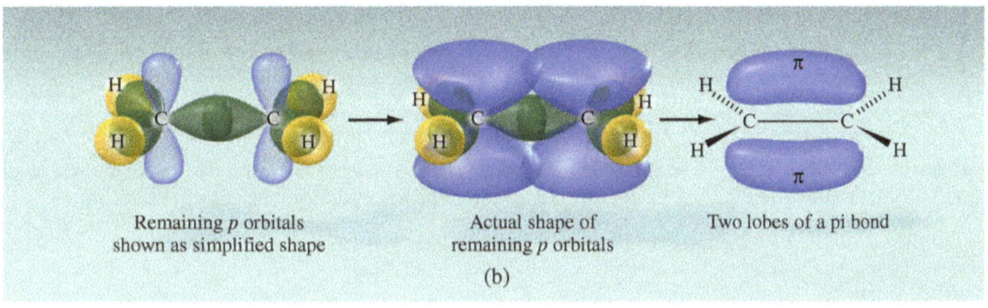

Figure 9.8 (a) A sigma bond forms when sp^2 hybrid orbitals on the C atoms overlap. Each C atom has one remaining unhybridized *p* orbital. (b) The remaining *p* orbitals overlap to form a pi bond.

the two C atoms (the result of the overlap of one of the sp^2 hybrid orbitals on each C atom) and four between the C and H atoms (each the result of the overlap of an sp^2 hybrid orbital on a C atom and the 1*s* orbital on an H atom). The leftover unhybridized *p* orbital is perpendicular to the plane in which the atoms of the molecule lie. Figure 9.8(a) illustrates the formation and the overlap of sp^2 hybrid orbitals and shows the positions of the remaining *p* orbital on each C atom.

Remember that the shapes used to represent the atomic and hybrid orbitals are simplified to make visualization of the molecules easier. The actual shapes of both atomic and hybrid orbitals are such that when the sp^2 hybrid orbitals overlap to form a sigma bond between the two C atoms, the remaining unhybridized *p* orbitals also overlap, although to a smaller extent. Figure 9.8(b) shows the overlap of the unhybridized *p* orbitals on the two C atoms in ethylene. The resulting regions of electron density are concentrated above and below the plane of the molecule, in contrast to a sigma bond in which the electron density is concentrated directly along the internuclear axis. Bonds that form from the sideways overlap of *p* orbitals are called ***pi* (*π*) *bonds*.** The two regions of overlap shown in Figure 9.8(b) together make up one pi bond. It is the formation of the pi bond that makes the ethylene molecule planar.

Animation
Chemical bonding—sigma and pi bonding in benzene.

A sigma bond and a pi bond together constitute a *double* bond. Because the sideways overlap of *p* orbitals is not as effective as the overlap of hybrid orbitals that point directly toward each other, the contribution of the pi bond to the overall strength of the bond is less than that of the sigma bond. The bond enthalpy of a carbon-carbon double bond (620 kJ/mol) is greater than that of a carbon-carbon single bond (347 kJ/mol), but it is not twice as large [⏮ Section 8.9].

Sample Problem 9.5 shows how to use a Lewis structure to determine the number of sigma and pi bonds in a molecule.

SAMPLE PROBLEM 9.5

Thalidomide ($C_{13}H_{10}N_2O_4$) is a sedative and antiemetic that was widely prescribed during the 1950s, although not in the United States, for pregnant women suffering from morning sickness. Its use was largely discontinued when it was determined to be responsible for thousands of devastating birth defects. Determine the number of carbon-carbon sigma bonds and the total number of pi bonds in thalidomide.

Student Note: For more on the history of thalidomide, see the Chapter 25 Chapter Opener and Standardized-Exam Practice Problems essay.

Thalidomide

Strategy Use the Lewis structure to determine the number of single and double bonds. Then, to convert the number of single and double bonds to the number of sigma and pi bonds, remember that a single bond typically is composed of a sigma bond, whereas a double bond is usually composed of one sigma bond and one pi bond.

Setup There are nine carbon-carbon single bonds and three carbon-carbon double bonds. Overall there are seven double bonds in the molecule (three C=C and four C=O).

Solution Thalidomide contains 12 carbon-carbon sigma bonds and a total of seven pi bonds (three in carbon-carbon double bonds and four in carbon-oxygen double bonds).

THINK ABOUT IT

The Lewis structure given for thalidomide is one of two possible resonance structures. Draw the other resonance structure, and count sigma and pi bonds again. Make sure you get the same answer.

Practice Problem ATTEMPT The active ingredient in Tylenol and a host of other over-the-counter pain relievers is acetaminophen ($C_8H_9NO_2$). Determine the total number of sigma and pi bonds in the acetaminophen molecule.

Acetaminophen

Practice Problem BUILD Determine the total number of sigma and pi bonds in a molecule of aspirin ($C_9H_8O_4$).

Aspirin

Practice Problem CONCEPTUALIZE In terms of valence bond theory and hybrid orbitals, explain why C_2H_2 and C_2H_4 contain pi bonds, whereas C_2H_6 does not.

Because the *p* orbitals that form pi bonds must be parallel to each other, pi bonds restrict the rotation of a molecule in a way that sigma bonds do not. For example, the molecule 1,2-dichloroethane exists as a single isomer. Although we can draw the molecule in several different ways, including the two shown in Figure 9.9(a), all of them are equivalent because the molecule can rotate freely about the sigma bond between the two carbon atoms.

Student Note: Remember that *isomers* are molecules with the same chemical formula but different structural arrangements of atoms (Section 9.2).

On the other hand, 1,2-dichloroethylene exists as two distinct isomers—*cis* and *trans*—as shown in Figure 9.9(b). The double bond between the carbon atoms consists of one sigma bond and one pi bond. The pi bond restricts rotation about the sigma bond, making the molecules rigid, planar, and not interchangeable. To change one isomer into the other, the pi bond would have to be broken and rotation would have to occur about the sigma bond and the pi bond. This process would require a significant input of energy.

The acetylene molecule (C_2H_2) is linear. Because each carbon atom has two electron domains around it in the Lewis structure, the carbon atoms are *sp* hybridized. As before, promotion of an electron first maximizes the number of unpaired electrons:

H—C≡C—H
C_2H_2

C $2s^2$ $2p^2$ —*promotion*→ C* $2s^1$ $2p^3$

The 2*s* orbital and one of the 2*p* orbitals then mix to form two *sp* hybrid orbitals:

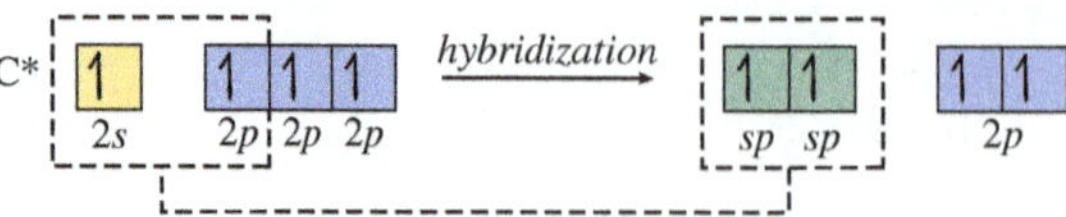

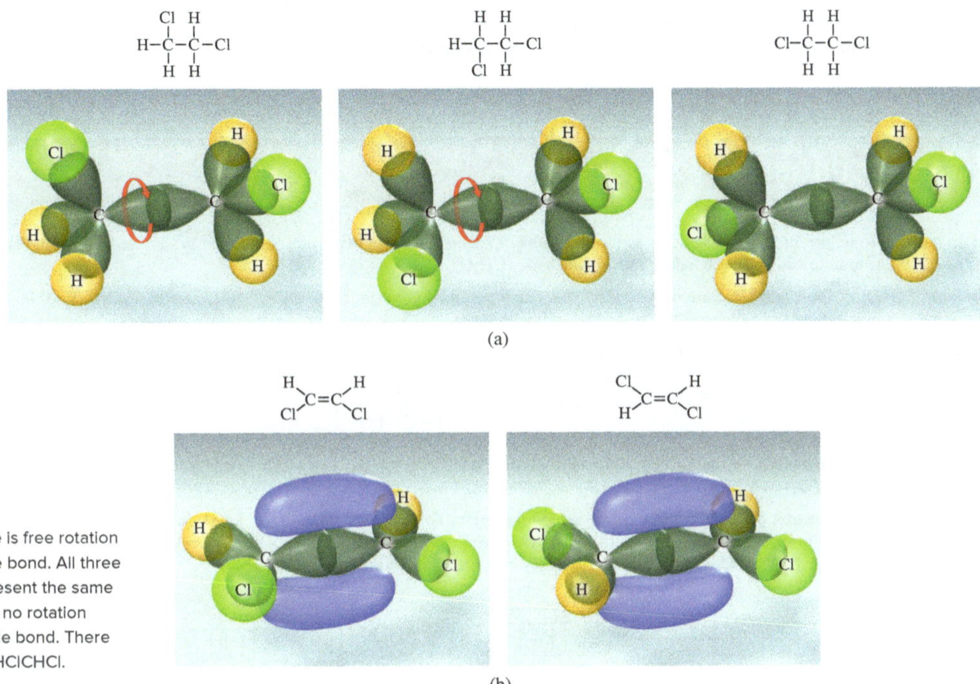

Figure 9.9 (a) There is free rotation about the C—C single bond. All three Lewis structures represent the same molecule. (b) There is no rotation about the C—C double bond. There are two isomers of CHClCHCl.

Animation
Figure 9.11, Formation of Pi Bonds in Ethylene and Acetylene.

This leaves two unhybridized *p* orbitals (each containing an electron) on each C atom. Figure 9.10 shows the sigma and pi bonds in the acetylene molecule (also known as *ethyne*). Just as one sigma bond and one pi bond make up a *double* bond, one sigma bond and two pi bonds make up a *triple* bond. Figure 9.11 summarizes the formation of bonds in ethane, ethylene, and acetylene.

Figure 9.10 (a) Formation of the sigma bond in acetylene. (b) Formation of the pi bonds in acetylene.

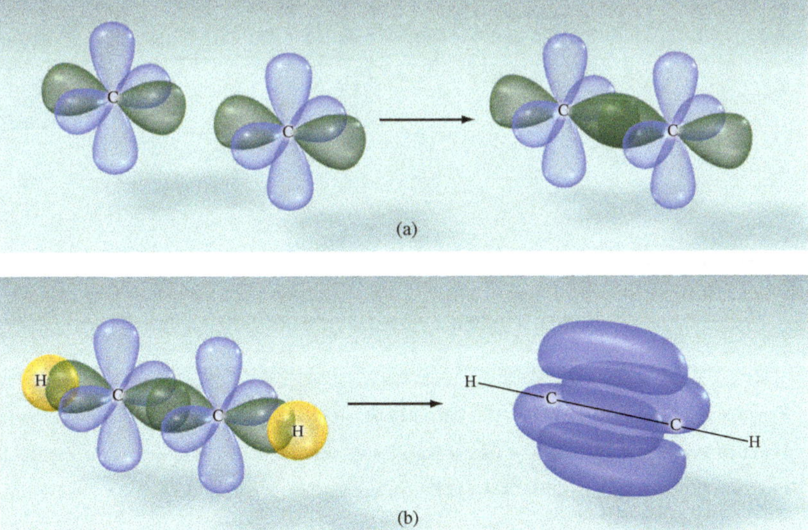

Sample Problem 9.6 shows how hybrid orbitals and pi bonds can be used to explain the bonding in formaldehyde, a molecule with a carbon-oxygen double bond.

SAMPLE PROBLEM 9.6

In addition to its use in aqueous solution as a preservative for laboratory specimens, formaldehyde gas is used as an antibacterial fumigant. Use hybridization to explain the bonding in formaldehyde (CH_2O).

Strategy Draw the Lewis structure of formaldehyde, determine the hybridization of the C and O atoms, and describe the formation of the sigma and pi bonds in the molecule.

Setup The Lewis structure of formaldehyde is

The C and O atoms each have three electron domains around them. [Carbon has two single bonds (C—H) and a double bond (C=O); oxygen has a double bond (O=C) and two lone pairs.]

Solution Three electron domains correspond to sp^2 hybridization. For carbon, promotion of an electron from the $2s$ orbital to the empty $2p$ orbital is necessary to maximize the number of unpaired electrons. For oxygen, no promotion is necessary. Each undergoes hybridization to produce sp^2 hybrid orbitals; and each is left with a singly occupied, unhybridized p orbital:

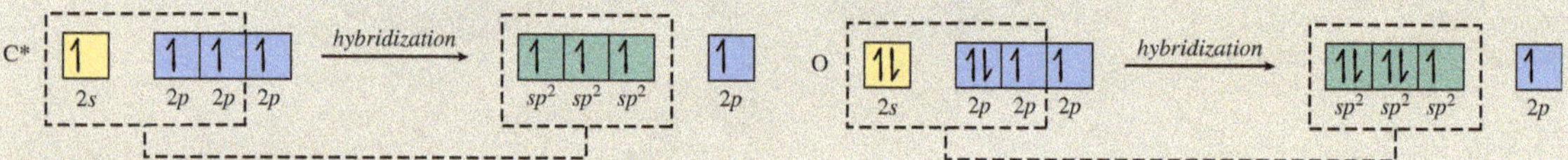

A sigma bond is formed between the C and O atoms by the overlap of one of the sp^2 hybrid orbitals from each of them. Two more sigma bonds form between the C atom and the H atoms by the overlap of carbon's remaining sp^2 hybrid orbitals with the $1s$ orbital on each H atom. Finally, the remaining p orbitals on C and O overlap to form a pi bond:

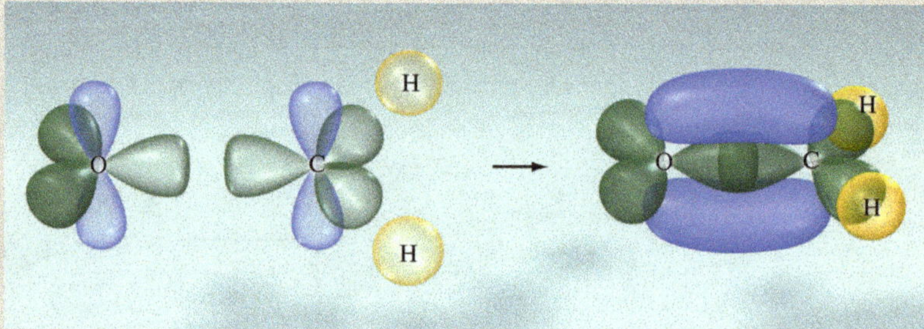

THINK ABOUT IT

Our analysis describes the formation of both a sigma bond and a pi bond between the C and O atoms. This corresponds correctly to the double bond predicted by the Lewis structure. (The two lone pairs on the O atom are the electrons in the doubly occupied sp^2 hybrid orbitals.)

Practice Problem ATTEMPT Use valence bond theory and hybrid orbitals to explain the bonding in hydrogen cyanide (HCN).

Practice Problem BUILD Use valence bond theory and hybrid orbitals to explain the bonding in diatomic nitrogen (N_2).

Practice Problem CONCEPTUALIZE Explain why hybrid orbitals are necessary to explain the bonding in N_2 and O_2, but not to explain the bonding in H_2 or Br_2.

Figure 9.11

Formation of Pi Bonds in Ethylene and Acetylene

2*s* and 2*p* atomic orbitals on two C atoms.

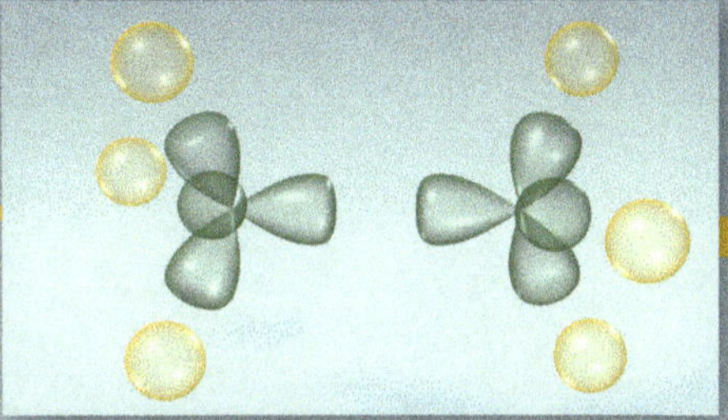

In ethane (C_2H_6) the C atoms are sp^3-hybridized.

In ethylene (C_2H_4) the C atoms are sp^2-hybridized.

Overlap of sp^2 hybrid orbitals forms a sigma bond between the two C atoms.

In acetylene (C_2H_2) the C atoms are *sp*-hybridized.

Overlap of *sp* hybrid orbitals forms a sigma bond between the two C atoms.

Overlap of sp^3 hybrid orbitals forms a sigma bond between the two C atoms.

Each C atom forms three sigma bonds with H atoms.

Each C atom forms two sigma bonds with H atoms. Each C atom has one leftover unhybridized *p* orbital.

The parallel, unhybridized *p* orbitals overlap to form a pi bond with two lobes.

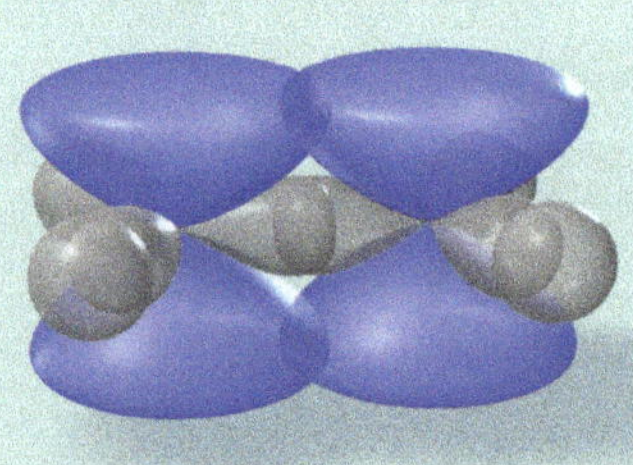

Using the more realistic shape shows how *p* orbitals overlap.

Each C atom forms one sigma bond with an H atom. Each C atom has two leftover unhybridized *p* orbitals.

Each pair of parallel, unhybridized *p* orbitals overlaps to form a pi bond with two lobes.

(See Visualizing Chemistry Questions VC 9.1–VC 9.4 on page 415.)

What's the point?

When carbon is sp^2- or *sp*-hybridized, parallel, unhybridized *p* orbitals interact to form pi bonds. Each pi bond consists of two lobes as a result of the overlap.

CHECKPOINT – SECTION 9.5 Hybridization in Molecules Containing Multiple Bonds

9.5.1 Which of the following molecules contain one or more pi bonds? (Select all that apply.)

a) N_2
b) Cl_2
c) CO_2
d) CH_3OH
e) CCl_4

9.5.2 From left to right, give the hybridization of each carbon atom in the allene molecule ($H_2C{=}C{=}CH_2$).

a) sp^2, sp^2, sp^2
b) sp^3, sp^2, sp^3
c) sp^2, sp, sp^2
d) sp^3, sp, sp^3
e) sp^3, sp^3, sp^3

9.5.3 Which of the following pairs of atomic orbitals on adjacent nuclei can overlap to form a sigma bond? Consider the *x* axis to be the internuclear axis.

a) 1*s* and 2*s*
b) 1*s* and $2p_x$
c) $2p_y$ and $2p_y$
d) $3p_y$ and $3p_z$
e) $2p_x$ and $3p_x$

9.5.4 Which of the following pairs of atomic orbitals on adjacent nuclei can overlap to form a pi bond? Consider the *x* axis to be the internuclear axis.

a) 1*s* and 2*s*
b) 2*s* and 2*s*
c) $2p_y$ and $2p_y$
d) $3p_y$ and $3p_z$
e) $2p_x$ and $3p_x$

9.6 Molecular Orbital Theory

Animation
Chemical bonding—paramagnetic liquid oxygen.

$\ddot{O}{=}\ddot{O}$

O_2

Although the bonding theories we have seen thus far provide simple and effective ways for us to visualize molecules and to predict their shapes and bond angles, Lewis structures and valence bond theory do not enable us to describe or predict some important properties of molecules. Diatomic oxygen, for example, exhibits a property called *paramagnetism. **Paramagnetic*** species are attracted by magnetic fields, whereas ***diamagnetic*** species are weakly repelled by them. Such magnetic properties are the result of a molecule's electron configuration. Species in which all the electrons are *paired* are diamagnetic, whereas species that contain one or more *unpaired* electrons are paramagnetic. Because O_2 exhibits paramagnetism, it must contain unpaired electrons. According to the Lewis structure of O_2 (shown in the left margin) and the valence bond theory description of O_2, however, all the electrons in O_2 are paired. Another bonding theory, called molecular orbital theory, is needed to describe the paramagnetism of O_2 and other important molecular properties.

According to ***molecular orbital theory,*** the atomic orbitals involved in bonding actually combine to form new orbitals that are the "property" of the entire molecule, rather than of the atoms forming the bonds. These new orbitals are called ***molecular orbitals.*** In molecular orbital theory, electrons shared by atoms in a molecule reside in the molecular orbitals.

Molecular orbitals are like atomic orbitals in several ways: they have specific shapes and specific energies, and they can each accommodate a maximum of two electrons. As was the case with atomic orbitals, two electrons residing in the same molecular orbital must have opposite spins, as required by the Pauli exclusion principle. And, like hybrid orbitals, the number of molecular orbitals we get is equal to the number of atomic orbitals we combine.

Our treatment of molecular orbital theory in this book will be limited to descriptions of bonding in diatomic molecules consisting of elements from the first two periods of the periodic table (H through Ne).

Bonding and Antibonding Molecular Orbitals

Student Note: Remember that the quantum mechanical approach treats atomic orbitals as *wave functions* [◀◀ Section 6.5], and that one of the properties of waves is their capacity for both constructive combination and destructive combination [◀◀ Section 6.1].

To begin our discussion, we consider H_2, the simplest homonuclear diatomic molecule. According to *valence* bond theory, an H_2 molecule forms when two H atoms are close enough for their 1*s* atomic orbitals to overlap. According to *molecular* orbital theory, two H atoms come together to form H_2 when their 1*s* atomic orbitals combine to give molecular orbitals. Figure 9.12 shows the 1*s* atomic orbitals of the isolated H atoms and the molecular orbitals that result from their constructive and destructive combinations. The *constructive* combination of the two 1*s* orbitals gives rise to a molecular orbital [Figure 9.12(b)] that lies along the internuclear axis, directly between the two H nuclei. Just as electron density shared between two nuclei in overlapping

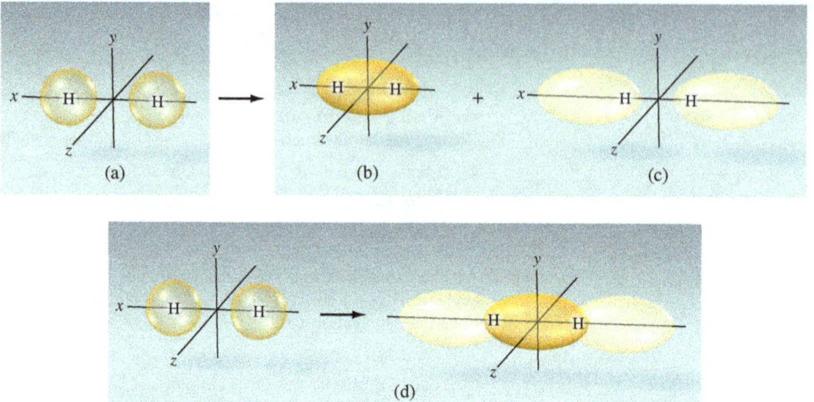

Figure 9.12 (a) Two *s* atomic orbitals combine to give two sigma molecular orbitals. (b) One of the molecular orbitals is lower in energy than the original atomic orbitals (darker), and (c) one is higher in energy (lighter). The two light yellow lobes make up one molecular orbital. (d) Atomic and molecular orbitals shown relative to the H nuclei.

atomic orbitals drew the nuclei together, electron density in a molecular orbital that lies between two nuclei will draw them together, too. Thus, this molecular orbital is referred to as a ***bonding molecular orbital.***

The *destructive* combination of the 1*s* atomic orbitals also gives rise to a molecular orbital that lies along the internuclear axis, but, as Figure 9.12(c) shows, this molecular orbital, which consists of two lobes, does not lie in between the two nuclei. Electron density in this molecular orbital would actually pull the two nuclei in opposite directions, rather than toward each other. This is referred to as an ***antibonding molecular orbital.***

σ Molecular Orbitals

Molecular orbitals that lie along the internuclear axis (such as the bonding and antibonding molecular orbitals in H_2) are referred to as σ molecular orbitals. Specifically, the *bonding* molecular orbital formed by the combination of two 1*s* atomic orbitals is designated σ_{1s} and the *antibonding* orbital is designated σ^*_{1s}, where the asterisk distinguishes an antibonding orbital from a bonding orbital. Figure 9.12(d) summarizes the combination of two 1*s* atomic orbitals to yield two molecular orbitals: one bonding and one antibonding.

Animation
Chemical bonding—formation of molecular orbitals.

Like atomic orbitals, molecular orbitals have specific energies. The combination of two atomic orbitals of equal energy, such as two 1*s* orbitals on two H atoms, yields one molecular orbital that is lower in energy (bonding) and one molecular orbital that is higher in energy (antibonding) than the original atomic orbitals. The bonding molecular orbital in H_2 is concentrated between the nuclei, along the internuclear axis. Electron density in this molecular orbital both attracts the nuclei and shields them from each other, stabilizing the molecule. Thus, the bonding molecular orbital is lower in energy than the isolated atomic orbitals. In contrast, the antibonding molecular orbital has most of its electron density outside the internuclear region. Electron density in this orbital does not shield one nucleus from the other, which increases the nuclear repulsions and makes the antibonding molecular orbital higher in energy than the isolated atomic orbitals.

Student Note: The designations σ and π are used in molecular orbital theory just as they are in valence bond theory: σ refers to electron density along the internuclear axis, and π refers to electron density that influences both nuclei but that does not lie directly along the internuclear axis.

Showing all the molecular orbitals in a molecule can make for a very complicated picture. Rather than represent molecules with pictures of their molecular orbitals, we generally use diagrams in which molecular orbitals are represented with boxes placed at the appropriate relative energy levels. Figure 9.13 shows the energies of the σ_{1s} and σ^*_{1s} molecular orbitals in H_2,

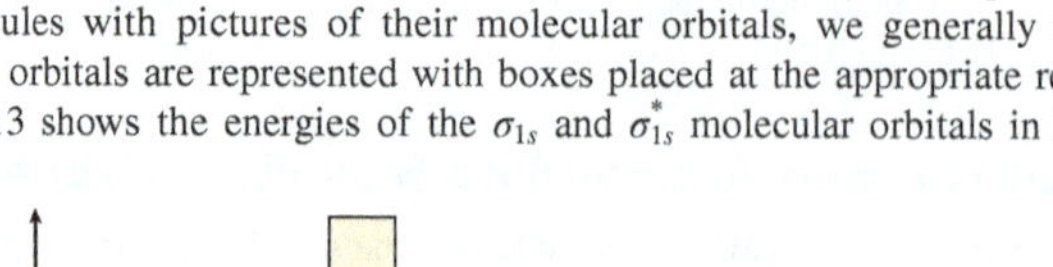

Figure 9.13 Relative energies of atomic orbitals in H and molecular orbitals in H_2.

Student Note: This molecular orbital diagram shows the two H atoms having electrons with paired spins (↑ and ↓). H atoms whose electrons have parallel spins (↑ and ↑ or ↓ and ↓) actually repel one another and will not bond to form H_2.

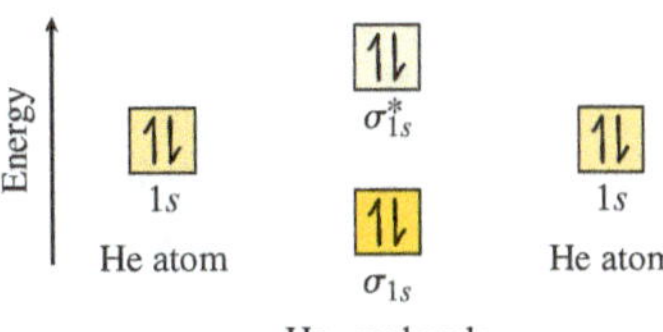

Figure 9.14 Relative energies of atomic orbitals in He and molecular orbitals in He_2.

relative to the energy of the original 1*s* orbitals on two isolated H atoms. Like atomic orbitals, molecular orbitals fill in order of increasing energy. Note that the electrons that originally resided in the atomic orbitals both occupy the lowest-energy molecular orbital, σ_{1s}, with opposite spins.

We can construct a similar molecular orbital diagram for the hypothetical molecule He_2. Like the H atom, the He atom has a 1*s* orbital (unlike H, though, the 1*s* orbital on He has two electrons, not one). The combination of 1*s* orbitals to form molecular orbitals in He_2 is essentially the same as what we have described for H_2. The placement of electrons is shown in Figure 9.14.

Bond Order

With molecular orbital diagrams such as those for H_2 and He_2, we can begin to see the power of molecular orbital theory. For a diatomic molecule described using molecular orbital theory, we can calculate the ***bond order.*** The value of the bond order indicates, qualitatively, how *stable* a molecule is. The higher the bond order, the more stable the molecule. Bond order is calculated in the following way:

Equation 9.1
$$\text{bond order} = \frac{\begin{pmatrix}\text{number of electrons}\\ \text{in bonding}\\ \text{molecular orbitals}\end{pmatrix} - \begin{pmatrix}\text{number of electrons}\\ \text{in antibonding}\\ \text{molecular orbitals}\end{pmatrix}}{2}$$

In the case of H_2, where both electrons reside in the σ_{1s} orbital, the bond order is $[(2 - 0)/2] = 1$. In the case of He_2, where the two additional electrons reside in the σ^*_{1s} orbital, the bond order is $[(2 - 2)/2] = 0$. Molecular orbital theory predicts that a molecule with a bond order of zero will not exist and He_2, in fact, does *not* exist under ordinary conditions.

We can do similar analyses of the molecules Li_2 and Be_2. (The Li and Be atoms have ground-state electron configurations of [He]$2s^1$ and [He]$2s^2$, respectively.) The 2*s* atomic orbitals also combine to form the corresponding σ and σ^* molecular orbitals. Figure 9.15 shows the molecular orbital diagrams and bond orders for Li_2 and Be_2.

Student Note: When we draw molecular orbital diagrams, we need only show the valence orbitals and electrons.

As predicted by molecular orbital theory, Li_2, with a bond order of 1, is a stable molecule, whereas Be_2, with a bond order of 0, does not exist.

π Molecular Orbitals

To consider diatomic molecules beyond Be_2, we must also consider the combination of *p* atomic orbitals. Like *s* orbitals, *p* orbitals combine both constructively, to give bonding

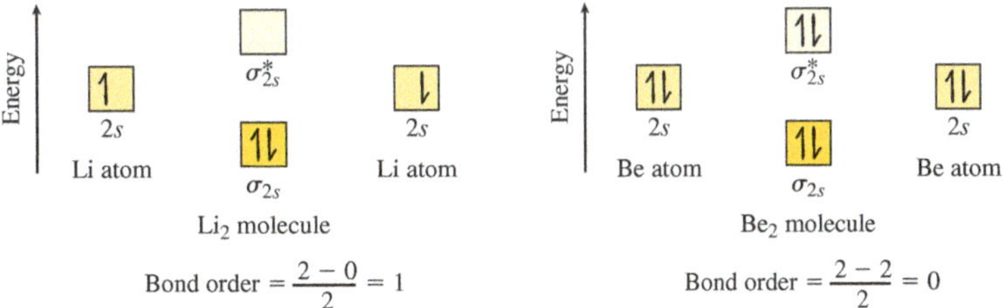

Figure 9.15 Bond order determination for Li_2 and Be_2.

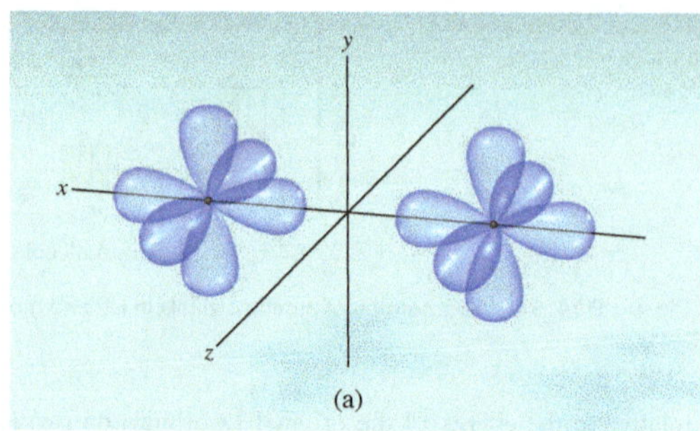

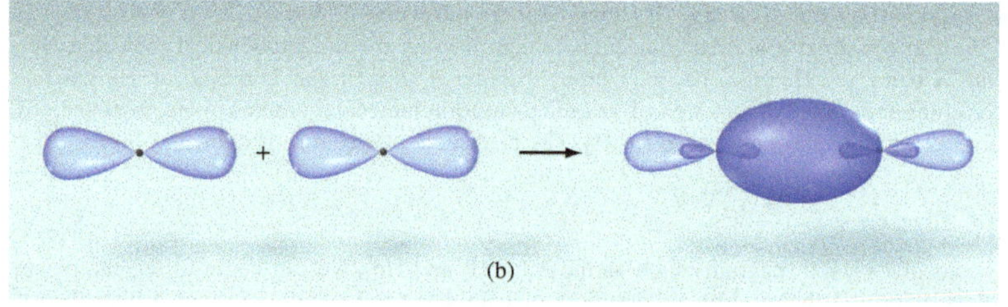

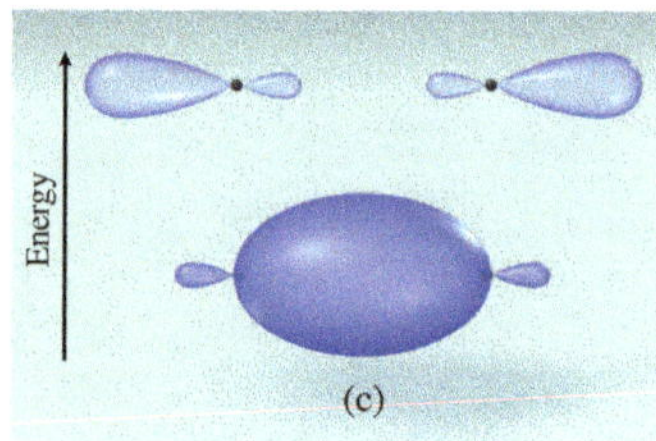

Figure 9.16 (a) Two sets of 2*p* orbitals. (b) The *p* atomic orbitals that point toward each other (p_x) combine to give bonding and antibonding σ molecular orbitals. (c) The antibonding σ molecular orbital is higher in energy than the corresponding bonding σ molecular orbital.

molecular orbitals that are lower in energy than the original atomic orbitals, and destructively, to give antibonding molecular orbitals that are higher in energy than the original atomic orbitals. However, the orientations of p_x, p_y, and p_z orbitals give rise to two different types of molecular orbitals: σ molecular orbitals, in which the regions of electron density in the bonding and antibonding molecular orbitals lie along the internuclear axis, and π molecular orbitals, in which the regions of electron density affect both nuclei but do *not* lie along the internuclear axis.

Orbitals that lie along the internuclear axis, as the $2p_x$ orbitals do in Figure 9.16(a), point directly toward each other and combine to form σ molecular orbitals. Figure 9.16(b) shows the combination of two $2p_x$ atomic orbitals to give two molecular orbitals designated σ_{2px} and $\sigma^*_{2p_x}$. Figure 9.16(c) shows the relative energies of these molecular orbitals.

Orbitals that are aligned parallel to each other, like the $2p_y$ and $2p_z$ orbitals shown in Figure 9.16(a), combine to form π molecular orbitals. These bonding molecular orbitals are designated π_{2p_y} and π_{2p_z}; the corresponding antibonding molecular orbitals are designated $\pi^*_{2p_y}$ and $\pi^*_{2p_z}$. Often we refer to the molecular orbitals collectively using the designations $\pi_{2p_{y,z}}$ and $\pi^*_{2p_{y,z}}$. Figure 9.17(a) shows the constructive and destructive combination of parallel *p* orbitals. Figure 9.17(b) shows the locations of the molecular orbitals resulting from the combination of p_y, and p_z orbitals relative to the two atomic nuclei. Again, electron density in the resulting *bonding* molecular orbitals serves to hold the nuclei together, whereas electron density in the *antibonding* molecular orbitals does not.

Just as the *p* atomic orbitals within a particular shell are higher in energy than the *s* orbital in the same shell, all the molecular orbitals resulting from the combination of *p* atomic orbitals are higher in energy than the molecular orbitals resulting from the combination of *s* atomic orbitals. To understand better the relative energy levels of the molecular orbitals resulting from *p*-orbital combinations, consider the fluorine molecule (F_2).

In general, molecular orbital theory predicts that the more effective the interaction or overlap of the atomic orbitals, the lower in energy will be the resulting bonding molecular orbital and the higher in energy will be the resulting antibonding molecular orbital. Thus, the relative energy levels of molecular orbitals in F_2 can be represented by the diagram in Figure 9.18(a). The p_x orbitals, which lie along the internuclear axis, overlap most effectively,

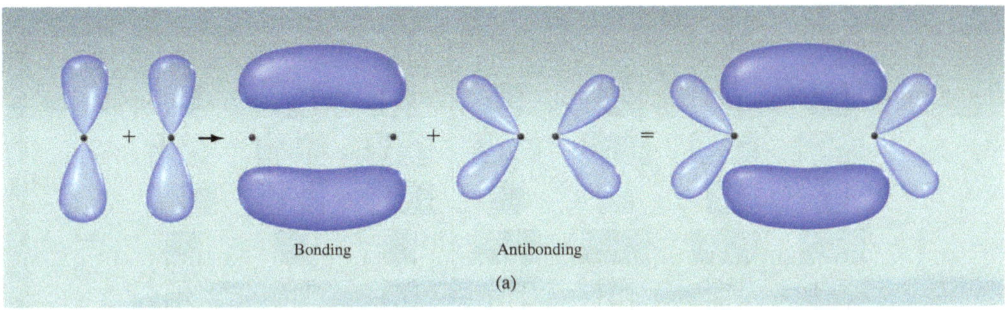

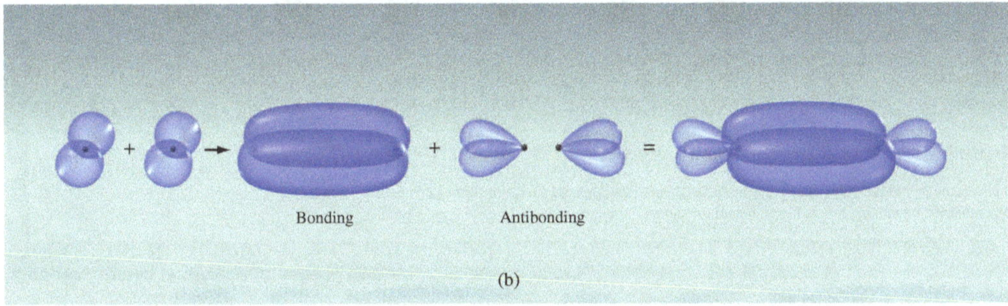

Figure 9.17 Parallel p atomic orbitals combine to give π molecular orbitals. (a) Bonding and antibonding molecular orbitals shown separately. (b) Bonding and antibonding molecular orbitals shown together relative to the two nuclei.

giving the lowest-energy bonding molecular orbital and the highest-energy antibonding molecular orbital.

The order of orbital energies shown in Figure 9.18(a) assumes that p orbitals interact only with other p orbitals and s orbitals interact only with other s orbitals—that there is no significant interaction between s and p orbitals. In fact, the relatively smaller nuclear charges of boron, carbon, and nitrogen atoms cause their atomic orbitals to be held less tightly than those of atoms with larger nuclear charges, and some s-p interaction does take place. This results in a change in the relative energies of the σ_{2p_x} and $\pi_{2p_{y,z}}$ molecular orbitals. Although energies of several of the resulting molecular orbitals change, the most important of these changes is the energy of the σ_{2p} orbital, making it higher than the $\pi_{2p_{y,z}}$ orbitals. The relative energy levels of molecular orbitals in the B_2, C_2, and N_2 molecules can be represented by the diagram in Figure 9.18(b).

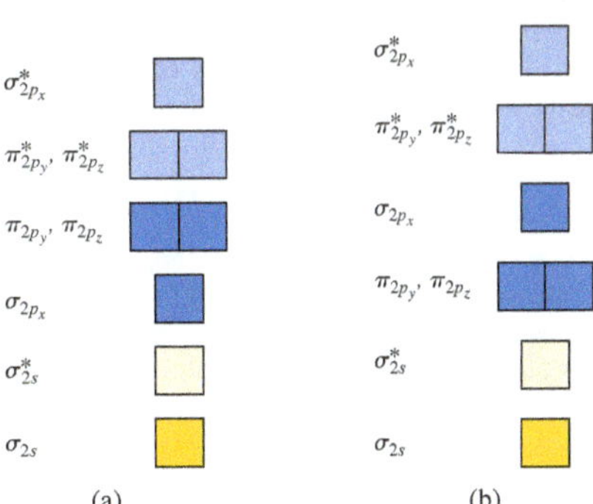

Figure 9.18 (a) Ordering of molecular orbital energies for O_2 and F_2. (b) Ordering of molecular orbital energies for Li_2, B_2, C_2, and N_2. Bonding orbitals are darker; antibonding orbitals are lighter.

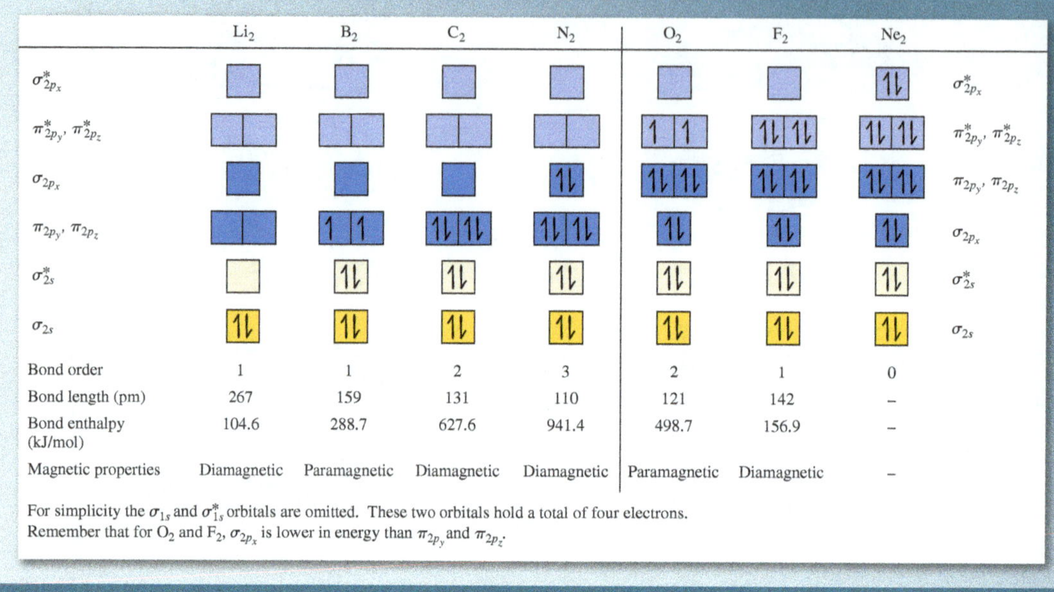

	Li$_2$	B$_2$	C$_2$	N$_2$	O$_2$	F$_2$	Ne$_2$
Bond order	1	1	2	3	2	1	0
Bond length (pm)	267	159	131	110	121	142	-
Bond enthalpy (kJ/mol)	104.6	288.7	627.6	941.4	498.7	156.9	-
Magnetic properties	Diamagnetic	Paramagnetic	Diamagnetic	Diamagnetic	Paramagnetic	Diamagnetic	-

For simplicity the σ_{1s} and σ^*_{1s} orbitals are omitted. These two orbitals hold a total of four electrons.
Remember that for O_2 and F_2, σ_{2p_x} is lower in energy than π_{2p_y} and π_{2p_z}.

Figure 9.19 Molecular orbital diagrams for second-period homonuclear diatomic molecules.

Molecular Orbital Diagrams

Beginning with oxygen, the nuclear charge is sufficiently large to prevent the interaction of *s* and *p* orbitals. Thus, for O_2 and Ne_2, the order of molecular orbital energies is the same as that for F_2, which is shown in Figure 9.18(a). Figure 9.19 gives the molecular orbital diagrams, magnetic properties, bond orders, and bond enthalpies for Li_2, B_2, C_2, N_2, O_2, F_2, and Ne_2. Note that the filling of molecular orbitals follows the same rules as the filling of atomic orbitals [|◀◀ Section 6.8]:

- Lower-energy orbitals fill first.
- Each orbital can accommodate a maximum of two electrons with opposite spins.
- Hund's rule is obeyed.

There are several important predictions made by the molecular orbital diagrams in Figure 9.19. First, molecular orbital theory correctly predicts that Ne_2, with a bond order of 0, does not exist. Second, it correctly predicts the magnetic properties of the molecules that do exist. Both B_2 and O_2 are known to be paramagnetic. Third, although bond order is only a qualitative measure of bond strength, the calculated bond orders of the molecules correlate well with the measured bond enthalpies. The N_2 molecule, with a bond order of 3, has the largest bond enthalpy of the five molecules. The B_2 and F_2 molecules, each with a bond order of 1, have the smallest bond enthalpies. Its ability to predict correctly the properties of molecules makes molecular orbital theory a powerful tool in the study of chemical bonding.

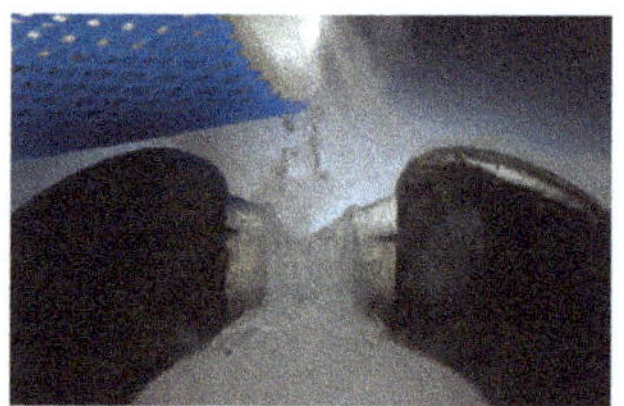

Liquid oxygen is attracted to the poles of a magnet because O_2 is paramagnetic.

Credit: © McGraw-Hill Education/Charles D. Winters, photographer.

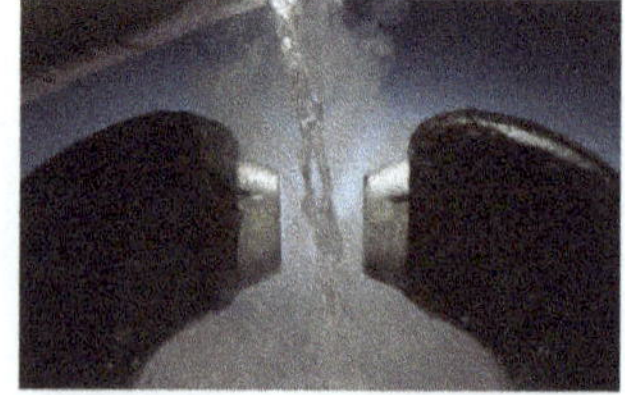

Liquid nitrogen is not attracted to the poles of a magnet because N_2 is diamagnetic.

Credit: © McGraw-Hill Education/Charles D. Winters, photographer.

Molecular Orbitals in Heteronuclear Diatomic Species

Our description of molecular orbital theory can also be applied to heteronuclear diatomic species, in which the two atoms are different, such as NO. In a case such as this, our description of the molecular orbitals involved in bonding must be modified slightly.

The atomic orbitals of a more electronegative atom are lower in energy than the corresponding atomic orbitals of a less electronegative atom. The 2*s* and 2*p* atomic orbitals are lower in energy for oxygen, which is more electronegative than nitrogen. (See Figure 8.9.) When atomic orbitals of different energies interact to form molecular orbitals, the lower-energy atomic orbital contributes more to the *bonding* molecular orbital; and the

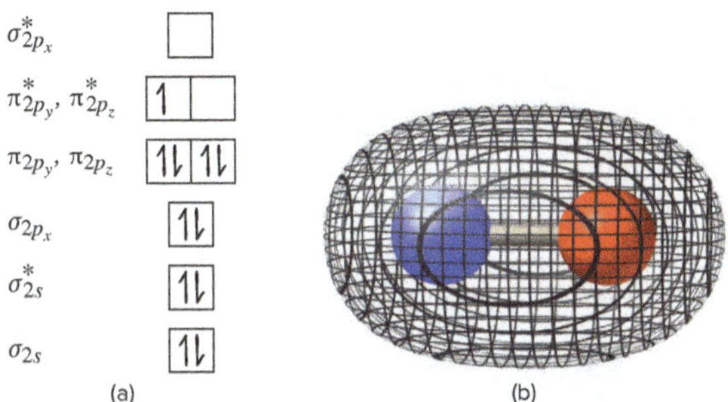

Figure 9.20 (a) Molecular orbital diagram for NO. (b) Molecular orbital representation of NO molecule.

higher-energy atomic orbital contributes more to the *antibonding* molecular orbital. The result is that the bonding molecular orbital more closely resembles the atomic orbital of the *more* electronegative atom, and the antibonding molecular orbital more closely resembles the atomic orbital of the *less* electronegative atom. The result of this is that electron density in the resulting bonding molecular orbital is *greater* in the vicinity of the *more* electronegative atom.

Recall that the order of energies of molecular orbitals in second-period homonuclear diatomic molecules is different for B_2, C_2, and N_2, than it is for O_2, F_2, and Ne_2 (Figure 9.19). For second-period heteronuclear diatomic species containing one atom from each group, such as NO and CO, there is no simple rule that can tell us which order the molecular orbitals follow. Note that the molecular orbitals in NO follow the same order as those in O_2.

Often the bond order determined from a molecular orbital diagram corresponds to the number of bonds in the Lewis structure of the molecule. In the case of NO, though, the Lewis structure contains a double bond whereas the molecular orbital approach gives a bond order of 2.5. In fact, the molecular orbital approach gives a bond order that is more consistent with experimental data. The experimentally determined strength of the bond in NO (631 kJ/mol) is greater than that of the average nitrogen-oxygen double bond (607 kJ/mol) [⏮ Table 8.6].

Sample Problem 9.7 shows how to use molecular orbital diagrams to determine the magnetic properties and bond order of the superoxide ion.

SAMPLE PROBLEM 9.7

The superoxide ion (O_2^-) has been implicated in a number of degenerative conditions, including aging and Alzheimer's disease. Using molecular orbital theory, determine whether O_2^- is paramagnetic or diamagnetic, and then calculate its bond order.

Strategy Start with the molecular orbital diagram for O_2, add an electron, and then use the resulting diagram to determine the magnetic properties and bond order.

Setup The molecular orbital diagram for O_2 is shown in Figure 9.19. The additional electron must be added to the lowest-energy molecular orbital available.

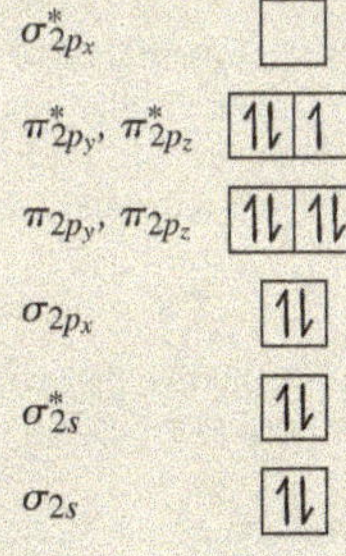

Molecular orbital diagram for O_2^-

Solution In this case, either of the two singly occupied π^*_{2p} orbitals can accommodate an additional electron. This gives a molecular orbital diagram in which there is one unpaired electron, making O_2^- paramagnetic. The new diagram has six electrons in bonding molecular orbitals and three in antibonding molecular orbitals. We can ignore the electrons in the σ_{2s} and σ^*_{2s} orbitals because their contributions to the bond order cancel each other. The bond order is $(6 - 3)/2 = 1.5$.

THINK ABOUT IT

Experiments confirm that the superoxide ion is paramagnetic. Also, any time we add one or more electrons to an antibonding molecular orbital, as we did in this problem, we should expect the bond order to decrease. Electrons in antibonding orbitals cause a bond to be less stable.

Practice Problem ATTEMPT Use molecular orbital theory to determine whether N_2^{2-} is paramagnetic or diamagnetic, and then calculate its bond order.

Practice Problem BUILD Use molecular orbital theory to determine whether F_2^{2+} is paramagnetic or diamagnetic, and then calculate its bond order.

Practice Problem CONCEPTUALIZE For most of the homonuclear diatomic species shown in Figure 9.19, *addition* and *removal* of one or more electrons (to form polyatomic ions) have opposite effects on the bond order. For some species, addition and removal of electrons have the *same* effect on bond order. Identify the species for which this is true and explain how it can be so.

CHECKPOINT – SECTION 9.6 Molecular Orbital Theory

9.6.1 Calculate the bond order of N_2^{2+}, and determine whether it is paramagnetic or diamagnetic.

a) 2, paramagnetic
b) 2, diamagnetic
c) 3, paramagnetic
d) 3, paramagnetic
e) 1, paramagnetic

9.6.2 Which of the following species is paramagnetic? (Select all that apply.)

a) C_2^{2-} b) O_2^{2+} c) F_2^{2+} d) F_2^{2-} e) C_2^{2+}

9.6.3 Calculate the bond order of He_2^+.

a) 0 b) 0.5 c) 1.0 d) 1.5 e) 2

9.6.4 Which if any of the following species has a bond order of 0? (Select all that apply.)

a) B_2^{2+} b) Ne_2^{2+} c) F_2^{2-} d) He_2^{2+} e) H_2^{2-}

9.7 Bonding Theories and Descriptions of Molecules with Delocalized Bonding

The progression of bonding theories in this chapter illustrates the importance of model development. Scientists use models to understand experimental results and to predict future observations. A model is useful as long as it agrees with observation. When it fails to do so, it must be replaced with a new model. What follows is a synopsis of the strengths and weaknesses of the bonding theories presented in Chapters 8 and 9:

Lewis Theory

Strength: The Lewis theory of bonding enables us to make qualitative predictions about bond strengths and bond lengths. Lewis structures are easy to draw and are widely used by chemists [⏮ Section 8.9].

Weakness: Lewis structures are two dimensional, whereas molecules are three dimensional. In addition, Lewis theory fails to account for the differences in bonds in compounds such as H_2, F_2, and HF. It also fails to explain *why* bonds form.

The Valence-Shell Electron-Pair Repulsion Model

Strength: The VSEPR model enables us to predict the shapes of many molecules and polyatomic ions.

Weakness: Because the VSEPR model is based on the Lewis theory of bonding, it also fails to explain why bonds form.

Valence Bond Theory

Strength: Valence bond theory describes the formation of covalent bonds as the overlap of atomic orbitals. Bonds form because the resulting molecule has a lower potential energy than the original, isolated atoms.

Weakness: Valence bond theory alone fails to explain the bonding in many molecules such as $BeCl_2$, BF_3, and CH_4, in which the central atom in its ground state does not have enough unpaired electrons to form the observed number of bonds.

Hybridization of Atomic Orbitals

Strength: The hybridization of atomic orbitals is not a separate bonding theory; rather, it is an *extension* of valence bond theory. Using hybrid orbitals, we can understand the bonding and geometry of more molecules, including $BeCl_2$, BF_3, and CH_4.

Weakness: Valence bond theory and hybrid orbitals fail to predict some of the important properties of molecules, such as the paramagnetism of O_2.

Molecular Orbital Theory

Strength: Molecular orbital theory enables us to predict accurately the magnetic and other properties of molecules and ions.

Weakness: Pictures of molecular orbitals can be very complex.

Although molecular orbital theory is in many ways the most powerful of the bonding models, it is also the most complex, so we continue to use the other models when they do an adequate job of explaining or predicting the properties of a molecule. For example, if you need to predict the three-dimensional shape of an AB_x molecule on an exam, you should draw its Lewis structure and apply the VSEPR model. Don't try to draw its molecular orbital diagram. On the other hand, if you need to determine the bond order of a diatomic molecule or ion, you should draw a molecular orbital diagram. In general chemistry, it is best to use the *simplest* theory that can answer a particular question.

Because they remain useful, we don't discard the old models when we develop new ones. In fact, the bonding in some molecules, such as benzene (C_6H_6), is best described using a combination of models. Benzene can be represented with two resonance structures [⏮ Section 8.7].

According to its Lewis structure and valence bond theory, the benzene molecule contains twelve σ bonds (six carbon-carbon, and six carbon-hydrogen) and three π bonds. From experimental evidence, however, we know that benzene does not have three single bonds and three double bonds between carbon atoms. Rather, there are six equivalent carbon-carbon bonds. This is precisely the reason that two different Lewis structures are necessary to represent the molecule. Neither one alone accurately depicts the nature of the carbon-carbon bonds. In fact, the π bonds in benzene are ***delocalized,*** meaning that they are spread out over the entire molecule, rather than confined between two specific atoms. (Bonds that are confined between two specific atoms are called ***localized*** bonds.) Valence bond theory does a good job of describing the localized σ bonds in benzene, but molecular orbital theory does a better job of using delocalized π bonds to describe the bonding scheme in benzene.

To describe the σ bonds in benzene, begin with a Lewis structure and count the electron domains on the carbon atoms. (Either resonance structure will give the same result.) Each C atom has three electron domains around it (two single bonds and one double bond). Recall from Table 9.4 that an atom that has three electron domains is sp^2 hybridized. To obtain the three unpaired electrons necessary on each C atom, one electron from each C atom must be promoted from the doubly occupied $2s$ orbital to an empty $2p$ orbital:

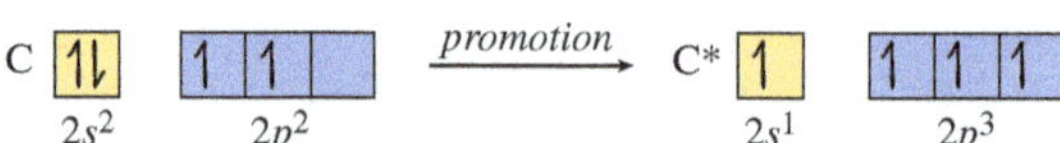

This actually creates four unpaired electrons. Next, the orbitals are sp^2 hybridized, leaving one singly occupied, unhybridized $2p$ orbital on each C atom:

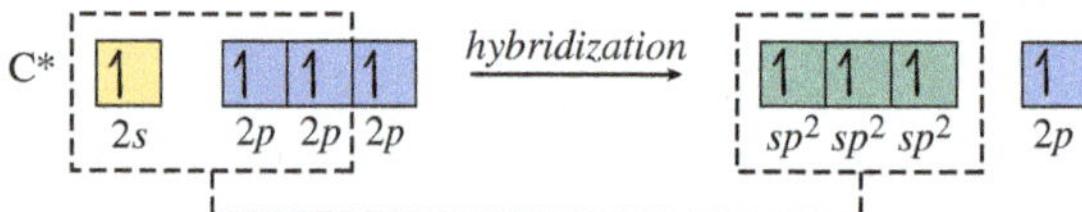

The sp^2 hybrid orbitals adopt a trigonal planar arrangement and overlap with one another (and with $1s$ orbitals on H atoms) to form the σ bonds in the molecule.

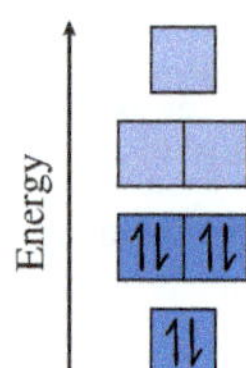

The remaining unhybridized $2p$ orbitals (one on each C atom) combine to form molecular orbitals. Because the p orbitals are all *parallel* to one another, only π_{2p} and π_{2p}^* molecular orbitals form. The combination of these six $2p$ atomic orbitals forms six molecular orbitals: three bonding and three antibonding. These molecular orbitals are delocalized over the entire benzene molecule:

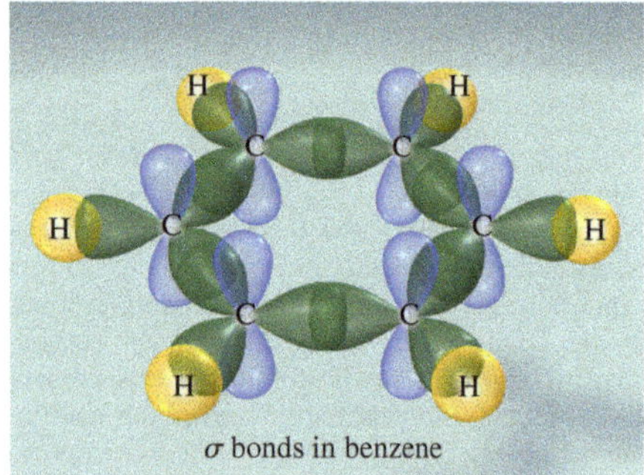

σ bonds in benzene

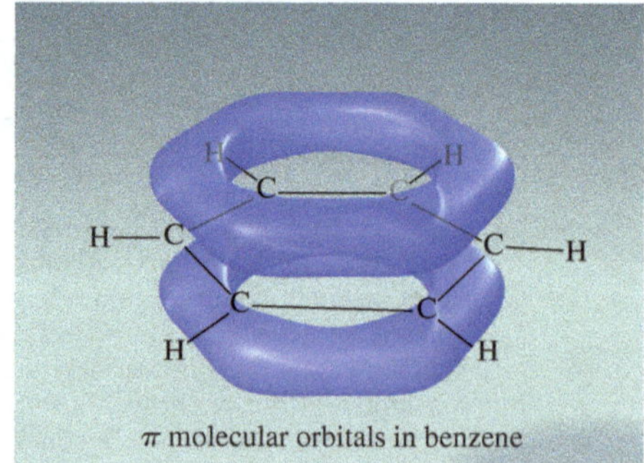

π molecular orbitals in benzene

In the ground state, the lower-energy *bonding* molecular orbitals contain all six electrons. The electron density in the delocalized π molecular orbitals lies above and below the plane that contains all the atoms and the σ bonds in the molecule.

Sample Problem 9.8 shows how to combine valence bond theory and molecular orbital theory to explain the bonding in the carbonate ion.

SAMPLE PROBLEM 9.8

It takes three resonance structures to represent the carbonate ion (CO_3^{2-}):

$$\left[\ \cdots\ \right]^{2-} \longleftrightarrow \left[\ \cdots\ \right]^{2-} \longleftrightarrow \left[\ \cdots\ \right]^{2-}$$

None of the three, though, is a completely accurate depiction. As with benzene, the bonds that are shown in the Lewis structure as one double and two single are actually three equivalent bonds. Use a combination of valence bond theory and molecular orbital theory to explain the bonding in CO_3^{2-}.

Strategy Starting with the Lewis structure, use valence bond theory and hybrid orbitals to describe the σ bonds. Then use molecular orbital theory to describe the delocalized π bonding.

Setup The Lewis structure of the carbonate ion shows three electron domains around the central C atom, so the carbon must be sp^2 hybridized.

Solution Each of the sp^2 hybrid orbitals on the C atom overlaps with a singly occupied p orbital on an O atom, forming the three σ bonds. Each O atom has an additional, singly occupied p orbital, perpendicular to the one involved in σ bonding. The unhybridized p orbital on C overlaps with the p orbitals on O to form π bonds, which have electron densities above and below the plane of the molecule. Because the species can be represented with resonance structures, we know that the π bonds are delocalized.

THINK ABOUT IT

Double bonds that appear in different places in different resonance structures represent delocalized π bonds.

Practice Problem ATTEMPT Use a combination of valence bond theory and molecular orbital theory to describe the bonding in ozone (O_3).

Practice Problem BUILD Use a combination of valence bond theory and molecular orbital theory to describe the bonding in the nitrite ion (NO_2^-).

Practice Problem CONCEPTUALIZE For which of the following species is the bonding best described as delocalized?

SO_3 $\quad SO_3^{2-}$ $\quad S_8$ $\quad O_2$

CHECKPOINT – SECTION 9.7 Bonding Theories and Descriptions of Molecules with Delocalized Bonding

9.7.1 Which of the following contain one or more delocalized π bonds? (Select all that apply.)

a) O_2

b) CO_2

c) NO_2^-

d) CH_4

e) CH_2Cl_2

9.7.2 Which of the atoms in BCl_3 need hybrid orbitals to describe the bonding in the molecule?

a) all four atoms

b) only the B atom

c) only the three Cl atoms

d) only the B atom and one Cl atom

e) only the B atom and two Cl atoms

9.7.3 Which of the following can hybrid orbitals be used for? (Select all that apply.)

a) to explain the geometry of a molecule

b) to explain how a central atom can form more bonds than the number of unpaired electrons in its ground-state configuration

c) to predict the geometry of a molecule

d) to explain the magnetic properties of a molecule

e) to predict the magnetic properties of a molecule

9.7.4 Which of the following enables us to explain the paramagnetism of O_2?

a) Lewis theory

b) valence bond theory

c) valence-shell electron-pair repulsion

d) hybridization of atomic orbitals

e) molecular orbital theory

Chapter Summary

Section 9.1

- According to the ***valence-shell electron-pair repulsion (VSEPR)*** model, electron pairs in the valence shell of an atom repel one another. An ***electron domain*** is a lone pair or a bond. Any bond (single, double, or triple) constitutes one electron domain.
- The arrangement of electron domains about a central atom, determined using the VSEPR model, is called the ***electron-domain geometry.*** The arrangement of *atoms* in a molecule is called the ***molecular geometry.*** The basic molecular geometries are linear, bent, trigonal planar, tetrahedral, trigonal pyramidal, trigonal bipyramidal, seesaw-shaped, T-shaped, octahedral, square pyramidal, and square planar.
- The ***bond angle*** is the angle between two adjacent bonds in a molecule or polyatomic ion. A trigonal bipyramid contains two types of bonds: ***axial*** and ***equatorial.***

Section 9.2

- The polarity of a molecule depends on the polarity of its individual bonds and on its molecular geometry. Even a molecule containing polar bonds may be nonpolar overall if the bonds are distributed symmetrically.
- ***Structural isomers*** are molecules with the same chemical formula, but different structural arrangements.

Section 9.3

- According to ***valence bond theory,*** bonds form between atoms when atomic orbitals overlap, thus allowing the atoms to share valence electrons. A bond forms, furthermore, when the resulting molecule is lower in energy than the original, isolated atoms.

Section 9.4

- To explain the bonding in some molecules, we need to employ the concept of ***hybridization,*** in which atomic orbitals mix to form hybrid orbitals.
- To use hybrid-orbital analysis, we must already know the molecular geometry and bond angles in a molecule. Hybrid orbitals are not used to predict molecular geometries.

Section 9.5

- ***Sigma (σ) bonds*** form when the region of orbital overlap lies directly between the two atoms. ***Pi (π) bonds*** form when parallel, unhybridized *p* orbitals interact. A *double* bond consists of one sigma bond and one pi bond. A *triple* bond consists of one sigma bond and two pi bonds.

Section 9.6

- A ***paramagnetic*** species is one that contains unpaired electrons. A ***diamagnetic*** species is one in which there are no unpaired electrons. Paramagnetic species are weakly attracted by a magnetic field, whereas diamagnetic species are weakly repelled by a magnetic field.
- According to ***molecular orbital theory,*** atomic orbitals combine to form new ***molecular orbitals*** that are associated with the molecule, rather than with individual atoms. Molecular orbitals may be *sigma,* if the orbital lies directly along the internuclear axis, or *pi,* if the orbital does not lie directly along the internuclear axis.
- Molecular orbitals may be *bonding* or *antibonding.* A ***bonding molecular orbital*** is lower in energy than the isolated atomic orbitals that combined to form it. The corresponding ***antibonding molecular orbital*** is higher in energy than the isolated atomic orbitals. ***Bond order*** is a measure of the *strength* of a bond and can be determined using a molecular orbital diagram.

Section 9.7

- It is generally best to use the bonding theory that most easily describes the bonding in a particular molecule or polyatomic ion. In species that can be represented by two or more resonance structures, the pi bonds are ***delocalized,*** meaning that they are spread out over the molecule and not constrained to just two atoms. ***Localized*** bonds are those constrained to two atoms. Many species are best described using a combination of valence bond theory and molecular orbital theory.

Key Words

Antibonding molecular orbital, 401
Axial, 374
Bond angle, 374
Bond order, 402
Bonding molecular orbital, 401
Delocalized, 408
Diamagnetic, 400
Electron domain, 372
Electron-domain geometry, 374
Equatorial, 374
Hybridization, 386
Localized, 408
Molecular geometry, 374
Molecular orbital, 400
Molecular orbital theory, 400
Paramagnetic, 400
Pi (π) bond, 394
Sigma (σ) bond, 393
Structural isomer, 380
Valence bond theory, 382
Valence-shell electron-pair repulsion (VSEPR), 372

Key Equation

9.1 $$\text{bond order} = \frac{\begin{pmatrix}\text{number of electrons}\\ \text{in bonding}\\ \text{molecular orbitals}\end{pmatrix} - \begin{pmatrix}\text{number of electrons}\\ \text{in antibonding}\\ \text{molecular orbitals}\end{pmatrix}}{2}$$

Bond order is calculated by subtracting the number of electrons in antibonding molecular orbitals from the number in bonding molecular orbitals, and dividing the result by 2.

KEY SKILLS Molecular Shape and Polarity

Molecular polarity is tremendously important in determining the physical and chemical properties of a substance. Indeed, molecular polarity is one of the most important consequences of molecular geometry. To determine the geometry or *shape* of a molecule or polyatomic ion, we use a stepwise procedure:

1. Draw a correct Lewis structure [◂◂ Chapter 8 Key Skills].
2. Count electron domains. Remember that an electron domain is a lone pair or a bond; and that a *bond* may be a single bond, a double bond, or a triple bond.
3. Apply the VSEPR model to determine electron-domain geometry.
4. Consider the positions of *atoms* to determine molecular geometry (shape), which may or may not be the same as the electron-domain geometry.

Consider the examples of SF_6, SF_4, and CH_2Cl_2. We determine the molecular geometry as follows:

Draw the Lewis structure	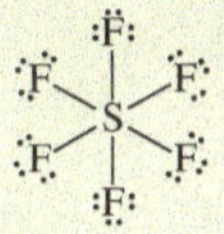	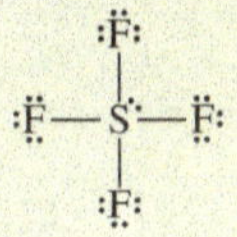	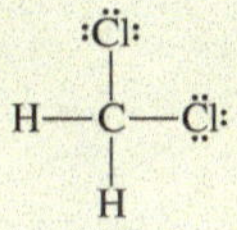
Count electron domains on the central atom	6 electron domains: • six bonds	5 electron domains: • four bonds • one lone pair	4 electron domains: • four bonds
Apply VSEPR to determine electron-domain geometry	6 electron domains arrange themselves in an octahedron.	5 electron domains arrange themselves in a trigonal bipyramid.	4 electron domains arrange themselves in a tetrahedron.
			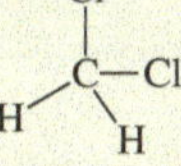
Consider positions of atoms to determine molecular geometry	With no lone pairs on the central atom, the molecular geometry is the same as the electron-domain geometry: Octahedral.	The lone pair occupies one of the equatorial positions, making the molecular geometry: See-saw shaped.	With no lone pairs on the central atom, the molecular geometry is the same as the electron-domain geometry: Tetrahedral.

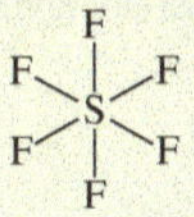

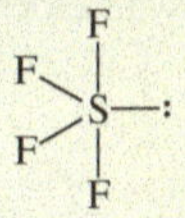

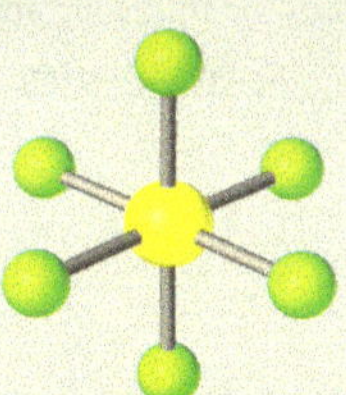

Having determined molecular geometry, we determine overall polarity of each molecule by examining the individual bond dipoles and their arrangement in three-dimensional space.

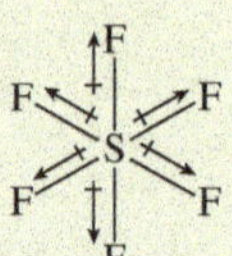

Determine whether or not the individual bonds are polar.

S and F have electronegativities of 2.5 and 4, respectively [⏮ Figure 8.9, page 338]. Therefore the individual bonds are polar and can be represented with arrows.

As in SF_6, the individual bonds in SF_4 are polar. The bond dipoles are represented with arrows.

C, H, and Cl have electronegativities of 2.5, 2.1, and 3.0, respectively. The individual bonds are polar. Bond dipoles are represented with arrows.

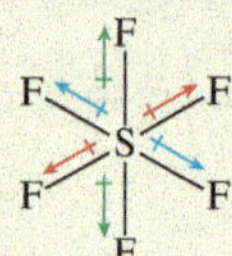

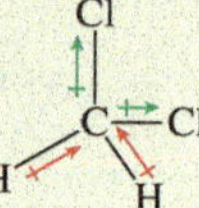

Consider the arrangement of bonds to determine which, if any, dipoles cancel one another.

The dipoles shown in red cancel each other; those shown in blue cancel each other; and those shown in green cancel each other. SF_6 is nonpolar.

The dipoles shown in green cancel each other; but the dipoles shown in red—because they are not directly across from each other—do not. SF_4 is polar.

Although the bonds are symmetrically distributed, they do not all have equivalent dipoles and therefore do not cancel each other. CH_2Cl_2 is polar.

Even with polar bonds, a molecule may be nonpolar if it consists of equivalent bonds that are distributed symmetrically. Molecules with equivalent bonds that are not distributed symmetrically—or with bonds that are not equivalent, are generally polar.

Key Skills Problems

9.1
What is the molecular geometry of PBr_3?

(a) trigonal planar (b) tetrahedral (c) trigonal pyramidal (d) bent (e) T-shaped

9.2
Which of the following species does not have tetrahedral molecular geometry?

(a) CCl_4 (b) SnH_4 (c) $AlCl_4^-$ (d) XeF_4 (e) PH_4^+

9.3
Which of the following species is polar?

(a) CF_4 (b) ClF_3 (c) PF_5 (d) AlF_3 (e) XeF_2

9.4
Which of the following species is nonpolar?

(a) ICl_2^- (b) SCl_4 (c) $SeCl_2$ (d) NCl_3 (e) $GeCl_4$

Questions and Problems

Applying What You've Learned

Recall from the beginning of the chapter that one of the fundamental scents for which we have olfactory receptors is musky. Musk is a familiar, provocative scent that has been part of the human experience for thousands of years. The primary odorous molecule in musk is muscone, a large, cyclic, organic molecule. The natural source of musk is a gland on the abdomen of the mature male musk deer, a small species of deer native to the Himalayas. The practice of harvesting musk nearly wiped out the musk deer population.

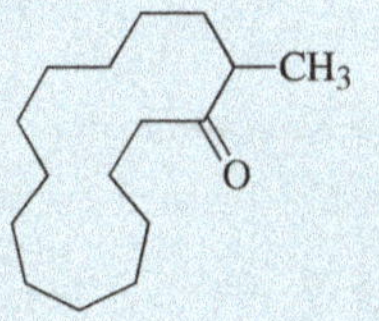

In 1888, Albert Baur, while experimenting with explosives, stumbled onto some synthetic compounds that smelled like musk. Two of these compounds are shown below. (Note the similarity between the structure of musk Baur and that of TNT.) Musk ketone, the molecule that smelled closest to muscone, was far cheaper than natural musk, but it was toxic and hazardous to prepare. In the 1950s, it was replaced with a series of synthetic musk compounds without the NO_2 groups. These new compounds are widely used in the fragrance industry.

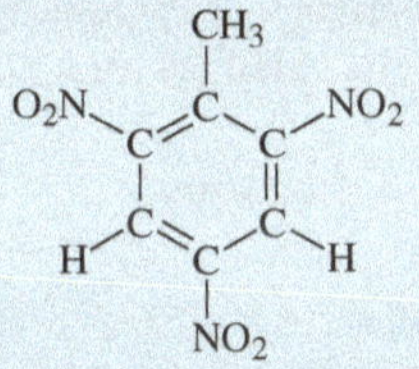

CH_3 O

Muscone

TNT

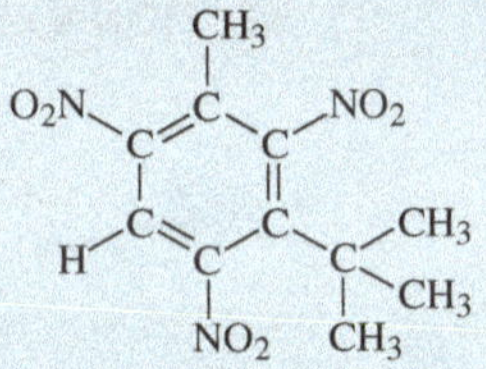

Musk Baur

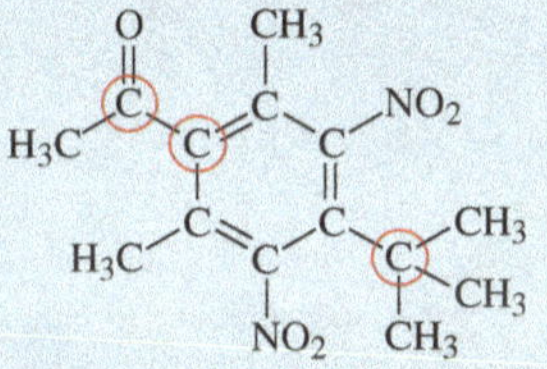

Musk Ketone

(a) How many carbon-carbon sigma bonds are there in musk ketone? How many carbon-carbon or carbon-oxygen pi bonds are there [⏮ Sample Problem 9.5]? (b) Determine the hybridization of the carbon atoms circled in red in the musk ketone molecule, and describe their geometries [⏮ Sample Problem 9.4]. (c) Of the pi bonds in the musk ketone molecule, which are localized and which are delocalized? (d) Describe the bonding in the six-membered ring portion of the musk ketone molecule using a combination of valence bond theory and molecular orbital theory [⏮ Sample Problem 9.8].

SECTION 9.1: MOLECULAR GEOMETRY

Review Questions

9.1 How is the geometry of a molecule defined, and why is the study of molecular geometry important?

9.2 Sketch the shape of a linear triatomic molecule, a trigonal planar molecule containing four atoms, a tetrahedral molecule, a trigonal bipyramidal molecule, and an octahedral molecule. Give the bond angles in each case.

9.3 How many atoms are directly bonded to the central atom in a tetrahedral molecule, a trigonal bipyramidal molecule, and an octahedral molecule?

9.4 Discuss the basic features of the VSEPR model. Explain why the magnitude of repulsion decreases in the following order: lone pair–lone pair > lone pair–bonding pair > bonding pair–bonding pair.

9.5 In the trigonal bipyramidal arrangement, why does a lone pair occupy an equatorial position rather than an axial position?

9.6 Explain why the CH_4 molecule is not square planar, although its Lewis structure makes it look as though it could be.

Conceptual Problems

9.7 Predict the geometries of the following species using the VSEPR method: (a) PCl_3, (b) $CHCl_3$, (c) SiH_4, (d) $TeCl_4$.

9.8 Predict the geometries of the following species: (a) $AlCl_3$, (b) $ZnCl_2$, (c) $HgBr_2$, (d) N_2O (arrangement of atoms is NNO).

9.9 Predict the geometry of the following molecules and ion using the VSEPR model: (a) CBr_4, (b) BCl_3, (c) NF_3, (d) H_2Se, (e) NO_2^-.

9.10 Predict the geometry of the following molecules and ion using the VSEPR model: (a) CH_3I, (b) ClF_3, (c) H_2S, (d) SO_3, (e) SO_4^{2-}.

9.11 Predict the geometry of the following ions using the VSEPR method: (a) SCN^- (arrangement of atoms is SCN), (b) AlH_4^-, (c) $SnCl_5^-$, (d) H_3O^+, (e) BeF_4^{2-}.

9.12 Predict the geometries of the following ions: (a) NH_4^+, (b) NH_2^-, (c) CO_3^{2-}, (d) ICl_2^-, (e) ICl_4^-.

9.13 Describe the geometry around each of the three central atoms in the CH_3COOH molecule.

9.14 Which of the following species are tetrahedral: $SiCl_4$, SeF_4, XeF_4, CI_4, $CdCl_4^{2-}$?

SECTION 9.2: MOLECULAR GEOMETRY AND POLARITY

Review Questions

9.15 Explain why an atom cannot have a permanent dipole moment.

9.16 The bonds in beryllium hydride (BeH_2) molecules are polar, and yet the dipole moment of the molecule is zero. Explain.

Conceptual Problems

9.17 Determine whether (a) BrF_5 and (b) BCl_3 are polar.

9.18 Determine whether (a) OCS and (b) XeF_4 are polar.

9.19 Which of the molecules shown is polar?

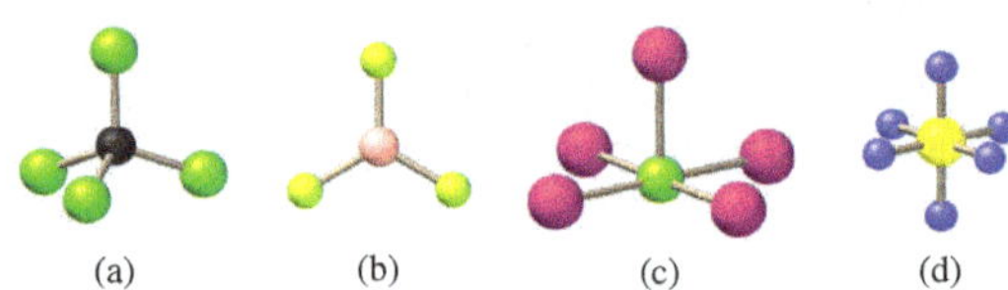

(a) (b) (c) (d)

9.20 Which of the molecules shown is polar?

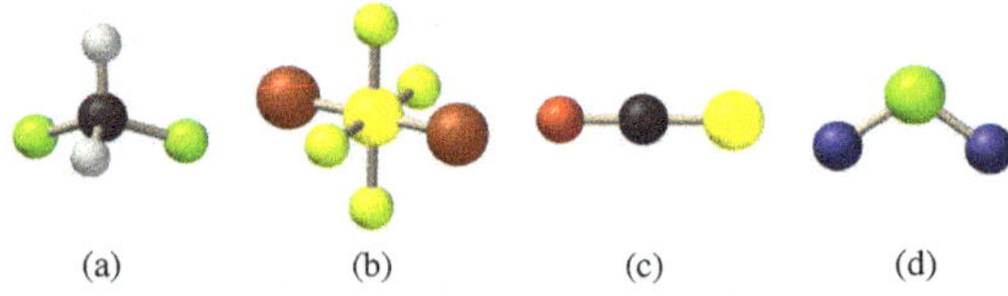

(a) (b) (c) (d)

SECTION 9.3: VALENCE BOND THEORY

Review Questions

9.21 What is valence bond theory? How does it differ from the Lewis concept of chemical bonding?

9.22 Use valence bond theory to explain the bonding in Cl_2 and HCl. Show how the atomic orbitals overlap when a bond is formed.

9.23 According to valence bond theory, how many bonds would you expect each of the following atoms (in the ground state) to form: Be, C?

9.24 According to valence bond theory, how many bonds would you expect each of the following atoms (in the ground state) to form: P, S?

SECTION 9.4: HYBRIDIZATION OF ATOMIC ORBITALS

Review Questions

9.25 What is the hybridization of atomic orbitals? Why is it impossible for an isolated atom to exist in the hybridized state?

9.26 How does a hybrid orbital differ from a pure atomic orbital? Can two 2*p* orbitals of an atom hybridize to give two hybridized orbitals?

9.27 What is the angle between the following two hybrid orbitals on the same atom: (a) *sp* and *sp* hybrid orbitals, (b) sp^2 and sp^2 hybrid orbitals, (c) sp^3 and sp^3 hybrid orbitals?

Conceptual Problems

9.28 Describe the bonding scheme of the AsH_3 molecule in terms of hybridization.

9.29 What is the hybridization state of Si (a) in SiH_4 and (b) in $H_3Si{-}SiH_3$?

9.30 Describe the change in hybridization (if any) of the Al atom in the following reaction:

$$AlCl_3 + Cl^- \longrightarrow AlCl_4^-$$

9.31 Consider the reaction

$$BF_3 + NH_3 \longrightarrow F_3B{-}NH_3$$

Describe the changes in hybridization (if any) of the B and N atoms as a result of this reaction.

9.32 What hybrid orbitals are used by nitrogen atoms in the following species: (a) NH_3, (b) $H_2N{-}NH_2$, (c) NO_3^-?

9.33 Describe the hybridization of phosphorus in PF_5.

SECTION 9.5: HYBRIDIZATION IN MOLECULES CONTAINING MULTIPLE BONDS

Visualizing Chemistry
Figure 9.11

VC 9.1 How is a sigma bond different from a pi bond?
a) A sigma bond is a bonding molecular orbital; a pi bond is an antibonding molecular orbital.
b) A sigma bond is a single bond, whereas a pi bond is a double bond.
c) The electron density in a sigma bond lies along the internuclear axis; that of a pi bond does not.

VC 9.2 Pi bonds form when __________ atomic orbitals on _________ atom(s) overlap.
a) perpendicular, adjacent
b) parallel, adjacent
c) parallel, the same

VC 9.3 Formation of two pi bonds requires the combination of ______ atomic orbitals.
a) two
b) four
c) six

VC 9.4 Why are there no pi bonds in ethane (C_2H_6)?
a) The remaining unhybridized *p* orbitals do not contain any electrons.
b) There are no unhybridized *p* orbitals remaining on either C atom.
c) The remaining unhybridized *p* orbitals are not parallel to each other.

Review Questions

9.34 How would you distinguish between a sigma bond and a pi bond?

9.35 Which of the following pairs of atomic orbitals of adjacent nuclei can overlap to form a sigma bond? Which overlap to form a pi bond? Which cannot overlap (no bond)? Consider the *x* axis to be the internuclear axis, that is, the line joining the nuclei of the two atoms. (a) 1*s* and 1*s*, (b) 1*s* and $2p_x$, (c) $2p_x$ and $2p_y$, (d) $3p_y$ and $3p_y$, (e) $2p_x$ and $2p_x$, (f) 1*s* and 2*s*.

Conceptual Problems

9.36 What are the hybrid orbitals of the carbon atoms in the following molecules?
(a) H_3C-CH_3
(b) $H_3C-CH=CH_2$
(c) $CH_3-C{\equiv}C-CH_2OH$
(d) $CH_3CH=O$
(e) CH_3COOH

9.37 Specify which hybrid orbitals are used by carbon atoms in the following species: (a) CO, (b) CO_2, (c) CN^-.

9.38 The allene molecule ($H_2C=C=CH_2$) is linear (the three C atoms lie on a straight line). What are the hybridization states of the carbon atoms? Draw diagrams to show the formation of sigma bonds and pi bonds in allene.

9.39 What is the hybridization of the central N atom in the azide ion (N_3^-)? (The arrangement of atoms is NNN.)

9.40 How many sigma bonds and pi bonds are there in each of the following molecules?

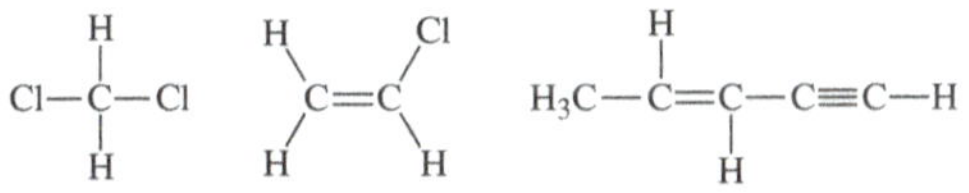

(a) (b) (c)

9.41 How many pi bonds and sigma bonds are there in the tetracyanoethylene molecule?

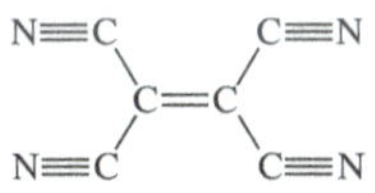

9.42 Tryptophan is one of the 20 amino acids in the human body. Describe the hybridization state of the C and N atoms, and determine the number of sigma and pi bonds in the molecule.

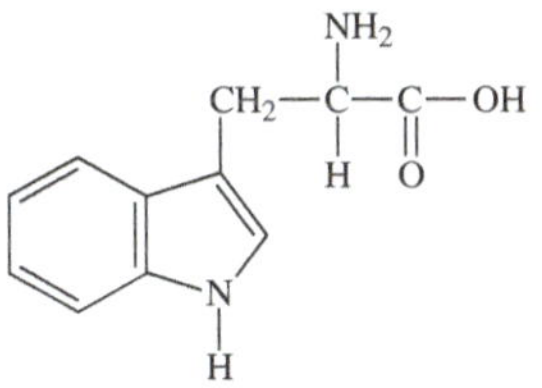

9.43 Benzo(*a*)pyrene is a potent carcinogen found in coal and cigarette smoke. Determine the number of sigma and pi bonds in the molecule.

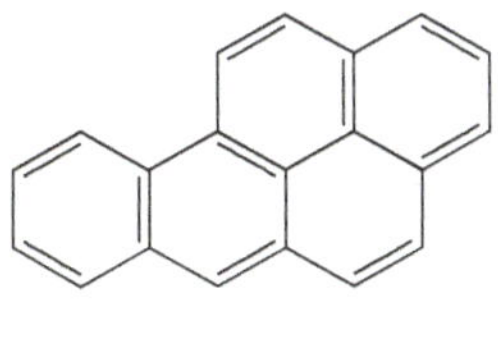

SECTION 9.6: MOLECULAR ORBITAL THEORY

Review Questions

9.44 What is molecular orbital theory? How does it differ from valence bond theory?

9.45 Define the following terms: bonding molecular orbital, antibonding molecular orbital, pi molecular orbital, sigma molecular orbital.

9.46 Sketch the shapes of the following molecular orbitals: σ_{1s}, $\sigma^{\star}_{1s}$, π_{2p}, $\pi^{\star}_{2p}$. How do their energies compare?

9.47 Explain the significance of bond order. Can bond order be used for quantitative comparisons of the strengths of chemical bonds?

Conceptual Problems

9.48 Explain in molecular orbital terms the changes in H−H internuclear distance that occur as the molecular H_2 is ionized first to H_2^+ and then to H_2^{2+}.

9.49 The formation of H_2 from two H atoms is an energetically favorable process. Yet statistically there is less than a 100 percent chance that any two H atoms will undergo the reaction. Apart from energy considerations, how would you account for this observation based on the electron spins in the two H atoms?

9.50 Draw a molecular orbital energy level diagram for each of the following species: He_2, HHe, He_2^+. Compare their relative stabilities in terms of bond orders. (Treat HHe as a diatomic molecule with three electrons.)

9.51 Arrange the following species in order of increasing stability: Li_2, Li_2^+, Li_2^-. Justify your choice with a molecular orbital energy level diagram.

9.52 Use molecular orbital theory to explain why the Be_2 molecule does not exist.

9.53 Which of these species has a longer bond, B_2 or B_2^+? Explain in terms of molecular orbital theory.

9.54 Acetylene (C_2H_2) has a tendency to lose two protons (H^+) and form the carbide ion (C_2^{2-}), which is present in a number of ionic compounds, such as CaC_2 and MgC_2. Describe the bonding scheme in the C_2^{2-} ion in terms of molecular orbital theory. Compare the bond order in C_2^{2-} with that in C_2.

9.55 Compare the Lewis and molecular orbital treatments of the oxygen molecule.

9.56 Explain why the bond order of N_2 is greater than that of N_2^+, but the bond order of O_2 is less than that of O_2^+.

9.57 Compare the relative bond orders of the following species and indicate their magnetic properties (i.e., diamagnetic or paramagnetic): O_2, O_2^+, O_2^- (superoxide ion), O_2^{2-} (peroxide ion).

9.58 Use molecular orbital theory to compare the relative stabilities of F_2 and F_2^+.

9.59 A single bond is almost always a sigma bond, and a double bond is almost always made up of a sigma bond and a pi bond. There are very few exceptions to this rule. Show that the B_2 and C_2 molecules are examples of the exceptions.

SECTION 9.7: BONDING THEORIES AND DESCRIPTIONS OF MOLECULES WITH DELOCALIZED BONDING

Review Questions

9.60 How does a delocalized molecular orbital differ from a molecular orbital such as that found in H_2 or C_2H_4? What do you think are the minimum conditions (e.g., number of atoms and types of orbitals) for forming a delocalized molecular orbital?

9.61 In Chapter 8, we saw that the resonance concept is useful for dealing with species such as the benzene molecule and the carbonate ion. How does molecular orbital theory deal with these species?

Conceptual Problems

9.62 Both ethylene (C_2H_4) and benzene (C_6H_6) contain the C=C bond. The reactivity of ethylene is greater than that of benzene. For example, ethylene readily reacts with molecular bromine, whereas benzene is normally quite inert toward molecular bromine and many other compounds. Explain this difference in reactivity.

9.63 Explain why the symbol on the left is a better representation of benzene molecules than that on the right.

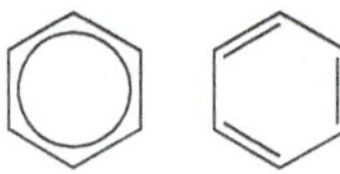

9.64 Determine which of these molecules has a more delocalized orbital, and justify your choice. (*Hint:* Both molecules contain two benzene rings. In naphthalene, the two rings are fused together. In biphenyl, the two rings are joined by a single bond, around which the two rings can rotate.)

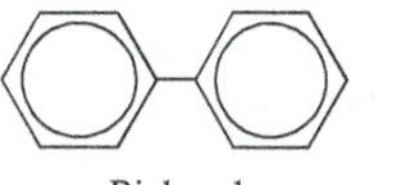

Biphenyl

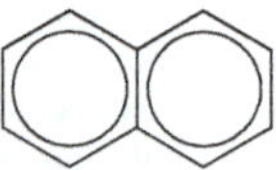

Naphthalene

9.65 Nitryl fluoride (FNO_2) is very reactive chemically. The fluorine and oxygen atoms are bonded to the nitrogen atom. (a) Write a Lewis structure for FNO_2. (b) Indicate the hybridization of the nitrogen atom. (c) Describe the bonding in terms of molecular orbital theory. Where would you expect delocalized molecular orbitals to form?

9.66 Describe the bonding in the nitrate ion NO_3^- in terms of delocalized molecular orbitals.

9.67 What is the state of hybridization of the central O atom in O_3? Describe the bonding in O_3 in terms of delocalized molecular orbitals.

ADDITIONAL PROBLEMS

9.68 Which of the following species is not likely to have a tetrahedral shape: (a) $SiBr_4$, (b) NF_4^+, (c) SF_4, (d) $BeCl_4^{2-}$, (e) BF_4^-, (f) $AlCl_4^-$?

9.69 Draw the Lewis structure of mercury(II) bromide. Is this molecule linear or bent? How would you establish its geometry?

9.70 Although both carbon and silicon are in Group 4A, very few Si=Si bonds are known. Account for the instability of silicon-to-silicon double bonds in general. (*Hint:* Compare the atomic radii of C and Si in Figure 7.6. What effect would the larger size have on pi bond formation?)

9.71 Predict the geometry of sulfur dichloride (SCl_2) and the hybridization of the sulfur atom.

9.72 Antimony pentafluoride (SbF_5) reacts with XeF_4 and XeF_6 to form ionic compounds, $XeF_3^+ SbF_6^-$ and $XeF_5^+ SbF_6^-$. Describe the geometries of the cations and anions in these two compounds.

9.73 Assume that the third-period element phosphorus forms a diatomic molecule, P_2, in an analogous way as nitrogen does to form N_2. (a) Write the electronic configuration for P_2. Use $[Ne_2]$ to represent the electron configuration for the first two periods. (b) Calculate its bond order. (c) What are its magnetic properties (diamagnetic or paramagnetic)?

9.74 The molecule benzyne (C_6H_4) is a very reactive species. It resembles benzene in that it has a six-membered ring of carbon atoms. Draw a Lewis structure of the molecule and account for the molecule's high reactivity.

9.75 Predict the bond angles for the following molecules: (a) $BeCl_2$, (b) BCl_3, (c) CCl_4, (d) CH_3Cl, (e) Hg_2Cl_2 (arrangement of atoms: ClHgHgCl), (f) $SnCl_2$, (g) H_2O_2, (h) SnH_4.

9.76 Briefly compare the VSEPR and hybridization approaches to the study of molecular geometry.

9.77 Draw Lewis structures and give the other information requested for the following molecules: (a) BF_3. Shape: planar or nonplanar? (b) ClO_3^-. Shape: planar or nonplanar? (c) HCN. Polar or nonpolar? (d) OF_2. Polar or nonpolar? (e) NO_2. Estimate the ONO bond angle.

9.78 Describe the hybridization state of arsenic in arsenic pentafluoride (AsF_5).

9.79 Determine whether (a) PCl_5 and (b) H_2CO (C double bonded to O) are polar.

9.80 Draw Lewis structures and give the other information requested for the following: (a) SO_3. Polar or nonpolar molecule? (b) PF_3. Polar or nonpolar? (c) F_3SiH. Polar or nonpolar? (d) SiH_3^-. Shape: planar or pyramidal? (e) Br_2CH_2. Polar or nonpolar molecule?

9.81 Which of the following molecules are linear: ICl_2^-, IF_2^+, OF_2, SnI_2, $CdBr_2$?

9.82 Draw the Lewis structure for the $BeCl_4^{2-}$ ion. Predict its geometry, and describe the hybridization state of the Be atom.

9.83 The N_2F_2 molecule can exist in either of the following two forms:

F–N=N–F (trans) and F–N=N–F (cis)

(a) What is the hybridization of N in the molecule?
(b) Which structure is polar?

9.84 Cyclopropane (C_3H_6) has the shape of a triangle in which a C atom is bonded to two H atoms and two other C atoms at each corner. Cubane (C_8H_8) has the shape of a cube in which a C atom is bonded to one H atom and three other C atoms at each corner. (a) Draw Lewis structures of these molecules. (b) Compare the CCC angles in these molecules with those predicted for an sp^3-hybridized C atom. (c) Would you expect these molecules to be easy to make?

9.85 Determine whether (a) CH_2Cl_2 and (b) XeF_4 are polar.

9.86 Does the following molecule have a dipole moment? Explain.

Cl(H)C=C=C(H)Cl

9.87 For which molecular geometries (linear, bent, trigonal planar, trigonal pyramidal, tetrahedral, square planar, T-shaped, seesaw-shaped, trigonal bipyramidal, square pyramidal, octahedral) can an AB_x molecule be nonpolar if there are (a) two different types of terminal atoms and (b) three different types of terminal atoms?

9.88 (a) From what group must the terminal atoms come in an AB_x molecule where the central atom is from Group 5A, for both the electron-domain geometry and the molecular geometry to be trigonal bipyramidal? (b) From what group must the terminal atoms come in an AB_x molecule where the central atom is from Group 6A, for the electron-domain geometry to be tetrahedral and the molecular geometry to be bent?

9.89 Carbon suboxide (C_3O_2) is a colorless pungent-smelling gas. Does this molecule possess a dipole moment? Explain.

9.90 The following molecules (AX_4Y_2) all have an octahedral geometry. Group the molecules that are equivalent to each other.

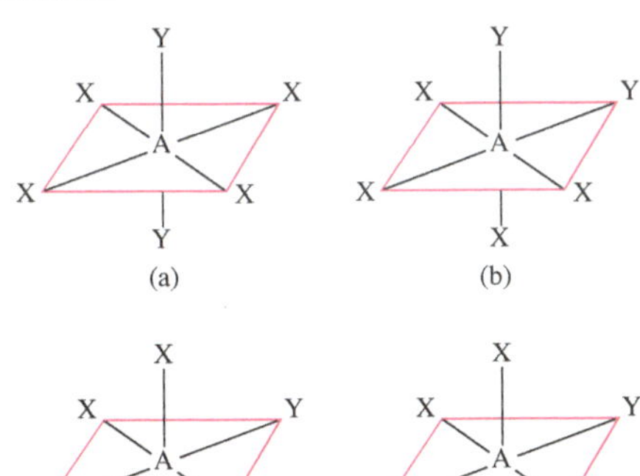

9.91 The compounds carbon tetrachloride (CCl_4) and silicon tetrachloride ($SiCl_4$) are similar in geometry and hybridization. However, CCl_4 does not react with water but $SiCl_4$ does. Explain the difference in their chemical reactivities. (*Hint:* The first step of the reaction is believed to be the addition of a water molecule to the Si atom in $SiCl_4$.)

9.92 Write the ground-state electron configuration for B_2. Is the molecule diamagnetic or paramagnetic?

9.93 What is the hybridization of C and of N in this molecule?

(ring structure: NH_2 on C; N=C; C–H; C=C–H; N–H; C=O)

9.94 The stable allotropic form of phosphorus is P_4, in which each P atom is bonded to three other P atoms. Draw a Lewis structure of this molecule and describe its geometry. At high temperatures, P_4 dissociates to form P_2 molecules containing a P=P bond. Explain why P_4 is more stable than P_2.

9.95 Use molecular orbital theory to explain the difference between the bond enthalpies of F_2 and F_2^-. (See Problem 8.125.)

9.96 Use molecular orbital theory to explain the bonding in the azide ion (N_3^-). (The arrangement of atoms is NNN.)

9.97 Carbon dioxide has a linear geometry and is nonpolar. Yet we know that the molecule exhibits bending and stretching motions that create a dipole moment. How would you reconcile these seemingly conflicting descriptions of CO_2?

9.98 Draw three Lewis structures for compounds with the formula $C_2H_2F_2$. Indicate which of the compounds are polar.

9.99 Write the electron configuration of the cyanide ion (CN^-). Name a stable molecule that is isoelectronic with the ion.

9.100 Aluminum trichloride ($AlCl_3$) is an electron-deficient molecule. It has a tendency to form a dimer (a molecule made up of two $AlCl_3$ units):

$$AlCl_3 + AlCl_3 \longrightarrow Al_2Cl_6$$

(a) Draw a Lewis structure for the dimer. (b) Describe the hybridization state of Al in $AlCl_3$ and Al_2Cl_6. (c) Sketch the geometry of the dimer. (d) Do these molecules possess a dipole moment?

9.101 The Lewis structure for O_2 is

:Ö=Ö:

Use molecular orbital theory to show that the structure actually corresponds to an excited state of the oxygen molecule.

9.102 Draw the Lewis structure of ketene (C_2H_2O) and describe the hybridization states of the C atoms. The molecule does not contain O–H bonds. On separate diagrams, sketch the formation of the sigma and pi bonds.

9.103 Which of the following geometries has a greater stability for tin(IV) hydride (SnH_4)?

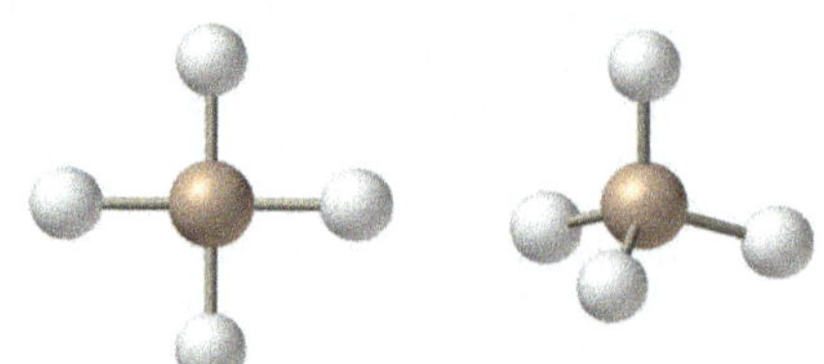

9.104 Which of the following ions possess a dipole moment: (a) ClF_2^+, (b) ClF_2^-, (c) IF_4^+, (d) IF_4^-?

Biological Problems

9.105 The molecular model of vitamin C is shown here. (a) Write the molecular formula of the compound. (b) What is the hybridization of each C and O atom? (c) Describe the geometry about each C and O atom.

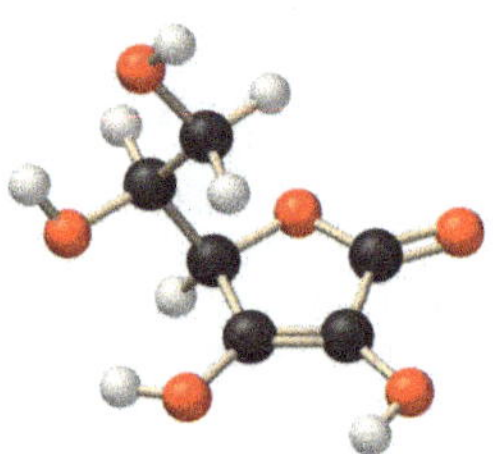

9.106 The molecular model of nicotine (a stimulant) is shown here. (a) Write the molecular formula of the compound. (b) What is the hybridization of each C and N atom? (c) Describe the geometry about each C and N atom.

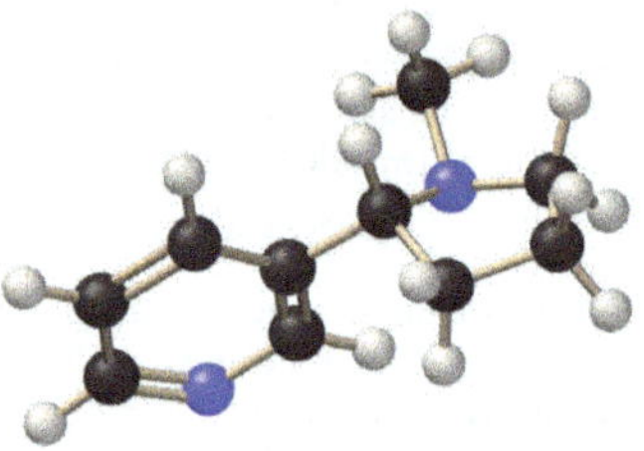

9.107 The compound TCDD, or 2,3,7,8-tetrachlorodibenzo-*p*-dioxin, is highly toxic:

It gained considerable notoriety in 2004 when it was implicated in the attempted murder of a Ukrainian politician. (a) Describe its geometry, and state whether the molecule has a dipole moment. (b) How many pi bonds and sigma bonds are there in the molecule?

9.108 Progesterone is a hormone responsible for female sex characteristics. In the usual shorthand structure, each point where lines meet represents a C atom, and most H atoms are not shown. Draw the complete structure of the molecule, showing all C and H atoms. Indicate which C atoms are sp^2- and sp^3-hybridized.

9.109 Carbon monoxide (CO) is a poisonous compound due to its ability to bind strongly to Fe^{2+} in the hemoglobin molecule. The molecular orbitals of CO have the same energy order as those of the N_2 molecule. (a) Draw a Lewis structure of CO and assign formal charges. Explain why CO has a rather small dipole moment of 0.12 D. (b) Compare the bond order of CO with that from molecular orbital theory. (c) Which of the atoms (C or O) is more likely to form bonds with the Fe^{2+} ion in hemoglobin?

9.110 The compound 3′-azido-3′-deoxythymidine, commonly known as AZT, is one of the drugs used to treat AIDS. What are the hybridization states of the C and N atoms in this molecule?

9.111 The disulfide bond, −S−S−, plays an important role in determining the three-dimensional structure of proteins. Describe the nature of the bond and the hybridization state of the S atoms.

Environmental Problems

9.112 Greenhouse gases absorb (and trap) outgoing infrared radiation (heat) from Earth and contribute to global warming. A molecule of a greenhouse gas either possesses a permanent dipole moment or has a changing dipole moment during its vibrational motions. Consider three of the vibrational modes of carbon dioxide

O=C=O   O=C=O   O=C=O

where the arrows indicate the movement of the atoms. (During a complete cycle of vibration, the atoms move toward one extreme position and then reverse their direction to the other extreme position.) Which of the preceding vibrations are responsible for CO_2 behaving as a greenhouse gas? Which of the following gases are greenhouse gases: N_2, O_2, O_3, CO, NO_2, N_2O, CH_4, $CFCl_3$?

9.113 The compound 1,2-dichloroethane ($C_2H_4Cl_2$) is nonpolar, while *cis*-dichloroethylene ($C_2H_2Cl_2$) has a dipole moment: The reason for the difference is that groups connected by a single bond can rotate with respect to each other, but no rotation occurs when a double bond connects the groups. On the basis of bonding considerations, explain why rotation occurs in 1,2-dichloroethane but not in *cis*-dichloroethylene.

1,2-dichloroethane   *cis*-dichloroethylene

Multiconcept Problems

9.114 Consider an N_2 molecule in its first excited electronic state, that is, when an electron in the highest occupied molecular orbital is promoted to the lowest empty molecular orbital. (a) Identify the molecular orbitals involved, and sketch a diagram to show the transition. (b) Compare the bond order and bond length of N_2^* with N_2, where the asterisk denotes the excited molecule. (c) Is N_2^* diamagnetic or paramagnetic? (d) When N_2^* loses its excess energy and converts to the ground state N_2, it emits a photon of wavelength 470 nm, which makes up part of the auroras' lights. Calculate the energy difference between these levels.

9.115 Imagine that sulfur dioxide can be prepared by combining elemental sulfur (S_8) and sulfur trioxide. Write a balanced equation for this hypothetical reaction and determine the oxidation state and the hybridization of sulfur in each species. What mass of sulfur trioxide would be required to combine with 1.00 kg of elemental sulfur; and what total mass of sulfur dioxide would result—assuming the reaction goes to completion?

Standardized-Exam Practice Problems

Physical Sciences

These questions are not based on a descriptive passage.

1. What is the shape of the ICl_3 molecule?

a) Trigonal planar c) T-shaped
b) Trigonal pyramidal d) Tetrahedral

2. Which of the following molecules is nonpolar?

a) NCl_3 c) PCl_3
b) BCl_3 d) $BrCl_3$

3. Which of the following has a bond order of 2?

N_2^{2-}	N_2	N_2^{2+}
I	II	III

a) I only c) III only
b) II only d) I and III

4. Which of the following are paramagnetic?

O_2^{2-}	O_2^-	O_2^+	O_2^{2+}
I	II	III	IV

a) I and II c) III and IV
b) II and III d) I and III

Answers to In-Chapter Materials

Answers to Practice Problems

9.1A (a) linear, (b) bent. **9.1B** (a) Group 6A, (b) Group 7A. **9.2A** Bent about O, tetrahedral about each C, trigonal pyramidal about N. All bond angles are ~109.5°. Angles labeled in blue are <109.5°.

```
      H  H
      |  |
H–Ö–C–C–N̈–H
      |  |  |
      H  H  H
```

9.2B SO_2 contains double bonds, which are not pushed together by the central atom's lone pair as easily as the single bonds in NH_3. **9.3A** Singly occupied 3*p* orbitals from the P atom overlap with *s* orbitals from H atoms. **9.3B** We cannot use valence bond theory to explain the bonding in SO_2 or CH_4. In the case of SO_2, although the central atom has two unpaired electrons and can form two bonds, the unpaired electrons on S are in 3*p* orbitals. Formation of two bonds by the overlap of two 3*p* orbitals on S would be expected to result in a bond angle of approximately 90°. In the case of CH_4, the central atom does not have enough unpaired electrons to form four bonds. **9.4A** Two of the 4*p* electrons in Br are promoted to empty *d* orbitals. The *s* orbital, all three *p* orbitals, and two of the *d* orbitals hybridize to form six sp^3d^2 hybrid orbitals. One of the hybrid orbitals contains the lone pair. Each of the remaining hybrid orbitals contains one electron and overlaps with a singly occupied 2*p* orbital on an F atom. The arrangement of hybrid orbitals is octahedral and the bond angles are ~90°. **9.4B** One of the 2*s* electrons in Be is promoted to an empty *p* orbital. The *s* orbital and one *p* orbital hybridize to form two *sp* hybrid orbitals. Each hybrid orbital contains one electron and overlaps with a singly occupied 2*p* orbital on an F atom. The arrangement is linear with a bond angle of ~180°. **9.5A** 16 σ bonds and 4 π bonds. **9.5B** 17 σ bonds and 5 π bonds. **9.6A** C and N atoms are *sp*-hybridized. The triple bond between C and N is composed of one sigma bond (from overlap of hybrid orbitals) and two pi bonds (from interaction of remaining *p* orbitals). The single bond between H and C is the result of an *sp* orbital from C overlapping with an *s* orbital from H. **9.6B** Each N atom is *sp*-hybridized. One *sp* orbital on each is singly occupied and one contains a lone pair. The singly occupied *sp* orbitals overlap to form a sigma bond between the N atoms. The remaining unhybridized *p* orbitals interact to form two pi bonds. **9.7A** paramagnetic; bond order = 2. **9.7B** paramagnetic; bond order = 2. **9.8A** Two different resonance structures are possible; therefore we consider all three atoms to be sp^2-hybridized. One of the hybrid orbitals on the central O atom contains the lone pair, the other two form sigma bonds to the terminal O atoms. Each atom has one remaining unhybridized *p* orbital. The *p* orbitals combine to form π molecular orbitals. **9.8B** Two different resonance structures are possible; therefore we consider all three atoms to be sp^2-hybridized. One of the hybrid orbitals on the central N atom contains the lone pair, the other two form sigma bonds to the terminal O atoms. Each atom has one remaining unhybridized *p* orbital. The *p* orbitals combine to form π molecular orbitals.

Answers to Checkpoints

9.1.1 d. **9.1.2** b. **9.1.3** e. **9.1.4** c. **9.1.5** d. **9.2.1** a. **9.2.2** c. **9.3.1** b, e. **9.3.2** b. **9.4.1** b. **9.4.2** d. **9.5.1** a, c. **9.5.2** c. **9.5.3** a, b, e. **9.5.4** c. **9.6.1** b. **9.6.2** c, e. **9.6.3** b. **9.6.4** a, c, e. **9.7.1** c. **9.7.2** b. **9.7.3** a, b. **9.7.4** e.

CHAPTER 18

Entropy, Free Energy, and Equilibrium

"Plotkin's Entropy" by artist Laurel Latto.

In This Chapter, You Will Learn

About the three laws of thermodynamics and how to use thermodynamic quantities to determine whether or not a process is expected to be spontaneous.

Before You Begin, Review These Skills

- System and surroundings [⏮ Section 5.1]
- Hess's law [⏮ Section 5.5]
- Equilibrium expressions [⏮ Section 15.3]

How Living Systems Must Obey the Laws of Thermodynamics

The painting "Plotkin's Entropy," by the artist Laurel Latto, is an abstract depiction of an extended polypeptide chain as it is synthesized in a biological cell. Under normal physiological conditions, the polypeptide spontaneously folds into a unique three-dimensional structure called the native protein, which performs various functions. Because the original chain can assume many possible configurations while the native protein can have only one specific arrangement, the folding process is accompanied by a decrease in entropy of the system. (Note that solvent molecules, water in this case, can also play a role in affecting the entropy change.) In accord with the second law of thermodynamics, any spontaneous process must result in an increase in the entropy of the universe. It follows, therefore, that there must be an increase in the entropy of the surroundings that outweighs the decrease in the entropy of the system. The intramolecular attractions between amino acid residues cause the folding of the polypeptide chain to be exothermic. The energy produced by the process spreads out, increasing molecular motion in the surroundings—thereby increasing the entropy of the surroundings.

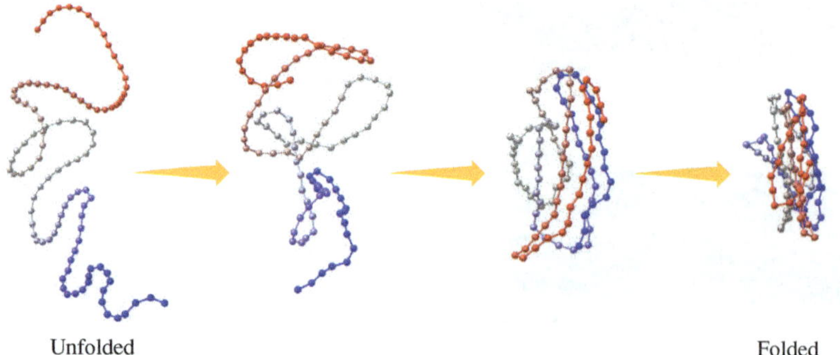

At the end of this chapter, you will be able to answer several questions regarding entropy changes [⏭ Applying What You've Learned, page 867].

18.1 Spontaneous Processes

Student Note: The conditions that most often are specified are *temperature*, *pressure*, and in the case of a solution, *concentration*.

An understanding of thermodynamics enables us to predict whether or not a reaction will occur when reactants are combined. This is important in the synthesis of new compounds in the laboratory, the manufacturing of chemicals on an industrial scale, and the understanding of natural processes such as cell function. A process that *does* occur under a specific set of conditions is called a ***spontaneous process***. One that does *not* occur under a specific set of conditions is called ***nonspontaneous***. Table 18.1 lists examples of familiar spontaneous processes and their nonspontaneous counterparts. These examples illustrate what we know intuitively: Under a given set of conditions, a process that occurs spontaneously in one direction does not also occur spontaneously in the opposite direction.

Processes that result in a decrease in the energy of a system often are spontaneous. For example, the combustion of methane is exothermic:

$$CH_4(g) + 2O_2(g) \longrightarrow CO_2(g) + 2H_2O(l) \qquad \Delta H° = -890.4 \text{ kJ/mol}$$

Thus, the energy of the system is lowered because heat is given off during the course of the reaction. Likewise, in the acid-base neutralization reaction:

$$H^+(aq) + OH^-(aq) \longrightarrow H_2O(l) \qquad \Delta H° = -56.2 \text{ kJ/mol}$$

heat is given off, lowering the energy of the system. Each of these processes is spontaneous, and each results in a lowering of the system's energy.

Now consider the melting of ice:

$$H_2O(s) \longrightarrow H_2O(l) \qquad \Delta H° = 6.01 \text{ kJ/mol}$$

In this case, the process is endothermic and yet it is also spontaneous at temperatures above 0°C. Conversely, the freezing of water is an *exothermic* process:

$$H_2O(l) \longrightarrow H_2O(s) \qquad \Delta H° = -6.01 \text{ kJ/mol}$$

Yet it is *not* spontaneous at temperatures above 0°C.

Based on the first two examples, and many others like them, we might conclude that exothermic processes tend to be spontaneous and, indeed, a negative ΔH does *favor* spontaneity. The last two examples, however, make it clear that the sign of ΔH *alone* is insufficient to predict spontaneity in every circumstance. For the remainder of this chapter, we will examine the *two* factors that determine whether or not a process is spontaneous under a given set of conditions.

18.2 Entropy

To predict the spontaneity of a chemical or physical process, we need to know both the change in *enthalpy* [⏮ Section 5.3] and the change in *entropy* associated with the process. We first encountered the concept of entropy in our discussion of solution formation [⏮ Section 13.2]. We will now look in more detail at what entropy is, and why it matters.

TABLE 18.1 Familiar Spontaneous and Nonspontaneous Processes

Spontaneous	Nonspontaneous
Ice melting at room temperature	Water freezing at room temperature
Sodium metal reacting violently with water to produce sodium hydroxide and hydrogen gas [⏮ Section 7.7]	Sodium hydroxide reacting with hydrogen gas to produce sodium metal and water
A ball rolling downhill	A ball rolling uphill
The rusting of iron at room temperature	The conversion of rust back to iron metal at room temperature
Water freezing at −10°C	Ice melting at −10°C

A Qualitative Description of Entropy

Qualitatively, the ***entropy*** **(*S*)** of a system is a measure of how *spread out* or how *dispersed* the system's energy is. The simplest interpretation of this is how spread out a system's energy is in *space*. In other words, for a given system, the greater the volume it occupies, the greater its entropy. This interpretation explains how the process in Figure 18.1 occurs spontaneously despite there being no enthalpy change. Because they are moving, the gas molecules that were originally confined to one side of the container possess *motional energy*. Motional energy includes *translational* energy, in which the entire molecule moves through space [⏮ Section 5.1]; *rotational energy*, in which the molecule spins about an axis running through its center of mass; and *vibrational energy*, in which atoms of a molecule move relative to one another. In the absence of a barrier preventing it, the motional energy of molecules will spread out to occupy a larger volume. The dispersal of a system's motional energy to occupy a larger volume when the barrier is removed constitutes an *increase* in the system's entropy. Just as spontaneity is favored by a process being exothermic, spontaneity is also favored by an increase in the system's entropy. Whether it is the enthalpy change, the entropy change, or both, for a process to be spontaneous, *something* must favor spontaneity.

A Quantitative Definition of Entropy

At this point, it is useful to introduce the mathematical definition of entropy proposed by Ludwig Boltzmann:

$$S = k \ln W \qquad \textbf{Equation 18.1}$$

where k is the Boltzmann constant (1.38×10^{-23} J/K) and W is the number of energetically equivalent different ways the molecules in a system can be arranged. To illustrate what this means, let's consider a simplified version of the process shown in Figure 18.1. Prior to the removal of the barrier between the left and right sides of the container, at any given instant, each molecule has a particular location, somewhere in the left side of the container. To narrow down the possible locations of the molecules, we imagine that each side of the container is divided into a number of equal smaller volumes called *cells*. In the simplest scenario, with just one molecule in the system, the number of possible locations of the molecule is equal to the number of cells. If the system contains *two* molecules, the number of possible arrangements is equal to the number of cells *squared*. (Note that a cell may contain more than one molecule.) Each time we increase the number of molecules by one, the number of possible arrangements increases by a factor equal to the number of cells. In general, for a volume consisting of X cells, and containing N molecules, the number of possible arrangements, W, is given by the equation:

Student Note: The Boltzmann constant is equal to the gas constant, R (in J/K · mol), divided by Avogadro's constant, N_A.

Student Note: The number of possible arrangements is sometimes called the number of *microstates*.

$$W = X^N \qquad \textbf{Equation 18.2}$$

Figure 18.2 illustrates this for a simple case involving just two molecules. We imagine the container is divided into four cells each with volume v. Initially, both molecules are confined to the left side, which consists of two cells. With two molecules in two cells, there are $2^2 = 4$ possible arrangements of the molecules [Figure 18.2(a)]. When the barrier is removed, doubling the volume available to the molecules, the number of cells also doubles. With four cells available, there are $4^2 = 16$ possible arrangements of the molecules. Eight of the sixteen arrangements have the molecules on

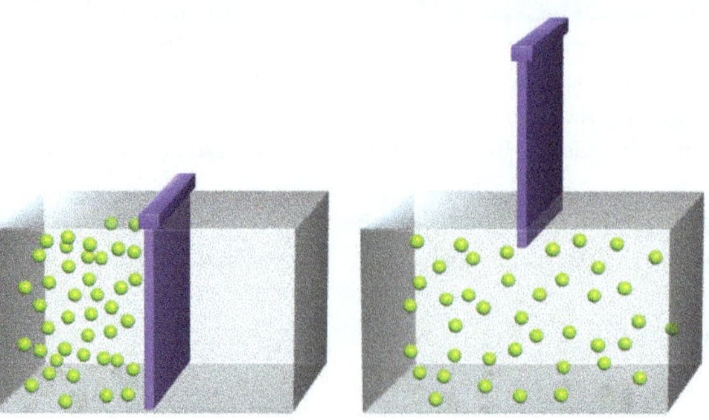

Figure 18.1 A spontaneous process. The rapidly moving gas molecules originally confined to one side of a container spread out to fill the whole container when the barrier is removed.

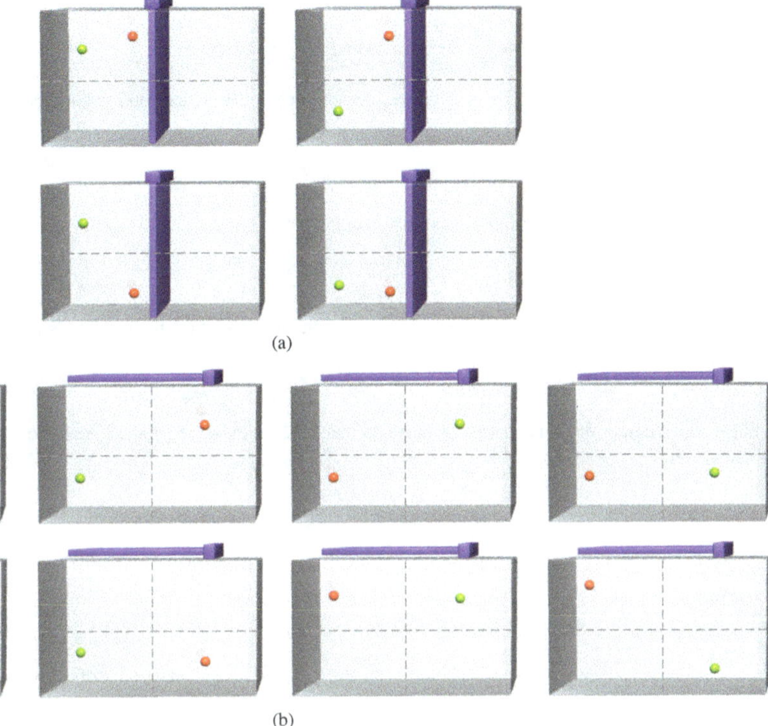

Figure 18.2 (a) Before the barrier is removed, the molecules are both in the left side of the container, which we imagine is divided into two cells of equal volume. There are four possible arrangements of two molecules in two cells. (b) When the barrier between the two sides of the container is removed, the volume (and the number of cells) available to the molecules doubles. The new number of possible arrangements is $4^2 = 16$, eight of which have the molecules in opposite sides of the container—the most probable outcome.

opposite sides of the container [Figure 18.2(b)]. Of the other eight arrangements, four have both molecules on the left side [as shown in Figure 18.2(a)], and four have both molecules on the right side (not shown). There are three different states possible for this system.

1. One molecule on each side (eight possible arrangements)
2. Both molecules on the left (four possible arrangements)
3. Both molecules on the right (four possible arrangements)

Student Note: The state with the largest number of possible arrangements has the greatest entropy.

The most *probable* state is the one with the *largest* number of possible arrangements. In this case, the most probable state is the one with one molecule on each side of the container. The same principle applies to systems with larger numbers of molecules. Increasing the number of molecules increases the number of possible arrangements, but the most probable state will be the one in which the gas molecules are divided evenly between the two sides of the container.

18.3 Entropy Changes in a System

Calculating ΔS_{sys}

The change in entropy of a system is the difference between the entropy of the final state and the entropy of the initial state:

Equation 18.3 $$\Delta S_{sys} = S_{final} - S_{initial}$$

Using Equation 18.1, we can write an expression for the entropy of each state:

$$\Delta S_{sys} = k \ln W_{final} - k \ln W_{initial} = k \ln \frac{W_{final}}{W_{initial}}$$

Student Note: Remember that the difference between two logs is equal to the log of the corresponding quotient:

$$\ln A - \ln B = \ln \frac{A}{B}$$

and that $\ln A^x = x \ln A$ [⏭ Appendix 1]

Combining this result with Equation 18.2 gives:

$$\Delta S_{sys} = k \ln \frac{(X_{final})^N}{(X_{initial})^N} = k \ln \left(\frac{X_{final}}{X_{initial}}\right)^N = k\,N \ln \left(\frac{X_{final}}{X_{initial}}\right)$$

Because X is the number of cells, and the volume of each cell is v, the total volume is related to the number of cells by:

$$V = Xv \qquad \text{or} \qquad X = \frac{V}{v}$$

We substitute V_{final}/v for X_{final} and $V_{initial}/v$ for $X_{initial}$ to get:

$$\Delta S_{sys} = k\,N \ln \frac{V_{final}/v}{V_{initial}/v} = k\,N \ln \frac{V_{final}}{V_{initial}}$$

Finally, because the Boltzmann constant, k, is the gas constant, R, divided by Avogadro's constant:

$$k = \frac{R}{N_A}$$

and because the number of molecules, N, is the product of the number of moles, n, and Avogadro's constant, N_A:

$$N = n \times N_A$$

$$kN = \left(\frac{R}{\cancel{N_A}} \times n \times \cancel{N_A}\right) = nR$$

the equation becomes:

$$\Delta S_{sys} = nR \ln \frac{V_{final}}{V_{initial}} \qquad \textbf{Equation 18.4}$$

Sample Problem 18.1 shows how to use Equation 18.4 to calculate the entropy change for a process like the one shown in Figure 18.1, the expansion of an ideal gas at constant temperature.

SAMPLE PROBLEM 18.1

Determine the change in entropy for 1.0 mole of an ideal gas originally confined to one-half of a 5.0-L container when the gas is allowed to expand to fill the entire container at constant temperature.

Strategy This is the isothermal expansion of an ideal gas. Because the molecules spread out to occupy a greater volume, we expect there to be an increase in the entropy of the system. Use Equation 18.3 to solve for ΔS_{sys}.

Setup $R = 8.314$ J/K · mol, $n = 1.0$ mole, $V_{final} = 5.0$ L, and $V_{initial} = 2.5$ L.

Solution

$$\Delta S_{sys} = nR \ln \frac{V_{final}}{V_{initial}} = 1.0\,\cancel{\text{mol}} \times \frac{8.314\text{ J}}{\text{K} \cdot \cancel{\text{mol}}} \times \ln \frac{5.0\text{ L}}{2.5\text{ L}} = 5.8\text{ J/K}$$

THINK ABOUT IT

Remember that for a process to be spontaneous, *something* must favor spontaneity. If the process is spontaneous but not exothermic (in this case, there is no enthalpy change), then we should expect ΔS_{sys} to be positive.

Practice Problem ATTEMPT Determine the change in entropy (ΔS_{sys}), for the expansion of 0.10 mole of an ideal gas from 2.0 L to 3.0 L at constant temperature.

Practice Problem BUILD To what fraction of its original volume must a 0.50-mol sample of ideal gas be compressed at constant temperature for ΔS_{sys} to be −6.7 J/K?

Practice Problem CONCEPTUALIZE Which equation is correct for calculating ΔS_{sys} for a gaseous reaction that occurs at constant volume?

(i) $\Delta S_{sys} = nR$

(ii) $\Delta S_{sys} = nRT \ln \frac{P_{initial}}{P_{final}}$

(iii) $\Delta S_{sys} = nR \ln \frac{P_{initial}}{P_{final}}$

(iv) $\Delta S_{sys} = \frac{nR}{T} \ln \frac{P_{final}}{P_{initial}}$

Student Note: For even the simplest of hypothetical systems, where there are only two possible positions for molecules ($X = 2$), most calculators cannot display a number as large as the result of Equation 18.2 for even as few as 500 molecules—much less for the *enormous number* of molecules present in any *real* sample. (If your calculator is like most, with $X = 2$, you can calculate the number of possible arrangements for $N \leq 332$ molecules. Try it: 2 ^ 332 = ? and 2 ^ 333 = ?)

Standard Entropy, $S°$

Although Equation 18.1 provides a quantitative definition of entropy, we seldom use it or Equation 18.3 to calculate the entropy change for a real process because of the difficulty involved in determining W, the number of different possible arrangements (Equation 18.2) in a macroscopic system. Instead, for processes other than isothermal expansion or compression of an ideal gas (for which we can use Equation 18.4), we routinely determine entropy changes using tabulated values.

Using calorimetry [⏮ Section 5.4], it is possible to determine the *absolute* value of the entropy of a substance, S; something we cannot do with either energy or enthalpy. (Recall that while we can determine ΔU and ΔH for a process that a system undergoes, we cannot determine the absolute values of either U or H for a system [⏮ Sections 5.2 and 5.3].) ***Standard entropy*** **($S°$)** is the absolute entropy of a substance at 1 atm. (Tables of standard entropy values typically are the values at 25°C because so many processes are carried out at room temperature—although temperature is *not* part of the standard state definition and therefore must be specified.) Table 18.2 lists standard entropies of a few elements and compounds. Appendix 2 provides a more extensive listing. The units of entropy are J/K · mol. We use joules rather than kilojoules because entropy values typically are quite small. The entropies of substances (elements and compounds) are always positive (i.e., $S > 0$), even for elements in their standard states. (Recall that the standard *enthalpy* of formation, $\Delta H_f°$, for elements in their standard states is arbitrarily defined as zero, and for compounds it may be either positive or negative [⏮ Section 5.6].) You will find that tables, including Appendix 2, contain *negative* absolute entropies for some aqueous ions. Unlike a substance, an individual ion cannot be studied experimentally. Therefore, standard entropies of ions are actually *relative* values, where a standard entropy of zero is arbitrarily assigned to the hydrated hydrogen (hydronium) ion. Depending on an ion's extent of hydration, its standard entropy may be positive or negative, relative to that of hydrogen (hydronium) ion.

Referring to Table 18.2, we can identify several important trends:

- For a given substance, the standard entropy is greater in the liquid phase than in the solid phase. [Compare the standard entropies of $Na(s)$ and $Na(l)$.] This results from there being greater molecular motion in a liquid, resulting in many possible arrangements of atoms in the liquid phase; whereas the positions of atoms in the solid are fixed.
- For a given substance, the standard entropy is greater in the gas phase than in the liquid phase. [Compare the standard entropies of $Na(l)$ and $Na(g)$ and those of $H_2O(l)$ and $H_2O(g)$.] This results from there being much greater molecular motion in a gas, resulting in many more possible arrangements of atoms in the gas phase than in the liquid phase—in part because the gas phase occupies a much greater volume than either of the condensed phases.
- For two monatomic species, the one with the larger molar mass has the greater standard entropy. [Compare the standard entropies of $He(g)$ and $Ne(g)$.]
- For two substances in the same phase, and with similar molar masses, the substance with the more complex molecular structure has the greater standard entropy. [Compare the standard entropies of $O_3(g)$ and $F_2(g)$.] The more complex a molecular structure, the more different types of motion the molecule can exhibit. A diatomic molecule such as F_2, for example, exhibits only one type of vibration, whereas a bent triatomic molecule such as O_3 exhibits three different types of vibrations. Each mode of motion contributes to the total number of available energy levels within which a system's energy can be dispersed. Figure 18.3 illustrates the ways in which the F_2 and O_3 molecules can rotate and vibrate.

TABLE 18.2 Standard Entropy Values ($S°$) for Some Substances at 25°C

Substance	$S°$ (J/K · mol)	Substance	$S°$ (J/K · mol)
$H_2O(l)$	69.9	C(diamond)	2.4
$H_2O(g)$	188.7	C(graphite)	5.69
$Na(s)$	51.05	$O_2(g)$	205.0
$Na(l)$	57.56	$O_3(g)$	237.6
$Na(g)$	153.7	$F_2(g)$	203.34
$He(g)$	126.1	$Au(s)$	47.7
$Ne(g)$	146.2	$Hg(l)$	77.4

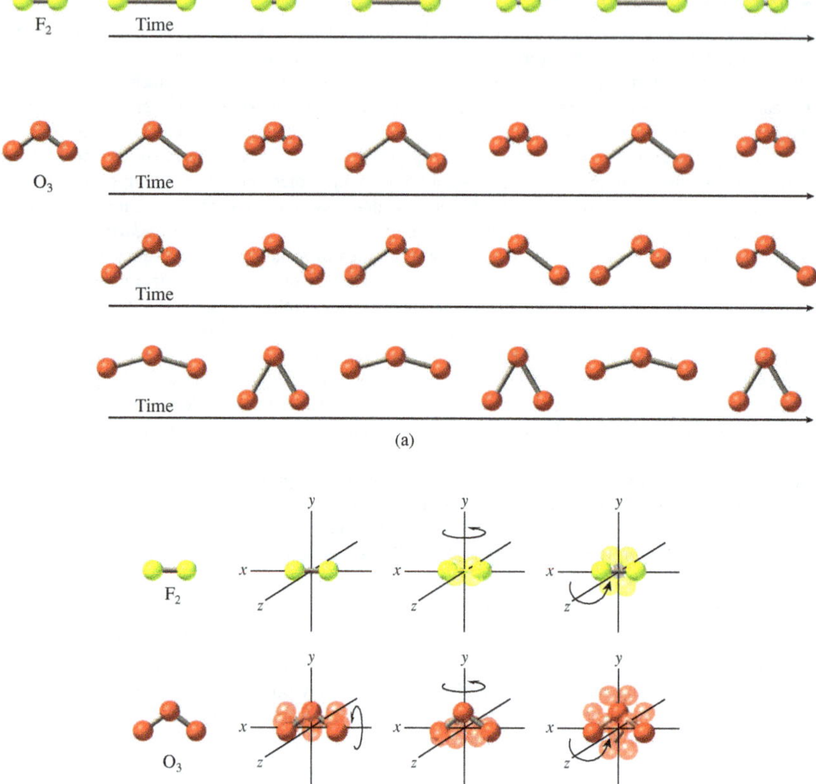

Figure 18.3 In addition to translational motion, molecules exhibit both *vibrations*, in which the atoms' positions relative to one another change, and *rotations*, in which the molecule rotates about its center of mass. (a) A diatomic molecule such as fluorine only exhibits one type of vibration. A bent, triatomic molecule such as ozone exhibits three types of vibration. (b) A diatomic molecule exhibits two different rotations, whereas a bent, triatomic molecule exhibits three different rotations. (Note that rotation of F_2 about the *x* axis would cause no change in the positions of either atom in the molecule.)

- In cases where an element exists in two or more allotropic forms, the form in which the atoms are more mobile has the greater entropy. [Compare the standard entropies of C(diamond) and C(graphite). In diamond, the carbon atoms occupy fixed positions in a three-dimensional array. In graphite, although the carbon atoms occupy fixed positions within the two-dimensional sheets (see Figure 11.27), the sheets are free to move with respect to one another, which increases the mobility and, therefore, total number of possible arrangements of atoms within the solid.]

Now let's consider a process represented by the following chemical equation:

$$a\text{A} + b\text{B} \longrightarrow c\text{C} + d\text{D}$$

Just as the enthalpy change of a reaction is the difference between the enthalpies of the products and reactants (Equation 5.12), the entropy change is the difference between the entropies of the products and reactants:

$$\Delta S^\circ_{\text{rxn}} = [cS^\circ(\text{C}) + dS^\circ(\text{D})] - [aS^\circ(\text{A}) + bS^\circ(B)] \qquad \textbf{Equation 18.5}$$

Or, using Σ to represent summation and m and n to represent the stoichiometric coefficients of the reactants and products, respectively, Equation 18.5 can be generalized as follows:

$$\Delta S^\circ_{\text{rxn}} = \Sigma nS^\circ(\text{products}) - \Sigma mS^\circ(\text{reactants}) \qquad \textbf{Equation 18.6}$$

The standard entropy values of a large number of substances have been measured in J/K · mol. To calculate the standard entropy change for a reaction ($\Delta S^\circ_{\text{rxn}}$), we look up the standard entropies of the products and reactants and use Equation 18.5.

Student Note: The *reaction* generally is the *system*. Therefore, $\Delta S^\circ_{\text{rxn}}$ is $\Delta S^\circ_{\text{sys}}$.

Sample Problem 18.2 demonstrates this approach.

SAMPLE PROBLEM 18.2

From the standard entropy values in Appendix 2, calculate the standard entropy changes for the following reactions at 25°C:

(a) $CaCO_3(s) \longrightarrow CaO(s) + CO_2(g)$

(b) $N_2(g) + 3H_2(g) \longrightarrow 2NH_3(g)$

(c) $H_2(g) + Cl_2(g) \longrightarrow 2HCl(g)$

Strategy Look up standard entropy values and use Equation 18.5 to calculate ΔS°_{rxn}. Just as we did when we calculated standard enthalpies of reaction, we consider stoichiometric coefficients to be dimensionless—giving ΔS°_{rxn} units of J/K · mol. Recall that here *per mole* means per mole of reaction as written [⏮ Section 5.6].

Setup From Appendix 2, $S^\circ[CaCO_3(s)] = 92.9$ J/K · mol, $S^\circ[CaO(s)] = 39.8$ J/K · mol, $S^\circ[CO_2(g)] = 213.6$ J/K · mol, $S^\circ[N_2(g)] = 191.5$ J/K · mol, $S^\circ[H_2(g)] = 131.0$ J/K · mol, $S^\circ[NH_3(g)] = 193.0$ J/K · mol, $S^\circ[Cl_2(g)] = 223.0$ J/K · mol, and $S^\circ[HCl(g)] = 187.0$ J/K · mol.

Solution

(a)
$$\begin{aligned}\Delta S^\circ_{rxn} &= [S^\circ(CaO) + S^\circ(CO_2)] - [S^\circ(CaCO_3)]\\ &= [(39.8\ J/K \cdot mol) + (213.6\ J/K \cdot mol)] - (92.9\ J/K \cdot mol)\\ &= 160.5\ J/K \cdot mol\end{aligned}$$

(b)
$$\begin{aligned}\Delta S^\circ_{rxn} &= [2S^\circ(NH_3)] - [S^\circ(N_2) + 3S^\circ(H_2)]\\ &= (2)(193.0\ J/K \cdot mol) - [(191.5\ J/K \cdot mol) + (3)(131.0\ J/K \cdot mol)]\\ &= -198.5\ J/K \cdot mol\end{aligned}$$

(c)
$$\begin{aligned}\Delta S^\circ_{rxn} &= [2S^\circ(HCl)] - [S^\circ(H_2) + S^\circ(Cl_2)]\\ &= (2)(187.0\ J/K \cdot mol) - [(131.0\ J/K \cdot mol) + (223.0\ J/K \cdot mol)]\\ &= 20.0\ J/K \cdot mol\end{aligned}$$

THINK ABOUT IT

Remember to multiply each standard entropy value by the correct stoichiometric coefficient. Like Equation 5.19, Equation 18.5 can only be used with a *balanced* chemical equation.

Practice Problem ATTEMPT Calculate the standard entropy change for the following reactions at 25°C. Predict first whether each one will be positive, negative, or too close to call.

(a) $2CO_2(g) \longrightarrow 2CO(g) + O_2(g)$

(b) $3O_2(g) \longrightarrow 2O_3(g)$

(c) $2NaHCO_3(s) \longrightarrow Na_2CO_3(s) + H_2O(l) + CO_2(g)$

Practice Problem BUILD In each of the following reactions, there is one species for which the standard entropy is not listed in Appendix 2. In each case, using the values that *are* in Appendix 2 and the ΔS°_{rxn} that is given, determine the value of the missing standard entropy at 25°C: (a) $K(s) \longrightarrow K(l)$, $\Delta S^\circ_{rxn} = 7.9$ J/K · mol, (b) $2S(\text{rhombic}) + Cl_2(g) \longrightarrow S_2Cl_2(g)$, $\Delta S^\circ_{rxn} = 44.74$ J/K · mol, (c) $O_2(g) + 2MgF_2(s) \longrightarrow 2MgO(s) + 2F_2(g)$, $\Delta S^\circ_{rxn} = 140.76$ J/K · mol.

Practice Problem CONCEPTUALIZE For each reaction shown in the diagrams, indicate whether ΔS°_{rxn} is positive, negative, or too close to call.

(i) + ⟶ +

(ii)

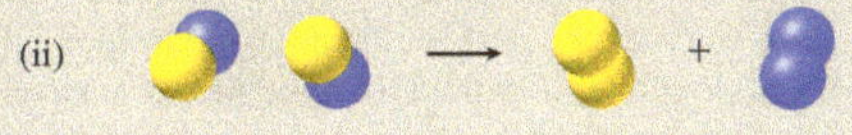

(iii)

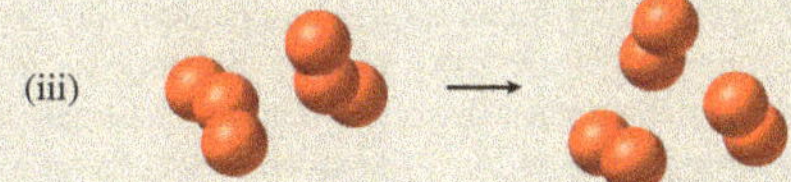

Qualitatively Predicting the Sign of ΔS°_{sys}

Equation 18.5 enables us to calculate ΔS°_{rxn} for a process when the standard entropies of the products and reactants are known. However, sometimes it's useful just to know the *sign* of ΔS°_{rxn}. Although multiple factors can influence the sign of ΔS°_{rxn}, the outcome is often dominated by a single factor, which can be used to make a qualitative prediction. Several processes that lead to an increase in entropy are

- Melting
- Vaporization or sublimation
- Temperature increase
- Reaction resulting in a greater number of gas molecules

When a solid is melted, the molecules have greater energy and are more mobile. They go from being in fixed positions in the solid, to being free to move about in the liquid. As we saw in the discussion of standard entropy, this leads to many more possible arrangements of the molecules and, therefore, greater entropy. The same rationale holds for the vaporization or sublimation of a substance. There is a dramatic increase in energy/mobility, and in the number of possible arrangements of a system's molecules when the molecules go from a condensed phase to the gas phase. Therefore, there is a much larger increase in the system's entropy, relative to the solid-to-liquid transition.

When the temperature of a system is increased, the energy of the system's molecules increases. To visualize this, recall from the discussion of kinetic molecular theory that increasing the temperature of a gas increases its average kinetic energy. This corresponds to an increase in the average speed of the gas molecules and a spreading out of the range of molecular speeds. [See Figure 10.20(a).] If we think of each of the possible molecular speeds within the range as a discrete energy level, we can see that at higher temperatures, there is a greater number of possible molecular speeds and, therefore, a greater number of energy levels available to the molecules in the system. With a greater number of available energy levels, there is a greater number of possible arrangements of molecules *within* those levels and, therefore, a greater entropy.

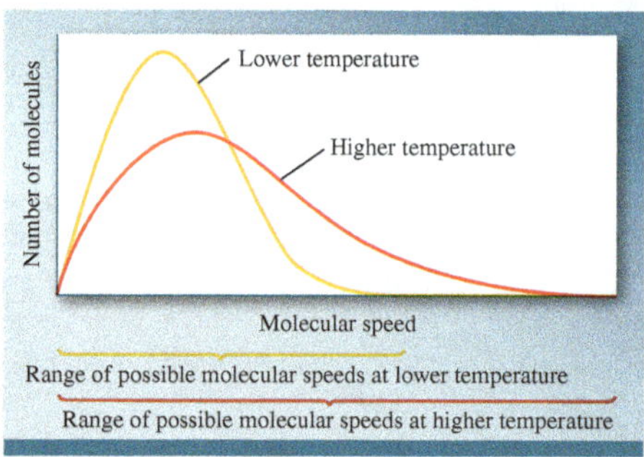

Because the entropy of a substance in the gas phase is always significantly greater than its entropy in either the liquid or solid phase, a reaction that results in an increase in the number of gas molecules causes an increase in the system's entropy. For reactions that do not involve gases, an increase in the number of solid, liquid, or aqueous molecules also usually causes an entropy increase.

By considering these factors, we can usually make a reasonably good prediction of the sign of ΔS°_{rxn} for a physical or chemical process, without having to look up the absolute entropy values for the species involved. Figure 18.4 summarizes the factors that can be used to compare entropies and illustrates several comparisons.

Animation
Figure 18.4, Factors That Influence the Entropy of a System.

In addition to melting, vaporization/sublimation, temperature increase, and reactions that increase the number of gas molecules, which can always be counted upon to result in an entropy increase, the process of *dissolving* a substance often leads to an increase in entropy. In the case of a molecular solute, such as sucrose (sugar), dissolving causes dispersal of the molecules (and consequently, of the system's *energy*) into a larger volume—resulting in an increase in entropy. In the case of an ionic solute, the analysis is slightly more complicated. We saw in our discussion of solution formation [⏮ Section 13.2] that the dissolution of ammonium nitrate (NH_4NO_3) is spontaneous, even though it is endothermic, because the system's entropy increases when the ionic solid dissociates and is dispersed throughout the solution. In general, this is the case for ionic solutes in which the charges on ions are small. (In the case of NH_4NO_3, they are +1 and −1.)

Figure 18.4

Factors That Influence the Entropy of a System

Volume Change

Quantum mechanical analysis shows that the spacing between translational energy levels is inversely proportional to the volume of the container. Thus, when the volume is increased, more energy levels become available within which the system's energy can be dispersed.

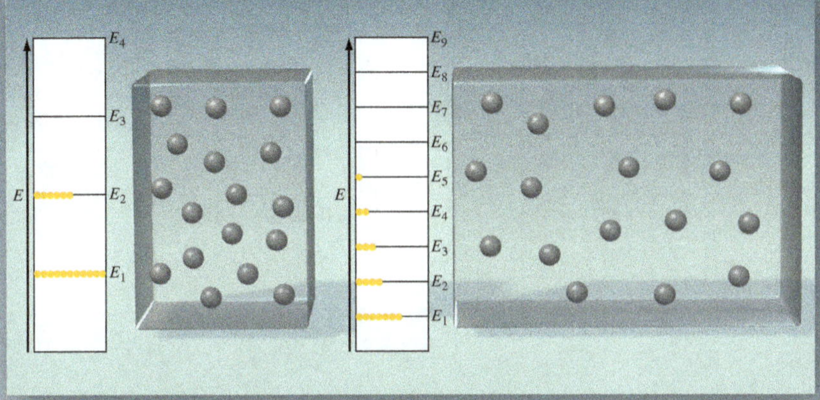

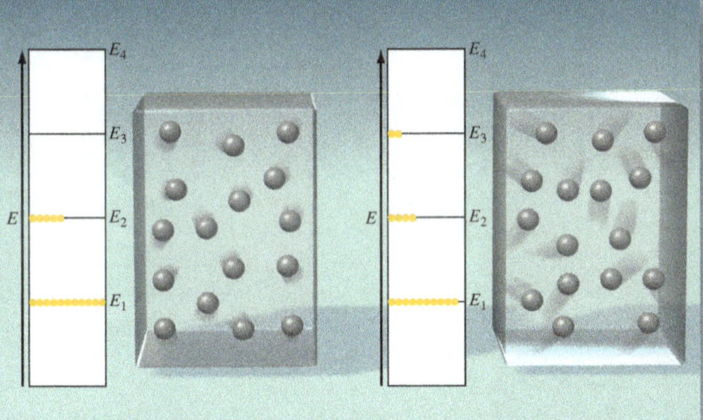

Temperature Change

At higher temperatures, molecules have greater kinetic energy—making more energy levels accessible. This increases the number of energy levels within which the system's energy can be dispersed, causing entropy to increase.

Molecular Complexity

Unlike atoms, which exhibit only translational motion, molecules can also exhibit rotational and vibrational motions. The greater a molecule's complexity, the greater the number of possible ways it can rotate and vibrate. The ozone molecule (O_3), for example, is more complex than the fluorine molecule (F_2) and exhibits more different kinds of vibrations and rotations. (See Figure 18.3.) This results in more energy levels within which the system's energy can be dispersed. The number and spacing of additional energy levels have been simplified to keep the illustration clear.

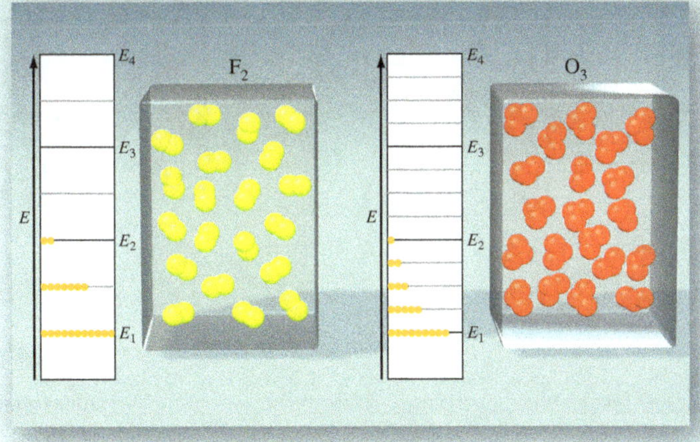

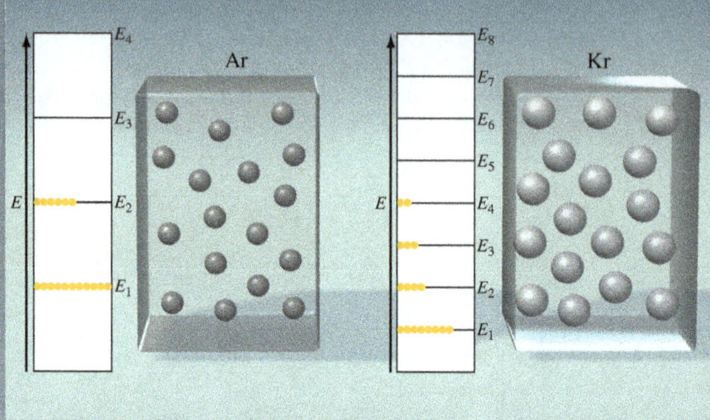

Molar Mass

The energy levels for a substance with a larger molar mass are more closely spaced. Kr, for example, has roughly twice the molar mass of Ar. Thus, Kr has roughly twice as many energy levels within which the system's energy can be dispersed.

Phase Change

Because of greater mobility, there are many more different possible arrangements (W) of molecules in the liquid phase than there are in the solid phase; and there are many, *many* more different possible arrangements of molecules in the gas phase than there are in the liquid phase. Entropy of a substance increases when it is melted ($s \longrightarrow l$), vaporized ($l \longrightarrow g$), or sublimed ($s \longrightarrow g$).

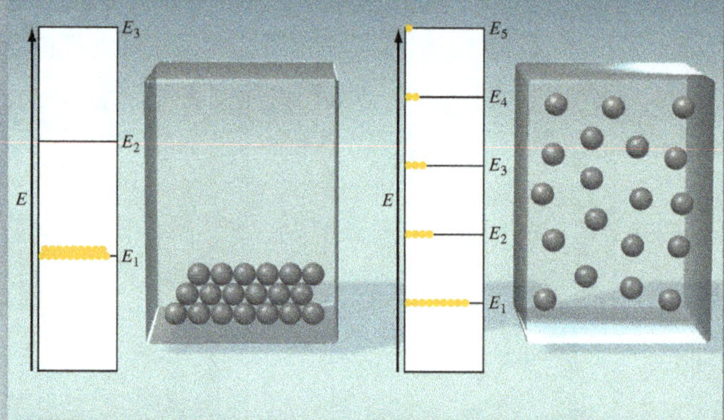

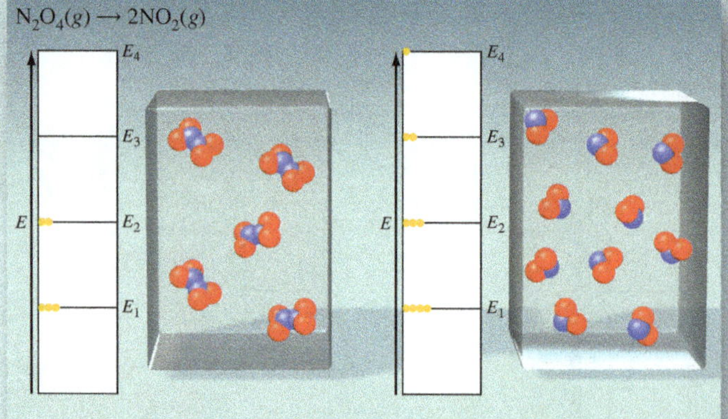

Chemical Reaction

When a chemical reaction produces more gas molecules than it consumes, the number of different possible arrangements of molecules (W) increases and entropy increases.

What's the point?

Although several factors can influence the entropy of a system or the entropy change associated with a process, often one factor dominates the outcome. Each of these comparisons shows a qualitative illustration of one of the important factors.

(See Visualizing Chemistry questions VC 18.1–VC 18.4 on pages 867–868.)

TABLE 18.3 Entropy Changes for the Dissolution (ΔS°_{soln})* of Some Ionic Solids at 25°C

Dissolution Equation	ΔS°_{soln} (J/K · mol)
$NH_4NO_3(s) \longrightarrow NH_4^+(aq) + NO_3^-(aq)$	108.1
$AlCl_3(s) \longrightarrow Al^{3+}(aq) + 3Cl^-(aq)$	−253.2
$FeCl_3(s) \longrightarrow Fe^{3+}(aq) + 3Cl^-(aq)$	−266.1

*Note that the subscript "rxn" in ΔS°_{rxn} changes to "soln" to refer specifically to the *solution* process.

However, when ions are dispersed in water, they become hydrated (surrounded by water molecules in a specific arrangement [◂◂ Figure 4.4]). This leads to a *decrease* in the entropy of the *water*, as hydration reduces the mobility of some of the water molecules by fixing them in positions around the dissolved ions. When the charges on ions are low, the increase in entropy of the solute typically outweighs the decrease in entropy of the water—resulting in an overall *increase* in the entropy of the system—as is the case with NH_4NO_3. By contrast, when highly charged ions such as Al^{3+} and Fe^{3+} are hydrated, the decrease in entropy of the water can actually outweigh the increase in entropy of the solute, leading to an overall *decrease* in entropy of the system. Table 18.3 lists the changes in entropy associated with the spontaneous dissolution of several ionic solids.

Sample Problem 18.3 lets you practice making qualitative predictions of the sign of ΔS°_{rxn}.

SAMPLE PROBLEM 18.3

For each process, determine the sign of ΔS for the system: (a) decomposition of $CaCO_3(s)$ to give $CaO(s)$ and $CO_2(g)$, (b) heating bromine vapor from 45°C to 80°C, (c) condensation of water vapor on a cold surface, (d) reaction of $NH_3(g)$ and $HCl(g)$ to give $NH_4Cl(s)$, and (e) dissolution of sugar in water.

Strategy Consider the change in energy/mobility of atoms and the resulting change in number of possible positions that each particle can occupy in each case. An increase in the number of arrangements corresponds to an increase in entropy and therefore a positive ΔS.

Setup Increases in entropy generally accompany solid-to-liquid, liquid-to-gas, and solid-to-gas transitions; the dissolving of one substance in another; a temperature increase; and reactions that increase the net number of moles of gas.

Solution ΔS is (a) positive, (b) positive, (c) negative, (d) negative, and (e) positive.

THINK ABOUT IT

For reactions involving only liquids and solids, predicting the sign of ΔS° can be more difficult, but in many such cases an increase in the total number of molecules and/or ions is accompanied by an increase in entropy.

Practice Problem ATTEMPT For each of the following processes, determine the sign of ΔS: (a) crystallization of sucrose from a supersaturated solution, (b) cooling water vapor from 150°C to 110°C, (c) sublimation of dry ice.

Practice Problem BUILD Make a qualitative prediction of the sign of ΔH°_{soln} for $AlCl_3(s)$ and the dissolution of $FeCl_3(s)$. See Table 18.3 (above). Explain your reasoning.

Practice Problem CONCEPTUALIZE Consider the gas-phase reaction of A_2 (blue) and B_2 (orange) to form AB_3. What are the correct balanced equation and the sign of ΔS for the reaction?

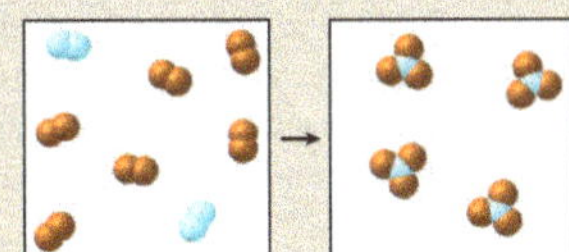

(a) $A_2 + B_2 \longrightarrow AB_3$, negative

(b) $2A_2 + 3B_2 \longrightarrow 4AB_3$, positive

(c) $2A_2 + 3B_2 \longrightarrow 4AB_3$, negative

(d) $A_2 + 3B_2 \longrightarrow 2AB_3$, negative

(e) $A_2 + 3B_2 \longrightarrow 2AB_3$, positive

CHECKPOINT – SECTION 18.3 Entropy Changes in a System

18.3.1 For which of the following physical processes is ΔS negative? (Select all that apply.)

a) Freezing ethanol
b) Evaporating water
c) Mixing carbon tetrachloride and benzene
d) Heating water
e) Condensing bromine vapor

18.3.2 For which of the following chemical reactions is ΔS negative? (Select all that apply.)

a) $2O_3(g) \longrightarrow 3O_2(g)$
b) $4Fe(s) + 3O_2(g) \longrightarrow 2Fe_2O_3(s)$
c) $2H_2O_2(aq) \longrightarrow 2H_2O(l) + O_2(g)$
d) $2Li(s) + 2H_2O(l) \longrightarrow 2LiOH(aq) + H_2(g)$
e) $2NH_3(g) \longrightarrow N_2(g) + 3H_2(g)$

18.3.3 Identify the correct balanced equation and the sign of ΔS for the reaction shown here.

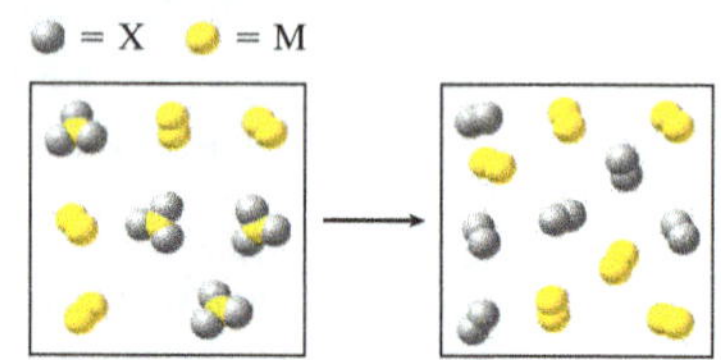

a) $4MX_3 + 4M_2 \longrightarrow 6X_2 + 6M_2$, ΔS positive
b) $4MX_3 \longrightarrow 6X_2 + 2M_2$, ΔS negative
c) $2MX_3 \longrightarrow 3X_2 + M_2$, ΔS positive
d) $2MX_3 + 2M_2 \longrightarrow 3X_2 + 3M_2$, ΔS positive
e) $2MX_3 + 2M_2 \longrightarrow 3X_2 + 3M_2$, ΔS negative

18.4 Entropy Changes in the Universe

Recall that the *system* typically is the part of the universe we are investigating (e.g., the reactants and products in a chemical reaction). The *surroundings* are everything else [|◀◀ Section 5.1]. Together, the system and surroundings make up the *universe*. We have seen that the *dispersal* or *spreading out* of a system's energy corresponds to an increase in the system's entropy. Moreover, an increase in the system's entropy is one of the factors that determines whether or not a process is spontaneous. However, correctly predicting the spontaneity of a process requires us to consider entropy changes in both the system and the surroundings.

Consider the following processes:

- An ice cube spontaneously melts in a room where the temperature is 25°C. In this case, the motional energy of the air molecules at 25°C is transferred to the ice cube (at 0°C), causing the ice to melt. There is no temperature change during a phase change. However, because the molecules are more mobile and there are many more different possible arrangements in *liquid* water than there are in ice, there is an increase in the entropy of the system. In this case, because the process of melting is endothermic, heat is transferred *from* the surroundings *to* the system and the temperature of the surroundings decreases. The slight decrease in temperature causes a small decrease in molecular motion and a decrease in the entropy of the surroundings.

  ΔS_{sys} is positive.
  ΔS_{surr} is negative.

- A cup of hot water spontaneously cools to room temperature as the motional energy of the water molecules spreads out to the cooler surrounding air. Although the loss of energy from the system and corresponding temperature decrease cause a *decrease* in the entropy of the *system*, the increased temperature of the surrounding air causes an *increase* in the entropy of the *surroundings*.

  ΔS_{sys} is negative.
  ΔS_{surr} is positive.

Thus, it is not just the entropy of the *system* that determines if a process is spontaneous, the entropy of the *surroundings* is also important. There are also examples of spontaneous processes in which ΔS_{sys} and ΔS_{surr} are *both* positive. The decomposition of hydrogen peroxide produces water and oxygen gas, $2H_2O_2(l) \longrightarrow 2H_2O(l) + O_2(g)$. Because the reaction results in an increase in the number of gas molecules, we know that there is an increase in the entropy of the system. However, this is an exothermic reaction, meaning that it also gives off heat to the surroundings. An increase in temperature of the surroundings causes an increase in the entropy of the surroundings as well. (Note that there are no spontaneous processes in which ΔS_{sys} and ΔS_{surr} are both negative, which will become clear shortly.)

Calculating ΔS_{surr}

When an exothermic process takes place, the heat transferred from the system to the surroundings increases the temperature of the molecules in the surroundings. Consequently, there is an increase in the number of energy levels accessible to the molecules in the surroundings and the entropy of the surroundings increases. Conversely, in an endothermic process, heat is transferred from the surroundings to the system, decreasing the entropy of the surroundings. Remember that for constant-pressure processes, the heat released or absorbed, q, is equal to the enthalpy change of the system, ΔH_{sys} [⏮ Section 5.3]. The change in entropy for the surroundings, ΔS_{surr}, is directly proportional to ΔH_{sys}:

$$\Delta S_{surr} \propto -\Delta H_{sys}$$

The minus sign indicates that a negative enthalpy change in the system (an *exothermic* process) corresponds to a positive entropy change in the surroundings. For an *endothermic* process, the enthalpy change in the system is a positive number and corresponds to a negative entropy change in the surroundings.

In addition to being directly proportional to ΔH_{sys}, ΔS_{surr} is inversely proportional to temperature:

$$\Delta S_{surr} \propto \frac{1}{T}$$

Combining the two expressions gives:

Equation 18.7 $$\Delta S_{surr} = \frac{-\Delta H_{sys}}{T}$$

The Second Law of Thermodynamics

We have seen that both the system and surroundings can undergo changes in entropy during a process. The sum of the entropy changes for the system and the surroundings is the entropy change for the universe overall:

Equation 18.8 $$\Delta S_{univ} = \Delta S_{sys} + \Delta S_{surr}$$

The ***second law of thermodynamics*** says that for a process to be spontaneous as written (in the forward direction), ΔS_{univ} must be positive. Therefore, the system may undergo a *decrease* in entropy, as long as the surroundings undergoes a larger *increase* in entropy, and vice versa. A process for which ΔS_{univ} is *negative* is not spontaneous as written.

In some cases, ΔS_{univ} is neither positive nor negative but is equal to zero. This happens when the entropy changes of the system and surroundings are equal in magnitude and opposite in sign and describes a specific type of process known as an *equilibrium* process. An ***equilibrium process*** is one that does not occur spontaneously in either the net forward or net reverse direction but can be made to occur by the addition or removal of energy to a system at equilibrium. An example of an equilibrium process is the melting of ice at 0°C. (Remember that at 0°C, ice and liquid water are in equilibrium with each other [⏮ Section 11.6].)

Student Note: Most of the equilibrium processes that we will encounter are phase changes.

With Equations 18.6 and 18.7, we can calculate the entropy changes for both the system and surroundings in a process. We can then use the second law of thermodynamics (Equation 18.8) to determine if the process is spontaneous or nonspontaneous as written or if it is an equilibrium process.

Consider the synthesis of ammonia at 25°C:

$$N_2(g) + 3H_2(g) \longrightarrow 2NH_3(g) \qquad \Delta H^\circ_{rxn} = -92.6 \text{ kJ/mol}$$

From Sample Problem 18.2(b), we have $\Delta S^\circ_{sys} = -199$ J/K · mol, and substituting ΔH°_{sys} (−92.6 kJ/mol) into Equation 18.7, we get:

$$\Delta S_{surr} = \frac{-(-92.6 \times 1000)\text{ J/mol}}{298\text{ K}} = 311\text{ J/K} \cdot \text{mol}$$

The entropy change for the universe is:

$$\begin{aligned}\Delta S^\circ_{univ} &= \Delta S^\circ_{sys} + \Delta S^\circ_{surr}\\ &= -199\text{ J/K} \cdot \text{mol} + 311\text{ J/K} \cdot \text{mol}\\ &= 112\text{ J/K} \cdot \text{mol}\end{aligned}$$

Because ΔS°_{univ} is positive, the reaction will be spontaneous at 25°C. Keep in mind, though, that just because a reaction is spontaneous does not mean that it will occur at an observable rate. The synthesis of ammonia is, in fact, extremely slow at room temperature. Thermodynamics can tell us whether or not a reaction will occur spontaneously under specific conditions, but it does not tell us how fast it will occur.

The spontaneity that we have seen as favored by a process being exothermic is due to the spreading out of energy from the system to the surroundings; thus, the negative ΔH_{sys} corresponds to a positive ΔS_{surr}. It is this positive contribution to the overall ΔS_{univ} that actually favors spontaneity.

Sample Problem 18.4 lets you practice identifying spontaneous, nonspontaneous, and equilibrium processes.

SAMPLE PROBLEM 18.4

Determine if each of the following is a spontaneous process, a nonspontaneous process, or an equilibrium process at the specified temperature: (a) $H_2(g) + I_2(g) \longrightarrow 2HI(g)$ at 0°C, (b) $CaCO_3(s) \longrightarrow CaO(s) + CO_2(g)$ at 200°C, (c) $CaCO_3(s) \longrightarrow CaO(s) + CO_2(g)$ at 1000°C, (d) $Na(s) \longrightarrow Na(l)$ at 98°C. (Assume that the thermodynamic data in Appendix 2 do not vary with temperature.)

Strategy For each process, use Equation 18.6 to determine ΔS°_{sys} and Equations 5.19 and 18.7 to determine ΔH°_{sys} and ΔS°_{surr}. At the specified temperature, the process is *spontaneous* if ΔS_{sys} and ΔS_{surr} sum to a positive number, *nonspontaneous* if they sum to a negative number, and an *equilibrium process* if they sum to zero. Note that because the *reaction* is the *system*, ΔS_{rxn} and ΔS_{sys} are used interchangeably.

Setup From Appendix 2:

(a) $S^\circ[H_2(g)] = 131.0$ J/K · mol, $S^\circ[I_2(g)] = 260.57$ J/K · mol, $S^\circ[HI(g)] = 206.3$ J/K · mol; $[H_2(g)] = 0$ kJ/mol, $\Delta H^\circ_f[I_2(g)] = 62.25$ kJ/mol, $\Delta H^\circ_f[HI(g)] = 25.9$ kJ/mol.

(b), (c) In Sample Problem 18.2(a), we determined that for this reaction, $\Delta S^\circ_{rxn} = 160.5$ J/K · mol, $\Delta H^\circ_f[CaCO_3(s)] = -1206.9$ kJ/mol, $\Delta H^\circ_f[CaO(s)] = -635.6$ kJ/mol, $\Delta H^\circ_f[CO_2(g)] = -393.5$ kJ/mol.

(d) $S^\circ[Na(s)] = 51.05$ J/K · mol, $S^\circ[Na(l)] = 57.56$ J/K · mol; $\Delta H^\circ_f[Na(s)] = 0$ kJ/mol, $\Delta H^\circ_f[Na(l)] = 2.41$ kJ/mol.

Solution

(a) $$\begin{aligned}\Delta S^\circ_{rxn} &= [2S^\circ(HI)] - [S^\circ(H_2) + S^\circ(I_2)]\\ &= (2)(206.3\text{ J/K} \cdot \text{mol}) - [131.0\text{ J/K} \cdot \text{mol} + 260.57\text{ J/K} \cdot \text{mol}] = 21.03\text{ J/K} \cdot \text{mol}\end{aligned}$$

$$\begin{aligned}\Delta H^\circ_{rxn} &= [2\Delta H^\circ_f(HI)] - [\Delta H^\circ_f(H_2) + \Delta H^\circ_f(I_2)]\\ &= (2)(25.9\text{ kJ/mol}) - [0\text{ kJ/mol} + 62.25\text{ kJ/mol}] = -10.5\text{ kJ/mol}\end{aligned}$$

$$\Delta S_{surr} = \frac{-\Delta H_{rxn}}{T} = \frac{-(-10.5\text{ kJ/mol})}{273\text{ K}} = 0.0385\text{ kJ/K} \cdot \text{mol} = 38.5\text{ J/K} \cdot \text{mol}$$

$$\Delta S_{univ} = \Delta S_{sys} + \Delta S_{surr} = 21.03\text{ J/K} \cdot \text{mol} + 38.5\text{ J/K} \cdot \text{mol} = 59.5\text{ J/K} \cdot \text{mol}$$

ΔS_{univ} is positive; therefore the reaction is spontaneous at 0°C.

(b), (c) $\Delta S^\circ_{rxn} = 160.5$ J/K · mol

$$\begin{aligned}\Delta H^\circ_{rxn} &= [\Delta H^\circ_f(CaO) + \Delta H^\circ_f(CO_2)] - [\Delta H^\circ_f(CaCO_3)]\\ &= [-635.6\text{ kJ/mol} + (-393.5\text{ kJ/mol})] - (-1206.9\text{ kJ/mol}) = 177.8\text{ kJ/mol}\end{aligned}$$

(b) $T = 200°C$ and

$$\Delta S_{surr} = \frac{-\Delta H_{sys}}{T} = \frac{-(177.8\text{ kJ/mol})}{473\text{ K}} = -0.376\text{ kJ/K} \cdot \text{mol} = -376\text{ J/K} \cdot \text{mol}$$

$$\Delta S_{univ} = \Delta S_{sys} + \Delta S_{surr} = 160.5\text{ J/K} \cdot \text{mol} + (-376\text{ J/K} \cdot \text{mol}) = -216\text{ J/K} \cdot \text{mol}$$

ΔS_{univ} is negative, therefore the reaction is nonspontaneous at 200°C.

(Continued on next page)

(c) $T = 1000°C$ and

$$\Delta S_{surr} = \frac{-\Delta H_{sys}}{T} = \frac{-(177.8 \text{ kJ/mol})}{1273 \text{ K}} = -0.1397 \text{ kJ/K} \cdot \text{mol} = -139.7 \text{ J/K} \cdot \text{mol}$$

$$\Delta S_{univ} = \Delta S_{sys} + \Delta S_{surr} = 160.5 \text{ J/K} \cdot \text{mol} + (-139.7 \text{ J/K} \cdot \text{mol}) = 20.8 \text{ J/K} \cdot \text{mol}$$

In this case, ΔS_{univ} is positive; therefore, the reaction is spontaneous at 1000°C.

(d) $\Delta S^{\circ}_{rxn} = S^{\circ}[Na(l)] - S^{\circ}[Na(s)] = 57.56 \text{ J/K} \cdot \text{mol} - 51.05 \text{ J/K} \cdot \text{mol} = 6.51 \text{ J/K} \cdot \text{mol}$

$\Delta H^{\circ}_{rxn} = \Delta H^{\circ}_f[Na(l)] - \Delta H^{\circ}_f[Na(s)] = 2.41 \text{ kJ/mol} - 0 \text{ kJ/mol} = 2.41 \text{ kJ/mol}$

$$\Delta S_{surr} = \frac{-\Delta H_{rxn}}{T} = \frac{-(2.41 \text{ kJ/mol})}{371 \text{ K}} = -0.0650 \text{ kJ/K} \cdot \text{mol} = -6.50 \text{ J/K} \cdot \text{mol}$$

$$\Delta S_{univ} = \Delta S_{sys} + \Delta S_{surr} = 6.51 \text{ J/K} \cdot \text{mol} + (-6.50 \text{ J/K} \cdot \text{mol}) = 0.01 \text{ J/K} \cdot \text{mol} \approx 0$$

ΔS_{univ} is zero; therefore, the reaction is an equilibrium process at 98°C. In fact, this is the melting point of sodium.

THINK ABOUT IT

Remember that standard enthalpies of formation have units of kJ/mol, whereas standard absolute entropies have units of J/K · mol. Make sure that you convert kilojoules to joules, or vice versa, before combining the terms. The small difference between the magnitudes of ΔS_{sys} and ΔS_{surr} in part (d) results from thermodynamic values not being entirely independent of temperature. The tabulated values of S° and ΔH°_f are for 25°C.

Practice Problem ATTEMPT For each of the following, calculate ΔS_{univ} and identify the process as a spontaneous process, a nonspontaneous process, or an equilibrium process at the specified temperature: (a) $CO_2(g) \longrightarrow CO_2(aq)$ at 25°C, (b) $N_2O_4(g) \longrightarrow 2NO_2(g)$ at 10.4°C, (c) $PCl_3(l) \longrightarrow PCl_3(g)$ at 61.2°C. (Assume that the thermodynamic data in Appendix 2 do not vary with temperature.)

Practice Problem BUILD (a) Calculate ΔS_{univ} and determine if the reaction $H_2O_2(l) \longrightarrow H_2O_2(g)$ is spontaneous, nonspontaneous, or an equilibrium process at 163°C. (b) The reaction $NH_3(g) + HCl(g) \longrightarrow NH_4Cl(s)$ is spontaneous in the forward direction at room temperature but, because it is exothermic, becomes less spontaneous with increasing temperature. Determine the temperature at which it is no longer spontaneous in the forward direction. (c) Determine the boiling point of Br_2. (Assume that the thermodynamic data in Appendix 2 do not vary with temperature.)

Practice Problem CONCEPTUALIZE The following table shows the signs of ΔS_{sys}, ΔS_{surr}, and ΔS_{univ} for four processes. Where possible, fill in the missing table entries. Indicate where it is not possible to determine the missing sign and explain.

Process	ΔS_{sys}	ΔS_{surr}	ΔS_{univ}
1	−	−	
2	+		+
3	−	+	
4		−	+

The Third Law of Thermodynamics

Finally, we consider the third law of thermodynamics briefly in connection with the determination of standard entropy. We have related the entropy of a system to the number of possible arrangements of the system's molecules. The larger the number of possible arrangements, the larger the entropy. Imagine a pure, perfect crystalline substance at absolute zero (0 K). Under these conditions, there is essentially no molecular motion and, because the molecules occupy fixed positions in the solid, there is only one way to arrange the molecules. From Equation 18.1, we write:

$$S = k \ln W = k \ln 1 = 0$$

According to the ***third law of thermodynamics***, the entropy of a perfect crystalline substance is *zero* at absolute zero. As temperature increases, molecular motion increases, causing an increase in the number of possible arrangements of the molecules and in the number of accessible energy levels, among which the system's energy can be dispersed. (See Figure 18.4.) This results in an increase in the system's entropy. Thus, the entropy of any substance at any temperature above 0 K is greater than zero. If the crystalline substance is impure or imperfect in any way, then its entropy is greater than zero even at 0 K because without perfect crystalline order, there is more than one possible arrangement of molecules.

The significance of the third law of thermodynamics is that it enables us to determine experimentally the *absolute* entropies of substances. Starting with the knowledge that the entropy of a pure crystalline substance is zero at 0 K, we can measure the increase in entropy of the

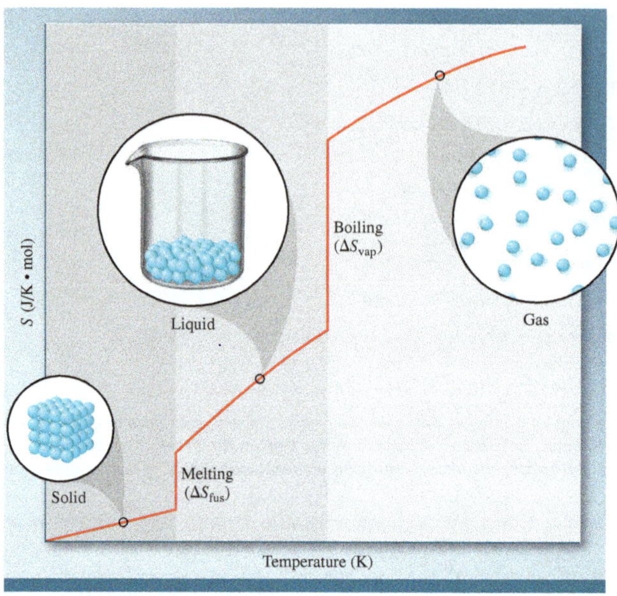

Figure 18.5 Entropy increases in a substance as temperature increases from absolute zero.

substance as it is heated. The change in entropy of a substance, ΔS, is the difference between the final and initial entropy values:

$$\Delta S = S_{\text{final}} - S_{\text{initial}}$$

where S_{initial} is zero if the substance starts at 0 K. Therefore, the measured *change* in entropy is equal to the *absolute* entropy of the substance at the final temperature:

$$\Delta S = S_{\text{final}}$$

The entropy values arrived at in this way are called *absolute* entropies because they are *true* values—unlike standard enthalpies of formation, which are derived using an arbitrary reference. Because the tabulated values are determined at 1 atm, we usually refer to absolute entropies as *standard* entropies, $S°$. Figure 18.5 shows the increase in entropy of a substance as temperature increases from absolute zero. At 0 K, it has a zero entropy value (assuming that it is a perfect crystalline substance). As it is heated, its entropy increases gradually at first because of greater molecular motion within the crystal. At the melting point, there is a large increase in entropy as the solid is transformed into the liquid. Further heating increases the entropy of the liquid again due to increased molecular motion. At the boiling point, there is a large increase in entropy as a result of the liquid-to-vapor transition. Beyond that temperature, the entropy of the gas continues to increase with increasing temperature.

CHECKPOINT – SECTION 18.4 Entropy Changes in the Universe

18.4.1 Using data from Appendix 2, calculate $\Delta S°$ (in J/K · mol) for the following reaction:

$$2NO(g) + O_2(g) \longrightarrow 2NO_2(g)$$

a) 145.3 J/K · mol

b) −145.3 J/K · mol

c) 59.7 J/K · mol

d) −59.7 J/K · mol

e) −421.2 J/K · mol

18.4.2 Using data from Appendix 2, calculate $\Delta S°$ (in J/K · mol) for the following reaction:

$$CH_4(g) + 2O_2(g) \longrightarrow CO_2(g) + 2H_2O(l)$$

a) 107.7 J/K · mol

b) −107.7 J/K · mol

c) 2.6 J/K · mol

d) 242.8 J/K · mol

e) −242.8 J/K · mol

18.4.3 The diagrams show a spontaneous chemical reaction. What can we deduce about ΔS_{surr} for this process?

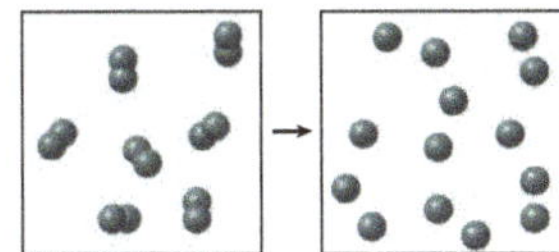

a) ΔS_{surr} is positive.

b) ΔS_{surr} is negative.

c) ΔS_{surr} is zero.

d) There is not enough information to deduce the sign of ΔS_{surr}.

18.4.4 The diagrams show a spontaneous chemical reaction. What can we deduce about ΔS_{surr} for this process?

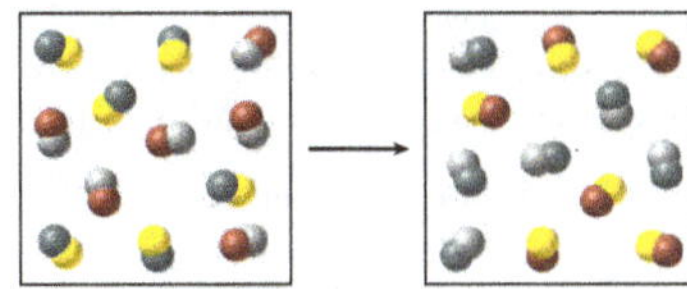

a) ΔS_{surr} is positive.

b) ΔS_{surr} is negative.

c) ΔS_{surr} is zero.

d) There is not enough information to deduce the sign of ΔS_{surr}.

18.5 Predicting Spontaneity

Gibbs Free-Energy Change, ΔG

According to the second law of thermodynamics, $\Delta S_{univ} > 0$ for a spontaneous process. What we are usually concerned with and usually *measure*, however, are the properties of the system rather than those of the surroundings or those of the universe overall. Therefore, it is convenient to have a thermodynamic function that enables us to determine whether or not a process is spontaneous by considering the system alone.

We begin with Equation 18.8. For a spontaneous process:

$$\Delta S_{univ} = \Delta S_{sys} + \Delta S_{surr} > 0$$

Substituting $-\Delta H_{sys}/T$ for ΔS_{surr}, we write:

$$\Delta S_{univ} = \Delta S_{sys} + \left(-\frac{\Delta H_{sys}}{T}\right) > 0$$

Multiplying both sides of the equation by T gives:

$$T\Delta S_{univ} = T\Delta S_{sys} - \Delta H_{sys} > 0$$

Now we have an equation that expresses the second law of thermodynamics (and predicts whether or not a process is spontaneous) in terms of only the *system*. We no longer need to consider the surroundings. For convenience, we can rearrange the preceding equation, multiply through by -1, and replace the $>$ sign with a $<$ sign:

$$-T\Delta S_{univ} = \Delta H_{sys} - T\Delta S_{sys} < 0$$

According to this equation, a process carried out at constant pressure and temperature is spontaneous if the changes in enthalpy and entropy of the system are such that $\Delta H_{sys} - T\Delta S_{sys}$ is less than zero.

To express the spontaneity of a process more directly, we introduce another thermodynamic function called the ***Gibbs***[1] ***free energy (G),*** or simply ***free energy:***

Equation 18.9 $$G = H - TS$$

Each of the terms in Equation 18.9 pertains to the system. G has units of energy just as H and TS do. Furthermore, like enthalpy and entropy, free energy is a state function. The change in free energy, ΔG, of a system for a process that occurs at constant temperature is:

Equation 18.10 $$\Delta G = \Delta H - T\Delta S$$

1. Josiah Willard Gibbs (1839–1903). American physicist. One of the founders of thermodynamics. Gibbs was a modest and private individual who spent almost all his professional life at Yale University. Because he published most of his work in obscure journals, Gibbs never gained the eminence that his contemporary and admirer James Maxwell did. Even today, very few people outside of chemistry and physics have ever heard of Gibbs.

Equation 18.10 enables us to predict the spontaneity of a process using the change in enthalpy, the change in entropy, and the absolute temperature. At constant temperature and pressure, for processes that are spontaneous as written (in the forward direction), ΔG is negative. For processes that are not spontaneous as written but that are spontaneous in the reverse direction, ΔG is positive. For systems at equilibrium, ΔG is zero.

Student Note: In this context, *free energy* is the energy available to do work. Thus, if a particular process is accompanied by a release of usable energy (i.e., if ΔG is negative), this fact alone guarantees that it is spontaneous, and there is no need to consider what happens to the rest of the universe.

- $\Delta G < 0$ The reaction is spontaneous in the forward direction (and nonspontaneous in the reverse direction).
- $\Delta G > 0$ The reaction is nonspontaneous in the forward direction (and spontaneous in the reverse direction).
- $\Delta G = 0$ The system is at equilibrium.

Often we can predict the sign of ΔG for a process if we know the signs of ΔH and ΔS. Table 18.4 shows how we can use Equation 18.10 to make such predictions.

Based on the information in Table 18.4, you may wonder what constitutes a "low" or a "high" temperature. For the example given in the table, 0°C is the temperature that divides high from low. Water freezes spontaneously at temperatures below 0°C, and ice melts spontaneously at temperatures above 0°C. At 0°C, a system of ice and water is at equilibrium. In general, though, the temperature that divides "high" from "low" depends on the individual reaction. To determine that temperature, we must set ΔG equal to 0 in Equation 18.10 (i.e., the equilibrium condition):

$$0 = \Delta H - T\Delta S$$

Rearranging to solve for T yields:

$$T = \frac{\Delta H}{\Delta S}$$

The temperature that divides high from low for a particular reaction can now be calculated if the values of ΔH and ΔS are known.

Sample Problem 18.5 demonstrates the use of this approach.

TABLE 18.4 Predicting the Sign of ΔG Using Equation 18.10 and the Signs of ΔH and ΔS

When ΔH Is	And ΔS Is	ΔG Will Be	And the Process Is	Example
Negative	Positive	Negative	Always spontaneous	$2H_2O_2(aq) \longrightarrow 2H_2O(l) + O_2(g)$
Positive	Negative	Positive	Always nonspontaneous	$3O_2(g) \longrightarrow 2O_3(g)$
Negative	Negative	Negative when $T\Delta S < \Delta H$ Positive when $T\Delta S > \Delta H$	Spontaneous at low T Nonspontaneous at high T	$H_2O(l) \longrightarrow H_2O(s)$ (freezing of water)
Positive	Positive	Negative when $T\Delta S > \Delta H$ Positive when $T\Delta S < \Delta H$	Spontaneous at high T Nonspontaneous at low T	$2HgO(s) \longrightarrow 2Hg(l) + O_2(g)$

SAMPLE PROBLEM 18.5

According to Table 18.4, a reaction will be spontaneous only at high temperatures if both ΔH and ΔS are positive. For a reaction in which $\Delta H = 199.5$ kJ/mol and $\Delta S = 476$ J/K · mol, determine the temperature (in °C) above which the reaction is spontaneous.

Strategy The temperature that divides high from low is the temperature at which $\Delta H = T\Delta S$ ($\Delta G = 0$). Therefore, we use Equation 18.10, substituting 0 for ΔG and solving for T to determine temperature in kelvins; we then convert to degrees Celsius.

Setup

$$\Delta S = \left(\frac{476\text{ J}}{\text{K}\cdot\text{mol}}\right)\left(\frac{1\text{ kJ}}{1000\text{ J}}\right) = 0.476\text{ kJ/K}\cdot\text{mol}$$

Solution

$$T = \frac{\Delta H}{\Delta S} = \frac{199.5\text{ kJ/mol}}{0.476\text{ kJ/K}\cdot\text{mol}} = 419\text{ K}$$

$$= (419 - 273) = 146°\text{C}$$

(Continued on next page)

THINK ABOUT IT

Spontaneity is favored by a release of energy (ΔH being negative) and by an increase in entropy (ΔS being positive). When both quantities are positive, as in this case, only the entropy change favors spontaneity. For an endothermic process such as this, which requires the input of heat, it should make sense that adding more heat by increasing the temperature will shift the equilibrium to the right, thus making it "more spontaneous."

Practice Problem ATTEMPT A reaction will be spontaneous only at low temperatures if both ΔH and ΔS are negative. For a reaction in which $\Delta H = -380.1$ kJ/mol and $\Delta S = -95.00$ J/K · mol, determine the temperature (in °C) below which the reaction is spontaneous.

Practice Problem BUILD Given that the reaction $4Fe(s) + 3O_2(g) + 6H_2O(l) \longrightarrow 4Fe(OH)_3(s)$ is spontaneous at temperatures below 1950°C, estimate the standard entropy of $Fe(OH)_3(s)$.

Practice Problem CONCEPTUALIZE Which of the following graphs best represents the relationship between ΔG and temperature for a process that is exothermic and for which ΔS is negative?

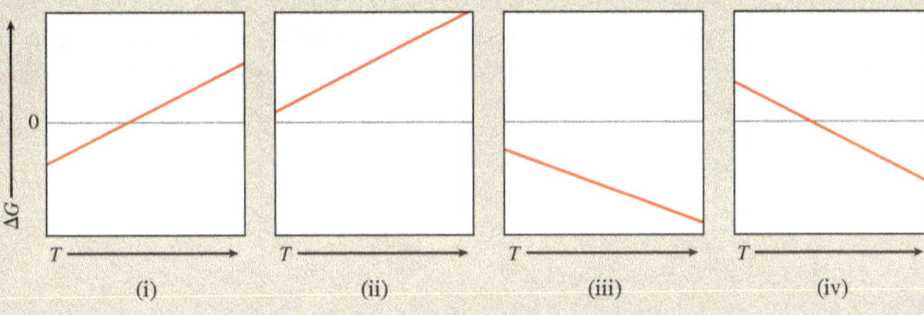

Student Note: The introduction of the term $\Delta G°$ enables us to write Equation 18.10 as $\Delta G° = \Delta H° - T\Delta S°$

Standard Free-Energy Changes, $\Delta G°$

The ***standard free energy of reaction*** **($\Delta G°_{rxn}$)** is the free-energy change for a reaction when it occurs under standard-state conditions—that is, when reactants in their standard states are converted to products in *their* standard states. The conventions used by chemists to define the standard states of pure substances and solutions are:

- Gases — 1 atm pressure
- Liquids — Pure liquid
- Solids — Pure solid
- Elements — The most stable allotropic form at 1 atm and 25°C
- Solutions — 1 molar concentration

To calculate $\Delta G°_{rxn}$, we start with the general equation:

$$aA + bB \longrightarrow cC + dD$$

The standard free-energy change for this reaction is given by:

Equation 18.11 $$\Delta G°_{rxn} = [c\Delta G°_f(C) + d\Delta G°_f(D)] - [a\Delta G°_f(A) + b\Delta G°_f(B)]$$

Equation 18.11 can be generalized as follows:

Equation 18.12 $$\Delta G°_{rxn} = \Sigma n\Delta G°_f \text{ (products)} - \Sigma m\Delta G°_f \text{ (reactants)}$$

where m and n are stoichiometric coefficients. The term **$\Delta G°_f$** is the ***standard free energy of formation*** of a compound—that is, the free-energy change that occurs when 1 mole of the compound is synthesized from its constituent elements, each in its standard state. For the combustion of graphite:

$$C(\text{graphite}) + O_2(g) \longrightarrow CO_2(g)$$

the standard free-energy change (from Equation 18.12) is:

$$\Delta G°_{rxn} = [\Delta G°_f(CO_2, \text{gas})] - [\Delta G°_f(C, \text{graphite}) + \Delta G°_f(O_2, \text{gas})]$$

As with standard enthalpy of formation, the standard free energy of formation of any element (in its most stable allotropic form at 1 atm) is defined as zero. Thus:

$$\Delta G°_f(C, \text{graphite}) = 0 \quad \text{and} \quad \Delta G°_f(O_2, \text{gas}) = 0$$

Therefore, the standard free-energy change for the reaction in this case is equal to the standard free energy of formation of CO_2:

$$\Delta G^\circ_{rxn} = \Delta G^\circ_f(CO_2)$$

Appendix 2 lists the values of ΔG°_f at 25°C for a number of compounds.

Sample Problem 18.6 demonstrates the calculation of standard free-energy changes.

SAMPLE PROBLEM 18.6

Calculate the standard free-energy changes for the following reactions at 25°C:

(a) $CH_4(g) + 2O_2(g) \longrightarrow CO_2(g) + 2H_2O(l)$

(b) $2MgO(s) \longrightarrow 2Mg(s) + O_2(g)$

Strategy Look up the ΔG°_f values for the reactants and products in each equation, and use Equation 18.12 to solve for ΔG°_{rxn}.

Setup From Appendix 2, we have the following values: $\Delta G^\circ_f[CH_4(g)] = -50.8$ kJ/mol, $\Delta G^\circ_f[CO_2(g)] = -394.4$ kJ/mol, $\Delta G^\circ_f[H_2O(l)] = -237.2$ kJ/mol, and $\Delta G^\circ_f[MgO(s)] = -569.6$ kJ/mol. All the other substances are elements in their standard states and have, by definition, $\Delta G^\circ_f = 0$.

Solution

(a) $\Delta G^\circ_{rxn} = (\Delta G^\circ_f[CO_2(g)] + 2\Delta G^\circ_f[H_2O(l)]) - (\Delta G^\circ_f[CH_4(g)] + 2\Delta G^\circ_f[O_2(g)])$

$= [(-394.4 \text{ kJ/mol}) + (2)(-237.2 \text{ kJ/mol})] - [(-50.8 \text{ kJ/mol}) + (2)(0 \text{ kJ/mol})]$

$= -818.0$ kJ/mol

(b) $\Delta G^\circ_{rxn} = (2\Delta G^\circ_f[Mg(s)] + \Delta G^\circ_f[O_2(g)]) - (2\Delta G^\circ_f[MgO(s)])$

$= [(2)(0 \text{ kJ/mol}) + (0 \text{ kJ/mol})] - [(2)(-569.6 \text{ kJ/mol})]$

$= 1139$ kJ/mol

THINK ABOUT IT

Note that, like standard enthalpies of formation (ΔH°_f), standard free energies of formation (ΔG°_f) depend on the *state* of matter. Using water as an example, $\Delta G^\circ_f[H_2O(l)] = -237.2$ kJ/mol and $\Delta G^\circ_f[H_2O(g)] = -228.6$ kJ/mol. Always double-check to make sure you have selected the right value from the table.

Practice Problem ATTEMPT Calculate the standard free-energy changes for the following reactions at 25°C:

(a) $H_2(g) + Br_2(l) \longrightarrow 2HBr(g)$

(b) $2C_2H_6(g) + 7O_2(g) \longrightarrow 4CO_2(g) + 6H_2O(l)$

Practice Problem BUILD For each reaction, determine the value of ΔG°_f that is not listed in Appendix 2:

(a) $Li_2O(s) + 2HCl(g) \longrightarrow 2LiCl(s) + H_2O(g)$ $\quad \Delta G^\circ_{rxn} = -244.88$ kJ/mol

(b) $Na_2O(s) + 2HI(g) \longrightarrow 2NaI(s) + H_2O(l)$ $\quad \Delta G^\circ_{rxn} = -435.44$ kJ/mol

Practice Problem CONCEPTUALIZE For which of the following species is $\Delta G^\circ_f = 0$?

$Br_2(l)$	$I_2(g)$	$CO_2(g)$	$Xe(g)$
(i)	(ii)	(iii)	(iv)

Using ΔG and ΔG° to Solve Problems

It is the sign of ΔG, the free-energy change, *not* the sign of ΔG°, the standard free-energy change, that indicates whether or not a process will occur spontaneously under a given set of conditions. What the *sign* of ΔG° tells us is the same thing that the magnitude of the equilibrium constant (K) tells us [◀◀ Section 15.2]. A negative ΔG° value corresponds to a large K value (products favored at equilibrium), whereas a positive ΔG° value corresponds to a small K value (reactants favored at equilibrium).

Student Note: The sign of ΔG° does indicate whether or not a process is spontaneous when *all* reactants and products are in their standard states, but this is very seldom the case.

Like equilibrium constants, ΔG° values change with temperature. One of the uses of Equation 18.10 is to determine the temperature at which a particular equilibrium will begin to favor a desired product. For example, calcium oxide (CaO), also called quicklime, is an extremely valuable inorganic substance with a variety of industrial uses, including water treatment and pollution control. It is prepared by heating limestone ($CaCO_3$), which decomposes at a high temperature:

$$CaCO_3(s) \rightleftharpoons CaO(s) + CO_2(g)$$

The reaction is reversible, and under the right conditions, CaO and CO_2 readily recombine to form $CaCO_3$ again. To prevent this from happening in the industrial preparation, the system is never maintained at equilibrium; rather, CO_2 is constantly removed as it forms, shifting the equilibrium from left to right, thus promoting the formation of calcium oxide.

An important piece of information for the chemist responsible for maximizing CaO production is the temperature at which the decomposition equilibrium of $CaCO_3$ begins to favor products. We can make a reliable estimate of that temperature as follows. First we calculate $\Delta H°$ and $\Delta S°$ for the reaction at 25°C, using the data in Appendix 2. To determine $\Delta H°$, we apply Equation 5.19:

$$\begin{aligned}\Delta H° &= [\Delta H_f°(\text{CaO}) + \Delta H_f°(\text{CO}_2)] - [\Delta H_f°(\text{CaCO}_3)]\\ &= [(-635.6 \text{ kJ/mol}) + (-393.5 \text{ kJ/mol})] - (-1206.9 \text{ kJ/mol})\\ &= 177.8 \text{ kJ/mol}\end{aligned}$$

Next we apply Equation 18.6 to find $\Delta S°$:

$$\begin{aligned}\Delta S° &= [S°(\text{CaO}) + S°\text{CO}_2)] - S°(\text{CaCO}_3)\\ &= [(39.8 \text{ J/K}\cdot\text{mol}) + (213.6 \text{ J/K}\cdot\text{mol})] - (92.9 \text{ J/K}\cdot\text{mol})\\ &= 160.5 \text{ J/K}\cdot\text{mol}\end{aligned}$$

From Equation 18.10, we can write:

$$\Delta G° = \Delta H° - T\Delta S°$$

and we obtain:

$$\begin{aligned}\Delta G° &= (177.8 \text{ kJ/mol}) - (298 \text{ K})(0.1605 \text{ kJ/K}\cdot\text{mol})\\ &= 130.0 \text{ kJ/mol}\end{aligned}$$

Student Note: Be careful with units in problems of this type. $S°$ values are tabulated using joules, whereas $\Delta H_f°$ values are tabulated using kilojoules.

Because $\Delta G°$ is a large positive number, the reaction does *not* favor product formation at 25°C (298 K). And, because $\Delta H°$ and $\Delta S°$ are both positive, we know that $\Delta G°$ will be negative (product formation will be favored) at high temperatures. We can determine what constitutes a high temperature for this reaction by calculating the temperature at which $\Delta G°$ is zero:

$$0 = \Delta H° - T\Delta S°$$

or:

$$\begin{aligned}T &= \frac{\Delta H°}{\Delta S°}\\ &= \frac{(177.8 \text{ kJ/mol})(1000 \text{ J/kJ})}{0.1605 \text{ kJ/K}\cdot\text{mol}}\\ &= 1108 \text{ K}(835°\text{C})\end{aligned}$$

At temperatures higher than 835°C, $\Delta G°$ becomes negative, indicating that the reaction would then favor the formation of CaO and CO_2. At 840°C (1113 K), for example:

$$\begin{aligned}\Delta G° &= \Delta H° - T\Delta S°\\ &= 177.8 \text{ kJ/mol} - (1113 \text{ K})(0.1605 \text{ kJ/K}\cdot\text{mol})\left(\frac{1 \text{ kJ}}{1000 \text{ J}}\right)\\ &= -0.8 \text{ kJ/mol}\end{aligned}$$

At still higher temperatures, $\Delta G°$ becomes increasingly negative, thus favoring product formation even more. Note that in this example we used the $\Delta H°$ and $\Delta S°$ values at 25°C to calculate changes to $\Delta G°$ at much higher temperatures. Because both $\Delta H°$ and $\Delta S°$ actually change with temperature, this approach does not give us a truly accurate value for $\Delta G°$, but it does give us a reasonably good estimate.

Equation 18.10 can also be used to calculate the change in entropy that accompanies a phase change. At the temperature at which a phase change occurs (i.e., the melting point or boiling point of a substance), the system is at equilibrium ($\Delta G = 0$). Therefore, Equation 18.10 becomes:

$$0 = \Delta H - T\Delta S$$

or:

$$\Delta S = \frac{\Delta H}{T}$$

Consider the ice-water equilibrium. For the ice-to-water transition, ΔH is the molar heat of fusion (see Table 11.8) and T is the melting point. The entropy change is therefore

$$\Delta S_{\text{ice}\rightarrow\text{water}} = \frac{6010 \text{ J/mol}}{273 \text{ K}} = 22.0 \text{ J/K} \cdot \text{mol}$$

Thus, when 1 mole of ice melts at 0°C, there is an increase in entropy of 22.0 J/K · mol. The increase in entropy is consistent with the increase in possible arrangements from solid to liquid. Conversely, for the water-to-ice transition, the decrease in entropy is given by:

$$\Delta S_{\text{water}\rightarrow\text{ice}} = \frac{-6010 \text{ J/mol}}{273 \text{ K}} = -22.0 \text{ J/K} \cdot \text{mol}$$

The same approach can be applied to the water-to-steam transition. In this case, ΔH is the heat of vaporization and T is the boiling point of water.

Sample Problem 18.7 examines the phase transitions in benzene.

SAMPLE PROBLEM 18.7

The molar heats of fusion and vaporization of benzene are 10.9 and 31.0 kJ/mol, respectively. Calculate the entropy changes for the solid-to-liquid and liquid-to-vapor transitions for benzene. At 1 atm pressure, benzene melts at 5.5°C and boils at 80.1°C.

Strategy The solid-liquid transition at the melting point and the liquid-vapor transition at the boiling point are *equilibrium* processes. Therefore, because ΔG is zero at equilibrium, in each case we can use Equation 18.10, substituting 0 for ΔG and solving for ΔS, to determine the entropy change associated with the process.

Setup The melting point of benzene is 5.5 + 273.15 = 278.7 K and the boiling point is 80.1 + 273.15 = 353.3 K.

Solution

$$\Delta S_{\text{fus}} = \frac{\Delta H_{\text{fus}}}{T_{\text{melting}}} = \frac{10.9 \text{ kJ/mol}}{278.7 \text{ K}}$$
$$= 0.0391 \text{ kJ/K} \cdot \text{mol} \quad \text{or} \quad 39.1 \text{ J/K} \cdot \text{mol}$$

$$\Delta S_{\text{vap}} = \frac{\Delta H_{\text{vap}}}{T_{\text{boiling}}} = \frac{31.0 \text{ kJ/mol}}{353.3 \text{ K}}$$
$$= 0.0877 \text{ kJ/K} \cdot \text{mol} \quad \text{or} \quad 87.7 \text{ J/K} \cdot \text{mol}$$

THINK ABOUT IT

For the same substance, ΔS_{vap} is always significantly larger than ΔS_{fus}. The change in number of possible arrangements is always bigger in a liquid-to-gas transition than in a solid-to-liquid transition.

Practice Problem ATTEMPT The molar heats of fusion and vaporization of argon are 1.3 and 6.3 kJ/mol, respectively, and argon's melting point and boiling point are −190°C and −186°C, respectively. Calculate the entropy changes for the fusion and vaporization of argon.

Practice Problem BUILD Using data from Appendix 2 and assuming that the tabulated values do not change with temperature, (a) calculate $\Delta H^\circ_{\text{fus}}$ and $\Delta S^\circ_{\text{fus}}$ for sodium metal and determine the melting temperature of sodium, and (b) calculate $\Delta H^\circ_{\text{vap}}$ and $\Delta S^\circ_{\text{vap}}$ for sodium metal and determine the boiling temperature of sodium.

Practice Problem CONCEPTUALIZE Explain why, in general, we can use the equation $\Delta S = \frac{\Delta H}{T}$ to calculate ΔS for a phase change but not for a chemical reaction?

CHECKPOINT – SECTION 18.5 Predicting Spontaneity

18.5.1 A reaction for which ΔH and ΔS are both negative is

a) nonspontaneous at all temperatures

b) spontaneous at all temperatures

c) spontaneous at high temperatures

d) spontaneous at low temperatures

e) at equilibrium

18.5.2 At what temperature (in °C) does a reaction go from being nonspontaneous to spontaneous if it has $\Delta H = 171$ kJ/mol and $\Delta S = 161$ J/K · mol?

a) 270°C

b) 670°C

c) 1100°C

d) 790°C

e) 28°C

18.5.3 Using data from Appendix 2, calculate $\Delta G°$ (in kJ/mol) at 25°C for the reaction:

$$CH_4(g) + 2O_2(g) \longrightarrow CO_2(g) + 2H_2O(g)$$

a) −580.8 kJ/mol

b) 580.8 kJ/mol

c) −572.0 kJ/mol

d) −800.8 kJ/mol

e) −818.0 kJ/mol

18.5.4 Calculate ΔS_{sub} (in J/K · mol) for the sublimation of iodine in a closed flask at 45°C:

$$I_2(s) \longrightarrow I_2(g)$$

$\Delta H_{sub} = 62.4$ kJ/mol.

a) 1.4 J/K · mol

b) 196 J/K · mol

c) 1387 J/K · mol

d) 0.196 J/K · mol

e) 721 J/K · mol

18.6 Free Energy and Chemical Equilibrium

Student Note: Even for a reaction that *starts* with all reactants and products in their standard states, as soon as the reaction begins, the concentrations of all species change and standard-state conditions no longer exist.

Reactants and products in a chemical reaction are almost always in something *other* than their standard states—that is, solutions usually have concentrations other than 1 *M* and gases usually have pressures other than 1 atm. To determine whether or not a reaction is spontaneous, therefore, we must take into account the actual concentrations and/or pressures of the species involved. And although we can determine $\Delta G°$ from tabulated values, we need to know ΔG to determine spontaneity.

Relationship Between ΔG and ΔG°

The relationship between ΔG and $\Delta G°$, which is derived from thermodynamics, is:

Equation 18.13 $$\Delta G = \Delta G° + RT \ln Q$$

Student Note: The *Q* used in Equation 18.13 can be either Q_c (for reactions that take place in solution) or Q_P (for reactions that take place in the gas phase).

where R is the gas constant (8.314 J/K · mol or 8.314×10^{-3} kJ/K · mol), T is the absolute temperature at which the reaction takes place, and Q is the reaction quotient [⏮ Section 15.2]. Thus, ΔG depends on two terms: $\Delta G°$ and $RT \ln Q$. For a given reaction at temperature T, the value of $\Delta G°$ is fixed but that of $RT \ln Q$ can vary because Q varies according to the composition of the reaction mixture.

Consider the following equilibrium:

$$H_2(g) + I_2(s) \rightleftharpoons 2HI(g)$$

Using Equation 18.12 and information from Appendix 2, we find that $\Delta G°$ for this reaction at 25°C is 2.60 kJ/mol. The value of ΔG, however, depends on the pressures of both gaseous species. If we start with a reaction mixture containing solid I_2, in which $P_{H_2} = 4.0$ atm and $P_{HI} = 3.0$ atm, the reaction quotient, Q_P, is:

$$Q_P = \frac{(P_{HI})^2}{(P_{H_2})} = \frac{(3.0)^2}{4.0} = \frac{9.0}{4.0}$$
$$= 2.25$$

Using this value in Equation 18.13 gives:

$$\Delta G = \frac{2.60 \text{ kJ}}{\text{mol}} + \left(\frac{8.314 \times 10^{-3} \text{ kJ}}{\text{K} \cdot \text{mol}}\right)(298 \text{ K})(\ln 2.25)$$
$$= 4.3 \text{ kJ/mol}$$

Because ΔG is positive, we conclude that, starting with these concentrations, the forward reaction will not occur spontaneously as written. Instead, the *reverse* reaction will occur spontaneously and the system will reach equilibrium by consuming part of the HI initially present and producing more H_2 and I_2.

If, on the other hand, we start with a mixture of gases in which $P_{H_2} = 4.0$ atm and $P_{HI} = 1.0$ atm, the reaction quotient, Q_P, is:

$$Q_P = \frac{(P_{HI})^2}{(P_{H_2})} = \frac{(1.0)^2}{(4.0)} = \frac{1}{4}$$
$$= 0.25$$

Using this value in Equation 18.13 gives:

$$\Delta G = \frac{2.60 \text{ kJ}}{\text{mol}} + \left(\frac{8.314 \times 10^{-3} \text{ kJ}}{\text{K} \cdot \text{mol}}\right)(298 \text{ K})(\ln 0.25)$$
$$= -0.8 \text{ kJ/mol}$$

With a negative value for ΔG, the reaction will be spontaneous as written—in the forward direction. In this case, the system will achieve equilibrium by consuming some of the H_2 and I_2 to produce more HI.

Sample Problem 18.8 uses $\Delta G°$ and the reaction quotient to determine in which direction a reaction is spontaneous.

SAMPLE PROBLEM 18.8

The equilibrium constant, K_P, for the reaction:

$$N_2O_4(g) \rightleftharpoons 2NO_2(g)$$

is 0.113 at 298 K, which corresponds to a standard free-energy change of 5.4 kJ/mol. In a certain experiment, the initial pressures are $P_{N_2O_4} = 0.453$ atm and $P_{NO_2} = 0.122$ atm. Calculate ΔG for the reaction at these pressures, and predict the direction in which the reaction will proceed spontaneously to establish equilibrium.

Strategy Use the partial pressures of N_2O_4 and NO_2 to calculate the reaction quotient Q_P, and then use Equation 18.13 to calculate ΔG.

Setup The reaction quotient expression is:

$$Q_P = \frac{(P_{NO_2})^2}{P_{N_2O_4}} = \frac{(0.122)^2}{0.453} = 0.0329$$

Solution

$$\Delta G = \Delta G° + RT \ln Q_P$$
$$= \frac{5.4 \text{ kJ}}{\text{mol}} + \left(\frac{8.314 \times 10^{-3} \text{ kJ}}{\text{K} \cdot \text{mol}}\right)(298 \text{ K})(\ln 0.0329)$$
$$= 5.4 \text{ kJ/mol} - 8.46 \text{ kJ/mol}$$
$$= -3.1 \text{ kJ/mol}$$

Because ΔG is negative, the reaction proceeds spontaneously from left to right to reach equilibrium.

THINK ABOUT IT

Remember, a reaction with a positive $\Delta G°$ value can be spontaneous if the starting concentrations of reactants and products are such that $Q < K$.

Practice Problem ATTEMPT $\Delta G°$ for the reaction:

$$H_2(g) + I_2(s) \rightleftharpoons 2HI(g)$$

is 2.60 kJ/mol at 25°C. Calculate ΔG, and predict the direction in which the reaction is spontaneous if the starting concentrations are $P_{H_2} = 5.25$ atm and $P_{HI} = 1.75$ atm.

Practice Problem BUILD What is the minimum partial pressure of H_2 required for the preceding reaction to be spontaneous in the forward direction at 25°C if the partial pressure of HI is 0.94?

(Continued on next page)

Practice Problem CONCEPTUALIZE Consider the reaction in Sample Problem 18.8. Which of the following graphs best shows what happens to ΔG as the partial pressure of N_2O_4 is increased?

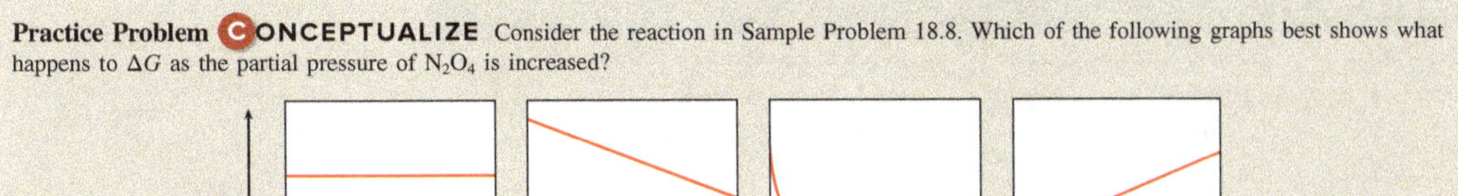
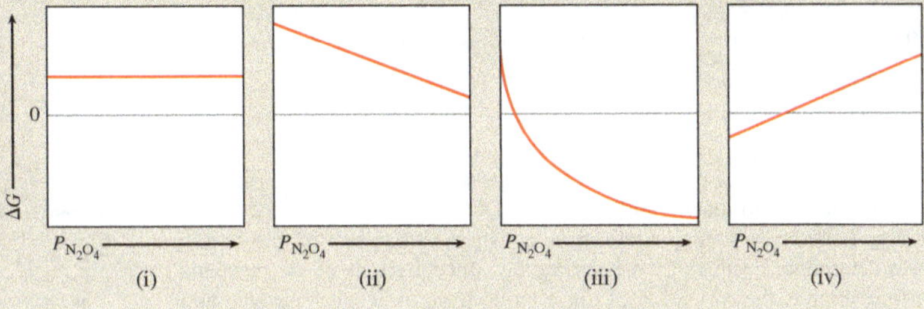

Relationship Between $\Delta G°$ and K

By definition, $\Delta G = 0$ and $Q = K$ at equilibrium, where K is the equilibrium constant. Thus, $\Delta G = \Delta G° + RT \ln Q$ (Equation 18.13) becomes:

$$0 = \Delta G° + RT \ln K$$

Student Note: In this equation and in Equation 18.14, K is K_c for reactions that take place in solution and K_P for reactions that take place in the gas phase.

or:

Equation 18.14 $$\Delta G° = -RT \ln K$$

According to Equation 18.14, then, the larger K is, the more negative $\Delta G°$ is. For chemists, Equation 18.14 is one of the most important equations in thermodynamics because it enables us to find the equilibrium constant of a reaction if we know the change in standard free energy, and vice versa.

Animation
Chemical Equilibrium—equilibrium.

It is significant that Equation 18.14 relates the equilibrium constant to the standard free-energy change, $\Delta G°$, rather than to the actual free-energy change, ΔG. The actual free-energy change of the system varies as the reaction progresses and becomes zero at equilibrium. On the other hand, $\Delta G°$, like K, is a constant for a particular reaction at a given temperature. Figure 18.6 shows plots of the free energy of a reacting system versus the extent of the reaction for two reactions. Table 18.5 summarizes the relationship between the *magnitude* of an equilibrium constant and the *sign* of the corresponding $\Delta G°$. Remember this important distinction: It is the sign of ΔG and not that of $\Delta G°$ that determines the direction of reaction spontaneity. The sign

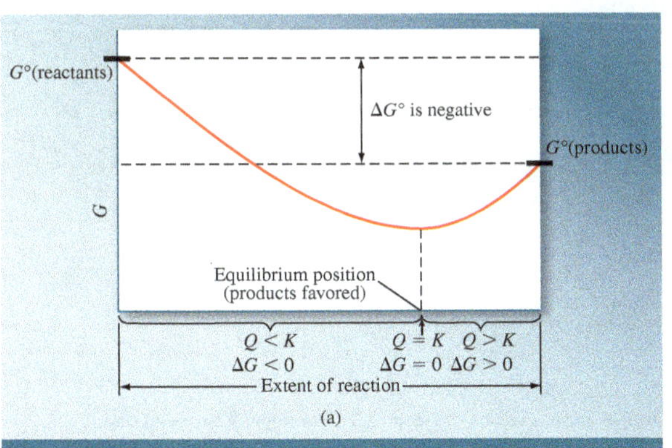

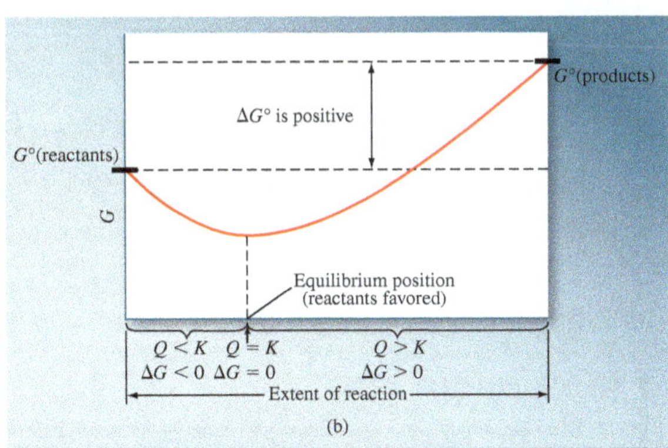

Figure 18.6 (a) $\Delta G° < 0$. At equilibrium, there is a significant conversion of reactants to products. (b) $\Delta G° > 0$. At equilibrium, reactants are favored over products. In both cases, the net reaction toward equilibrium is from left to right (reactants to products) if $Q < K$ and right to left (products to reactants) if $Q > K$. At equilibrium, $Q = K$.

TABLE 18.5 Relationship Between K and $\Delta G°$ as Predicted by Equation 18.14

K	ln K	$\Delta G°$	Result at Equilibrium
> 1	Positive	Negative	Products are favored.
= 1	0	0	Neither products nor reactants are favored.
< 1	Negative	Positive	Reactants are favored.

of $\Delta G°$ only tells us the relative amounts of products and reactants when equilibrium is reached, *not* the direction the reaction must go to reach equilibrium.

Student Note: The sign of $\Delta G°$ tells us the same thing that the magnitude of K tells us. The sign of ΔG tells us the same thing as the comparison of Q and K values [Section 15.4].

For reactions with very large or very small equilibrium constants, it can be very difficult—sometimes impossible—to determine K values by measuring the concentrations of the reactants and products. Consider, for example, the formation of nitric oxide from molecular nitrogen and molecular oxygen:

$$N_2(g) + O_2(g) \rightleftharpoons 2NO(g)$$

At 25°C, the equilibrium constant, K_P, is:

$$K_P = \frac{(P_{NO})^2}{(P_{N_2})(P_{O_2})} = 4.0 \times 10^{-31}$$

The very small value of K_P means that the concentration of NO at equilibrium will be exceedingly low and, for all intents and purposes, impossible to measure directly. In such a case, the equilibrium constant is more conveniently determined using $\Delta G°$, which can be calculated either from tabulated $\Delta G_f°$ values or from $\Delta H°$ and $\Delta S°$.

Sample Problems 18.9 and 18.10 show how to use $\Delta G°$ to calculate K and how to use K to calculate $\Delta G°$, respectively.

SAMPLE PROBLEM 18.9

Using data from Appendix 2, calculate the equilibrium constant, K_P, for the following reaction at 25°C:

$$2H_2O(l) \rightleftharpoons 2H_2(g) + O_2(g)$$

Strategy Use data from Appendix 2 and Equation 18.12 to calculate $\Delta G°$ for the reaction. Then use Equation 18.14 to solve for K_P.

Setup

$$\begin{aligned}\Delta G° &= (2\Delta G_f°[H_2(g)] + \Delta G_f°[O_2(g)]) - (2\Delta G_f°[H_2O(l)])\\ &= [(2)(0 \text{ kJ/mol}) + (2)(0 \text{ kJ/mol})] - [(2)(-237.2 \text{ kJ/mol})]\\ &= 474.4 \text{ kJ/mol}\end{aligned}$$

Solution

$$\Delta G° = -RT \ln K_P$$

$$\frac{474.4 \text{ kJ}}{\text{mol}} = -\left(\frac{8.314 \times 10^{-3} \text{ kJ}}{\text{K} \cdot \text{mol}}\right)(298 \text{ K}) \ln K_P$$

$$-191.5 = \ln K_P$$

$$K_P = e^{-191.5}$$

$$= 7 \times 10^{-84}$$

THINK ABOUT IT

This is an extremely small equilibrium constant, which is consistent with the large, positive value of $\Delta G°$. We know from everyday experience that water does not decompose spontaneously into its constituent elements at 25°C.

(*Continued on next page*)

Practice Problem ATTEMPT Using data from Appendix 2, calculate the equilibrium constant, K_P, for the following reaction at 25°C:

$$2O_3(g) \rightleftharpoons 3O_2(g)$$

Practice Problem BUILD K_f for the complex ion $Ag(NH_3)_2^+$ is 1.5×10^7 at 25°C. Using this and data from Appendix 2, calculate the value of ΔG_f° for $Ag(NH_3)_2^+(aq)$.

Practice Problem CONCEPTUALIZE Which of the following graphs best shows the relationship between ΔG° and equilibrium constant (K)?

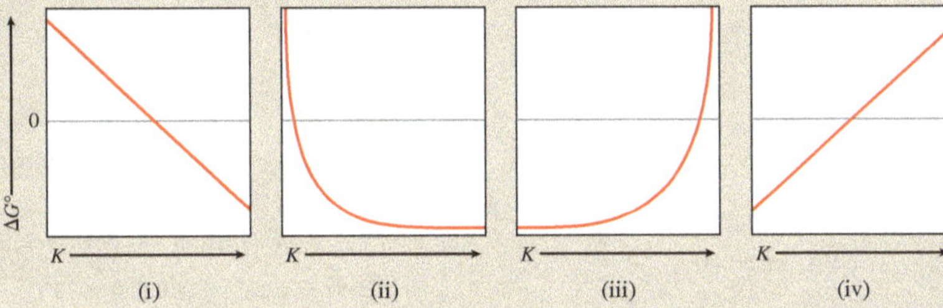

SAMPLE PROBLEM 18.10

The equilibrium constant, K_{sp}, for the dissolution of silver chloride in water at 25°C:

$$AgCl(s) \rightleftharpoons Ag^+(aq) + Cl^-(aq)$$

is 1.6×10^{-10}. Calculate ΔG° for the process.

Strategy Use Equation 18.14 to calculate ΔG°.

Setup $R = 8.314 \times 10^{-3}$ kJ/K · mol and $T = (25 + 273) = 298$ K.

Solution

$$\Delta G^\circ = -RT \ln K_{sp}$$

$$= -\left(\frac{8.314 \times 10^{-3}\text{ kJ}}{\text{K} \cdot \text{mol}}\right)(298\text{ K}) \ln (1.6 \times 10^{-10})$$

$$= 55.9\text{ kJ/mol}$$

THINK ABOUT IT

The relatively large, positive ΔG°, like the very small K value, corresponds to a process that lies very far to the left. Note that the K in Equation 18.14 can be any type of K_c (K_a, K_b, K_{sp}, etc.) or K_P.

Practice Problem ATTEMPT Calculate ΔG° for the process:

$$BaF_2(s) \rightleftharpoons Ba^{2+}(aq) + 2F^-(aq)$$

The K_{sp} of BaF_2 at 25°C is 1.7×10^{-6}.

Practice Problem BUILD K_{sp} for $Co(OH)_2$ at 25°C is 3.3×10^{-16}. Using this and data from Appendix 2, calculate the value of ΔG_f° for $Co(OH)_2(s)$.

Practice Problem CONCEPTUALIZE Fill in each blank with *positive, negative, zero,* or *impossible to determine*:

When $Q = K$, ΔG is ___________. When $Q < K$, ΔG is ___________. When $Q > K$, ΔG is ___________.

CHECKPOINT – SECTION 18.6 Free Energy and Chemical Equilibrium

18.6.1 For the reaction:

$$A(aq) + B(aq) \rightleftharpoons C(aq)$$

$\Delta G° = -1.95$ kJ/mol at 25°C. What is ΔG (in kJ/mol) at 25°C when the concentrations are [A] = [B] = 0.315 *M* and [C] = 0.405 *M*?

a) −1.95 kJ/mol

b) 1.53 kJ/mol

c) −5.43 kJ/mol

d) 8.16 kJ/mol

e) −12.1 kJ/mol

18.6.2 Consider the reaction:

$$X(g) + Y(g) \rightleftharpoons Z(g)$$

for which $\Delta G° = -1.2$ kJ/mol at 25°C. Without knowing the pressures of the reactants and the product, we know that this reaction

a) is spontaneous as written.

b) is nonspontaneous as written.

c) favors reactants at equilibrium.

d) favors products at equilibrium.

e) is at equilibrium.

18.6.3 The $\Delta G°$ for the reaction:

$$N_2(g) + 3H_2(g) \rightleftharpoons 2NH_3(g)$$

is −33.3 kJ/mol at 25°C. What is the value of K_P?

a) 2×10^{-6}

b) 7×10^{5}

c) 3×10^{-70}

d) 1

e) 1×10^{-1}

18.6.4 The K_{sp} for iron(III) hydroxide [$Fe(OH)_3$] is 1.1×10^{-36} at 25°C. For the process:

$$Fe(OH)_3(s) \rightleftharpoons Fe^{3+}(aq) + 3OH^-(aq)$$

determine $\Delta G°$ (in kJ/mol) at 25°C.

a) -2.73×10^{-36} kJ/mol

b) −17.2 kJ/mol

c) 17.2 kJ/mol

d) −205 kJ/mol

e) 205 kJ/mol

18.7 Thermodynamics in Living Systems

Many biochemical reactions have a positive $\Delta G°$ value, yet they are essential to the maintenance of life. In living systems, these reactions are coupled to an energetically favorable process, one that has a negative $\Delta G°$ value. The principle of coupled reactions is based on a simple concept: we can use a thermodynamically favorable reaction to drive an unfavorable one. Suppose, for example, that we want to extract zinc from a zinc sulfide (ZnS). The following reaction will not work because it has a large positive $\Delta G°$ value:

$$ZnS(s) \longrightarrow Zn(s) + S(s) \qquad \Delta G° = 198.3 \text{ kJ/mol}$$

On the other hand, the combustion of sulfur to form sulfur dioxide is favored because of its large negative $\Delta G°$ value:

$$S(s) + O_2(g) \longrightarrow SO_2(g) \qquad \Delta G° = -300.1 \text{ kJ/mol}$$

By coupling the two processes, we can bring about the separation of zinc from zinc sulfide. In practice, this means heating ZnS in air so that the tendency of S to form SO_2 will promote the decomposition of ZnS:

$$ZnS(s) \longrightarrow Zn(s) + \cancel{S(s)} \qquad \Delta G° = 198.3 \text{ kJ/mol}$$

$$\cancel{S(s)} + O_2(g) \longrightarrow SO_2(g) \qquad \Delta G° = -300.1 \text{ kJ/mol}$$

$$ZnS(s) + O_2(g) \longrightarrow Zn(s) + SO_2(g) \qquad \Delta G° = -101.8 \text{ kJ/mol}$$

Coupled reactions play a crucial role in our survival. In biological systems, enzymes facilitate a wide variety of nonspontaneous reactions. In the human body, for example, food molecules, represented by glucose ($C_6H_{12}O_6$), are converted to carbon dioxide and water during metabolism, resulting in a substantial release of free energy:

$$C_6H_{12}O_6(s) + 6O_2(g) \longrightarrow 6CO_2(g) + 6H_2O(l) \qquad \Delta G° = -2880 \text{ kJ/mol}$$

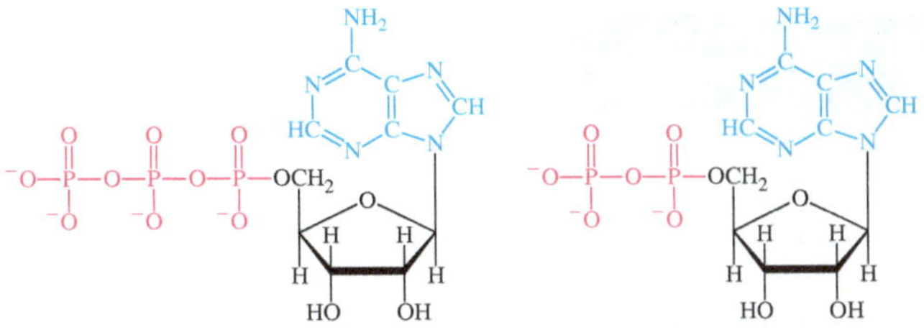

Figure 18.7 Structures of ATP and ADP.

Adenosine triphosphate (ATP)

Adenosine diphosphate (ADP)

In a living cell, this reaction does not take place in a single step; rather, the glucose molecule is broken down with the aid of enzymes in a series of steps. Much of the free energy released along the way is used to synthesize adenosine triphosphate (ATP) from adenosine diphosphate (ADP) and phosphoric acid (Figure 18.7):

$$ADP + H_3PO_4 \longrightarrow ATP + H_2O \qquad \Delta G^\circ = 31 \text{ kJ/mol}$$

The function of ATP is to store free energy until it is needed by cells. Under appropriate conditions, ATP undergoes hydrolysis to give ADP and phosphoric acid, with a release of 31 kJ/mol of free energy, which can be used to drive energetically unfavorable reactions, such as protein synthesis.

Proteins are polymers made of amino acids. The stepwise synthesis of a protein molecule involves the joining of individual amino acids. Consider the formation of the dipeptide (a unit composed of two amino acids) alanylglycine from alanine and glycine. This reaction represents the first step in the synthesis of a protein molecule:

$$\text{alanine} + \text{glycine} \longrightarrow \text{alanylglycine} \qquad \Delta G^\circ = 29 \text{ kJ/mol}$$

The positive ΔG° value means this reaction does not favor the formation of product, so only a little of the dipeptide would be formed at equilibrium. With the aid of an enzyme, however, the reaction is coupled to the hydrolysis of ATP as follows:

$$ATP + H_2O + \text{alanine} + \text{glycine} \longrightarrow ADP + H_3PO_4 + \text{alanylglycine}$$

The overall free-energy change is given by $\Delta G^\circ = -31 \text{ kJ/mol} + 29 \text{ kJ/mol} = -2 \text{ kJ/mol}$, which means that the coupled reaction now favors the formation of product and an appreciable amount of alanylglycine will be formed under these conditions. Figure 18.8 shows the ATP-ADP interconversions that act as energy storage (from metabolism) and free-energy release (from ATP hydrolysis) to drive essential reactions.

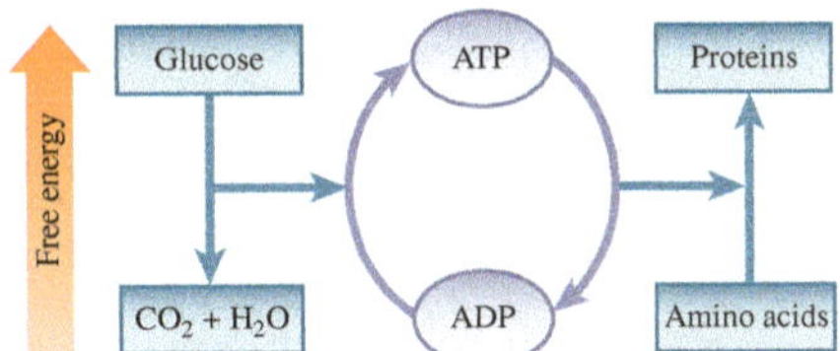

Figure 18.8 Schematic representation of ATP synthesis and coupled reactions in living systems. The conversion of glucose to carbon dioxide and water during metabolism releases free energy. The released free energy is used to convert ADP to ATP. The ATP molecules are then used as an energy source to drive unfavorable reactions such as protein synthesis from amino acids.

Chapter Summary

Section 18.1

- A ***spontaneous process*** is one that occurs under a specified set of conditions.
- A ***nonspontaneous process*** is one that does *not* occur under a specified set of conditions.
- Spontaneous processes do not necessarily happen quickly.

Section 18.2

- ***Entropy (S)*** is a thermodynamic state function that measures how dispersed or spread out a system's energy is.

Section 18.3

- Entropy change for a process can be calculated using standard entropy values or can be predicted qualitatively based on factors such as temperature, phase, and number of molecules.
- Whether or not a process is spontaneous depends on the change in *enthalpy* and the change in *entropy* of the system.
- Tabulated ***standard entropy*** values are absolute values.

Section 18.4

- According to the ***second law of thermodynamics,*** the entropy change for the universe is positive for a *spontaneous* process and zero for an ***equilibrium process.***
- According to the ***third law of thermodynamics,*** the entropy of a perfectly crystalline substance at 0 K is zero.

Section 18.5

- The ***Gibbs free energy (G)*** or simply the ***free energy*** of a system is the energy available to do work.
- The ***standard free energy of reaction*** (ΔG°_{rxn}) for a reaction tells us whether the equilibrium lies to the right (negative ΔG°_{rxn}) or to the left (positive ΔG°_{rxn}).
- ***Standard free energies of formation*** (ΔG°_f) can be used to calculate standard free energies of reaction.

Section 18.6

- The free-energy change (ΔG) is determined using the standard free-energy change (ΔG°) and the reaction quotient (Q).
- The sign of ΔG tells us whether the reaction is spontaneous under the conditions described.
- ΔG° is related to the equilibrium constant, K. A negative ΔG° corresponds to a large K; a positive ΔG° corresponds to a small K.

Section 18.7

- In living systems, thermodynamically favorable reactions provide the free energy needed to drive necessary but thermodynamically unfavorable reactions.

Key Words

Entropy (S), 835
Equilibrium process, 846
Free energy, 850
Gibbs free energy (G), 850
Nonspontaneous process, 834
Second law of thermodynamics, 846
Spontaneous process, 834
Standard entropy (S°), 838
Standard free energy of formation (ΔG°_f), 852
Standard free energy of reaction (ΔG°_{rxn}), 852
Third law of thermodynamics, 848

Key Equations

18.1 $S = k \ln W$	The entropy S of a system is equal to the product of the Boltzmann constant (k) and ln of W, the number of possible arrangements of molecules in the system.
18.2 $W = X^N$	The number of possible arrangements W is equal to the number of possible locations of molecules X raised to the number of molecules in the system N.
18.3 $\Delta S_{sys} = S_{final} - S_{initial}$	The entropy change in a system ΔS_{sys}, is equal to final entropy, S_{final}, minus initial entropy, $S_{initial}$.
18.4 $\Delta S_{sys} = nR \ln \frac{V_{final}}{V_{initial}}$	For a gaseous process involving a volume change, entropy change is calculated as the product of the number of moles (n), the gas constant (R), and ln of the ratio of final volume to initial volume [$\ln(V_{final}/V_{initial})$].
18.5 $\Delta S^\circ_{rxn} = [cS^\circ(C) + dS^\circ(D)] - [aS^\circ(A) + bS^\circ(B)]$	Standard entropy change for a reaction (ΔS°_{rxn}) can be calculated using tabulated values of absolute entropies (S°) for products and reactants.
18.6 $\Delta S^\circ_{rxn} = \Sigma nS^\circ(\text{products}) - \Sigma mS^\circ(\text{reactants})$	ΔS°_{rxn} is calculated as the sum of absolute entropies for products minus the sum of absolute entropies for reactants. Each species in a chemical equation must be multiplied by its coefficient.
18.7 $\Delta S_{surr} = \frac{-\Delta H_{sys}}{T}$	Entropy change in the surroundings (ΔS_{surr}) is calculated as the ratio of minus the enthalpy change in the system ($-\Delta H_{sys}$) to absolute temperature (T).
18.8 $\Delta S_{univ} = \Delta S_{sys} + \Delta S_{surr}$	Entropy change in the universe (ΔS_{univ}) is equal to the sum of entropy change for the system (ΔS_{sys}) and entropy change for the surroundings (ΔS_{surr}).
18.9 $G = H - TS$	Gibbs free energy (G) is the difference between enthalpy (H) and the product of absolute temperature and entropy (TS).
18.10 $\Delta G = \Delta H - T\Delta S$	The change in free energy (ΔG) is calculated as the difference between change in enthalpy (ΔH) and the product of absolute temperature and change in entropy ($T\Delta S$).
18.11 $\Delta G^\circ_{rxn} = [c\Delta G^\circ_f(C) + d\Delta G^\circ_f(D)] - [a\Delta G^\circ_f(A) + b\Delta G^\circ_f(B)]$	Standard free-energy change for a reaction (ΔG°_{rxn}) can be calculated using tabulated values of free energies of formation (ΔG°_f) for products and reactants.
18.12 $\Delta G^\circ_{rxn} = \Sigma n\Delta G^\circ_f(\text{products}) - \Sigma m\Delta G^\circ_f(\text{reactants})$	ΔG°_{rxn} is calculated as the sum of free energies of formation for products minus the sum of free energies of formation for reactants. Each species in a chemical equation must be multiplied by its coefficient.
18.13 $\Delta G = \Delta G^\circ + RT \ln Q$	ΔG is calculated as the sum of ΔG° and the product of R, T, and ln of the reaction quotient, Q.
18.14 $\Delta G^\circ = -RT \ln K$	At equilibrium, Q is equal to K and ΔG is equal to zero. ΔG° is equal to minus the product of R, T, and ln of K.

KEY SKILLS

Free Energy and Equilibrium

In Chapter 15, you learned how to calculate reaction quotient (Q), and that by comparing its magnitude to that of the equilibrium constant (K), you can determine whether a reaction must proceed to the right or to the left in order to reach equilibrium. In Chapter 18, you learned how to calculate free-energy change (ΔG) and standard free-energy change ($\Delta G°$), and that the sign of ΔG indicates whether or not a process is spontaneous. It is useful to consider how these parameters are related, and that they give us essentially the same information. Equation 18.14 relates K and $\Delta G°$:

$$\Delta G° = -RT \ln K$$

According to this equation, if $\Delta G°$ is negative, the value of K is greater than 1. Recall that $K > 1$ tells us the equilibrium *lies to the right*. This is different from saying that it will *proceed* to the right. A reaction may have a large equilibrium constant and yet proceed to the *left* in order to reach equilibrium—if the starting conditions are such that Q is greater than K. For example, $K = 54.3$ for the following reaction at 430°C:

$$H_2(g) + I_2(g) \rightleftharpoons 2HI(g)$$

Because K is greater than 1, we would say that this equilibrium lies to the right. However, if we initially have only HI and no H_2 or I_2, clearly the reaction must *proceed* to the *left* in order for equilibrium to be established.

So what does it mean for an equilibrium to *lie* to the right? It means that if the starting conditions are *standard* conditions (the concentrations of all aqueous species are 1 M and the pressures of all gaseous species are 1 atm), products will predominate at equilibrium. This corresponds to the reaction proceeding to the right from *standard* initial conditions. Likewise, for an equilibrium that lies to the *left*, if the starting conditions are standard conditions, reactants will predominate at equilibrium—the reaction will proceed to the *left* from standard initial conditions.

We can also describe this in terms of spontaneity. The sign of $\Delta G°$ tells us whether or not a process or reaction is *spontaneous* under *standard* conditions. This is simply another way to say that the equilibrium lies to the right. ΔG, on the other hand, tells us whether or not a process is spontaneous under *actual* conditions—which usually are not standard. ΔG and $\Delta G°$ are related by Equation 18.13:

$$\Delta G = \Delta G° + RT \ln Q$$

The sign of ΔG depends not only on the sign of $\Delta G°$, but also on the value of Q. The information provided by the sign of ΔG is the same as that provided by the *comparison* of Q and K.

By combining Equations 18.13 and 18.14, we can write an expression that indicates at a glance the relationship between the sign of ΔG and the relative magnitudes of Q and K.

Substituting the right side of Equation 18.14 for the ΔG° term in Equation 18.13 gives:

$$\Delta G = -RT \ln K + RT \ln Q$$

or:

$$\Delta G = -RT \ln Q - RT \ln K$$

And because $\ln A - \ln B = \ln \frac{A}{B}$ [▸▸| Appendix 1], we get:

$$\Delta G = RT \ln \frac{Q}{K}$$

This enables us to simply compare Q and K. When Q is greater than K, the log term in our final equation will be positive and ΔG will be positive—meaning that the reaction must proceed to the left to establish equilibrium. When Q is *less* than K, the log term and ΔG will both be *negative*—meaning that the reaction will proceed to the *right*.

Key Skills Problems

18.1
Which of the following must be negative for a process to be spontaneous as written?

(a) ΔG° (b) ΔG (c) K (d) Q (e) R

18.2
ΔG for a reaction is always negative when

(a) ΔG° is negative (b) $K < 1$ (c) $K > 1$
(d) $Q < K$ (e) $Q > K$

18.3
The diagram shown here depicts a system at equilibrium for the reaction $A_2 + B_2 \rightleftharpoons 2AB$.

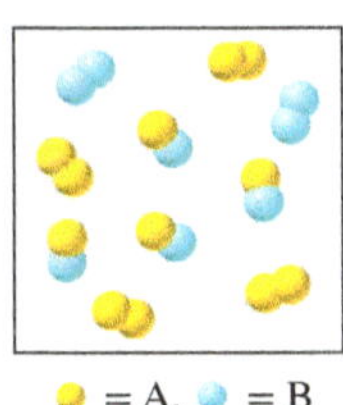

● = A, ● = B

ΔG° for the reaction is −4.8 kJ/mol. Which of the following diagrams depicts starting conditions for the reaction where ΔG is negative? Select all that apply.

(a)
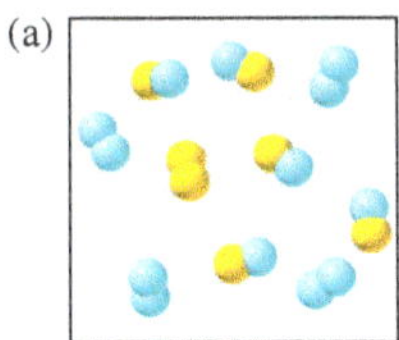

(c)
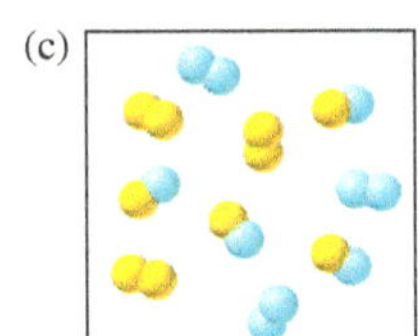

(b)
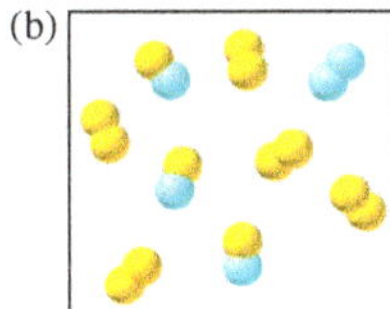

(d)
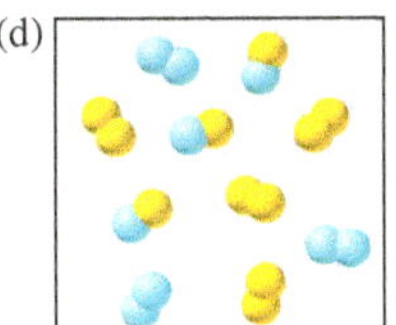

18.4
The reaction shown here has $\Delta G^\circ = -1.83$ kJ/mol at 25°C. Under the starting conditions shown in the diagram, the reaction will proceed to the right. Which of the following could be the reaction depicted in the diagram? Select all that apply.

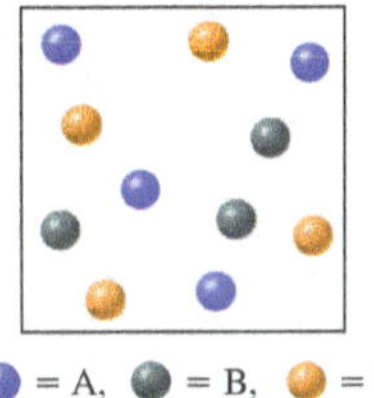

● = A, ● = B, ● = C

(a) $C \rightleftharpoons A + B$
(b) $2C \rightleftharpoons A + B$
(c) $2C \rightleftharpoons 2A + B$
(d) $2C \rightleftharpoons A + 2B$

Questions and Problems

Applying What You've Learned

When a folded protein in solution is heated to a high enough temperature, its polypeptide chain will unfold to become the denatured protein—a process known as "denaturation." The temperature at which most of the protein unfolds is called the "melting" temperature. The melting temperature of a certain protein is found to be 63°C, and the enthalpy of denaturation is 510 kJ/mol.

Problems:

(a) Estimate the entropy of denaturation, assuming that the denaturation is a single-step equilibrium process; that is, folded protein $\rightleftharpoons$ denatured protein [⏮ Sample Problem 18.7]. The single polypeptide protein chain has 98 amino acids. Calculate the entropy of denaturation per amino acid. (b) Assuming that ΔH and ΔS do not change with temperature, determine ΔG for the denaturation at 20°C [⏮ Sample Problem 18.5]. (c) Assuming that the ΔG value from part (b) is $\Delta G°$ for the denaturation, determine the value of the equilibrium constant for the process at 20°C [⏮ Sample Problem 18.9].

SECTION 18.1: SPONTANEOUS PROCESSES

Review Questions

18.1 Explain what is meant by a *spontaneous process*. Give two examples each of spontaneous and nonspontaneous processes.

18.2 Which of the following processes are spontaneous and which are nonspontaneous: (a) dissolving table salt (NaCl) in hot soup, (b) climbing Mt. Everest, (c) spreading fragrance in a room by removing the cap from a perfume bottle, (d) separating helium and neon from a mixture of the gases?

18.3 Which of the following processes are spontaneous and which are nonspontaneous at a given temperature?
(a) $NaNO_3(s) \xrightarrow{H_2O} NaNO_3(aq)$ saturated soln
(b) $NaNO_3(s) \xrightarrow{H_2O} NaNO_3(aq)$ unsaturated soln
(c) $NaNO_3(s) \xrightarrow{H_2O} NaNO_3(aq)$ supersaturated soln

SECTION 18.2: ENTROPY

Review Questions

18.4 Describe what is meant by the term *entropy*. What are the units of entropy?

18.5 What is the relationship between entropy and the number of possible arrangements of molecules in a system?

Conceptual Problems

18.6 Referring to the setup in Figure 18.2, determine the number of possible arrangements, W, and calculate the entropy before and after removal of the barrier if the number of molecules is (a) 10, (b) 50, (c) 100.

18.7 In the setup shown, a container is divided into eight cells and contains two molecules. Initially, both molecules are confined to the left side of the container. (a) Determine the number of possible arrangements before and after removal of the central barrier. (b) After the removal of the barrier, how many of the arrangements correspond to the state in which both molecules are in the left side of the container? How many correspond to the state in which both molecules are in the right side of the container? How many correspond to the state in which the molecules are in opposite sides of the container? Calculate the entropy for each state and comment on the most probable state of the system after removal of the barrier.

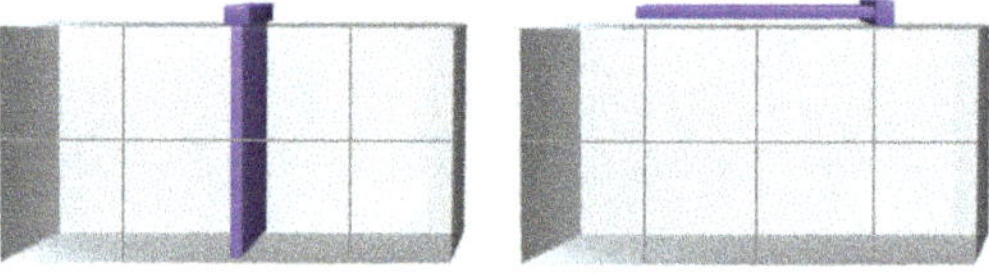

SECTION 18.3: ENTROPY CHANGES IN A SYSTEM

Visualizing Chemistry
Figure 18.4

VC 18.1 Consider two gas samples at STP: one consisting of a mole of F_2 gas ($S° = 203.34$ J/K · mol) and one consisting of a mole of F gas ($S° = 158.7$ J/K · mol). What factors account for the difference in standard entropies of these two species?

Volume	Molar mass	Temperature	Phase	Molecular complexity
(i)	(ii)	(iii)	(iv)	(v)

a) i, ii, iii, and iv
b) ii and v
c) ii, iv, and v

VC 18.2 Now consider the reaction $F_2(g) \longrightarrow 2F(g)$ at constant temperature and pressure. What factors contribute to the entropy increase associated with the reaction?

Volume increase	Molar mass increase	Increased number of molecules	Phase change	Increase in molecular complexity
(i)	(ii)	(iii)	(iv)	(v)

a) i and iii
b) i, ii, and iii
c) i, iv, and v

VC 18.3 Which of the following best describes why entropy always increases with temperature?
a) As temperature increases, the number of molecules increases.
b) As temperature increases, energy levels become more closely spaced.
c) As temperature increases, the molecules become more energetic and can access more energy levels.

VC 18.4 Which of the following best explains why entropy typically increases with molar mass?
a) As molar mass increases, the number of molecules increases.
b) As molar mass increases, energy levels become more closely spaced.
c) As molar mass increases, the molecules become more energetic and can access more energy levels.

Review Questions

18.8 How does the entropy of a system change for each of the following processes?
(a) A solid melts.
(b) A liquid freezes.
(c) A liquid boils.
(d) A vapor is converted to a solid.
(e) A vapor condenses to a liquid.
(f) A solid sublimes.
(g) A solid dissolves in water.

18.9 How does the entropy of a system change for each of the following processes?
(a) Bromine liquid vaporizes.
(b) Water freezes to form ice.
(c) Naphthalene, the key component of mothballs, sublimes.
(d) Sugar crystals form from a supersaturated solution.
(e) A block of lead melts.
(f) Iodine vapor condenses to form solid iodine.
(g) Carbon tetrachloride dissolves in liquid benzene.

18.10 Predict whether the entropy change is positive or negative for each of the following reactions. Give reasons for your predictions.
(a) $2KClO_4(s) \longrightarrow 2KClO_3(s) + O_2(g)$
(b) $H_2O(g) \longrightarrow H_2O(l)$
(c) $2Na(s) + 2H_2O(l) \longrightarrow 2NaOH(aq) + H_2(g)$
(d) $N_2(g) \longrightarrow 2N(g)$

18.11 State whether the sign of the entropy change expected for each of the following processes will be positive or negative, and explain your predictions.
(a) $PCl_3(l) + Cl_2(g) \longrightarrow PCl_5(g)$
(b) $2HgO(s) \longrightarrow 2Hg(l) + O_2(g)$
(c) $H_2(g) \longrightarrow 2H(g)$
(d) $U(s) + 3F_2(g) \longrightarrow UF_6(s)$

Computational Problems

18.12 Calculate ΔS_{sys} for (a) the isothermal expansion of 2.0 mol of an ideal gas from 10.0 L to 15.0 L, (b) the isothermal expansion of 1.5 mol of an ideal gas from 20.0 L to 22.5 L, and (c) the isothermal compression of 5.0 mol of an ideal gas from 100.0 L to 75.0 L.

18.13 Calculate ΔS_{sys} for (a) the isothermal compression of 0.0050 mol of an ideal gas from 112 mL to 52.5 mL, (b) the isothermal compression of 0.015 mol of an ideal gas from 225 mL to 22.5 mL, and (c) the isothermal expansion of 22.1 mol of an ideal gas from 122 L to 275 L.

18.14 Using the data in Appendix 2, calculate the standard entropy changes for the following reactions at 25°C:
(a) $S(\text{rhombic}) + O_2(g) \longrightarrow SO_2(g)$
(b) $MgCO_3(s) \longrightarrow MgO(s) + CO_2(g)$
(c) $2C_2H_6(g) + 7O_2(g) \longrightarrow 4CO_2(g) + 6H_2O(l)$

18.15 Using the data in Appendix 2, calculate the standard entropy changes for the following reactions at 25°C:
(a) $H_2(g) + CuO(s) \longrightarrow Cu(s) + H_2O(g)$
(b) $2Al(s) + 3ZnO(s) \longrightarrow Al_2O_3(s) + 3Zn(s)$
(c) $CH_4(g) + 2O_2(g) \longrightarrow CO_2(g) + 2H_2O(l)$

Conceptual Problems

18.16 For each pair of substances listed here, choose the one having the larger standard entropy value at 25°C. The same molar amount is used in the comparison. Explain the basis for your choice. (a) $Li(s)$ or $Li(l)$, (b) $C_2H_5OH(l)$ or $CH_3OCH_3(l)$ (*Hint:* Which molecule can hydrogen-bond?), (c) $Ar(g)$ or $Xe(g)$, (d) $CO(g)$ or $CO_2(g)$, (e) $O_2(g)$ or $O_3(g)$, (f) $NO_2(g)$ or $N_2O_4(g)$.

18.17 Arrange the following substances (1 mole each) in order of increasing entropy at 25°C: (a) $Ne(g)$, (b) $SO_2(g)$, (c) $Na(s)$, (d) $NaCl(s)$, (e) $H_2(g)$. Give the reasons for your arrangement.

SECTION 18.4: ENTROPY CHANGES IN THE UNIVERSE

Review Questions

18.18 State the second law of thermodynamics in words, and express it mathematically.

18.19 State the third law of thermodynamics in words, and explain its usefulness in calculating entropy values.

Computational Problems

18.20 Calculate ΔS_{surr} for each of the reactions in Problem 18.14 and determine if each reaction is spontaneous at 25°C.

18.21 Calculate ΔS_{surr} for each of the reactions in Problem 18.15 and determine if each reaction is spontaneous at 25°C.

18.22 Using data from Appendix 2, calculate ΔS°_{rxn} and ΔS_{surr} for each of the reactions in Problem 18.10 and determine if each reaction is spontaneous at 25°C.

18.23 Using data from Appendix 2, calculate ΔS°_{rxn} and ΔS_{surr} for each of the reactions in Problem 18.11 and determine if each reaction is spontaneous at 25°C.

SECTION 18.5: PREDICTING SPONTANEITY

Review Questions

18.24 Define *free energy*. What are its units?

18.25 Why is it more convenient to predict the direction of a reaction in terms of ΔG_{sys} instead of ΔS_{univ}? Under what conditions can ΔG_{sys} be used to predict the spontaneity of a reaction?

18.26 What is the significance of the sign of ΔG_{sys}?

18.27 From the following combinations of ΔH and ΔS, predict if a process will be spontaneous at a high or low temperature: (a) both ΔH and ΔS are negative, (b) ΔH is negative and ΔS is positive, (c) both ΔH and ΔS are positive, (d) ΔH is positive and ΔS is negative.

Computational Problems

18.28 Calculate ΔG° for the following reactions at 25°C:
(a) $N_2(g) + O_2(g) \longrightarrow 2NO(g)$
(b) $H_2O(l) \longrightarrow H_2O(g)$
(c) $2C_2H_2(g) + 5O_2(g) \longrightarrow 4CO_2(g) + 2H_2O(l)$
(*Hint:* Look up the standard free energies of formation of the reactants and products in Appendix 2.)

18.29 Calculate ΔG° for the following reactions at 25°C:
(a) $2Mg(s) + O_2(g) \longrightarrow 2MgO(s)$
(b) $2SO_2(g) + O_2(g) \longrightarrow 2SO_3(g)$
(c) $2C_2H_6(g) + 7O_2(g) \longrightarrow 4CO_2(g) + 6H_2O(l)$
(See Appendix 2 for thermodynamic data.)

18.30 From the values of ΔH and ΔS, predict which of the following reactions would be spontaneous at 25°C: reaction A: $\Delta H = 10.5$ kJ/mol, $\Delta S = 30$ J/K · mol; reaction B: $\Delta H = 1.8$ kJ/mol, $\Delta S = -113$ J/K · mol. If either of the reactions is nonspontaneous at 25°C, at what temperature might it become spontaneous?

18.31 Find the temperatures at which reactions with the following ΔH and ΔS values would become spontaneous: (a) $\Delta H = -126$ kJ/mol, $\Delta S = 84$ J/K · mol; (b) $\Delta H = -11.7$ kJ/mol, $\Delta S = -105$ J/K · mol.

18.32 The molar heats of fusion and vaporization of ethanol are 7.61 and 26.0 kJ/mol, respectively. Calculate the molar entropy changes for the solid-liquid and liquid-vapor transitions for ethanol. At 1 atm pressure, ethanol melts at −117.3°C and boils at 78.3°C.

18.33 The molar heats of fusion and vaporization of mercury are 23.4 and 59.0 kJ/mol, respectively. Calculate the molar entropy changes for the solid-liquid and liquid-vapor transitions for mercury. At 1 atm pressure, mercury melts at −38.9°C and boils at 357°C.

18.34 Consider the formation of a dimeric protein:

$$2P \longrightarrow P_2$$

At 25°C, we have $\Delta H^\circ = 17$ kJ/mol and $\Delta S^\circ = 65$ J/K · mol. Is the dimerization favored at this temperature? Comment on the effect of lowering the temperature. Does your result explain why some enzymes lose their activities under cold conditions?

18.35 Use the values listed in Appendix 2 to calculate ΔG° for the following alcohol fermentation:

$$C_6H_{12}O_6(s) \longrightarrow 2C_2H_5OH(l) + 2CO_2(g)$$

18.36 As an approximation, we can assume that proteins exist either in the native (physiologically functioning) state or the denatured state. The standard molar enthalpy and entropy of the denaturation of a certain protein are 512 kJ/mol and 1.60 kJ/K · mol, respectively. Comment on the signs and magnitudes of these quantities, and calculate the temperature at which the denaturation becomes spontaneous.

18.37 Certain bacteria in the soil obtain the necessary energy for growth by oxidizing nitrites to nitrates:

$$2NO_2^- + O_2 \longrightarrow 2NO_3^-$$

Given that the standard Gibbs free energies of formation of NO_2^- and NO_3^- are −34.6 and −110.5 kJ/mol, respectively, calculate the amount of Gibbs free energy released when 1 mole of NO_2^- is oxidized to 1 mole of NO_3^-.

SECTION 18.6: FREE ENERGY AND CHEMICAL EQUILIBRIUM

Review Questions

18.38 Explain the difference between ΔG and ΔG°.

18.39 Explain why Equation 18.14 is of great importance in chemistry.

18.40 Fill in the missing entries in the following table:

K	$\ln K$	ΔG°	Result at equilibrium
< 1			
		0	
			Products are favored

Computational Problems

18.41 Calculate K_P for the following reaction at 25°C:

$$H_2(g) + I_2(g) \rightleftharpoons 2HI(g) \qquad \Delta G^\circ = 2.60 \text{ kJ/mol}$$

18.42 For the autoionization of water at 25°C:

$$H_2O(l) \rightleftharpoons H^+(aq) + OH^-(aq)$$

K_w is 1.0×10^{-14}. What is ΔG° for the process?

18.43 Consider the following reaction at 25°C:

$$Fe(OH)_2(s) \rightleftharpoons Fe^{2+}(aq) + 2OH^-(aq)$$

Calculate ΔG° for the reaction. K_{sp} for $Fe(OH)_2$ is 1.6×10^{-14}.

18.44 Calculate ΔG° and K_P for the following equilibrium reaction at 25°C:

$$2H_2O(g) \rightleftharpoons 2H_2(g) + O_2(g)$$

18.45 (a) Calculate ΔG° and K_P for the following equilibrium reaction at 25°C:

$$PCl_5(g) \rightleftharpoons PCl_3(g) + Cl_2(g)$$

(b) Calculate ΔG for the reaction if the partial pressures of the initial mixture are $P_{PCl_5} = 0.0029$ atm, $P_{PCl_3} = 0.27$ atm, and $P_{Cl_2} = 0.40$ atm.

18.46 The equilibrium constant (K_P) for the reaction:

$$H_2(g) + CO_2(g) \rightleftharpoons H_2O(g) + CO(g)$$

is 4.40 at 2000 K. (a) Calculate $\Delta G°$ for the reaction. (b) Calculate ΔG for the reaction when the partial pressures are $P_{H_2} = 0.25$ atm, $P_{CO_2} = 0.78$ atm, $P_{H_2O} = 0.66$ atm, and $P_{CO} = 1.20$ atm.

18.47 Consider the decomposition of calcium carbonate:

$$CaCO_3(s) \rightleftharpoons CaO(s) + CO_2(g)$$

Calculate the pressure in atm of CO_2 in an equilibrium process (a) at 25°C and (b) at 800°C. Assume that $\Delta H° = 177.8$ kJ/mol and $\Delta S° = 160.5$ J/K · mol for the temperature range.

18.48 The equilibrium constant K_P for the reaction:

$$CO(g) + Cl_2(g) \rightleftharpoons COCl_2(g)$$

is 5.62×10^{35} at 25°C. Calculate $\Delta G_f°$ for $COCl_2$ at 25°C.

18.49 At 25°C, $\Delta G°$ for the process:

$$H_2O(l) \rightleftharpoons H_2O(g)$$

is 8.6 kJ/mol. Calculate the vapor pressure of water at this temperature.

18.50 Calculate $\Delta G°$ for the process

$$C(\text{diamond}) \rightleftharpoons C(\text{graphite})$$

Is the formation of graphite from diamond favored at 25°C? If so, why is it that diamonds do not become graphite on standing?

SECTION 18.7: THERMODYNAMICS IN LIVING SYSTEMS

Review Questions

18.51 What is a coupled reaction? What is its importance in biological reactions?

18.52 What is the role of ATP in biological reactions?

Computational Problems

18.53 Referring to the metabolic process involving glucose in Section 18.7, calculate the maximum number of moles of ATP that can be synthesized from ADP from the breakdown of 1 mole of glucose.

18.54 In the metabolism of glucose, the first step is the conversion of glucose to glucose 6-phosphate:

$$\text{glucose} + H_3PO_4 \longrightarrow \text{glucose 6-phosphate} + H_2O \qquad \Delta G° = 13.4 \text{ kJ/mol}$$

Because $\Delta G°$ is positive, this reaction does not favor the formation of products. Show how this reaction can be made to proceed by coupling it with the hydrolysis of ATP. Write an equation for the coupled reaction, and estimate the equilibrium constant for the coupled process.

ADDITIONAL PROBLEMS

18.55 Predict the signs of ΔH, ΔS, and ΔG of the system for the following processes at 1 atm: (a) ammonia melts at −60°C, (b) ammonia melts at −77.7°C, (c) ammonia melts at −100°C. (The normal melting point of ammonia is −77.7°C.)

18.56 Calculate ΔG for the reaction:

$$H_2O(l) \rightleftharpoons H^+(aq) + OH^-(aq)$$

at 25°C for the following initial concentrations:
(a) $[H^+] = 1.0 \times 10^{-7}$ *M*, $[OH^-] = 1.0 \times 10^{-7}$ *M*
(b) $[H^+] = 1.0 \times 10^{-3}$ *M*, $[OH^-] = 1.0 \times 10^{-4}$ *M*
(c) $[H^+] = 1.0 \times 10^{-12}$ *M*, $[OH^-] = 2.0 \times 10^{-8}$ *M*
(d) $[H^+] = 3.5$ *M*, $[OH^-] = 4.8 \times 10^{-4}$ *M*

18.57 Which of the following thermodynamic functions are associated only with the first law of thermodynamics: *S*, *U*, *G*, and *H*?

18.58 A student placed 1 g of each of three compounds A, B, and C in a container and found that after 1 week no change had occurred. Offer some possible explanations for the fact that no reactions took place. Assume that A, B, and C are totally miscible liquids.

18.59 Consider the following reaction at 298 K:

$$2H_2(g) + O_2(g) \longrightarrow 2H_2O(l) \qquad \Delta H° = -571.6 \text{ kJ/mol}$$

Calculate ΔS_{sys}, ΔS_{surr}, and ΔS_{univ} for the reaction.

18.60 Consider the following facts: Water freezes spontaneously at −5°C and 1 atm, and ice has a lower entropy than liquid water. Explain how a spontaneous process can lead to a decrease in entropy.

18.61 Ammonium nitrate (NH_4NO_3) dissolves spontaneously and endothermically in water. What can you deduce about the sign of ΔS for the solution process?

18.62 Calculate the equilibrium pressure of CO_2 due to the decomposition of barium carbonate ($BaCO_3$) at 25°C.

18.63 (a) Trouton's rule states that the ratio of the molar heat of vaporization of a liquid (ΔH_{vap}) to its boiling point in kelvins is approximately 90 J/K · mol. Use the following data to show that this is the case and explain why Trouton's rule holds true:

	T_{bp} (°C)	ΔH_{vap}(kJ/mol)
Benzene	80.1	31.0
Hexane	68.7	30.8
Mercury	357	59.0
Toluene	110.6	35.2

(b) Use the values in Table 11.6 to calculate the same ratio for ethanol and water. Explain why Trouton's rule does not apply to these two substances as well as it does to other liquids.

18.64 Referring to Problem 18.63, explain why the ratio is considerably smaller than 90 J/K · mol for liquid HF.

18.65 Which of the following are not state functions: *S*, *H*, *q*, *w*, *T*?

18.66 For reactions carried out under standard-state conditions, Equation 18.10 takes the form $\Delta G° = \Delta H° - T\Delta S°$. (a) Assuming $\Delta H°$ and $\Delta S°$ are independent of temperature, derive the equation:

$$\ln \frac{K_2}{K_1} = \frac{\Delta H°}{R}\left(\frac{T_2 - T_1}{T_1 T_2}\right)$$

where K_1 and K_2 are the equilibrium constants at T_1 and T_2, respectively. (b) Given that at 25°C K_c is 4.63×10^{-3} for the reaction:

$$N_2O_4(g) \rightleftharpoons 2NO_2(g) \qquad \Delta H° = 58.0 \text{ kJ/mol}$$

calculate the equilibrium constant at 65°C.

18.67 The sublimation of carbon dioxide at −78°C is given by:

$$CO_2(s) \longrightarrow CO_2(g) \qquad \Delta H_{sub} = 25.2 \text{ kJ/mol}$$

Calculate ΔS_{sub} when 84.8 g of CO_2 sublimes at this temperature.

18.68 Under what conditions does a substance have a standard entropy of zero? Can an element or a compound ever have a negative standard entropy?

18.69 A student looked up the $\Delta G_f°$, $\Delta H_f°$, and $\Delta S°$ values for CO_2 in Appendix 2. Plugging these values into Equation 18.10, the student found that $\Delta G_f° \neq \Delta H_f° - T\Delta S°$ at 298 K. What is wrong with this approach?

18.70 Consider the following Brønsted acid-base reaction at 25°C:

$$HF(aq) + Cl^-(aq) \rightleftharpoons HCl(aq) + F^-(aq)$$

(a) Predict whether K will be greater or smaller than 1. (b) Does $\Delta S°$ or $\Delta H°$ make a greater contribution to $\Delta G°$? (c) Is $\Delta H°$ likely to be positive or negative?

18.71 At 0 K, the entropy of carbon monoxide crystal is not zero but has a value of 4.2 J/K · mol, called the residual entropy. According to the third law of thermodynamics, this means that the crystal does not have a perfect arrangement of the CO molecules. (a) What would be the residual entropy if the arrangement were totally random? (b) Comment on the difference between the result in part (a) and 4.2 J/K · mol. (*Hint:* Assume that each CO molecule has two choices for orientation, and use Equation 18.1 to calculate the residual entropy.)

18.72 Crystallization of sodium acetate from a supersaturated solution occurs spontaneously (see Figure 13.2). Based on this, what can you deduce about the signs of ΔS and ΔH?

18.73 Consider the thermal decomposition of $CaCO_3$:

$$CaCO_3(s) \rightleftharpoons CaO(s) + CO_2(g)$$

The equilibrium vapor pressures of CO_2 are 22.6 mmHg at 700°C and 1829 mmHg at 950°C. Calculate the standard enthalpy of the reaction. [*Hint:* See Problem 18.66(a).]

18.74 A certain reaction is spontaneous at 72°C. If the enthalpy change for the reaction is 19 kJ/mol, what is the *minimum* value of ΔS (in J/K · mol) for the reaction?

18.75 Predict whether the entropy change is positive or negative for each of these reactions:
(a) $Zn(s) + 2HCl(aq) \rightleftharpoons ZnCl_2(aq) + H_2(g)$
(b) $O(g) + O(g) \rightleftharpoons O_2(g)$
(c) $NH_4NO_3(s) \rightleftharpoons N_2O(g) + 2H_2O(g)$
(d) $2H_2O_2(l) \rightleftharpoons 2H_2O(l) + O_2(g)$

18.76 The reaction $NH_3(g) + HCl(g) \longrightarrow NH_4Cl(s)$ proceeds spontaneously at 25°C even though there is a decrease in entropy in the system (gases are converted to a solid). Explain.

18.77 Use the following data to determine the normal boiling point, in kelvins, of mercury. What assumptions must you make to do the calculation?
Hg(l): $\Delta H_f° = 0$ (by definition)
$S° = 77.4$ J/K · mol
Hg(g): $\Delta H_f° = 60.78$ kJ/mol
$S° = 174.7$ J/K · mol

18.78 The molar heat of vaporization of ethanol is 39.3 kJ/mol, and the boiling point of ethanol is 78.3°C. Calculate ΔS for the vaporization of 0.50 mole of ethanol.

18.79 A certain reaction is known to have a $\Delta G°$ value of −122 kJ/mol. Will the reaction necessarily occur if the reactants are mixed together?

18.80 Derive the equation:

$$\Delta G = RT \ln \frac{Q}{K}$$

where Q is the reaction quotient, and describe how you would use it to predict the spontaneity of a reaction.

18.81 Calculate $\Delta G°$ and K_P for the following processes at 25°C:
(a) $H_2(g) + Br_2(l) \rightleftharpoons 2HBr(g)$
(b) $\frac{1}{2}H_2(g) + \frac{1}{2}Br_2(l) \rightleftharpoons HBr(g)$
Account for the differences in $\Delta G°$ and K_P obtained for parts (a) and (b).

18.82 Calculate the pressure of O_2 (in atm) over a sample of NiO at 25°C if $\Delta G° = 212$ kJ/mol for the reaction:

$$NiO(s) \rightleftharpoons Ni(s) + \tfrac{1}{2}O_2(g)$$

18.83 Comment on the statement: "Just talking about entropy increases its value in the universe."

18.84 For a reaction with a negative $\Delta G°$ value, which of the following statements is false? (a) The equilibrium constant K is greater than one. (b) The reaction is spontaneous when all the reactants and products are in their standard states. (c) The reaction is always exothermic.

18.85 Consider the reaction:

$$N_2(g) + O_2(g) \rightleftharpoons 2NO(g)$$

Given that $\Delta G°$ for the reaction at 25°C is 173.4 kJ/mol, (a) calculate the standard free energy of formation of NO and (b) calculate K_P of the reaction. (c) One of the starting substances in smog formation is NO. Assuming that the temperature in a running automobile engine is 1100°C, estimate K_P for the given reaction. (d) As farmers know, lightning helps to produce a better crop. Why?

18.86 Heating copper(II) oxide at 400°C does not produce any appreciable amount of Cu:

$$CuO(s) \rightleftharpoons Cu(s) + \frac{1}{2}O_2(g) \qquad \Delta G° = 127.2 \text{ kJ/mol}$$

However, if this reaction is coupled to the conversion of graphite to carbon monoxide, it becomes spontaneous. Write an equation for the coupled process, and calculate the equilibrium constant for the coupled reaction.

18.87 Consider the decomposition of magnesium carbonate:

$$MgCO_3(s) \rightleftharpoons MgO(s) + CO_2(g)$$

Calculate the temperature at which the decomposition begins to favor products. Assume that both $\Delta H°$ and $\Delta S°$ are independent of temperature.

18.88 Describe two ways that you could determine $\Delta G°$ of a reaction.

18.89 The activity series in Table 4.6 shows that reaction (a) is spontaneous whereas reaction (b) is nonspontaneous at 25°C:
(a) $Fe(s) + 2H^+(aq) \longrightarrow Fe^{2+}(aq) + H_2(g)$
(b) $Cu(s) + 2H^+(aq) \longrightarrow Cu^{2+}(aq) + H_2(g)$
Use the data in Appendix 2 to calculate the equilibrium constant for these reactions and hence confirm that the activity series is correct.

18.90 The rate constant for the elementary reaction:

$$2NO(g) + O_2(g) \longrightarrow 2NO_2(g)$$

is $7.1 \times 10^9/M^2 \cdot s$ at 25°C. What is the rate constant for the reverse reaction at the same temperature?

18.91 A 74.6-g ice cube floats in the Arctic Sea. The pressure and temperature of the system and surroundings are at 1 atm and 0°C, respectively. Calculate ΔS_{sys}, ΔS_{surr}, and ΔS_{univ} for the melting of the ice cube. What can you conclude about the nature of the process from the value of ΔS_{univ}? (The molar heat of fusion of water is 6.01 kJ/mol.)

18.92 Which of the following is not accompanied by an increase in the entropy of the system: (a) mixing of two gases at the same temperature and pressure, (b) mixing of ethanol and water, (c) discharging a battery, (d) expansion of a gas followed by compression to its original temperature, pressure, and volume?

18.93 Hydrogenation reactions (e.g., the process of converting C=C bonds to C—C bonds in the food industry) are facilitated by the use of a transition metal catalyst, such as Ni or Pt. The initial step is the adsorption, or binding, of hydrogen gas onto the metal surface. Predict the signs of ΔH, ΔS, and ΔG when hydrogen gas is adsorbed onto the surface of Ni metal.

18.94 Give a detailed example of each of the following, with an explanation: (a) a thermodynamically spontaneous process, (b) a process that would violate the first law of thermodynamics, (c) a process that would violate the second law of thermodynamics, (d) an irreversible process, (e) an equilibrium process.

18.95 The following diagram shows the variation of the equilibrium constant with temperature for the reaction:

$$I_2(g) \rightleftharpoons 2I(g)$$

Calculate $\Delta G°$, $\Delta H°$, and $\Delta S°$ for the reaction at 872 K. (*Hint:* See Problem 18.66.)

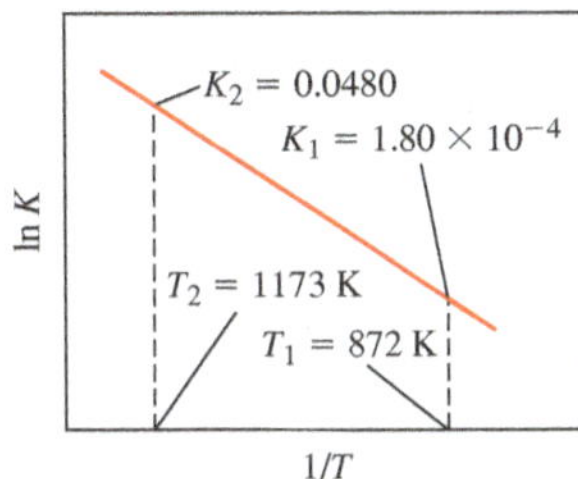

18.96 The standard enthalpy of formation and the standard entropy of gaseous benzene are 82.93 kJ/mol and 269.2 J/K · mol, respectively. Calculate $\Delta H°$, $\Delta S°$, and $\Delta G°$ for the given process at 25°C. Comment on your answers:

$$C_6H_6(l) \longrightarrow C_6H_6(g)$$

18.97 Consider the gas-phase reaction between A_2 (green) and B_2 (red) to form AB at 298 K:

$$A_2(g) + B_2(g) \rightleftharpoons 2AB(g) \qquad \Delta G° = -3.43 \text{ kJ/mol}$$

(a) Which of the following reaction mixtures is at equilibrium?
(b) Which of the following reaction mixtures has a negative ΔG value?
(c) Which of the following reaction mixtures has a positive ΔG value?
The partial pressures of the gases in each frame are equal to the number of A_2, B_2, and AB molecules times 0.10 atm. Round your results to two significant figures.

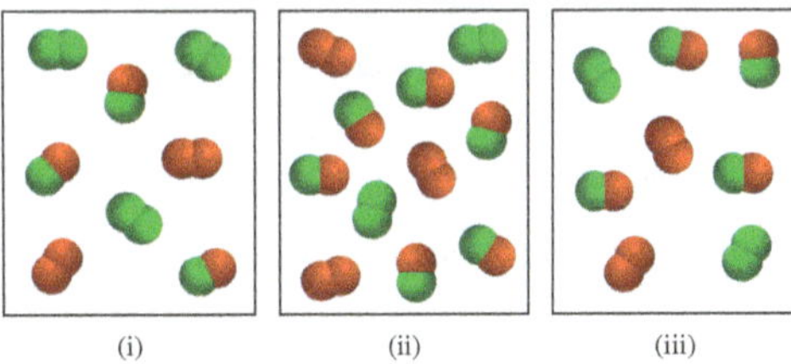

Engineering Problems

18.98 Large quantities of hydrogen are needed for the synthesis of ammonia. One preparation of hydrogen involves the reaction between carbon monoxide and steam at 300°C in the presence of a copper-zinc catalyst:

$$CO(g) + H_2O(g) \rightleftharpoons CO_2(g) + H_2(g)$$

Calculate the equilibrium constant (K_P) for the reaction and the temperature at which the reaction favors the formation of CO and H_2O. Will a larger K_P be attained at the same temperature if a more efficient catalyst is used?

18.99 The following reaction was described as the cause of sulfur deposits formed at volcanic sites:

$$2H_2S(g) + SO_2(g) \rightleftharpoons 3S(s) + 2H_2O(g)$$

It may also be used to remove SO_2 from powerplant stack gases. (a) Identify the type of redox reaction it is.

(b) Calculate the equilibrium constant (K_P) at 25°C, and comment on whether this method is feasible for removing SO_2. (c) Would this procedure become more effective or less effective at a higher temperature?

18.100 Comment on the feasibility of extracting copper from its ore chalcocite (Cu_2S) by heating:

$$Cu_2S(s) \longrightarrow 2Cu(s) + S(s)$$

Calculate the $\Delta G°$ for the overall reaction if this process is coupled to the conversion of sulfur to sulfur dioxide and $\Delta G_f°(Cu_2S) = -86.1$ kJ/mol.

18.101 One of the steps in the extraction of iron from its ore (FeO) is the reduction of iron(II) oxide by carbon monoxide at 900°C:

$$FeO(s) + CO(g) \rightleftharpoons Fe(s) + CO_2(g)$$

If CO is allowed to react with an excess of FeO, calculate the mole fractions of CO and CO_2 at equilibrium. State any assumptions.

18.102 In the Mond process for the purification of nickel, carbon monoxide is combined with heated nickel to produce $Ni(CO)_4$, which is a gas and can therefore be separated from solid impurities:

$$Ni(s) + 4CO(g) \rightleftharpoons Ni(CO)_4(g)$$

Given that the standard free energies of formation of $CO(g)$ and $Ni(CO)_4(g)$ are -137.3 and -587.4 kJ/mol, respectively, calculate the equilibrium constant of the reaction at 80°C. Assume that $\Delta G_f°$ is temperature independent.

18.103 Water gas, a mixture of H_2 and CO, is a fuel made by combining steam with red-hot coke (a by-product of coal distillation):

$$H_2O(g) + C(s) \rightleftharpoons CO(g) + H_2(g)$$

From the data in Appendix 2, estimate the temperature at which the reaction begins to favor the formation of products.

18.104 (a) Over the years, there have been numerous claims about "perpetual motion machines," machines that will produce useful work with no input of energy. Explain why the first law of thermodynamics prohibits the possibility of such a machine existing. (b) Another kind of machine, sometimes called a "perpetual motion of the second kind," operates as follows. Suppose an ocean liner sails by scooping up water from the ocean and then extracting heat from the water, converting the heat to electric power to run the ship, and dumping the water back into the ocean. This process does not violate the first law of thermodynamics, for no energy is created—energy from the ocean is just converted to electric energy. Show that the second law of thermodynamics prohibits the existence of such a machine.

Biological Problems

18.105 The enthalpy change in the denaturation of a certain protein is 125 kJ/mol. If the entropy change is 397 J/K · mol, calculate the minimum temperature at which the protein would denature spontaneously.

18.106 The concentration of glucose inside a cell is 0.12 mM and that outside a cell is 12.3 mM. Calculate the Gibbs free-energy change for the transport of 3 moles of glucose into the cell at 37°C.

18.107 Active transport is the process in which a substance is transferred from a region of lower concentration to one of higher concentration. This is a nonspontaneous process and must be coupled to a spontaneous process, such as the hydrolysis of ATP. The concentrations of K^+ ions in the blood plasma and in nerve cells are 15 mM and 400 mM, respectively (1 mM = 1×10^{-3} M). Use Equation 18.13 to calculate ΔG for the process at the physiological temperature of 37°C:

$$K^+(15\ mM) \longrightarrow K^+(400\ mM)$$

In this calculation, the $\Delta G°$ term can be set to zero. What is the justification for this step?

18.108 When a native protein in solution is heated to a high enough temperature, its polypeptide chain will unfold to become the denatured protein. The temperature at which a large portion of the protein unfolds is called the melting temperature. The melting temperature of a certain protein is found to be 46°C, and the enthalpy of denaturation is 382 kJ/mol. Estimate the entropy of denaturation, assuming that the denaturation is a two-state process; that is, native protein $\longrightarrow$ denatured protein. The single polypeptide protein chain has 122 amino acids. Calculate the entropy of denaturation per amino acid. Comment on your result.

18.109 The pH of gastric juice is about 1.00 and that of blood plasma is 7.40. Calculate the Gibbs free energy required to secrete a mole of H^+ ions from blood plasma to the stomach at 37°C.

18.110 The following reaction represents the removal of ozone in the stratosphere:

$$2O_3(g) \rightleftharpoons 3O_2(g)$$

Calculate the equilibrium constant (K_P) for this reaction. In view of the magnitude of the equilibrium constant, explain why this reaction is not considered a major cause of ozone depletion in the absence of human-made pollutants such as the nitrogen oxides and CFCs. Assume the temperature of the stratosphere is -30°C and $\Delta G_f°$ is temperature independent.

18.111 Carbon monoxide (CO) and nitric oxide (NO) are polluting gases contained in automobile exhaust. Under suitable conditions, these gases can be made to react to form nitrogen (N_2) and the less harmful carbon dioxide (CO_2). (a) Write an equation for this reaction. (b) Identify the oxidizing and reducing agents. (c) Calculate the K_P for the reaction at 25°C. (d) Under normal atmospheric conditions, the partial pressures are $P_{N_2} = 0.80$ atm, $P_{CO_2} = 3.0 \times 10^{-4}$ atm, $P_{CO} = 5.0 \times 10^{-5}$ atm, and $P_{NO} = 5.0 \times 10^{-7}$ atm. Calculate Q_P, and predict the direction toward which the reaction will proceed.
(e) Will raising the temperature favor the formation of N_2 and CO_2?

Multiconcept Problems

18.112 The K_{sp} of AgCl is given in Table 17.4. What is its value at 60°C? [*Hint:* You need the result of Problem 18.66(a) and the data in Appendix 2 to calculate $\Delta H°$.]

18.113 Consider two carboxylic acids (acids that contain the −COOH group): CH_3COOH (acetic acid, $K_a = 1.8 \times 10^{-5}$) and $CH_2ClCOOH$ (chloroacetic acid, $K_a = 1.4 \times 10^{-3}$). (a) Calculate $\Delta G°$ for the ionization of these acids at 25°C. (b) From the equation $\Delta G° = \Delta H° - T\Delta S°$, we see that the contributions to the $\Delta G°$ term are an enthalpy term ($\Delta H°$) and a temperature times entropy term ($T\Delta S°$). These contributions are listed here for the two acids:

	$\Delta H°$ (kJ/mol)	$T\Delta S°$ (kJ/mol)
CH_3COOH	−0.57	−27.6
$CH_2ClCOOH$	−4.7	−21.1

Which is the dominant term in determining the value of $\Delta G°$ (and hence K_a of the acid)? (c) What processes contribute to $\Delta H°$? (Consider the ionization of the acids as a Brønsted acid-base reaction.) (d) Explain why the $T\Delta S°$ term is more negative for CH_3COOH.

18.114 Many hydrocarbons exist as structural isomers, which are compounds that have the same molecular formula but different structures. For example, both butane and isobutane have the same molecular formula of C_4H_{10}. Calculate the mole percent of these molecules in an equilibrium mixture at 25°C, given that the standard free energy of formation of butane is −15.9 kJ/mol and that of isobutane is −18.0 kJ/mol. Does your result support the notion that straight-chain hydrocarbons (i.e., hydrocarbons in which the C atoms are joined along a line) are less stable than branch-chain hydrocarbons?

Standardized-Exam Practice Problems

Physical and Biological Sciences

In chemistry, the standard state for a solution is 1 *M*. This means that each solute concentration expressed in molarity is divided by 1 *M*. In biological systems, however, we define the standard state for the H^+ ions to be 1×10^{-7} *M*. Consequently, the change in the standard Gibbs free energy according to these two conventions will be different for reactions involving the uptake or release of H^+ ions, depending on which convention is used. We will therefore replace $\Delta G°$ with $\Delta G°'$, where the prime denotes that it is the standard Gibbs free-energy change for a biological process.

1. What is most likely the reason for the standard state for H^+ being defined as 1×10^{-7} *M* in biological systems?

    a) It is an arbitrary convention.
    b) Physiological conditions reduce all concentrations by a factor of 10^7.
    c) Physiological conditions increase all concentrations by a factor of 10^7.
    d) Biological systems have a pH of about 7.

2. Consider the reaction:

$$A + B \longrightarrow C + xH^+$$

where x is a stoichiometric coefficient. Using the equation:

$$\Delta G = \Delta G° + RT \ln Q$$

determine the relationship between $\Delta G°'$ and $\Delta G°$, keeping in mind that ΔG is the same for a process regardless of which convention is used.

    a) $\Delta G°' = \Delta G° + RT \ln (1 \times 10^7)$
    b) $\Delta G°' = \Delta G° + RT \ln (1 \times 10^{-7})^x$
    c) $\Delta G°' = \Delta G° + e^{-7/RT}$
    d) $\Delta G°' = \Delta G° + RT \times e^{-7}$

3. Determine the relationship between $\Delta G°'$ and $\Delta G°$ for the reverse process:

$$C + xH^+ \longrightarrow A + B$$

    a) $\Delta G°' = \Delta G° + RT \ln (1 \times 10^7)^x$
    b) $\Delta G°' = \Delta G° + RT \ln (1 \times 10^{-7})$
    c) $\Delta G°' = \Delta G° + e^{7/RT}$
    d) $\Delta G°' = \Delta G° + RT \times e^7$

4. NAD^+ and NADH are the oxidized and reduced forms of nicotinamide adenine dinucleotide, two key compounds in the metabolic pathways. For the oxidation of NADH:

$$NADH + H^+ \longrightarrow NAD^+ + H_2$$

$\Delta G°$ is −21.8 kJ/mol at 298 K. Calculate $\Delta G°'$ for this process, and calculate ΔG using either the chemical or the biological convention when [NADH] = 1.5×10^{-2} *M*, [H^+] = 3.0×10^{-5} *M*, [NAD] = 4.6×10^{-3} *M*, and P_{H_2} = 0.010 atm.

    a) 18.1 kJ/mol, 50.3 kJ/mol
    b) 21.8 kJ/mol, 29.6 kJ/mol
    c) 18.1 kJ/mol, 10.3 kJ/mol
    d) 21.8 kJ/mol, 10.3 kJ/mol

Answers to In-Chapter Materials

Answers to Practice Problems

18.1A 0.34 J/K. **18.1B** $\frac{1}{5}$. **18.2A** (a) 173.6 J/K · mol, (b) −139.8 J/K · mol, (c) 215.3 J/K · mol. **18.2B** (a) $S°[K(l)]$ = 71.5 J/K · mol, (b) $S°[S_2Cl_2(g)]$ = 331.5 J/K · mol, (c) $S°[MgF_2(s)]$ = 57.24 J/K · mol. **18.3A** (a) Negative, (b) negative, (c) positive. **18.3B** The sign of $\Delta H°$ for both dissolution processes is negative. Something must favor spontaneity; if not entropy change, then enthalpy change. Because these processes both involve decreases in the system's entropy, they must be exothermic, or they could not be spontaneous. **18.4A** (a) ΔS_{univ} = −27.2 J/K · mol, nonspontaneous, (b) ΔS_{univ} = −28.1 J/K · mol, nonspontaneous, (c) ΔS_{univ} = 0 J/K · mol, equilibrium. **18.4B** (a) ΔS_{univ} = 5.2 J/K · mol, spontaneous, (b) 346°C, (c) 58°C. **18.5A** 3728°C. **18.5B** 108 J/K · mol. **18.6A** (a) −106 kJ/mol, (b) −2935 kJ/mol. **18.6B** (a) $\Delta G_f°[Li_2O(s)]$ = −561.2 kJ/mol, (b) $\Delta G_f°[NaI(s)]$ = −286.1 kJ/mol. **18.7A** ΔS_{fus} = 16 J/K · mol, ΔS_{vap} = 72 J/K · mol. **18.7B** (a) $\Delta H°_{fus}$ = 2.41 kJ/mol, $\Delta S°_{fus}$ = 6.51 J/K · mol, $T_{melting}$ = 97°C. (b) $\Delta H°_{vap}$ = 105.3 kJ/mol, $\Delta S°_{vap}$ = 96.1 J/K · mol, $T_{boiling}$ = 823°C. **18.8A** 1.3 kJ/mol, reverse reaction is spontaneous. **18.8B** 2.5 atm. **18.9A** 2×10^{57}. **18.9B** −16.8 kJ/mol. **18.10A** 32.9 kJ/mol. **18.10B** −454.4 kJ/mol.

Answers to Checkpoints

18.3.1 a, e. **18.3.2** b. **18.3.3** c. **18.4.1** b. **18.4.2** e. **18.4.3** b. **18.4.4** d. **18.5.1** d. **18.5.2** d. **18.5.3** d. **18.5.4** b. **18.6.1** b. **18.6.2** d. **18.6.3** b. **18.6.4** e.

Structure and Bonding

1

L-Dopa, also called levodopa, was first isolated from seeds of the broad bean plant *Vicia faba* in 1913. Since 1967 it has been the drug of choice for the treatment of Parkinson's disease, a debilitating illness that results from the degeneration of neurons that produce the neurotransmitter dopamine in the brain. L-Dopa is an oral medication that is transported to the brain by the bloodstream, where it is converted to dopamine. Since L-dopa must be taken in large doses with some serious side effects, today it is often given with other drugs that lessen its negative impact on an individual. In Chapter 1, we learn about the structure, bonding, and properties of organic molecules like L-dopa.

Before examining organic molecules in detail, we must review some important features about structure and bonding learned in previous chemistry courses. We will discuss these concepts primarily from an organic chemist's perspective, and spend time on only the particulars needed to understand organic compounds.

Important topics in Chapter 1 include drawing Lewis structures, predicting the shape of molecules, determining what orbitals are used to form bonds, and how electronegativity affects bond polarity. Equally important is Section 1.8 on drawing organic molecules, both shorthand methods routinely used for simple and complex compounds, as well as three-dimensional representations that allow us to more clearly visualize them.

1.1 The Periodic Table

All matter is composed of the same building blocks called **atoms.** There are two main components of an atom.

- The **nucleus** contains positively charged **protons** and uncharged **neutrons.** Most of the mass of the atom is contained in the nucleus.
- The **electron cloud** is composed of negatively charged **electrons.** The electron cloud comprises most of the volume of the atom.

Schematic of an atom

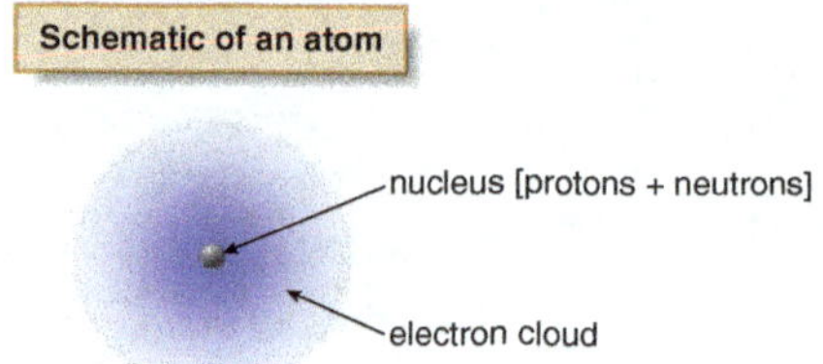

The charge on a proton is equal in magnitude but opposite in sign to the charge on an electron. In a neutral atom, the **number of protons in the nucleus equals the number of electrons.** This quantity, called the **atomic number,** is unique to a particular element. For example, every neutral carbon atom has an atomic number of six, meaning it has six protons in its nucleus and six electrons surrounding the nucleus.

In addition to neutral atoms, we will also encounter **charged ions.**

- A *cation* is positively charged and has fewer electrons than its neutral form.
- An *anion* is negatively charged and has more electrons than its neutral form.

The number of neutrons in the nucleus of a particular element can vary. **Isotopes** are two atoms of the same element having a different number of neutrons. The **mass number** of an atom is the total number of protons and neutrons in the nucleus. Isotopes have different mass numbers.

Isotopes of carbon and hydrogen are sometimes used in organic chemistry, as we will see in Chapter 14.

- The most common isotope of hydrogen has one proton and no neutrons in the nucleus, but 0.02% of hydrogen atoms have one proton and one neutron. This isotope of hydrogen is called **deuterium,** and is sometimes symbolized by the letter **D.**
- Most carbon atoms have six protons and six neutrons in the nucleus, but 1.1% have six protons and seven neutrons.

The **atomic weight** is the weighted average of the mass of all isotopes of a particular element, reported in atomic mass units (amu).

Figure 1.1
A comparison of two isotopes of the element carbon

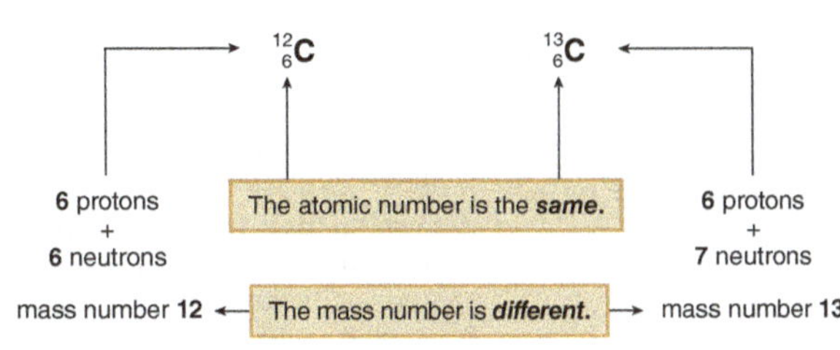

Each atom is identified by a one- or two-letter abbreviation that is the characteristic symbol for that element. Carbon is identified by the single letter **C.** Sometimes the atomic number is indicated as a subscript to the left of the element symbol, and the mass number is indicated as a superscript, as shown in Figure 1.1.

A **row** in the periodic table is also called a ***period,*** and a **column** is also called a ***group.*** A periodic table is located on the inside front cover for your reference.

Long ago it was realized that groups of elements have similar properties, and that these atoms could be arranged in a schematic way called the **periodic table.** There are more than 100 known elements, arranged in the periodic table in order of increasing atomic number. The periodic table is composed of rows and columns.

- Elements in the same row are similar in *size.*
- Elements in the same column have similar *electronic and chemical properties.*

Each column in the periodic table is identified by a **group number,** an Arabic (1 to 8) or Roman (I to VIII) numeral followed by the letter A or B. For example, carbon is located in group **4A** in the periodic table in this text.

Carbon's entry in the periodic table:

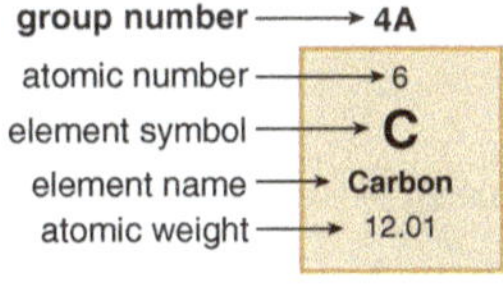

Although more than 100 elements exist, most are not common in organic compounds. Figure 1.2 contains a truncated periodic table, indicating the handful of elements that are routinely seen in this text. **Most of these elements are located in the first and second rows of the periodic table.**

Across each row of the periodic table, electrons are added to a particular shell of orbitals around the nucleus. The shells are numbered 1, 2, 3, and so on. Adding electrons to the first shell forms the first row. Adding electrons to the second shell forms the second row. **Electrons are first added to the shells closest to the nucleus.** These electrons are held most tightly.

Each shell contains a certain number of subshells called **orbitals.** An orbital is a region of space that is high in electron density. There are four different kinds of orbitals, called ***s, p, d,*** and ***f.*** The first shell has only one orbital, called an *s* orbital. The second shell has two kinds of orbitals, *s* and *p,* and so on. Each type of orbital occupies a certain space and has a particular shape.

Figure 1.2
A periodic table of the common elements seen in organic chemistry

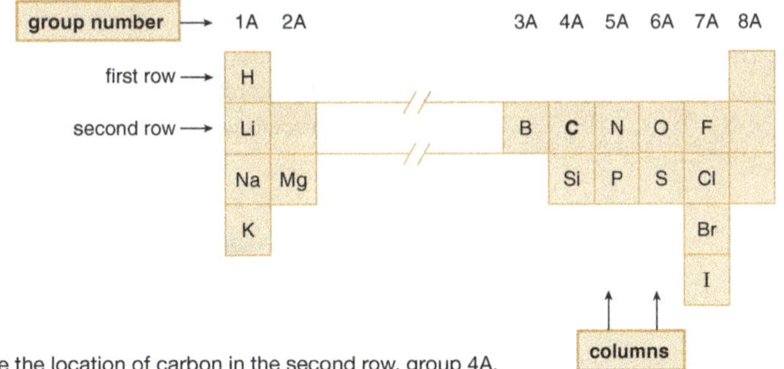

- Note the location of carbon in the second row, group 4A.

For the first- and second-row elements, we must deal with only ***s* orbitals** and ***p* orbitals.**

- An ***s* orbital** has a **sphere of electron density.** It is ***lower in energy*** than other orbitals of the same shell, because electrons are kept close to the positively charged nucleus. An *s* orbital is filled with electrons before a *p* orbital in the same shell.
- A ***p* orbital** has a **dumbbell shape.** It contains a **node of electron density** at the nucleus. A node means there is no electron density in this region. A *p* orbital is ***higher in energy*** than an *s* orbital (in the same shell) because its electron density is farther away from the nucleus. A *p* orbital is filled with electrons only after an *s* orbital of the same shell is full.

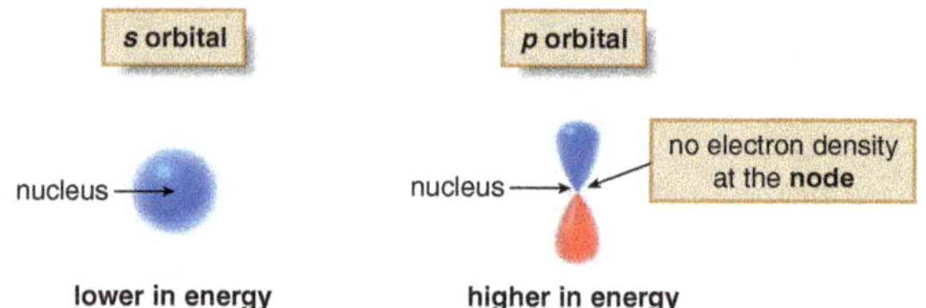

Let's now look at the elements in the first and second rows of the periodic table.

1.1A The First Row

The first row of the periodic table is formed by adding electrons to the first shell of orbitals around the nucleus. There is only one orbital in the first shell, called the **1*s* orbital.**

- **Remember: Each orbital can have a maximum of two electrons.**

As a result, there are **two elements in the first row,** one having one electron added to the 1*s* orbital, and one having two. The element hydrogen (H) has what is called a $1s^1$ configuration with one electron in the 1*s* orbital, and helium (He) has a $1s^2$ configuration with two electrons in the 1*s* orbital.

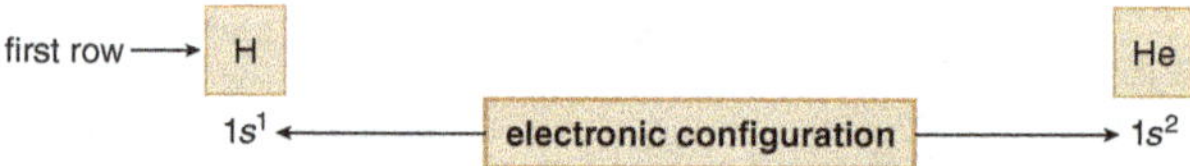

1.1B The Second Row

Every element in the second row has a filled first shell of electrons. Thus, all second-row elements have a $1s^2$ configuration. These electrons in the inner shell of orbitals are called **core electrons,** and are not usually involved in the chemistry of a particular element.

Each element in the second row of the periodic table has four orbitals available to accept additional electrons:

- **one 2*s* orbital,** the *s* orbital in the second shell
- **three 2*p* orbitals,** all dumbbell-shaped and perpendicular to each other along the *x*, *y*, and *z* axes

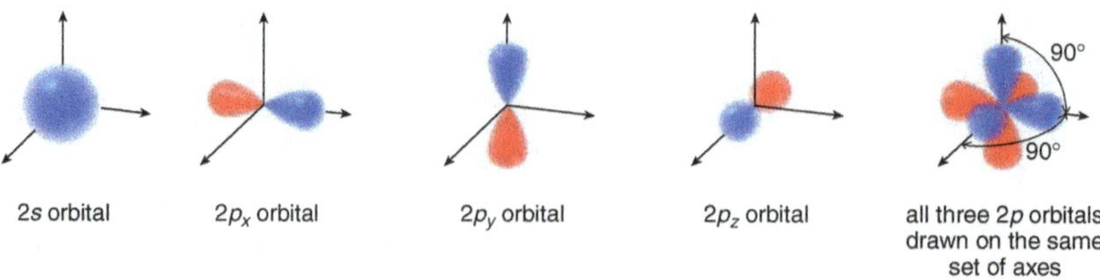

Because each of the four orbitals in the second shell can hold two electrons, there is a **maximum capacity of *eight* electrons** for elements in the second row. The second row of the periodic table consists of eight elements, obtained by adding electrons to the 2*s* and three 2*p* orbitals.

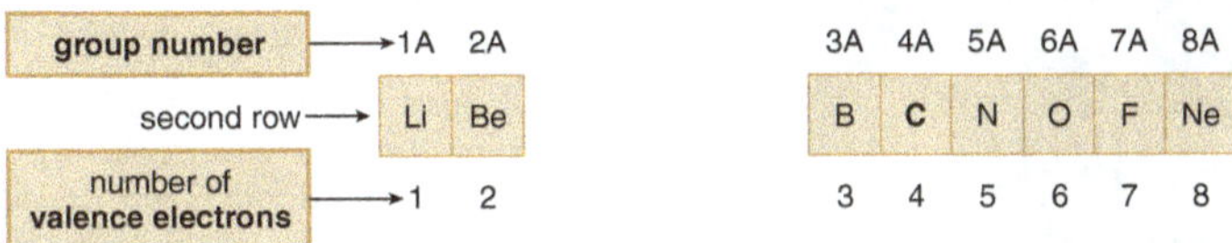

The outermost electrons are called **valence electrons.** The valence electrons are more loosely held than the electrons closer to the nucleus, and as such, they participate in chemical reactions. **The group number of a second-row element reveals its number of valence electrons.** For example, carbon in group 4A has four valence electrons, and oxygen in group 6A has six.

Problem 1.1 While the most common isotope of nitrogen has a mass number of 14 (nitrogen-14), a radioactive isotope of nitrogen has a mass number of 13 (nitrogen-13). Nitrogen-13 is used in PET (positron emission tomography) scans by physicians to monitor brain activity and diagnose dementia. For each isotope, give the following information: (a) the number of protons; (b) the number of neutrons; (c) the number of electrons in the neutral atom; and (d) the group number.

1.2 Bonding

Until now our discussion has centered on individual atoms, but it is more common in nature to find two or more atoms joined together.

- *Bonding* is the joining of two atoms in a stable arrangement.

Bonding may occur between atoms of the same or different elements. Bonding is a favorable process because it always leads to ***lowered energy and increased stability.*** Joining two or more elements forms **compounds.** Although only about 100 elements exist, more than 50 million compounds are known. Examples of compounds include hydrogen gas (H_2), formed by joining two hydrogen atoms, and methane (CH_4), the simplest organic compound, formed by joining a carbon atom with four hydrogen atoms.

One general rule governs the bonding process.

- Through bonding, atoms attain a complete outer shell of valence electrons.

Alternatively, because the noble gases in group 8A of the periodic table are especially stable as atoms having a filled shell of valence electrons, the general rule can be restated.

- Through bonding, atoms attain a stable noble gas configuration of electrons.

What does this mean for first- and second-row elements? **A first-row element like hydrogen can accommodate *two electrons* around it.** This would make it like the noble gas helium at the end of the same row. **A second-row element is most stable with *eight valence electrons* around it** like neon. Elements that behave in this manner are said to follow the **octet rule.**

There are two different kinds of bonding: **ionic bonding** and **covalent bonding.**

- *Ionic bonds* result from the *transfer* of electrons from one element to another.
- *Covalent bonds* result from the *sharing* of electrons between two nuclei.

The type of bonding is determined by the location of an element in the periodic table. An ionic bond generally occurs when elements on the **far left** side of the periodic table combine with elements on the **far right** side, ignoring the noble gases, which form bonds only rarely. **The resulting ions**

Atoms readily form ionic bonds when they can attain a noble gas configuration by gaining or losing just one or two electrons. NaCl and KI are ionic compounds.

are held together by extremely strong electrostatic interactions. A positively charged cation formed from the element on the left side attracts a negatively charged anion formed from the element on the right side. The resulting **salts** are seen in many of the inorganic compounds with which you are familiar. Sodium chloride (NaCl) is common table salt, and potassium iodide (KI) is an essential nutrient added to make iodized salt.

Ionic compounds form extended crystal lattices that maximize the positive and negative electrostatic interactions. In NaCl, each positively charged Na^+ ion is surrounded by six negatively charged Cl^- ions, and each Cl^- ion is surrounded by six Na^+ ions.

NaCl—An ionic crystalline lattice

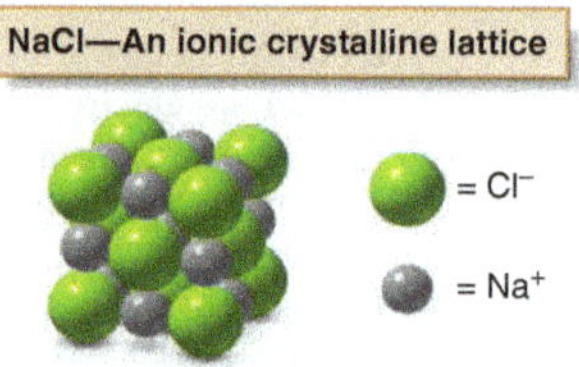

Lithium fluoride, LiF, is an example of an ionic compound.

- The element **lithium,** located in group 1A of the periodic table, has just one valence electron in its second shell. If this electron is lost, lithium forms the cation Li^+ having no electrons in the second shell. However, it will have a stable electronic arrangement with two electrons in the first shell like helium.
- The element **fluorine,** located in group 7A of the periodic table, has seven valence electrons. By gaining one it forms the anion F^-, which has a filled valence shell (an octet of electrons), like neon.
- Thus, lithium fluoride is a stable ionic compound.

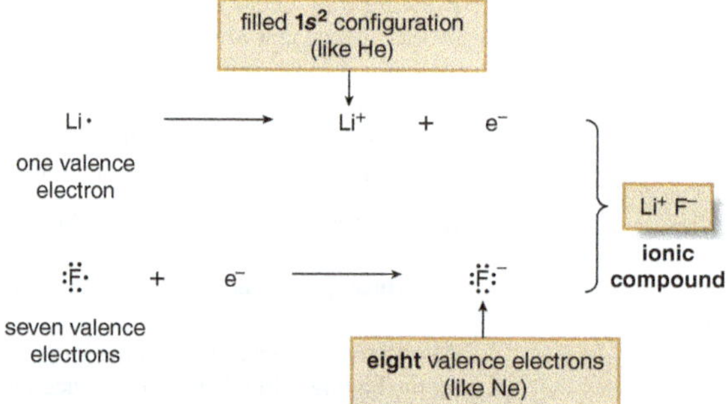

- **The transfer of electrons forms stable salts composed of cations and anions.**

A **compound** may have either ionic or covalent bonds. A **molecule** has only covalent bonds.

The second type of bonding, **covalent bonding,** occurs with elements like carbon in the middle of the periodic table, which would otherwise have to gain or lose several electrons to form an ion with a complete valence shell. **A covalent bond is a two-electron bond,** and a compound with covalent bonds is called a **molecule.** Covalent bonds also form between two elements from the same side of the table, such as two hydrogen atoms or two chlorine atoms. H_2, Cl_2, and CH_4 are all examples of covalent molecules.

Problem 1.2 Label each bond in the following compounds as ionic or covalent.

a. F_2 b. LiBr c. CH_3CH_3 d. $NaNH_2$

How many covalent bonds will a particular atom typically form? As you might expect, it depends on the location of the atom in the periodic table. In the first row, **hydrogen forms one covalent bond** using its one valence electron. When two hydrogen atoms are joined in a bond, each has a filled valence shell of two electrons.

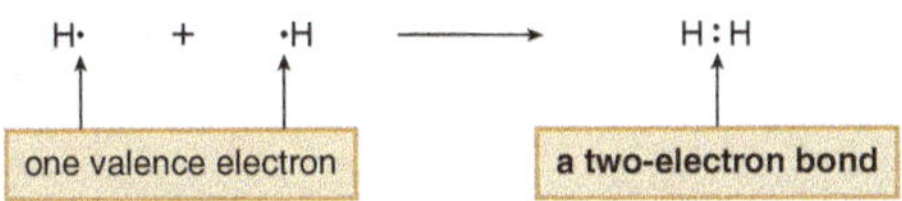

Second-row elements can have no more than eight valence electrons around them. For neutral molecules, two consequences result.

- **Atoms with one, two, three, or four valence electrons form one, two, three, or four bonds,** respectively, in neutral molecules.
- **Atoms with five or more valence electrons** form enough bonds to give an octet. This results in the following simple equation:

predicted number of bonds = **8 – number of valence electrons**

For example, B has three valence electrons, so it forms three bonds, as in BF_3. N has five valence electrons, so it also forms three bonds (8 – 5 = 3 bonds), as in NH_3.

Nonbonded pair of electrons = unshared pair of electrons = lone pair

These guidelines are used in Figure 1.3 to summarize the usual number of bonds formed by the common atoms in organic compounds. Notice that when second-row elements form fewer than four bonds their octets consist of both **bonding (shared) electrons** and **nonbonding (unshared) electrons.** Unshared electrons are also called **lone pairs.**

Problem 1.3 How many covalent bonds are predicted for each atom?

a. O b. Al c. Br d. Si

Figure 1.3

Summary: The usual number of bonds of common neutral atoms

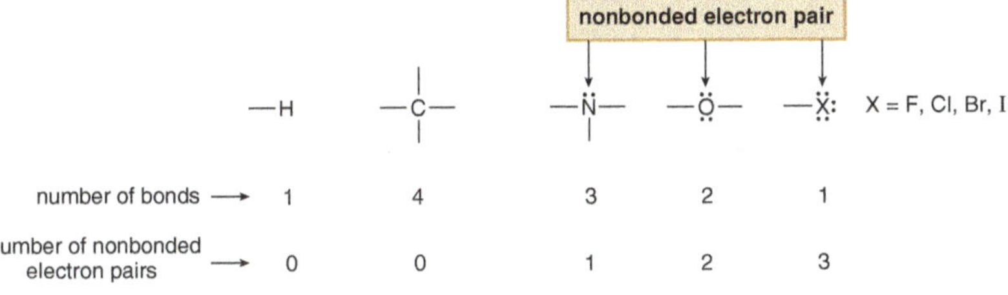

1.3 Lewis Structures

Lewis structures **are electron dot representations for molecules.** There are three general rules for drawing Lewis structures.

1. Draw only the valence electrons.
2. Give every second-row element no more than eight electrons.
3. Give each hydrogen two electrons.

The letter **X** is often used to represent one of the halogens in group 7A: F, Cl, Br, or I.

To draw a Lewis structure for a diatomic molecule like **HF,** recall that hydrogen has one valence electron and fluorine has seven. H and F each donate one electron to form a two-electron bond. The resulting molecule gives both H and F a filled valence shell.

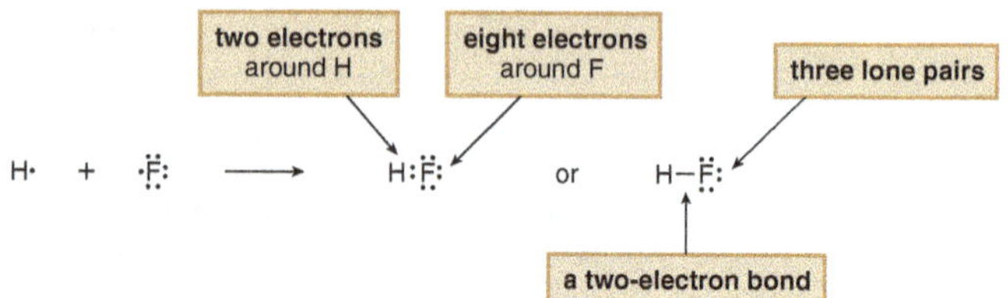

In a Lewis structure, **a *solid line* indicates a two-electron covalent bond.**

1.3A A Procedure for Drawing Lewis Structures

Drawing a Lewis structure for larger molecules is easier if you follow a stepwise procedure.

How To Draw a Lewis Structure

Step [1] Arrange atoms next to each other that you think are bonded together.

- Always place hydrogen atoms and halogen atoms on the periphery because H and X (X = F, Cl, Br, and I) form only one bond each.

- As a first approximation, place no more atoms around an atom than the number of bonds it usually forms.

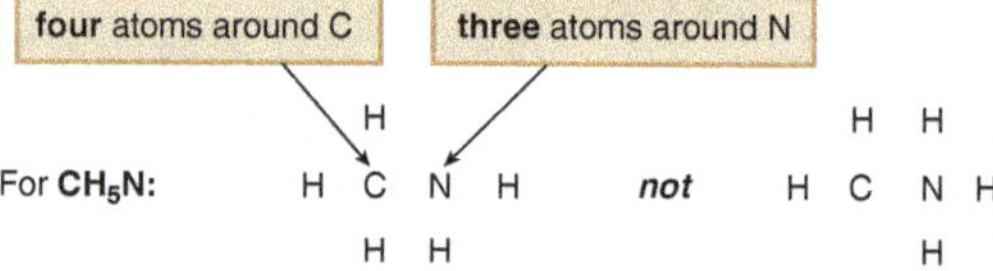

- In truth, the proper arrangement of atoms may not be obvious, or more than one arrangement may be possible (Section 1.4). Even in many simple molecules, the connectivity between atoms must be determined experimentally.

Step [2] Count the electrons.

- Count the number of valence electrons from all atoms.
- Add one electron for each negative charge.
- Subtract one electron for each positive charge.
- This sum gives the total number of electrons that must be used in drawing the Lewis structure.

Step [3] Arrange the electrons around the atoms.

- Place a bond between every two atoms, giving two electrons to each H and no more than eight to any second-row atom.
- Use all remaining electrons to fill octets with lone pairs.
- If all valence electrons are used and an atom does not have an octet, form multiple bonds, as shown in Sample Problem 1.2.

Step [4] Assign formal charges to all atoms.

- Formal charges are discussed in Section 1.3C.

Sample Problem 1.1 illustrates how to draw a Lewis structure of a simple organic molecule.

Sample Problem 1.1 Draw a Lewis structure for methanol, a compound with molecular formula CH_4O.

Solution

Step [1] Arrange the atoms.

```
    H
H   C   O   H
    H
```

- Place the second-row elements, C and O, in the middle.
- Place three H's around C to surround C by four atoms.
- Place one H next to O to surround O by two atoms.

Step [2] Count the electrons.

$$1\ C \times 4\ e^- = 4\ e^-$$
$$1\ O \times 6\ e^- = 6\ e^-$$
$$4\ H \times 1\ e^- = 4\ e^-$$
$$\mathbf{14\ e^-\ total}$$

Step [3] Add the bonds and lone pairs.

- Add five two-electron bonds to form the C–H, C–O, and O–H bonds, using 10 of the 14 electrons.
- Place two lone pairs on the O atom to use the remaining four electrons and give the O atom an octet.

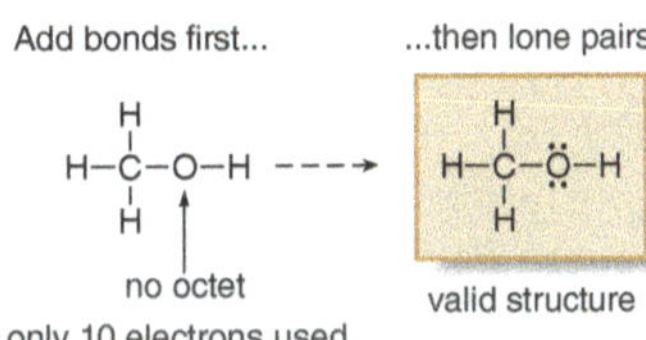

To check if a Lewis structure is valid, we must answer YES to three questions.

- Have all the electrons been used?
- Is each H surrounded by two electrons?
- Is each second-row element surrounded by no more than eight electrons?

Since the answer to all three questions is YES, we have drawn a valid Lewis structure for CH_4O.

Problem 1.4 Draw a valid Lewis structure for each species:

a. CH_3CH_3 b. CH_5N c. CH_3^- d. CH_3Cl

1.3B Multiple Bonds

Sample Problem 1.2 illustrates two examples of when multiple bonds are needed in Lewis structures.

Sample Problem 1.2 Draw a Lewis structure for each compound. Assume the atoms are arranged as follows:

a. ethylene, C_2H_4

```
H C C H
  H H
```

b. acetylene, C_2H_2

```
H C C H
```

Solution

a. **Ethylene, C_2H_4:** Follow Steps [1] to [3] to draw a Lewis structure. After placing five bonds between the atoms and adding the two remaining electrons as a lone pair, one C still has no octet.

Step [2] Count the electrons.

$$2\ C \times 4\ e^- = 8\ e^-$$
$$4\ H \times 1\ e^- = 4\ e^-$$
$$\mathbf{12\ e^-\ total}$$

Step [3] Add the bonds and lone pairs.

Add bonds first... ...then lone pairs.

```
H–C–C–H  ------>  H–C–C̈–H
  H H               H H
```

no octet

To give both C's an octet, **change *one* lone pair into *one* bonding pair of electrons between the two C's, forming a double bond.**

H–C–C̈–H (with H below each C) ---Move a lone pair.---> H–C=C–H (with H below each C) ← **Each C now has four bonds.**

ethylene
a valid Lewis structure

This uses all 12 electrons, each C has an octet, and each H has two electrons. The Lewis structure is valid. **Ethylene contains a carbon–carbon double bond.**

b. **Acetylene, C_2H_2:** A similar phenomenon occurs with acetylene. Placing the 10 valence electrons gives a Lewis structure in which one or both of the C's lack an octet.

Step [2] Count the electrons.

$2\ C \times 4\ e^- = 8\ e^-$
$2\ H \times 1\ e^- = 2\ e^-$
10 e⁻ total

Step [3] Add the bonds and lone pairs.

Add bonds first... ...then lone pairs.

H–C–C–H -----> H–C–C̈–H (no octet) or H–C̈–C̈–H (no octets)

Carbon always forms four bonds in stable organic molecules. Carbon forms single, double, and triple bonds to itself and other elements.

For a second-row element in a stable molecule:

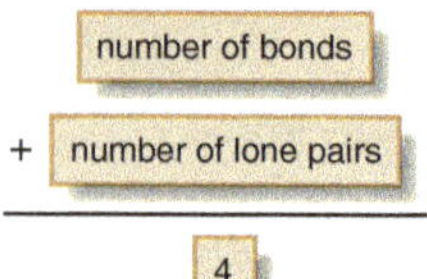

In this case, **change *two* lone pairs into *two* bonding pairs of electrons, forming a triple bond.**

H–C–C̈–H ---> H–C=C̈–H (no octet) ---> H–C≡C–H ← **Each C now has four bonds.**

acetylene
a valid Lewis structure

This uses all 10 electrons, each C has an octet, and each H has two electrons. The Lewis structure is valid. **Acetylene contains a carbon–carbon triple bond.**

- **After placing all electrons in bonds and lone pairs, use a lone pair to form a multiple bond if an atom does not have an octet.**

You must change *one* lone pair into *one* new bond for each *two* electrons needed to complete an octet. In acetylene, for example, four electrons were needed to complete an octet, so two lone pairs were used to form two new bonds, forming a triple bond.

Problem 1.5 Draw an acceptable Lewis structure for each compound, assuming the atoms are connected as arranged. Hydrogen cyanide (HCN) is a poison, formaldehyde (H_2CO) is a preservative, and glycolic acid ($HOCH_2CO_2H$) is used to make dissolving sutures.

a. HCN H C N

b. H_2CO H C O (H below C)

c. $HOCH_2CO_2H$ H O C C O H (H above and below the first C; O above the second C)

1.3C Formal Charge

To manage electron bookkeeping in a Lewis structure, organic chemists use **formal charge.**

- *Formal charge* **is the charge assigned to individual atoms in a Lewis structure.**

By calculating formal charge, we determine how the number of electrons around a particular atom compares to its number of valence electrons. Formal charge is calculated as follows:

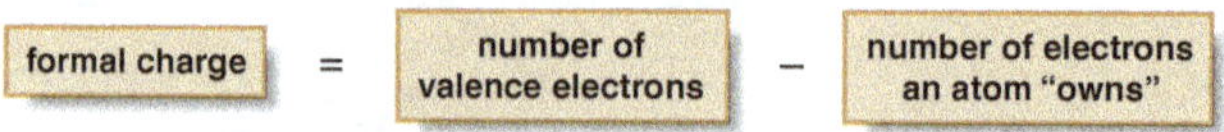

The number of electrons "owned" by an atom is determined by its number of bonds and lone pairs.

- **An atom "owns" *all* of its unshared electrons and *half* of its shared electrons.**

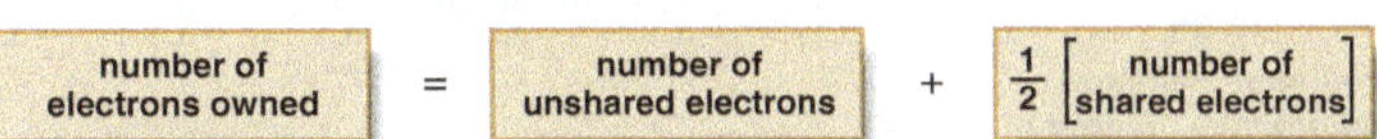

The number of electrons "owned" by different carbon atoms is indicated in the following examples:

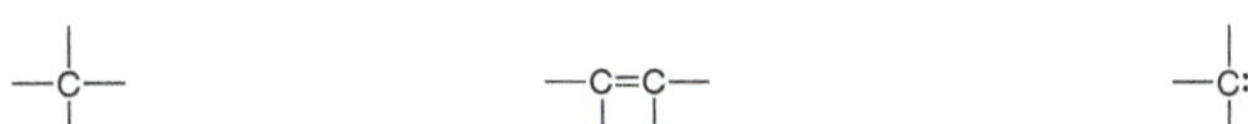

• C shares eight electrons. • C "owns" **four** electrons.	• Each C shares eight electrons. • Each C "owns" **four** electrons.	• C shares six electrons. • C has two unshared electrons. • C "owns" **five** electrons.

Sample Problem 1.3 illustrates how formal charge is calculated on the atoms of a polyatomic ion. **The sum of the formal charges on the individual atoms equals the net charge on the molecule or ion.**

Sample Problem 1.3 Determine the formal charge on each atom in the ion H_3O^+.

$$\left[\mathrm{H{-}\ddot{O}{-}H} \atop \mathrm{H}\right]^+$$

Solution

For each atom, two steps are needed:

Step [1] Determine the number of electrons an atom "owns."

Step [2] Subtract this sum from its number of valence electrons.

O atom	H atoms
[1] number of electrons "owned" by O $2 + \frac{1}{2}(6) = 5$	[1] number of electrons "owned" by each H $0 + \frac{1}{2}(2) = 1$
[2] formal charge on O $6 - 5 =$ **+1**	[2] formal charge on each H $1 - 1 =$ **0**

The formal charge on each H is 0. The formal charge on oxygen is +1. The overall charge on the ion is the sum of all of the formal charges; 0 + 0 + 0 + 1 = **+1.**

Problem 1.6 Calculate the formal charge on each second-row atom:

a. $[NH_4]^+$ (H–N–H with H above and below N)   b. $CH_3{-}N{\equiv}C{:}$ c. $:\ddot{O}{=}\ddot{O}{-}\ddot{\underset{..}{O}}:$

Problem 1.7 Draw a Lewis structure for each ion:

a. CH_3O^- b. HC_2^- c. $(CH_3NH_3)^+$ d. $(CH_3NH)^-$

Sometimes it is easier to count bonds, rather than shared electrons when determining formal charge.

$\frac{1}{2}$[number of shared electrons] = number of bonds

A shortcut method to determine the number of bonds in a Lewis structure is given in the *Student Study Guide/ Solutions Manual* (page 1–4).

When you first add formal charges to Lewis structures, use the procedure in Sample Problem 1.3. With practice, you will notice that certain bonding patterns always result in the same formal charge. For example, any N atom with four bonds (and, thus no lone pairs) has a +1 formal charge. Table 1.1 lists the bonding patterns and resulting formal charges for carbon, nitrogen, and oxygen.

Table 1.1 Formal Charge Observed with Common Bonding Patterns for C, N, and O

Atom	Number of valence electrons	Formal charge +1	Formal charge 0	Formal charge −1
C	4	 $-\overset{+}{C}-$ (three bonds)	 $-\overset{\vert}{\underset{\vert}{C}}-$ (four bonds)	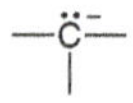 $-\ddot{C}^{-}-$ (three bonds, one lone pair)
N	5	 $-\overset{\vert}{\underset{\vert}{N}}{}^{+}-$ (four bonds)	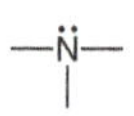 $-\ddot{N}-$ (three bonds, one lone pair)	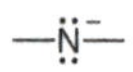 $-\ddot{\underset{..}{N}}{}^{-}-$ (two bonds, two lone pairs)
O	6	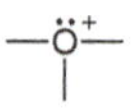 $-\ddot{O}^{+}-$ (three bonds, one lone pair)	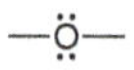 $-\ddot{\underset{..}{O}}-$ (two bonds, two lone pairs)	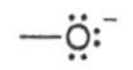 $-\ddot{\underset{..}{O}}{:}^{-}$ (one bond, three lone pairs)

1.4 Isomers

In drawing a Lewis structure for a molecule with several atoms, sometimes more than one arrangement of atoms is possible for a given molecular formula. For example, there are two acceptable arrangements of atoms for the molecular formula C_2H_6O.

```
  H H                          H   H
  | |                          |   |
H-C-C-Ö-H  <-- isomers -->  H-C-Ö-C-H
  | |   ..                     | .. |
  H H                          H   H
```

ethanol dimethyl ether

same molecular formula
C_2H_6O

Both are valid Lewis structures, and both molecules exist. One is called ethanol, and the other, dimethyl ether. These two compounds are called **isomers.**

- *Isomers* are different molecules having the same molecular formula.

Ethanol and dimethyl ether are **constitutional isomers** because they have the same molecular formula, but the ***connectivity of their atoms is different.*** For example, ethanol has one C–C bond and one O–H bond, whereas dimethyl ether has two C–O bonds. A second class of isomers, called **stereoisomers,** is introduced in Section 4.13B.

Problem 1.8 Draw Lewis structures for each molecular formula.

a. $C_2H_4Cl_2$ (two isomers) b. C_3H_8O (three isomers) c. C_3H_6 (two isomers)

1.5 Exceptions to the Octet Rule

Most of the common elements in organic compounds—**C, N, O, and the halogens**—follow the octet rule. Hydrogen is a notable exception, because it accommodates only two electrons in bonding. Additional exceptions include boron and beryllium (second-row elements in groups 3A and 2A, respectively), and elements in the third row (particularly phosphorus and sulfur).

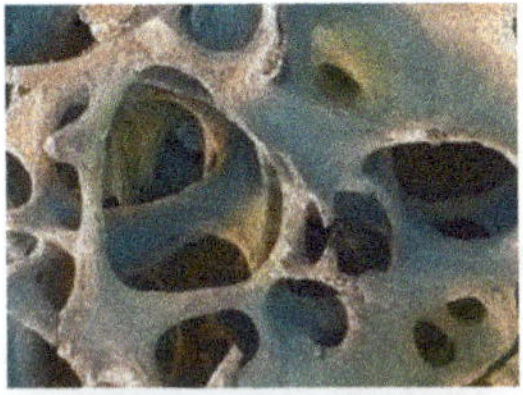

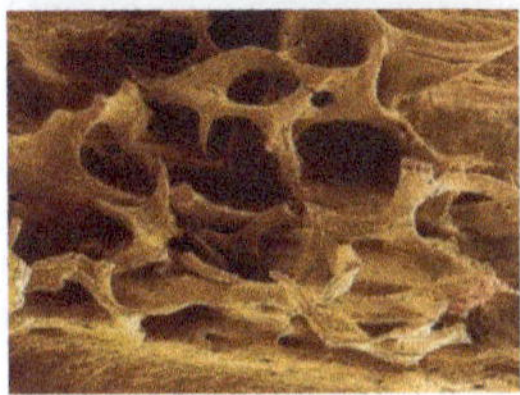

Alendronic acid, sold as a sodium salt under the trade name of **Fosamax,** is used to prevent osteoporosis in women. Osteoporosis decreases bone density, as shown by comparing normal bone (top) with brittle bone (bottom).

Elements in groups 2A and 3A of the periodic table, such as beryllium and boron, do not have enough valence electrons to form an octet in a neutral molecule. Lewis structures for BeH_2 and BF_3 show that these atoms have only four and six electrons, respectively, around the central atom. There is nothing we can do about this! There simply aren't enough electrons to form an octet.

H–Be–H

four electrons around Be

:F̤:
|
:F̤–B–F̤:

six electrons around B

Because the Be and B atoms each have less than an octet of electrons, these molecules are highly reactive.

A second exception to the octet rule occurs with some elements located in the third row and later in the periodic table. These elements have empty *d* orbitals available to accept electrons, and thus they may have ***more than eight*** **electrons** around them. For organic chemists, the two most common elements in this category are **phosphorus** and **sulfur,** which can have 10 or even 12 electrons around them. Examples of these phosphorus and sulfur compounds include the following:

10 electrons around S

CH_3–S(=O)–CH_3

dimethyl sulfoxide
(abbreviated as DMSO)

12 electrons around S

HO–S(=O)(=O)–OH

sulfuric acid

10 electrons around each P

HO–P(=O)(OH)–C(OH)(CH₂CH₂CH₂NH₂)–P(=O)(OH)–OH

alendronic acid

1.6 Resonance

Some molecules can't be adequately represented by a single Lewis structure. For example, two valid Lewis structures can be drawn for the anion $(HCONH)^-$. One structure has a negatively charged N atom and a C–O double bond; the other has a negatively charged O atom and a C–N double bond. These structures are called **resonance structures** or **resonance forms.** A **double-headed arrow** is used to separate two resonance structures.

H–C(=Ö:)–N̈⁻–H ⟷ H–C(–Ö:⁻)=N̈–H

double-headed arrow

- *Resonance structures* are two Lewis structures having the *same* placement of atoms but a *different* arrangement of electrons.

Which resonance structure is an accurate representation for $(HCONH)^-$? **The answer is *neither* of them.** The true structure is a composite of both resonance forms, and is called a **resonance hybrid.** The hybrid shows characteristics of ***both*** resonance structures.

Each resonance structure implies that electron pairs are localized in bonds or on atoms. In actuality, resonance allows certain electron pairs to be ***delocalized*** over two or more atoms, and this delocalization of electron density adds stability. **A molecule with two or more resonance structures is said to be *resonance stabilized.*** We will return to the resonance hybrid in Section 1.6C. First, however, we examine the general principles of resonance theory and learn how to interconvert two or more resonance structures.

1.6A An Introduction to Resonance Theory

Keep in mind the following basic principles of resonance theory.

- **Resonance structures are not real. An individual resonance structure does not accurately represent the structure of a molecule or ion. Only the hybrid does.**
- **Resonance structures are *not* in equilibrium with each other. There is no movement of electrons from one form to another.**
- **Resonance structures are *not* isomers. Two isomers differ in the arrangement of *both* atoms and electrons, whereas resonance structures differ *only* in the *arrangement of electrons.***

Resonance structures are different ways of drawing the same compound. Two resonance structures are *not* different compounds.

For example, ions **A** and **B** are resonance structures because the atom position is the same in both compounds, but the location of an electron pair is different. In contrast, compounds **C** and **D** are isomers since the atom placement is different; **C** has an O–H bond, and **D** has an additional C–H bond.

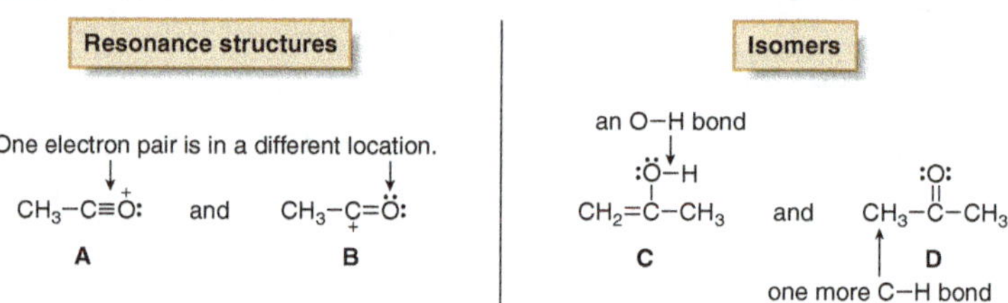

Problem 1.9 Classify each pair of compounds as isomers or resonance structures.

a. $^{-}:\ddot{N}=C=\ddot{O}:$ and $^{-}:C≡\overset{+}{N}–\ddot{O}:^{-}$ b. $H\ddot{O}–C(=O:)–\ddot{O}:^{-}$ and $H\ddot{O}–C(–\ddot{O}:^{-})=\ddot{O}:$

Problem 1.10 Considering structures **A–D,** classify each pair of compounds as isomers, resonance structures, or neither: (a) **A** and **B;** (b) **A** and **C;** (c) **A** and **D;** (d) **B** and **D.**

A: $CH_3–C(=O:)–\ddot{O}H$

B: $CH_3–C(–\ddot{O}:^{-})=\overset{+}{O}H$

C: $CH_3–CH(–\ddot{O}:^{-})–\ddot{O}H$

D: $H–C(=O:)–CH_2\ddot{O}H$

1.6B Drawing Resonance Structures

To draw resonance structures, use the three rules that follow:

Rule [1] **Two resonance structures differ in the position of multiple bonds and nonbonded electrons. The placement of atoms and single bonds always stays the same.**

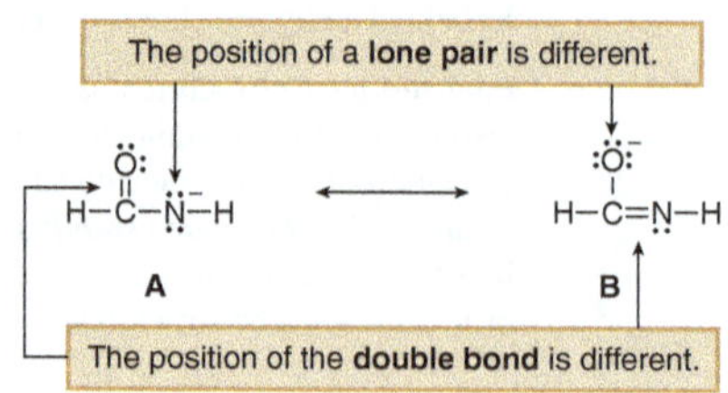

Rule [2] Two resonance structures must have the same number of unpaired electrons.

two unpaired electrons

C

- **A** and **B** have no unpaired electrons.
- **C** is *not* a resonance structure of **A** and **B**.

Rule [3] Resonance structures must be valid Lewis structures. Hydrogen must have two electrons and no second-row element can have more than eight electrons.

10 electrons around C
not a valid Lewis structure

Curved arrow notation is a convention that shows how electron position differs between the two resonance forms.

- *Curved arrow notation shows the movement of an electron pair.* The tail of the arrow always begins at an electron pair, either in a bond or lone pair. The head points to where the electron pair "moves."

A curved arrow always begins at an electron pair. It ends at an atom or a bond.

Move an electron pair to O.

A ⟷ B

Use this electron pair to form a double bond.

Resonance structures **A** and **B** differ in the location of two electron pairs, so two curved arrows are needed. To convert **A** to **B,** take the lone pair on N and form a double bond between C and N. Then, move an electron pair in the C – O double bond to form a lone pair on O. Curved arrows thus show how to reposition the electrons in converting one resonance form to another. **The electrons themselves do not actually move.** Sample Problem 1.4 illustrates the use of curved arrows to convert one resonance structure to another.

Sample Problem 1.4

Follow the curved arrows to draw a second resonance structure for each ion.

a. $CH_2{=}\overset{|}{C}H{-}\overset{+}{C}H_2$ b. $H{-}\overset{-}{C}H{-}C({=}O){-}CH_3$

Solution

a. The curved arrow tells us to move **one** electron pair in the double bond to the adjacent C – C bond. Then determine the formal charge on any atom whose bonding is different.

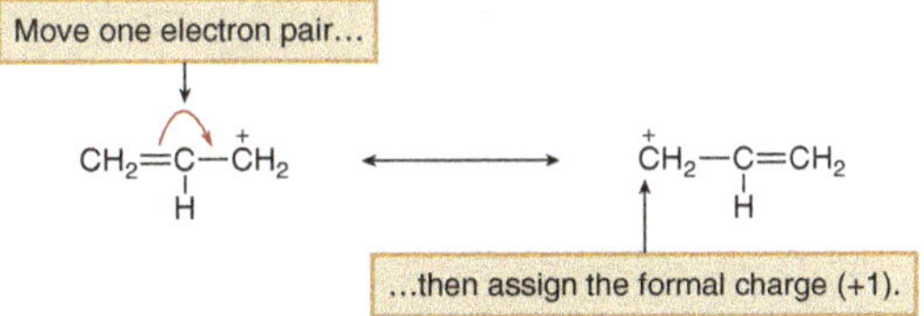

Positively charged carbon atoms are called **carbocations.** Carbocations are unstable intermediates because they contain a carbon atom that is lacking an octet of electrons.

b. **Two** curved arrows tell us to move **two** electron pairs. The second resonance structure has a formal charge of (–1) on O.

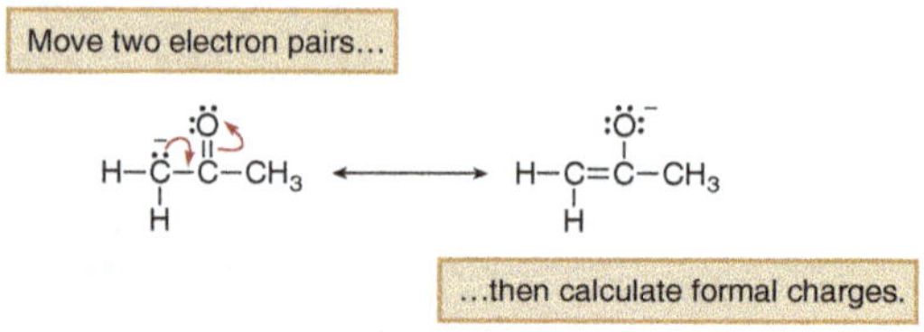

Move two electron pairs...

H–C–C(=O)–CH$_3$ ⟷ H–C=C(–O$^-$)–CH$_3$

...then calculate formal charges.

This type of resonance-stabilized anion is called an **enolate anion.** Enolates are important intermediates in many organic reactions, and all of Chapters 23 and 24 is devoted to their preparation and reactions.

Problem 1.11 Follow the curved arrows to draw a second resonance structure for each species.

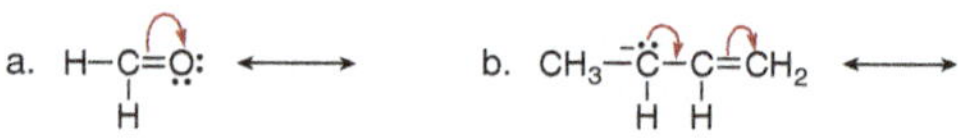

a. H–C(H)=O: ⟷ b. CH$_3$–C(H)–C(H)=CH$_2$ ⟷

Problem 1.12 Use curved arrow notation to show how the first resonance structure can be converted to the second.

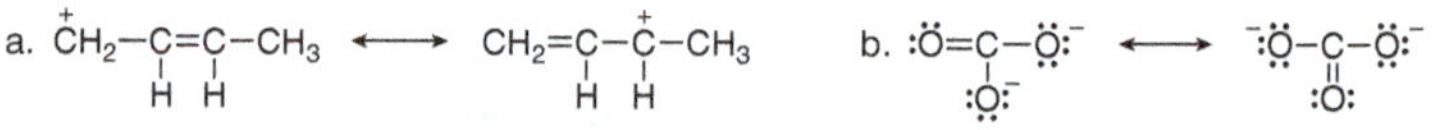

a. $\overset{+}{C}H_2$–C(H)=C(H)–CH$_3$ ⟷ CH$_2$=C(H)–$\overset{+}{C}$(H)–CH$_3$ b. :Ö=C(–O:$^-$)–Ö:$^-$ ⟷ $^-$:Ö–C(=O:)–Ö:$^-$

Two resonance structures can have exactly the same kinds of bonds, as they do in the carbocation in Sample Problem 1.4a, or they may have different types of bonds, as they do in the enolate in Sample Problem 1.4b. Either possibility is fine as long as the individual resonance structures are valid Lewis structures.

A resonance structure can have an atom with *fewer* than eight electrons around it. **B** is a resonance structure of **A** even though the carbon atom is surrounded by only six electrons.

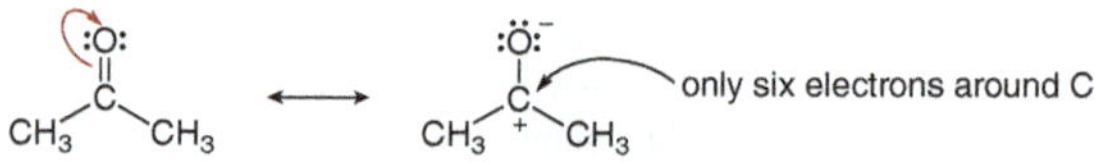

only six electrons around C

A ⟷ **B**

a valid resonance structure

In contrast, a resonance structure can *never* have a second-row element with more than eight electrons. **C** is *not* a resonance structure of **A** because the carbon atom is now surrounded by 10 electrons.

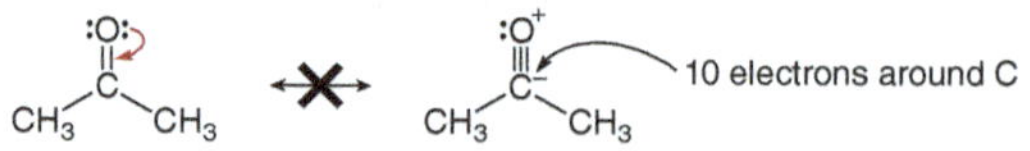

10 electrons around C

A ✕ **C**

not a valid resonance structure

We will learn much more about resonance in Chapter 16.

The ability to draw and manipulate resonance structures is an important skill that will be needed throughout your study of organic chemistry. With practice, you will begin to recognize certain common bonding patterns for which more than one Lewis structure can be drawn. For now, notice that:

- **A second resonance structure can be drawn when a lone pair is located on an atom directly bonded to a multiple bond.**

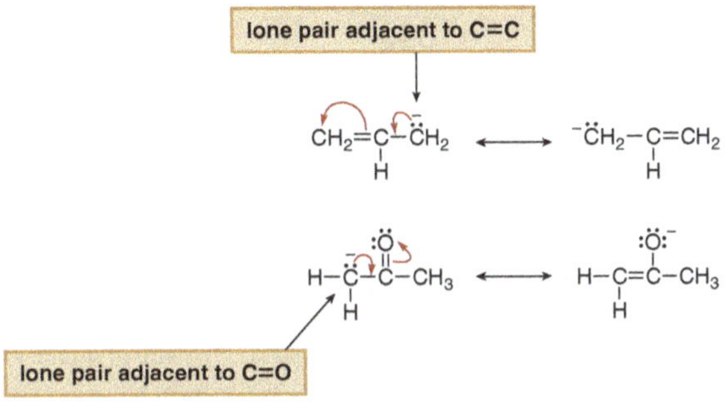

lone pair adjacent to C=C

$CH_2{=}C(H){-}\ddot{C}H_2^- \longleftrightarrow {}^-\ddot{C}H_2{-}C(H){=}CH_2$

$H{-}\ddot{C}^-(H){-}C({=}\ddot{O}){-}CH_3 \longleftrightarrow H{-}C(H){=}C(\ddot{O}:^-){-}CH_3$

lone pair adjacent to C=O

- **A second resonance structure can be drawn when an atom bearing a (+) charge is bonded to either a multiple bond or an atom with a lone pair.**

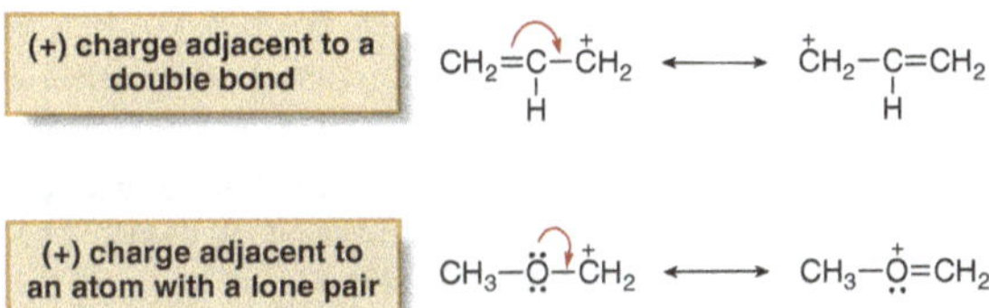

Problem 1.13 Draw a second resonance structure for each species.

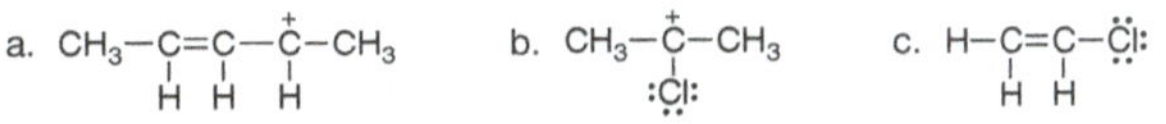

a. $CH_3{-}CH{=}CH{-}\overset{+}{C}H{-}CH_3$

b. $CH_3{-}\overset{+}{C}(:\ddot{Cl}:){-}CH_3$

c. $H{-}CH{=}CH{-}\ddot{Cl}:$

1.6C The Resonance Hybrid

The **resonance hybrid** is the composite of all possible resonance structures. In the resonance hybrid, the electron pairs drawn in different locations in individual resonance structures are ***delocalized.***

- **The resonance hybrid is more stable than any resonance structure because it delocalizes electron density over a larger volume.**

What does the hybrid look like? When all resonance forms are identical, as they were in the carbocation in Sample Problem 1.4a, each resonance form contributes **equally** to the hybrid.

When two resonance structures are different, the hybrid looks more like the "better" resonance structure. The "better" resonance structure is called the **major contributor** to the hybrid, and all others are **minor contributors.** The hybrid is the weighted average of the contributing resonance structures. What makes one resonance structure "better" than another? There are many factors, but for now, we will learn just two.

- A "better" resonance structure is one that has *more bonds* and *fewer charges.*

CH3–C(=Ö:)–H ⟷ CH3–C(+)(–Ö:⁻)–H

X more bonds fewer charges — **major contributor**

Y — **minor contributor**

Comparing resonance structures **X** and **Y, X** is the major contributor because it has more bonds and fewer charges. Thus, the hybrid looks more like **X** than **Y.**

How can we draw a hybrid, which has delocalized electron density? First, we must determine what is different in the resonance structures. Two differences commonly seen are the **position of a multiple bond** and the **site of a charge.** The anion $(HCONH)^-$ illustrates two conventions for drawing resonance hybrids.

H–C(=Ö:)–N̈⁻–H ⟷ H–C(–Ö:⁻)=N̈–H

A **B**

individual resonance structures

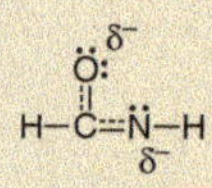

resonance *hybrid*

- **Double bond position.** Structure **A** has a C–O double bond, whereas structure **B** has a C–N double bond. A dashed line in the hybrid indicates partial double bond character between these atoms.
- **Location of the charge.** A negative charge resides on different atoms in **A** and **B.** The symbol δ^- (for a partial negative charge) indicates that the charge is delocalized on the N and O atoms in the hybrid.

Common symbols and conventions used in organic chemistry are listed on the inside back cover.

This discussion of resonance is meant to serve as an introduction only. You will learn many more facets of resonance theory in later chapters. In Chapter 2, for example, the enormous effect of resonance on acidity is discussed.

Problem 1.14 Label the resonance structures in each pair as major, minor, or equal contributors to the hybrid. Then draw the hybrid.

a.  CH_3–C(H)(+)–N̈(H)–CH_3 ⟷ CH_3–C(H)=N(H)(+)–CH_3

b.  CH_2=C(H)–C̈H$_2^-$ ⟷ $^-$C̈H$_2$–C(H)=CH_2

Problem 1.15 Draw a second resonance structure for nitrous acid. Label each resonance structure as a major, minor, or equal contributor to the hybrid. Then draw the resonance hybrid.

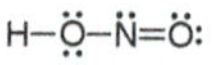

H–Ö–N̈=Ö:

nitrous acid

1.7 Determining Molecular Shape

We can now use Lewis structures to determine the shape around a particular atom in a molecule. Consider the H_2O molecule. The Lewis structure tells us only which atoms are connected to each other, but it implies nothing about the geometry. What does the overall molecule look like? Is H_2O a bent or linear molecule? Two variables define a molecule's structure: **bond length** and **bond angle.**

1.7A Bond Length

Although the SI unit for bond length is the picometer (pm), the angstrom (Å) is still widely used in the chemical literature; 1 Å = 10^{-10} m. As a result, 1 pm = 10^{-2} Å, and 95.8 pm = 0.958 Å.

***Bond length* is the average distance between the centers of two bonded nuclei.** Bond lengths are typically reported in picometers (pm), where 1 pm = 10^{-12} m. For example, the O–H bond length in H_2O is 95.8 pm. Average bond lengths for common bonds are listed in Table 1.2.

- **Bond length *decreases* across a row of the periodic table as the size of the atom *decreases*.**

—C—H > —N—H > —O—H

Increasing bond length

- **Bond length *increases* down a column of the periodic table as the size of an atom *increases*.**

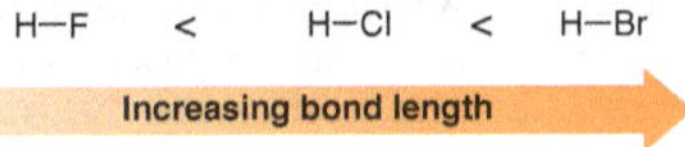

Learn these general trends. Often, knowing such trends is more useful than learning a set of exact numbers, because we are usually interested in comparisons rather than absolute values.

Table 1.2 Average Bond Lengths

Bond	Length (pm)	Bond	Length (pm)	Bond	Length (pm)
H–H	74	H–F	92	C–F	133
C–H	109	H–Cl	127	C–Cl	177
N–H	101	H–Br	141	C–Br	194
O–H	96	H–I	161	C–I	213

1.7B Bond Angle

Bond angle determines the shape around any atom bonded to two other atoms. To determine the bond angle and shape around a given atom, we must first determine how many groups surround the atom. **A group is either an atom or a lone pair of electrons.** Then we use the **valence shell electron pair repulsion (VSEPR) theory** to determine the shape. VSEPR is based on the fact that electron pairs repel each other; thus:

- **The most stable arrangement keeps these groups as far away from each other as possible.**

A second-row element has only three possible arrangements, defined by the number of groups surrounding it.

To determine geometry: [1] Draw a valid Lewis structure; [2] count groups around a given atom.

Number of groups	Geometry	Bond angle
• two groups	linear	180°
• three groups	trigonal planar	120°
• four groups	tetrahedral	109.5°

Let's examine several molecules to illustrate this phenomenon. In each example, we first need a valid Lewis structure, and then we merely count groups around a given atom to predict its geometry.

Two Groups Around an Atom

Any atom surrounded by only two groups is linear and has a bond angle of 180°. Two examples illustrating this geometry are $\mathbf{BeH_2}$ (beryllium hydride) and **HC≡CH** (acetylene). We consider each carbon atom in acetylene *separately*. Because each C is surrounded by two atoms and no lone pairs, each H–C–C bond angle in acetylene is 180°, and therefore all four atoms are linear.

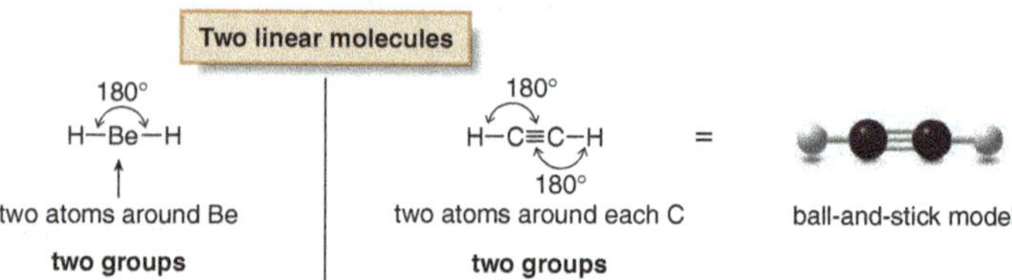

Most students in organic chemistry find that building models helps them visualize the shape of molecules. Invest in a set of models *now*.

Common element colors are also shown on the inside back cover.

Acetylene illustrates another important feature: ***ignore multiple bonds in predicting geometry.* Count only atoms and lone pairs.**

We will begin to represent molecules with models having balls for atoms and sticks for bonds, as in the ball-and-stick model of acetylene just shown. These representations are analogous to a set of molecular models. Balls are color-coded using accepted conventions: carbon (black), hydrogen (white or gray), oxygen (red), and so forth, as shown.

Three Groups Around an Atom

Any atom surrounded by three groups is trigonal planar and has bond angles of 120°. Two examples illustrating this geometry are $\mathbf{BF_3}$ (boron trifluoride) and $\mathbf{CH_2{=}CH_2}$ (ethylene). *Each* carbon atom of ethylene is surrounded by three atoms and no lone pairs, making *each* H–C–C bond angle 120°.

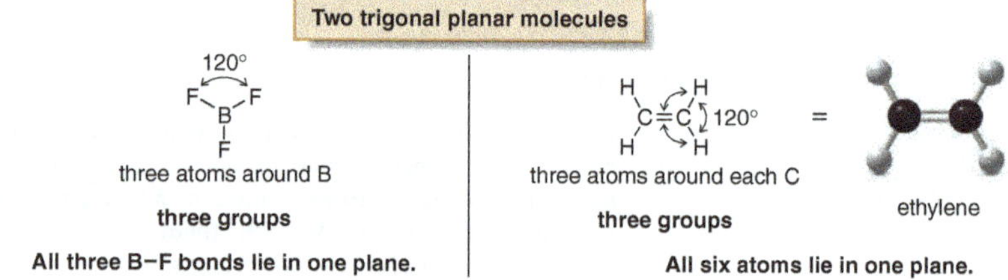

Four Groups Around an Atom

Any atom surrounded by four groups is tetrahedral and has bond angles of approximately 109.5°. For example, the simple organic compound methane, $\mathbf{CH_4}$, has a central carbon atom with bonds to four hydrogen atoms, each pointing to a corner of a tetrahedron. This arrangement keeps four groups farther apart than a square planar arrangement in which all bond angles would be only 90°.

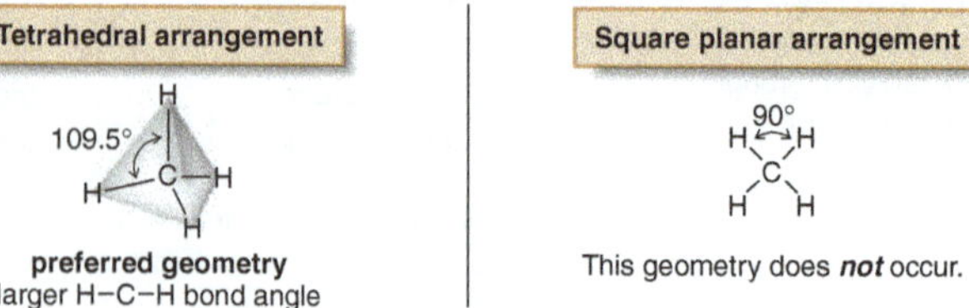

How can we represent the three-dimensional geometry of a tetrahedron on a two-dimensional piece of paper? **Place two of the bonds in the plane of the paper, one bond in front and one bond behind,** using the following conventions:

- A *solid line* is used for a bond *in* the plane.
- A *wedge* is used for a bond in *front* of the plane.
- A *dashed line* is used for a bond *behind* the plane.

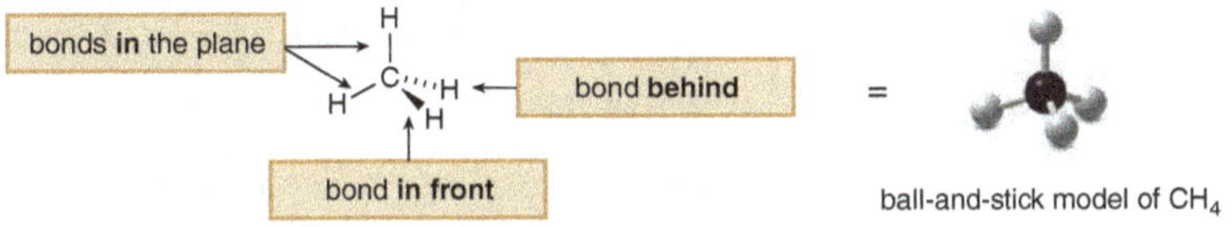

bonds **in** the plane

bond **behind**

bond **in front**

ball-and-stick model of CH_4

This is just one way to draw a tetrahedron for CH_4. We can turn the molecule in many different ways, generating many equivalent representations. For example, all of the following are acceptable drawings for CH_4.

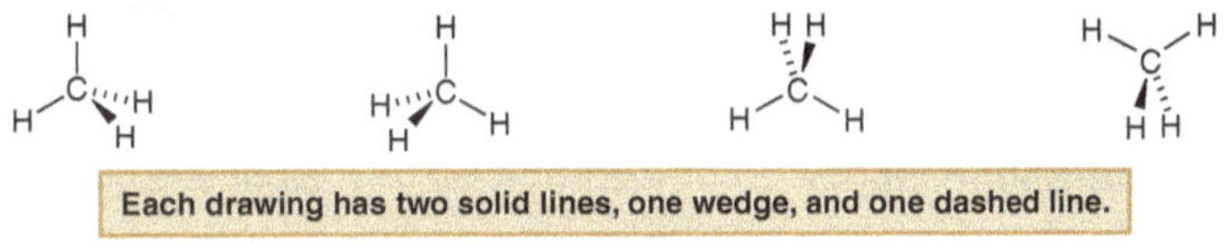

Each drawing has two solid lines, one wedge, and one dashed line.

Finally, **wedges and dashes are used for groups that are really *aligned one behind another.*** It does not matter in the following two drawings whether the wedge or dash is skewed to the left or right, because the two H atoms are really aligned as shown in the three-dimensional model.

All carbons in stable molecules are ***tetravalent,*** but the geometry varies with the number of groups around the particular carbon.

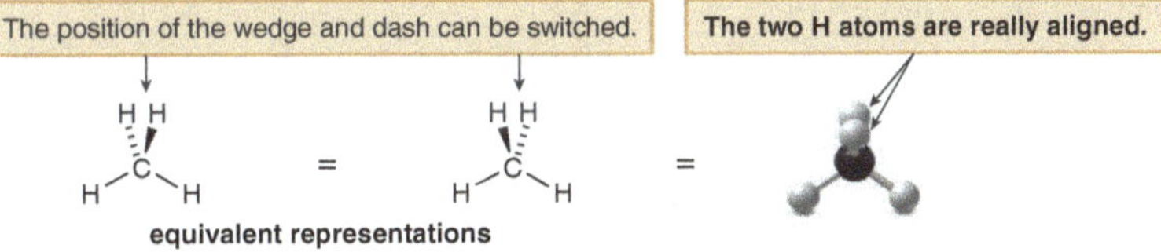

Problem 1.16 Draw two different three-dimensional representations for CH_2Cl_2 (dichloromethane) using wedges and dashes.

Ammonia (NH_3) and water (H_2O) both have atoms surrounded by four groups, some of which are lone pairs. In $\mathbf{NH_3}$**,** the three H atoms and one lone pair around N point to the corners of a tetrahedron. The H–N–H bond angle of 107° is close to the theoretical tetrahedral bond angle of 109.5°. This molecular shape is referred to as a **trigonal pyramid,** because one of the groups around the N is a nonbonded electron pair, not another atom.

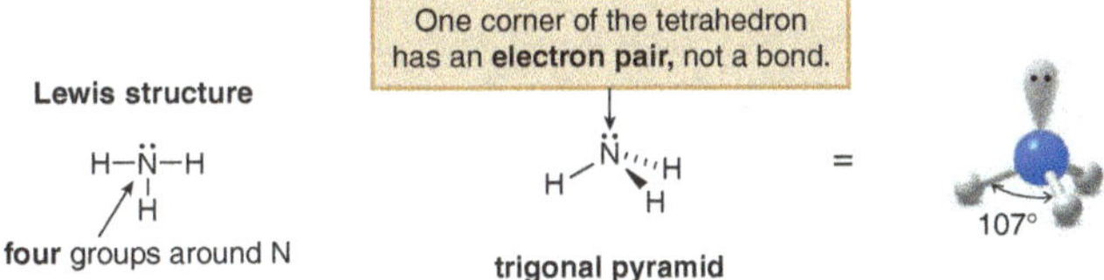

In $\mathbf{H_2O}$**,** the two H atoms and two lone pairs around O point to the corners of a tetrahedron. The H–O–H bond angle of 105° is close to the theoretical tetrahedral bond angle of 109.5°. Water has a **bent** molecular shape, because two of the groups around oxygen are lone pairs of electrons.

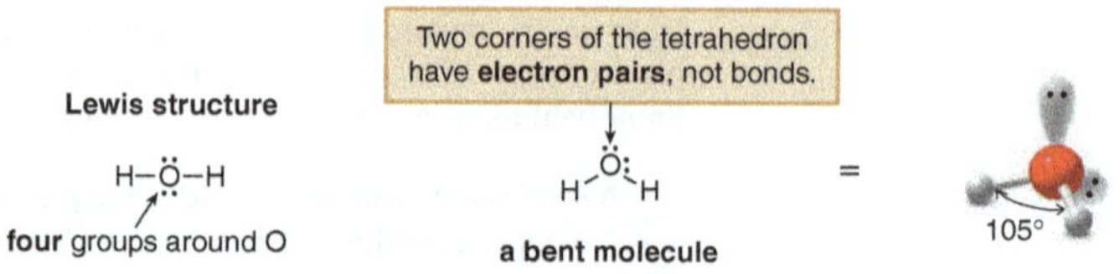

Two corners of the tetrahedron have **electron pairs**, not bonds.

Lewis structure

H–Ö–H

four groups around O

a bent molecule

105°

In both NH_3 and H_2O the bond angle is somewhat smaller than the theoretical tetrahedral bond angle because of repulsion of the lone pairs of electrons. The bonded atoms are compressed into a smaller space with a smaller bond angle.

Predicting geometry based on counting groups is summarized in Table 1.3.

Table 1.3 Summary: Determining Geometry Based on the Number of Groups

Number of groups around an atom	Geometry	Bond angle	Examples
2	linear	180°	BeH_2, HC≡CH
3	trigonal planar	120°	BF_3, $CH_2=CH_2$
4	tetrahedral	109.5°	CH_4, NH_3, H_2O

Sample Problem 1.5 Determine the geometry around the indicated atom in each species.

a. :Ö=C=Ö: b. $H–N^+(H)_2–H$

Solution

a. :Ö=C=Ö: 180°

two atoms around C
no lone pairs

two groups

linear

b. $H–N^+(H)_2–H$ → 109.5°

four atoms around N
no lone pairs

four groups → **tetrahedral**

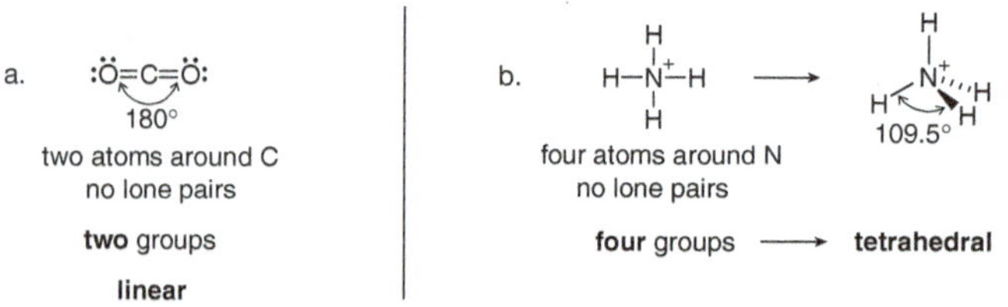

Problem 1.17 Determine the geometry around all second-row elements in each compound.

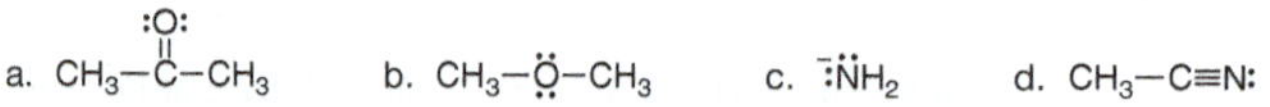

a. $CH_3–C(=O:)–CH_3$ b. $CH_3–\ddot{O}–CH_3$ c. $^{-}$:NH$_2$ d. $CH_3–C≡N:$

Problem 1.18 Predict the indicated bond angles in each compound.

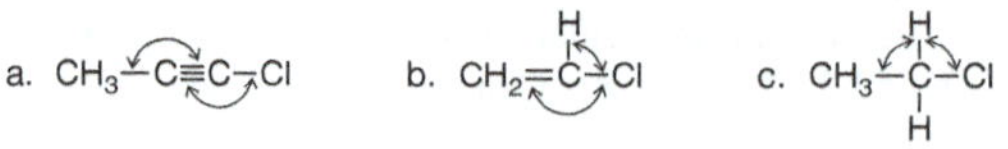

a. $CH_3–C≡C–Cl$ b. $CH_2=CH–Cl$ c. $CH_3–CH_2–Cl$

Problem 1.19 Using the principles of VSEPR theory, you can predict the geometry around any atom in any molecule, no matter how complex. Enanthotoxin is a poisonous compound isolated from a common variety of hemlock grown in England. Predict the geometry around the indicated atoms in enanthotoxin.

enanthotoxin

The hemlock water dropwort is the source of the poison enanthotoxin.

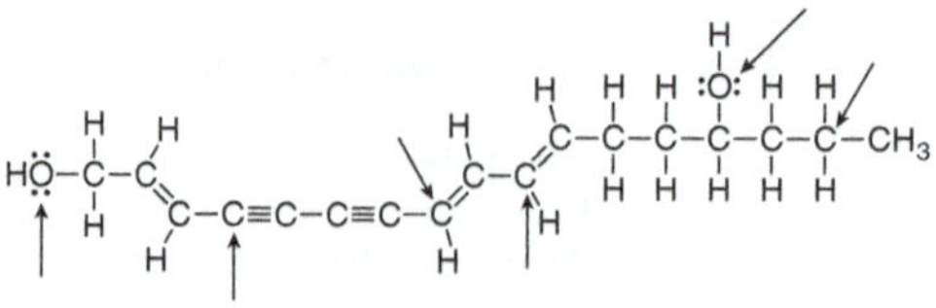

1.8 Drawing Organic Structures

Drawing organic molecules presents a special challenge. Because they often contain many atoms, we need shorthand methods to simplify their structures. The two main types of shorthand representations used for organic compounds are **condensed structures** and **skeletal structures.**

1.8A Condensed Structures

Condensed structures are most often used for compounds having a chain of atoms bonded together, rather than a ring. The following conventions are used:

- All of the atoms are drawn in, but the two-electron bond lines are generally omitted.
- Atoms are usually drawn next to the atoms to which they are bonded.
- Parentheses are used around similar groups bonded to the same atom.
- Lone pairs are omitted.

To interpret a condensed formula, it is usually best to start at the **left side** of the molecule and remember that the ***carbon atoms must be tetravalent.*** A carbon bonded to three H atoms becomes CH_3; a carbon bonded to two H atoms becomes CH_2, and so forth. Two examples of condensed formulas for compounds having only carbon and hydrogen are given below.

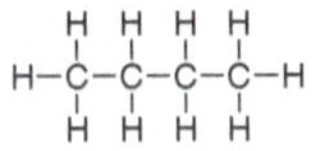

= $CH_3CH_2CH_2CH_3$ or $CH_3(CH_2)_2CH_3$

2 CH_2 groups bonded together

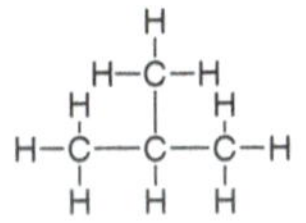

= $(CH_3)_3CH$

Recall from the prologue: A **heteroatom** is any atom that is not C or H.

Other examples of condensed structures with heteroatoms and carbon–carbon multiple bonds are given in Figure 1.4. You must learn how to convert a Lewis structure to a condensed structure, and vice versa.

Figure 1.4 Examples of condensed structures

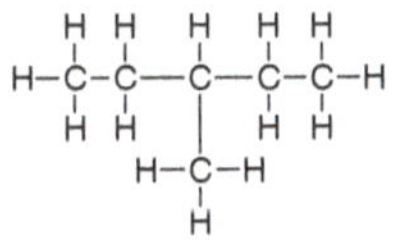

= $CH_3CH_2CHCH_2CH_3$ (with CH_3 below) or $CH_3CH_2CH(CH_3)CH_2CH_3$

Parentheses indicate the CH_3 is bonded to the carbon chain.

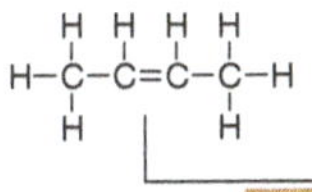

= $CH_3CH{=}CHCH_3$

Keep the double bond.

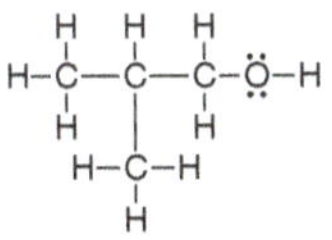

= $(CH_3)_2CHCH_2OH$

Draw the heteroatoms without lone pairs.

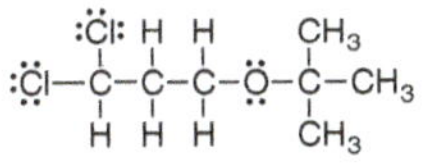

= $Cl_2CHCH_2CH_2OC(CH_3)_3$ or $Cl_2CH(CH_2)_2OC(CH_3)_3$

Figure 1.5
Condensed structures containing a C–O double bond

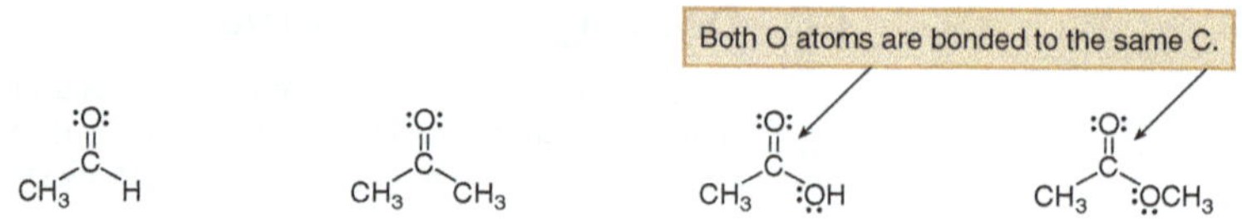

Both O atoms are bonded to the same C.

CH_3CHO CH_3COCH_3 CH_3CO_2H $CH_3CO_2CH_3$

- Take special note of condensed structures with oxygen atoms. In these examples, the only way for all atoms to have an octet is by having a carbon–oxygen double bond.

Translating some condensed formulas is not obvious, and it will come only with practice. This is especially true for compounds containing a carbon–oxygen double bond. Some noteworthy examples in this category are given in Figure 1.5. While carbon–carbon double bonds are generally drawn in condensed structures, carbon–oxygen double bonds are usually omitted.

Sample Problem 1.6

Convert each condensed formula to a Lewis structure.

a. $(CH_3)_2CHOCH_2CH_2CH_2OH$ b. $CH_3(CH_2)_2CO_2C(CH_3)_3$

Solution

Start at the left and proceed to the right, making sure that each carbon has four bonds. Give any heteroatom enough lone pairs to have an octet.

a. $(CH_3)_2CHOCH_2CH_2CH_2OH$

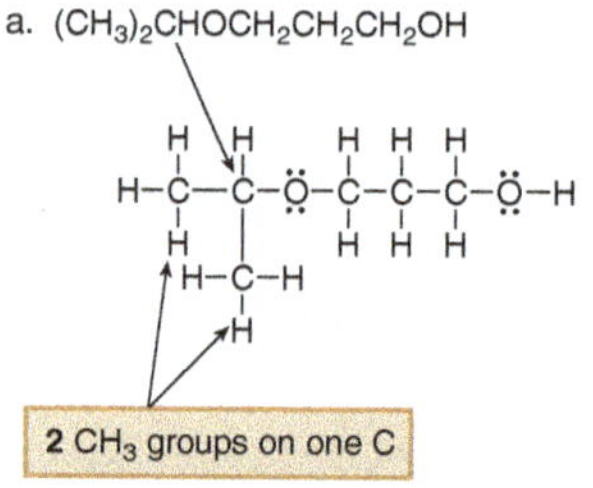

2 CH_3 groups on one C

b. $CH_3(CH_2)_2CO_2C(CH_3)_3$

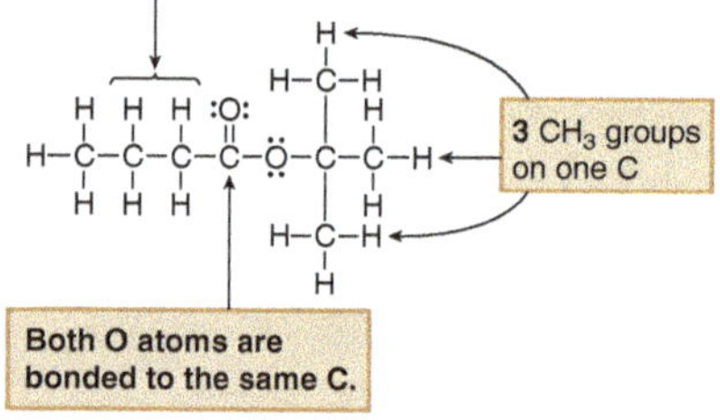

3 CH_3 groups on one C

Both O atoms are bonded to the same C.

Problem 1.20

Convert each condensed formula to a Lewis structure.

a. $CH_3(CH_2)_4CH(CH_3)_2$
b. $(CH_3)_2CHCH(CH_2CH_3)_2$
c. $(CH_3)_3CCH(OH)CH_2CH_3$
d. $(CH_3)_2CHCHO$

Problem 1.21

Simplify each condensed structure by using parentheses around similar groups.

a. $CH_3CH_2CH_2CH_2CH_2Cl$

b. 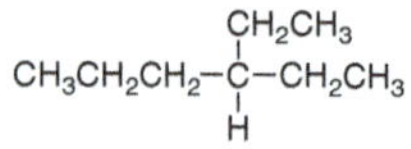 $CH_3CH_2CH_2\text{–}C(H)(CH_2CH_3)\text{–}CH_2CH_3$

c. $HOCH_2\text{–}C(H)(CH_2OH)\text{–}CH_2CH_2CH_2\text{–}C(CH_3)_2\text{–}CH_2\text{–}C(CH_3)_2\text{–}CH_3$

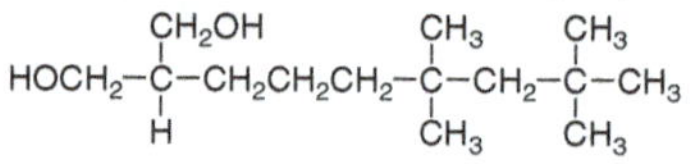

Problem 1.22

During periods of strenuous exercise, the buildup of lactic acid [$CH_3CH(OH)CO_2H$] causes the aching feeling in sore muscles. Convert this condensed structure to a Lewis structure of lactic acid.

1.8B Skeletal Structures

Skeletal structures are used for organic compounds containing both rings and chains of atoms. Three important rules are used to draw them:

- **Assume there is a carbon atom at the junction of any two lines or at the end of any line.**
- **Assume there are enough hydrogens around each carbon to make it tetravalent.**
- **Draw in all heteroatoms and the hydrogens directly bonded to them.**

Carbon chains are drawn in a zigzag fashion, and rings are drawn as polygons, as shown for hexane and cyclohexane.

Hexane

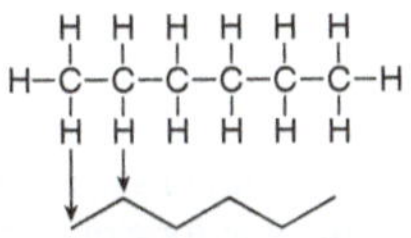

skeletal structure

$CH_3CH_2CH_2CH_2CH_2CH_3$
condensed structure

Cyclohexane

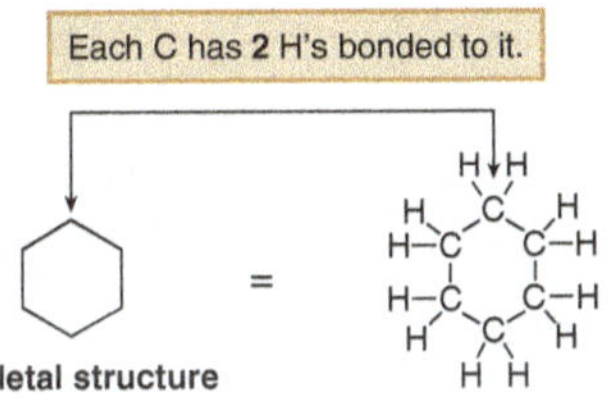

Each C has **2** H's bonded to it.

skeletal structure

How To Interpret a Skeletal Structure

Example Draw in all C atoms, H atoms, and lone pairs in the following molecule:

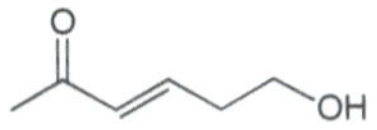

Step [1] **Place a C atom at the intersection of any two lines and at the end of any line.**

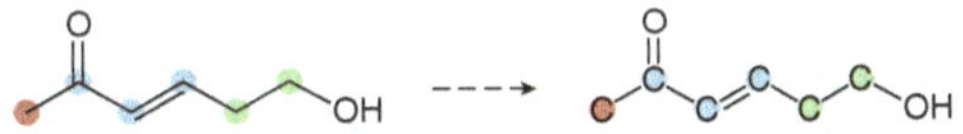

- This molecule has six carbons, including the C labeled in red at the left end of the chain.
- There are two C's (labeled in green) between the C = C and the OH group.

Step [2] **Add enough H's to make each C tetravalent.**

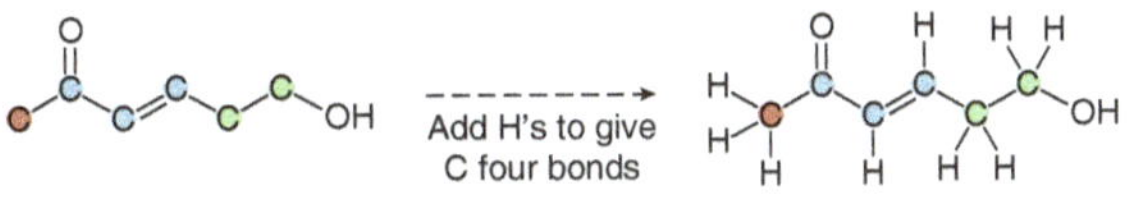

Add H's to give C four bonds

- The end C labeled in red needs three H's to be tetravalent.
- Each C on the C = C has three bonds already, so only one H must be drawn.
- There are two CH_2 groups between the C = C and the OH group.

Step [3] **Add lone pairs to give each heteroatom an octet.**

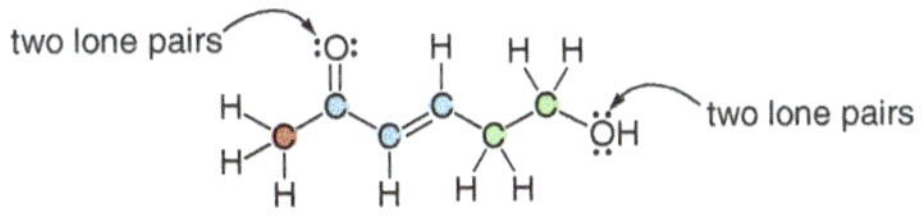

two lone pairs

two lone pairs

Each O needs **two** lone pairs for an octet.

Figure 1.6
Interpreting skeletal structures

Figure 1.6 shows other examples of skeletal structures, and Sample Problem 1.7 illustrates how to interpret the skeletal structure for a more complex cyclic compound.

Sample Problem 1.7 Draw a complete structure for vanillin showing all C atoms, H atoms, and lone pairs. Vanillin is the principal component of the extract of the vanilla bean.

vanillin

vanilla bean

Solution

Skeletal structures have a C atom at the junction of any two lines and at the end of any line. Each C must have enough H's to make it tetravalent. In structures that contain a –CHO group, the C atom is doubly bonded to the O atom and singly bonded to H. Each O atom needs two lone pairs to have a complete octet.

$C_8H_8O_3$

Problem 1.23 How many hydrogen atoms are present around each indicated carbon atom in the following molecules? What is the molecular formula for each molecule? Both compounds are active ingredients in some common sunscreens.

a. octinoxate (2-ethylhexyl 4-methoxycinnamate)

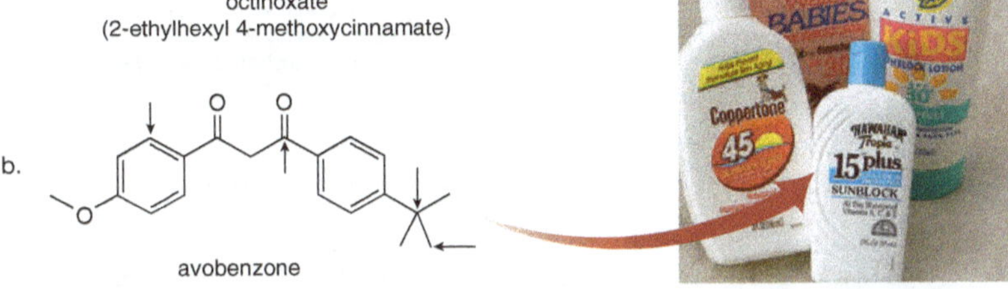

b. avobenzone

Problem 1.24 Convert each skeletal structure to a complete structure with all C's, H's, and lone pairs drawn in.

a. b. c. d.

When heteroatoms are bonded to a carbon skeleton, the heteroatom is joined *directly* to the carbon to which it is bonded, with no H atoms in between. Thus, an OH group is drawn as OH or HO depending on where the OH is located. In contrast, when carbon appendages are bonded to a carbon skeleton, the H atoms will be drawn to the *right* of the carbon to which they are bonded regardless of the location.

Place the O and N atoms directly joined to the ring.

This C is bonded to the C=O.

This C is bonded to the ring.

1.8C Skeletal Structures with Charged Carbon Atoms

Take care in interpreting skeletal structures for positively and negatively charged carbon atoms, because *both* the hydrogen atoms *and* the lone pairs are omitted. Keep in mind the following:

- **A charge on a carbon atom takes the place of one hydrogen atom.**
- **The charge determines the number of lone pairs. Negatively charged carbon atoms have one lone pair and positively charged carbon atoms have none.**

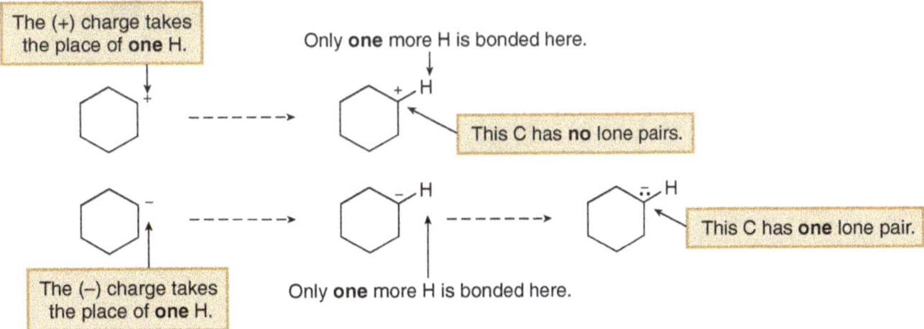

Skeletal structures often leave out lone pairs on heteroatoms, but *don't forget about them.* Use the formal charge on an atom to determine the number of lone pairs. For example, a neutral O atom with two bonds needs two additional lone pairs, and a positively charged O atom with three bonds needs only one lone pair.

A neutral O atom "owns" six electrons:
- two bonds (four bonding electrons)
- **two** lone pairs (four unshared electrons).

A positively charged O atom "owns" five electrons, one fewer than its group number of six:
- three bonds (six bonding electrons)
- **one** lone pair (two unshared electrons).

Problem 1.25 Draw in all hydrogens and lone pairs on the charged carbons in each ion.

a. b. 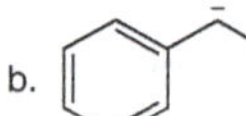 c. d.

Problem 1.26 Draw a skeletal structure for the molecules in parts (a) and (b), and a condensed structure for the molecules in parts (c) and (d).

a. $(CH_3)_2C{=}CH(CH_2)_4CH_3$

b. (four-membered ring: CH_3–C(H)–C(H)–CH_2CH_2Cl bonded to H_2N–C(H)–C(H)–H)

c.

d. HO

1.9 Hybridization

What orbitals do the first- and second-row atoms use to form bonds? Let's begin with hydrogen and then examine the orbitals used for bonding by atoms in the second row.

1.9A Hydrogen

Recall from Section 1.2 that two hydrogen atoms share each of their electrons to form H_2. Thus, the 1*s* orbital on one H overlaps with the 1*s* orbital on the other H to form a bond that concentrates electron density between the two nuclei. This type of bond, called a **σ** (sigma) **bond,** is cylindrically symmetrical because the electrons forming the bond are distributed symmetrically about an imaginary line connecting the two nuclei.

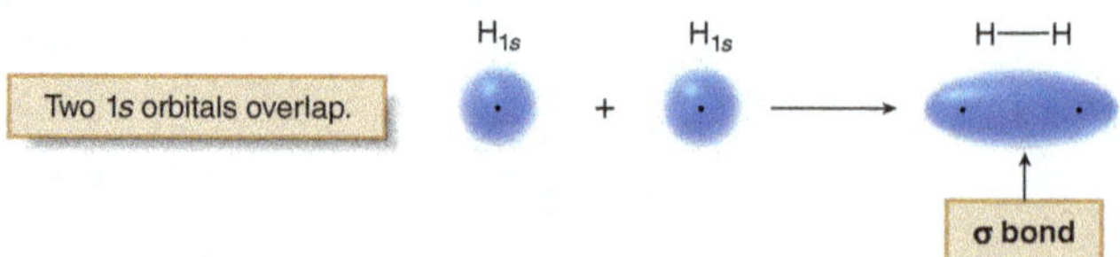

- **A σ bond concentrates electron density on the axis that joins two nuclei. All single bonds are σ bonds.**

1.9B Bonding in Methane

To account for the bonding patterns observed in more complex molecules, we must take a closer look at the 2*s* and 2*p* orbitals of atoms of the second row. Let's illustrate this with methane, CH_4.

Carbon has two core electrons, plus **four valence electrons.** To fill atomic orbitals in the most stable arrangement, electrons are placed in the orbitals of lowest energy. For carbon, this places two in the 2*s* orbital and one each in two 2*p* orbitals.

C ($1s^2$) + 4 valence electrons

2*p* ↑ ↑ —

2*s* ↑↓

- This lowest energy arrangement of electrons for an atom is called its **ground state.**

In this description, **carbon should form *only two bonds*** because it has only two unpaired valence electrons, and CH_2 should be a stable molecule. In reality, however, CH_2 is a highly reactive species that cannot be isolated under typical laboratory conditions. In CH_2, carbon would not have an octet of electrons.

H–C̈–H

no octet (unstable)

There is a second possibility. Promotion of an electron from a 2*s* to a vacant 2*p* orbital would form **four** unpaired electrons for bonding. This process requires energy because it moves an electron to a higher energy orbital. This higher energy electron configuration is called an electronically **excited state.**

2*p* ↑ ↑ — → energy → 2*p* ↑ ↑ ↑

2*s* ↑↓ → 2*s* ↑ (**four** unpaired electrons)

ground state for carbon — excited state for carbon

This description is still not adequate. Carbon would form two different types of bonds: three with 2*p* orbitals and one with a 2*s* orbital. **But experimental evidence points to carbon forming *four identical bonds* in methane.**

To solve this dilemma, chemists have proposed that atoms like carbon do not use pure *s* and pure *p* orbitals in forming bonds. Instead, atoms use a set of new orbitals called **hybrid orbitals.** The mathematical process by which these orbitals are formed is called **hybridization.**

- *Hybridization* is the combination of two or more atomic orbitals to form the same number of hybrid orbitals, each having the same shape and energy.

Hybridization of one 2*s* orbital and three 2*p* orbitals for carbon forms four hybrid orbitals, each with one electron. These new hybrid orbitals are intermediate in energy between the 2*s* and 2*p* orbitals.

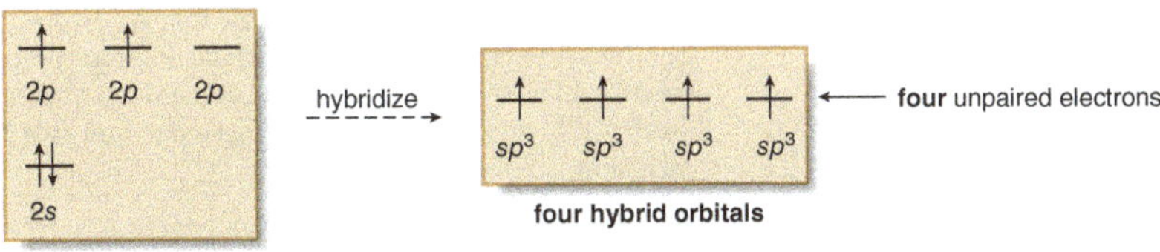

- These hybrid orbitals are called *sp^3 hybrids* because they are formed from *one s* orbital and *three p* orbitals.

p orbital

sp^3 hybrid orbital

What do these new hybrid orbitals look like? Mixing a spherical 2*s* orbital and three dumbbell-shaped 2*p* orbitals together produces four orbitals having one large lobe and one small lobe, oriented toward the corners of a tetrahedron. Each large lobe concentrates electron density in the bonding direction between two nuclei. This makes bonds formed from hybrid orbitals ***stronger*** than bonds formed from pure *p* orbitals.

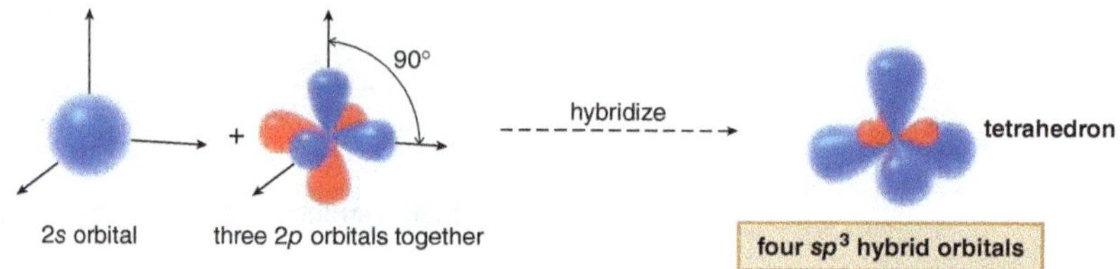

The four hybrid orbitals form four equivalent bonds. We can now explain the observed bonding in CH_4.

- **Each bond in CH_4 is formed by overlap of an sp^3 hybrid orbital of carbon with a 1*s* orbital of hydrogen. These four bonds point to the corners of a tetrahedron.**

All four C–H bonds in methane are **σ bonds,** because the electron density is concentrated on the axis joining C and H. An orbital picture of the bonding in CH_4 is given in Figure 1.7.

Problem 1.27 What orbitals are used to form each of the C–C and C–H bonds in $CH_3CH_2CH_3$ (propane)? How many σ bonds are present in this molecule?

Figure 1.7 Bonding in CH_4 using sp^3 hybrid orbitals

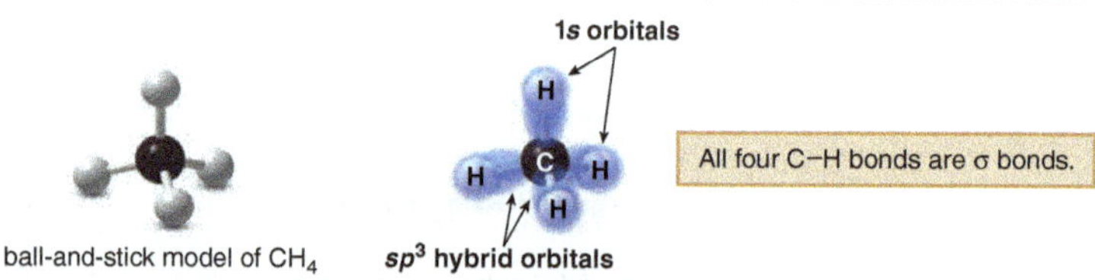

1.9C Other Hybridization Patterns—*sp* and *sp^2* Hybrid Orbitals

Forming sp^3 hybrid orbitals is just one way that 2*s* and 2*p* orbitals can hybridize. In fact, three common modes of hybridization are seen in organic molecules. The number of orbitals is always conserved in hybridization; that is, a **given number of atomic orbitals hybridize to form an equivalent number of hybrid orbitals.**

- *One* 2*s* orbital and *three* 2*p* orbitals form *four* sp^3 hybrid orbitals.
- *One* 2*s* orbital and *two* 2*p* orbitals form *three* sp^2 hybrid orbitals.
- *One* 2*s* orbital and *one* 2*p* orbital form *two sp* hybrid orbitals.

We have already seen pictorially how four sp^3 hybrid orbitals are formed from one 2*s* and three 2*p* orbitals. Figures 1.8 and 1.9 illustrate the same process for *sp* and sp^2 hybrids. Each *sp* and sp^2 hybrid orbital has one large and one small lobe, much like an sp^3 hybrid orbital. Note, however, that both sp^2 and *sp* hybridization **leave one and two 2*p* orbitals *unhybridized,*** respectively, on each atom.

To determine the hybridization of an atom in a molecule, we count groups around the atom, just as we did in determining geometry. **The number of groups (atoms and nonbonded electron**

Figure 1.8
Forming two *sp* hybrid orbitals

2p 2p 2p
2s
two atomic orbitals
hybridize
sp sp
two hybrid orbitals
2p 2p
Two 2*p* orbitals remain unhybridized.

- Forming **two *sp* hybrid orbitals** uses **one 2s** and **one 2p orbital,** leaving **two 2p orbitals *unhybridized.***

Figure 1.9
Forming three sp^2 hybrid orbitals

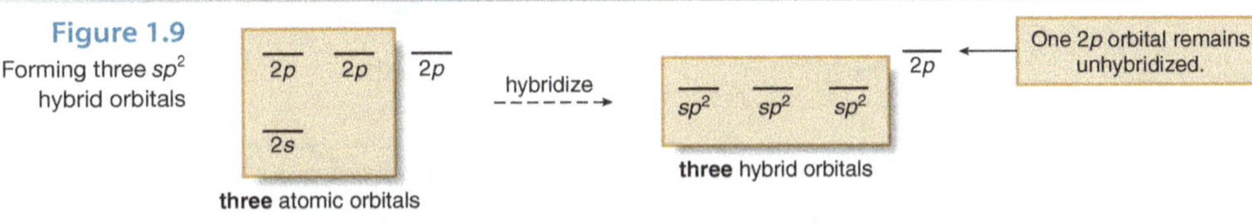

- Forming **three sp^2 hybrid orbitals** uses **one 2s** and **two 2p orbitals,** leaving **one 2p orbital *unhybridized.***

The **superscripts** for hybrid orbitals correspond to the **number of atomic orbitals** used to form them. The number "1" is understood.

For example: $sp^3 = s^1p^3$

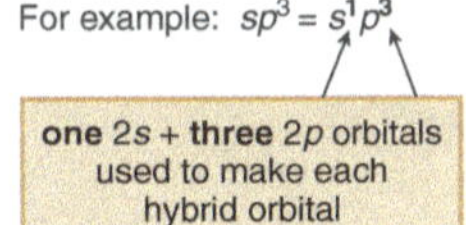

pairs) corresponds to the number of atomic orbitals that must be hybridized to form the hybrid orbitals.

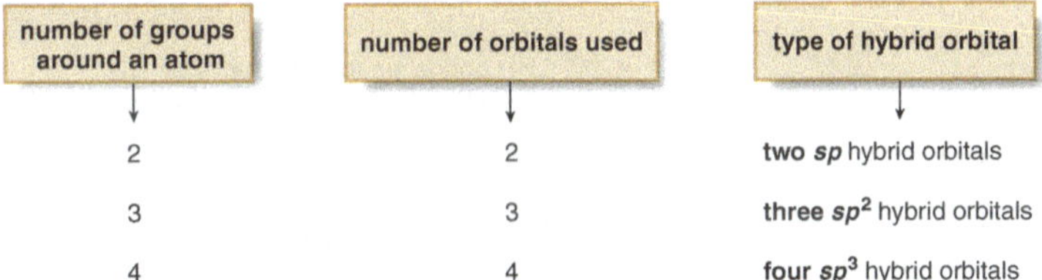

number of groups around an atom	number of orbitals used	type of hybrid orbital
2	2	**two *sp*** hybrid orbitals
3	3	**three sp^2** hybrid orbitals
4	4	**four sp^3** hybrid orbitals

Let's illustrate this phenomenon with BeH_2, BF_3, NH_3, and H_2O. We already determined the geometry in these molecules by counting groups in Section 1.7.

In **BeH_2,** the **Be atom is *sp* hybridized** because it is surrounded by two groups (two H atoms). Each Be–H bond is formed by overlap of an *sp* hybrid orbital from Be and a 1*s* orbital from H. The *sp* hybrid orbitals are oriented 180° away from each other.

In **BF_3,** the **B atom is sp^2 hybridized** because it is surrounded by three groups (three F atoms). Each B–F bond is formed by overlap of an sp^2 hybrid orbital from B and a 2*p* orbital from F. The sp^2 hybrid orbitals all lie in a plane, and are oriented 120° apart. The B atom also has a vacant unhybridized 2*p* orbital. This orbital is located ***above and below the plane*** of the BF_3 molecule.

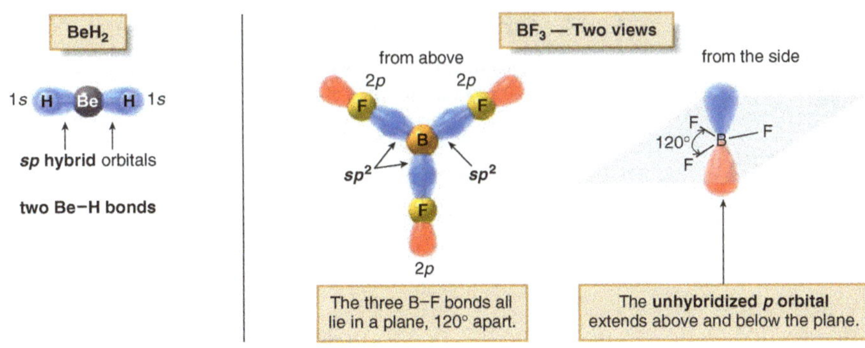

The N atom in $\mathbf{NH_3}$ and the O atom in $\mathbf{H_2O}$ are both surrounded by four groups, making them sp^3 hybridized. Each N–H and O–H bond in these molecules is formed by overlap of an sp^3 hybrid orbital with a 1*s* orbital from H. The lone pairs of electrons on N and O also occupy sp^3 hybrid orbitals, as shown in Figure 1.10.

Figure 1.10
Hybrid orbitals of NH_3 and H_2O

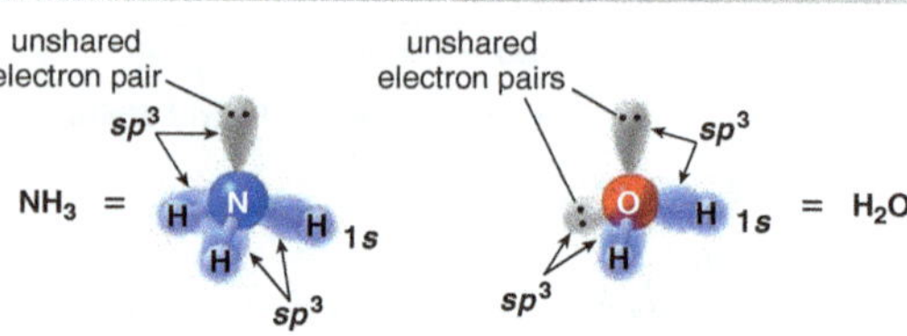

Sample Problem 1.8 What orbitals are used to form each bond in methanol, CH_3OH?

Solution

To solve this problem, draw a valid Lewis structure and count groups around each atom. Then, use the rule to determine hybridization: two groups = *sp*, three groups = sp^2, and four groups = sp^3.

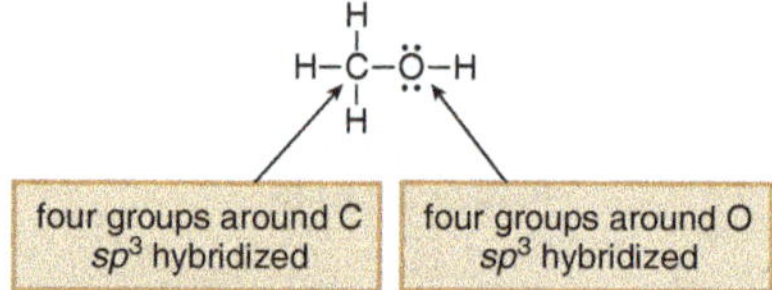

- All C–H bonds are formed from $C_{sp^3}–H_{1s}$.
- The C–O bond is formed from $C_{sp^3}–O_{sp^3}$.
- The O–H bond is formed from $O_{sp^3}–H_{1s}$.

Problem 1.28 What orbitals are used to form each bond in the following molecules?

a. CH_3BeH b. $(CH_3)_3B$ c. CH_3OCH_3

1.10 Ethane, Ethylene, and Acetylene

Let's now use the principles of hybridization to determine the type of bonds in **ethane, ethylene,** and **acetylene.**

ethane ethylene acetylene

H–C≡C–H

1.10A Ethane—CH_3CH_3

According to the Lewis structure for **ethane, CH_3CH_3,** each carbon atom is singly bonded to four other atoms. As a result:

- **Each carbon is tetrahedral.**
- **Each carbon is sp^3 hybridized.**

CH_3CH_3 = =

ethane tetrahedral C atoms

Ethane is a constituent of natural gas.

All of the bonds in ethane are σ bonds. The C–H bonds are formed from the overlap of one of the three sp^3 hybrid orbitals on each carbon atom with the 1*s* orbital on hydrogen. The C–C bond is formed from the overlap of an sp^3 hybrid orbital on each carbon atom.

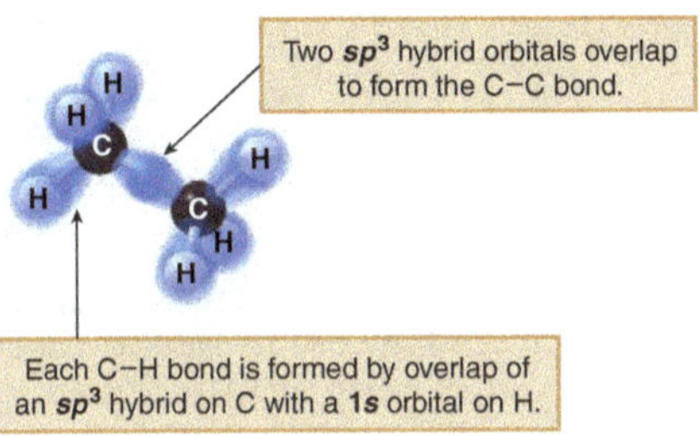

Finally, a model of ethane shows that **rotation can occur around the central C–C σ bond.** Note how the relative position of the H atoms on the adjacent CH_3 groups changes from one representation to another. This process is discussed in greater detail in Chapter 4.

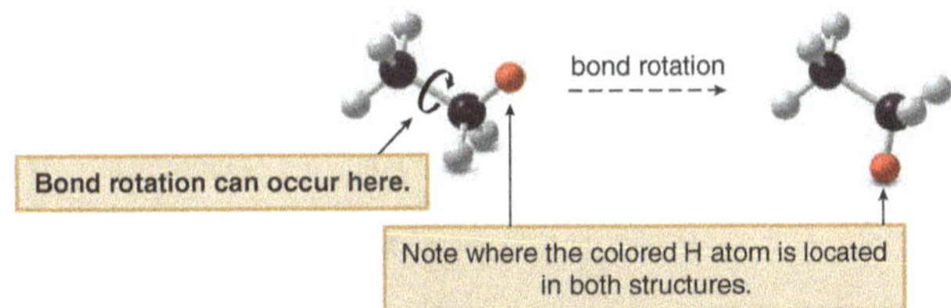

1.10B Ethylene—C_2H_4

Ethylene is an important starting material in the preparation of the plastic polyethylene.

Based on the Lewis structure of **ethylene, $CH_2{=}CH_2$,** each carbon atom is singly bonded to two H atoms and doubly bonded to the other C atom, so each C is surrounded by three groups. As a result:

- **Each carbon is trigonal planar (Section 1.7B).**
- **Each carbon is sp^2 hybridized.**

$CH_2{=}CH_2$ = (H)(H)C=C(H)(H) 120°

ethylene three groups around C

What orbitals are used to form the two bonds of the C–C double bond? Recall from Section 1.9 that **sp^2 hybrid orbitals** are formed from **one 2*s* and two 2*p* orbitals,** leaving one **2*p* orbital unhybridized.** Because carbon has four valence electrons, **each of these orbitals has one electron** that can be used to form a bond.

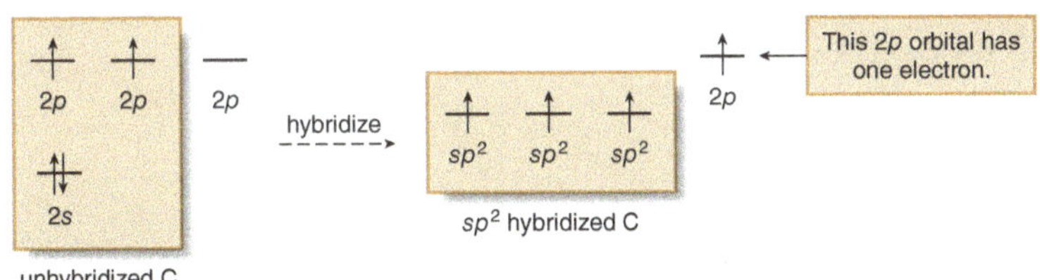

Each C−H bond results from the end-on overlap of an sp^2 hybrid orbital on carbon and the 1*s* orbital on hydrogen. Similarly, one of the C−C bonds results from the end-on overlap of an sp^2 hybrid orbital on each carbon atom. Each of these bonds is a **σ bond.**

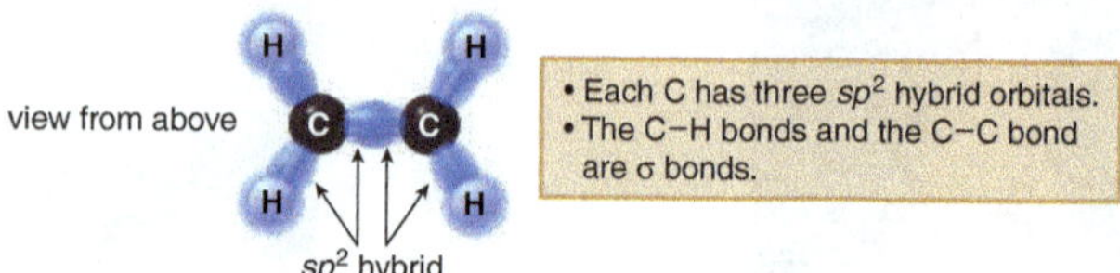

The second C−C bond results from the side-by-side overlap of the 2*p* orbitals on each carbon. Side-by-side overlap creates an area of electron density above and below the plane containing the sp^2 hybrid orbitals (that is, the plane containing the six atoms in the σ bonding system).

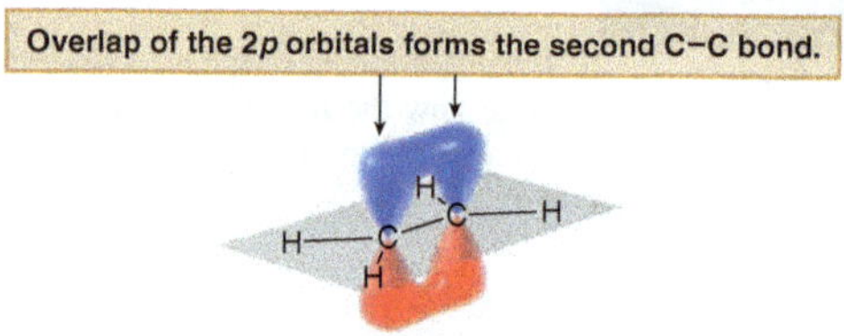

In this second bond, the electron density is ***not*** concentrated on the axis joining the two nuclei. This new type of bond is called a **π bond.** Because the electron density in a π bond is farther from the two nuclei, **π bonds are usually weaker and therefore more easily broken than σ bonds.**

Thus, a carbon–carbon double bond has two components:

- a σ bond, formed by end-on overlap of two sp^2 hybrid orbitals;
- a π bond, formed by side-by-side overlap of two 2*p* orbitals.

Figure 1.11 summarizes the bonding observed in ethylene.

Unlike the C−C single bond in ethane, rotation about the C−C double bond in ethylene is **restricted.** It can occur only if the π bond first breaks and then re-forms, a process that requires considerable energy.

All double bonds are composed of one σ and one π bond.

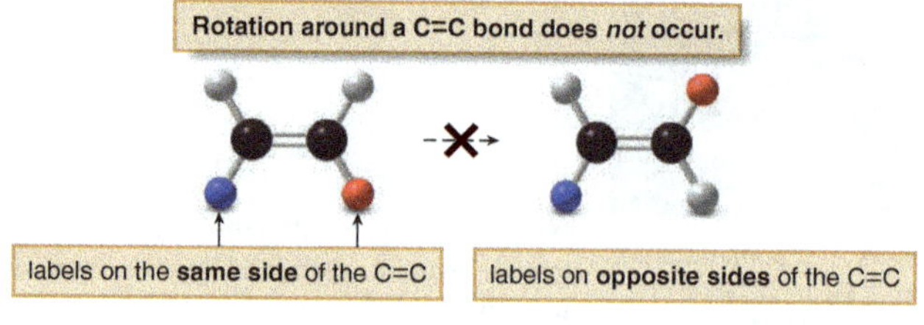

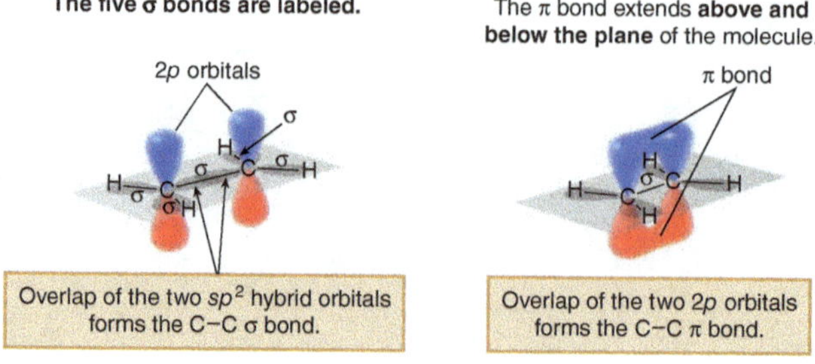

Figure 1.11 Summary: The σ and π bonds in ethylene

1.10C Acetylene—C_2H_2

Based on the Lewis structure of **acetylene, HC≡CH,** each carbon atom is singly bonded to one hydrogen atom and triply bonded to the other carbon atom, so each carbon atom is surrounded by two groups. As a result:

- **Each carbon is linear (Section 1.7B).**
- **Each carbon is *sp* hybridized.**

180°
H–C≡C–H
two groups around C
acetylene

Because acetylene produces a very hot flame on burning, it is often used in welding torches. The fire is very bright, too, so it was once used in the lamps worn by spelunkers—people who study and explore caves.

What orbitals are used to form the bonds of the C–C triple bond? Recall from Section 1.9 that ***sp* hybrid orbitals** are formed from **one 2*s* and one 2*p* orbital,** leaving **two 2*p* orbitals unhybridized.** Because carbon has four valence electrons, **each of these orbitals has one electron** that can be used to form a bond.

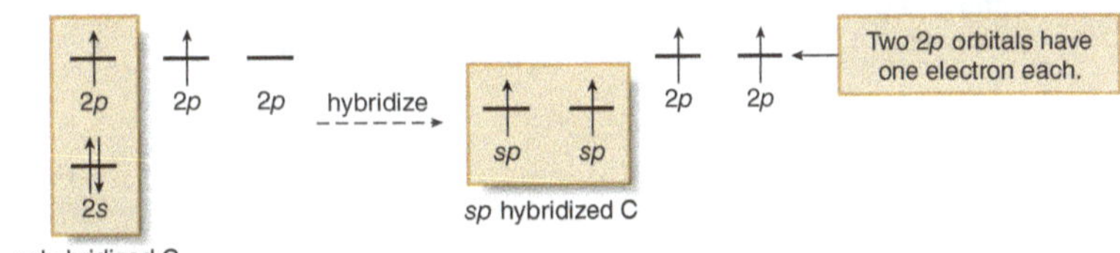

Each C–H bond results from the end-on overlap of an *sp* hybrid orbital on carbon and the 1*s* orbital on hydrogen. Similarly, one of the C–C bonds results from the end-on overlap of an *sp* hybrid orbital on each carbon atom. Each of these bonds is a **σ bond.**

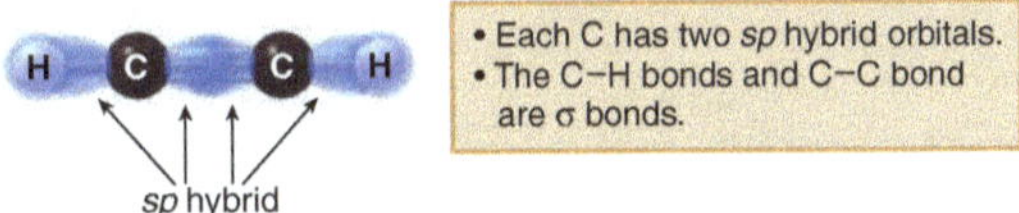

Each carbon atom also has two **unhybridized 2*p* orbitals** that are perpendicular to each other and to the *sp* hybrid orbitals. Side-by-side overlap between the two 2*p* orbitals on one carbon with the two 2*p* orbitals on the other carbon creates the second and third bonds of the C–C triple bond. The electron density from one of these two bonds is above and below the axis joining the two nuclei, and the electron density from the second of these two bonds is in front of and behind the axis, so both of these bonds are **π bonds.**

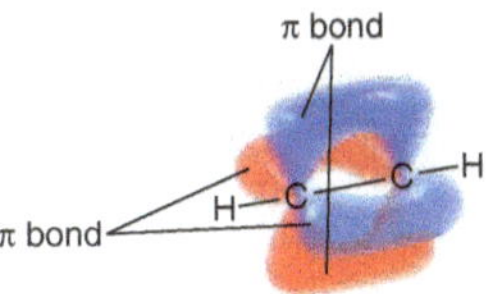

The side-by-side overlap of two *p* orbitals always forms a π bond.

Thus, a carbon–carbon triple bond has three components:

- **a σ bond, formed by end-on overlap of two *sp* hybrid orbitals;**
- **two π bonds, formed by side-by-side overlap of two sets of 2*p* orbitals.**

All triple bonds are composed of one σ and two π bonds.

Figure 1.12 summarizes the bonding observed in acetylene. Table 1.4 summarizes the three possible types of bonding in carbon compounds.

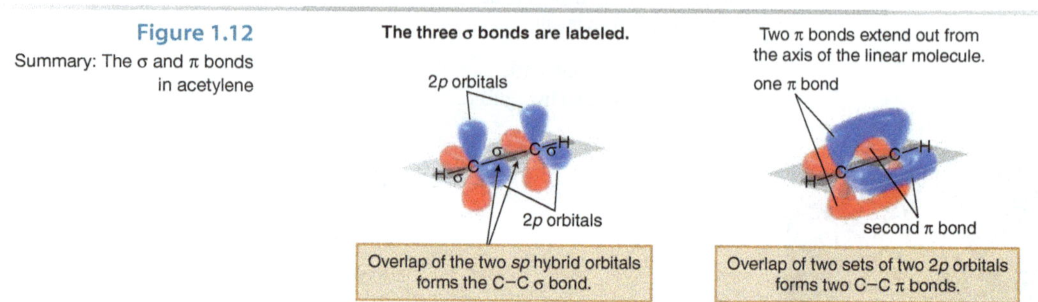

Figure 1.12
Summary: The σ and π bonds in acetylene

Table 1.4 A Summary of Covalent Bonding Seen in Carbon Compounds

Number of groups bonded to C	Hybridization	Bond angle	Example	Observed bonding
4	sp^3	109.5°	CH_3CH_3 ethane	one σ bond $C_{sp^3}-C_{sp^3}$
3	sp^2	120°	$CH_2=CH_2$ ethylene	one σ bond + one π bond $C_{sp^2}-C_{sp^2}$ $C_{2p}-C_{2p}$
2	sp	180°	$HC\equiv CH$ acetylene	one σ bond + two π bonds $C_{sp}-C_{sp}$ $C_{2p}-C_{2p}$ $C_{2p}-C_{2p}$

Sample Problem 1.9 Answer each question for the molecule acetone, drawn below:

CH_3
→C=Ö←
→CH_3
acetone

a. Determine the hybridization of the indicated atoms.
b. What orbitals are used to form the C–O double bond?
c. In what type of orbital does each lone pair reside?

Solution

a.

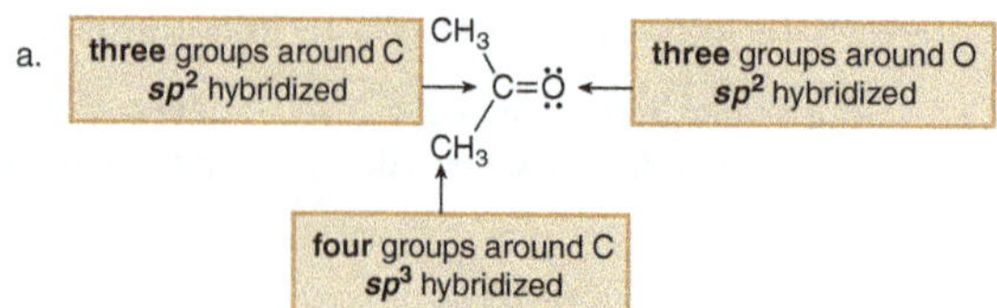

b. • The σ bond is formed from the end-on overlap of $C_{sp^2}-O_{sp^2}$.
• The π bond is formed from the side-by-side overlap of $C_{2p}-O_{2p}$.

c. The O atom has three sp^2 hybrid orbitals.
• One is used for the σ bond of the double bond.
• The remaining two sp^2 hybrids are occupied by the lone pairs.

Problem 1.29 Determine the hybridization around the indicated atoms in the following molecules:

a. $CH_3-C\equiv CH$ b. (cyclohexylidene)$=\ddot{N}-CH_3$ c. $CH_2=C=CH_2$

Problem 1.30 Classify each bond in the following molecules as σ or π:

a. CH_3CHO b. $CH_3-C\equiv N$ c. $HCOOCH_3$

1.11 Bond Length and Bond Strength

Let's now examine the relative bond length and bond strength of the C–C and C–H bonds in ethane, ethylene, and acetylene.

1.11A A Comparison of Carbon–Carbon Bonds

While the SI unit of energy is the **joule** (J), organic chemists often report energy values in **calories** (cal). For this reason, energy values in the tables in this text are reported in joules, followed by the number of calories in parentheses.
1 cal = 4.18 J

Note the inverse relationship between bond length and bond strength. The shorter the bond, the closer the electron density is kept to the nucleus, and the harder the bond is to break. ***Shorter* bonds are *stronger* bonds.**

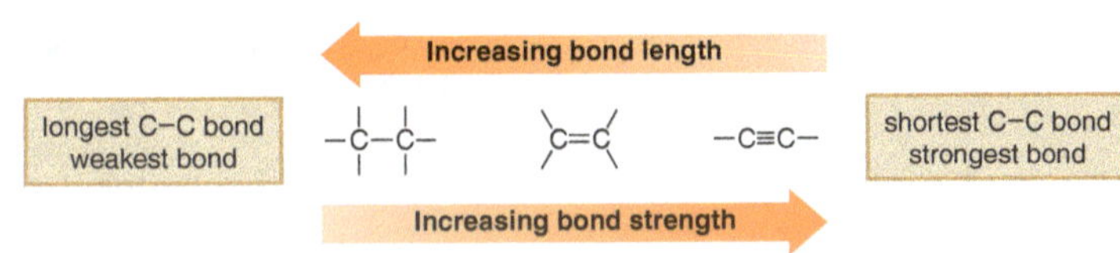

- **As the number of electrons between two nuclei *increases*, bonds become shorter and stronger.**
- **Thus, triple bonds are shorter and stronger than double bonds, which are shorter and stronger than single bonds.**

Values for bond lengths and bond strengths for CH_3CH_3, $CH_2=CH_2$, and $HC\equiv CH$ are listed in Table 1.5. Be careful not to confuse two related but different principles regarding multiple bonds such as C–C double bonds. **Double bonds, consisting of both a σ and a π bond, are *strong*.** The **π component** of the double bond, however, is usually much ***weaker* than the σ component.** This is a particularly important consideration when studying alkenes in Chapter 10.

1.11B A Comparison of Carbon–Hydrogen Bonds

The length and strength of a C–H bond vary slightly depending on the hybridization of the carbon atom.

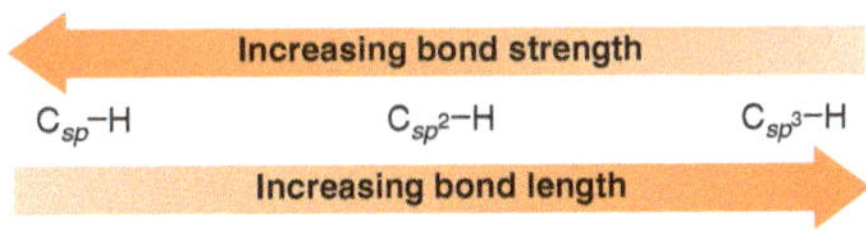

To understand why this is so, we must look at the atomic orbitals used to form each type of hybrid orbital. A single 2*s* orbital is always used, but the number of 2*p* orbitals varies with the type of

Table 1.5 Bond Lengths and Bond Strengths for Ethane, Ethylene, and Acetylene

Compound	C–C bond length (pm)	Bond strength kJ/mol (kcal/mol)
CH_3–CH_3	153	368 (88)
CH_2=CH_2	134	635 (152)
HC≡CH	121	837 (200)

Increasing bond length (upward); Increasing bond strength (downward)

Compound	C–H bond length (pm)	Bond strength kJ/mol (kcal/mol)
CH_3CH_2–H	111	410 (98)
CH_2=C(H)–H	110	435 (104)
HC≡C–H	109	523 (125)

Increasing bond length (upward); Increasing bond strength (downward)

hybridization. A quantity called **percent *s*-character** indicates the fraction of a hybrid orbital due to the 2*s* orbital used to form it.

sp hybrid $\frac{\text{one 2s orbital}}{\text{two hybrid orbitals}}$ = 50% *s*-character

sp^2 hybrid $\frac{\text{one 2s orbital}}{\text{three hybrid orbitals}}$ = 33% *s*-character

sp^3 hybrid $\frac{\text{one 2s orbital}}{\text{four hybrid orbitals}}$ = 25% *s*-character

Why should the percent *s*-character of a hybrid orbital affect the length of a C–H bond? A 2*s* orbital keeps electron density closer to a nucleus compared to a 2*p* orbital. As the **percent *s*-character increases,** a hybrid orbital holds its electrons closer to the nucleus, and the **bond becomes shorter and stronger.**

Increased percent *s*-character ---→ Increased bond strength ---→ Decreased bond length

Problem 1.31 Which of the indicated bonds in each pair of compounds is shorter?

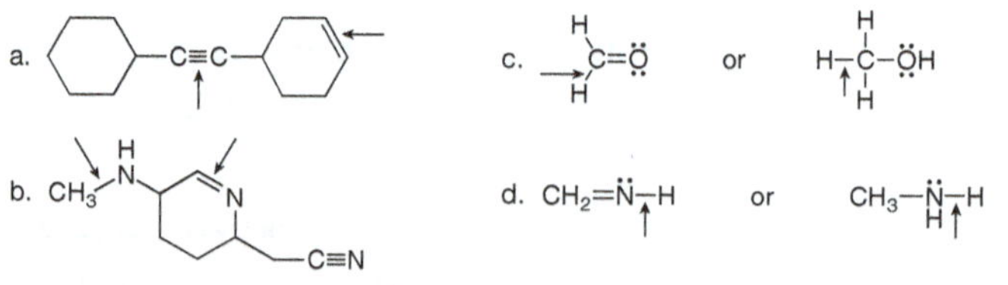

a. (cyclohexyl)–C≡C–(cyclohexenyl)

b. CH_3–NH–(ring with C=N)–C≡N

c. H_2C=Ö or H_3C–ÖH

d. CH_2=N̈–H or CH_3–N̈H–H

1.12 Electronegativity and Bond Polarity

***Electronegativity* is a measure of an atom's attraction for electrons in a bond.** Thus, electronegativity indicates how much a particular atom "*wants*" electrons. The following trends in electronegativity are observed in the periodic table:

- Electronegativity *increases* across a row of the periodic table as the nuclear charge increases (excluding the noble gases).
- Electronegativity *decreases* down a column of the periodic table as the atomic radius increases, pushing the valence electrons farther from the nucleus.

As a result, the *most* electronegative elements are located at the **upper right-hand corner** of the periodic table, and the *least* electronegative elements in the **lower left-hand corner.** A scale has been established to represent electronegativity values arbitrarily, from 0 to 4, as shown in Figure 1.13.

Electronegativity values are relative, so they can be used for comparison purposes only. When comparing two different elements, one is **more electronegative** than the other if it attracts electron density toward itself. One is less electronegative—**more electro*positive***—if it gives up electron density to the other element.

Problem 1.32 Rank the following atoms in order of increasing electronegativity. Label the most electronegative and most electropositive atom in each group.

a. Se, O, S b. P, Na, Cl c. Cl, S, F d. O, P, N

Electronegativity values are used as a guideline to indicate whether the electrons in a bond are **equally shared** or **unequally shared** between two atoms. For example, whenever two identical atoms are bonded together, each atom attracts the electrons in the bond to the same extent. The electrons are equally shared, and the **bond is *nonpolar.*** Thus, a **carbon–carbon bond is nonpolar.** The same is true whenever two different atoms having similar electronegativities are bonded together. **C−H bonds are considered to be nonpolar,** because the electronegativity difference between C (2.5) and H (2.2) is small.

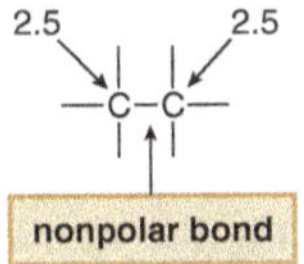

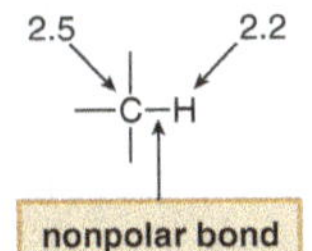

The small electronegativity difference between C and H is ignored.

Figure 1.13 Electronegativity values for some common elements

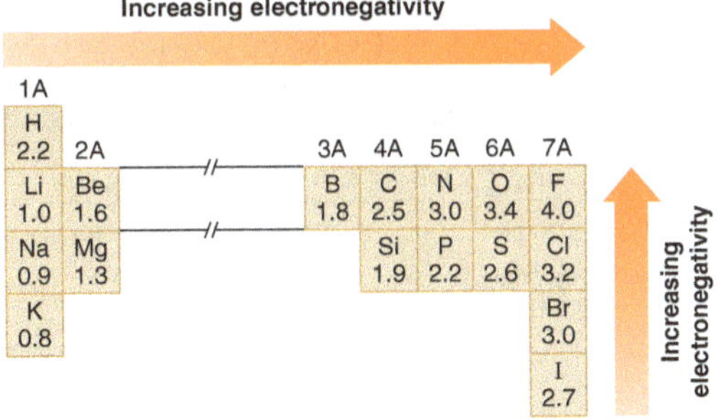

Bonding between atoms of different electronegativity values results in the **unequal sharing** of electrons. For example, in a C–O bond, the electrons are pulled away from C (2.5) toward O (3.4), the element of higher electronegativity. **The bond is *polar,*** or ***polar covalent.*** The bond is said to have a **dipole;** that is, **a separation of charge.**

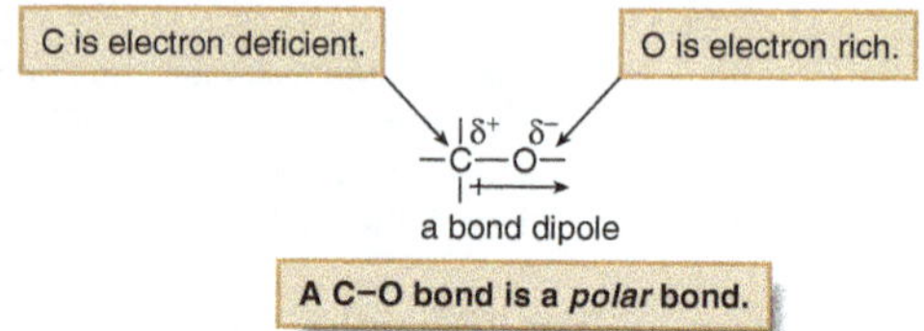

The direction of polarity in a bond is often indicated by an arrow, with the head of the arrow pointing toward the more electronegative element. The tail of the arrow, with a perpendicular line drawn through it, is positioned at the less electronegative element. Alternatively, the symbols δ^+ and δ^- indicate this unequal sharing of electron density.

- δ^+ means an atom is electron deficient (has a partial positive charge).
- δ^- means the atom is electron rich (has a partial negative charge).

Problem 1.33 Show the direction of the dipole in each bond. Label the atoms with δ^+ and δ^-.

a. H–F b. –B–C– c. –C–Li d. –C–Cl

Students often wonder how large an electronegativity difference must be to consider a bond polar. That's hard to say. We will set an arbitrary value for this difference and use it as an *approximation.* **Usually, a polar bond will be one in which the electronegativity difference between two atoms is ≥ 0.5 units.**

The distribution of electron density in a molecule can be shown using an **electrostatic potential map.** These maps are color coded to illustrate areas of high and low electron density. Electron-rich regions are indicated in red, and electron-deficient sites are indicated in blue. Regions of intermediate electron density are shown in orange, yellow, and green.

For example, an electrostatic potential map of CH_3Cl clearly indicates the polar nature of the C–Cl bond (Figure 1.14). The more electronegative Cl atom pulls electron density toward it, making it electron rich. This is indicated by the red around the Cl in the plot. The carbon is electron deficient, and this is shown with blue.

When comparing two maps, the comparison is useful only if they are plotted *using the same scale* of color gradation. For this reason, whenever we compare two plots in this text, they will be drawn side by side using the same scale. It will be difficult to compare two plots in different parts of the book, because the scale may be different. Despite this limitation, an electrostatic potential plot is a useful tool for visually evaluating the distribution of electron density in a molecule, and with care, comparing the electron density in two different molecules.

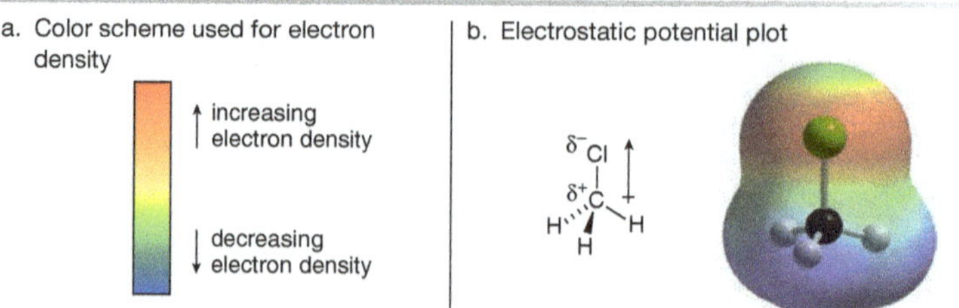

Figure 1.14 Electrostatic potential plot of CH_3Cl

1.13 Polarity of Molecules

A **polar molecule** has either one polar bond, or two or more bond dipoles that reinforce. A **nonpolar molecule** has either no polar bonds, or two or more bond dipoles that cancel.

Whenever C or H is bonded to N, O, and all halogens, the bond is ***polar.*** Thus, the C–I bond is considered polar even though the electronegativity difference between C and I is small. Remember, electronegativity is just an approximation.

Thus far, we have been concerned with the polarity of one bond. To determine whether a molecule has a net dipole, use the following two-step procedure:

[1] Use electronegativity differences to **identify all of the polar bonds and the directions of the bond dipoles.**

[2] **Determine the geometry** around individual atoms by counting groups, and decide if individual dipoles **cancel** or **reinforce each other in space.**

The two molecules $\mathbf{H_2O}$ and $\mathbf{CO_2}$ illustrate different outcomes of this process. In H_2O, each O–H bond is polar because the electronegativity difference between O (3.4) and H (2.2) is large. Since H_2O is a **bent** molecule, the two dipoles reinforce (both point *up*). Thus, **$\mathbf{H_2O}$ has a net dipole, making it a polar molecule.** CO_2 also has polar C–O bonds because the electronegativity difference between O (3.4) and C (2.5) is large. However, CO_2 is a **linear** molecule, so the two dipoles, which are equal and opposite in direction, **cancel.** Thus, CO_2 is a **nonpolar molecule** with **no net dipole.**

δ⁻ net dipole δ⁺

The net dipole bisects the H–O–H bond angle.
The two individual dipoles reinforce.

H_2O is a polar molecule.

δ⁻ δ⁺ δ⁻

no net dipole
The two dipoles cancel.

CO_2 is a nonpolar molecule.

Electrostatic potential plots for H_2O and CO_2 appear in Figure 1.15. Additional examples of polar and nonpolar molecules are given in Figure 1.16.

Problem 1.34 Indicate which of the following molecules is polar because it possesses a net dipole. Show the direction of the net dipole if one exists.

a. CH_3Br b. CH_2Br_2 c. CF_4 d. (Cl)(H)C=C(Cl)(H), cis e. (Cl)(H)C=C(H)(Cl), trans

Figure 1.15 Electrostatic potential plots for H_2O and CO_2

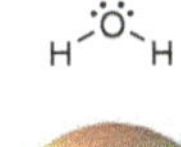

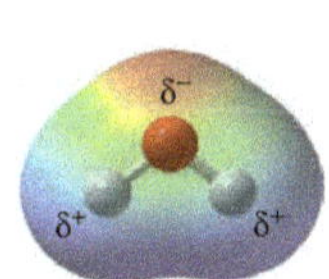

- The electron-rich (red) region is concentrated on the electronegative O atom. Both H atoms are electron deficient (blue-green).

:O=C=O:

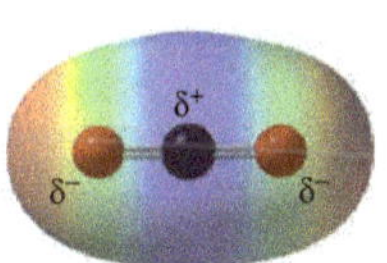

- Both electronegative O atoms are electron rich (red) and the central C atom is electron deficient (blue).

Figure 1.16 More examples of polar and nonpolar molecules

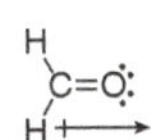

one polar bond

a **polar** molecule

three polar bonds
All dipoles cancel.
no net dipole

a **nonpolar** molecule

net dipole

three polar bonds
All dipoles reinforce.

a **polar** molecule

net dipole

two polar bonds
Both dipoles reinforce.

a **polar** molecule

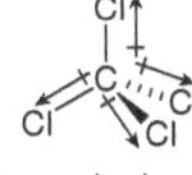

four polar bonds
All dipoles cancel.
no net dipole

a **nonpolar** molecule

1.14 L-Dopa—A Representative Organic Molecule

The principles learned in this chapter apply to all organic molecules regardless of size or complexity. For example, we now know a great deal about the structure of **L-dopa,** a drug used to treat Parkinson's disease described in the chapter opener.

Sample Problem 1.10

Answer each question about L-dopa.

L-dopa

a. Convert the skeletal structure to a Lewis structure.
b. What is the hybridization and geometry around each labeled atom?
c. Label three polar bonds.

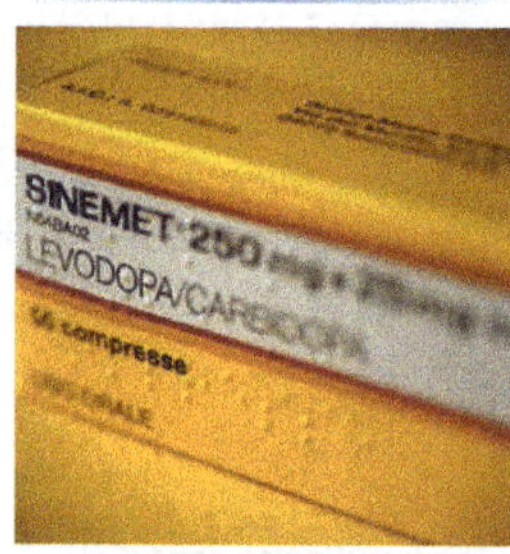

Sinemet, the trade name of a drug used to treat Parkinson's disease, contains a combination of L-dopa (Sample Problem 1.10) and carbidopa (Problem 1.36). Carbidopa increases the effectiveness of L-dopa by inhibiting its metabolism prior to crossing the blood–brain barrier and entering the brain.

Solution

a, b. Convert the skeletal structure to a Lewis structure as in Sample Problem 1.7. Each O atom needs two lone pairs and each N atom needs one for an octet. Count groups around an atom to determine hybridization and shape.

sp^2 hybridized and trigonal planar

sp^3 hybridized and tetrahedral

sp^3 hybridized and trigonal pyramidal

Lewis structure of L-dopa

c. All C–O, O–H, N–H, and C–N bonds are polar because the electronegativity difference between the atoms is large.

Problem 1.35

Provide the following information about L-dopa.

a. Label all sp^3 hybridized C atoms.
b. Label all H atoms that bear a partial positive charge (δ^+).
c. Draw another resonance structure.

Sample Problem 1.11 illustrates how to derive structural information from a ball-and-stick model.

Sample Problem 1.11

Use the ball-and-stick model of vitamin B_6 to answer each question.

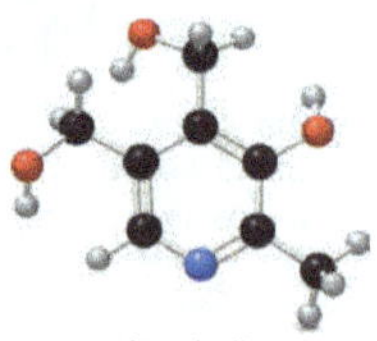

vitamin B_6

a. Draw a skeletal structure of vitamin B_6.
b. How many sp^2 hybridized carbons are present?
c. What is the hybridization of the N atom in the ring?

Solution

Use the element colors shown in Section 1.7B to convert the 3-D model to a skeletal structure [black (C), gray (H), red (O), blue (N)]. H atoms on carbon are omitted, but H atoms on heteroatoms are drawn in. Count groups to determine hybridization. Each O atom needs two lone pairs and each N needs one to give an octet of electrons.

a. 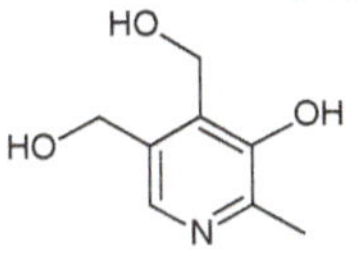

skeletal structure of vitamin B_6

b, c. 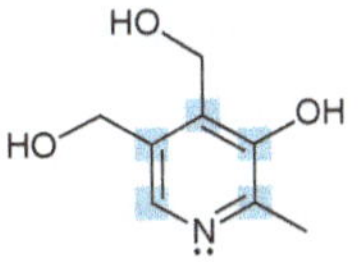

- Each C labeled in blue is sp^2 hybridized.
- The N atom is surrounded by three groups (two atoms and one lone pair), making it sp^2 hybridized.

Problem 1.36 Use the ball-and-stick model to answer each question about carbidopa, a drug used in combination with L-dopa to treat Parkinson's disease

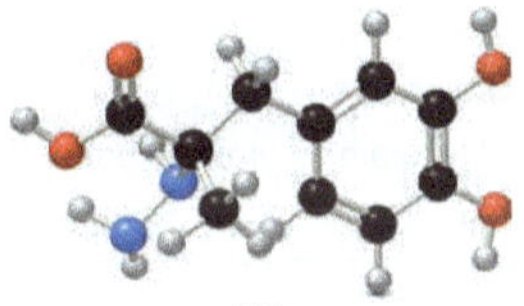

carbidopa

a. Draw a skeletal structure of carbidopa.
b. Determine the hybridization around each carbon atom.
c. What is the hybridization and geometry around each N atom?
d. How many polar bonds are present?

KEY CONCEPTS

Structure and Bonding

Important Facts

- **The general rule of bonding:** Atoms "strive" to attain a complete outer shell of valence electrons (Section 1.2). H "wants" two electrons. Second-row elements "want" eight electrons.
- **Formal charge** is the difference between the number of valence electrons of an atom and the number of electrons it "owns" (Section 1.3C). See Sample Problem 1.3 for a stepwise example.
- **Curved arrow notation** shows the movement of an electron pair. The tail of the arrow always begins at an electron pair, either in a bond or a lone pair. The head points to where the electron pair "moves" (Section 1.6).
- **Electrostatic potential plots** are color-coded maps of electron density, indicating electron-rich and electron-deficient regions (Section 1.12).

The Importance of Lewis Structures (Sections 1.3–1.5)

A properly drawn Lewis structure shows the number of bonds and lone pairs present around each atom in a molecule. In a valid Lewis structure, each H has two electrons, and each second-row element has no more than eight. This is the first step needed to determine many properties of a molecule.

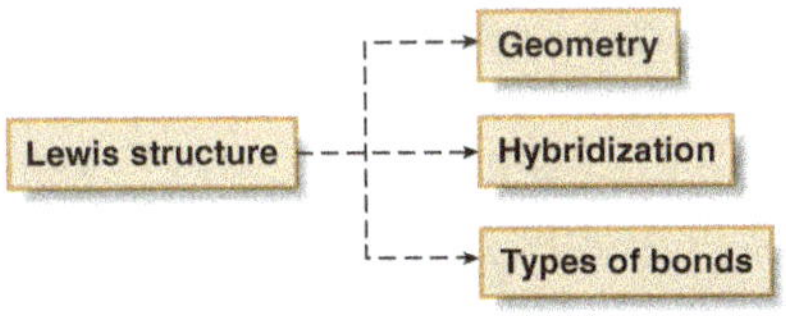

Geometry: [linear, trigonal planar, or tetrahedral] (Section 1.7)

Hybridization: [*sp*, sp^2, or sp^3] (Section 1.9)

Types of bonds: [single, double, or triple] (Sections 1.3, 1.10)

Resonance (Section 1.6)

The basic principles:

- Resonance exists when a compound cannot be represented by a single Lewis structure.
- Resonance structures differ in the position of only nonbonded electrons and π bonds, not atoms.
- The resonance hybrid is the only accurate representation for a resonance-stabilized compound. A hybrid represents the compound better than any single resonance structure because electron density is delocalized.

The difference between resonance structures and isomers:

- Two **isomers** differ in the arrangement of *both* atoms and electrons.
- **Resonance structures** differ *only* in the arrangement of electrons.

$CH_3-C(=O)-O-CH_3$ $CH_3CH_2-C(=O)-O-H$ ⟷ $CH_3CH_2-C(-O^-)=O^+-H$

isomers

resonance structures

Geometry and Hybridization

The number of groups around an atom determines both its geometry (Section 1.7) and hybridization (Section 1.9).

Number of groups	Geometry	Bond angle (°)	Hybridization
2	linear	180	*sp*
3	trigonal planar	120	sp^2
4	tetrahedral	109.5	sp^3

Drawing Organic Molecules (Section 1.8)

- Shorthand methods are used to abbreviate the structure of organic molecules.

skeletal structure = $CH_3-C(CH_3)(H)-C(H)(H)-C(CH_3)(CH_3)-CH_3$ = $(CH_3)_2CHCH_2C(CH_3)_3$

skeletal structure | isooctane | condensed structure

- A carbon bonded to four atoms is tetrahedral. The best way to represent a tetrahedron is to draw two bonds in the plane, one bond in front, and one bond behind.

Bond Length

- Bond length decreases across a row and increases down a column of the periodic table (Section 1.7A).
- Bond length decreases as the number of electrons between two nuclei increases (Section 1.11A).
- Bond length decreases as the percent *s*-character increases (Section 1.11B).
- Bond length and bond strength are inversely related. In general, shorter bonds are stronger bonds (Section 1.11).
- Sigma (σ) bonds are generally stronger than π bonds (Section 1.10).

Electronegativity and Polarity (Sections 1.12, 1.13)

- Electronegativity increases across a row and decreases down a column of the periodic table.
- A polar bond results when two atoms of different electronegativity values are bonded together. Whenever C or H is bonded to N, O, or any halogen, the bond is polar.
- A polar molecule has either one polar bond, or two or more bond dipoles that reinforce.

PROBLEMS

Problems Using Three-Dimensional Models

1.37 Citric acid is responsible for the tartness of citrus fruits, especially lemons and limes.

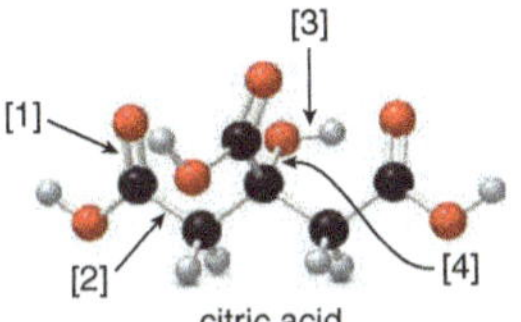

citric acid

a. What is the molecular formula for citric acid?
b. How many lone pairs are present?
c. Draw a skeletal structure.
d. How many sp^2 hybridized carbons are present?
e. What orbitals are used to form each indicated bond ([1]–[4])?

1.38 Zingerone gives ginger its pungent taste.

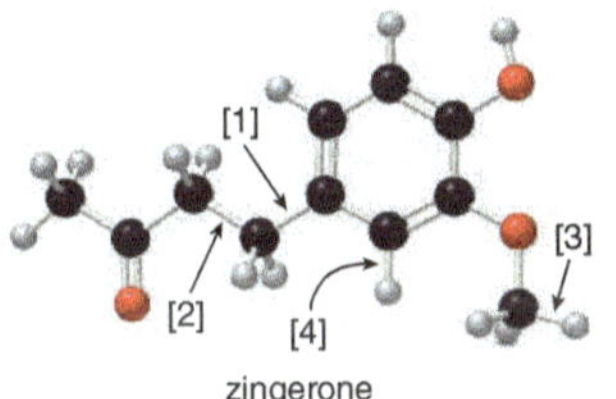

zingerone

a. What is the molecular formula for zingerone?
b. How many lone pairs are present?
c. Draw a skeletal structure.
d. How many sp^2 hybridized carbons are present?
e. What orbitals are used to form each indicated bond ([1]–[4])?

Atomic Structure, Ionic Bonding, and Covalent Bonding

1.39 Two radioactive isotopes of iodine used for the diagnosis and treatment of thyroid disease have mass numbers of 123 and 131. For each isotope, give the following information: (a) the number of protons; (b) the number of neutrons; (c) the number of electrons in the neutral atom; (d) the group number.

1.40 Label each bond in the following compounds as ionic or covalent.

a. NaI b. BrCl c. HCl d. CH_3NH_2 e. $NaOCH_3$

Lewis Structures and Formal Charge

1.41 Assign formal charges to each carbon atom in the given species. All lone pairs have been drawn in.

a. $CH_2{=}\ddot{C}H$ b. $H{-}\ddot{C}{-}H$ c. $H{-}\dot{C}{-}H$ (with H below C) d. $H{-}C{-}C$ (with H above and H below each C)

1.42 Assign formal charges to each N and O atom in the given molecules. All lone pairs have been drawn in.

a. $CH_3{-}\ddot{N}{-}CH_3$ (lone pair below N) c. $CH_3{-}N{\equiv}N{:}$ e. $CH_3{-}\ddot{O}{\cdot}$ (lone pair below O)

b. $:\ddot{N}{=}N{=}\ddot{N}:$ d. $CH_3{-}C({=}\ddot{O}H){-}CH_3$ (:OH above C) f. $CH_3{-}\ddot{N}{=}\ddot{O}$ (lone pair below O)

1.43 Draw one valid Lewis structure for each compound. Assume the atoms are arranged as drawn.

a. CH_2N_2 H C N N (H below C)

b. CH_3NO_2 H C N O (H above C; H below C, O below N)

c. CH_3CNO H C C N O (H above and below first C)

d. HCO_2^- H C O (O above C)

e. HCO_3^- H O C O (O above C)

f. $^-CH_2CN$ H C C N (H below first C)

1.44 Draw a valid Lewis structure for each compound.

a. N_2 b. $(CH_3OH_2)^+$ c. $(CH_3CH_2)^-$ d. HNNH e. H_6BN

1.45 Draw an acceptable Lewis structure from each condensed structure, such that all atoms have zero formal charge.

a. diethyl ether, $(CH_3CH_2)_2O$, the first general anesthetic used in medical procedures
b. acrylonitrile, CH_2CHCN, starting material used to manufacture synthetic Orlon fibers
c. dihydroxyacetone, $(HOCH_2)_2CO$, an ingredient in sunless tanning products
d. acetic anhydride, $(CH_3CO)_2O$, a reagent used to synthesize aspirin

Isomers and Resonance Structures

1.46 Creatine is a dietary supplement used by some athletes to boost their athletic performance. (a) Draw in all lone pairs in creatine. (b) Draw two additional resonance structures showing all lone pairs and formal charges.

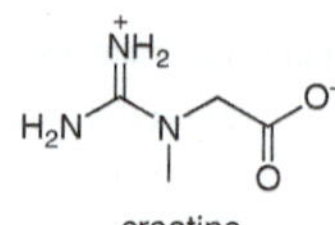

creatine

1.47 Draw all possible isomers for each molecular formula.

a. C_3H_7Cl (two isomers) b. C_2H_4O (three isomers) c. C_3H_9N (four isomers)

1.48 Draw Lewis structures for the nine isomers having molecular formula C_3H_6O, with all atoms having a zero formal charge.

1.49 With reference to compound **A** drawn below, label each compound as an isomer, a resonance structure, or neither.

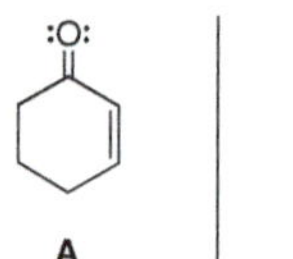

A

a. 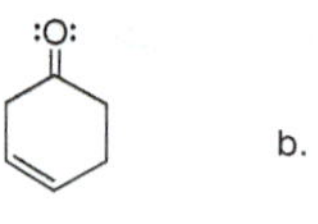 b. c. 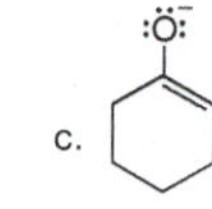 d. 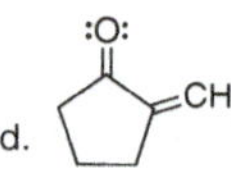

1.50 With reference to cation **B,** label each species as an isomer, a resonance structure, or neither.

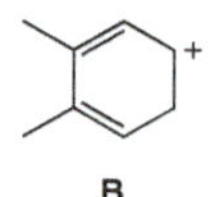

B

a. 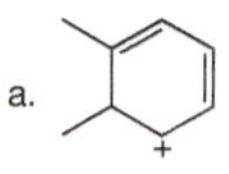 b. 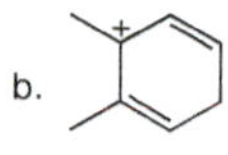 c. 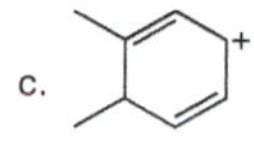 d.

1.51 How are the molecules or ions in each pair related? Classify them as resonance structures, isomers, or neither.

a. $CH_3-\ddot{O}-CH_2CH_3$ and $CH_3-C(CH_3)(H)-\ddot{O}H$

b. □ and $CH_3-C(H)=C(H)-CH_3$

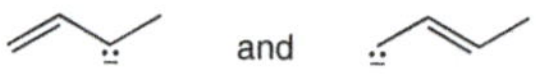

c. and

d. $CH_3CH_2CH_3$ and $CH_3CH_2\ddot{C}H_2$

1.52 Add curved arrows to show how the first resonance structure can be converted into the second.

a. $CH_3-\overset{+}{C}(H)-\ddot{N}(H)-CH_3 \longleftrightarrow CH_3-C(H)=\overset{+}{N}(H)-CH_3$

b. $H-C(=\ddot{O})-\ddot{N}H_2 \longleftrightarrow H-C(-\ddot{O}:^-)=\overset{+}{N}H_2$

c. $\longleftrightarrow$

1.53 Follow the curved arrows to draw a second resonance structure for each species.

a. $CH_3-\overset{+}{N}\equiv N: \longleftrightarrow$

b. $CH_3-C(-\ddot{O}:^-)=CH-\overset{+}{C}H_2 \longleftrightarrow$

c. 

d. 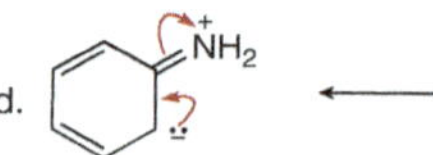

1.54 Draw a second resonance structure for each ion. Then, draw the resonance hybrid.

a. $CH_3-C(=O)-\ddot{O}:^-$

b. $CH_2=\overset{+}{N}H_2$

c.

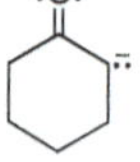

d. $H-C(=\overset{+}{O}H)-H$

1.55 Draw all reasonable resonance structures for each species.

a. O_3

b. NO_3^- (a central N atom)

c. N_3^-

d. $CH_3-C(=O)-\ddot{C}H-C(=O)-CH_3$

e.

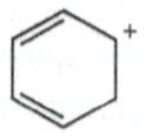

f. $CH_2=CH-\ddot{C}H-CH=CH_2$

g.

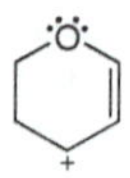

1.56 Draw four additional resonance structures for the following cation. Then draw the resonance hybrid.

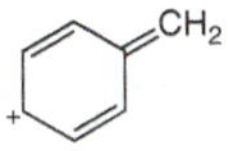

1.57 Rank the resonance structures in each group in order of increasing contribution to the resonance hybrid. Label the resonance structure that contributes the most as **3** and the resonance structure that contributes the least as **1.** Label the intermediate contributor as **2.**

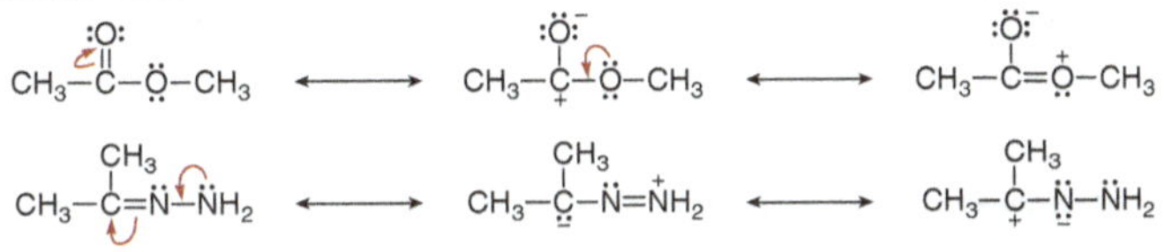

a. $CH_3-C(=O)-\ddot{O}-CH_3 \longleftrightarrow CH_3-\overset{+}{C}(-O^-)-\ddot{O}-CH_3 \longleftrightarrow CH_3-C(-O^-)=\overset{+}{O}-CH_3$

b. $CH_3-C(CH_3)=\ddot{N}-\ddot{N}H_2 \longleftrightarrow CH_3-\overset{-}{C}(CH_3)-\ddot{N}=\overset{+}{N}H_2 \longleftrightarrow CH_3-\overset{+}{C}(CH_3)-\overset{-}{\ddot{N}}-\ddot{N}H_2$

1.58 Consider the compounds and ions with curved arrows drawn below. When the curved arrows give a second valid resonance structure, draw the resonance structure. When the curved arrows generate an invalid Lewis structure, explain why the structure is unacceptable.

a.

b.

c. 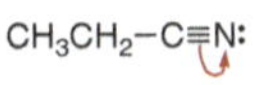 $CH_3CH_2-C\equiv N:$

d.

Geometry

1.59 Predict all bond angles in each compound.

a. CH_3Cl b. NH_2OH c. $CH_2{=}NCH_3$ d. $HC\equiv CCH_2OH$ e.

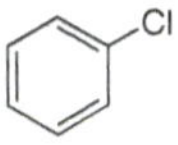

1.60 Predict the geometry around each indicated atom.

a. $CH_3CH_2CH_2CH_3$

b. $(CH_3)_2N^-$

c.

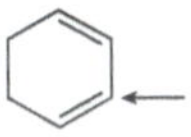

d. BF_4^-

e. 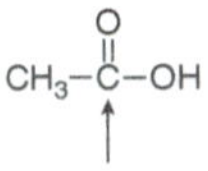 $CH_3-C(=O)-OH$

f. $(CH_3)_3N$

1.61 Draw the structure of halothane, $CF_3CHClBr$, in three dimensions, using solid lines, wedges, and dashes to illustrate the position of atoms. Halothane is a safe and widely used general anesthetic.

Drawing Organic Molecules

1.62 How many hydrogens are present around each carbon atom in the following molecules?

a. 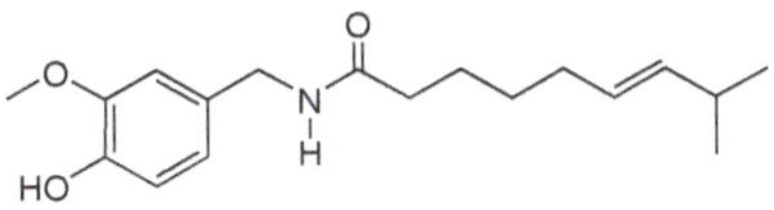 capsaicin
(spicy component of hot peppers)

b. 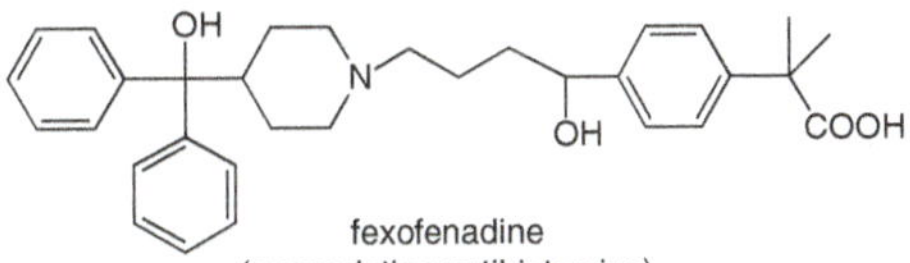 fexofenadine
(nonsedating antihistamine)

1.63 Draw in all the carbon and hydrogen atoms in each molecule.

a. 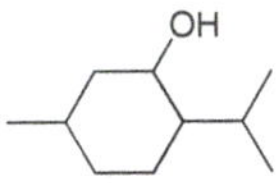 menthol
(isolated from peppermint oil)

b. 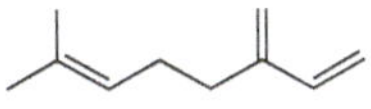 myrcene
(isolated from bayberry)

c. 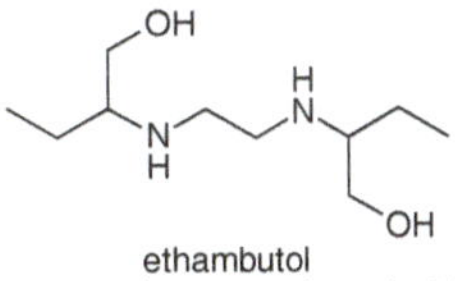 ethambutol
(drug used to treat tuberculosis)

d. 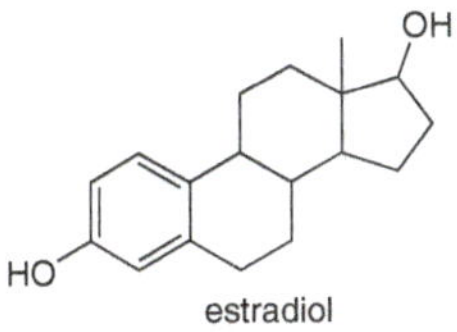 estradiol
(a female sex hormone)

1.64 Convert each molecule into a skeletal structure.

a. $(CH_3)_2CHCH_2CH_2CH(CH_3)_2$

b. $CH_3CH(Cl)CH(OH)CH_3$

c. $(CH_3)_3C(CH_2)_5CH_3$

d. 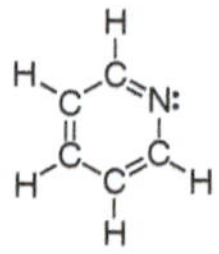 (ring structure with H, C, N:)

e. 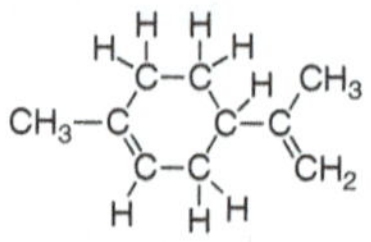 CH_3–(ring structure)

limonene
(oil of lemon)

f. $CH_3(CH_2)_2C(CH_3)_2CH(CH_3)CH(CH_3)CH(Br)CH_3$

1.65 Convert the following condensed formulas into Lewis structures.

a. CH_3CH_2COOH
b. $CH_3CONHCH_3$
c. CH_3COCH_2Br
d. $(CH_3)_3COH$
e. $(CH_3)_3CCHO$
f. CH_3COCl
g. $CH_3COCH_2CO_2H$
h. $HO_2CCH(OH)CO_2H$

1.66 Draw in all the hydrogen atoms and nonbonded electron pairs in each ion.

a. 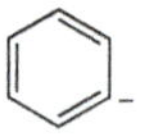 b. 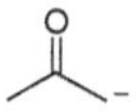 c. 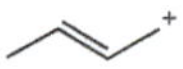 d.

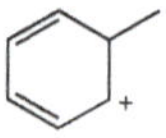

1.67 Draw in all H atoms and lone pairs in each ion.

a. 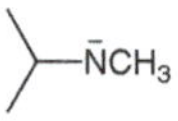 $\bar{N}CH_3$ b.  $\equiv\overset{+}{N}H$ c. 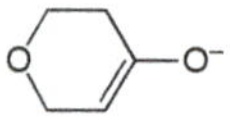 O ... O^- d. 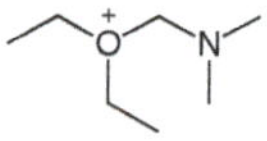 $\overset{+}{O}$ ... N

1.68 Each of the following condensed or skeletal structures is an incorrect representation of a molecule or ion. Explain what is wrong in each structure.

a. $CH_3CH=CH=CHCH_3$
b. $(CH_3)_3CHCH_2CH_2CH_3$
c. 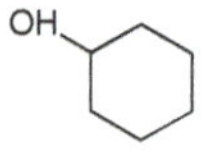 OH
d. 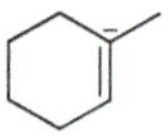
e. 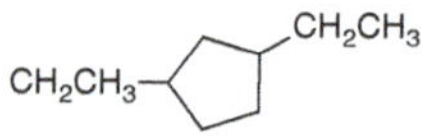 CH_2CH_3– ... CH_2CH_3

Hybridization

1.69 Predict the hybridization and geometry around each indicated atom.

a. $CH_3\bar{C}H_2$
b. 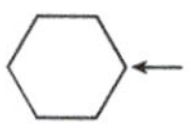
c. $(CH_3)_3O^+$
d. 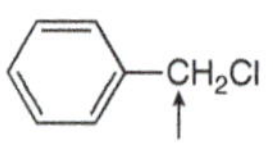 CH_2Cl
e. $CH_3-C\equiv C-H$
f. $CH_2=NOCH_3$
g. $CH_3CH=C=CH_2$

1.70 What orbitals are used to form each indicated bond? For multiple bonds, indicate the orbitals used in individual bonds.

a.

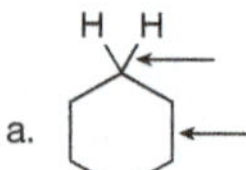

b. 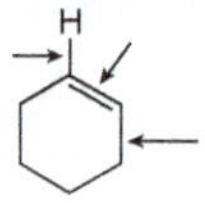 H

c. 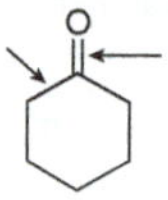 O

d. 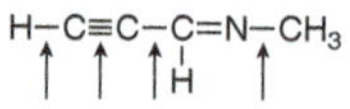 $H-C\equiv C-C=N-CH_3$ (H)

1.71 Ketene, $CH_2=C=O$, is an unusual organic molecule that has a single carbon atom doubly bonded to two different atoms. Determine the hybridization of both C atoms and the O in ketene. Then, draw a diagram showing what orbitals are used to form each bond (similar to Figures 1.11 and 1.12).

1.72 Consider the unstable cation and anion drawn below.

$CH_2=\overset{+}{C}H$ $\qquad$ $CH_2=\bar{\ddot{C}}H$

a. What is the hybridization of each carbon atom in these ions?
b. What orbitals are used to form both bonds in each carbon–carbon double bond?

Bond Length and Strength

1.73 Rank the following bonds in order of *increasing* bond length.

a. cyclohexyl–C≡C–H   cyclohexyl–C(H)=C(H)–H   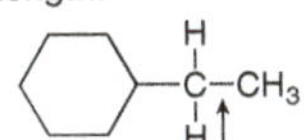

b. 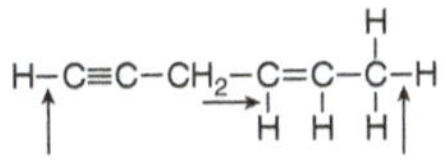 H–C≡C–CH2–C(H)=C(H)–C(H)(H)–H

1.74 Answer the following questions about compound **A**.

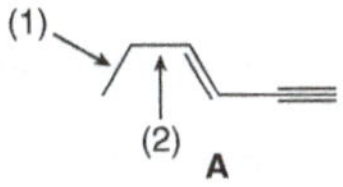

(1) (2) **A**

a. Label the shortest C–C single bond.
b. Label the longest C–C single bond.
c. Considering all the bonds, label the shortest C–C bond.
d. Label the weakest C–C bond.
e. Label the strongest C–H bond.
f. Explain why bond (1) and bond (2) are different in length, even though they are both C–C single bonds.

1.75 Two useful organic compounds that contain Cl atoms are drawn below. Vinyl chloride is the starting material used to prepare poly(vinyl chloride), a plastic used in insulation, pipes, and bottles. Chloroethane (ethyl chloride) is a local anesthetic. Why is the C–Cl bond of vinyl chloride stronger than the C–Cl bond in chloroethane?

$CH_2{=}CHCl$	CH_3CH_2Cl
vinyl chloride	chloroethane (ethyl chloride)

Bond Polarity

1.76 Use the symbols δ⁺ and δ⁻ to indicate the polarity of the labeled bonds.

a. 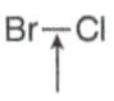 Br–Cl   b. 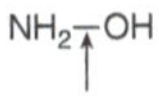 NH_2–OH   c. 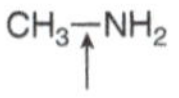 CH_3–NH_2 d. 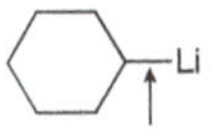 cyclohexyl–Li

1.77 Label the polar bonds in each molecule. Indicate the direction of the net dipole (if there is one).

a. $CHBr_3$
b. $CH_3CH_2OCH_2CH_3$
c. CBr_4
d. 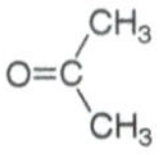 $O{=}C(CH_3)_2$
e. 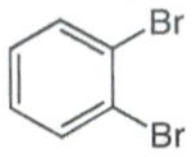 1,2-dibromobenzene (Br, Br)
f. 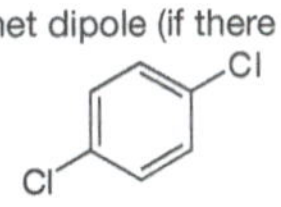 1,4-dichlorobenzene (Cl, Cl)

General Questions

1.78 Answer the following questions about acetonitrile ($CH_3C{\equiv}N$:).

a. Determine the hybridization of both C atoms and the N atom.
b. Label all bonds as σ or π.
c. In what type of orbital does the lone pair on N reside?
d. Label all bonds as polar or nonpolar.

1.79 Benzene is the simplest member of a whole class of compounds called aromatic hydrocarbons.

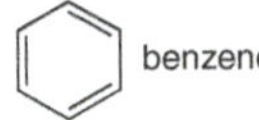
benzene

a. How is each carbon atom hybridized?
b. What is the geometry around each carbon atom? What is the overall geometry of the benzene ring?
c. Draw a diagram showing the orbitals used to join the carbon atoms of the ring.
d. Follow the indicated curved arrow notation to draw a second resonance structure.

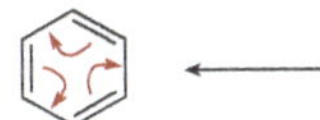

e. Benzene and other aromatic hydrocarbons are shown in Chapter 17 to be very stable. Offer an explanation.

1.80 The principles of this chapter can be applied to organic molecules of any size. Answer the following questions about amoxicillin, an antibiotic from the penicillin family.

amoxicillin

a. Predict the hybridization and geometry around each indicated atom.
b. Label five polar bonds using the symbols δ^+ and δ^-.
c. Draw a skeletal structure.
d. How many π bonds does amoxicillin have? Label them.
e. Find a C–H bond containing a carbon atom having a hybrid orbital with 33% *s*-character.

1.81

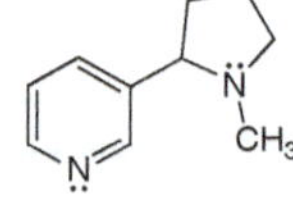

nicotine

a. What is the hybridization of each N atom in nicotine?
b. What is the geometry around each N atom?
c. In what type of orbital does the lone pair on each N atom reside?
d. Draw a constitutional isomer of nicotine.
e. Draw a resonance structure of nicotine.

1.82 Stalevo is the trade name for a medication used for Parkinson's disease, which contains L-dopa, carbidopa, and entacapone.

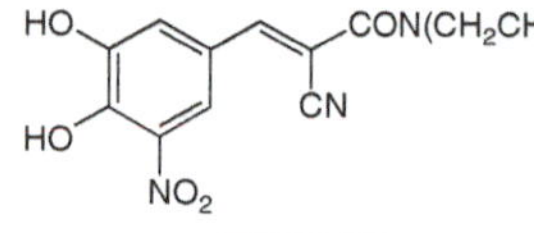

entacapone

a. Draw a Lewis structure for entacapone.
b. Which C–C bond in entacapone is the longest?
c. Which C–C single bond is the shortest?
d. Which C–N bond is the longest?
e. Which C–N bond is the shortest?
f. Use curved arrows to draw a resonance structure that is an equal contributor to the resonance hybrid.
g. Use curved arrows to draw a resonance structure that is a minor contributor to the resonance hybrid.

1.83 CH_3^+ and CH_3^- are two highly reactive carbon species.
a. What is the predicted hybridization and geometry around each carbon atom?
b. Two electrostatic potential plots are drawn for these species. Which ion corresponds to which diagram and why?

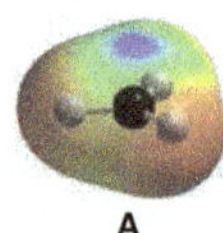

A

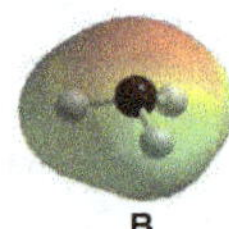

B

Challenge Problems

1.84 The N atom in CH_3CONH_2 (acetamide) is sp^2 hybridized, even though it is surrounded by four groups. Using this information, draw a diagram that shows the orbitals used by the atoms in the $-CONH_2$ portion of acetamide, and offer an explanation as to the observed hybridization.

1.85 Use the observed bond lengths to answer each question. (a) Why is bond [1] longer than bond [2]? (b) Why are bonds [3] and [4] equal in length, and shorter than bond [2]?

CH_3–OH: [1] 143 pm

CH_3COOH: [2] 136 pm

CH_3COO^-: [3] 127 pm, [4] 127 pm

1.86 Draw at least 10 more resonance structures for acetaminophen, the active pain reliever in Tylenol.

acetaminophen

1.87 When two carbons having different hybridization are bonded together, the C–C bond contains a slight dipole. In a $C_{sp^2}-C_{sp^3}$ bond, what is the direction of the dipole? Which carbon is considered more electronegative?

1.88 Draw all possible isomers having molecular formula C_4H_8 that contain one π bond.

1.89 Use the principles of resonance theory to explain why carbocation **A** is more stable than carbocation **B.**

A B

1.90 The curved arrow notation introduced in Section 1.6 is a powerful method used by organic chemists to show the movement of electrons not only in resonance structures, but also in chemical reactions. Since each curved arrow shows the movement of two electrons, following the curved arrows illustrates what bonds are broken and formed in a reaction. Consider the following three-step process. (a) Add curved arrows in Step [1] to show the movement of electrons. (b) Use the curved arrows drawn in Step [2] to identify the structure of **X. X** is converted in Step [3] to phenol and HCl.

+ H–Cl: →[1] + :Cl:⁻

→[2] X →[3] (:Cl:⁻) + H–Cl:

phenol

3 Introduction to Organic Molecules and Functional Groups

Vitamin C, or **ascorbic acid,** is important in the formation of collagen, a protein that holds together the connective tissues of skin, muscle, and blood vessels. Citrus fruits—oranges, grapefruit, and lemons—are well known sources of vitamin C. A deficiency of vitamin C causes scurvy, a common disease of sailors in the 1600s when they had no access to fresh fruits on long voyages. Early Arctic explorers, however, were able to obtain vitamin C by eating scurvy grass, a low-growing plant found along the coasts of the Arctic region of the northern hemisphere. In Chapter 3, we learn why some vitamins like vitamin A can be stored in the fat cells in the body, whereas others like vitamin C are excreted in urine.

Having learned some basic concepts about structure, bonding, and acid–base chemistry in Chapters 1 and 2, we will now concentrate on organic molecules.

- What are the characteristic features of an organic compound?
- What determines the properties of an organic compound?

After these questions are answered, we can understand some important phenomena. For example, why do we store some vitamins in the body and readily excrete others? How does soap clean away dirt? We will also use the properties of organic molecules to explain some basic biological phenomena, such as the structure of cell membranes and the transport of species across these membranes.

3.1 Functional Groups

What are the characteristic features of an organic compound? Most organic molecules have C–C and C–H σ bonds. These bonds are strong, nonpolar, and not readily broken. Organic molecules may have the following structural features as well:

- ***Heteroatoms*—atoms other than carbon or hydrogen.** Common heteroatoms are nitrogen, oxygen, sulfur, phosphorus, and the halogens.
- ***π Bonds.*** The most common π bonds occur in C–C and C–O double bonds.

These structural features distinguish one organic molecule from another. They determine a molecule's geometry, physical properties, and reactivity, and comprise what is called a **functional group.**

- A *functional group* is an atom or a group of atoms with characteristic chemical and physical properties. It is the *reactive part* of the molecule.

Why do heteroatoms and π bonds confer reactivity on a particular molecule?

- Heteroatoms have lone pairs and create electron-deficient sites on carbon.
- π Bonds are easily broken in chemical reactions. A π bond makes a molecule a base and a nucleophile.

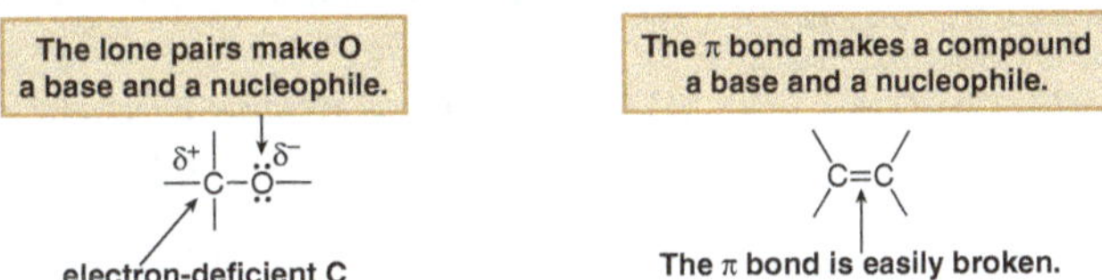

Don't think, though, that the C–C and C–H σ bonds are unimportant. They form the **carbon backbone** or **skeleton** to which the functional groups are bonded. A functional group usually behaves the same whether it is bonded to a carbon skeleton having as few as two or as many as 20 carbons. For this reason, we often abbreviate the carbon and hydrogen portion of the molecule by a capital letter **R,** and draw the **R** bonded to a particular functional group.

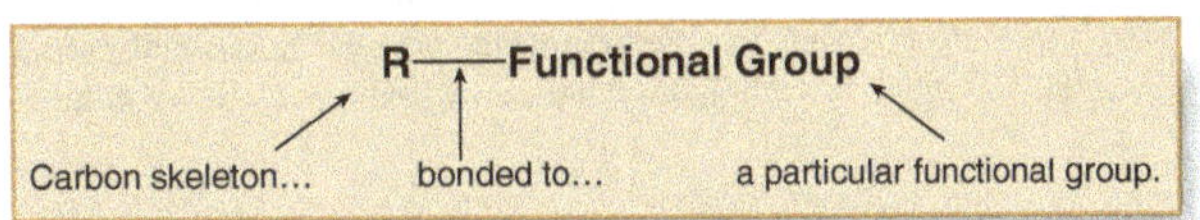

Ethane, for example, has only C–C and C–H σ bonds, so it has *no* functional group. Ethane has no polar bonds, no lone pairs, and no π bonds, so it has **no reactive sites.** Because of this, ethane and molecules like it are very unreactive.

Ethanol, on the other hand, has two carbons and five hydrogens in its carbon backbone, as well as an OH group, a functional group called a **hydroxy** group. Ethanol has lone pairs and polar bonds that make it reactive with a variety of reagents, including the acids and bases discussed in Chapter 2. The hydroxy group makes the properties of ethanol very different from the properties of ethane. Moreover, any organic molecule containing a hydroxy group has properties similar to ethanol.

Ethane

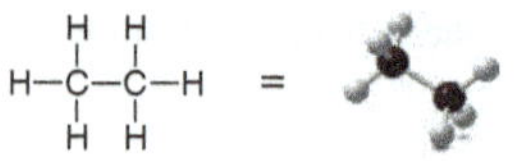

• all C–C and C–H σ bonds
• no functional group

Ethanol

carbon backbone

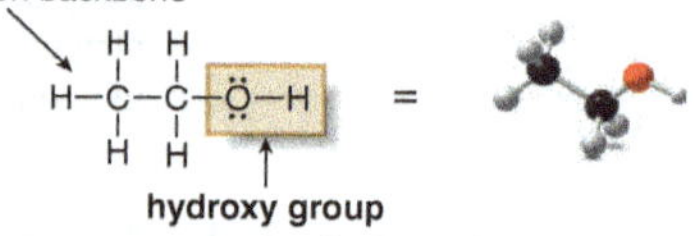

hydroxy group
• polar C–O and O–H bonds
• two lone pairs

Most organic compounds can be grouped into a relatively small number of categories, based on the structure of their functional group. Ethane, for example, is an **alkane,** whereas ethanol is a simple **alcohol.**

Problem 3.1 What reaction occurs when CH_3CH_2OH is treated with (a) H_2SO_4? (b) NaH? What happens when CH_3CH_3 is treated with these same reagents?

3.2 An Overview of Functional Groups

We can subdivide the most common functional groups into three types. A more complete list of functional groups is presented on the inside front cover.

- **Hydrocarbons**
- **Compounds containing a C–Z σ bond** where Z = an electronegative element
- **Compounds containing a C=O group**

3.2A Hydrocarbons

To review the structure and bonding of the simple aliphatic hydrocarbons, return to Section 1.10.

The word *aliphatic* is derived from the Greek word *aleiphas* meaning *fat.* Aliphatic compounds have physical properties similar to fats.

***Hydrocarbons* are compounds made up of only the elements carbon and hydrogen.** They may be **aliphatic** or **aromatic.**

[1] **Aliphatic hydrocarbons.** Aliphatic hydrocarbons can be divided into three subgroups.
- ***Alkanes*** have only C–C σ bonds and no functional group. Ethane, $\mathbf{CH_3CH_3}$, is a simple alkane.
- ***Alkenes*** have a C–C double bond as a functional group. Ethylene, $\mathbf{CH_2{=}CH_2}$, is a simple alkene.
- ***Alkynes*** have a C–C triple bond as a functional group. Acetylene, $\mathbf{HC{\equiv}CH}$, is a simple alkyne.

[2] **Aromatic hydrocarbons.** This class of hydrocarbons was so named because many of the earliest known aromatic compounds had strong, characteristic odors.

The simplest aromatic hydrocarbon is **benzene.** The six-membered ring and three π bonds of benzene comprise a *single* functional group. Benzene is a component of the **BTX** mixture (**B** for **b**enzene) added to gasoline to boost octane ratings.

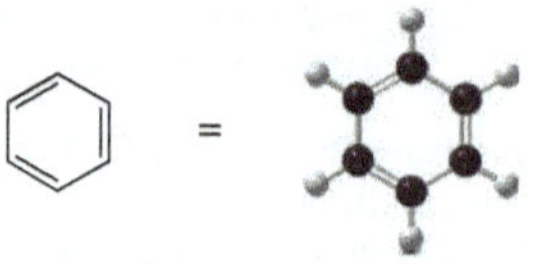

benzene
molecular formula C_6H_6

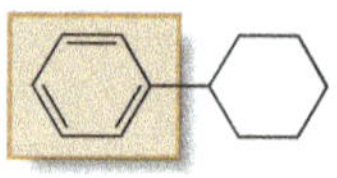

phenyl group
C_6H_5-
phenylcyclohexane

Table 3.1 Hydrocarbons

Type of compound	General structure	Example	Functional group
Alkane	R–H	CH_3CH_3	—
Alkene	C=C	$H_2C=CH_2$	double bond
Alkyne	—C≡C—	H–C≡C–H	triple bond
Aromatic compound	 benzene ring	 benzene ring	phenyl group

Polyethylene is a synthetic plastic first produced in the 1930s, and initially used as insulating material for radar during World War II. It is now a plastic used in milk containers, sandwich bags, and plastic wrapping. Over 100 billion pounds of polyethylene are manufactured each year.

When a benzene ring is bonded to another group, it is called a **phenyl group.** In phenylcyclohexane, for example, a phenyl group is bonded to the six-membered cyclohexane ring. Table 3.1 summarizes the four different types of hydrocarbons.

Alkanes, which have no functional groups, are notoriously unreactive except under very drastic conditions. For example, **polyethylene** is a synthetic plastic and high molecular weight alkane, consisting of chains of $-CH_2-$ groups bonded together, hundreds or even thousands of atoms long. Because it is an alkane with no reactive sites, it is a very stable compound that does not readily degrade and thus persists for years in landfills.

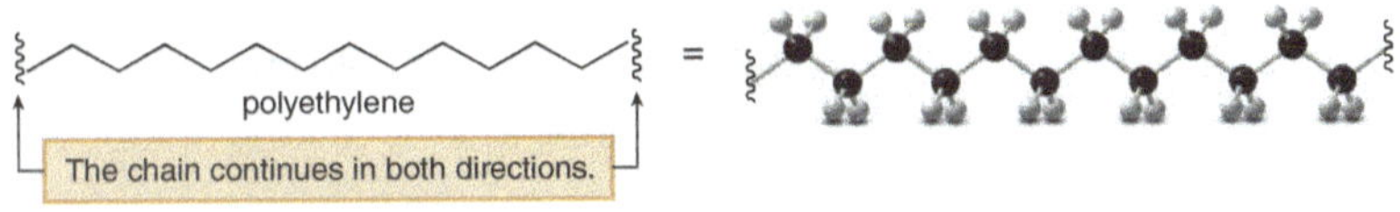

polyethylene

The chain continues in both directions.

3.2B Compounds Containing C–Z σ Bonds

Several types of functional groups that contain C–Z σ bonds are listed in Table 3.2. The electronegative heteroatom Z creates a polar bond, making carbon electron deficient. The lone pairs on Z are available for reaction with protons and other electrophiles, especially when Z = N or O.

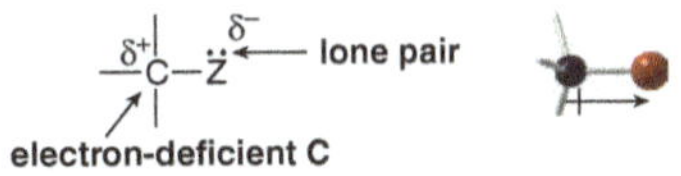

δ+ δ− lone pair

electron-deficient C

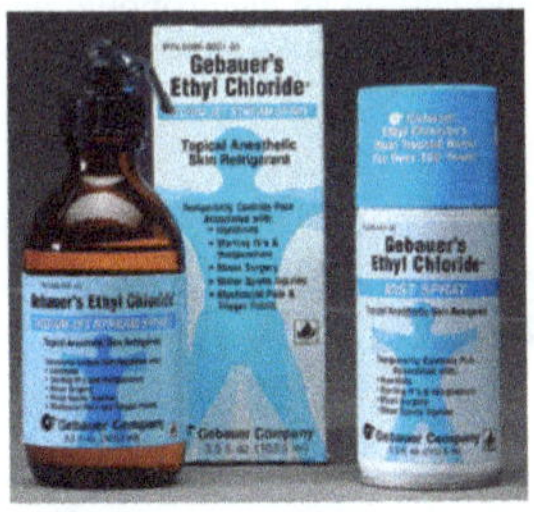

Chloroethane, CH_3CH_2Cl, is a local anesthetic.

Several simple compounds in this category are widely used. As an example, chloroethane (CH_3CH_2Cl, commonly called ethyl chloride) is an alkyl halide used as a local anesthetic. Chloroethane quickly evaporates when sprayed on a wound, causing a cooling sensation that numbs the site of an injury.

Hemibrevetoxin B is a neurotoxin produced by algal blooms referred to as "red tides," because of the color often seen in shallow ocean waters when these algae proliferate.

Molecules containing these functional groups may be simple or very complex. Diethyl ether, the first common general anesthetic, is a simple ether because it contains a single O atom, depicted in red, bonded to two C atoms. Hemibrevetoxin B, on the other hand, contains four ether groups, in addition to other functional groups.

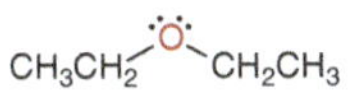

$CH_3CH_2-O-CH_2CH_3$

diethyl ether

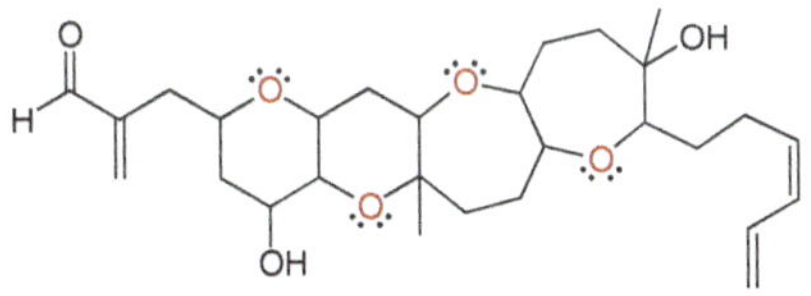

hemibrevetoxin B

Table 3.2 Compounds Containing C–Z σ Bonds

Type of compound	General structure	Example	3-D structure	Functional group
Alkyl halide	R–X: (X = F, Cl, Br, I)	CH_3–Br:		**–X** halo group
Alcohol	R–OH	CH_3–OH		**–OH** hydroxy group
Ether	R–O–R	CH_3–O–CH_3		**–OR** alkoxy group
Amine	R–NH_2 or R_2NH or R_3N	CH_3–NH_2		**–NH_2** amino group
Thiol	R–SH	CH_3–SH		**–SH** mercapto group
Sulfide	R–S–R	CH_3–S–CH_3		**–SR** alkylthio group

3.2C Compounds Containing a C=O Group

Many different types of functional groups possess a C–O double bond (a **carbonyl group**), as shown in Table 3.3. The polar C–O bond makes the carbonyl carbon an electrophile, while the lone pairs on O allow it to react as a nucleophile and base. The carbonyl group also contains a π bond that is more easily broken than a C–O σ bond.

Reactive features of a carbonyl group

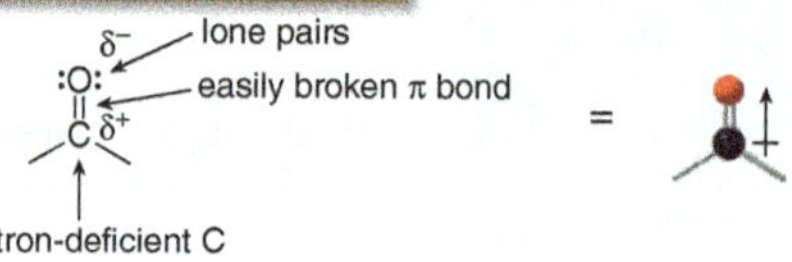

δ− lone pairs
:O:
easily broken π bond
C δ+
=
electron-deficient C

Atenolol and donepezil are examples of useful drugs that contain a variety of functional groups. **Atenolol is a β blocker,** a group of drugs used to treat hypertension. Donepezil, sold under the trade name Aricept, is used to treat mild to moderate dementia associated with Alzheimer's disease.

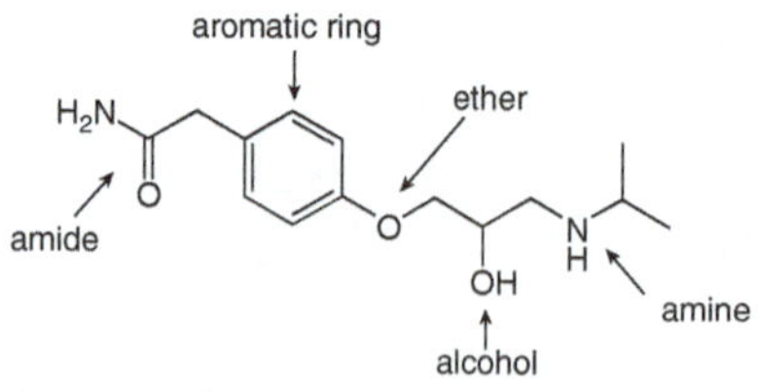

aromatic ring
H_2N
ether
amide
O
N
H
OH
amine
alcohol

atenolol
(used to treat high blood pressure)

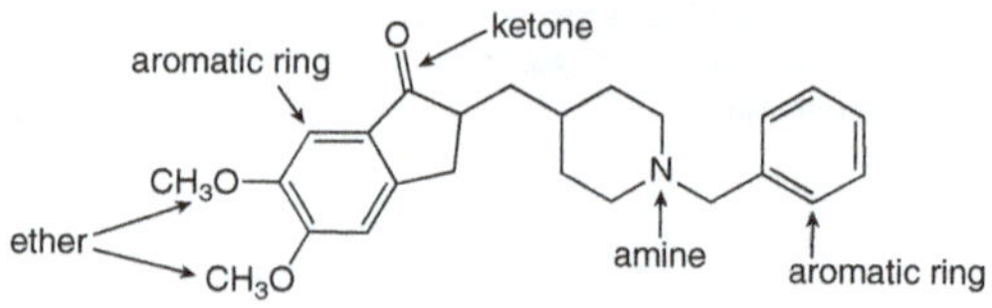

O
ketone
aromatic ring
CH_3O
ether
CH_3O
N
amine
aromatic ring

donepezil
(used to treat Alzheimer's disease)

Table 3.3 Compounds Containing a C=O Group

Type of compound	General structure	Example	3-D structure	Functional group
Aldehyde	RCHO	CH_3CHO		C=O carbonyl group
Ketone	RCOR	CH_3COCH_3		C=O carbonyl group
Carboxylic acid	RCOOH	CH_3COOH		–COOH carboxy group
Ester	RCOOR	CH_3COOCH_3		–COOR
Amide	$RCON$–H (or R), –H (or R)	CH_3CONH_2		$-CONH_2$, –CONHR, or $-CONR_2$
Acid chloride	RCOCl	CH_3COCl		–COCl

The importance of a functional group cannot be overstated. A functional group determines all the following properties of a molecule:

- bonding and shape
- type and strength of intermolecular forces
- physical properties
- nomenclature
- chemical reactivity

Problem 3.2 Oseltamivir can be prepared in 10 steps from shikimic acid. Identify the functional groups in oseltamivir and shikimic acid.

shikimic acid —10 steps→ oseltamivir

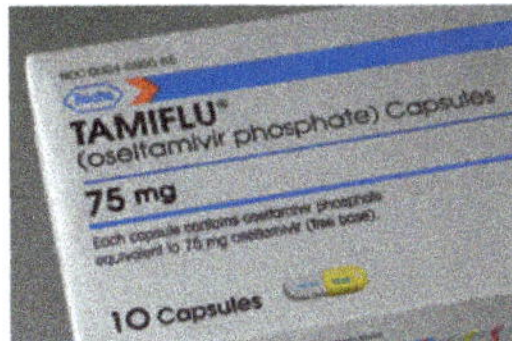

Tamiflu is the trade name for oseltamivir, an antiviral drug used to treat influenza.

Problem 3.3 Draw the structure of a compound fitting each description:

a. an aldehyde with molecular formula C_4H_8O
b. a ketone with molecular formula C_4H_8O
c. a carboxylic acid with molecular formula $C_4H_8O_2$
d. an ester with molecular formula $C_4H_8O_2$

Problem 3.4 Draw structures that fit each description and name the functional group in each molecule: (a) two constitutional isomers with molecular formula $C_5H_{10}O$ that contain different functional groups; (b) two constitutional isomers with molecular formula $C_6H_{10}O$ that contain the same functional group.

3.3 Intermolecular Forces

Intermolecular forces are the interactions that exist *between* molecules. A functional group determines the type and strength of these interactions.

3.3A Ionic Compounds

Intermolecular forces are also referred to as **noncovalent interactions** or **nonbonded interactions.**

Ionic compounds contain oppositely charged particles held together by **extremely strong electrostatic interactions.** These ionic interactions are much stronger than the intermolecular forces present between covalent molecules, so it takes a great deal of energy to separate oppositely charged ions from each other.

3.3B Covalent Compounds

Covalent compounds are composed of discrete molecules. The nature of the forces between the molecules depends on the functional group present. There are three different types of interactions, presented here in order of *increasing strength:*

- **van der Waals forces**
- **dipole–dipole interactions**
- **hydrogen bonding**

Van der Waals Forces

Van der Waals forces, also called **London forces,** are very weak interactions caused by the **momentary changes in electron density in a molecule.** Van der Waals forces are the only attractive forces present in nonpolar compounds.

Although any single van der Waals interaction is weak, a large number of van der Waals interactions creates a strong force. For example, geckos stick to walls and ceilings by van der Waals interactions of the surfaces with the 500,000 tiny hairs on each foot.

For example, although a nonpolar CH_4 molecule has no net dipole, at any one instant its electron density may not be completely symmetrical, creating a *temporary* dipole. This can induce a temporary dipole in another CH_4 molecule, with the partial positive and negative charges arranged close to each other. **The weak interaction of these temporary dipoles constitutes van der Waals forces.** All compounds exhibit van der Waals forces.

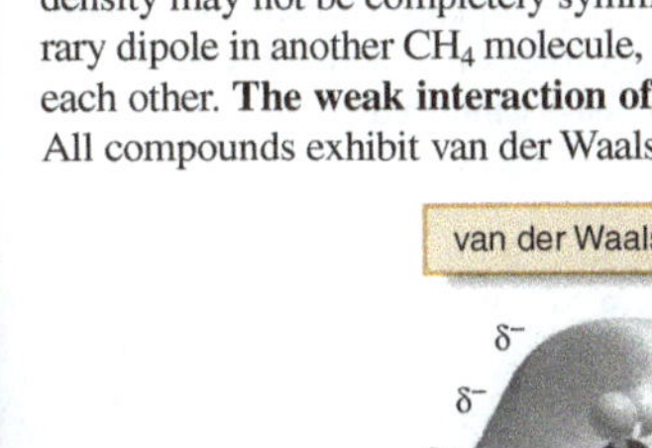

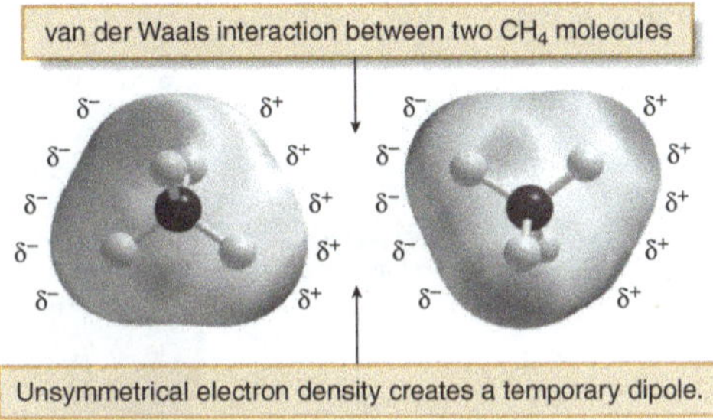

The surface area of a molecule determines the strength of the van der Waals interactions. **The larger the surface area, the larger the attractive force between two molecules, and the stronger the intermolecular forces.** Long, sausage-shaped molecules such as $CH_3CH_2CH_2CH_2CH_3$ (pentane) have stronger van der Waals interactions than compact spherical ones like $C(CH_3)_4$ (2,2-dimethylpropane), as shown in Figure 3.1.

Another factor affecting the strength of van der Waals forces is **polarizability.**

- *Polarizability* is a measure of how the electron cloud around an atom responds to changes in its electronic environment.

Larger atoms like iodine, which have more loosely held valence electrons, are more polarizable than smaller atoms like fluorine, which have more tightly held electrons. Because larger atoms have more easily induced dipoles, compounds containing them possess stronger intermolecular interactions.

Figure 3.1
Surface area and van der Waals forces

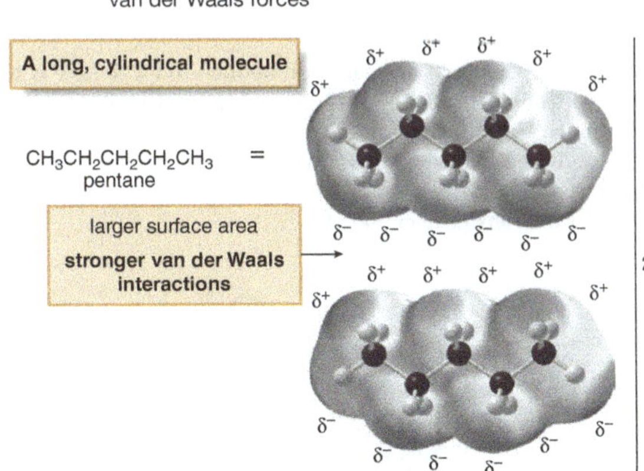

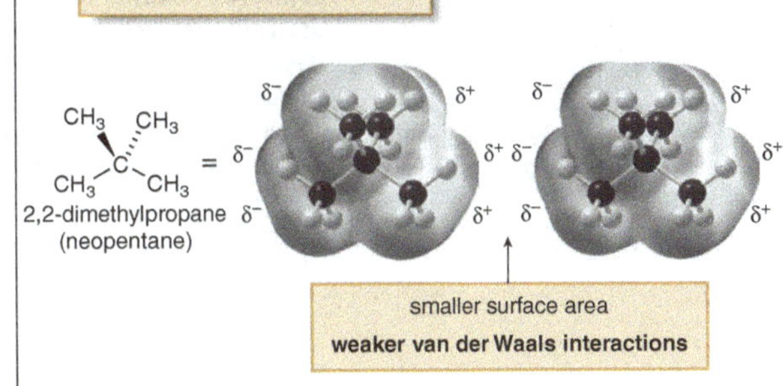

Thus, two F_2 molecules have little force of attraction between them, because the electrons are held very tightly and temporary dipoles are difficult to induce. On the other hand, two I_2 molecules exhibit a much stronger force of attraction because the electrons are held much more loosely and temporary dipoles are easily induced.

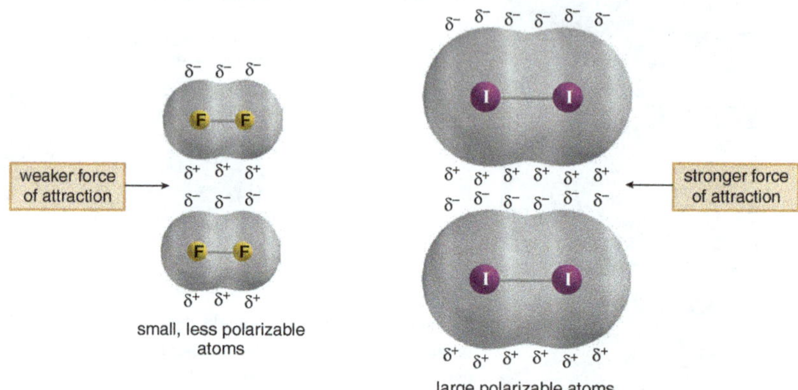

Dipole–Dipole Interactions

Dipole–dipole interactions **are the attractive forces between the permanent dipoles of two polar molecules.** In acetone, $(CH_3)_2C{=}O$, for example, the dipoles in adjacent molecules align so that the partial positive and partial negative charges are in close proximity. These attractive forces caused by permanent dipoles are much stronger than weak van der Waals forces.

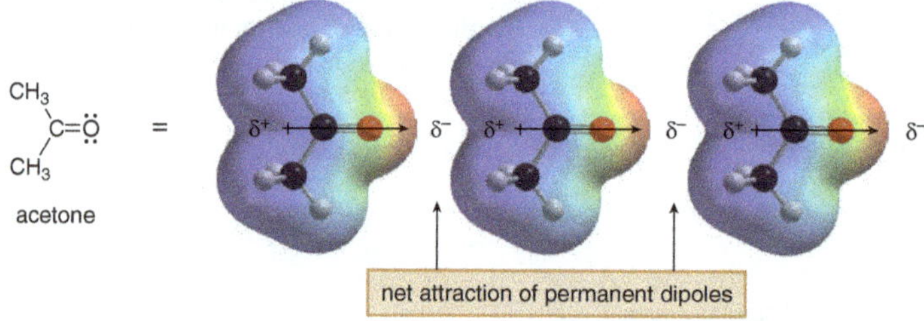

Hydrogen Bonding

Hydrogen bonding helps determine the three-dimensional shape of large biomolecules such as carbohydrates and proteins. See Chapters 28 and 29 for details.

Hydrogen bonding **typically occurs when a hydrogen atom bonded to O, N, or F is electrostatically attracted to a lone pair of electrons on an O, N, or F atom in another molecule.** Thus, H_2O molecules can hydrogen bond to each other. When they do, an H atom covalently bonded to O in one water molecule is attracted to a lone pair of electrons on the O in another water molecule. Hydrogen bonds are the strongest of the three types of intermolecular forces, though they are still much weaker than any covalent bond.

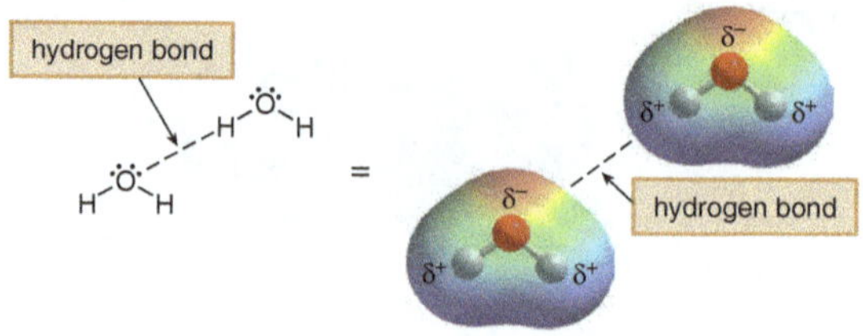

hydrogen bond

hydrogen bond

Sample Problem 3.1 illustrates how to determine the relative strength of intermolecular forces for a group of compounds. Table 3.4 summarizes the four types of interactions that affect the properties of all compounds.

Sample Problem 3.1 Rank the following compounds in order of increasing strength of intermolecular forces: $CH_3CH_2CH_2CH_2CH_3$ (pentane), $CH_3CH_2CH_2CH_2OH$ (1-butanol), and $CH_3CH_2CH_2CHO$ (butanal).

Solution

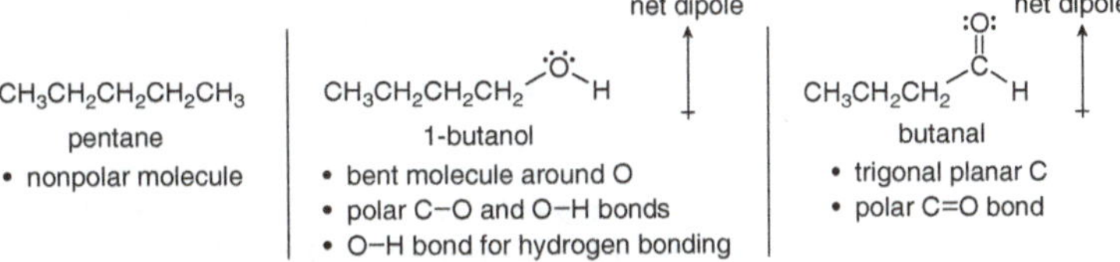

- Pentane has only nonpolar C–C and C–H bonds, so its molecules are held together by only **van der Waals** forces.
- 1-Butanol is a polar bent molecule, so it can have **dipole–dipole** interactions in addition to **van der Waals** forces. Because it has an O–H bond, 1-butanol molecules are held together by intermolecular **hydrogen bonds** as well.
- Butanal has a trigonal planar carbon with a polar C=O bond, so it exhibits **dipole–dipole** interactions in addition to **van der Waals** forces. There is *no* H atom bonded to O, so two butanal molecules *cannot* hydrogen bond to each other.

$CH_3CH_2CH_2CH_2CH_3$ $CH_3CH_2CH_2CHO$ $CH_3CH_2CH_2CH_2OH$

Increasing strength of intermolecular forces →

Table 3.4 Summary of Types of Intermolecular Forces

Type of force	Relative strength	Exhibited by	Example
van der Waals	weak	all molecules	$CH_3CH_2CH_2CH_2CH_3$ $CH_3CH_2CH_2CHO$ $CH_3CH_2CH_2CH_2OH$
dipole–dipole	moderate	molecules with a net dipole	$CH_3CH_2CH_2CHO$ $CH_3CH_2CH_2CH_2OH$
hydrogen bonding	strong	molecules with an O–H, N–H, or H–F bond	$CH_3CH_2CH_2CH_2OH$
ion–ion	very strong	ionic compounds	NaCl, LiF

Problem 3.5 What types of intermolecular forces are present in each compound?

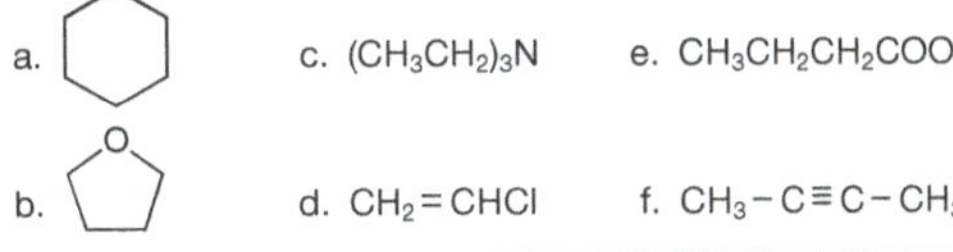

a. (cyclohexane)
c. $(CH_3CH_2)_3N$
e. $CH_3CH_2CH_2COOH$
b. (tetrahydrofuran)
d. $CH_2=CHCl$
f. $CH_3-C\equiv C-CH_3$

3.4 Physical Properties

The strength of a compound's intermolecular forces determines many of its physical properties, including its boiling point, melting point, and solubility.

3.4A Boiling Point (bp)

The *boiling point* of a compound is the temperature at which a liquid is converted to a gas. In boiling, energy is needed to overcome the attractive forces in the more ordered liquid state.

- The stronger the intermolecular forces, the higher the boiling point.

Because **ionic compounds** are held together by extremely strong interactions, they have **very high boiling points.** The boiling point of NaCl, for example, is 1413 °C. **With covalent molecules, the boiling point depends on the identity of the functional group.** For compounds of approximately the same molecular weight:

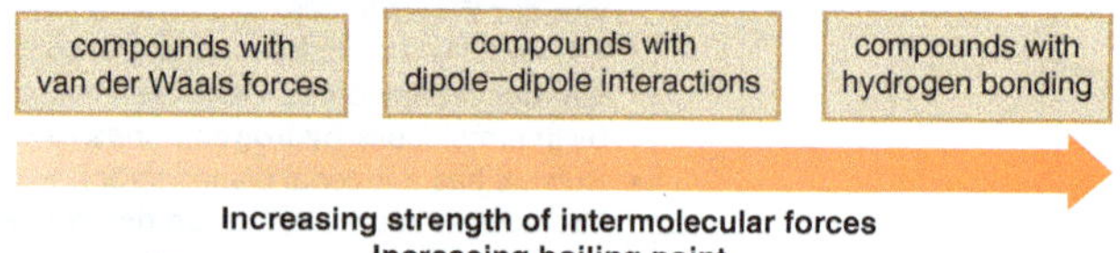

Recall from Sample Problem 3.1, for example, that the relative strength of the intermolecular forces increases from pentane to butanal to 1-butanol. The boiling points of these compounds increase in the same order.

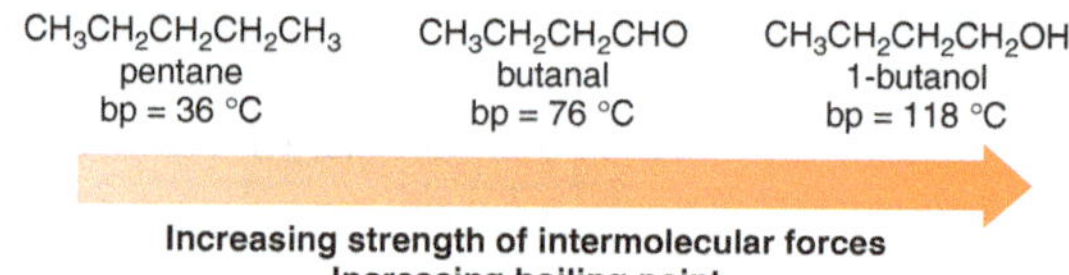

Because surface area and polarizability affect the strength of intermolecular forces, they also affect the boiling point. For two compounds with similar functional groups:

- The larger the surface area, the higher the boiling point.
- The more polarizable the atoms, the higher the boiling point.

Examples of each phenomenon are illustrated in Figure 3.2. In comparing two ketones that differ in size, 3-pentanone has a higher boiling point than acetone because it has a greater molecular weight and larger surface area. In comparing two alkyl halides having the same number of carbon atoms, CH_3I has a higher boiling point than CH_3F because I is more polarizable than F.

Sample Problem 3.2 Which compound in each pair has the higher boiling point?

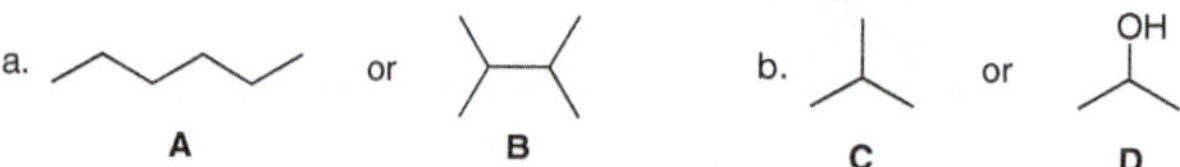

Solution

a. Isomers **A** and **B** have only nonpolar C–C and C–H bonds, so they exhibit only van der Waals forces. Because **B** is more compact, it has less surface area and a lower boiling point.

b. Compounds **C** and **D** have approximately the same molecular weight but different functional groups. **C** is a nonpolar alkane, exhibiting only van der Waals forces. **D** is an alcohol with an O–H group available for hydrogen bonding, so it has stronger intermolecular forces and a higher boiling point.

Figure 3.2 Effect of surface area and polarizability on boiling point

a. Effect of surface area

$CH_3CH_2C(=O)CH_2CH_3$ =

3-pentanone

larger surface
higher boiling point
bp = 102 °C

$CH_3C(=O)CH_3$ =

acetone

smaller surface
lower boiling point
bp = 56 °C

b. Effect of polarizability

CH_3—I =

iodomethane

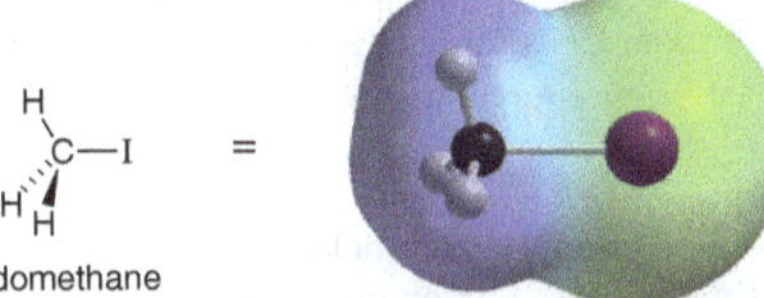

more polarizable I atom
higher boiling point
bp = 42 °C

CH_3—F =

fluoromethane

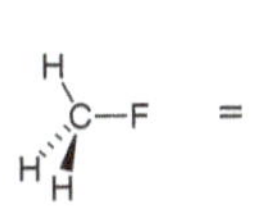

less polarizable F atom
lower boiling point
bp = −78 °C

Problem 3.6 Which compound in each pair has the higher boiling point?

a. $(CH_3)_2C{=}CH_2$ or $(CH_3)_2C{=}O$
b. CH_3CH_2COOH or CH_3COOCH_3
c. $CH_3(CH_2)_4CH_3$ or $CH_3(CH_2)_5CH_3$
d. $CH_2{=}CHCl$ or $CH_2{=}CHI$

Problem 3.7 Explain why the boiling point of propanamide, $CH_3CH_2CONH_2$, is considerably higher than the boiling point of *N,N*-dimethylformamide, $HCON(CH_3)_2$ (213 °C vs. 153 °C), even though both compounds are isomeric amides.

Liquids having different boiling points can be separated in the laboratory using a *distillation* apparatus (Figure 3.3). When a mixture of two liquids is heated in the distilling flask,

Figure 3.3 Schematic of a distillation apparatus

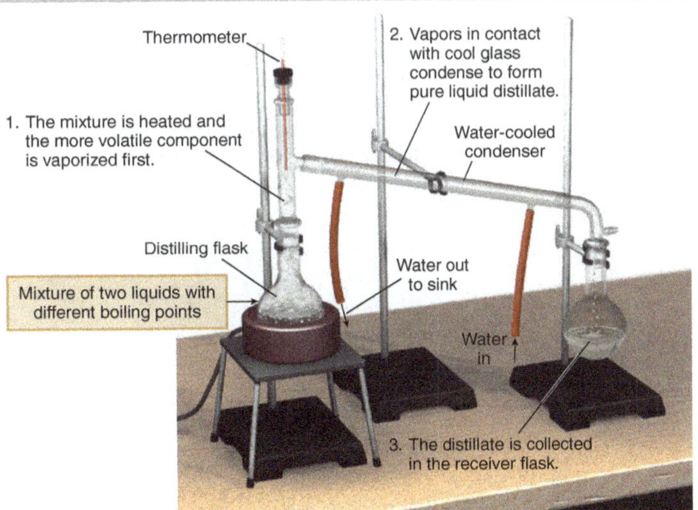

the lower boiling compound, the **more volatile component,** distills first, followed by the **less volatile, higher boiling component.** By collecting the distillate in a series of receiver flasks, the two liquids can usually be separated from each other. The best separations are generally achieved when the liquids in the mixture have widely different boiling points.

3.4B Melting Point (mp)

The *melting point* is the temperature at which a solid is converted to its liquid phase. In melting, energy is needed to overcome the attractive forces in the more ordered crystalline solid. Two factors determine the melting point of a compound.

- The stronger the intermolecular forces, the higher the melting point.
- Given the same functional group, the more symmetrical the compound, the higher the melting point.

Because **ionic compounds** are held together by extremely strong interactions, they have **very high melting points.** For example, the melting point of NaCl is 801 °C. With covalent molecules, the melting point once again depends on the identity of the functional group. For compounds of approximately the same molecular weight:

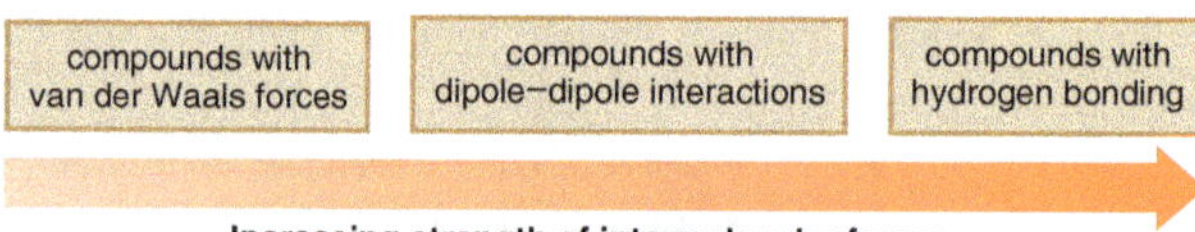

The trend in the melting points of pentane, butanal, and 1-butanol parallels the trend observed in their boiling points.

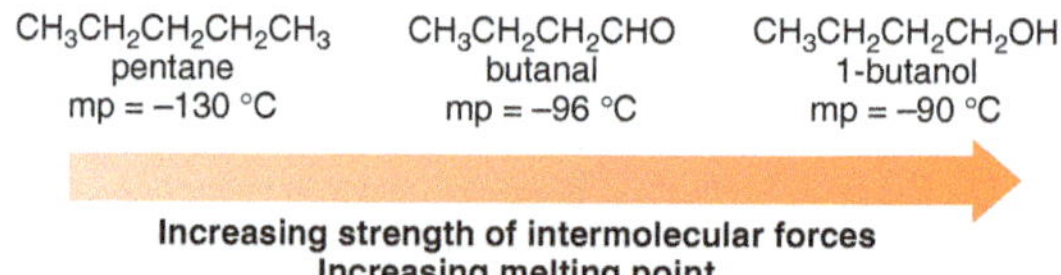

Symmetry also plays a role in determining the melting points of compounds having the same functional group and similar molecular weights, but very different shapes. A compact symmetrical molecule like 2,2-dimethylpropane packs well into a crystalline lattice whereas 2-methylbutane, which has a CH_3 group dangling from a four-carbon chain, does not. Thus, 2,2-dimethylpropane has a much higher melting point.

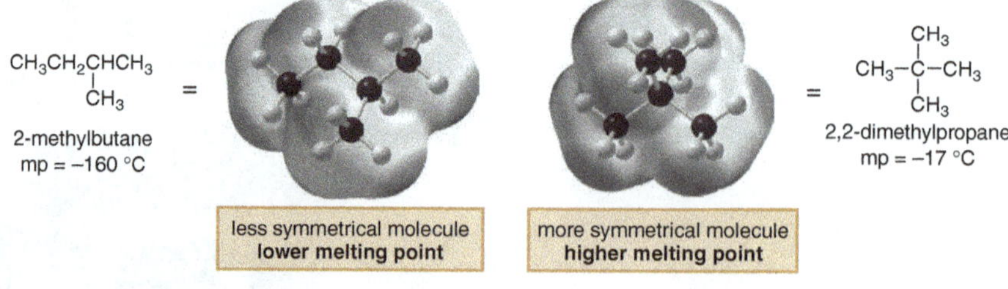

Problem 3.8 Predict which compound in each pair has the higher melting point.

a. or NH_2 b. or

Problem 3.9 Consider acetic acid (CH_3CO_2H) and its conjugate base, sodium acetate (CH_3CO_2Na). (a) What intermolecular forces are present in each compound? (b) Explain why the melting point of sodium acetate (324 °C) is considerably higher than the melting point of acetic acid (17 °C).

3.4C Solubility

Quantitatively, a compound may be considered soluble when 3 g of solute dissolves in 100 mL of solvent.

Solubility* is the extent to which a compound, called the *solute,* dissolves in a liquid, called the *solvent. In dissolving a compound, the energy needed to break up the interactions between the molecules or ions of the solute comes from new interactions between the solute and the solvent.

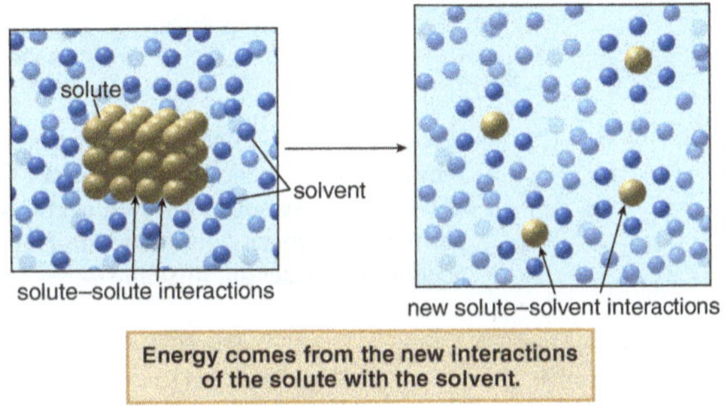

Energy comes from the new interactions of the solute with the solvent.

Compounds dissolve in solvents having similar kinds of intermolecular forces.

- "Like dissolves like."
- Polar compounds dissolve in polar solvents. Nonpolar or weakly polar compounds dissolve in nonpolar or weakly polar solvents.

Water and **organic liquids** are two different kinds of solvents. Water is very polar because it is capable of hydrogen bonding with a solute. Many organic solvents are either nonpolar, like carbon tetrachloride (CCl_4) and hexane [$CH_3(CH_2)_4CH_3$], or weakly polar like diethyl ether ($CH_3CH_2OCH_2CH_3$).

Ionic compounds are held together by strong electrostatic forces, so they need very polar solvents to dissolve. **Most ionic compounds are soluble in water, but are insoluble in organic solvents.** To dissolve an ionic compound, the strong ion–ion interactions must be replaced by many weaker **ion–dipole interactions,** as illustrated in Figure 3.4.

Figure 3.4 Dissolving an ionic compound in H_2O

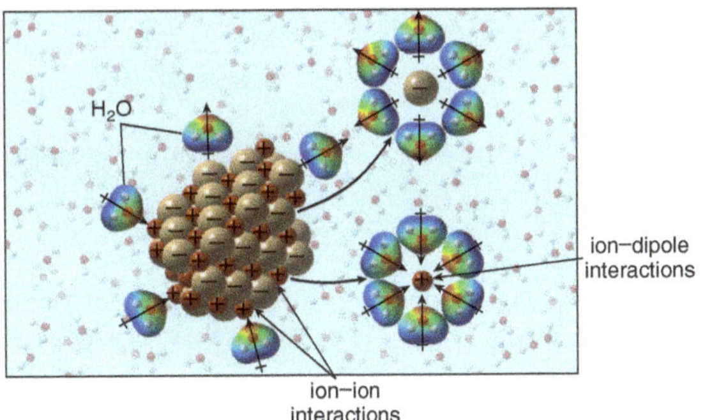

- When an ionic solid is dissolved in H_2O, the ion–ion interactions are replaced by ion–dipole interactions. Though these forces are weaker, there are so many of them that they compensate for the stronger ionic bonds.

Most organic compounds are soluble in organic solvents (remember, *like dissolves like*). **An organic compound is water soluble only if it contains one polar functional group capable of hydrogen bonding with the solvent for every five C atoms it contains.** In other words, a water-soluble organic compound has an O- or N-containing functional group that solubilizes its nonpolar carbon backbone.

Compare, for example, the solubility of butane and acetone in H_2O and CCl_4.

$CH_3CH_2CH_2CH_3$ (butane): CCl_4 → **soluble**; H_2O → **insoluble**

$(CH_3)_2C{=}O$ (acetone): CCl_4 → **soluble**; H_2O → **soluble**

Because butane and acetone are both organic compounds having a C–C and C–H backbone, they are soluble in the organic solvent CCl_4. Butane, a nonpolar molecule, is insoluble in the polar solvent H_2O. Acetone, however, is H_2O soluble because it contains only three C atoms and its O atom can hydrogen bond with one H atom of H_2O. In fact, acetone is so soluble in water that acetone and water are **miscible**—they form solutions in all proportions with each other.

$(CH_3)_2C{=}O$ molecules cannot hydrogen bond to each other because they have no OH group. However, $(CH_3)_2C{=}O$ can hydrogen bond to H_2O because its O atom can hydrogen bond to one of the H atoms of H_2O.

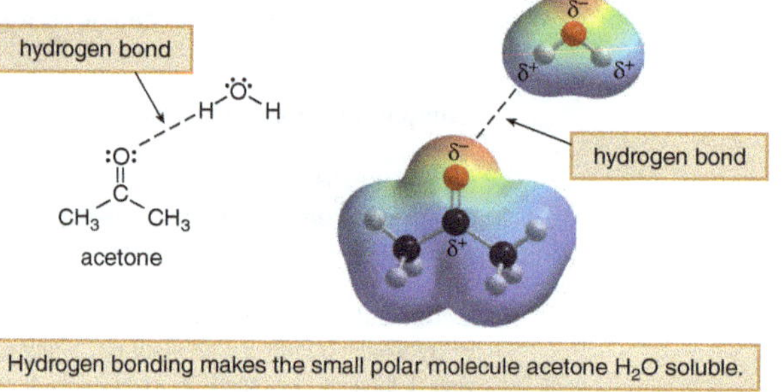

Hydrogen bonding makes the small polar molecule acetone H_2O soluble.

For an organic compound with one functional group, **a compound is water soluble only if it has ≤ five C atoms and contains an O or N atom.**

The size of an organic molecule with a polar functional group determines its water solubility. A low molecular weight alcohol like **ethanol is water soluble** because it has a small carbon skeleton (≤ five C atoms) compared to the size of its polar OH group. Cholesterol, on the other hand, has 27 carbon atoms and only one OH group. Its carbon skeleton is too large for the OH group to solubilize by hydrogen bonding, so **cholesterol is insoluble in water.**

$CH_3CH_2–OH$ (ethanol) — one OH group that can hydrogen bond — **H_2O soluble**

cholesterol — too large to be H_2O soluble — **H_2O insoluble**

Hydrophobic = afraid of H_2O; hydrophilic = H_2O loving.

- The nonpolar part of a molecule that is not attracted to H_2O is said to be *hydrophobic.*
- The polar part of a molecule that can hydrogen bond to H_2O is said to be *hydrophilic.*

In cholesterol, for example, the **hydroxy group is hydrophilic,** whereas the **carbon skeleton is hydrophobic.**

MTBE (*tert*-butyl methyl ether) and 4,4'-dichlorobiphenyl (a polychlorinated biphenyl, abbreviated as PCB) demonstrate that solubility properties can help determine the fate of organic compounds in the environment.

$CH_3-O-C(CH_3)_3$ — MTBE, *tert*-butyl methyl ether

Cl–C₆H₄–C₆H₄–Cl — 4,4'-dichlorobiphenyl (a polychlorinated biphenyl, PCB)

Using **MTBE** as a high-octane additive in unleaded gasoline has had a negative environmental impact. Although MTBE is not toxic or carcinogenic, it has a distinctive, nauseating odor, and **it is water soluble.** Small amounts of MTBE have contaminated the drinking water in several communities, making it unfit for consumption. For this reason, the use of MTBE as a gasoline additive has steadily declined in the United States since 1999.

4,4'-Dichlorobiphenyl is a polychlorinated biphenyl **(PCB),** a compound that contains two benzene rings joined by a C–C bond, and substituted by one or more chlorine atoms on each ring. PCBs have been used as plasticizers in polystyrene coffee cups and coolants in transformers. They have been released into the environment during production, use, storage, and disposal, making them one of the most widespread organic pollutants. **PCBs are insoluble in H_2O, but very soluble in organic media,** so they are soluble in fatty tissue, including that found in all types of fish and birds around the world. Although PCBs are not acutely toxic, frequently ingesting large quantities of fish contaminated with PCBs has been shown to retard growth and memory retention in children.

Solubility properties of some representative compounds are summarized in Figure 3.5.

Figure 3.5 Solubility summary

Type of compound	Solubility in H_2O	Solubility in organic solvents (such as CCl_4)
• Ionic		
$NaCl$	**soluble**	**insoluble**
• Covalent		
$CH_3CH_2CH_2CH_3$	**insoluble** (no N or O atom to hydrogen bond to H_2O)	**soluble**
$CH_3CH_2CH_2OH$	**soluble** (≤ 5 C's and an O atom for hydrogen bonding to H_2O)	**soluble**
$CH_3(CH_2)_{10}OH$	**insoluble** (> 5 C's; too large to be soluble even though it has an O atom for hydrogen bonding to H_2O)	**soluble**

Sample Problem 3.3 Which compounds are water soluble?

chlorocyclopentane **A** (cyclopentane–Cl)

cyclopentanol **B** (cyclopentane–OH)

Solution

A has five C atoms and a polar C–Cl bond, but it is incapable of hydrogen bonding with H_2O, making it H_2O insoluble. **B** has five C atoms and a polar OH group able to hydrogen bond to H_2O, making it H_2O soluble.

Problem 3.10 Which compounds are water soluble?

a. $CH_3CH_2OCH_2CH_3$ b. $CH_3CH_2CH_2CH_2CH_3$ c. $(CH_3CH_2CH_2CH_2)_3N$

Sample Problem 3.4 a. Which of the following molecules can hydrogen bond to another molecule like itself?
b. Which of the following molecules can hydrogen bond with water?

$CH_3CH_2-\ddot{O}-CH_2CH_3$ — diethyl ether — **A**

$CH_3CH_2-\ddot{N}(H)-CH_2CH_3$ — diethylamine — **B**

Solution

a. Compounds **A** and **B** have polar C–O and C–N bonds but **A** does not have an O–H bond, so it cannot hydrogen bond to another **A** molecule. **B,** on the other hand, has an N–H bond, so two **B** molecules can hydrogen bond with each other.

b. Because **A** has an electronegative O atom and **B** has an electronegative N atom, both **A** and **B** can hydrogen bond with a hydrogen atom in water.

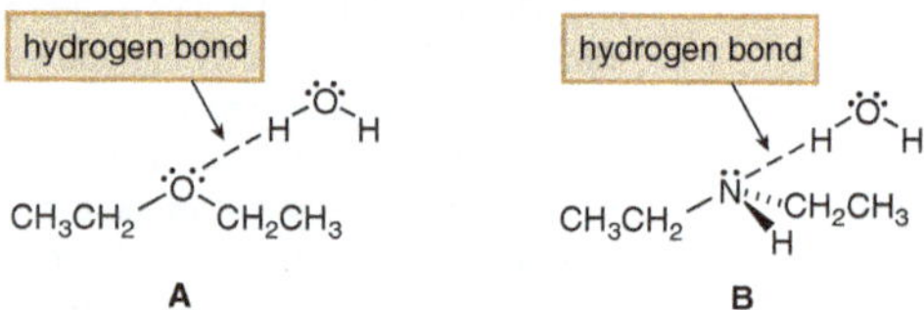

Problem 3.11 Label the hydrophobic and hydrophilic portions of each molecule:

a. norethindrone (oral contraceptive component) [structure with OH, C≡C–H, O]

b. arachidonic acid (fatty acid) [structure with COOH]

3.5 Application: Vitamins

Vitamins **are organic compounds needed in small amounts for normal cell function.** Our bodies cannot synthesize these compounds, so they must be obtained in the diet. Most vitamins are identified by a letter, such as A, C, D, E, and K. There are several different B vitamins, though, so a subscript is added to distinguish them: for example, B_1, B_2, and B_{12}.

Whether a vitamin is **fat soluble** (it dissolves in organic media) or **water soluble** can be determined by applying the solubility principles discussed in Section 3.4C. Vitamins A and C illustrate the differences between fat-soluble and water-soluble vitamins.

3.5A Vitamin A

The name **vitamin** was first used in 1912 by the Polish chemist Casimir Funk, who called them *vitamines,* because he thought that they all contained an *amine* functional group. Later the word was shortened to vitamin, because some are amines but others, like vitamins A and C, are not.

Vitamin A, or **retinol,** is an essential component of the vision receptors in the eyes. It also helps to maintain the health of mucous membranes and the skin, so many anti-aging creams contain vitamin A. A deficiency of this vitamin leads to a loss of night vision.

[structure with OH] vitamin A

Vitamin A contains 20 carbons and a single OH group, making it **water insoluble.** Because it is organic, it is **soluble in any organic medium.** To understand the consequences of these solubility characteristics, we must learn about the chemical environment of the body.

Vitamin A is synthesized from β-carotene, the orange pigment in carrots.

About 70% of the body is composed of water. Fluids such as blood, gastric juices in the stomach, and urine are largely water with dissolved ions such as Na^+ and K^+. Vitamin A is insoluble in these fluids. There are also fat cells composed of organic compounds having C−C and C−H bonds. Vitamin A is soluble in this organic environment, and thus it is readily stored in these fat cells, particularly in the liver.

Vitamin A may be obtained directly from the diet. In addition, β-carotene, the orange pigment found in many plants including carrots, is readily converted to vitamin A in our bodies.

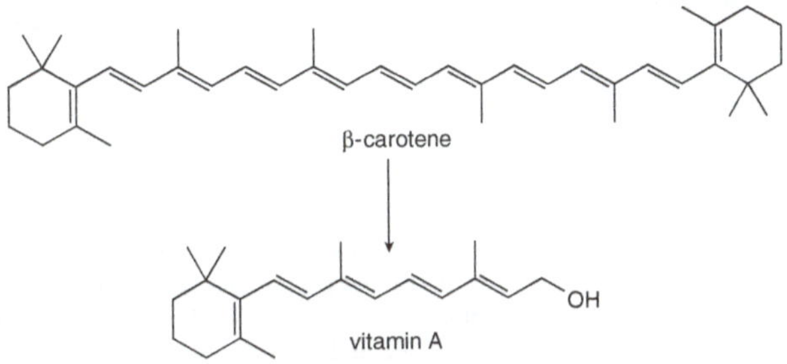

β-carotene

OH

vitamin A

Eating too many carrots does not result in an excess of stored vitamin A. If you consume more β-carotene than you need, your body stores this precursor until it needs more vitamin A. Some β-carotene reaches the surface tissues of the skin and eyes, giving them an orange color. This phenomenon may look odd, but it is harmless and reversible. When stored β-carotene is converted to vitamin A and is no longer in excess, these tissues will return to their normal hue.

Vitamin C is obtained by eating citrus fruits and a wide variety of other fruits and vegetables. Individuals can also obtain the recommended daily dose of vitamin C by taking tablets that contain vitamin C prepared in the laboratory. Both the "natural" vitamin C in oranges and the "synthetic" vitamin C in vitamin supplements are identical.

3.5B Vitamin C

Although most animal species can synthesize vitamin C, humans, guinea pigs, the Indian fruit bat, and the bulbul bird must obtain this vitamin from dietary sources. Citrus fruits, strawberries, tomatoes, and sweet potatoes are all excellent sources of vitamin C.

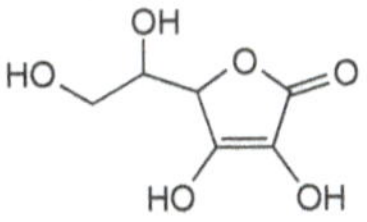

OH HO O O HO OH

vitamin C
(ascorbic acid)

Vitamin C has six carbon atoms, each bonded to an oxygen atom that is capable of hydrogen bonding, making it **water soluble.** Vitamin C thus dissolves in urine. Although it has been acclaimed as a deterrent for all kinds of diseases, from the common cold to cancer, the consequences of taking large amounts of vitamin C are not really known, because any excess of the minimum daily requirement is excreted in the urine.

Problem 3.12 Predict the water solubility of each vitamin.

a. 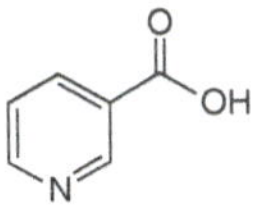 O OH N

vitamin B_3
(niacin)

b. 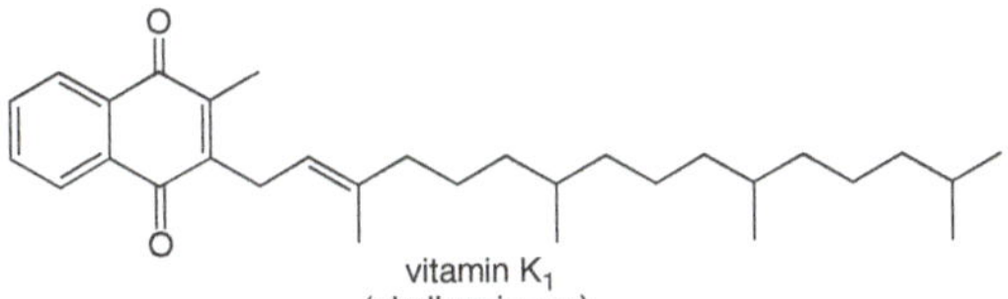 O O

vitamin K_1
(phylloquinone)

Problem 3.13 (a) Identify the functional groups in the ball-and-stick model of pantothenic acid, vitamin B_5. (b) At which sites can pantothenic acid hydrogen bond to water? (c) Predict the water solubility of pantothenic acid.

Avocados are an excellent dietary source of pantothenic acid, vitamin B_5.

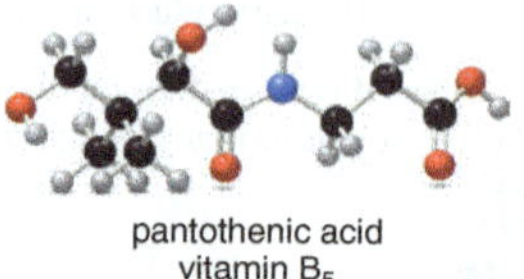

pantothenic acid
vitamin B_5

3.6 Application of Solubility: Soap

Soap has been used by humankind for some 2000 years. Historical records describe its manufacture in the first century and document the presence of a soap factory in Pompeii. Before this time clothes were cleaned by rubbing them on rocks in water, or by forming soapy lathers from the roots, bark, and leaves of certain plants. These plants produced natural materials called *saponins,* which act in much the same way as modern soaps.

On a molecular level, soap has two distinct parts:

- **a hydrophilic portion composed of ions called the *polar head***
- **a hydrophobic carbon chain of nonpolar C–C and C–H bonds, called the *nonpolar tail***

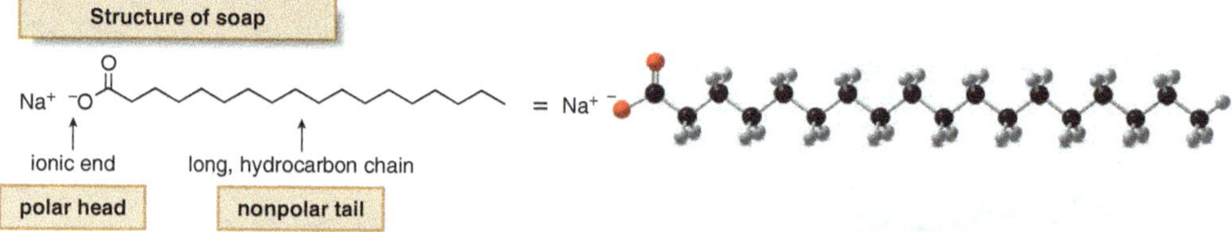

Dissolving soap in water forms ***micelles,* spherical droplets having the ionic heads on the surface and the nonpolar tails packed together in the interior,** as shown in Figure 3.6. In this arrangement, the ionic heads are solvated by the polar solvent water, thus solubilizing the nonpolar, "greasy" hydrocarbon portion of the soap.

How does soap dissolve grease and oil? Water alone cannot dissolve dirt, which is composed largely of nonpolar hydrocarbons. When soap is mixed with water, however, the nonpolar hydrocarbon tails dissolve the dirt in the interior of the micelle. The polar head of the soap remains on the surface of the micelle to interact with water. The nonpolar tails of the soap molecules are so well sealed off from the water by the polar head groups that the micelles are water soluble, allowing them to separate from the fibers of our clothes and be washed down the drain with water. In this way, soaps do a seemingly impossible task: they remove nonpolar hydrocarbon material from skin and clothes, by solubilizing it in the polar solvent water.

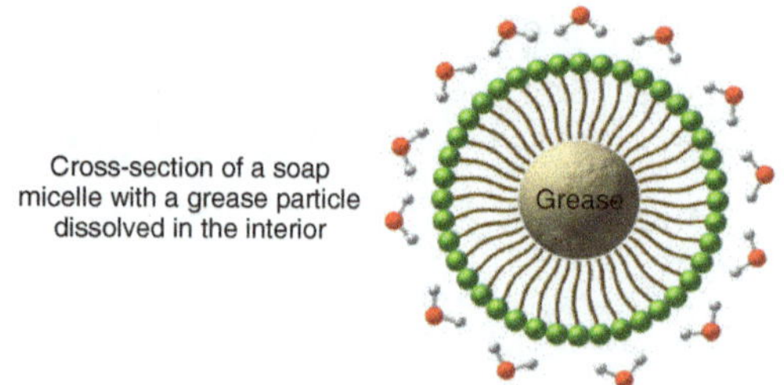

Cross-section of a soap micelle with a grease particle dissolved in the interior

Figure 3.6
Dissolving soap in water

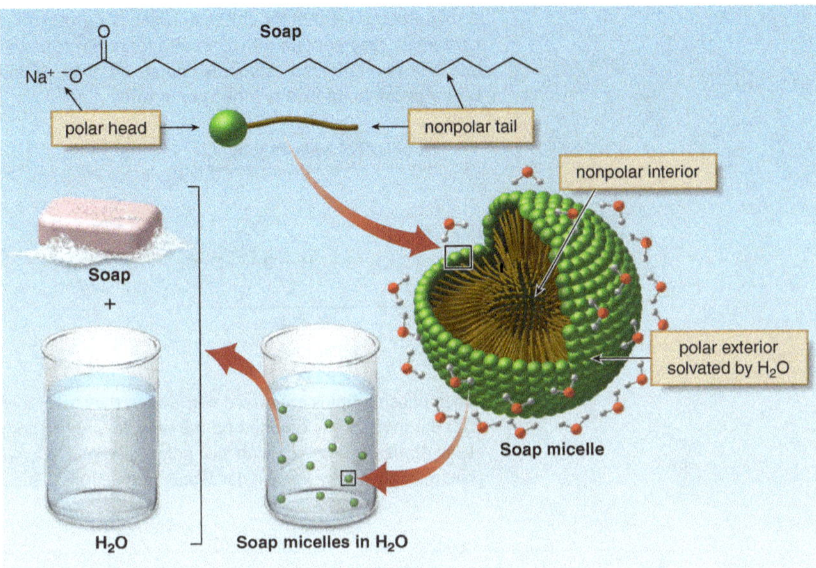

- When soap is dissolved in H_2O, it forms micelles with the nonpolar tails in the interior and the polar heads on the surface. The polar heads are solvated by ion–dipole interactions with H_2O molecules.

Problem 3.14 Which of the following structures represent soaps? Explain your answers.

a. $CH_3CO_2^-Na^+$ b. $CH_3(CH_2)_{14}CO_2^-Na^+$ c. $CH_3(CH_2)_{12}COOH$ d. $CH_3(CH_2)_9CO_2^-Na^+$

Problem 3.15 Today, synthetic detergents like the compound drawn here, not soaps, are used to clean clothes. Explain how this detergent cleans away dirt.

SO_3^- Na^+

a detergent

3.7 Application: The Cell Membrane

The cell membrane is a beautifully complex example of how the principles of organic chemistry come into play in a biological system.

3.7A Structure of the Cell Membrane

The basic unit of living organisms is the **cell.** The cytoplasm is the aqueous medium inside the cell, separated from water outside the cell by the **cell membrane.** The cell membrane serves two apparently contradictory functions. It acts as a barrier to the passage of ions, water, and other molecules into and out of the cell, and it is also selectively permeable, letting nutrients in and waste out.

A major component of the cell membrane is a group of organic compounds called **phospholipids.** Like soap, they contain a hydrophilic ionic portion, and a hydrophobic hydrocarbon portion, in this case two long carbon chains composed of C−C and C−H bonds. **Phospholipids thus contain a polar head and *two* nonpolar tails.**

Structure of a phospholipid

$(CH_3)_3\overset{+}{N}CH_2CH_2{-}O{-}P(=O)(O^-){-}O{-}CH_2$

ionic end

polar head

two long hydrocarbon chains

nonpolar tails

When phospholipids are mixed with water, they assemble in an arrangement called a **lipid bilayer,** with the ionic heads oriented on the outside and the nonpolar tails on the inside. The polar heads electrostatically interact with the polar solvent H_2O, while the nonpolar tails are held in close proximity by numerous van der Waals interactions. This is schematically illustrated in Figure 3.7.

Figure 3.7
The cell membrane

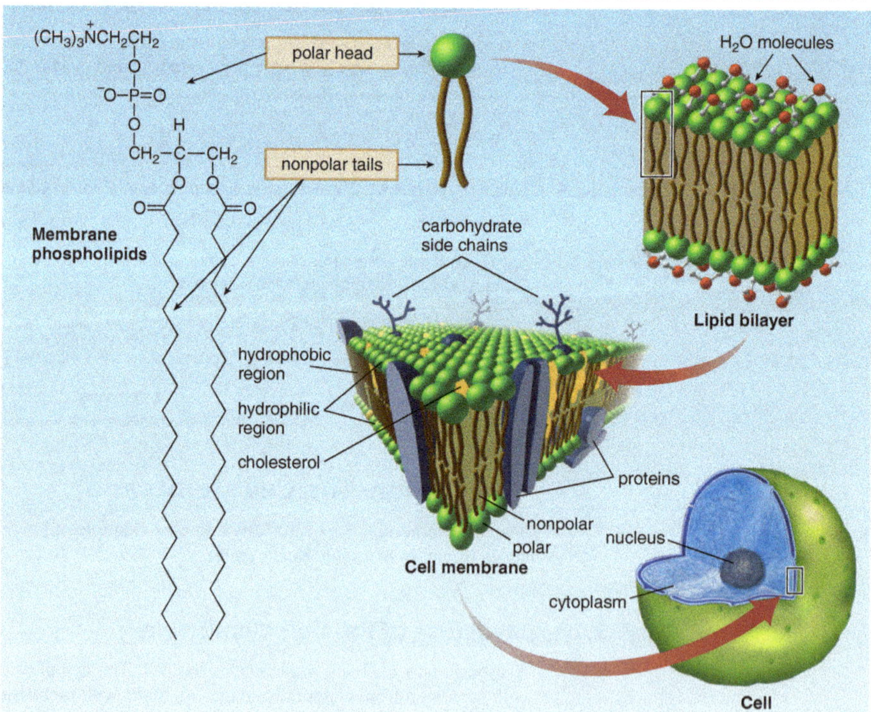

- Phospholipids contain an ionic or polar head, and two long nonpolar hydrocarbon tails. In an aqueous environment, phospholipids form a lipid bilayer, with the polar heads oriented toward the aqueous exterior and the nonpolar tails forming a hydrophobic interior. Cell membranes are composed largely of this lipid bilayer.

Cell membranes are composed of these lipid bilayers. The charged heads of the phospholipids are oriented toward the aqueous interior and exterior of the cell. The nonpolar tails form the hydrophobic interior of the membrane, thus serving as an insoluble barrier that protects the cell from the outside.

3.7B Transport Across a Cell Membrane

How does a polar molecule or ion in the water outside a cell pass through the nonpolar interior of the cell membrane and enter the cell? Some nonpolar molecules like O_2 are small enough to enter and exit the cell by diffusion. Polar molecules and ions, on the other hand, may be too large or too polar to diffuse efficiently. Some ions are transported across the membrane with the help of molecules called **ionophores.**

Ionophores **are organic molecules that complex cations.** They have a hydrophobic exterior that makes them soluble in the nonpolar interior of the cell membrane, and a central cavity with several oxygen atoms whose lone pairs complex with a given ion. The size of the cavity determines the identity of the cation with which the ionophore complexes. Two naturally occurring antibiotics that act as ionophores are **nonactin** and **valinomycin.**

Naturally occurring antibiotic ionophores

nonactin

valinomycin

Each molecule contains a large central cavity to hold a cation.

Several synthetic ionophores have also been prepared, including one group called **crown ethers.** ***Crown ethers*** **are cyclic ethers containing several oxygen atoms that bind specific cations depending on the size of their cavity.** Crown ethers are named according to the general format ***x*-crown-*y*,** where ***x*** is the total number of atoms in the ring and ***y*** is the number of oxygen atoms. For example, 18-crown-6 contains 18 atoms in the ring, including 6 O atoms. This crown ether binds potassium ions. Sodium ions are too small to form a tight complex with the O atoms, and larger cations do not fit in the cavity.

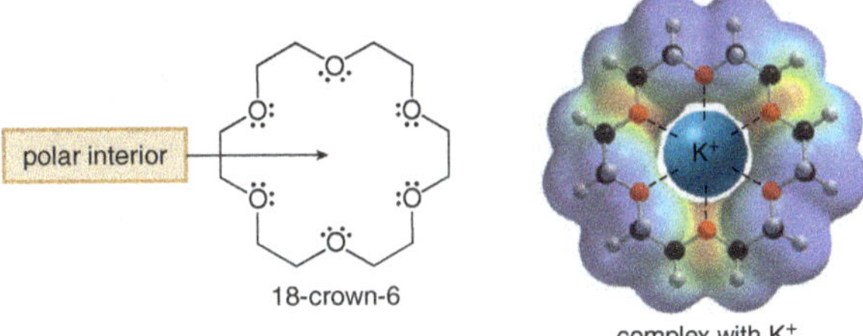

18-crown-6

complex with K^+

Figure 3.8
Transport of ions across a cell membrane

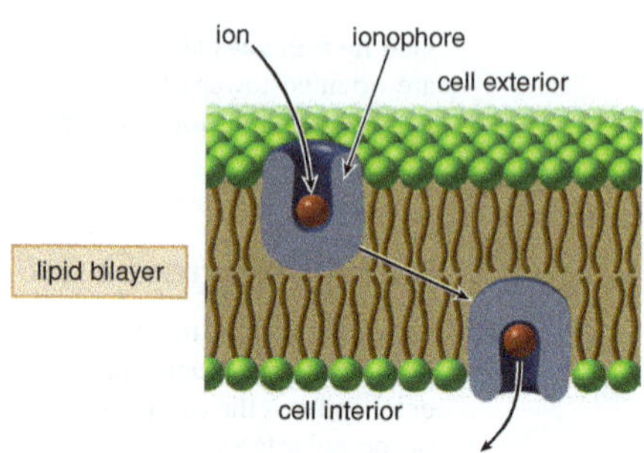

- By binding an ion on one side of a lipid bilayer (where the concentration of the ion is high) and releasing it on the other side of the bilayer (where the concentration of the ion is low), an ionophore transports an ion across a cell membrane.

How does an ionophore transfer an ion across a membrane? The ionophore binds the ion on one side of the membrane in its polar interior. It can then move across the membrane because its hydrophobic exterior interacts with the hydrophobic tails of the phospholipid. The ionophore then releases the ion on the other side of the membrane. This ion-transfer role is essential for normal cell function. This process is illustrated in Figure 3.8.

In this manner, antibiotic ionophores like nonactin transport ions across a cell membrane of bacteria. This disrupts the normal ionic balance in the cell, thus interfering with cell function and causing the bacteria to die.

Problem 3.16 Nonactin and valinomycin each contain only two different types of functional groups. What two functional groups are present in nonactin? In valinomycin?

Problem 3.17 Now that you have learned about solubility, explain why aspirin (Section 2.7) crosses a cell membrane as a neutral carboxylic acid rather than an ionic conjugate base.

3.8 Functional Groups and Reactivity

Much of Chapter 3 has been devoted to how a functional group determines the strength of intermolecular forces and, consequently, the physical properties of molecules. A functional group also determines reactivity. What type of reaction does a particular kind of organic compound undergo? Begin by recalling two fundamental concepts.

- **Functional groups create reactive sites in molecules.**
- **Electron-rich sites react with electron-poor sites.**

All functional groups contain a heteroatom, a π bond, or both, and these features make electron-deficient (or electrophilic) sites and electron-rich (or nucleophilic) sites in a molecule. Molecules react at these sites. To predict reactivity, first locate the functional group and then determine the resulting electron-rich or electron-deficient sites it creates. Keep three guidelines in mind.

- **An electronegative heteroatom like N, O, or X makes a carbon atom *electrophilic.***

CH_3CH_2—Cl (δ^+ on C) cyclohexyl—OH (δ^+ on C)

electrophilic site

- **A lone pair on a heteroatom makes it basic and nucleophilic.**

$CH_3-\ddot{O}-CH_3$ $CH_3-\ddot{N}(CH_3)-CH_3$

basic and nucleophilic site

- **π Bonds create *nucleophilic* sites and are more easily broken than σ bonds.**

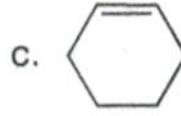

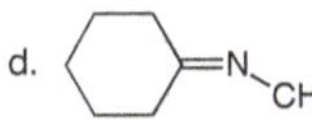

Problem 3.18 Label the electrophilic and nucleophilic sites in each molecule.

a. (cyclohexyl)—Br b. H_2O c. (cyclohexene) d. (cyclohexylidene)=N—CH_3

By identifying the nucleophilic and electrophilic sites in a compound you can begin to understand how it will react. In general, electron-rich sites react with electron-deficient sites:

:Nu⁻ = a nucleophile; E^+ = an electrophile.

- **An electron-deficient carbon atom reacts with a nucleophile, symbolized as $:Nu^-$.**
- **An electron-rich carbon reacts with an electrophile, symbolized as E^+.**

At this point we don't know enough organic chemistry to draw the products of many reactions with confidence. We do know enough, however, to begin to predict if two compounds might react together based solely on electron density arguments, and at what atoms that reaction is most likely to occur.

For example, alkenes contain an electron-rich C–C double bond and so they react with electrophiles, E^+. On the other hand, alkyl halides possess an electrophilic carbon atom, so they react with electron-rich nucleophiles.

Alkenes react with electrophiles.

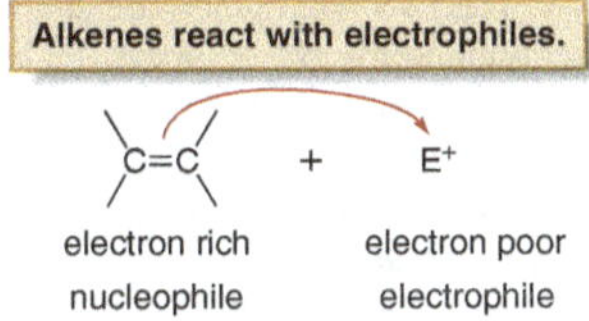

Alkyl halides react with nucleophiles.

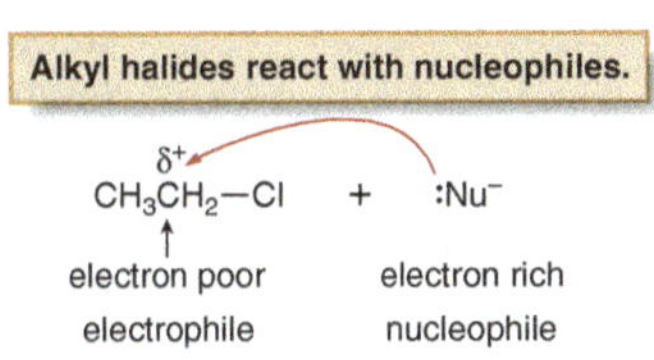

For now, you don't need to worry about the products of these reactions. At this point you should only be able to find reactive sites in molecules and begin to understand why a reaction might occur at these sites. After you learn more about the structure of organic molecules in Chapters 4 and 5, we will begin a detailed discussion of organic reactions in Chapter 6.

Problem 3.19 Considering only electron density, state whether the following reactions will occur:

a. $CH_3CH_2-Br + {}^-OH \longrightarrow$

b. $CH_3-C\equiv C-CH_3 + Br^- \longrightarrow$

c. $CH_3C(=O)Cl + {}^-OCH_3 \longrightarrow$

d. $CH_3-C\equiv C-CH_3 + Br^+ \longrightarrow$

3.9 Biomolecules

***Biomolecules* are organic compounds found in biological systems.** Many are relatively small, with molecular weights of less than 1000 g/mol. There are four main families of these small molecules—simple sugars, amino acids, lipids, and nucleotides. Many simple biomolecules are used to synthesize larger compounds that have important cellular functions.

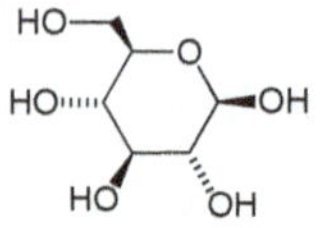

glucose
a simple sugar

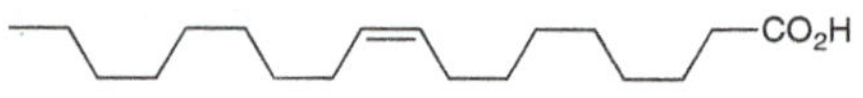

oleic acid
a fatty acid

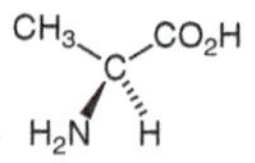

alanine
an amino acid

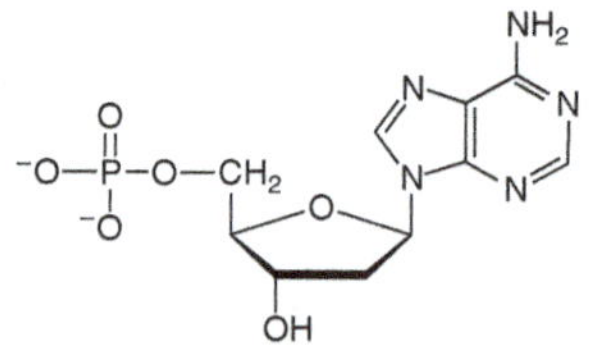

deoxyadenosine 5'-monophosphate
a nucleotide

Simple sugars such as glucose combine to form the complex carbohydrates starch and cellulose, as described in Chapter 28. Alanine is an amino acid used to synthesize proteins, the subject of Chapter 29. Fatty acids such as oleic acid react with alcohols to form triacylglycerols, the most prevalent lipids, first mentioned in Chapter 10, and discussed in more detail in Chapters 22 and 30. While these biomolecules all contain more than one functional group, their properties and reactions are explained by the principles of basic organic chemistry.

Finally, deoxyadenosine 5'-monophosphate is a nucleotide that combines with thousands of other nucleotides to form DNA, deoxyribonucleic acid, the high molecular weight polynucleotide that stores the genetic information of an organism. DNA consists of two polynucleotide chains that wind together in a double helix. Figure 3.9 illustrates the importance of hydrogen bonding in the structure of DNA. The two polynucleotide chains are held together by an extensive network of hydrogen bonds in which the N–H groups on one chain intermolecularly hydrogen bond to an oxygen or nitrogen atom on the adjacent chain.

Problem 3.20 The fact that sweet-tasting carbohydrates like table sugar are also high in calories has prompted the development of sweet, low-calorie alternatives. (a) Identify the functional groups in aspartame, the artificial sweetener in Equal. (b) Label all of the sites that can hydrogen bond to the oxygen atom of water. (c) Label all of the sites that can hydrogen bond with a hydrogen atom of water.

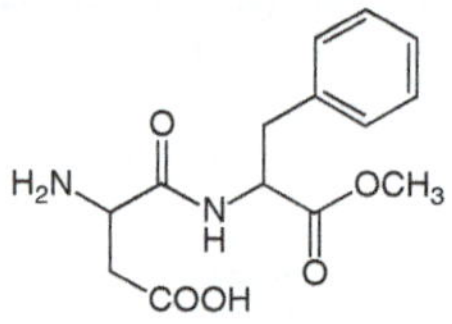

aspartame

Figure 3.9
The double helix of DNA

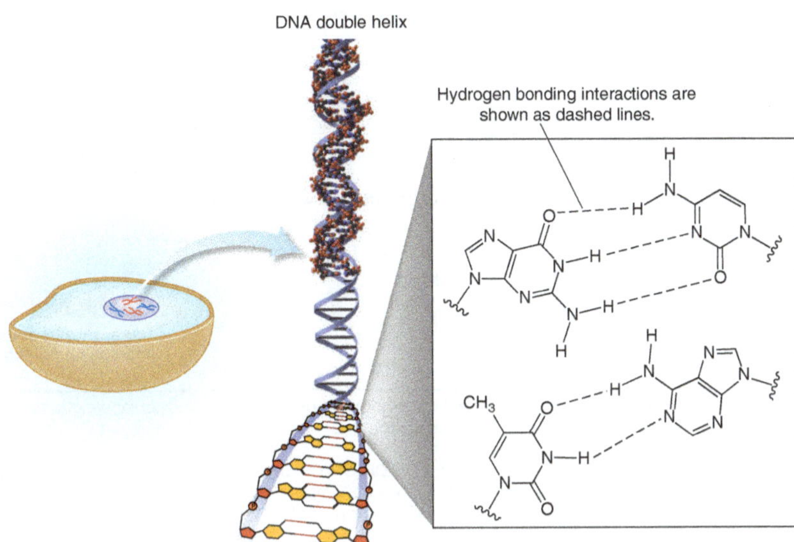

- DNA, which is contained in the chromosomes of the nucleus of a cell, stores all of the genetic information in an organism. DNA consists of two long strands of polynucleotides held together by hydrogen bonding.

KEY CONCEPTS

Introduction to Organic Molecules and Functional Groups

Types of Intermolecular Forces (3.3)

Increasing strength ↓

Type of force	Cause
van der Waals	Caused by the interaction of temporary dipoles • Larger surface area, stronger forces • Larger, more polarizable atoms, stronger forces
dipole–dipole	Caused by the interaction of permanent dipoles
hydrogen bonding	Caused by the electrostatic interaction of a H atom in an O–H, N–H, or H–F bond with the lone pair of another N, O, or F atom
ion–ion	Caused by the charge attraction of two ions

Physical Properties

Property	Observation
Boiling point (3.4A)	• For compounds of comparable molecular weight, the stronger the intermolecular forces the higher the bp.

$CH_3CH_2CH_2CH_2CH_3$	$CH_3CH_2CH_2CHO$	$CH_3CH_2CH_2CH_2OH$
VDW	VDW, DD	VDW, DD, HB
bp = 36 °C	bp = 76 °C	bp = 118 °C

Increasing strength of intermolecular forces
Increasing boiling point

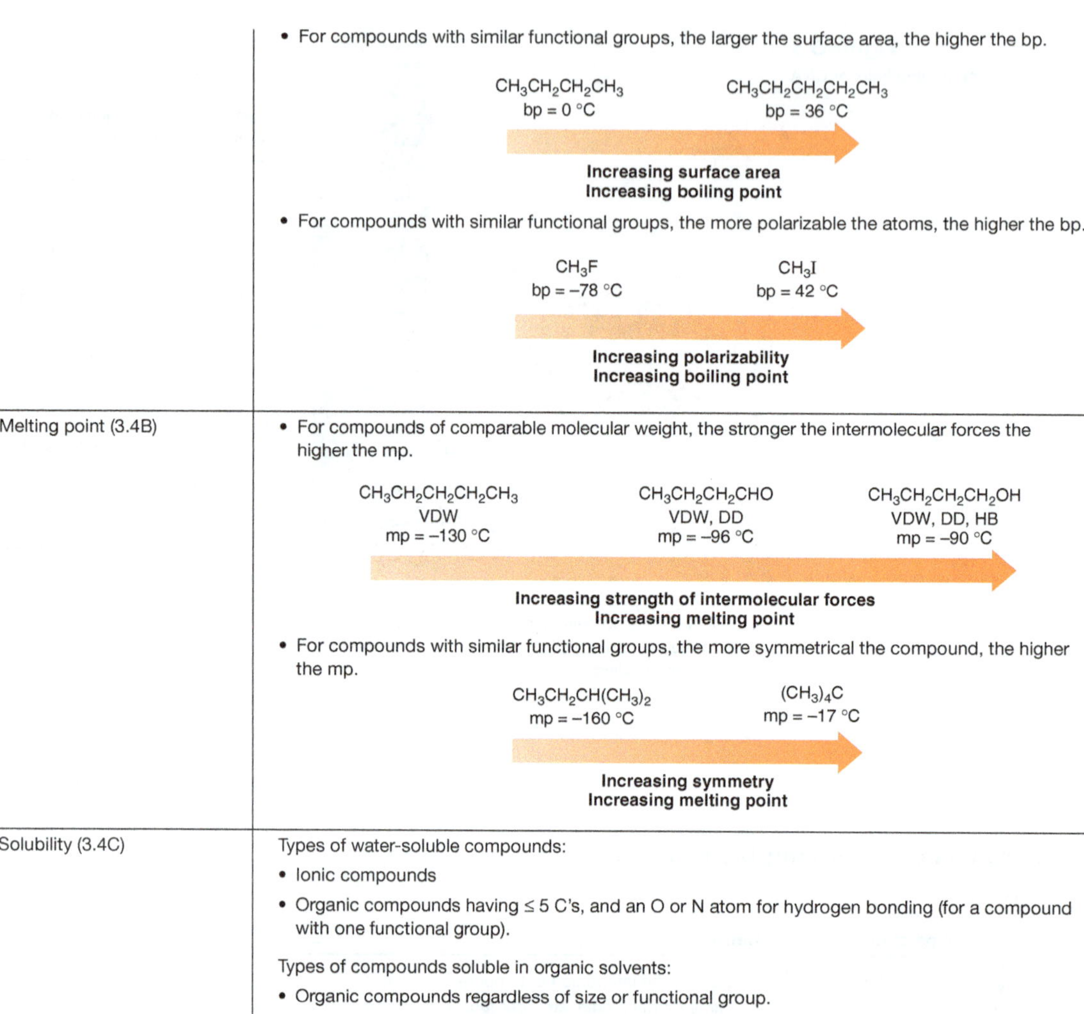

	• For compounds with similar functional groups, the larger the surface area, the higher the bp. $CH_3CH_2CH_2CH_3$ bp = 0 °C; $CH_3CH_2CH_2CH_2CH_3$ bp = 36 °C **Increasing surface area** **Increasing boiling point** • For compounds with similar functional groups, the more polarizable the atoms, the higher the bp. CH_3F bp = –78 °C; CH_3I bp = 42 °C **Increasing polarizability** **Increasing boiling point**
Melting point (3.4B)	• For compounds of comparable molecular weight, the stronger the intermolecular forces the higher the mp. $CH_3CH_2CH_2CH_2CH_3$ VDW mp = –130 °C; $CH_3CH_2CH_2CHO$ VDW, DD mp = –96 °C; $CH_3CH_2CH_2CH_2OH$ VDW, DD, HB mp = –90 °C **Increasing strength of intermolecular forces** **Increasing melting point** • For compounds with similar functional groups, the more symmetrical the compound, the higher the mp. $CH_3CH_2CH(CH_3)_2$ mp = –160 °C; $(CH_3)_4C$ mp = –17 °C **Increasing symmetry** **Increasing melting point**
Solubility (3.4C)	Types of water-soluble compounds: • Ionic compounds • Organic compounds having ≤ 5 C's, and an O or N atom for hydrogen bonding (for a compound with one functional group). Types of compounds soluble in organic solvents: • Organic compounds regardless of size or functional group.

Key: VDW = van der Waals, DD = dipole–dipole, HB = hydrogen bonding

Reactivity (3.8)

- **Nucleophiles react with electrophiles.**
- Electronegative heteroatoms create electrophilic carbon atoms, which tend to react with nucleophiles.
- Lone pairs and π bonds are nucleophilic sites that tend to react with electrophiles.

PROBLEMS

Problems with Three-Dimensional Models

3.21 (a) Identify the functional groups in the ball-and-stick model of elemicin, a compound partly responsible for the flavor and fragrance of nutmeg. (b) Draw a skeletal structure of a constitutional isomer of elemicin that should have a higher boiling point and melting point. (c) Label all electrophilic carbon atoms.

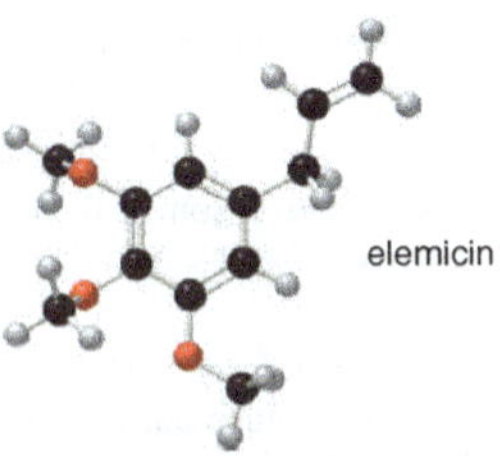

elemicin

3.22 (a) Identify the functional groups in the ball-and-stick model of neral, a compound with a lemony odor isolated from lemon grass. (b) Draw a skeletal structure of a constitutional isomer of neral that should be more water soluble. (c) Label the most electrophilic carbon atom.

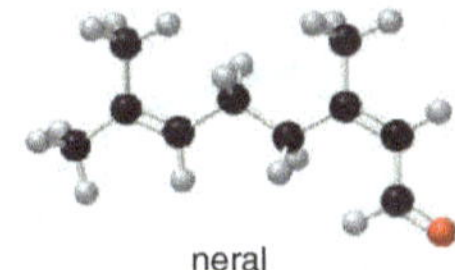

neral

Functional Groups

3.23 Identify the functional groups in each molecule.

a. 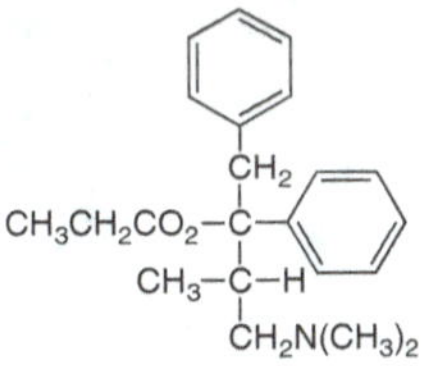 $CH_3CH_2CO_2$–C(–CH_2–C_6H_5)(C_6H_5)–CH_3–C–H–$CH_2N(CH_3)_2$

Darvon (analgesic)

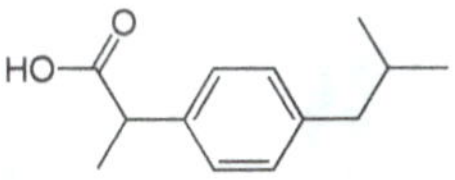

b. 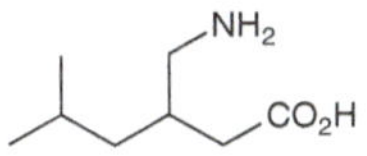 pregabalin Trade name: Lyrica (used in treating chronic pain)

c. ibuprofen (analgesic)

d. 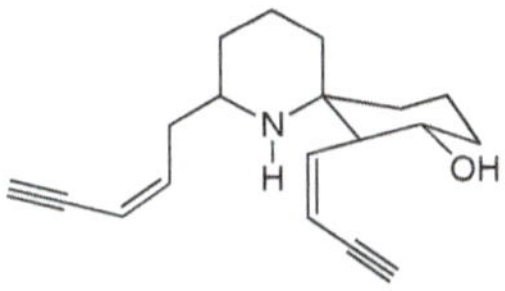 histrionicotoxin (poison secreted by a South American frog)

e. penicillin G (an antibiotic)

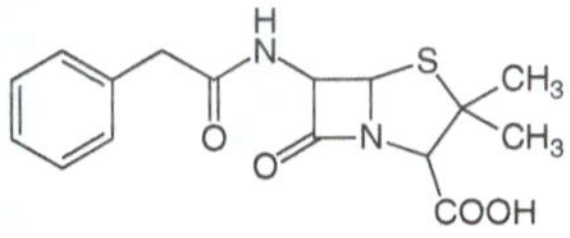

f. 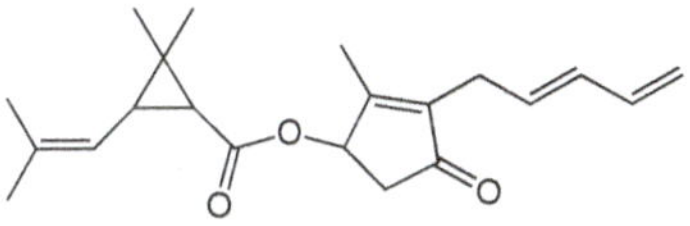 pyrethrin I (potent insecticide from chrysanthemums)

3.24 Draw the seven constitutional isomers having molecular formula $C_4H_{10}O$. Identify the functional group in each isomer.

3.25 Identify each functional group located in the following rings. Which structure represents a lactone—a cyclic ester—and which represents a lactam—a cyclic amide?

a. 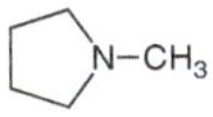 N–CH_3 b. 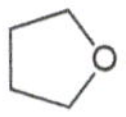 O c. 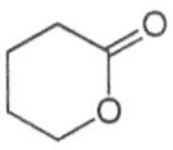 d. 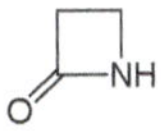 NH

3.26 Draw seven constitutional isomers with molecular formula $C_3H_6O_2$ that contain a carbonyl group. Identify the functional group(s) in each isomer.

Intermolecular Forces

3.27 What types of intermolecular forces are exhibited by each compound?

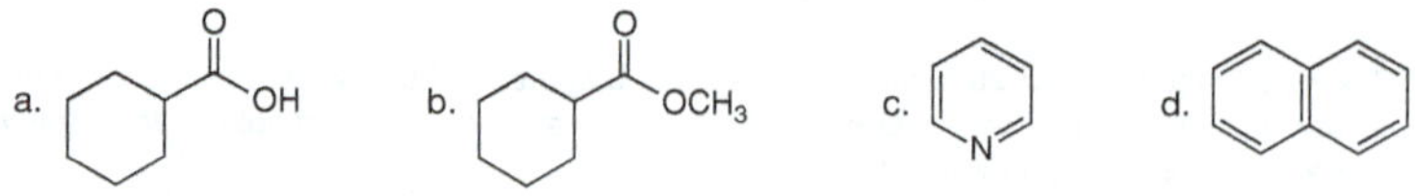

a. OH b. OCH_3 c. N d.

3.28 Rank the compounds in each group in order of increasing strength of intermolecular forces:

a. CH_3NH_2, CH_3CH_3, CH_3Cl

b. CH_3Br, CH_3I, CH_3Cl

c. $(CH_3)_2C=C(CH_3)_2$, $(CH_3)_2CHCOOH$, $(CH_3)_2CHCOCH_3$

d. NaCl, CH_3OH, CH_3Cl

3.29 Carboxylic acids (RCOOH) can exist as dimers in some situations, with two molecules held together by two intermolecular hydrogen bonds. Show how two molecules of acetic acid (CH_3CO_2H), the carboxylic acid present in vinegar, can hydrogen bond to each other.

3.30 Intramolecular forces of attraction are often important in holding large molecules together. For example, some proteins fold into compact shapes, held together by attractive forces between nearby functional groups. A schematic of a folded protein is drawn here, with the protein backbone indicated by a blue-green ribbon, and various appendages drawn dangling from the chain. What types of intramolecular forces occur at each labeled site (**A–F**)?

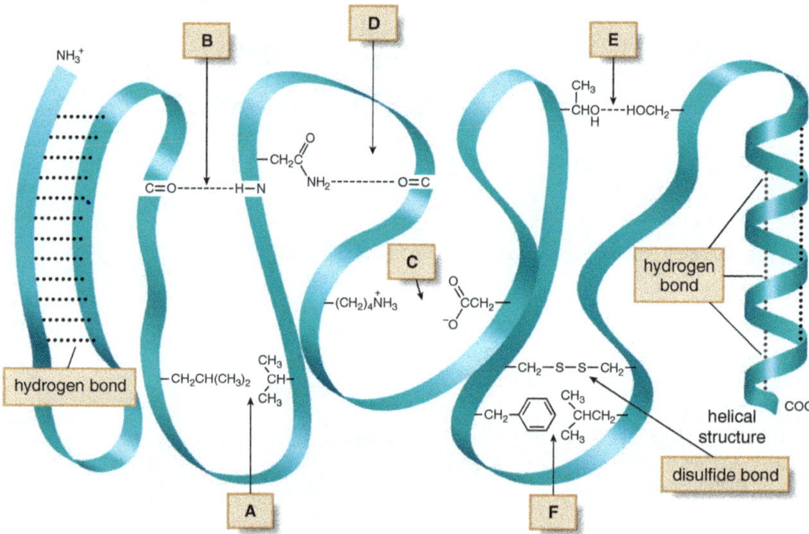

Physical Properties

3.31 (a) Draw four compounds with molecular formula $C_6H_{12}O$, each containing at least one different functional group. (b) Predict which compound has the highest boiling point, and explain your reasoning.

3.32 Rank the compounds in each group in order of increasing boiling point.

a. $CH_3(CH_2)_4I$, $CH_3(CH_2)_5I$, $CH_3(CH_2)_6I$

b. $CH_3CH_2CH_2NH_2$, $(CH_3)_3N$, $CH_3CH_2CH_2CH_3$

c. $(CH_3)_3COC(CH_3)_3$, $CH_3(CH_2)_3O(CH_2)_3CH_3$, $CH_3(CH_2)_7OH$

d.  OH Br OH

e.

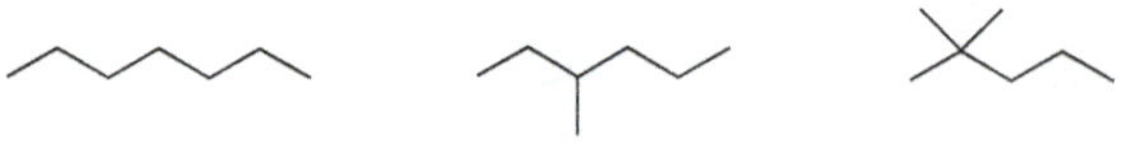

f. 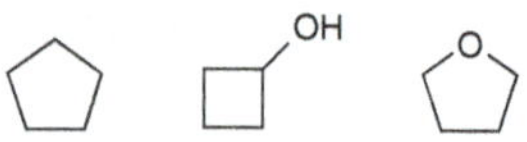 OH O

3.33 Explain why $CH_3CH_2NHCH_3$ has a higher boiling point than $(CH_3)_3N$, even though they have the same molecular weight.

3.34 Menthone and menthol are both isolated from mint. Explain why menthol is a solid at room temperature but menthone is a liquid.

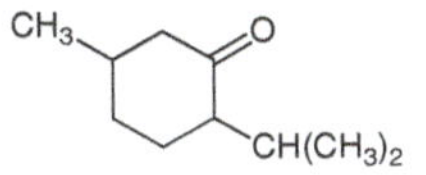

menthone 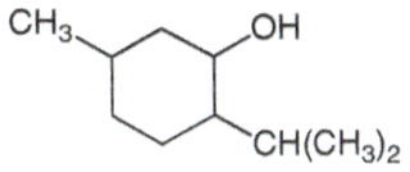 menthol

3.35 Rank the compounds in each group in order of increasing melting point.

a. $(CH_3)_2CHOH$, $(CH_3)_3CH$, $(CH_3)_2C{=}O$

b. CH_3F, CH_3Cl, CH_3I

c.

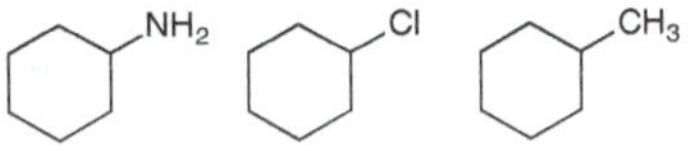

3.36 Explain the observed trend in the melting points for four isomers of molecular formula C_7H_{16}.

	mp (°C)
$CH_3CH_2CH_2CH(CH_3)CH_2CH_3$	−119
$CH_3CH_2CH_2CH_2CH(CH_3)_2$	−118
$CH_3CH_2CH_2CH_2CH_2CH_2CH_3$	−91
$(CH_3)_2CHC(CH_3)_3$	−25

3.37 Explain why benzene has a lower boiling point but much higher melting point than toluene.

benzene bp = 80 °C mp = 5 °C and toluene bp = 111 °C mp = −93 °C

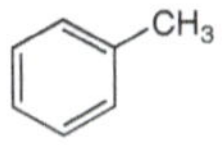

3.38 Rank the compounds in each group in order of increasing water solubility.

a. $(CH_3)_3CH$, $CH_3OCH_2CH_3$, $CH_3CH_2CH_2CH_3$, $CH_3CH_2CH_2OH$

b.

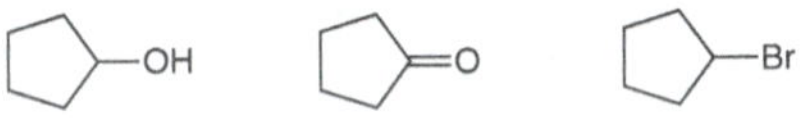

3.39 Which of the following molecules can hydrogen bond to another molecule of itself? Which can hydrogen bond with water?

a. CBr_3CH_3

b. CH_3NH_2

c. CH_3OCH_3

d. $(CH_3CH_2)_3N$

e. $CH_3CH_2CH_2CONH_2$

f. $CH_3CH_2CH_2Cl$

g. CH_3SOCH_3

h. $CH_3CH_2COOCH_3$

3.40 Explain why diethyl ether ($CH_3CH_2OCH_2CH_3$) and 1-butanol ($CH_3CH_2CH_2CH_2OH$) have similar solubility properties in water, but 1-butanol has a much higher boiling point.

3.41 Predict the water solubility of each of the following organic molecules:

a. 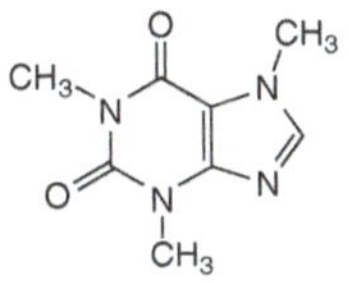 caffeine (stimulant in coffee, tea, and many soft drinks)

b. 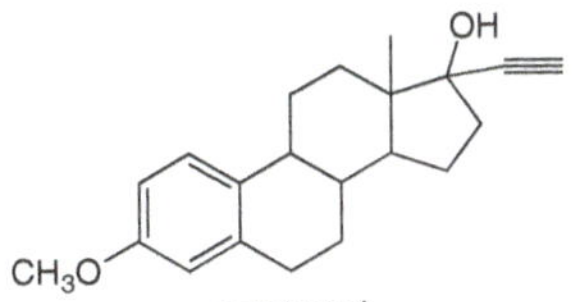 mestranol (component in oral contraceptives)

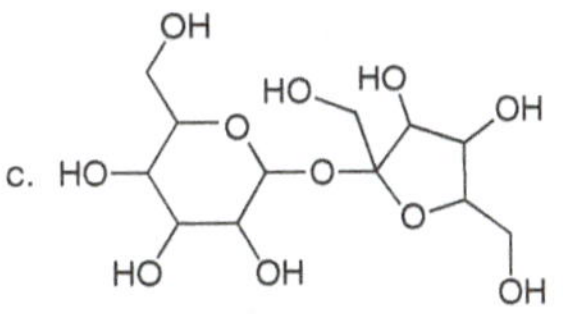

c. sucrose (table sugar)

d. 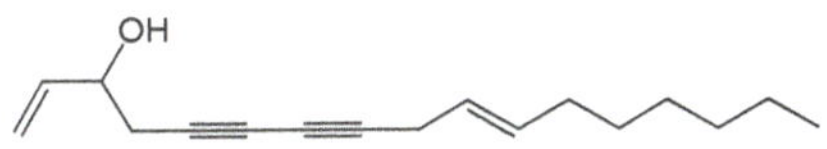 carotatoxin (neurotoxin isolated from carrots)

Applications

3.42 Predict the solubility of each of the following vitamins in water and in organic solvents:

a. 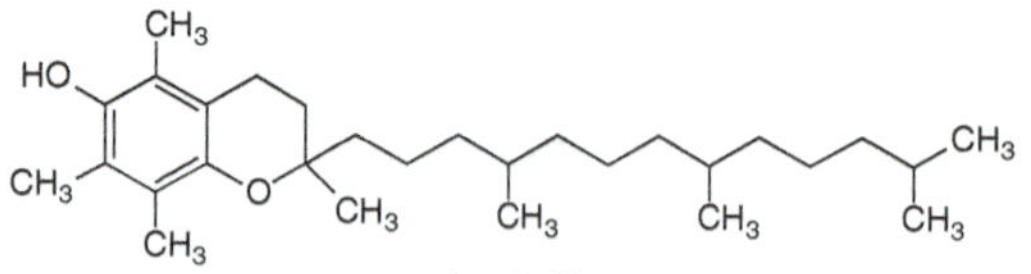 vitamin E

b. 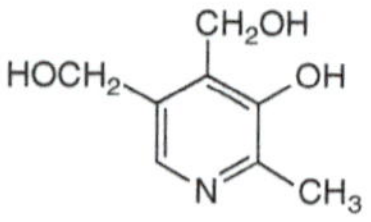 pyridoxine
vitamin B_6

3.43 Avobenzone and dioxybenzone are two commercial sunscreens. Using the principles of solubility, predict which sunscreen is more readily washed off when an individual goes swimming. Explain your choice.

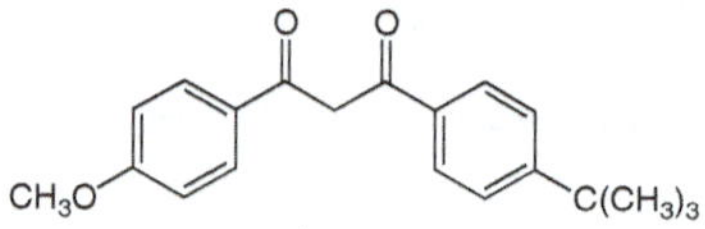

avobenzone

dioxybenzone

3.44 Poly(ethylene glycol) (PEG) and poly(vinyl chloride) (PVC) are examples of polymers, large organic molecules composed of repeating smaller units covalently bonded together. Polymers have very different properties depending (in part) on their functional groups. Discuss the water solubility of each polymer and suggest why PEG is used in shampoos while PVC is used to make garden hoses and pipes. Synthetic polymers are discussed in detail in Chapters 15 and 31.

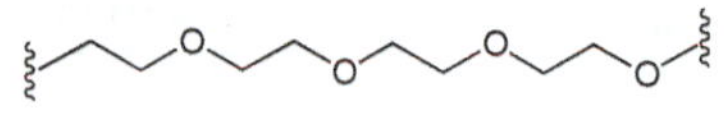

poly(ethylene glycol)
PEG

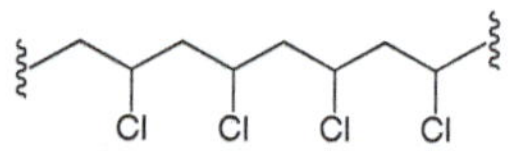

poly(vinyl chloride)
PVC

3.45 THC is the active component in marijuana, and ethanol is the alcohol in alcoholic beverages. Explain why drug screenings are able to detect the presence of THC but not ethanol weeks after these substances have been introduced into the body.

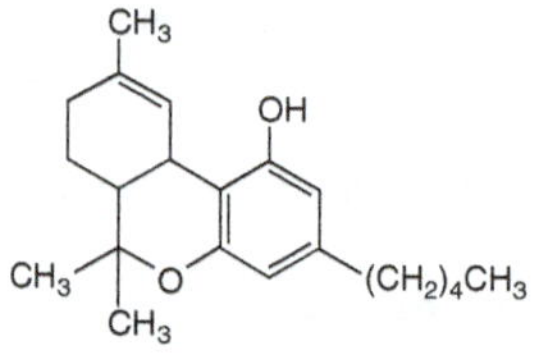

tetrahydrocannabinol
THC

$CH_3CH_2—OH$
ethanol

3.46 Cocaine is a widely abused, addicting drug. Cocaine is usually obtained as its hydrochloride salt (cocaine hydrochloride) but can be converted to crack (the neutral molecule) by treatment with base. Which of the two compounds here has a higher boiling point? Which is more soluble in water? How does the relative solubility explain why crack is usually smoked but cocaine hydrochloride is injected directly into the bloodstream?

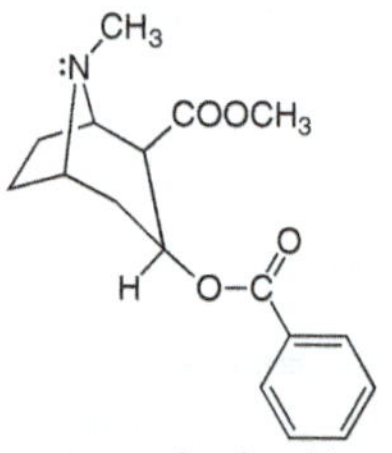

cocaine (crack)
neutral organic molecule

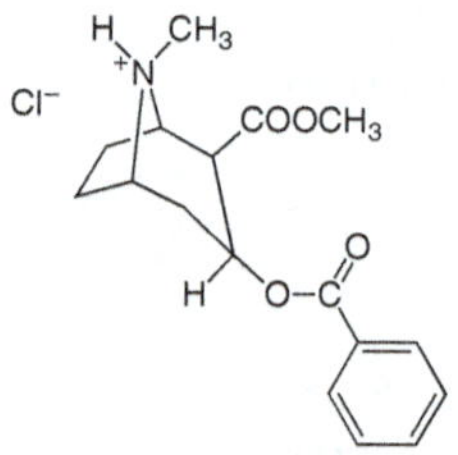

cocaine hydrochloride
a salt

3.47 Unlike soap, which is ionic, some liquid laundry detergents are neutral molecules. Explain how each of the following molecules behaves like soap and cleans away dirt.

a. 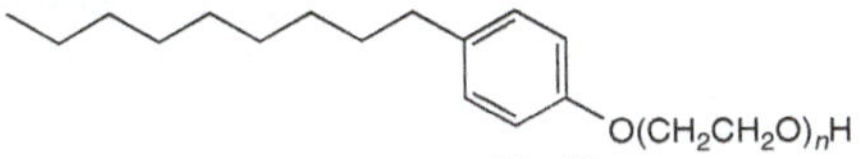 $O(CH_2CH_2O)_nH$

$n = 7$ to 13

b. 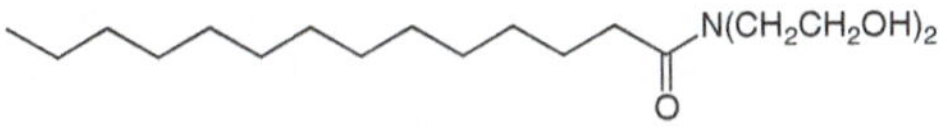 $N(CH_2CH_2OH)_2$

3.48 Many drugs are sold as their hydrochloride salts ($R_2NH_2^+$ Cl^-), formed by reaction of an amine (R_2NH) with HCl.

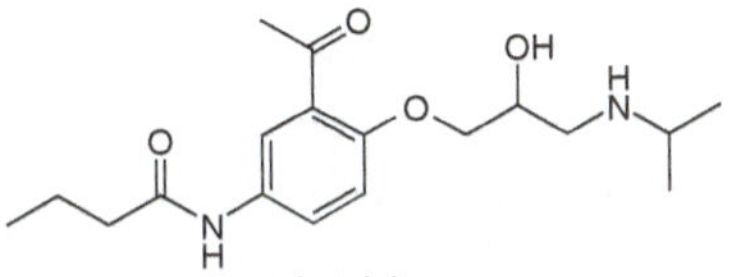

acebutolol

a. Draw the product (a hydrochloride salt) formed by reaction of acebutolol with HCl. Acebutolol is a β blocker used to treat high blood pressure.
b. Discuss the solubility of acebutolol and its hydrochloride salt in water.
c. Offer a reason as to why the drug is marketed as a hydrochloride salt rather than a neutral amine.

Reactivity of Organic Molecules

3.49 Label the electrophilic and nucleophilic sites in each molecule.

a. 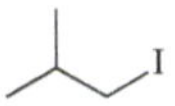 I
b. 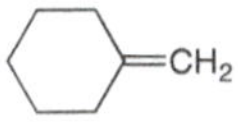 $=CH_2$
c. 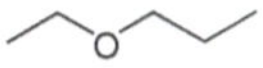 O
d.
e. CH_3OH
f. 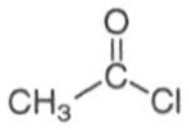 CH_3COCl

3.50 By using only electron density arguments, determine whether the following reactions will occur:

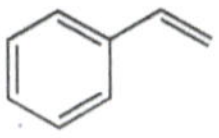

a. 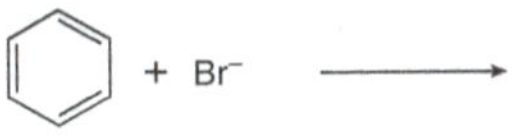 + Br^- ⟶
b. 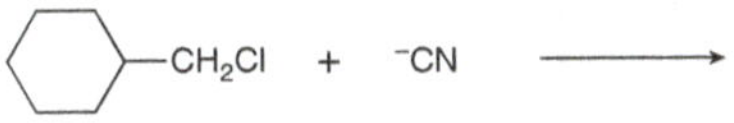 $-CH_2Cl$ + ^-CN ⟶
c. 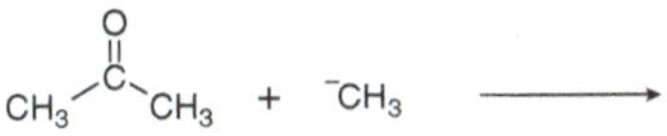 CH_3COCH_3 + $^-CH_3$ ⟶
d. + ^-OH ⟶
e. + H_3O^+ ⟶

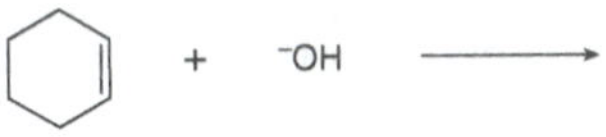

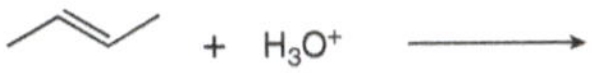

Cell Membrane

3.51 The composition of a cell membrane is not uniform for all types of cells. Some cell membranes are more rigid than others. Rigidity is determined by a variety of factors, one of which is the structure of the carbon chains in the phospholipids that comprise the membrane. One example of a phospholipid was drawn in Section 3.7A, and another, having C–C double bonds in its carbon chains, is drawn here. Which phospholipid would be present in the more rigid cell membrane and why?

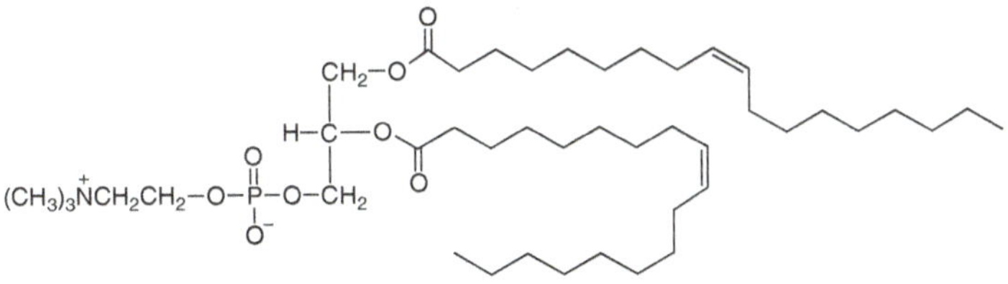

$(CH_3)_3\overset{+}{N}CH_2CH_2-O-P(=O)(O^-)-O-CH_2$, H–C–O, CH_2-O

a phospholipid with π bonds in the long hydrocarbon chains

General Question

3.52 Vancomycin is a useful antibiotic for treating infections in cancer patients on chemotherapy and renal patients on dialysis. Unlike mammalian cells, bacterial cells are surrounded by a fairly rigid cell wall, which is crucial to the bacterium's survival. Vancomycin kills bacteria by interfering with their cell wall synthesis.

a. How many amide functional groups are present in vancomycin?
b. Which OH groups are bonded to sp^3 hybridized carbon atoms and which are bonded to sp^2 hybridized carbons?
c. Would you expect vancomycin to be water soluble? Explain.
d. Which proton is the most acidic?
e. Label three *different* functional groups capable of hydrogen bonding.

vancomycin

Challenge Problems

3.53 Although diethyl ether and tetrahydrofuran are both four-carbon ethers, one compound is much more water soluble than the other. Predict which compound has higher water solubility and offer an explanation.

diethyl ether tetrahydrofuran

3.54 Answer the following questions by referring to the ball-and-stick model of fentanyl, a potent narcotic analgesic used in surgical procedures.

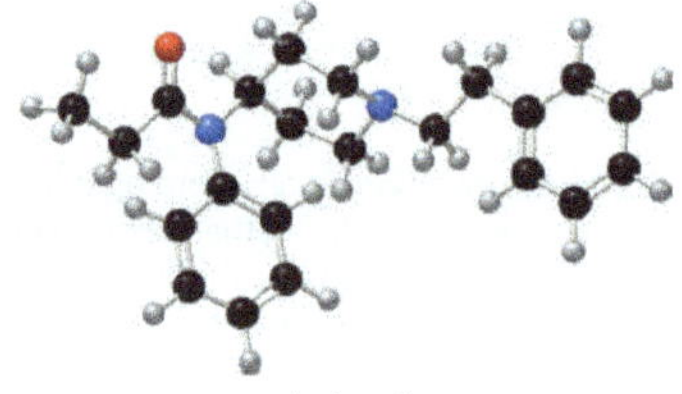

fentanyl

a. Identify the functional groups.
b. Label the most acidic proton.
c. Label the most basic atom.
d. What types of intermolecular forces are present between two molecules of fentanyl?
e. Draw an isomer predicted to have a higher boiling point.
f. Which sites in the molecule can hydrogen bond to water?
g. Label all electrophilic carbons.

3.55 Explain why **A** is less water soluble than **B,** even though both compounds have the same functional groups.

OH, CHO = **A** OH, CHO = **B**

3.56 Recall from Section 1.10B that there is restricted rotation around carbon–carbon double bonds. Maleic acid and fumaric acid are two isomers with vastly different physical properties and pK_a values for loss of both protons. Explain why each of these differences occurs.

	maleic acid	fumaric acid
mp (°C)	130	286
solubility (g/L) in H_2O at 25 °C	788	7
pK_{a1}	1.9	3.0
pK_{a2}	6.5	4.5

Appendix 1
Mathematical Operations

Scientific Notation

Chemists often deal with numbers that are either extremely large or extremely small. For example, in 1 g of the element hydrogen there are roughly

$$602{,}200{,}000{,}000{,}000{,}000{,}000{,}000$$

hydrogen atoms. Each hydrogen atom has a mass of only

$$0.00000000000000000000000166 \text{ g}$$

These numbers are cumbersome to handle, and it is easy to make mistakes when using them in arithmetic computations. Consider the following multiplication:

$$0.0000000056 \times 0.00000000048 = 0.000000000000000002688$$

It would be easy for us to miss one zero or add one more zero after the decimal point. Consequently, when working with very large and very small numbers, we use a system called *scientific notation*. Regardless of their magnitude, all numbers can be expressed in the form

$$N \times 10^n$$

where N is a number between 1 and 10 and n, the exponent, is a positive or negative integer (whole number). Any number expressed in this way is said to be written in scientific notation.

Suppose that we are given a certain number and asked to express it in scientific notation. Basically, this assignment calls for us to find n. We count the number of places that the decimal point must be moved to give the number N (which is between 1 and 10). If the decimal point has to be moved to the left, then n is a positive integer; if it has to be moved to the right, n is a negative integer. The following examples illustrate the use of scientific notation:

1. Express 568.762 in scientific notation:

$$568.762 = 5.68762 \times 10^2$$

Note that the decimal point is moved to the left by two places and $n = 2$.

2. Express 0.00000772 in scientific notation:

$$0.00000772 = 7.72 \times 10^{-6}$$

Here the decimal point is moved to the right by six places and $n = -6$.

Keep in mind the following two points. First, $n = 0$ is used for numbers that are not expressed in scientific notation. For example, 74.6×10^0 ($n = 0$) is equivalent to 74.6. Second, the usual practice is to omit the superscript when $n = 1$. Thus, the scientific notation for 74.6 is 7.46×10 and not 7.46×10^1.

Next, we consider how scientific notation is handled in arithmetic operations.

Addition and Subtraction

To add or subtract using scientific notation, we first write each quantity—say N_1 and N_2—with the same exponent n. Then we combine N_1 and N_2; the exponents remain the same. Consider the following examples:

$$(7.4 \times 10^3) + (2.1 \times 10^3) = 9.5 \times 10^3$$

$$\begin{aligned}(4.31 \times 10^4) + (3.9 \times 10^3) &= (4.31 \times 10^4) + (0.39 \times 10^4)\\ &= 4.70 \times 10^4\end{aligned}$$

$$\begin{aligned}(2.22 \times 10^{-2}) - (4.10 \times 10^{-3}) &= (2.22 \times 10^{-2}) - (0.41 \times 10^{-2})\\ &= 1.81 \times 10^{-2}\end{aligned}$$

Multiplication and Division

To multiply numbers expressed in scientific notation, we multiply N_1 and N_2 in the usual way, but *add* the exponents together. To divide using scientific notation, we divide N_1 and N_2 as usual and subtract the exponents. The following examples show how these operations are performed:

$$\begin{aligned}(8.0 \times 10^4) \times (5.0 \times 10^2) &= (8.0 \times 5.0)(10^{4+2})\\ &= 40 \times 10^6\\ &= 4.0 \times 10^7\end{aligned}$$

$$\begin{aligned}(4.0 \times 10^{-5}) \times (7.0 \times 10^3) &= (4.0 \times 7.0)(10^{-5+3})\\ &= 28 \times 10^{-2}\\ &= 2.8 \times 10^{-1}\end{aligned}$$

$$\begin{aligned}\frac{6.9 \times 10^7}{3.0 \times 10^{-5}} &= \frac{6.9}{3.0} \times 10^{7-(-5)}\\ &= 2.3 \times 10^{12}\end{aligned}$$

$$\begin{aligned}\frac{8.5 \times 10^4}{5.0 \times 10^9} &= \frac{8.5}{5.0} \times 10^{4-9}\\ &= 1.7 \times 10^{-5}\end{aligned}$$

Basic Trigonometry

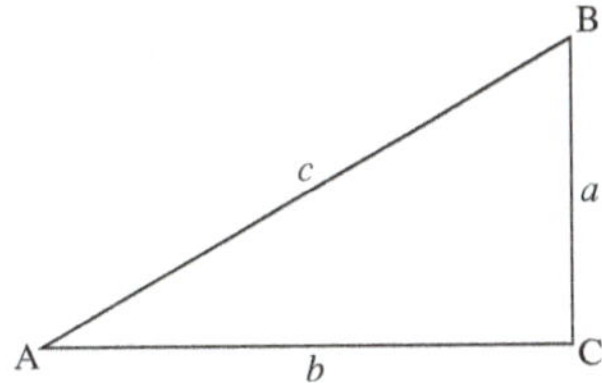

In the triangle shown, A, B, and C are angles (C = 90°) and a, b, and c are side lengths. To calculate unknown angles or sides, we use the following relationships:

$$a^2 + b^2 = c^2$$

$$\sin A = \frac{a}{c}$$

$$\cos A = \frac{b}{c}$$

$$\tan A = \frac{a}{b}$$

Logarithms

Common Logarithms

The concept of logarithms is an extension of the concept of exponents, which is discussed on page A-1. The common, or base-10, logarithm of any number is the power to which 10 must be raised to equal the number. The following examples illustrate this relationship:

Logarithm	Exponent
$\log 1 = 0$	$10^0 = 1$
$\log 10 = 1$	$10^1 = 10$
$\log 100 = 2$	$10^2 = 100$
$\log 10^{-1} = -1$	$10^{-1} = 0.1$
$\log 10^{-2} = -2$	$10^{-2} = 0.01$

In each case, the logarithm of the number can be obtained by inspection.

Because the logarithms of numbers are exponents, they have the same properties as exponents. Thus, we have

Logarithm	Exponent
$\log AB = \log A + \log B$	$10^A \times 10^B = 10^{A+B}$
$\log \frac{A}{B} = \log A - \log B$	$\frac{10^A}{10^B} = 10^{A-B}$

Furthermore, $\log A^n = n \log A$.

Now suppose we want to find the common logarithm of 6.7×10^{-4}. On most electronic calculators, the number is entered first and then the log key is pressed. This operation gives us

$$\log 6.7 \times 10^{-4} = -3.17$$

Note that there are as many digits *after* the decimal point as there are significant figures in the original number. The original number has two significant figures, and the "17" in −3.17 tells us that the log has two significant figures. The "3" in −3.17 serves only to locate the decimal point in the number 6.7×10^{-4}. Other examples are

Number	Common Logarithm
62	1.79
0.872	−0.0595
1.0×10^{-7}	−7.00

Sometimes (as in the case of pH calculations) it is necessary to obtain the number whose logarithm is known. This procedure is known as taking the antilogarithm; it is simply the reverse of taking the logarithm of a number. Suppose in a certain calculation we have pH = 1.46 and are asked to calculate $[H^+]$. From the definition of pH ($pH = -\log [H^+]$). We can write

$$[H^+] = 10^{-1.46}$$

Many calculators have a key labeled $\log^{-1}$ or INV log to obtain antilogs. Other calculators have a 10^x or y^x key (where x corresponds to −1.46 in our example and y is 10 for base-10 logarithm). Therefore, we find that $[H^+] = 0.035\ M$.

Natural Logarithms

Logarithms taken to the base e instead of 10 are known as natural logarithms (denoted by ln or $\log_e$); e is equal to 2.7183. The relationship between common logarithms and natural logarithms is as follows:

$$\log 10 = 1 \qquad 10^1 = 10$$

$$\ln 10 = 2.303 \qquad e^{2.303} = 10$$

Thus,

$$\ln x = 2.303 \log x$$

To find the natural logarithm of 2.27, say, we use ln key to get

$$\ln 2.27 = 0.820$$

For a calculator that does not have an ln key, we can proceed as follows:

$$\begin{aligned} 2.303 \log 2.27 &= 2.303 \times 0.356 \\ &= 0.820 \end{aligned}$$

Sometimes we may be given the natural logarithm and asked to find the number it represents. For example,

$$\ln x = 59.7$$

On many calculators, we simply use the e key:

$$e^{59.7} = 8 \times 10^{25}$$

The Quadratic Equation

A quadratic equation takes the form

$$ax^2 + bx + c = 0$$

If coefficients a, b, and c are known, then x is given by

$$x = \frac{-b \pm \sqrt{b^2 - 4ac}}{2a}$$

Suppose we have the following quadratic equation:

$$2x^2 + 5x - 12 = 0$$

Solving for x, we write

$$\begin{aligned} x &= \frac{-5 \pm \sqrt{(5)^2 - 4(2)(-12)}}{2(2)} \\ &= \frac{-5 \pm \sqrt{25 + 96}}{4} \end{aligned}$$

Therefore,

$$x = \frac{-5 + 11}{4} = \frac{3}{2}$$

and

$$x = \frac{-5 - 11}{4} = -4$$

Successive Approximation

In the determination of hydrogen ion concentration in a weak acid solution, use of the quadratic equation can sometimes be avoided using a method known as *successive approximation*. Consider the example of a 0.0150 M solution of hydrofluoric acid (HF). The K_a for HF is 7.10×10^{-4}. To determine the hydrogen ion concentration in this solution, we construct an equilibrium table and enter the initial concentrations, the expected change in concentrations, and the equilibrium concentrations of all species:

	$HF(aq)$ $\rightleftharpoons$	$H^+(aq)$ +	$F^-(aq)$
Initial concentration (M):	0.0150	0	0
Change in concentration (M):	$-x$	$+x$	$+x$
Equilibrium concentration (M):	$0.0150 - x$	x	x

Using the rule that x can be neglected if the initial acid concentration divided by the K_a is greater than 100, we find that in this case x cannot be neglected ($0.0150/7.1 \times 10^{-4} \approx 21$). Successive approximation involves first neglecting x with respect to initial acid concentration

$$\frac{x^2}{0.0150 - x} \approx \frac{x^2}{0.0150} = 7.10 \times 10^{-4}$$

and solving for x:

$$x^2 = (0.0150)(7.10 \times 10^{-4}) = 1.07 \times 10^{-5}$$

$$x = \sqrt{1.07 \times 10^{-5}} = 0.00326\ M$$

We then solve for x again, this time using the calculated value of x on the bottom of the fraction:

$$\frac{x^2}{0.0150 - x} = \frac{x^2}{0.0150 - 0.00326} = 7.10 \times 10^{-4}$$

$$x^2 = (0.0150 - 0.00326)(7.10 \times 10^{-4}) = 8.33 \times 10^{-6}$$

$$x = \sqrt{8.33 \times 10^{-6}} = 0.00289\ M$$

Note that the calculated value of x decreased from 0.00326 to 0.00289. We now use the new calculated value of x on the bottom of the fraction and solve for x again:

$$\frac{x^2}{0.0150 - x} = \frac{x^2}{0.0150 - 0.00289} = 7.10 \times 10^{-4}$$

$$x^2 = (0.0150 - 0.00289)(7.10 \times 10^{-4}) = 8.60 \times 10^{-6}$$

$$x = \sqrt{8.60 \times 10^{-6}} = 0.00293\ M$$

This time the value of x increased slightly. We use the new calculated value and solve for x again.

$$\frac{x^2}{0.0150 - x} = \frac{x^2}{0.0150 - 0.00293} = 7.10 \times 10^{-4}$$

$$x^2 = (0.0150 - 0.00293)(7.10 \times 10^{-4}) = 8.57 \times 10^{-6}$$

$$x = \sqrt{8.57 \times 10^{-6}} = 0.00293\ M$$

This time we find that the answer is still 0.00293 *M*, so there is no need to repeat the process. In general, we apply the method of successive approximation until the value of x obtained does not differ from the value obtained in the previous step. The value of x determined using successive approximation is the same value we would get if we were to use the quadratic equation.

Appendix 2

Thermodynamic Data at 1 ATM and 25°C*

Inorganic Substances			
Substance	**ΔH_f° (kJ/mol)**	**ΔG_f° (kJ/mol)**	**S° (J/K · mol)**
$Ag(s)$	0	0	42.7
$Ag^+(aq)$	105.9	77.1	73.9
$AgCl(s)$	−127.0	−109.7	96.1
$AgBr(s)$	−99.5	−95.9	107.1
$AgI(s)$	−62.4	−66.3	114.2
$AgNO_3(s)$	−123.1	−32.2	140.9
$Al(s)$	0	0	28.3
$Al^{3+}(aq)$	−524.7	−481.2	−313.38
$Al_2O_3(s)$	−1669.8	−1576.4	50.99
$As(s)$	0	0	35.15
$AsO_4^{3-}(aq)$	−870.3	−635.97	−144.77
$AsH_3(g)$	171.5		
$H_3AsO_4(s)$	−900.4		
$Au(s)$	0	0	47.7
$Au_2O_3(s)$	80.8	163.2	125.5
$AuCl(s)$	−35.2		
$AuCl_3(s)$	−118.4		
$B(s)$	0	0	6.5
$B_2O_3(s)$	−1263.6	−1184.1	54.0
$H_3BO_3(s)$	−1087.9	−963.16	89.58
$H_3BO_3(aq)$	−1067.8	−963.3	159.8
$Ba(s)$	0	0	66.9
$Ba^{2+}(aq)$	−538.4	−560.66	12.55
$BaO(s)$	−558.2	−528.4	70.3
$BaCl_2(s)$	−860.1	−810.66	125.5
$BaSO_4(s)$	−1464.4	−1353.1	132.2
$BaCO_3(s)$	−1218.8	−1138.9	112.1
$Be(s)$	0	0	9.5
$BeO(s)$	−610.9	−581.58	14.1
$Br_2(l)$	0	0	152.3
$Br_2(g)$	30.7	3.14	245.13
$Br^-(aq)$	−120.9	−102.8	80.7

* The thermodynamic quantities of ions are based on the reference states that $\Delta H_f^\circ[H^+(aq)] = 0$, $\Delta G_f^\circ[H^+(aq)] = 0$, and $S^\circ[H^+(aq)] = 0$.

Substance	ΔH_f° (kJ/mol)	ΔG_f° (kJ/mol)	S° (J/K · mol)
$HBr(g)$	−36.2	−53.2	198.48
C(graphite)	0	0	5.69
C(diamond)	1.90	2.87	2.4
$CCl_4(g)$	−95.7	−62.3	309.7
$CCl_4(l)$	−128.2	−66.4	216.4
$CO(g)$	−110.5	−137.3	197.9
$CO_2(g)$	−393.5	−394.4	213.6
$CO_2(aq)$	−412.9	−386.2	121.3
$CO_3^{2-}(aq)$	−676.3	−528.1	−53.1
$HCO_3^-(aq)$	−691.1	−587.1	94.98
$H_2CO_3(aq)$	−699.7	−623.2	187.4
$CS_2(g)$	115.3	65.1	237.8
$CS_2(l)$	87.3	63.6	151.0
$HCN(aq)$	105.4	112.1	128.9
$CN^-(aq)$	151.0	165.69	117.99
$(NH_2)_2CO(s)$	−333.19	−197.15	104.6
$(NH_2)_2CO(aq)$	−319.2	−203.84	173.85
$Ca(s)$	0	0	41.6
$Ca(g)$	179.3	145.5	154.8
$Ca^{2+}(aq)$	−542.96	−553.0	−55.2
$CaO(s)$	−635.6	−604.2	39.8
$Ca(OH)_2(s)$	−986.6	−896.8	83.4
$CaF_2(s)$	−1214.6	−1161.9	68.87
$CaCl_2(s)$	−794.96	−750.19	113.8
$CaSO_4(s)$	−1432.69	−1320.3	106.69
$CaCO_3(s)$	−1206.9	−1128.8	92.9
$Cd(s)$	0	0	51.46
$Cd^{2+}(aq)$	−72.38	−77.7	−61.09
$CdO(s)$	−254.6	−225.06	54.8
$CdCl_2(s)$	−389.1	−342.59	118.4
$CdSO_4(s)$	−926.17	−820.2	137.2
$Cl_2(g)$	0	0	223.0
$Cl(g)$	121.7	105.7	165.2
$Cl^-(aq)$	−167.2	−131.2	56.5
$HCl(g)$	−92.3	−95.27	187.0
$Co(s)$	0	0	28.45
$Co^{2+}(aq)$	−67.36	−51.46	155.2
$CoO(s)$	−239.3	−213.38	43.9
$Cr(s)$	0	0	23.77
$Cr^{2+}(aq)$	−138.9		
$Cr_2O_3(s)$	−1128.4	−1046.8	81.17
$CrO_4^{2-}(aq)$	−863.16	−706.26	38.49
$Cr_2O_7^{2-}(aq)$	−1460.6	−1257.29	213.8

(Continued)

Substance	ΔH_f° (kJ/mol)	ΔG_f° (kJ/mol)	S° (J/K · mol)
$Cs(s)$	0	0	82.8
$Cs(g)$	76.50	49.53	175.6
$Cs^+(aq)$	−247.69	−282.0	133.05
$CsCl(s)$	−442.8	−414.4	101.2
$Cu(s)$	0	0	33.3
$Cu^+(aq)$	51.88	50.2	−26.4
$Cu^{2+}(aq)$	64.39	64.98	−99.6
$CuO(s)$	−155.2	−127.2	43.5
$Cu_2O(s)$	−166.69	−146.36	100.8
$CuCl(s)$	−134.7	−118.8	91.6
$CuCl_2(s)$	−205.85		
$CuS(s)$	−48.5	−49.0	66.5
$CuSO_4(s)$	−769.86	−661.9	113.39
$F_2(g)$	0	0	203.34
$F(g)$	80.0	61.9	158.7
$F^-(aq)$	−329.1	−276.48	−9.6
$HF(g)$	−271.6	−270.7	173.5
$Fe(s)$	0	0	27.2
$Fe^{2+}(aq)$	−87.86	−84.9	−113.39
$Fe^{3+}(aq)$	−47.7	−10.5	−293.3
$FeO(s)$	−272.0	−255.2	60.8
$Fe_2O_3(s)$	−822.2	−741.0	90.0
$Fe(OH)_2(s)$	−568.19	−483.55	79.5
$Fe(OH)_3(s)$	−824.25		
$H(g)$	218.2	203.2	114.6
$H_2(g)$	0	0	131.0
$H^+(aq)$	0	0	0
$OH^-(aq)$	−229.94	−157.30	−10.5
$H_2O(g)$	−241.8	−228.6	188.7
$H_2O(l)$	−285.8	−237.2	69.9
$H_2O_2(g)$	−136.1	−105.5	232.9
$H_2O_2(l)$	−187.6	−118.1	109.6
$Hg(l)$	0	0	77.4
$Hg^{2+}(aq)$		−164.38	
$HgO(s)$	−90.7	−58.5	72.0
$HgCl_2(s)$	−230.1		
$Hg_2Cl_2(s)$	−264.9	−210.66	196.2
$HgS(s)$	−58.16	−48.8	77.8
$HgSO_4(s)$	−704.17		
$Hg_2SO_4(s)$	−741.99	−623.92	200.75
$I_2(g)$	62.25	19.37	260.57
$I_2(s)$	0	0	116.7
$I(g)$	106.8	70.21	180.67
$I^-(aq)$	−55.9	−51.67	109.37

Substance	ΔH_f° (kJ/mol)	ΔG_f° (kJ/mol)	S° (J/K · mol)
$HI(g)$	25.9	1.30	206.3
$K(s)$	0	0	63.6
$K^+(aq)$	−251.2	−282.28	102.5
$KOH(s)$	−425.85		
$KCl(s)$	−435.87	−408.3	82.68
$KClO_3(s)$	−391.20	−289.9	142.97
$KClO_4(s)$	−433.46	−304.18	151.0
$KBr(s)$	−392.17	−379.2	96.4
$KI(s)$	−327.65	−322.29	104.35
$KNO_3(s)$	−492.7	−393.1	132.9
$Li(s)$	0	0	28.0
$Li(g)$	159.3	126.6	138.8
$Li^+(aq)$	−278.46	−293.8	14.2
$LiCl(s)$	−408.3	−384.0	59.30
$Li_2O(s)$	−595.8		
$LiOH(s)$	−487.2	−443.9	50.2
$Mg(s)$	0	0	32.5
$Mg(g)$	150	115	148.55
$Mg^{2+}(aq)$	−461.96	−456.0	−117.99
$MgO(s)$	−601.8	−569.6	26.78
$Mg(OH)_2(s)$	−924.66	−833.75	63.1
$MgCl_2(s)$	−641.8	−592.3	89.5
$MgSO_4(s)$	−1278.2	−1173.6	91.6
$MgCO_3(s)$	−1112.9	−1029.3	65.69
$Mn(s)$	0	0	31.76
$Mn^{2+}(aq)$	−218.8	−223.4	−83.68
$MnO_2(s)$	−520.9	−466.1	53.1
$N(g)$	470.4	455.5	153.3
$N_2(g)$	0	0	191.5
$N_3^-(aq)$	245.18		
$NH_3(g)$	−46.3	−16.6	193.0
$NH_4^+(aq)$	−132.80	−79.5	112.8
$NH_4Cl(s)$	−315.39	−203.89	94.56
$NH_3(aq)$	−80.3	−26.5	111.3
$N_2H_4(l)$	50.4		
$NO(g)$	90.4	86.7	210.6
$NO_2(g)$	33.85	51.8	240.46
$N_2O_4(g)$	9.66	98.29	304.3
$N_2O(g)$	81.56	103.6	219.99
$HNO_2(aq)$	−118.8	−53.6	
$HNO_3(l)$	−173.2	−79.9	155.6
$NO_3^-(aq)$	−206.57	−110.5	146.4
$Na(s)$	0	0	51.05
$Na(l)$	2.41	0.50	57.56

(Continued)

Substance	ΔH_f° (kJ/mol)	ΔG_f° (kJ/mol)	S° (J/K · mol)
Na(*g*)	107.7	77.3	153.7
Na^+(*aq*)	−239.66	−261.87	60.25
NaOH(*aq*)	−469.6	−419.2	49.8
Na_2O(*s*)	−415.9	−376.56	72.8
NaCl(*s*)	−410.9	−384.0	72.38
NaI(*s*)	−288.0		
Na_2SO_4(*s*)	−1384.49	−1266.8	149.49
$NaNO_3$(*s*)	−466.68	−365.89	116.3
Na_2CO_3(*s*)	−1130.9	−1047.67	135.98
$NaHCO_3$(*s*)	−947.68	−851.86	102.09
Ni(*s*)	0	0	30.1
Ni^{2+}(*aq*)	−64.0	−46.4	−159.4
NiO(*s*)	−244.35	−216.3	38.58
$Ni(OH)_2$(*s*)	−538.06	−453.1	79.5
O(*g*)	249.4	230.1	160.95
O_2(*g*)	0	0	205.0
O_3(*aq*)	−12.09	16.3	110.88
O_3(*g*)	142.2	163.4	237.6
P(white)	0	0	44.0
P(red)	−18.4	13.8	29.3
PCl_3(*l*)	−319.7	−272.3	217.1
PCl_3(*g*)	−288.07	−269.6	311.7
PCl_5(*g*)	−374.9	−305.0	364.5
PO_4^{3-}(*aq*)	−1284.07	−1025.59	−217.57
P_4O_{10}(*s*)	−3012.48		
PH_3(*g*)	9.25	18.2	210.0
HPO_4^{2-}(*aq*)	−1298.7	−1094.1	−35.98
$H_2PO_4^-$(*aq*)	−1302.48	−1135.1	89.1
Pb(*s*)	0	0	64.89
Pb^{2+}(*aq*)	1.6	−24.3	21.3
PbO(*s*)	−217.86	−188.49	69.45
PbO_2(*s*)	−276.65	−218.99	76.57
$PbCl_2$(*s*)	−359.2	−313.97	136.4
PbS(*s*)	−94.3	−92.68	91.2
$PbSO_4$(*s*)	−918.4	−811.2	147.28
Pt(*s*)	0	0	41.84
$PtCl_4^{2-}$(*aq*)	−516.3	−384.5	175.7
Rb(*s*)	0	0	69.45
Rb(*g*)	85.8	55.8	170.0
Rb^+(*aq*)	−246.4	−282.2	124.27
RbBr(*s*)	−389.2	−378.1	108.3
RbCl(*s*)	−435.35	−407.8	95.90
RbI(*s*)	−328	−326	118.0
S(rhombic)	0	0	31.88

Substance	ΔH_f° (kJ/mol)	ΔG_f° (kJ/mol)	S° (J/K · mol)
S(monoclinic)	0.30	0.10	32.55
$SO(g)$	5.01	−19.9	221.8
$SO_2(g)$	−296.4	−300.4	248.5
$SO_3(g)$	−395.2	−370.4	256.2
$SO_3^{2-}(aq)$	−624.25	−497.06	43.5
$SO_4^{2-}(aq)$	−907.5	−741.99	17.15
$H_2S(g)$	−20.15	−33.0	205.64
$HSO_3^-(aq)$	−627.98	−527.3	132.38
$HSO_4^-(aq)$	−885.75	−752.87	126.86
$H_2SO_4(l)$	−811.3		
$SF_6(g)$	−1096.2		
$Si(s)$	0	0	18.70
$SiO_2(s)$	−859.3	−805.0	41.84
$Sr(s)$	0	0	54.39
$Sr^{2+}(aq)$	−545.5	−557.3	−39.33
$SrCl_2(s)$	−828.4	−781.15	117.15
$SrSO_4(s)$	−1444.74	−1334.28	121.75
$SrCO_3(s)$	−1218.38	−1137.6	97.07
$U(s)$	0	0	50.21
$UF_6(g)$	−2147	−2064	378
$Zn(s)$	0	0	41.6
$Zn^{2+}(aq)$	−152.4	−147.2	−106.48
$ZnO(s)$	−348.0	−318.2	43.9
$ZnCl_2(s)$	−415.89	−369.26	108.37
$ZnS(s)$	−202.9	−198.3	57.7
$ZnSO_4(s)$	−978.6	−871.6	124.7

Organic Substances

Substance	Formula	ΔH_f° (kJ/mol)	ΔG_f° (kJ/mol)	S° (J/K · mol)
Acetic acid(l)	CH_3COOH	−484.2	−389.45	159.8
Acetaldehyde(g)	CH_3CHO	−166.35	−139.08	264.2
Acetone(l)	CH_3COCH_3	−246.8	−153.55	198.7
Acetylene(g)	C_2H_2	226.6	209.2	200.8
Benzene(l)	C_6H_6	49.04	124.5	172.8
Butane(g)	C_4H_{10}	−124.7	−15.7	310.0
Ethanol(l)	C_2H_5OH	−276.98	−174.18	161.0
Ethane(g)	C_2H_6	−84.7	−32.89	229.5
Ethylene(g)	C_2H_4	52.3	68.1	219.5
Formic acid(l)	$HCOOH$	−409.2	−346.0	129.0
Glucose(s)	$C_6H_{12}O_6$	−1274.5	−910.56	212.1
Methane(g)	CH_4	−74.85	−50.8	186.2
Methanol(l)	CH_3OH	−238.7	−166.3	126.8
Propane(g)	C_3H_8	−103.9	−23.5	269.9
Sucrose(s)	$C_{12}H_{22}O_{11}$	−2221.7	−1544.3	360.2

Appendix 3

Solubility Product Constants at 25°C

Compound	Dissolution Equilibrium	K_{sp}
Bromides		
Copper(I) bromide	$CuBr(s) \rightleftharpoons Cu^+(aq) + Br^-(aq)$	4.2×10^{-8}
Lead(II) bromide	$PbBr_2(s) \rightleftharpoons Pb^{2+}(aq) + 2Br^-(aq)$	6.6×10^{-6}
Mercury(I) bromide	$Hg_2Br_2(s) \rightleftharpoons Hg_2^{2+}(aq) + 2Br^-(aq)$	6.4×10^{-23}
Silver bromide	$AgBr(s) \rightleftharpoons Ag^+(aq) + Br^-(aq)$	7.7×10^{-13}
Carbonates		
Barium carbonate	$BaCO_3(s) \rightleftharpoons Ba^{2+}(aq) + CO_3^{2}(aq)$	8.1×10^{-9}
Calcium carbonate	$CaCO_3(s) \rightleftharpoons Ca^{2+}(aq) + CO_3^{2-}(aq)$	8.7×10^{-9}
Lead(II) carbonate	$PbCO_3(s) \rightleftharpoons Pb^{2+}(aq) + CO_3^{2-}(aq)$	3.3×10^{-14}
Magnesium carbonate	$MgCO_3(s) \rightleftharpoons Mg^{2+}(aq) + CO_3^{2-}(aq)$	4.0×10^{-5}
Silver carbonate	$Ag_2CO_3(s) \rightleftharpoons 2Ag^+(aq) + CO_3^{2-}(aq)$	8.1×10^{-12}
Strontium carbonate	$SrCO_3(s) \rightleftharpoons Sr^{2+}(aq) + CO_3^{2-}(aq)$	1.6×10^{-9}
Chlorides		
Lead(II) chloride	$PbCl_2(s) \rightleftharpoons Pb^{2+}(aq) + 2Cl^-(aq)$	2.4×10^{-4}
Mercury(I) chloride	$Hg_2Cl_2(s) \rightleftharpoons Hg_2^{2+}(aq) + 2Cl^-(aq)$	3.5×10^{-18}
Silver chloride	$AgCl(s) \rightleftharpoons Ag^+(aq) + Cl^-(aq)$	1.6×10^{-10}
Chromates		
Lead(II) chromate	$PbCrO_4(s) \rightleftharpoons Pb^{2+}(aq) + CrO_4^{2-}(aq)$	2.0×10^{-14}
Silver(I) chromate	$Ag_2CrO_4(s) \rightleftharpoons 2Ag^+(aq) + CrO_4^{2-}(aq)$	1.2×10^{-12}
Fluorides		
Barium fluoride	$BaF_2(s) \rightleftharpoons Ba^{2+}(aq) + 2F^-(aq)$	1.7×10^{-6}
Calcium fluoride	$CaF_2(s) \rightleftharpoons Ca^{2+}(aq) + 2F^-(aq)$	4.0×10^{-11}
Lead(II) fluoride	$PbF_2(s) \rightleftharpoons Pb^{2+}(aq) + 2F^-(aq)$	4.0×10^{-8}

Compound	Dissolution Equilibrium	K_{sp}
Hydroxides		
Aluminum hydroxide	$Al(OH)_3(s) \rightleftharpoons Al^{3+}(aq) + 3OH^-(aq)$	1.8×10^{-33}
Calcium hydroxide	$Ca(OH)_2(s) \rightleftharpoons Ca^{2+}(aq) + 2OH^-(aq)$	8.0×10^{-6}
Chromium(III) hydroxide	$Cr(OH)_3(s) \rightleftharpoons Cr^{3+}(aq) + 3OH^-(aq)$	3.0×10^{-29}
Copper(II) hydroxide	$Cu(OH)_2(s) \rightleftharpoons Cu^{2+}(aq) + 2OH^-(aq)$	2.2×10^{-20}
Iron(II) hydroxide	$Fe(OH)_2(s) \rightleftharpoons Fe^{2+}(aq) + 2OH^-(aq)$	1.6×10^{-14}
Iron(III) hydroxide	$Fe(OH)_3(s) \rightleftharpoons Fe^{3+}(aq) + 3OH^-(aq)$	1.1×10^{-36}
Magnesium hydroxide	$Mg(OH)_2(s) \rightleftharpoons Mg^{2+}(aq) + 2OH^-(aq)$	1.2×10^{-11}
Strontium hydroxide	$Sr(OH)_2(s) \rightleftharpoons Sr^{2+}(aq) + 2OH^-(aq)$	3.2×10^{-4}
Zinc hydroxide	$Zn(OH)_2(s) \rightleftharpoons Zn^{2+}(aq) + 2OH^-(aq)$	1.8×10^{-14}
Iodides		
Copper(I) iodide	$CuI(s) \rightleftharpoons Cu^+(aq) + I^-(aq)$	5.1×10^{-12}
Lead(II) iodide	$PbI_2(s) \rightleftharpoons Pb^{2+}(aq) + 2I^-(aq)$	1.4×10^{-8}
Silver iodide	$AgI(s) \rightleftharpoons Ag^+(aq) + I^-(aq)$	8.3×10^{-17}
Phosphates		
Calcium phosphate	$Ca_3(PO_4)_2(s) \rightleftharpoons 3Ca^{2+}(aq) + 2PO_4^{3-}(aq)$	1.2×10^{-26}
Iron(III) phosphate	$FePO_4(s) \rightleftharpoons Fe^{3+}(aq) + PO_4^{3-}(aq)$	1.3×10^{-22}
Sulfates		
Barium sulfate	$BaSO_4(s) \rightleftharpoons Ba^{2+}(aq) + SO_4^{2-}(aq)$	1.1×10^{-10}
Calcium sulfate	$CaSO_4(s) \rightleftharpoons Ca^{2+}(aq) + SO_4^{2-}(aq)$	2.4×10^{-5}
Lead(II) sulfate	$PbSO_4(s) \rightleftharpoons Pb^{2+}(aq) + SO_4^{2-}(aq)$	1.8×10^{-8}
Mercury(I) sulfate	$Hg_2SO_4(s) \rightleftharpoons Hg_2^{2+}(aq) + SO_4^{2-}(aq)$	6.5×10^{-7}
Silver sulfate	$Ag_2SO_4(s) \rightleftharpoons 2Ag^+(aq) + SO_4^{2-}(aq)$	1.5×10^{-5}
Strontium sulfate	$SrSO_4(s) \rightleftharpoons Sr^{2+}(aq) + SO_4^{2-}(aq)$	3.8×10^{-7}
Sulfides		
Bismuth sulfide	$Bi_2S_3(s) \rightleftharpoons 2Bi^{3+}(aq) + 3S^{2-}(aq)$	1.6×10^{-72}
Cadmium sulfide	$CdS(s) \rightleftharpoons Cd^{2+}(aq) + S^{2-}(aq)$	8.0×10^{-28}
Cobalt(II) sulfide	$CoS(s) \rightleftharpoons Co^{2+}(aq) + S^{2-}(aq)$	4.0×10^{-21}
Copper(II) sulfide	$CuS(s) \rightleftharpoons Cu^{2+}(aq) + S^{2-}(aq)$	6.0×10^{-37}
Iron(II) sulfide	$FeS(s) \rightleftharpoons Fe^{2+}(aq) + S^{2-}(aq)$	6.0×10^{-19}
Lead(II) sulfide	$PbS(s) \rightleftharpoons Pb^{2+}(aq) + S^{2-}(aq)$	3.4×10^{-28}
Manganese(II) sulfide	$MnS(s) \rightleftharpoons Mn^{2+}(aq) + S^{2-}(aq)$	3.0×10^{-14}
Mercury(II) sulfide	$HgS(s) \rightleftharpoons Hg^{2+}(aq) + S^{2-}(aq)$	4.0×10^{-54}
Nickel(II) sulfide	$NiS(s) \rightleftharpoons Ni^{2+}(aq) + S^{2-}(aq)$	1.4×10^{-24}
Silver sulfide	$Ag_2S(s) \rightleftharpoons 2Ag^+(aq) + S^{2-}(aq)$	6.0×10^{-51}
Tin(II) sulfide	$SnS(s) \rightleftharpoons Sn^{2+}(aq) + S^{2-}(aq)$	1.0×10^{-26}
Zinc sulfide	$ZnS(s) \rightleftharpoons Zn^{2+}(aq) + S^{2-}(aq)$	3.0×10^{-23}